THE VOLATILE OILS

BY

E. GILDEMEISTER AND FR. HOFFMANN.

LEIPZIG BERLIN

Written under the auspices of the firm of

SCHIMMEL & CO.,

LEIPZIG.

Authorized translation by

EDWARD KREMERS

MADISON, WIS.

With four maps and numerous illustrations.

British Library Cataloguing-in-Publication Data
A catalogue record for this book is available from the
British Library

Essential Oils

Essential oils are also known as volatile oils, ethereal oils, aetherolea, or simply as the 'oil of' the plant from which they are extracted, such as the oil of clove. An oil is 'essential' in the sense that it contains the characteristic fragrance of the plant that it is taken from. Essential oils do not form a distinctive category for any medicinal, pharmacological, or culinary purpose - and they are not essential for health, although they have been used medicinally in history. Although some are suspicious or dismissive towards the use of essential oils in healthcare or pharmacology, essential oils retain considerable popular use, partly in fringe medicine and partly in popular remedies. Therefore it is difficult to obtain reliable references concerning their pharmacological merits.

Medicinal applications proposed by those who sell or use medical oils range from skin treatments to remedies from cancer - and are generally based on historical efficacy. Having said this, some essential oils such as those of juniper and agathosma are valued for their diuretic effects. Other oils, such as clove oil or eugenol were popular for many hundreds of years in dentistry and as antiseptics and local anaesthetics. However as the use of essential oils has declined in evidence based medicine, older text-books are frequently our only sources for information! Modern works are less inclined to generalise; rather than referring to 'essential oils' as a class at all, they prefer to discuss specific compounds, such as methyl salicylate, rather than 'oil of wintergreen.'

Nevertheless, interest in essential oils has considerably revived in recent decades, with the popularity of aromatherapy, alternative health stores and massage. Generally, the oils are volatized or diluted with a carrier oil to be used in massage, or diffused in the air by a nebulizer, heated over a candle flame, or burned as incense. Their usage goes way back, and the earliest recorded mention of such methods used to produce essential oils was made by Ibn al-Baitar (1188-1248), an Andalusian physician, pharmacist and chemist. Different oils were claimed to have differing properties; some to have an uplifting and energizing effect on the mind such as grapefruit and jasmine, whilst others such as rose lavender have a reputation as de-stressing and relaxing - and also, usefully, as an insect repellent.

The oils themselves are usually extracted by 'distillation', often by using steam -but some other processes include 'expression' or 'solvent extraction'. Distillation involves raw plant material (be that flowers, leaves, wood, bark, roots, seeds or peel) put into an alembic (distillation apparatus) over water. As the water is heated, the steam passes through the plant material, vaporizing the volatile compounds. The vapours flow through a coil, where they condense back to liquid, which is then collected in the receiving vessel. 'Expression' differs in that it usually merely uses a mechanical or cold press to extract the oil. Most citrus peel oils are made in this way, and due to the relatively large quantities of oil in citrus peel and low cost to grow and harvest the raw materials, citrus-fruit oils are cheaper than most other essential oils. 'Solvent extraction' is perhaps the most difficult of the three methods, and is generally used for flowers, which contain too little volatile oil to undergo expression. Instead, a solvent such as hexane or supercritical carbon dioxide is used to extract the oils.

These techniques have allowed essential oils to be used in all manner of products; from perfumes to cosmetics, soaps - and as flavourings for food and drinks as well as adding scent to incense and household cleaning products. The science, history and folkloric tradition of essential oils is incredibly fascinating - and a still much debated area. We hope the reader is inspired by this book to find out more.

THE VOLATILE OILS.

PREFACE.

It is only within the last few decades that the former empiric manufacture of volatile oils has been placed on a scientific basis, which has enabled it to develop into an independent branch of chemical industry. During the period of transition in which this branch still finds itself, those factories which have done pioneer work, both scientifically and technically, as well as those which use the oils in various manufactures, are often compelled to suffer from the competition of inferior and adulterated products. As a matter of fact, the proper understanding of the estimation and appreciation of quality and purity of the much used volatile oils is not as common as is desirable for industry, commerce and the trades. The principal cause for this condition is the fact that the recent chemical investigations and their application to the arts have not yet been generally offered in suitable form.

The want of a work which from a modern standpoint treats in an exhaustive and critical manner the entire subject of volatile oils, has caused the firm of Schimmel & Co. of Leipzig to commission the authors with the preparation of a treatise that would meet the present demands. The successful completion of this task was greatly facilitated by having placed at their disposal the observation records of many years of manufacturing on a large scale.

Special stress has been laid on the description of properties and on tried methods of testing the commercially more important oils. Thus the consumer is placed in a position to distinguish pure from adulterated oils, good oils from those of inferior quality. Inasmuch as rational methods of examination are dependent on a knowledge of the physical properties and chemical composition of the oils, it became necessary to thoroughly discuss the results of the scientific investigations of, the subject. Those investigations, however, that cannot claim permanent value as well as antiquated methods, such as color reactions, etc., have not been included.

Fully realizing the importance of an historical basis for such a work, this has received special attention. Detailed references to the original sources enable the reader to make farther investigations in this direction.

Although the object and aim of this book are primarily of a practical nature, the authors venture to hope that it also offers to the scientific investigator a complete summary, with numerous references to original literature, of everything that has been accomplished in this line.

The authors acknowledge with thanks the cooperation of Dr. C. von Rechenberg who wrote the chapter "Theoretical foundation for the preparation of volatile oils by steam distillation;" also of Dr. J. Hellé who contributed "The more common constituents of volatile oils;" finally that of Dr. J. Bertram who kindly assisted in reading proof and who repeatedly made valuable suggestions.

With regard to the share in the work by the two authors whose names appear on the title page: the "Historical introduction," also everything that pertains to the history of the volatile oils and the crude materials, as well as the description of the methods of production of the American products, and in part the statements regarding origin and production of the drugs, are from the pen of Dr. Friedrich Hoffmann; the entire chemical text and those parts not specially enumerated have been written by Dr. Eduard Gildemeister.

Leipzig and Berlin, June 1899.

PREFACE TO THE AMERICAN EDITION.

In the work of Doctors Gildemeister and Hoffmann we find a happy blending of history with chemical science and technology that is quite unique in modern chemical literature. Moreover it covers a chapter of organic chemistry of which Professor Emil Fischer of Berlin recently remarked that it had undergone more rapid development within the past fifteen years than any other.

The translation of such a work should indeed prove of value. To make such a translation and to take cognizance of the numerous contributions on volatile oils and related subjects that have appeared since the original German edition was issued a year ago, has proved a severe task for one whose time was already divided between instructional duties and editorial labors. In judging the English edition, it is hoped that the limitations as to time and the difficulties encountered will be taken into consideration.

Owing to the impossibility of satisfactorily translating many of the numerous quotations in the historical introduction, chapters two and three have been condensed. Inasmuch as but very few if any readers are in position to consult the numerous historical works quoted by Dr. Hoffmann, all bibliographic information has been placed in an appendix.

To the special part a few oils were added, and changes rendered necessary by recent investigations were made as far as time permitted.

The writer desires to acknowledge with thanks his indebtedness to Mr. Carl Fritzsche and to Dr. Hoffmann for suggestions, as to the scope of the translation and as to minor details. Mr. O. Schreiner has assisted in the translation and proof reading and Dr. C. Kleber of Garfield, N. J., has kindly read one proof. His long experience as chemical expert on the subject of volatile oils rendered his cooperation especially valuable.

EDWARD KREMERS.

MADISON, WIS., July 1900.

TABLE OF CONTENTS.

HISTORICAL INTRODUCTION.

1. THE SPICE-TRADE IN ANTIQUITY AND DURING THE MIDDLE AGES.

(With two maps.)

Parts of plants as well as natural plant products which have been used since antiquity on account of their agreeable odor, their pleasant taste, or their medicinal virtue, enter the world's commerce up to the present time in their original form, being either previously dried or prepared in some other expedient manner. The essential constituents of these crude materials (drugs), the aromatic volatile oils, the resins, gum resins, bitter principles, alkaloids and glucosides, have been recognized in the course of development of the natural sciences. With the improvements in technology they have gradually been prepared in a purer and better condition.

Of these various products of the plant world, the spices and aromatics have from the very beginning ministered to the needs and welfare of man, and have, therefore, been appreciated by him in a special degree. As a result, they have always been a prominent and influential factor in the intercourse of nations as well as in the world's commerce. After several thousand years of knowledge and actual use of the spices in their original form, their essential constituents, the volatile oils, have since the middle ages and more particularly in modern times been successfully isolated in their natural freshness and entire efficiency.

In a treatise on volatile oils, a brief historical retrospect of the origin of and commerce in the bearers of these products, viz. the spices and aromatics, may be regarded as eminently proper. This all the more, since in this branch of knowledge as well as in others the historical element constitutes a valuable basis for a proper understanding and investigation.

All investigation in the realm of the history of civilization, that considers not merely a single people but mankind in general, and that goes back to the earliest historic documents, invariably leads to the wonderful orient so rich in legends—to central Asia, the traditional cradle of mankind. This is also true of the history of the trade of the oldest peoples, and especially of the source and distribution of the useful spices and aromatics.

Its geographical position and topographical configuration make Asia a very highly favored continent. Broad as it is, it extends from pole to equator. Favored by mighty mountain chains and rivers, its most beautiful and richest countries lie in latitudes where soil and climate afford all conditions for a profusion of luxuriant subtropical vegetation. The eastern and southern coastlands are cleft by large bays which penetrate far inland. Many navigable rivers which flow into these bays have their origin in distant highlands. The mainland is bordered by a wreath of islands extending from the Japanese island realm through the Malay archipelago to Ceylon. These islands abound in tropical vegetation. The entire continent, therefore, reveals a diversity and richness of plant life such as no other possesses.

These advantages have made southern Asia and the islands bordering on its shore the oldest and principal scene of international traffic and commerce, spices and aromatics constituting the main articles of exchange. They not only found general use on account of their agreeable odor and aromatic taste, but were employed by most peoples in religious rites and sacrificial customs, and thus acquired symbolic meaning. With the increase of prosperity and luxury, also with the development of the sense of cleanliness and of physical well-being, spices and aromatics not only became more valuable, but their consumption increased.

According to documents discovered in recent years, the territory between the Indus and the Oxus was the starting point of the early commerce between the oldest peoples of central and southern Asia. Attock, Cabura, Bactra and Maracanda seem to have been the first larger centres for storage and exchange of oriental products. These were spices and aromatics, the noble metals, silk, and jewelry. To Attock were brought the products of the eastern Chinese empire, which, at an early date, closed its markets to the rest of the world.

From Attock, at the junction of the Kabul river with the Indus, the caravan road led via Cabura (the present capital Kabul of Afghanistan) to the north via Bactra, Bokhara and Maracanda

(Samarkand) to the countries of the Oxus and to the Scythian tribes. Also from Cabura southward to Kandahar, thence in a western direction through the realm of the Parthians to the *Pylae Caspiae* (Caspian gate), and to Ecbatana in Media. Thence the land route crossed the Tigris to Babylon on the Euphrates. In a later period, after the traffic along the water routes had developed, a round-about way via Susa to the mouth of the Tigris was taken and the caravan freight shipped up the Euphrates to Babylon. Between Attock and the ports on the Black and Mediterranean seas, Babylon—existing 3000 years B. C. — was in early antiquity the most important place of traffic and commerce for westward bound Chinese and Indian merchandise. To the northward the caravan roads led out of Babylon through Assyria and Armenia to the Black sea (*Pontus Euxinus*) and westward through Syria to the Mediterranean sea (*Mare Internum*), thence through Palestine to Egypt. In spite of their highly developed industry the Egyptians, as is well known, closed their doors to foreign peoples as did the Chinese. As a result commercial centres were wanting in Egypt that were open to foreign merchants and to transitory commerce.

During the prime of the Babylonian empire, about 2000 to 1000 B. C., a lively caravan trade was developed which extended from China, India and Arabia to Egypt, Palestine, Syria and the Black sea.

During this period, Arabia acquired a special importance by means of the sea traffic of her southern coast, which was favored by the Persian gulf and the Red sea. At an early date, the Arabian population conducted a lively intermediate trade with Indian and Egyptian goods which were brought to the Arabian ports. By means of caravans these were carried northward to Babylonia, Syria and other countries. The principal route from southwestern Arabia to Babylon, Damascus and Egypt led from Cane on the Arabian gulf (Erythraean sea) via Saba, Macoraba, Hippos and Onne to Elath (the present Akabah) at the north-eastern end of the Red sea. From this point the eastern route crossed the Jordan via Petra, Kir Moab, Ammonitis and Dan to Damascus; the western route to Egypt via Azab, Axomis and Meroë.

About 15 centuries before the Christian era, the world's commerce was gradually and, in the course of time, very greatly expanded by the Phoenicians, who lived on the narrow Syrian coast district. In the industrial and commercial field they acquired a prominent position; as mariners, however, a dominating position among the nations of their time. Besides having practical control of sea navigation, the Phoenicians were the first extensive and successful colonizing nation of antiquity. They established or extended commerce with the peoples living along the coast of the Mediterranean, they ventured through the "Pillars of Hercules" (Gibraltar) into the ocean and made accessible the products of the Madeira and Canary islands, the western coasts of Spain and France, the British islands, and the northland as far as the amber coasts of the Baltic sea.

For almost a thousand years, during which time they held their prominent position in marine traffic, the Phoenicians were the principal commercial agents between the nations of the orient and the occident. Sidon and, since the ninth century B. C., Tyre became prominent centres of the world's commerce of that time.[1]

The Phoenicians also extended their navigation to the Red sea and the Arabian gulf and from these to the Persian gulf. In the latter they established the colonies on Arados and Tylos, islands belonging to the present Bahrein group. From the twelfth century up to their decline in the fifth century B. C., these cities carried on a large transit-trade with goods from India and Ceylon to Babylon, Damascus, Tyre and Sidon, and to Egypt. A caravan route led from Gerra via Salma, Thaema and Madiana to Elath. From Elath the older routes to the north, to Damascus, Tyre and Sidon were followed, also westward to Egypt. To Babylon, on the other hand, the water route up the Euphrates or Tigris was taken from Arados and Tylos.

Carthage, a Phoenician colony established in 846 B. C., soon flourished and developed such power that it became the greatest rival of the mother country in the following century.

Owing to the rise of the Persian empire, the inland commerce of western Asia was somewhat shifted during the period from the sixth to the fourth century B. C. The old routes of traffic passing through countries controlled by the Persians were not only kept in good condition but extensions were also made. These old highways of transcontinental commerce underwent further changes at the time of the Greek conquests under Alexander the Great at the close of the fourth century B. C. Still greater, however, were the changes brought about by the migration of nations during the fourth and fifth centuries of the Christian era. Wars and other disturbances of the commercial intercourse along the old caravan routes frequently restricted traffic to the

1) As is well known, the Phoenicians supplied King Solomon with the material for the building of the temple at Jerusalem about 1000 B. C. (1. Kings, 5: 9—10; also 2. Chron., 2: 8, 9; Ezekiel, 27.)

rivers and seas. Upon the reestablishment of peace, however, commerce always seems to have found its way back to the traditional caravan routes.

In the course of time, however, and especially during the sixth and seventh centuries still other changes were made. Thus, e. g. the products of the Chinese and Indian coast districts and of the Indian islands were brought in part by ship over the Bay of Bengal and by way of Ceylon to the commercial centres of the Persian gulf and the Red sea. From these they were distributed by coast-wise trade, by river navigation up the Tigris and Euphrates, or by caravan to the north and west. From the more northern Chinese and Indian districts the caravans passed through the present East-Turkestan following the older routes mentioned on p. 8 through the countries of the Oxus to the Araxes. The goods, instead of being carried by river to Phasis and the Black sea, were taken as far as Artaxata and then by caravan through Persia to the ports of Asia Minor. The old route from Kandahar along the northern border of the Iranian plateau, which likewise led through Persian territory, was also followed by caravans.

During the reign of the East Roman emperor Justinian, in the sixth century, when the world empire of the Romans was broken up by the migration of nations, Persia experienced a new rise to power under the Sassanidæ. The Persians ruled the entire territory from the Caspian to the Arabian seas and from Afghanistan of to-day to Syria and Armenia. They improved the old high-ways and caravansaries, kept them in repair and promoted commerce and traffic, directing both over routes leading through their own territory. Owing to the wealth and luxury of the Roman empire, the commerce in oriental spices, aromatics etc. had risen to an unusual height. The East Roman empire which at that time was the principal western state with its capital at Constantinople, was forced by the Persians to procure such oriental goods as were not shipped by water, from and through Persia and to pay a heavy duty on them. The principal places of storage and for the collection of revenue at that time were Artaxata on the Araxes, Nisibis, south of the Tigris, and Callinicum (Rakka) on the Euphrates. To Artaxata were brought the goods from the countries of the Oxus over the Caspian sea. Those that were conveyed along the caravan routes south of the Caspian sea, and those that came from the coastlands of the Persian bay up the Tigris or Euphrates centered at Nisibis. For nearly five centuries, the aromatics of China and India, of the Malayan archipelago which came via Ceylon, and in part those of Arabia were transported over the two last mentioned routes to the western countries. About this time also, the levant commerce, which became so important in a future period, had its beginning.

During the life of the Persian empire (up to the middle of the seventh century A. D.) all attempts, made by Justinian and his successors, to divert the transmarine commerce from its course through Persian territory remained unsuccessful. They did not even succeed in opening up marine traffic between India and Ethiopia, because Persian merchants visited the Indian markets and persuaded the Indians and Chinese not to sell their goods to new customers. In the course of time, however, the Greeks succeeded in obtaining larger consignments by water from the ports of India and Ceylon and more particularly from the coast-lands of the Arabian sea, which were rich in spices. These were delivered directly to their own ports, Kolsum, and Akabah and Berenice near the entrance to the Red sea.

About this time there existed three great caravan routes from China westward. They began in the territory of the Hoang Ho and the Yangtsekiang and passed through the Gobi desert. The northern route then took its course through the oasis of Chami, then northward along the Thian-Shan mountains through the present Dsungarai, past the Balkash sea, and via Talas. It then followed the Syr-Darya river to the Aral sea and the Caspian sea.

The middle route passed to the south of the Thian-Shan mountains through the northern part of East Turkestan via Chami, Turfan, Karaschar, Kutscha, and Aksu to Kashgar; thence over the Terek pass to Ferghana via Samarkand, Buchara and Merw to Persia.

From the Gobi desert, the southern route passed through the southern part of East Turkestan via Chotan and Yarkand, then over the Pamir plateau and through Afghanistan to the Pandschab (India) crossing the Bamian and Gazna passes to Multan. Goods intended for the west, coming via this route, were taken down the Indus river to Daybal. From this port they were shipped by sea with other goods from India and Ceylon.

During the seventh and eighth centuries A. D. the Arabs carried on an extensive marine commerce with India and China, especially in spices and aromatics. These were supplied in large quantities to the luxurious courts of the caliphs and the Byzantine emperors. The principal centres between China and Arabia were at that time on the Malay peninsula, to which were also brought the products of Java and other Sunda islands. Later commerce concentrated itself in Kalah, a city on the east shore of the Malay peninsula. In the tenth century there existed between Kalah and Siraf, a city on the east coast of the Persian gulf, regular commercial intercourse between the Arabians and Chinese. From the northern point of Sumatra the Chinese crossed the bay of Bengal to Ceylon.

Arabian merchants also settled along the Malabar coast, in Ceylon and in the Indian sea ports. From the eighth to the tenth century, Daybal at the mouth of the Indus was the most important commercial centre and seaport of India. It was the principal emporium for the products of the Indus valley and the Pandschab on the one side and of Mesopotamia, Persia and Arabia on the other. For the products of northern India, Multan on the Dschelam river in the Pandschab, was the first larger rallying point. It was also a place of pilgrimage that was much revered and visited by the Hindoos.

From the eighth century on, Suhar and Muscat, near the entrance of the Persian bay, developed as rival ports for Indian and Chinese commerce with occidental countries. At the same time Aden, at the entrance to the Red sea, became the principal port and commercial centre for the products from Yemen, Hedsjaz, Ethiopia and Egypt.

In addition there were caravan routes: one from India to Persia through Seistan; the other via Gazna and Kabul to Afghanistan.

Seti I. and Rameses II., Egyptian pharaohs, had during the first quarter of the fourteenth century B. C. connected the Red sea with the Mediterranean sea. In order to reestablish this sea-route, Pharaoh Necho toward the end of the seventh century B. C. tried to have a new canal constructed from Bubastis on the Nile to Patumos on the Red sea. This was not completed, however, until 500 B. C. by Darius Hystaspes and was widened and improved by the Ptolemies. Before the beginning of the Christian era it was again choked up with sand. Under the caliph Omar in the seventh century A. D. the canal from Cairo to the Red sea was again restored, but did not exist longer than a century.

From the seventh to the twelfth century there also existed several land routes across the Suez isthmus. One of these followed the course of the old canal (choked up with sand) from the Red sea to Cairo, whence the goods were shipped down the Nile and then by sea. If the passage of the goods through Alexandria was not necessary, the shorter route over the isthmus from Kolsum to Pelusium (Faramiah) was preferred. At this time Damascus and Jerusalem were also important commercial centres. Here also oriental goods were exchanged by the merchants of Mecca on the one hand and those of Tripoli, Beirut, Tyre and Acre on the other.

From the seventh to the twelfth century there existed an active coastwise trade along the north African coast, having its centres in Syria and Egypt and extending as far as Morrocco and Spain. This commerce acquired a special importance for spices and aromatics although for a time it was limited by Mohammedan laws against intercourse with Christians. Commerce also soon flourished among the Greeks, who obtained spices and aromatics, possibly also rose water and aromatized fatty oils from Antioch, Alexandria and Trapezunt and brought them to Constantinople, Thessalionica and Cherson. Already in the tenth century Trapezunt was an important emporium for the spices and aromatics of India and Arabia and for Persian perfumes. The Greeks, however, purchased these luxuries only for home consumption, which was large, without distributing them farther to other nations.

From the tenth to the fifteenth centuries, the commerce of the Mediterranean was conducted principally by Italian cities. In the tenth and eleventh centuries Bari, Salerno, Naples, Gaëta, and above all Amalfi, Pisa and Venice were the principal commercial centres. The levant commerce, which was in its prime from the twelfth to the fifteenth century centered in Venice and Genoa. In the levant itself, at the time of the crusades during the twelfth and thirteenth centuries, Acre on the Palestine coast was the most important commercial port. When this city, the last held by the Christians, likewise fell into the hands of the Mohammedans in 1291, Famagusta on Cyprus, and, for a longer period, Lajazza on the bay of Alexandretta, became the commercial centres of the Levant up to the fifteenth century. The latter port was the junction for the western merchants with those coming from Asia.

Toward the end of the thirteenth century, the cities of Bagdad and Basra on the Euphrates, lost their commercial supremacy which they held for several centuries to Tebriz, the new rising capital of Persia near the Caspian sea. When Egypt in the thirteenth and fourteenth centuries began to levy a high toll on the finer Indian spices and aromatics, the land transportation was deviated more and more through Persia via Bagdad and Tebriz to Lajazzo and Trapezunt.

During the fourteenth century, the island port Ormuz in the Persian gulf became an emporium for the westward bound goods from India and Ceylon. It maintained this position until its capture by the Portugese at the beginning of the sixteenth century. The more important ports along the western coast of India at that time were Mangalore, Calicut and Quilon. Ginger, cinnamon, cardamoms, pepper, cloves, nutmegs, sandalwood, lignaloes, indigo, etc., were brought to these ports from the interior. From the Chinese ports and the East Indian islands large importations of these and other spices and aromatics were received.

Toward the close of the thirteenth and at the beginning of the fourteenth century the direct traffic between Europe and China became very active, more particularly over the land route. Under the protection of the Mongols the caravan routes through central Asia were generally safe. The greater part of the Chinese empire was also accessible to Europeans. About this time also Marco Polo, the first European world-traveler, visited China, India and the islands of the Indian ocean.

Owing to disturbances and invasions of central Asia, the overland traffic diminished after the middle of the fourteenth century. Up to the discovery of the sea route around Africa at the close of the fifteenth century, Tebriz, however, remained an important centre for transitory trade. The capture of Constantinople by the Turks in 1453 enabled them to interrupt the trade of the Italians via Trapezunt and the Crimea and soon to cut it off altogether. Cyprus also lost its former importance for the levant trade about this time.

Egyptian commerce, however, once more increased considerably toward the close of the fourteenth and in the course of the fifteenth century. In place of Aden, Dschidda, the sea-port of Mecca, became the principal junction of commerce between the Indian seas and the occident. The heavier goods were transported by water, the lighter by pilgrim caravans to Tor on the Sinai peninsula. The high toll levied by the Egyptians caused some of this trade to be deviated for the time being to Syria. Owing to the occupation of Lajazzo by the Turks in 1347, and the conquest of the Crimea in the fifteenth century, this traffic became very prosperous for a short time.

Thus in the course of several thousand years, the commercial intercourse between the nations of Asia, later also those of Africa, and of Europe underwent a variety of changes as did also the various routes of traffic. The circumnavigation of Africa by the Portugese in 1498, their conquest of Ormuz, the key to the Persian gulf, and their extended marine traffic brought about important changes in the traditional channels of traffic. The transport by means of caravans was gradually diminished; the highways, formerly well kept, got out of repair, the ocean ship displaced the "ship of the desert," the camel of the caravans.

From the sixteenth century on, the ocean became the preferred highway of international commerce. Hence the levant commerce, which had flourished for several centuries and had enriched the commercial centres of Italy and of other Mediterranean countries, lost its importance and finally ceased.

Numerous ruins of architecturally magnificent structures in cities once large and powerful and in fortified markets, well built high-ways covered with the sand and alluvium of centuries, also caravansaries on the plateau and desert lands of western Asia and the Arabian peninsula reveal to these after-days the former greatness, the artistic skill and the commercial prosperity of peoples who still live in history but principally in name only.

The spices and aromatics of southern Asia and the Asiatic islands, which constituted the first foundation of international commerce and which have played an important role in the commerce of all ages, have retained their original value in spite of all changes in the world's history and in spite of all mismanagement and waste on the part of foreign nations. The same spicy cinnamon, cloves, nutmegs and cardamoms, pepper and ginger and other spices used and highly appreciated since antiquity; frankincense and myrrh, benzoin and other incenses, camphor, sandalwood and lignaloes and other plant products acquiring use in ever increasing numbers thrive after thousands of years in the sunny countries and islands of the orient in the primitive freshness and profusion.

However, they are no longer brought to the occident on the backs of camels stretched out into almost endless chains over the plateaus and deserts of Asia, but in trim sailing vessels and speedy steamships crossing the ocean; or they are transported in freight cars that hasten over steel tracks encircling the continents. They are used either in their original state, or in a concentrated form, purified as it were by the giant stills of modern chemical industry. They still serve man in the humble hut as well as in the palace.

---

2. HISTORY OF VOLATILE OILS.

In the course of centuries various more or less independent branches have separated from the common trunk of natural science and have made a history of their own. This is true also of the subject presented in this treatise. To be sure a true knowledge and application of those plant products, which are termed volatile oils, has been gained only in recent periods. Nevertheless the nature and value of these substances, that so materially enhance the attractiveness of plant life, does not appear to have escaped the observation of the oldest peoples. It seems almost certain that not only the grace and vivid coloring of the flowers but also the fragrance of the vegetation in southern Asia and the islands encircling the mainland, must have aroused the curiosity of man, as well as the use of the plants for the purpose of food, clothing and other necessaries. Indeed, those plants and plant products that were conspicuous on account of their aromatic odor and taste, seem to have attracted the attention of man in a special degree. These very properties seem to have induced him to use them and to seek proper methods for their cultivation and preservation.

It is true that the oldest documents pertaining to the history of the beginning of human industry have recorded only the most primitive methods for the preparation of implements that were used in the chase, the cultivation of the soil, and for the collection and preparation of foodstuffs and other useful products. Nevertheless it may not be amiss to suppose that the exigencies of self-preservation, of protection, and of wellbeing at an early period caused man to utilize the heat of the sun and of fire for the preparation of foodstuffs as well as for various other purposes in the production and preparation of a variety of natural products. It may have required long periods of time before fire was used for the preserving of perishable foodstuffs, for the separation of more valuable substances from those of lesser value, or of the pleasant from disagreeable ones: so e. g., the distillation of the "spirit of wine" from the wine itself; the separation of the "subtle principle," the aroma, from spices and balsams. The earliest documents, however, show that the spices were among the earliest and most highly prized articles of barter and commerce during antiquity; that, as products of nature agreeable to the gods, they were offered as sacrifices in religious ceremonies, and were also used in the embalming of the dead preparatory to their transit into the realm of the gods.

It is especially this use of spices and of aromatic plant products by the priests, those promotors and supporters of natural science during antiquity, that renders it probable that their knowledge was early applied to the production, preparation and improvement of the spices which were used in the sacrifices and in embalming. Whether a beginning in the preparation of the aromatic principles of plants, our modern volatile oils, was made previous to early Hindoo and Egyptian civilization, does not become apparent from the oldest documents. Even the Bible, which gives so much information concerning the customs of the Jews, makes no other statements than those pertaining to the spices and aromatics used in various countries. The early preparation of the metals and the manufacture of metallic utensils would indicate that furnaces and other apparatus for heating were early used in a variety of ways. One may suppose, therefore, that they were gradually used in primitive attempts to separate the spirit from wine, from other fermented fruit juices and honey; also the subtle aromatic principles from spices, balsams and oleoresins. These crude experiments may be considered as constituting the first stages in the art of distillation which later became of such great importance.

The Egyptians, owing to the continuity of their early and highly developed civilization and to the preservation of their monuments and literary productions, are generally considered as standing at the portals of history. With reference to time, however, the Chinese and Aryans are probably the oldest peoples. These races, with whom civilization seems to have had its beginning, lived in central and southern Asia, a mountainous district favored with a mild climate and a luxuriant vegetation rich in useful and spicy products. Our knowledge of the peoples who first occupied this wide territory is but legendary. Concerning their industrial and technical accomplishments as well as their literary productions, but little definite information has come down to us. These Chinese and Indians may have developed considerable dexterity in industrial pursuits and may even have accomplished much in the scientific realm. Their attitude of exclusiveness, however, toward the outside world and their secrecy have prevented them from exerting a lasting influence on other nations. The oldest documents that throw light on their early scientific accomplishments are the Ayur-Veda (book of the science of life) by Charaka and Susruta.[1]) As is the case with so many writings of early antiquity, nothing definite is known with regard to the age of these documents. It is possible that they are traditions reduced to writing at a rather late time. From this work it becomes apparent that the Indians were acquainted with primitive apparatus for distillation, with fermentation and the products obtained by distillation. Of "distilled oils," those of rose, schoenus (andropogon), and calamus are mentioned.[2]). Whether these oils are "distilled" in the modern sense of this term can not be ascertained.

From the likewise legendary documents of the old Persians it would seem that they also were acquainted with the process of distillation and distilling apparatus.[3])

With regard to the Egyptians, however, whose history goes back as far as 4000 B. C., we have definite information concerning the early development of industry and art, the acquirement of scientific knowledge and its practical application. Their commerce, which extended as far as India, Babylonia, Syria, Ethiopia and other countries, as well as their industry and art undoubtedly developed slowly before it reached that height which we now admire. The application of their scientific knowledge finds manifold expression. Thus, they were acquainted with the preparation of the more common and useful metals, with furnaces and distilling apparatus, with the distillation of wine, of cedar resin,[4]) the preparation of soda, alum, vinegar,[5]) soap, leather, with the preparation and use of colors and the manufacture of glass. The Egyptians were acquainted with and used cedar (turpentine) oil[6]) and colophonium,[7]) and probably knew how to obtain plant aromatics in purified form, possibly as distilled oils.

---

1) See under Bibliography.
2) Susruta's Ayur-veda, pp. 111 & 130.
3) Gebri de alchimia; Schmieder, Gesch. d. Alch., p. 34.
4) Aetii medici graeci, fol. 10.
5) Numbers 6: 3.
6) Herodoti historiae II, 85.
Dioscoridis, De mat. med., lib. 1, cap. 34, 39, 80, 95, 97.
Plinii sec.; Nat. hist., lib. 15, cap. 6 & 7, and lib. 16, cap. 22.
Scrib. Largi, Comp. med., p. 328.
Theophrasti opera: Hist. plant., lib. 9, cap. 3.
7) Dioscoridis, De mat. med., vol. 1, p. 660; and vol. 2, p. 689.

The height which Egyptian civilization reached is revealed better by the architectural monuments and their contents, in the case of the pyramids, than by the few written documents that have come down to us. Like their writings, many of their trades were lost at least in part to the civilized nations succeeding them and had to be rediscovered. In judging the relationship between their scientific knowledge and their accomplishments in the arts and trades, it should be remembered that the manufacture of metals, of glass, and even dyeing were based on a rather crude empiricism and seem to have been almost wholly independent of the theoretical consideration of the age. Thus, with but little scientific knowledge of permanent value, these most ancient peoples, the Hindoos, Egyptians, Assyrians, Babylonians and Phoenicians, like the Chinese, who stand at the dawn of civilization, have in the course of centuries of development accomplished much. They, and among them the Egyptians more particularly, served as the teachers of the classical nations of Greece and Rome.

The scientific knowledge as well as the industrial and artistic accomplishments of the Hebrews and the Greeks, and indirectly also of the Romans, had their root in Egyptian civilization. However, the Greeks like the Hebrews tended toward the ideal rather than the practical in their conception of nature. They did not experiment and were not bent on applying their scientific knowledge. Their philosophers and writers collected and systematized the information that had come down to them and thus aided in preserving it, without, however, putting it to practical use or adding anything new to it.

The Greeks, however, were well informed as to the Egyptian arts: they understood the preparation and working of the metals and the manufacture of glass and other industrial arts. Their commerce, however, was mostly barter in natural products. The oriental spices were highly prized by them for incenses, cosmetic and sanitary purposes. Whether the primitive method of distillation practiced by the Egyptians and Persians was known to the Greeks does not appear from their literature. It is not improbable, however, for medicine and the cosmetics were hardly less thought of by the Greeks than by the Egyptians. Owing to the luxury of the later Greeks, perfumes and spices were extensively used. The much praised oriental perfumes, especially the sandal wood (ξύλα ἰνδικὰ) were considered a necessity at all festivities.

When Greek culture spread westward and became the basis of Roman civilization, Greek views concerning nature and Greek knowledge of it were likewise transmitted. In their numerous conquests, the Romans increased their knowledge of oriental natural products. These were brought by the old caravan routes and then by sea to Rome. Among them were the finest spices and aromatics for the kitchen, perfumes and ointments for the toilet, balsams and incenses. Whether aromatized fats, or distilled oils as well, were used cannot be ascertained definitely from Roman literature. That the Romans themselves were adept in the preparation of toilet articles seems highly probable. That the natural sciences, including the science of drugs, were well cultivated is shown by the writings of Dioscorides,*) Pliny*) and Galen.*) Yet, although the Romans were good observers of natural objects and phenomena, and equally good compilers of the knowledge of their own period and of previous periods, they did not, in general, penetrate into the secrets of nature. Neither did they add much to the stock of knowledge handed down to them. The natural sciences and medicine were, therefore, but little advanced by them.

With the decay of Greek and Roman culture and the long night in the history of civilization that followed, many of the earlier discoveries in the arts and sciences were lost. At the close of that period which we now designate as antiquity, birth was given to a new civilization, of which the Arabians were the forerunners. They themselves, however, contributed but little to this Mohammedan or Arabian period of civilization. This had its roots in the Alexandrian school, the spirit of which was imparted to the later Mohammedan conglomerate of peoples through the Syrians and Persians and their languages, also through the Greeks of Asia Minor.

Permeated by the conceptions of the Alexandrian school, the Arabs revived the study of the natural sciences in the ninth century. Mathematics, astronomy and medicine were rapidly developed. With their tendency to faith in the miraculous, alchemy and magic developed with the natural sciences and played an important role in the theory of transmutation of the metals and in medicine. The philosopher's stone and a universal medicine were to banish all misery and disease from this world.

Above all it was Geber[1]) (Dschabir), one of the most influential and prominent scholars of his time, who developed this theory and established a firm faith in it which lasted for centuries. During the period when

*) See Bibliography.

1) Summa perfectionis etc., lib. 4, pp. 156—178 Alchimiae Gebri etc., lib. 2, cap. 12; Bergmann, De prim. chem., § 3 D & § 4 C.

Bagdad, Bassora and Damascus were the principal centres of commerce, no people was more skilled and more productive in the arts and trades and also in natural science. Their commercial relations extended to almost all known countries, and the use and knowledge of Eastern spices and aromatics was greatly fostered by them.

With the development of the traditional knowledge, the Arabians also fostered the process of distillation in connection with the hermetic art.[1]) As early as the fourth century, the Greek scholars Synesios of Ptolomais[2]) and Zosimos of Panopolis[3]) had described the distilling apparatus and methods of the Egyptians. Aëtius of Amida, a physician and writer who lived in Constantinople during the beginning of the sixth century, also described the preparation of empyreumatic oils by downward distillation (*Destillatio per descensum*).[4]) This as well as the upward distillation (*Destillatio per ascensum*) are reported by Geber. According to Porta[5]) and other writers of the sixteenth century the Arabian physicians and alchemists introduced the condensing tube (*Serpentina*) for the better cooling of the distillate, and a kind of fractionation for the distillation of wine. Geber is indeed the first, since the Egyptians, who possessed a thorough knowledge of the subject of distillation: of dry distillation and distillation with water vapor, of utensils of glass and glazed earthenware used in the process.

Next to the writings of Geber, those of Mesuë, who lived some time between the eighth and tenth centuries, reveal what knowledge the Arabians had on the subject of distillation and volatile oils. In the chapter on oils[6]) their method of preparation is described. Most of them are aromatised fatty oils, only juniper oil and oil from asphaltum were prepared by destructive distillation. According to Bergmann,[7]) Mesuë is supposed to have been acquainted with distilled rose oil and oil of amber.

Other Arabian physicians also give information about distilled waters and oils of that period. Thus Ibn Khaldun[8]) of the ninth century mentions that distilled rose water was an important article of commerce of the Persians during his and the previous century. Nonus Theophanes[9]) who in the tenth century was physician of Emperor Michael VIII in Constantinople, recommended rose water as a medica-

1) Conringius, De herm. Aegypt., lib. II, cap. 4; Bergmann, Hist. chem., vol. 4; Schmieder's Gesch. d. Alch., p. 85.

2) Synesii Tract. chimicus.

3) See Bibliography; also Hoefer, Hist. de la chim., I, pp. 261—270.

4) Aëtius, Libri medicinales, fol. 10.

5) De destillatione.

6) Antidotarium, fol. 80.

7) Historiae chemiae etc., p. 7.

8) Notices et extraits, 19, p. 364.

9) Synesius de febribus, cap. 28, p. 112.

ment. The Syrian physician Serapion (Janus Dàmascenus) who lived in the ninth century; also Avenzoar[1]) who lived about a century later and who was physician to the Caliph Ebn Attafin of Morocco, used rose water as an eye-remedy and rose oil sugar as an internal remedy. In the medical writings of Abn Dschafar Achmed, an Arabian physician of the eleventh century, which were translated into Greek by Synesius of Constantinople, rose water, rose oil and camphor are mentioned among the current remedies.[2])

While Geber was the first, and also the most important Arabian writer who was acquainted with distillation, the writings of Albucasis who lived three centuries later, reveal such an exact knowledge of the subject, that the thought seems justified that distillation was largely practiced by the Arabians. The "Liber servitoris"[3]) of Albucasis contains a very clear description of the process of distillation, also specific directions for the distillation e. g. of water, acetic acid and alcohol. Torbert Bergmann,[4]) the Swedish chemical historian, regards this description as one of the first and best.

During the period from the eighth to the twelfth centuries, many Arabian scientists were especially active in furthering the sciences of medicine and materia medica. Distillation was frequently resorted to as a means of securing the active constituents of drugs. In spite of the crude methods often employed, it seems probable, therefore, that the volatile oils which, no doubt, separated at times must have attracted the attention of the curious experimenter. Owing to the complete lack of knowledge regarding the nature of the oils as well as on account of the idea that the distilled waters contained the subtle active ingredients of the drugs, the oils which separated may have received but little consideration, being regarded as crude separations. They were, therefore, not valued highly and, as the literature of that period shows, only a few of them found application.

The desire to transmute the baser metals into gold, which began to predominate in the eleventh century, brought about the decay of Arabian science. From the conquest of Bagdad by the Mongols in 1258 dates the political downfall of the Arabs. In Spain and also in Italy, through their school at Salerno, Arabian science still held sway

---

1) Liber Theisier etc., Liber 7, fol. 1; Lib. 5, cap. 9, fol. 44.
2) Synesius de febribus, pp. 58 & 240.
3) See Arabian writer under Bibliography.
4) Bergmann, Historiae chimiae. Comp. also Liber Servitoris, fol. 339 b, 341b & 342.

for some time. In these countries also, through the crusades, Europeans became acquainted with Arabian scientific ideas, as well as with their methods, among others with the process of distillation and the utensils used in this process. In the European countries alchemistic speculations were welcomed in the monasteries where mysticism and faith in miracles were at home. Though distillation did not become a lost art by any means, it was employed principally in the search for the philosopher's stone and the panacea for all human woes.

The distillation of alcohol also was given much attention during this period. Thus the Cardinal Vitalis de Furno at the beginning of the fourteenth century declared it a true panacea. Albertus Magnus (1193—1280), Bishop of Regensburg, carefully described the distillation of alcohol in his works. Arnoldus Villanovus (1235—1312) seems first to have introduced the Arabian term alcohol into German literature, and on account of its properties, designated it *aqua vitae.*[1]) The latter was also acquainted with the distillation of the oils of turpentine,[2]) rosemary,[3]) and sage. His *oleum mirabile* consisted principally of an alcoholic solution of two of these oils. This mixture was used as an external remedy and later, with the omission of the turpentine oil, as an aromatic perfume. For centuries it remained a much praised specialty under the name of Hungarian water.

Alcohol was also used for the extraction of spices and other plant products. In this manner alcoholic solutions of volatile oils, of aromatic resins and balsams were obtained.[4])

From the thirteenth century on distilled aromatic waters were more extensively used as medicaments. The separation of oils both at the surface and beneath the aqueous distillate was observed, but apparently received but little attention. Owing to the practice of using alcohol in the preparation of many of these aromatic waters, the oil must frequently have remained in solution wholly or in part. Thus, e. g. the plants or plant products to be distilled were moistened with wine or *aqua vitae* before distillation; or steeped in water they were first allowed to undergo fermentation. Both alcohol and volatile oil were lost, in part at least, by submitting the plant products to a process known as circulation, a preliminary operation consisting of more or less prolonged digestion. In this manner inferior distilled aromatic waters were obtained.

---

1) Opera omnia, Lib. de vinis, p. 558.
2) Berviarum practicae, p. 1055.
3) Opera omnia, Lib. de vinis, p. 589.
4) Ph. Ulstadii, Coelum philosophorum.

Nevertheless, several of the more important experimenters and writers of that period knew and described volatile oils. In addition to the oils of turpentine and rosemary already mentioned, Arnoldus Villanovus[1]) and Raymundus Lullus[2]) describe the distillation of oil of sage; Sancto Amando[3]) that of bitter almond oil, oil of rue and oil of cinnamon; Saladinus of Aesculo[4]) that of oil of rose and oil of sandalwood. The writings of their contemporaries also reveal a knowledge of these and other distilled oils without, however, making mention of their use in medicine or the arts.

The epoch-making inventions and discoveries of the fourteenth and fifteenth centuries wrought great changes in the natural sciences and their application. The discovery of the new world and the circumnavigation of Africa to the East Indies widened the horizon of the people. The period of the Renaissance and the Reformation assisted in doing away with the blind faith in authority, not only in theology but in the natural sciences, as well. The founding of universities and the invention of printing assisted largely in the spreading of knowledge. Thus information and skill, previously the secret of the few, now soon became common property of many.

With the rise of Paracelsus, who taught that the object of chemistry was to make remedies and not gold, and the establishment of the iatrochemical school, the art of distillation was once more directed into its more proper course. More particularly, with the separation of pharmacy from medicine and the establishment of apothecary shops, the distillation of aromatic waters was carried on in the laboratories of these pharmacies and was here developed until the distillation of volatile oils became an independent industry.

As was largely the case with the Arabians, the progress of the art of distillation again finds expression in medical literature. Though medical books became much more numerous with the invention of printing, they cease largely to be a source of information with regard to methods of preparation of volatile oils and their introduction. Nevertheless, they give information about the introduction of aromatic drugs and aromatic waters. The numerous works that come under consideration may be classed in three groups of equal importance: the antidotaries and later dispensatories; the treatises on distillation, which were prominent from the close of the fifteenth to the close of the

---

1) Opera omnia, Lib. de vinis, fol. 589.

2) Experimenta nova, vol. 5, fol. 829.

3) Expositio supra Antidot., fol 328.

4) Compendium aromatariorum, fol. 349b.

sixteenth centuries; and the price ordinances for spices and drugs of various cities, which came into use about the same time.

Before discussing these works, attention should once more be called to the fact that the term "distilled" as used in ancient and mediaeval writings is not always employed in the same sense as to-day. In fact, up to and including the middle ages it was a collective term implying the preparation of vegetable and animal extracts according to the rules of the art, or rectification and separation: it involved such processes as maceration, digestion, expression, straining, filtering and even processes of fermentation and decay.

Aside from turpentine or cedar oil, the term distilled oil, as it is used in older literature, applies as a rule to fatty oils which had been aromatised with the respective plants or parts of plants. That many seeds and fruits often yielded aromatic oils upon hot or cold expression, or when boiled with water was also known in early antiquity. These aromatic and aromatised fats and fatty oils were used medicinally as ointments and for cosmetic purposes.

Whether the oils of rose, andropogon, and calamus, mentioned in the Ayur-Veda as distilled oils, were such in the modern sense of the term, cannot be decided. The same is true of the oils of spike, rosemary and sage, as well as of other oils of later writers. Although the Indians, the Babylonians, and especially the Egyptians were acquainted with the art of distillation, and also with volatile oils, a sharp distinction between true distilled oils and aromatised fatty oils does not seem to have existed at the beginning of the Christian era. Inasmuch as the aromatised oils were used principally in religious rites and for purposes of toilet, it seems natural that they were given preference over the distilled volatile oils. Indeed, the process of preparation of "distilled" oils described by Dioscorides[1]) and copied by Pliny[2]) is one of aromatization. It may well be doubted that volatile oils escaped observation by Arabian naturalists and others who distilled aromatic waters, although, as has already been pointed out, the presence of alcohol in the distillate may frequently have kept the oil in solution. Though a number of distilled oils are mentioned in various treatises and were evidently known, yet one of the oldest known lists of current drugs and spices, that of the city of Frankfort-on-the-Main of the

---

1) Matthioli Opera quae et., lib. 1, cap. 53: Germ. translation of Dioscorides' works in Trommsdorff's Jour. d. Pharm., 11, pp. 112.

2) Naturalis historiae, lib. 13, cap. 2.

Liber de arte distillandi. de Simplicibus.

Das buch der rechten kunst zů distilieren die eintzigē ding

von Hieronymo Brunschwygk/bürtig vñ wund artzot der keiserlichē fryē statt straßburg.

un getruckt durch den wohlgeachte Johannem grueninger zu Strassburg in den achte tag des meyen als man zelt von der geburt Christi funfzehnhundert. Lob sy got. Anno 1500.

Fig. 1.

year 1450, does not mention any distilled oils. However, a similar list of the same city for 1582 mentions forty-two, and another of the year 1587 enumerates fifty-nine such oils.[1])

To return, however, to the literary documents of the sixteenth century. Among the most interesting, if not the most valuable, are the treatises on distillation, the "Destillirbücher". The first larger work of this class was written by the Strassburg physician, Hieronymus Brunschwig (1450—1534), the two volumes being published in 1500 and 1507 respectively.[2]) The work describes principally the preparation and use of the much lauded distilled waters (*gebrannte Wässer*), distilled wines (*gebrannte Weine*), life elixirs, simple and mixed oils and balsams. How little attention was given to distilled oils is shown by the fact that but four distilled oils are mentioned and described, viz., the oils of spike,[3]) turpentine,[4]) juniper wood[5]) and rosemary.[6]) Directions for the rectification of turpentine oil are also given, viz., by shaking first with water, then with rose water or wine, and by final distillation. An *oleum benedictum compositum*[7]) consists of a distillate of rosemary, turpentine, olibanum, mastic, ammoniac, galbanum, opopanax, cloves and cinnamon. Directions are further given for the preparation of a number of aromatic balsams, mixtures of volatile oils, by the distillation of mixtures of oleo resins and spices with the addition of turpentine oil.[8])

How little the nature of the distillates was understood becomes apparent from Brunschwig's definition of the process of distillation. He states that it consists merely of the separation of the subtle from the crude, to make that which is fragile and destructible indestructible, to render immaterial that which is material, spiritual that which is corporeal, handsome that which is not handsome. Nevertheless he displays a considerable knowledge of the technique of distillation. For this very reason, however, it seems strange that no mention is made of the observation of oils when such aromatic plant products as the umbelliferous fruits, the labiate leaves, juniper berries, cloves, cinnamon and other spices were subjected to distillation with water. This is all the more remarkable since the volatile and even empyreumatic oils and other products of distillation such as acetic acid were regarded, like alcohol, as the quintessences of the crude materials from which they were obtained.

---

1) See the price ordinances for the corresponding years under Bibliography.

2) See fig. 1 and 2, pp. 23 and 25.

3) Hieron. Brunschwig, Liber de arte distillandi, Vol. 1, fol. 72.

4) ibidem, Vol. 1, fol. 88.

5) do Vol. 2, fol. 289.

6) do Vol. 1, fol. 52.

7) do Vol. 1, fol. 58.

8) do Vol. 2, fol. 271.

Liber de arte Distil
landi de Compositis.
Das büch der waren kunst zü distillieren die
Composita vñ simplicia/ vnd dz Büch thesaurus pauperū/ Ein schatz d armē ge/
nāt Micariū/ die brösamlin gefallen võ dē büchern d Artzny/ vnd durch Experimēt
võ mir Jheronimo brüschwick vff geclubt vñ geoffenbart zü trost denē die es begerē.

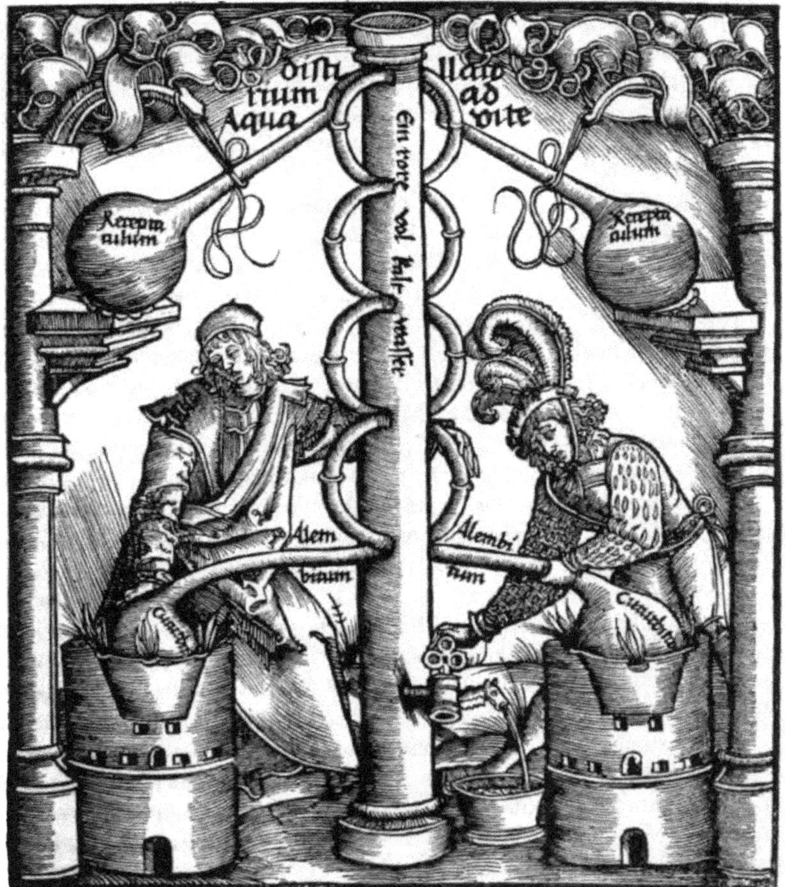

getruckt un gendigt in die keisserliche frye statt Strassburg uff sanct Mathis abent in dem jar 1507.

Fig. 2.

Brunschwig's "Destillirbuch" was followed by a number of similar treatises, all of which reveal the important position which the distilled waters held in the materia medica of the sixteenth century. To some extent they contribute to the history of the volatile oils themselves. About twenty-five years after the publication of Brunschwig's treatise, the smaller work of Philipp Ulstad, physician and professor of medicine in Nürnberg, appeared[1]) and became widely known. Ulstad held notions similar to those of Brunschwig with regard to the nature of distillates, and since his quintessences contained more or less alcohol, no mention is made of the oils themselves. The principal value which these works have at the present time as far as the history of volatile oils is concerned, lies in the thorough and careful description and figurative reproduction of the methods of distillation and utensils employed at that time.

About 56 years after the appearance of Brunschwig's "Destillirbuch" and 28 years after that of the first edition of Ulstad's "Coelum Philosophorum," Walther Hermann Reiff (Gualtherus H. Ryff), who was surgeon in Strassburg during the first half of the sixteenth century, published a third treatise[2]) of this kind which for a long time was held in high repute. His definition of distillation does not differ materially from that of Brunschwig. In the last part of the book he also describes "the correct method of preparing by means of artificial distillation several precious oils." They are distilled, some of them with wine, from myrrh, liquid storax, sagapenum, opopanax, ammoniac, storax calamita, sacocolla, benzoin, labdanum, galbanum, turpentine, mastic, sandarac, guaiac wood, rosemary, spike, anise, cloves, cinnamon, mace, safron, and from various mixtures of spices (balsams).

Under spike and lavender oil (fol. 186) he mentions that these oils are commonly imported from France in small bottles and sold at a high price. In his "Reformirte Apothek" which was published in 1563 he states (fol. 191) that "when lavender flowers are distilled a fragrant oil usually floats on the surface of the distillate. In France, in the province about Narbonne, where the plant grows abundantly it is especially distilled; likewise oils from other useful and fragrant herbs, flowers, fruits and roots." This statement is of special interest inasmuch as it seems to be the earliest reference in German literature to the French volatile oil industry which evidently dates back as far as the early part of the sixteenth century.

---

1) See Bibliography.

2) Gualth. Ryff, Neu gross Destillirbuch. See Bibliography.

How little Ryff knew about the nature of volatile oils becomes apparent from fol. 187 and 188 where he describes "how from several strong and good spices precious oils can be distilled." To prepare specially good oils from cloves, nutmeg, mace and safron these spices are to be comminuted and distilled with rectified spirit. When the "spirits" have been distilled off and oil begins to come over, the mass is to be taken out and pressed between warm plates. The oil thus obtained is to be rectified by "circulation" until it is clear.

Brunschwig's "Destillirbuch" seems to have stimulated the distillation of aromatic waters and of spirituous aromatic distillates, and their introduction into medicine, as well as the art of distillation itself. It would appear that his and other treatises of a like nature, in a measure at least, displaced the older antidotaries. However, in the course of the sixteenth century the latter were also reprinted and revised in various European cities. Among the principal treatises in this group are those of Valerius Cordus and Conrad Gesner, whose writings acquired great reputation and served as the standard for others. Valerius Cordus was born in 1515 in Simshausen. His father was professor of medicine in Marburg. Here he studied medicine, receiving the bachelor's degree in 1531. In the same year he went to Wittenberg to attend the lectures of Melanchthon, and soon received permission to deliver a course of lectures on the materia medica of Dioscorides. He died 1544 in Rome. His commentaries[1]) on Dioscorides and other scientific writings were published after his death by Conrad Gesner (1516—1565) of Zürich, a talented medical writer, who probably made additions of his own to the text of Cordus.

These "Annotationes" of Cordus are of special importance in the history of volatile oils, partly on account of the reputation of the author, partly because of his knowledge of the subject and also because they appeared in a century that was so productive of literature. Whereas Brunschwig's book reveals a retrogression in the technique of distillation as compared with the Arabian period, Ulstad, Ryff, Matthiolus, Lonicer and others made many improvements themselves and also made known many of the older pieces of apparatus that had been forgotten. Cordus and Gesner utilized their discoveries and went even farther.

In the chapter on the distillation of oils,*) Cordus discussed the nature of the "extracts" of plants obtained by expression and distillation. Concerning the oily plant extracts, Cordus distinguished

1) See Bibliography.

*) Liber de artificiosis extractionibus, fol. 226 of the Annotationes.

between the viscid, fatty oils (*oleum crassum, viscosum, terrestre*) obtained by expression, e. g. of seeds, and those of a spirituous nature (*aerea*) which can be separated from the "terrestrial" substances by distillation. As illustrations of the first class, he mentions a number of the common fatty oils, as illustrations of the second class the oils of carpobalsam,[1] cardamom, cubeb, pepper, cloves, cinnamon, mace, nutmeg, lignaloes and those of some of the common umbelliferous fruits, such as anise, fennel, caraway, cumin, angelica, *Ligustrum, Libanotus, Pastinaca, Apium, Petroselinum, Pimpinella* and *Anethum.* In his description of the properties of volatile oils, Cordus makes mention of the remarkable property of the oils of anise and fennel to congeal to a butyraceous or spermaceti-like mass; also of the property of the oils of cinnamon and cloves to sink under water. The method of distillation of volatile oils has been carefully described by Cordus, the description being accompanied by a cut of a primitive glass still constructed by himself.[2])

Of still greater value than the "Annotationes" of Cordus is the "Thesaurus Euonymi Philiatri" of Conrad Gesner. The Latin edition was possibly published as early as 1550; a German edition in 1555 under the title of "Ein köstlicher theurer Schatz des Euonymus Philiatrus." Compared with Brunschwig's treatise it not only reveals a decided advance in the technique of distillation, but the first German edition of 1555 also contains several chapters on distilled oils (pp. 212—249) and on "balsams" and other mixed oils (pp. 249—273). The distillation of a number of oils, viz.: of the oils of lavender, rosemary, rue, cinnamon, cloves, nutmeg and others, is described, and the description accompanied with cuts of the apparatus to be used; also the distillation of juniper berries and juniper wood by *destillatio per ascensum* and *destillatio per descensum.* The oils of galbanum, labdanum, myrrh, opopanax, liquid storax and *styrax calamita*, mastic and turpentine are described. The oils from guaiac wood and sandal wood and from several other woods and barks are mentioned (pp. 244—247) and their distillation described.

Gesner's notion about the nature of the volatile oils does not differ much from the traditional one mentioned in connection with Brunschwig. His practical conception of a volatile oil is also rather confused, for in describing the several methods according to which one and the same

---

[1]) The fruits of *Balsamea meccanensis* Gleditsch (*Balsamodendron opobalsamum* Kunth) were formerly known as *carpobalsamum* and used medicinally.

[2]) De artificiosis extractionibus, vol. 2, fol. 226.

"distilled" oil can be obtained, he not only enumerates distillation proper, but also the digestion with fatty oils, e. g. for rose oil (pp. 224 and 236), lavender oil (p. 337), marjoram, myrrh and other oils (p. 332). Another method applied to nutmegs, mace, etc., is to moisten with alcohol and distill until oil begins to come over. The process is then interrupted and the oil expressed with the aid of warm plates. The aromatized fatty oil thus obtained is then distilled.[1]) He also frequently directs the plant material to be moistened with alcohol. Such flowers as have a delicate fragrance are directed to be placed in layers in the still which are separated by similar layers of odorless flowers or leaves. These are to absorb part of the aroma and thus assist in imparting it to the distillate. The oil of lavender flowers he directs to be made by first distilling the aromatic water from a glass retort. This water is set aside in a warm place during the entire summer when the oil and some water will distill over spontaneously. The oil thus obtained is to be carefully separated from the water and preserved in a glass vial.[2])

Of this period, one important author remains to be mentioned, viz.: Giovanni Battista della Porta (1537—1615), a broadly educated Neapolitan nobleman. Of his works, published in twenty books,[3]) two are of special importance for the history of volatile oils, his "Liber de destillatione," and his "Liber de vinis." Porta has a clearer conception of the process and of the products of distillation than any of his contemporaries. Both books, published about 1560, reveal a more comprehensive knowledge of facts and of literature than any of their predecessors. They are further characterized by originality of investigation and presentation.

Porta distinguishes clearly between expressed fatty and distilled oils and describes the methods according to which they are prepared. He also describes the preparation of aromatic waters and the apparatus by means of which the volatile oils can be separated from the aqueous distillate. However, in spite of his practical insight, Porta still adheres to the traditional conception of the nature of volatile oils.[4]) He also retains the term oil for hygroscopic substances, such as *oleum ex salibus, ol. ex tartaro, ol. ex soda*, etc.

These sixteenth century treatises on herbs and their distillation, as well as a number of less important ones, were the principal handbooks for the preparation of medicaments, especially of distilled waters,

---

1) Ein köstlicher theurer Schatz, fol. 215—217.

2) Ibidem, fol. 222.

3) Portae, Magiae naturális. See Bibl.

4) De destillatione, p. 367.

oils, and vinous distillates. At first they supplemented the older antidotaries, later they replaced them. The gradual change from these treatises on distillation to the so-called dispensatories is marked by the appearance of several works classed with the latter. Those of Ortolff Meydenberger,[1]) and the later ones of Otto Brunfels[2]) (1488—1534), Leonhard Fuchs[3]) (1501—1566) and of W. H. Ryff[4]) (first half of sixteenth century) may here be mentioned.

With the appearance of Paracelsus (1493—1541) and the spread of his iatrochemical ideas in medicine, vegetable remedies lost their former importance, their place being taken more or less by chemical products. Thus the distilled waters had to surrender their supremacy. This resulted in a lessening of appreciation of the "Kräuter-" and "Destillirbücher." The volatile oils, however, gradually taking the place of the distilled waters, gained in importance. This change was contemporaneous with the transition from the "Destillirbücher" to the dispensatories. Although this change was brought about gradually, it is usually identified historically with the appearance of the "Dispensatorium Noricum" of Valerius Cordus in the year 1546.

While in Wittenberg, Cordus was in the habit of visiting with his uncle who from 1532 to 1560 was proprietor of the "Salomo-Apotheke" in Leipzig. Here he seems to have taken an active interest in the art of distillation and in making chemico-pharmaceutical preparations. At Ralla's instigation and with his assistance, Cordus collected tried formulas for the preparation of distilled waters and other current pharmaceutical preparations. These were published by Ralla.

This compilation, and still more his lectures on the materia medica of Dioscorides as well as his "Historia Plantarum" published in 1540, had established the fame of the young scholar. On one of his botanical excursions, Cordus appears to have stopped at Nürnberg where he received due attention in medical circles. In 1542 the council of that city charged him with the preparation of a dispensatory for the guidance of physicians and apothecaries of that municipality. This task Cordus accomplished with the aid of his uncle Ralla and of Caspar Pfreund, a friend and able apothecary at Torgau. The book was favorably received by the council of Nürnberg and was published in 1546, two years after the death of its author. Several editions appeared in rapid succession, the third Nürnberg edition bearing the

1) Arzneibuch.

2) Spiegel der Arznei; and Reformation der Apotheken.

3) Annotationes de simplicibus.

4) Reformirte deutsche Apothek.

date 1548. As an authoritative treatise, the book seems to have found general recognition. It was frequently reprinted, both in its original form and with the additions made by Conrad Gesner.

The long title of this work was abbreviated to "Dispensatorium Noricum"[1]) and it is commonly regarded as the first German pharmacopoeia, though this is not quite true.[2]) It was recognized as standard up to the close of the seventeenth century, although it had to share honors with the Augsburg Pharmacopoeia of Adolph Occo, and several other local pharmacopoeias.[3])

Notwithstanding the want of a clear understanding of the nature of the distilled oils during the whole of the sixteenth and part of the seventeenth centuries, their preparation was fostered and their use in medicine, the arts and in the household increased. Among the medical experimenters and writers of this period, Johann Winther,[3]) professor of medicine in Strassburg, seems to have distilled a large number of the more common volatile oils with great care. Moreover, the distillation of aromatic waters and volatile oils was being conducted principally in the pharmaceutical laboratories where both the process and the utensils were variously improved in the course of time.

In addition to the Nürnberg and Augsburg pharmacopoeias and similar authoritative works, the municipal price ordinances, which since the sixteenth century were issued in various cities to regulate the sale of drugs and spices, are reliable sources of information concerning the introduction of distilled oils into medicine and the arts. As documents they are of similar importance to the price lists of modern wholesale merchants and manufacturers.

The following list has been prepared with the aid of the previously discussed historical documents.[4]) It should, however, be definitely understood that the dates given are not necessarily those of the first introduction or use, but those of their legal recognition as articles of commerce.

---

1) Comp. Bibliography.

2) See Ortloff v. Bayrland, Arzneibuch.

3) See Bibliography: Pharmacopoeias of Augsburg, Antwerp, Cologne and Metz.

4) In addition to the "Destillirbücher" previously mentioned, the following pharmacopoeial works have been used in the compilation of this list: Of the "Dispensatorium Noricum" the editions of 1546, 1552, 1559, 1568, 1580, 1589, 1592 and 1612; of the "Pharmacopoea Augustana" the editions of 1580, 1597 and 1640; and the "Dispensatorium Brandenburgicum" of 1698.

Of the large number of municipal price ordinances the following were consulted: Frankfort-on-the-Main, for 1582, 1587, 1668, 1710; Nürnberg, for 1552, 1618, 1624, 1644, 1652; Worms, 1582; Strassburg, 1586; Wittenberg, 1599, 1682; Halberstadt, 1607, 1697; Halle, 1643, 1700; Ulm, 1649; Bremen, 1644, 1664; Dresden, 1652; Leipzig, 1669, 1689, 1694; Berlin, 1574; Cologne, 1628.

Oils of animal origin are not included in this list.

Distilled oils known and in use:

Up to the beginning of the sixteenth century:

The oils of benzoin, calamus, cedarwood, costus root, mastix, rose, rosemary, sage, spike, turpentine, juniperwood, frankincense, cinnamon.

To these were added:

From 1500 to 1540:

The oils of lignaloes, angelica, anise, cardamom, carpobalsam,[1]) cubeb, wild caraway, fennel, caraway, libanotis, lovage, mace, nutmeg, *Pastinaca sativa* L., pimpinella, pepper (from *Piper nigrum*), celery, sandal wood, juniper berries, juniper tar (*Oleum cadinum*), mastix.

From 1540 to 1589:

The oils of elecampane, ammoniac, horehound, anime, asafetida, basilicum, bdellium, mountain melissa (*Calaminta montana*), mountain thyme (*Thymus acinos*), amber, citrus, coriander, "costiver", dill, origanum, sweet marjoram, elemi, galbanum, galangal, guaiac, chamomile, Roman chamomile, spearmint, labdanum, lavender, lemon, spoonwort, laurel, "marum verum", marjorum, balm, mints, carrot seeds, feverfew, cumin, myrrh, cloves, opopanax, parsley, pepper (from *Piper longum*), summer savoy (*Satureja hortensis*), European penny-royal, orange peel, tansy, wild thyme, rue, rhodium, sagpenum, sandarac, sassafras, false cumin, storax, tacamahac, tar, thyme, iris, wormwood, hyssop, zedoary (root), saffron, grains of paradise.

From 1589 to 1607:

The oils of *Chaerophyllum bulbosum*, peppermint, savin, white mustard, seseli, zedoary (flowers).

From 1607 to 1652:

The oils of ginger, arbor vitae, costmary (*Tanacetum balsamita*).

From 1652 to 1672:

The oil of cow-parsnip (*Heracleum sphondylium*), cascarilla, cypress, *Anthriscus cerefolium*, *Eupatorium cannabinum*, black mustard.

From 1672 to 1708:

The oils of valerian, bergamot, mugwort, box-tree, masterwort, neroli, *Oleum templinum* (from *Pinus pumilo*).

From 1708 to 1730:

Bitter almond oil[2]) and oil of cajeput.

---

[1]) See footnote, p. 28.

[2]) Bitter almond oil and several other poisonous oils, such as cherry laurel oil were excluded from general commerce on account of their poisonous properties. Hence they do not appear in the price ordinances. Inasmuch as they were not used medicinally when they first became known, they do not appear in the pharmacopoeial treatises. Both of the above mentioned oils were known before the middle of the sixteenth century, bitter almond oil even during the middle ages.

As has already been shown, the art of distillation, particularly as applied to volatile oils, made considerable progress during the sixteenth century. The knowledge of the nature of the oils themselves, however, made but little genuine advance. During the seventeenth century the conditions for advancement were even less favorable than during the sixteenth. The thirty years' war, which affected so many European countries, well nigh crushed Germany and for almost a century stagnated the scientific and industrial life of the nation. In many instances, traditional knowledge of the arts and trades became lost. Superstition flourished and with it alchemy. During the seventeenth century there were more alchemists in Germany than during the two previous ones. The war had produced a lack of funds at the courts so that these became a productive field for cultivation by the adepts. But few of these attained practical results of any kind. Among the few was Boettger (1685—1719) who instead of transmuting baser metals into gold succeeded in making porcelain. That under such conditions science could make but little progress is self-evident. Indeed, scientists and physicians as well as all classes of intelligent society were either open or secret adherents of the theory of transmutation.

While the desire to convert the baser metals into gold once more became well nigh universal, there were a few who, away from the seat of war, cultivated science and with it the art of distillation. Among these are Joh. Baptista van Helmont in Brussels (1577—1644), Johann Rudolph Glauber in Amsterdam (1604—1668), Nicolas Lemery in Paris (1645—1715) and Wilhelm Homberg in Paris (1652—1715).

About this time salts were added to the water in the still, e. g. common salt, potash, alum, tartar. The idea was to make the water heavier and thus prevent the plants from settling to the bottom and being burnt. Possibly it was found that in some instances a larger yield of oil was obtained. Glauber also made similar use of muriatic acid.

These presumptive improvements, however, did not raise the art of distillation in general above the basis of empirical experimentation, and the seventeenth century closes without having made any material contribution to the history of volatile oils. Neither did the eighteenth century have much to add. The phlogistic theory conceived by J. J. Becher (1635—1681) and logicaly worked out by G. E. Stahl (1660—1734) failed altogether in throwing new light on the organic world and thus could not assist in a better understanding of the composition of volatile oils.

Renewed progress, however, in the manufacture and use of volatile oils is to be recorded during the eighteenth century. The technique of distillation was improved in the laboratories of the apothecary shops where the oils were largely distilled and a better product was prepared. The oils found application not only in medicine, but also in the arts and in the household. The number of oils mentioned in municipal price ordinances and other literature up to 1500 had been only thirteen; in 1540 the number had increased to thirty-four and in 1589 to one hundred and eight oils. The "Dispensatorium Noricum" of Cordus mentions only three oils in 1543; the edition of 1552 mentions five; that of 1563 six; and that of 1589 fifty-six oils. In 1708 one hundred and twenty oils are mentioned in the price ordinances of that time.

The distillation of pure volatile oils and the skill to mix them so as to produce agreeably fragrant mixtures, not only stimulated the improvement of methods of preparation, but also their use for purposes of comfort and luxury. As has already been indicated [1]) a volatile oil industry seems to have developed from small beginnings in southern France during the fifteenth and sixteenth centuries with the distillation of lavender and rosemary oils. In like manner, the perfume industry seems to have had its origin. The preparation of "Hungarian water" in the sixteenth century by making an alcoholic distillate from fresh rosemary has already been referred to.[2]) During the seventeenth century, a "Karmeliter Geist," an alcoholic distillate from balm and lavender,[3]) was introduced. In 1725 Johann Maria Farina of Cologne introduced his famous *Eau de Cologne*. The successful mixture of several odors and the prime quality of the oils used proved an important stimulus to the manufacture of these oils.[4]) From these small beginnings the perfume industry gradually developed into the important position it has held since the middle of this century.

With the increased importance of the volatile oils, more attention was bestowed upon their nature and composition. Boerhaave, who at the beginning of the eighteenth century was professor of medicine, botany and chemistry at the University of Leyden, in his treatise on chemistry states that volatile oils consist of two elements: the one cruder and resinous, insoluble in water (*mater*); the other more subtle, ethereal, which can scarcely be weighed and which by itself is possibly gaseous (*spiritus rector*). The first part he considered to be common to all oils and a unit by itself. The characteristic odor and taste,

[1]) See p. 26.
[2]) See p. 20.
[3]) See Oil of Balm.
[4]) See Oil of Spike.

however, of the various oils were due to the *spiritus rector* which was peculiar to each oil. It was water soluble and therefore gave to the distilled waters their odor, taste and medicinal virtue. The changes produced in volatile oils upon exposure to air and light were attributed, in harmony with this theory, to the escape of the *spiritus rector.*[1])

This conception was perfectly in harmony with the belief in the subtle properties and medicinal virtues of aromatic plant substances and their aqueous distillates. With the assumption of the water solubility of the *spiritus rector* the distilled waters were naturally regarded as being charged in the highest degree with the medicinal properties of the crude drugs Boerhaave's dualistic theory was therefore received as the most rational explanation of the firmly established belief in the efficacy of distilled waters, and was also accepted as a further argument for their retention in medicine. Even after the antiphlogistic nomenclature came into vogue after 1787, the *spiritus rector* was not discarded, being rebaptized as *arôme.*

The first chemists who discarded the dualistic theory of the volatile oils in their writings, and claimed that odor and taste are due to the oil as such, are T. A. C. Gren,[2]) Professor of Medicine in Halle, and the French chemist Ant. François de Fourcroy[3]) of Paris. The former exposed the untenability of Boerhaave's theory in 1796, the latter in 1798. Indeed Fr. Hoffmann (1660—1742), a contemporary of Boerhaave and professor at Halle, had not accepted the latter's theory without reserve. A many-sided investigator and writer, he had prepared and studied the volatile oils with great care.[4]) Yet he had no clearer conception concerning the preparation, yield and properties of the oils than his contemporaries. He distinguished between oils obtained by expression, by *destillatio per ascensum* and *per descensum.*[5]) He regarded sulphur as a fundamental principle of all oils, the bituminous and empyreumatic oils containing a relatively large amount of sulphur.[6]) He also believed that the color and odor of oils was influenced by their larger or lesser sulphur content.

---

1) Boerhaave; Elementa chemiae, vol 2, p. 124.

2) Grundriss d. Chemie, vol. 2, p. 217.

3) Ann. de chimie, 25, p. 282; also Systeme des connaissances chimique.

4) Opera omnia. See Bibliography.

5) The *destillatio per ascensum* corresponds to the method now generally used, allowing the vapors to pass upwards in the still and removing them from above. In the *destillatio per descensum* the vapors were forced downward through the material and collected in a receptacle underneath the still. An incomplete extraction was thus effected. (Comp. chapter 8.)

6) Opera omnia, tom 4, liber 1, p. 449—451.

It should be of interest to note that camphor which had been regarded as a volatile organic salt, was pronounced by Hoffmann to be a congealed volatile oil.[1]) He also made the observation that most of the commercial oils of his time were adulterated with turpentine oil, *oleum vini,* alcohol and fatty oils.[2]) Further he determined the yield[3]) and specific gravity[4]) of many oils. Glauber's suggestion that oils which had become colored by age be rectified with dilute hydrochloric acid[5]) was opposed by Hoffmann. He declared the employment of *spiritus salis,*[6]) dilute sulphuric acid,[7]) potash, tartar and alum[8]) in the distillation of volatile oils as useless, but consented to the use of common salt. He argued that the addition of salt facilitated the separation of the oil particles and prevented decay; that it made the water "heavier" and thus prevented the settling and burning of the plant material; that it also purified the distillate.[9])

In some instances recourse was again taken to the process of fermentation before distillation which was in vogue during the fifteenth and sixteenth centuries. This was done e. g. with juniper berries, wormwood, sage and other herbs, honey and yeast[10]) occasionally being added. The old practice of previously moistening the plant material with alcohol[11]) was also resorted to. In this manner a larger yield of oil was obtained but it would seem that the dilution of the oil with alcohol was not recognized. Downward distillation (*destillatio per descensum*) was applied by Hoffmann[12]) in the preparation of oils with high specific gravity such as the oils of cinnamon and cloves. This method was evidently regarded as being better because the dark colored oils thus obtained were supposed to contain more sulphur.

With the increased use of volatile oils during the first half of the eighteenth century it became more and more desirable not only to prepare oils of good quality but to obtain the largest possible yield as well. As guides there appeared new treatises on distillation which had

---

1) Opera omnia. Liber 72. Observatio 18, p. 44—50.
2) Ibidem. Liber 67. Observatio 2, p. 9—11.
3) " " 65. " 1, p. 1—9.
4) " " 72. " 8, p. 27—30.
5) Glauberii, Furni novi philosophici, pars 1, pp 35, 36 et 41; et pars 3, p. 30.
6) Ibidem. Pars 1, p. 36. Crude hydrochloric acid prepared by distillation of salt with alum or sulphuric acid.
7) Crell's Chem. Journ., 3, p. 30.—Pfaff, System der Materia medica (1815), vol. 4, p. 50.
8) Glauberii, Furni novi philosophici, pars 1, p. 38; et pars 3, p. 31.
9) Hoffmann, Opera omnia. Supplementum secundum. Pars 1, p. 730.
10) Berl. Jahrb. f. Pharm., 1804, p. 380.
11) Demachy, Laborant im Grossen, p. 238.
12) Hoffmann, Opera omnia, tom. 4, lib. 1, p. 449—451.—Supplementum secundum, pars 1, p. 730.

little more than the title in common with the older *Destillirbücher.* Of these the works of Burghart,[1]) Dejean and Demachy[1]) may here be mentioned.

Following the lead of Winther,[2]) Boerhaave and Hoffmann, a number of investigators of the eighteenth century ascertained the yield of oil obtainable from the more common aromatic plant products. Of these the following deserve special mention: Joh. Fr. Cartheuser[3]) (1704—1769), Professor of Medicine, Botany and Chemistry at the University of Frankfurt-on-the-Oder; Caspar Neumann[4]) (1683—1737), a Berlin apothecary; Claude Joseph Geoffroy,[5]) a Parisian apothecary; and François Rouelle[5]) (1703—1770). Their experiments were conducted on a small scale and with simple apparatus. Their results, however, published in their works and in journals,[6]) were regarded as standard and were quite generally introduced into the literature on the subject. Through the dispensatory of the English physician and chemist, William Lewis,[7]) the results of the above mentioned investigators and of others found their way into English literature.

Aside from the publications already mentioned, the interest shown in the study of volatile oils toward the close of the seventeenth and during the course of the eighteenth century is possibly best shown by the number of dissertations on the subject which were written at German universities under the stimulus of a number of university teachers. The more important ones are herewith enumerated:

1670. "De oleorum destillatiorum natura et usu in genere." Dissertatio ab David Kellner. Helmstadii.

1696. "De oleis destillatis." Dissertatio ab Henrico Rosenberg. Jenae.

1744. "De oleis destillatis empyreumaticis." Dissertatio ab Christian Lindner. Francofurti ad Viadrum.

---

1) See Bibliography.

2) See p. 31.

3) See works enumerated under Bibliography.

4) In the second volume of his Chymia medica, etc.

5) Mémoires de l'Academie Royale des Sciences de Paris, 1730—1760.

6) In 1789, Remler of Erfurt collected and tabulated the observations relative to the yield and properties of volatile oils published up to that year. A similar tabular compilation taking into consideration also the origin of the oils was published in the "Journal de pharmacie" for August 1884 by Raybaud of Paris in connection with the industrial exposition of the previous year. A German translation appeared in Buchner's "Repert. d. Pharm." for 1885, vol. 51, p. 54. Two further treatises on this subject appeared by G. H. Zeller in 1850 and 1855 respectively in the "Jahrbuch für praktische Pharmacie und verwandte Fächer." The former appeared also as a separate under the title of "Studien über ätherische Oele," Landau, 1850; the latter under the title "Ausbeute und Darstellung der ätherischen Oele." Stuttgart, 1855.

7) See Bibliography.

1744. "De sale volatili oleoso solido in oleis aethereis nonnunquam reperto." Dissertatio ab Fr. Günther. Francofurti ad Viadrum.

1745. "De oleis vegetabilium essentialibus." Dissertatio ab A. Fr. Walther. Lipsiae.

1746. "De spiritu rectore in regno animali, vegetabili et fossili, atmosphaerico." Dissertatio ab Gottfried de Xhora. Leydae.

1747. "De oleorum destillatorum usu multiplice principue in castris." Dissertatio ab Joh. Paul Ziegler. Altorfii.

1748. "Dissertatio chemica inauguralis sistens Dosimasiam concretionum in nonnullis oleis aethereis observatum" ab F. Hagen. Regiomontanae.

1752. "De oleis essentialibus aethereis eorumque modo operandi et usu." Dissertatio ab Johann Friedr. Vangerow. Hallae.

1759. "De oleis destillatis aethereis." Dissertatio ab Fr. W. Eiken. Helmstadii.

1765. "De partibus oleorum aethereorum constitutivis." Dissertatio ab Johannes Christ. Schmidtius. Jenae.

1765. "De partibus oleorum aethereorum constitutivis." Dissertatio ab J. Fr. Faselius. Jenae.

1765. "De oleis vegetabilium essentialibus, eorumque partibus constitutivis." Dissertatio ab W. B. Trommsdorff. Erfurti.

1778. "De adulterationibus oleum aethereorum. Dissertatio ab K. W. Chr. Müller. Goettingen.

The investigations reported in these dissertations, however, rest on false premises and, therefore, produced no valuable results. Research based on the phlogistic theory and the doctrines of Boerhaave and Hoffmann concerning the constitution of volatile oils, could hardly be expected to yield results of any importance. How crude the notions concerning the chemical nature of volatile oils were even at the time of Scheele, is shown in a dissertation[1]) of the year 1765 accepted by the University of Jena. From it the following propositions or conclusions are quoted:

"The essential constituents of volatile oils are of two kinds, solid and liquid. To the first class belong sulphur, phlogiston, earth and salts; to the second class air, fire and water. The presence of the first is revealed by the inflammability of the oils, for every object that burns with a flame contains much sulphur or phlogiston. The color as well as the coloration of the oil likewise argue in favor of their presence. Some oils are yellow, others green or blue; with age, the colors become darker. As is known, all coloration is due to particles of sulphur or phlogiston. Such oils have a penetrating odor, which is caused by their content of volatile saline sulphur particles. They, therefore, contain sulphur or phlogiston in sufficiently large quantities.

"In the course of time these oils are converted into a resinous mass, a change that is not conceivable without phlogiston.

---

[1]) "De partibus oleorum aethereorum constitutivis." Dissertatio inauguralis per Johannes Christianus Schmidtius. Jenae d. 30. Maerz 1765.

"Volatile oils always burn with a smoking flame. All soot, however, consists of earth, salt, water and phlogiston. When the oils are treated with nitric acid, a residue of earth and carbon remains.

"Some volatile oils have a higher specific gravity than water. This is due to their larger content of earthy constituents and salts."

The crystalline deposits formed in some oils upon standing, also the congealing of certain oils at lower temperatures, which had been observed by Valerius Cordus in 1539, by Kunkel in 1685, by J. H. Link in 1717, by Friedr. Hoffmann in 1701, by Caspar Neumann in 1719 and by others, were studied. The crystalline parts were regarded as a volatile salt, later as a camphor peculiar to each oil, at times also as benzoic acid.[1]) Hoffmann explained the congealing of oils of rose, anise and fennel by assuming the formation of a curdled modification of the oil. Neumann in 1719 and Geoffroy in 1726 regarded the crystals formed upon standing as camphor.[2]) The formation of such crystals was observed in the oils of thyme, cardamom and marjoram by Neumann;[3]) in peppermint oil by Gaubius[4]) of Leyden in 1770; in oil of mace by Wiegleb[5]) in 1774; in the oils of lavender, rosemary, sage and marjoram by Arezula[6]) in 1785. They regarded these separations as varieties of camphor, only Wiegleb thought them to be peculiar combustible salts.[7])

In 1793 and 1794 Margueran studied the action of frost on volatile oils and observed the formation of crystals and congealing in connection with a number of the more common oils.[8])

The study of the action of various reagents on volatile oils, which was begun about the middle of the seventeenth century, yielded but a superficial insight into their nature. The repeated distillation of oils over chalk or burnt lime,[9]) conducted by the excellent chemist Homberg about the year 1700, produced no results whatever. The action of strong acids had been observed by Glauber[10]) as early as 1663. The

---

1) Hagen, "Dissertatio chemica inauguralis sistens dosimasiam, concretionum in nonnullis oleis aethereis observatorum." Regiomontanae 1748.

P. J. Macquer's "Dictionnaire de Chymie." Germ. transl. by J. G. Leonhardi. Vol. 4, p. 465, foot note 9.

2) Mem. de l'Acad., 1726, p. 95.

3) De salibus alcalino fixis et camphora. Berolini 1727, p. 105.

4) Adversariorum varii argumenti liber unus. Leidae 1771. Sectio 7, p. 99—112

5) Vogel's Lehrsätze der Chemie, § 842.

6) Resultato de las experiencas hachas sobre alcanfor de Murcia con licencia. En Segovia 1789.

7) Comp. Vogel's Lehrsätze der Chemie, edited by Wiegleb.

8) Jour. de chim. et de phys., 2, p. 178; Crell's Chem. Ann., 2, pp. 195, 310 and 480.

9) Mém. de l'Acad., 1700, p. 298; and 1701, p. 129; also Chem. u. botan. Abhandl., vol. 3, p. 155—157.

10) Prosperitas Germaniae.

effect of strong nitric acid on a number of distilled oils was studied by Borrichius[1]) in 1671, by Tournefort[2]) in 1698, by Hasse[3]) in 1783; that of sulphuric acid by Kunkel[4]) in 1700 and by Homberg[5]) in 1701. A more detailed study of the action of strong acids on volatile oils was made by Hoffmann[6]) and by Geoffroy[7]) in 1726 and by Rouelle[8]) in 1747. Upon distillation of oils with strong hydrochloric acid, especially if the acid was generated in an almost anhydrous condition in the experiment, it was supposed that compounds of the oil with the acid were obtained. Such a supposed compound was known to Homberg[9]) as early as 1709. The preparation, however, of such a compound of definite chemical composition was first accomplished by Kind,[10]) an apothecary in Eutin, in 1803 by the action of hydrogen chloride gas on turpentine oil.

The solubility and color of distilled oils also received attention during the eighteenth century. Thus Macquer[11]) in 1745 published his investigations on the solubility of distilled oils in alcohol, which were the most extensive on this subject. The color of oils and the changes in color were studied by Homberg[12]) in 1707 and by Bindheim[13]) of Moscow in 1788. The latter arrived at the conclusion that the color depends on a larger or lesser amount of resin carried over in the process of distillation, hence the darker colored oils are apt to contain considerable resin.

As has already been pointed out, the phlogistic theory afforded no satisfactory basis for the study of organic substances and consequently of volatile oils. With the discovery of oxygen by Scheele and Priestley during the years 1771[14]) to 1774 and the ingenious interpretation of this and other discoveries by Lavoisier with the aid of the balance, a reaction against the phlogistic theory set in which resulted in the inauguration of the present chemical period. The study of the chemical constitution of substances was placed on a rational scientific basis. Inorganic chemistry, having to deal with the simpler substances,

---

1) Acta med. et phil. Hafn., 1671, p. 188.

2) Hist. reg. scient. acad., p. 495.

3) Crell's Neueste Entd. in d. Chem., 9, p. 88; Crell's Chem. Ann., 1, p. 417.

4) Laboratorium chymicum, p. 847.

5) Chem. bot. Abh., 1. p. 720.

6) Observat. phys.-chim., lib. 3, p. 123.

7) Mém. de l'Acad., 1726, p. 95.

8) Ibidem, 1747, p. 45.

9) Chem. bot. Abh., 3, p. 155.

10) Trommsdorff's Journ. d. Pharm. 11, p. 132.

11) Mém. de l'Acad., 1745, p. 4.

12) Chem. botan. Abh., 3. p. 155.

13) Crell's Chem. Ann., 1788 II., pp. 219 and 248.

14) A. E. von Nordenskiöld, Scheele's Nachgelassene Briefe und Aufzeichnungen, pp. xxi, 86, 408, 458 and 466; Pharm. Rundschau, 11, pp. 28 and 48.

profited first by the new theories of the opponents of the phlogistic school. Organic chemistry, and with it the study of the volatile oils, were benefited somewhat later.

Though of little consequence, the experiments of the Dutch chemists Deimann, Troostwyck, Bond and Lauwerenburg[1]) should here be mentioned. They passed the vapors of volatile oils through red hot iron tubes and examined the resulting gases. At the same time they made the bold attempt to synthesize oils by the action of gaseous hydrogen chloride on olefiant gas.

The first investigation suggested by the new theories that was of positive value, was the elementary analysis of turpentine oil made by Houtton-Labilliadière.[2]) He found the ratio of carbon to hydrogen to be five to eight, the same that was later established for all hemiterpenes, terpenes, sesquiterpenes and polyterpenes.

Attention has already been called to the crystalline deposits that had been observed in the course of several centuries. These were mostly considered as identical with ordinary camphor because like it they were volatile, soluble in alcohol and fatty oils, and burned with a smoky flame. Only in a few instances, however, had these deposits been proven to be identical with camphor. Berzelius,[3]) therefore, argued against the indiscriminate generic use of the term camphor. In its place he suggested the use of the term stearoptene (from στέαρ, tallow, and πτήνον, volatile). He pointed out the analogy existing between volatile and fatty oils in so far as they can be a mixture of several oils having different congealing points. Thus oils may, under favorable circumstances, be separated into an oil which is solid at ordinary temperature, the stearoptene, and one which is liquid at low temperatures, the elaoptene (from ἔλαιον, oil, and πτήνον, volatile). The result of this was that the solid deposits from volatile oils were thereafter designated alternately stearoptene as well as camphor. Up to this day the older abuse of the term camphor has not ceased as becomes apparent from such words as cedar camphor, cubeb camphor, juniper camphor, etc. Soubeiran and Capitaine[4]) even made things worse by applying the term "liquid camphor" to the liquid hydrogen-chloride addition products of the terpenes. After it had been shown that true camphor contained oxygen, the term camphor in its generic sense was also applied to other oxygenated constituents of volatile oils though they were liquid.

---

1) Journ. de chim. et de phys., 2, p. 178; Crell's Chem. Ann., 2, pp. 195, 310 and 430.

2) Journ. de pharm., 4, p. 5.

3) Lehrb. d. Chemie [3], vol. 6, p. 580.

4) Liebig's Annalen, 34, p. 311.

In 1833 Dumas published an article entitled "Ueber die vegetabilischen Substanzen, welche sich dem Campher nähern und über einige ätherische Oele."[1] Although a number of important observations of rather striking properties of individual oils had been made, the systematic study of the volatile oils may be said to have begun with the analysis of a number of stearoptenes by Dumas. He suggested the following classification of volatile oils:

1.) Those that consist of carbon and hydrogen only, like turpentine oil and oil of lemon;

2.) Those that contain oxygen, like camphor and anise oil;

3.) Those that contain sulphur,[2] like mustard oil, or nitrogen, like oil of bitter almonds.

The elementary analysis of solid peppermint oil, camphor and solid anise oil revealed the composition $C_5H_{10}\frac{1}{2}O$, $C_5H_8\frac{1}{2}O$ and $C_5H_6\frac{1}{2}O$. By doubling these formulas of Dumas the modern formulas for the respective substances are obtained. Of the oxygen free oils, he analyzed turpentine oil and the hydrocarbons of lemon oil, verifying the earlier results of Houtton Labilliardière. During the years 1833—1835, Dumas published further contributions on the subject of volatile oils, several jointly with Pelouze and Peligot. They pertain to artificial camphor (pinene hydrochloride), mustard oil, cinnamon oil, terpin hydrate, orris oil, pepper oil, oil of juniper berries and others.

Almost simultaneously with the first publications by Dumas, Blanchet and Sell[3] published the results of their investigation which had been carried out in Liebig's laboratory and which involve in large part the same substances studied by Dumas. The most noteworthy result of these investigations is the recognition of the identity of the stearoptene from fennel oil and that from anise oil.

Several years later, in 1837, the highly important and very interesting results of Liebig and Woehler's work on bitter almond oil were published.[4] As early as 1802 Schrader and Vauquelin had discovered hydrocyanic acid in the distillate of bitter almonds. In 1822 Robiquet showed that no volatile oil preexisted in the almonds, and with Boutron-Charlard he had prepared amygdalin in 1830. They had not succeeded,

[1] Liebig's Annalen, 6, p. 245.

[2] The fact that mustard oil contains sulphur was recognized by Thibierge in 1819 (Journ. de pharm., 5, pp. 20, 439 and 446; Trommsdorff's Neues Journ. d. Pharm., 4^{II}, p. 250.) That sulphuretted hydrogen is given off during the distillation of several umbelliferous fruits, such as caraway, dill, fennel, etc., was pointed out by L. A. Planche of Paris in 1820. (Trommsdorff's Neues Journ. d. Pharm., 7^{I}, p. 356.)

[3] Liebig's Annalen, 7, p. 154.

[4] Ibidem, 22, p. 1.

however, in preparing bitter almond oil from amygdalin. That this is decomposed by emulsin into benzaldehyde, hydrocyanic acid and sugar was demonstrated by Liebig and Woehler. They also point out that the manner of formation of mustard oil must be closely related to that of bitter almond oil, for the mustard seeds deprived of their fatty oil possess no odor, this being produced only when water is present. The investigation of mustard oil by Will[1]) in 1844 substantiated this supposition.

Chemists now became especially interested in the action of hydrogen chloride on various terpenes and in the resulting hydrochlorides, some of which were solid, others liquid; also in the study of terpin hydrate and its decomposition products. The study of the literature pertaining to these subjects is rendered difficult by the error of regarding mixtures of several substances as chemical individuals and describing them as such; further by the fact that almost every author, irrespective of the work of others, coined a nomenclature of his own. This confusion continued until very recently when Wallach and his disciples created order.[2])

Crystalline pinene monohydrochloride had been discovered by Kindt,[3]) an apothecary, in 1802. He regarded it as artificial camphor, a view shared by Trommsdorff.[4]) The true composition of this compound was ascertained by Dumas in 1833. Crystalline dipentene dihydrochloride was discovered by Thenard in 1807. It is the "salzsaures Citronenöl," muriate of lemon oil, of Blanchet and Sell, the artificial lemon camphor of Dumas. These and similar substances were investigated by Soubeiran and Capitaine (turpentine oil), Deville (turpentine oil and elemi oil), Schweizer (carvene) and Berthelot (turpentine oil).[5]) The formation of terpin hydrate and the action of acids on this substance was studied principally by Wiggers, List, Deville and Berthelot.[6])

---

1) Liebig's Annalen, 52, p. 1. A more complete insight into the mechanism of the reaction by which mustard oil is produced was supplied by the later investigations of Will and Koerner in 1863. (Liebig's Annalen, 125, p. 257.) See Oil of Mustard.

2) The historical development of this chapter of the chemistry of the terpenes is described in "Terpene und Terpenderivate, ein Beitrag zur Geschichte der ätherischen Oele" by E. Kremers (Pharm. Rundschau, 9, pp. 55, 110, 159, 217, 237; and 10, pp. 10, 31, 60; also Proc. Wisc. Acad. Sc. Arts and Letters, 8, pp. 312—362.

3) Trommsdorff's Journ. d. Pharm., 11 II, p. 132.

4) Ibidem, p. 135.

5) Of later investigators of this subject Oppenheim (1864), Hell and Ritter (1884), Bouchardat and Lafont (1886), and finally Wallach (1884—1887) may be mentioned.

6) The same subject was later investigated by Oppenheim (1884), Flawitzky (1879), Tilden (1878—79), Bouchardat and Voiry (1887). Here also Wallach's exact investigations revealed the fact that different acids, as well as the same acid in different degrees of concentration, produce different results.

A paper published about this time (1841) by Gerhardt and Cahours[1]) is of special interest in so far as it contains a definition of a volatile oil which in a general way holds good to-day. It also makes known new methods of investigation. About oils in general the authors state:

"There are, indeed, but very few oils which can be crystallized; most oils are liquid and consist of a mixture of two and even three peculiar substances, which rarely are obtained by themselves when distilled at different temperatures."

The separation of the individual substances is effected by first allowing any solid constituent to crystallize out, then the lower boiling hydrocarbon is isolated by distillation at a temperature 20—30° below the boiling point of the crude oil.[2]) Inasmuch, however, as the hydrocarbon cannot be completely freed from oxygenated constituents in this manner it is treated with fused alkali. The oxygenated constituents also are subjected to like treatment with fused alkali, and cumin oil is thus made to yield cuminic acid, oil of valerian valerianic acid.

Strong reagents are also employed by Rochleder, Persoz, Laurent and Gerhardt in order to obtain an insight into the nature of volatile oils. They oxidized either the entire oil or fractions thereof with chromic acid or nitric acid. Their investigations included the oils of valerian, sage, anise, staranise, fennel, cumin, cinnamon, tansy and estragon. The conclusions drawn from these oxidation experiments were in part correct, in part wrong. Thus e. g. Gerhardt pointed out the identity of dragonic acid, obtained from estragon oil, with anisic acid, and claimed that estragon oil and anise oil were absolutely identical. This conclusion was wrong, for the anethol of anise oil is paramethoxypropenylbenzene, whereas the formation of anisic acid from estragon oil is due to the presence of paramethoxyallylbenzene.[3])

This method, however, rendered it impossible to decide whether a substance obtained after the oxidation preexisted in the oil or not. Thus camphor was found in several oxidized oils and was regarded as an original constituent although, as was the case in the oils of valerian and sage, it had resulted from borneol. Persoz, however, seems to have

---

[1]) Liebig's Annalen, 38, p. 67.

[2]) Fractional distillation, however, was previously employed in the examination of volatile oils. As early as 1888 Walter had subjected peppermint oil to interrupted distillation, "gebrochene Destillation" (Gmelin, Handbuch d. Chem. [4], vol. 7a, p. 404). In 1840 Völckel (Liebig's Annalen, 35, p. 306) speaks of "fractional distillation." Already Blanchet and Sell in 1833 had applied fractionation with water vapor as a means of separation and had found that the first fraction of lemon oil boiled at 167°, the last fraction at 178°.

[3]) This difference was first ascertained in the laboratory of Schimmel & Co. (Bericht S. & Co., April 1892, p. 17) and verified by Grimaux in 1893 (Compt. rend., 117, p. 1189).

had doubts as to the reliability of these conclusions, for he leaves it undecided whether the camphor obtained from oil of tansy was contained in the oil or not. As a matter of fact, tansy oil contains camphor as an original constituent,[1]) this being less readily attacked by the oxidizing agents than the other constituents of the oil.

Of considerable importance in the farther development of the chemistry of volatile oils are the investigations of Berthelot from 1852 to 1863, which involve principally the hydrocarbons contained in these oils. He studied first of all the hydrocarbon of turpentine oil[2]) and its isomers and polymers obtained from its hydrochloride. By heating pinene hydrochloride with barium stearate or sodium benzoate he obtained a new hydrocarbon which he regarded as "camphene proper"[3]) and which is identical with the camphene of to-day. This new camphene was either dextrogyrate, laevogyrate or optically inactive according to the turpentine oil employed. Berthelot, therefore, distinguished between the following hydrocarbons:

1.) *Terebentene* (l-pinene) from French turpentine oil, laevogyrate,[4]) b. p. 161°. It yields a laevogyrate monhydrochloride, also under proper conditions an inactive dihydrochloride (dipentene dihydrochloride).

2.) *Terecamphene* (l-camphene) from terebentene hydrochloride, laevogyrate, m. p. 45°, b. p. 160°. With hydrogen chloride it forms a dextrogyrate hydrochloride.

3.) *Australene* (d-pinene) from American turpentine oil, b. p. 161°, dextrogyrate like its hydrochloride. Its behavior to hydrogen chloride is analogous to that of terebentene.

4.) *Austracamphene* (d-camphene) from australene hydrochloride. It corresponds to terecamphene.

5.) *Inactive camphene* (i-camphene) can be obtained by proper treatment from the hydrochloride of terebentene as well as that of australene.

6. *Terebene*,[5]) b. p. 160°.

---

1) Bericht von S. & Co., Oct. 1895, p. 84.

2) Compt. rend. 55, pp. 496 and 544; also Liebig's Annalen, Suppl. II, p. 226.

3) Soubeiran and Capitaine in 1840 had applied the term camphene to all hydrocarbons C_5H_8 (Liebig's Annalen. 84, p. 311).

4) The rotatory power of volatile oils was first observed by Biot in 1817 in connection with French oil of turpentine (Mém. d. l'Acad. des Sc., 18), later also with oil of lemon. The turpentine oil was shown to be laevogyrate, the oil of lemon dextrogyrate. In 1848, Leeson of London found that American turpentine oil possessed a rotatory power opposite to that of the French oil. This observation was soon after verified by Pereira and Guibourt. Pareira introduced the terms laevo-gyrate and dextro-gyrate. (Pharm. Journ., 5, p. 70.)

5) This substance which was considered a chemical unit by Berthelot was shown by Riban to be a mixture of a terpene, cymene and camphor. Power and Kleber in 1894 found camphene, dipentene, terpinene and cymene in terebene. (Pharm. Rundsch., 12, p. 16.)

These six hydrocarbons are isomeric and have the formula $C_{10}H_{16}$. With these the following are polymeric:

1.) A liquid hydrocarbon boiling at 250° which is probably a sesquiterebene, $C_{15}H_{24}$.

2.) Diterebene (Deville's colophene) $C_{20}H_{32}$, an inactive liquid boiling at about 300°.

3.) Several polyterebenes $C_{10n}H_{16n}$, optically inactive liquids becoming more and more viscid, which boil between 360° and a dark red heat.

After a discussion of the methods of formation of the individual hydrocarbons, Berthelot continues:

In accordance with the known facts, the hydrocarbon $C_{10}H_{16}$—e. g. terebentene—may be regarded as the starting point of two series:

1.) Of a monatomic or camphol[1]) series (monohydrochlorides or chlorine esters of camphol, $C_{10}H_{17}Cl$; camphene, $C_{10}H_{16}$: camphol alcohols, $C_{10}H_{18}O$);

2.) Of a diatomic or terpil series (dihydrochlorides, $C_{10}H_{18}Cl_2$; terpilene, $C_{10}H_{16}$; hydrate $C_{10}H_{20}O_2$).

Each of these two series constitutes a larger group, which can be divided into secondary series (australene, terebentene, etc.) the parallel and isomeric members of which occur in twos; each has as type an inactive hydrocarbon, namely camphene in the first group, terpilene in the second.

A similar, but much less detailed classification was attempted by Gladstone[2]) in 1864, after having determined the specific gravity, the index of refraction and the optical rotation of a number of oils. By means of fractional distillation he isolated the hydrocarbons of various oils, rectified them by distillation over sodium and arranged them into three large groups:

1.) Hydrocarbons of the formula $C_{10}H_{16}$ which boil between 160—170°;
2.) Hydrocarbons of the formula $C_{15}H_{24}$ which boil between 249—260°;
3.) Colophene, $C_{20}H_{32}$, b. p. 315°, representing the third group.

About this time the word terpene was introduced, evidently by Kekulé.[3]) His investigations, preceded by those of Barbier and Oppenheim were of considerable importance, inasmuch as by revealing the relations between the terpenes and cymene, they threw new light on the molecular structure of these hydrocarbons. Almost simultaneously Barbier[4]) (1872) and Oppenheim[5]) (1872) obtained cymene by heating the dibromide of terpin either by itself or with aniline. By the action

---

[1]) Berthelot in 1858 changed the name borneol to camphol. Liebig's Annalen, 110, p. 368; from Compt. rend., 47, p. 266.

[2]) Jour. Chem. Soc., 17, p. 1. A second contribution appeared eight years later. Ibidem, 25, p. 1.

[3]) Lehrbuch der organischen Chemie (1866), vol. 2, p. 487: ".... on the other hand the oil of turpentine and numerous isomeric hydrocarbons, which may be designated by the generic term terpenes."

[4]) Compt. rend., 74, p. 194.

[5]) Berichte, 5, p. 94.

of iodine on turpentine oil, Kekulé[1]) (1873) obtained the same hydrocarbon. He, therefore, thought himself justified in supposing six atoms of the turpentine oil to be arranged in similar manner as in benzene. Further that the methyl and propyl groups in the turpentine oil occupy the same relative positions as in cymene.[2]) This view of the constitution of the terpenes was the predominant one for a long time. With this the question of the constitution of the terpenes had its origin. Important in this direction was also the synthesis of a terpene—the polymerization of isoprene to dipentene—by Bouchardat[3]) in 1875.

In the same year, Tilden[4]) found that the hydrocarbon of turpentine oil combines with nitrosylchloride to form a well crystallizing compound. Together with Stenhouse, he applied this reaction to the terpenes from the oils of sage, orange, lemon and bergamot. On the behavior of these hydrocarbons to nitrosylchloride he based a new classification, concerning which he makes the following statement:

The natural terpenes are colorless limpid liquids which vary in specific gravity from about 0.84 to about 0.86. They are divisible into two groups as follows:—

1. Turpentine group: b. p. 156° to 160°; m. p. of nitroso-derivative 129°; form solid crystalline hydrated terpin $C_{10}H_{20}O_2H_2O$.
2. Orange group: b. p. 174° to 176°; m. p. of nitroso-derivative 71°; form (by Wigger's process) no solid crystalline terpin hydrate.[5])

The liquids included in each group are allotropic modifications of the same hydrocarbon distinguished one from another by their varions rotatory action on the polarized ray. It will, however, be found I believe that the terpenes from several different plants will on further examination be conclusively proved to be really identical and not simply isomeric. This, I believe, to be the case with the terpenes from French turpentine and sage, also with the terpenes from orange peel, bergamot and lemon."

Tilden's prediction, that the number of terpenes would be shown to be much smaller than assumed in his days, has proven itself true. His classification, however, was insufficient, for it included only a small number of terpenes. Indeed the material at hand was not sufficiently sifted for an attempt of that kind. It consisted of a large number of disconnected observations, the study of which was rendered difficult by an arbitrary nomenclature. Only by a systematic exploration of this disorderly realm could a clear insight into the subject be gained.

---

1) Berichte, 6, p. 437.
2) Kekulé's formula for camphor was based on the same consideration.
3) Compt. rend., 80, p. 1446.
4) Jour. Chem. Soc., 28, p. 514; Ibidem, 31, p. 554; Pharm. Journ., 1877, p. 191.
5) This statement is incorrect for dipentene and limonene likewise produce terpin hydrate. Comp. Flückiger, Arch. d. Pharm., 222, p. 362; also Kremers, Am. Chem. Journ., 17, p. 695.

That we are able to-day to distinguish sharply between so many terpenes and their derivatives is due primarily to the excellent experimental researches of Otto Wallach, the founder of modern terpene chemistry.

Inasmuch as it was impossible to isolate the numerous terpenes boiling between 155 and 185° by fractional distillation, methods had to be sought which enabled the characterization of these hydrocarbons, even in mixtures, by means of crystalline derivatives. Only after the characterization of the numerous isomers was accomplished was it possible to study successfully the relation of one terpene to another, the relation of the terpenes to their oxygenated derivatives, and the problem of their constitution.

These problems have been solved in so far that it is now possible to identify many if not most terpenes without great difficulty. The inversions, or changes from one to the other, are also better understood. The problem of their constitution, however, is still far from a satisfactory solution, though structural formulas have been proposed for a number the terpenes.

In 1884 Wallach[1]) began his researches on this subject with the investigation of the oil of wormseed (*Oleum cinae*). Three years later he was in a position to characterize eight terpenes by means of crystalline derivatives (tetrabromides, hydrochlorides, hydrobromides, nitrosites etc.), viz. pinene, camphene, limonene, dipentene, sylvestrene, terpinolene, terpinene and phellandrene. To these fenchene was added later. The sesquiterpenes were also included in his investigations as well as the oxygen derivatives of the terpenes with their greater capacity for reaction, viz.: cineol, terpin hydrate, terpineol, pinol, fenchone, thujone, pulegone, menthone, carvone, methyl heptenone, patchouly alcohol, caryophyllene hydrate and guajol.

After Wallach had removed the principal difficulties in the investigation of volatile oils, other chemists also developed a successful activity. A. v. Baeyer's valuable investigations into the constitution of the terpenes and related compounds appeared in the Proceedings of the German Chemical Society since 1893. Whereas Wallach and v. Baeyer

---

1) Wallach's contributions are to be found in the following volumes of Liebig's Annalen: 225, 227, 230, 238, 239, 241, 245, 246, 252, 253, 258, 259, 263, 264, 268, 269, 270, 271, 272, 275, 276, 277, 278, 279, 281, 284, 286, 287, 289, 291, 296, 300, 302, 305, 306, 309. Several papers from his pen also appeared in the Proceedings of the German Chemical Society since 1890. Of the latter, his lecture before this society in 1891 in which he presented a general survey of his work up to that date should receive special mention (Berichte, 24, p. 1525).

investigated primarily the cyclic compounds, Semmler paid special attention to chain compounds. He showed that the alcohols geraniol and linalool and the aldehydes citral and citronellal, which occur frequently in volatile oils, are chain compounds; also that they, like the more or less closely related cyclic compounds, can be converted into cymene.[1]) Other chapters of the chemistry of volatile oils were included in his researches,[2]) some of which were conducted jointly with Tiemann.[3]) The investigations by the latter and Krüger of orris oil and irone, led to the discovery of ionone, the artificial perfume of violets.

In addition to these, numerous other chemists have investigated the composition of volatile oils and the constitution of their constituents. In so far as their results pertain to the composition of volatile oils, their results are included in the special part of this treatise.

The constitutional problems involved in the study of the terpenes and their derivatives are among the most difficult of organic chemistry because of the readiness with which rearrangements within the molecule take place under the influence of ordinary physical and chemical reagents. A typical example of some of the difficulties involved may be had in ordinary camphor. Although investigators had an almost unlimited supply at their disposal and in spite of the fact that many chemists have been incessantly investigating this compound for several decades in order to obtain an insight into its constitution, the views held with regard to its structure[4]) are still divided.

This brief history of volatile oils would not be complete without some reference to the literature on the subject since the beginning of this century. So long as the volatile oils were principally prepared in the laboratories of apothecary shops, the description of the oils and of the methods of their preparation was found in pharmaceutical reference works and pharmacopoeial commentaries. The results of scientific and technical studies were also published in pharmaceutical rather than in purely chemical journals. With the early forties the preparation of volatile oils was taken out of the pharmaceutical laboratories and a separate literature was created.

---

1) The reverse process, the conversion of the cyclic menthone into chain compounds of the citronellal series was accomplished by Wallach in 1897. (Liebig's Annalen, 296).

2) The oils of *Allium ursinum*, asafetida, garlic, onion; also tanacetone (thujone), menthone, pulegone, terpineol, etc.

3) Bergamot oil, lavender oil, citral, pinene, methyl-heptenone, tanacetone, etc.

4) Not less than two dozen structural formulas have been proposed for camphor in the course of these investigations.

Zeller's*) publications (1850—1855), already mentioned on p. 37, contain a compilation of the yield, also a very meagre description of the physical properties of the oils and their behavior toward reagents. Maier's*) treatise (1867) also takes into consideration the scientific investigations. The methods of preparation and the subject of distillation are described in detail by Mieržinski*) (1872). A similar work was written by Askinson*) in 1876. "Die Toiletten-chemie" of Hirzel*) which passed through four editions, also "The art of perfumery" by Piesse,*) which was translated into several languages, may here be mentioned.

The results of the earlier papers by Wallach are contained in the excellent work of Bornemann*) (1891), whereas the "Odorographia" of Sawer*) emphasizes the botanical side of the subject. Finally the monograph by Heusler*) on the terpenes should be mentioned for it has proven itself well nigh indispensable for the scientific investigation of terpenes and their derivatives.

Although remarkable progress has been made in the chemistry of the volatile oils during the past fifteen years, the investigations of this branch of science are far from being completed. On the contrary, the newly acquired knowledge has caused an ever-increasing number of problems to present themselves for solution. It is quite remarkable, however, that the interest bestowed upon purely theoretical questions has caused a falling off in the systematic analytical study of volatile oils. There are still a considerable number of oils that are much used and readily accessible, the composition of which is but imperfectly known. Even many oils that have been repeatedly examined will be made to yield still other constituents when carefully reexamined. Plant physiological problems also have scarcely been touched upon, and the assay of volatile oils is just beginning to assume a modern scientific aspect in harmony with recent theoretical results. Hence there still remains a large field for profitable work.

It may well be said, therefore, that in spite of the unprecedented progress in recent years, the chemistry of volatile oils is but in its initial stages. Following the paths of Wallach and of other reformers, chemical investigators of all civilized countries are busily at work opening up new fields for investigation. Science and industry are working hand in hand and it may reasonably be expected that both will profit greatly by the results of the incessant labor thus fostered and stimulated.

---

*) See Bibliography.

3. HISTORY OF THE METHODS OF DISTILLATION AND OF DISTILLING APPARATUS.

The literary documents considered in the previous chapter, reveal the slow process of evolution leading up to a better understanding of the subject of volatile oils. In like manner, a short historical retrospect of the methods of distillation and distilling apparatus may result in an insight into the gradual development of the art of distillation and the methods of preparation of distilled oils. The history of the evolution from the primitive *Cucurbita*, the *Alembic* and the *Berchile* to the steam and vacuum appåratus of our own time reveals a long and varied course which had to be followed by this apparently modern branch of industry in order to bring it to its present technical and scientific perfection.

Fig. 8.

The first definite statement found in ancient writings which indicates a kind of primitive distillation, although probably not pictured until the middle ages, is the mention of the method for obtaining oil of cedar (*πισσέλαιον*) in the writings of Herodotus,[1] Dioscorides[2] and Pliny.[3] This oil is said to have been obtained from the oleoresin by boiling with water in an open earthen kettle. The oil either collected at the surface of the liquid and was removed, or its vapors were condensed in layers of wool spread over sticks of wood laid crosswise

1) Historiae, lib. 2, p. 85.
2) De mat. med., lib. I, 84, 89, 80.
3) Hist. nat., lib. 15, cap. 6—7; lib. 16, cap. 22.

on the kettle as in figure 3. The wool was replaced from time to time by fresh portions and the saturated wool expressed with the hands.

Authentic representations of the distilling vessels used by the Egyptians do not exist. Some of their forms of apparatus were undoubtedly adopted by the Arabians and improved by them. To the oldest known writings which give information on methods of distillation and distilling apparatus belong those of the Greek physician Dioscorides[1]) and of the Greek philosopher Zosimos.[1])

Fig. 4.

Fig. 5.

In a manuscript Arabian translation of Dioscorides' Materia Medica in the library at Leyden distilling ranges and apparatus are mentioned and described. These descriptions probably occur in the original Greek text. Among them are found the *cucurbita* and the *alembic*.[2])

Fig. 6. Fig 7.

Just as pictures of animals have served as symbols in the oldest mythology and as characters in writing of the earliest peoples, the forms of animals were used by the ancients as prototypes in the making of jewelry, and of all kinds of useful articles and apparatus. The same

[1]) Compare Bibliography.

[2]) Extracts from this as well as the much later Arabic writings of Rhases and an unimportant illustration of an Arabic distilling apparatus were published in 1878 by Prof. Wiedemann in the Ztsch. d. deutsch. morgenländ. Gesell., 82, p. 575.

seems to have occured in the preparation of primitive digestion and distilling vessels. Such pictorial representations have been carried over from the writings of Zosimos and probably also of others into the writings of the Arabians and from these into other alchemical works of the middle ages.[1])

As prototype of a common flask the figure of an ostrich is given (fig. 4); as that of a retort a goose (fig. 5) or a pelican (fig. 6). The shape of a bear served for a still (*cucurbita*) and head (*alembicus*) (fig. 7). An improved form of this simple distilling apparatus is found in the writings of Geber[2]) and Albucasis.[3]) The latter not only described glass distilling vessels but

Fig. 8.

Fig. 9.

also those prepared of glazed earthenware (fig. 8) and a kind of fractional distillation for the purpose of a better condensation and separation of subtle spirits by placing several *alembices*[4]) on top of one another (fig. 9).

From the writings of Geber and Albucasis, also from those of the excellent physician and writer Rhases (El RAzi), it becomes apparent that the Arabians distinguished as early as the eighth century between distillation over open fire and from a water bath and ash bath.[5]) Geber described both methods in detail.[6])

For the purpose of better condensation, Costaeus of Lodi,[7]) a physician and alchemist of Bologna, recommended that the beak of the

1) Joannis Rhenani. See Bibliography.
2) Summa perfectionis magisterii.
3) See Bibliography.
4) Albucasis, Liber servitoris, lib. 27, p. 247.
5) Rhases, Das Buch der Geheimnisse.
6) Summa perfectionis mag., cap. 50.
7) See Mesuë, Simplicia et composita.

alembic be cooled by water (figs. 10 and 11), also that the distillate be improved by the use of the water bath (*balneum Mariae*) (fig. 12) and the sand bath (*balneum arenae*) (fig. 13). Among the writings left by the Arabians, the work of Albucasis previously mentioned probably contains one of the first and most striking descriptions of the manner of distillation and distilling apparatus. From the fourteenth century on the practice of making distilled liquors increased very considerably. As a result the methods of distillation and the

Fig. 10.

Fig. 11.

distilling apparatus, especially those parts employed for the condensation of the vapors, were greatly improved. The method of con-

Fig. 12.

Fig. 13.

densation of the vapors, already well known to the Arabians, of passing the straight or bent tube of the alembic, or an elongation of the same wound into a spiral (wormtube, *serpentina*) through cold water was already in general use at that time for the distillation of wine and fermented plant juices. As examples of such distilling apparatus and methods, "die mancherley Kühlungen der Teutschen und Welsche Weinbrenner" are described and illustrated in treatises on distillation of the first half of the sixteenth century, namely in those of Brunschwig, Ulstad, Ryff and Lonicer. In these a distilling apparatus constructed with considerable skill is described. The helm of the still and the outer condenser jacket were made of sheet copper. The form of the headlike expansion of the helm with the outer jacket, the lower open rim of which was tightly luted to the still, gave rise to the name of "Mohrenkopf." The condensation was effected by a continuous flow of cold water through the outer jacket (fig. 14).

Fig. 14.

The method of condensation derived from the Arabians was considered as the most perfect for the distillation of spirit of wine (*aqua vitae*). The illustration of this apparatus was selected for the title page of the second volume of Brunschwig's Destillirbuch published in the year 1507 and is reproduced on page 25.

Where the two upright serpentine connecting tubes (*serpentinae*) between the retorts (*cucurbitae*) and the receivers (*receptacula*) cross each other they pass through a condensing tube filled with cold water. The cooling effect thus produced is not sufficient for the condensation of all the vapor. The worm acts therefore, as a dephlegmator and increases the alcoholic strength of the distillate. This is emphasized correctly by Brunschwig.[1])

The perfection of the apparatus for the preparation of distilled liquors, also distilled oils, which up to that time had received but little attention, seems to have progressed much more slowly and with greater difficulty.

In comparison with the ready volatility of the alcohol, water was considered as that product of distillation most closely related to it, whereas the oil was regarded as the "obese and fatty substance that had to be driven over with a stronger and more violent heat." This had led to the firmly established belief that in the process of purification,

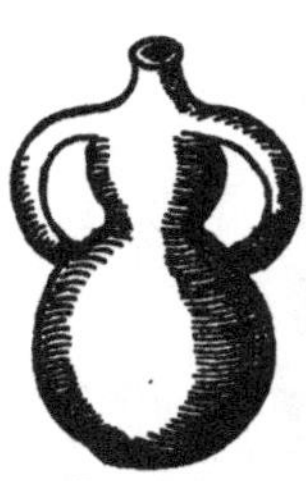

Fig. 15.

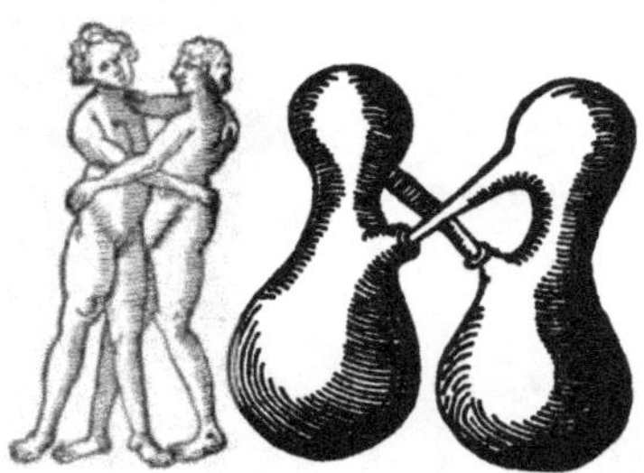

Fig. 16.

the volatile and subtle part must penetrate and exhaust the material as much as possible. As a result all sorts of queer apparatus and sources of heat were invented. They all resulted in a prolonged digestion and an unintentional loss of the alcohol ofttimes formed by fermentation, and of the aroma.

Circulation was therefore considered not only as the essence of distillation but also as an important preparatory part of it. It was believed that the plant and animal material finally to be distilled was thereby prepared for the refinement and purification of the "geistige Wesen" contained in them and for their better and easier separation and purification. A large variety of vessels, usually constructed after some symbolic prototype was used for this purpose. The simple *Circulatoria* were ordinary glass flasks, retorts with the tubes bent in a variety of ways, also so-called urine glasses used by physicians for diagnosis.

---

1) De arte distillandi, vol. 2, lib. 1.

The operation performed in the pelican (fig. 15) and double or twin circulatoria (fig. 16) provided with reflux tubes were considered as the most perfect kinds of circulation, especially for refining the "spirits."

Fig. 17.

Fig. 18.

Still more peculiar than the form of the circulatoria was the source of heat used for the purpose of circulation, usually accompanied by fermentation, and even decomposition processes. Not only was the water bath (*balneum Mariae*) (fig. 17) and the ash bath (*balneum per cinerem*) (fig. 18) employed, but also the sun bath (*destillatio solis*) (fig. 19). The circulation vessels were also immersed in fermenting dough, and heated with this in an oven (*destillatio panis*); or they were imbedded in decomposing well wetted horse manure which was placed in a layer above unslaked lime in pits (*destillatio per ventrem equinum*) (fig. 20).

Fig. 19.

Fig. 20.

With the introduction of aromatic waters as one of the principal forms of medication, the condensation of the vapors gave rise to difficulties, because a greater degree of heat was necessary for their distillation. Plant material lying on the bottom of the still was also easily burnt, and the distillate received therefrom an empyreumatic odor and taste. With a strong heat a serious overheating of the helm and tube, which were usually constructed of lead or tin, took place, while with the employment of a moderate heat the yield of the distillate remained unsatisfactory. In order to overcome these disadvantages and to prevent the flowing back of the distillate condensed in the helm, as well as to increase the cooling effect of the air, the helm known as the "Rosenhut" (fig. 1, p. 23) and (fig. 21) was constructed as early

Fig. 21.

as the fifteenth century. Near the base, at about the height of the outlet tube, this had a groove extending around the inside of the helm, and through which the water condensing on the upper wall of the helm and running down was conducted into the outlet tube and from there into the receiver. The "Rosenhut" was therefore in itself an inefficient air condenser, which served its purpose with much less efficiency than did the "Mohrenkopf" in the alcohol distillation (fig. 14).

The first step toward a better condensation with cold water, in the preparation of the distilled waters, consisted in surrounding the head of the helm (*alembic*) with an oxbladder. It was securely fastened and provided with a wooden stop-cock. The hood-like basin thus formed (fig. 22) was kept cold by means of flowing water. In a similar manner the helm was also surrounded by a basin-like metallic addition which was either fastened by luting or soldering. Thus the helm could be well cooled by running water (fig. 23). By means of an inner horizontal groove like that in the "Rosenhut" (fig. 21) the distillate which condensed on the walls of the helm was conducted into the receiver.

Walter Ryff in his treatise on distillation describes and illustrates distilling apparatus with condensing tubes which are passed through vessels of cold water. The first apparatus has two tubes connected with the helm (fig. 24) which are passed through a barrel of water.

Fig. 22.

Fig. 23.

However, Ryff declares this method of condensation as insufficient and recommends a worm tube for the shape of which he gives two forms (figs. 25 and 26) and concerning the use of which he gives a detailed account.

A peculiarly constructed apparatus for the distillation of aromatic waters and oils was recommended by Adam Lonicer in his "Kräuter- und Destillirbuch" published in 1573. The construction of the apparatus becomes readily apparent from the accompanying cut (fig. 27).

Finally, the Arabians and probably others before them practiced "downward distillation" which corresponds on the whole to our modern dry distillation for obtaining empyreumatic and tar oils. At the time of the revival of the distilling art this method was also used in the preparation of the oils of certain woods, barks and spices. Juniper

Fig. 24.

Fig. 25.

wood especially had been submitted to this *destillatio per descensum* since antiquity, later also guaiac wood, cinnamon, cloves, mace and other spices were distilled in this manner. The furnace contained a division in the middle, with a central opening into which a pot provided with a beak-like opening at the bottom was either hung, or plastered in. On top of the opening extending into the upper part of the furnace

Fig. 26.

Fig. 27.

was placed a wire gauze and a second pot filled with the dry substance to be distilled was luted with its opening on the top of the lower pot. The heating was then effected by building a fire around the upper pot (fig. 28). Sometimes the lower pot was buried in the earth and a fire built about the upper pot fastened on top of this one in the same manner.

For the *destillatio per descensum* on a small scale glass vessels heated from the side (fig. 29) were also employed and even for some easily distilled substances the heat of the sun (*destillatio solis*) (fig. 30) was used. At present the preparation of empyreumatic oils as well as of the finer tars is effected in cast iron or earthenware cylinders.

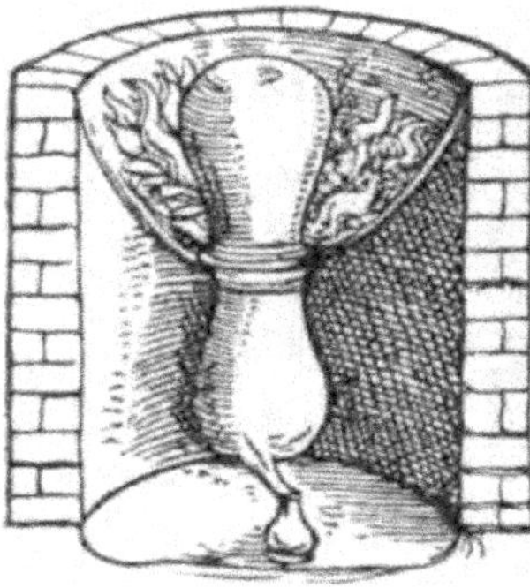
Fig. 28.

The treatises on distillation published during the sixteenth century reveal by both text and illustration that about that time more attention was again given to the construction of furnaces and implements used for the purpose of distillation.

Besides the distilling apparatus most generally used at that time and reproduced in figs. 10, 11, 14, 17, 21, 23, 24 and 25, the socalled "faule Heinz" or *Athanor* (from ἀθάνατος, imperishable) called by Ulstad *furnus Acediae* (fig. 31) was much in favor and was used to a great extent for the distillation of waters and oils. Above a common

Fig. 29.

Fig. 30.

fireplace were placed three or more distilling retorts with "Rosenhuthelm" (p. 104, fig. 21). The fireplace ended in a central iron, copper or earthenware pipe the opening of which could be closed by a cover. By means of slides at the sides of the fireplace the heat could be conducted under any one of the stills or retorts as desired and the distillation was thus regulated.

For the distillation of large quantities in a large number of single retorts or stills, larger cupil furnaces after the manner of the so-called "galley furnaces" appear also to have been in use. The illustrations and descriptions of these in the treatises on distillation of the sixteenth century represent no doubt more the possibility than the realization of perfection. The illustrations of these furnaces were transferred from one distilling book to another, but probably have not been generally used in practice. Among others they are thoroughly described in the text and reproduced in illustration in the works of Matthiolus and of Lonicer previously mentioned. They are built either in the shape of a terrace (fig. 32) or of a bee hive (fig. 33).

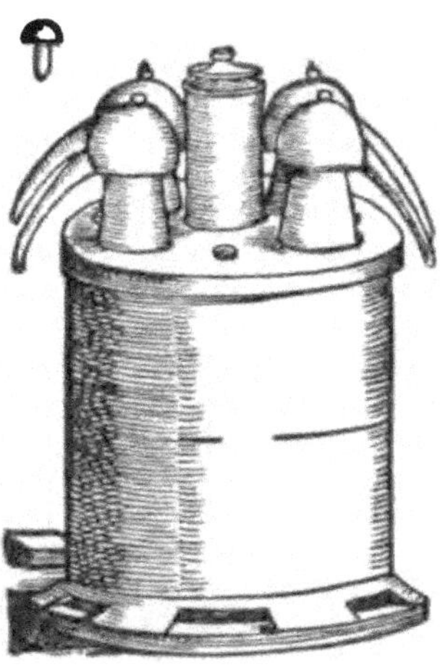

Fig. 31.

Although the compilers of the distilling books of the sixteenth century have in succession followed the pioneer work of Brunschwig, especially in regard to illustrations, their writings nevertheless quite often show considerable differences in views, practical skill, and experience, and also in the originality of their knowledge and ability.

Fig. 32.

With but little public intercourse these secluded workers and writers toiled mostly far from one another, each in his own sphere and manner, often with but a slight knowledge of the older writings and of the work of his contemporaries. With regard to the manner of distillation of the aromatic waters and oils this is shown in an unmistakable manner in works compiled in the course of the first half of the sixteenth century by Philipp Ulstad, Walter Ryff, Adam Lonicer, Valerius Cordus

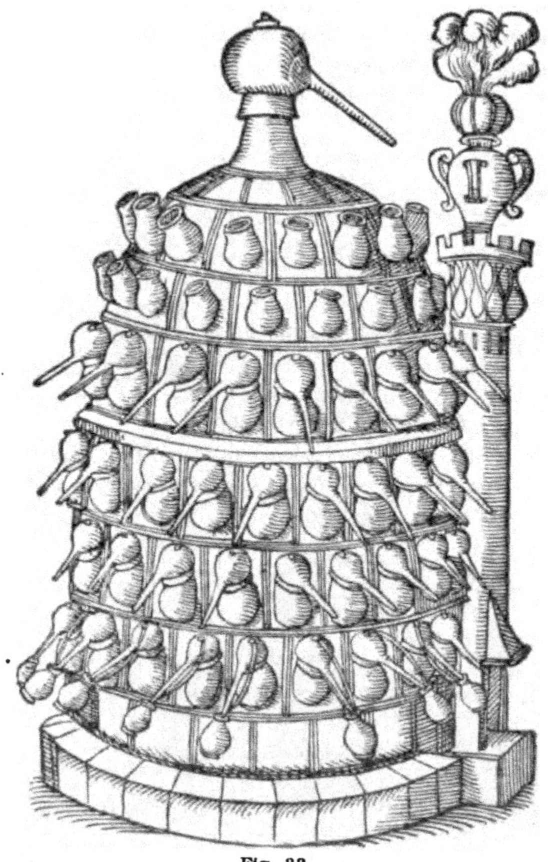

Fig. 38.

and Conrad Gesner. All of these were mainly based on the writings of Hieronymus Brunschwig. Their views, however, as to the nature of distillation itself and of the distilling methods and apparatus are neither in accord with those of Brunschwig, nor with those of their contemporaries.

How little personal skill, practical experience and familiarity with the literature on the subject may be found in the writings and methods of working of the most prominent experimenters of that time becomes apparent e. g. from the construction of and preference for the distilling vessels employed. Thus for instance, Valerius Cordus, profound in theoretical science, but ignoring the rationally constructed distilling apparatus then well known, used and recommended "ein Kolb mit einem angeschmelztem Helm" (fig. 34) as an efficient apparatus meeting all the requirements of the art. At the same time Conrad Gesner, his contemporary, used for the same purpose a distilling furnace (fig. 35), which had been used for some time.

As has already been mentioned, the seventeenth century, crippled as it was by the destructive storms of the Thirty Years war, added but little to the further development of the art of distillation and other technical scientific industries. The few active experimenters, however, favored with a better understanding, endeavored to perfect not only the apparatus used but the processes as well.

Fig. 34.

Fig. 35.

As the "Destillirbuch" of Brunschwig and similar treatises of his successors reflect the practical and theoretical knowledge of the sixteenth century with its mistakes and imperfections, so Glauber's treatise on distillation reflects the condition of the art and science of distillation during the second half of the seventeenth century. Although Glauber's laboratory work and the character of his writings was of a wider scope than that covered by the older "Destillirbücher," yet he also paid special attention to the distillation of aromatic plants and spices. In this, he and his contemporaries seem to have paid special attention to the improvement of the methods of distillation for the purpose of relatively increasing the products of distillation. For this purpose as has already been mentioned on page 33, a very rational expedient of increasing the specific gravity and thus raising the boiling point of the water used for the distillation, was resorted to. This was effected by the addition of salts. The use of muriatic acid (*spiritus salis*) recommended by Glauber for the distillation of oil of cinnamon and other expensive oils has also been alluded to (p. 33). His idea was that the acid penetrated the material and thus drove out the oil. A small amount of muriatic acid was supposed to take the place of a large quantity of water thereby avoiding loss of oil. He also recommends that dark colored and resinified oils be rectified with *spiritus salis*.

Glauber's authority was recognized until the middle of the eighteenth century, and the methods of distillation recommended by him in his several writings were employed by his contemporaries and their successors. Boerhaave, therefore, and Hoffmann, their contemporaries, and later investigators prepared the volatile oils by using common and other salts or hydrochloric acid.

It is perhaps due to the observation that metal was present in an oil or a distilled water, especially if an acid had been used in its preparation, that in the course of the eighteenth century more attention was again bestowed upon the material from which the still was constructed. In consequence glass and glazed earthenware were substituted for metal. As a matter of fact it seems that as early as the sixteenth century the presence of metals in the distillates obtained from metallic stills did not escape the notice of some of the experimenters. Among others Joh. Krafft[1]) (1519—1585) cautions against the use of copper distilling vessels. The famous Parisian physician Ambroise Paré[2]) (1510—1590) warns against the use of lead helms and condenser tubes "which ofttimes cause the distilled water to be milky." The Bologna physician and professor Benedetto Vettori of Faenza (1481—1561) declared about the year 1555, that water on being conducted through lead pipes dissolves lead and thus becomes poisonous.[3])

However, these observations like so many others made in the art of distillation appear either to have been known to but a few or else were unheeded and again forgotten, for even during the seventeenth and eighteenth centuries when oils were distilled with acids, lead and tin heads and condensers were in general use in connection with copper stills or glass and earthenware retorts.

---

1) Conciliorum et epist., 1, fol. 190.
2) Les oeuvres, p. 746.
3) Practicae maguae, 1, cap. 21, fol. 144.

As already mentioned in the preceding chapter, the distillation of the volatile oils and the construction of the distilling apparatus received more attention and underwent a more rapid development with their general introduction into the laboratories of apothecaries. In these the volatile oils used in medicine and the arts were prepared up to the first decades of the nineteenth century. Only a few oils, such as the oils of lavender, rosemary and rose which could be readily produced in some countries and which were largely used in the perfume and soap industries, have been obtained since the sixteenth century in larger quantities by means of primitive portable distilling apparatus.[1])

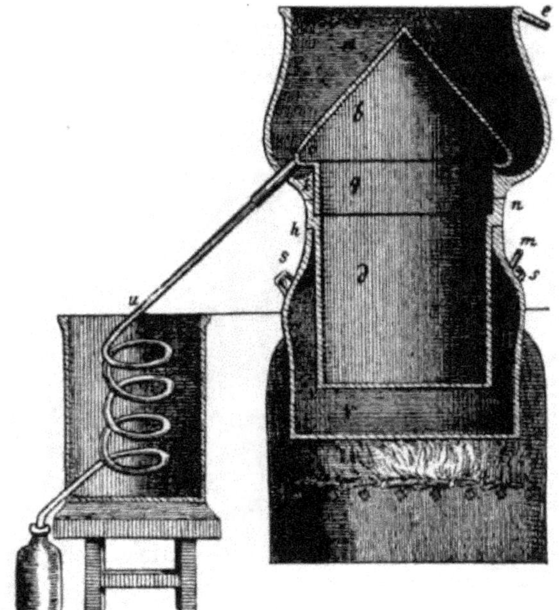

Fig. 86.

The distilling vessels used in the apothecary laboratories and the itinerant stills (*Wanderdestillirgeräthe*), or *alambics voyageants* used in France, Spain, Italy and Bulgaria, consisted of copper stills with a copper or tin head and tin condensing tubes of various shapes.

One of the better distilling apparatus used for the distillation of volatile oils in the eighteenth century consisted of a tin or copper body

1) Comp. p. 26; also under Oil of Lavender, Oil of Rose and Oil of Rosemary.

suspended in a water bath, and provided with a "Mohrenkopf" (fig. 14, p. 55), a "Rosenhut" (fig. 21, p. 58) and a spiral tube for condensation. An illustration (fig. 36) and description of this distilling apparatus was published in 1784 by François Demachy,[1]) director of the apothecary laboratories of the civil hospitals of Paris.

The copper kettle *v* serves as a waterbath which can be turned by the handles *s s* and refilled with a fresh supply of water through the side tube *m*. The tin still *d* rests with the upper ring *n* on the rim *h* of the kettle. The lower neck *q* of the head of the still *b* rests at *n* on the upper rim of the still. Around the lower edge of the head runs the trough *c* in which the distillate that has been condensed in the cone collects and passes with uncondensed vapors through the tube *c—u* into the spiral condenser.

The "Mohrenkopf" *a* serving as a cooler for the "Rosenhut" *b* is soldered to the neck *k* of the condensing cone. The water in the cooler warmed during the process of distillation, runs off through the upper tube *e* as fast as cold water is added.

Fig. 37.

Since the beginning of the nineteenth century attempts have been made to simplify and to improve the construction of the distilling apparatus, more especially of the cooler, also to prevent the burning of the plants on the bottom of the still with the use of direct heat. Such improvements were made especially by Joh. Gottfr. Dingler, the apothecary, in Augsburg[2]) during the years 1815—1820, by Smithson Tennant[3]) in 1815 and by Henry Tritton[4]) in 1818, both of England.

---

[1]) L'art du destillateur. Germ. transl. by Hahnemann, vol. 1, p. 192, and plate 2, fig. 1.

[2]) Trommsdorff's Journ. d. Pharm., 11[1], p. 241; also Buchner's Repert., 3, p. 137, and 6, p. 142.

[3]) Phil. Trans. 1815; Repertory of Arts, Sept. 1815.

[4]) Annals of Phil., June 1818; Buchner's Repert., 6, p. 98.

The latter attempted to carry on the distillation at a lower temperature by putting the apparatus in connection with an air pump. The distilling apparatus more commonly in use at that time for the distillation of volatile oils was the one shown in the accompanying cut (fig. 37).

Steam distillation was recommended in 1826 by H. Zeise[1]) and especially for volatile oils by van Dyk in Utrecht,[2]) who thereby materially aided in its introduction. He demonstrated that the volatile

Fig. 38.

oils which were obtained by steam alone from the vegetable material, distinguished themselves from those obtained by distillation over open fire, by a lighter color and purer odor. Clove oil distilled with steam is nearly colorless, cinnamon oil light straw yellow and orange peel oil completely colorless.[3]) The first steam distillation on a larger scale in a pharmaceutical laboratory appears to have been that in the old Apothecary's Hall in London.[4])

---

1) Beiträge zur Nutzanwendung der Wasserdämpfe; Arch. d. Pharm., 16, p. 69.
2) Buchner's Repert., 29, p. 94.
3) Ibidem, 29, p. 110.
4) do., 29, pp. 112 & 188.

In Germany steam distillation for the preparation of volatile oils in apothecary laboratories was also made possible by the introduction of a steam distilling apparatus,[1] constructed about the year 1826 by Johann Beindorff, mechanic and tin founder in Frankfurt a. M. (fig. 38). With this apparatus, soon perfected in many ways, the distillation of volatile oils with steam under pressure was possible. The condensing arrangement also had the advantage over the spiral tube of being made up of separable parts, and thus allowing it to be readily cleaned even on the inside.

For the preparation of volatile oils on a small scale, the arrangements based on the original steam distilling apparatus of Beindorff remained, until the industry conducted on a large scale became dominant also in this field and prepared products of a quality and at prices with which the preparation on a small scale could not compete.

Of the arrangements used for a long time for the separation and removal of the oils from the water the Florentine flask in various forms and sizes has shown its utility and has been in continual use. It probably came into use in the middle ages. A method of separation of oil and water which in its principles corresponds to those of the Florentine flask has, it appears been described for the first time by Porta[2]) in the latter half of the sixteenth century.

The Florentine flask like many other facts and improvements pertaining to the art of distillation which were not generally known, was soon forgotten. As a result it was rediscovered several times from the beginning of the seventeenth century to the year 1823.

Thus the flask was again described and introduced by Homberg[3]) at the end of the seventeenth century about one hundred years after Porta's description—only, however, to be again forgotten for a considerable period of time. A century later, in the year 1803, the Florentine flask was again recommended for the distillation of volatile oils by the Augsburg apothecary Johann Gottfried Dingler[4]) and later in 1823 once more introduced as something new by the apothecary Samuel Peetz in Pesth.[5])

The Florentine flask of older construction as described by Porta, has been in use for a long time. The oil was siphoned off by means of a porous siphon consisting of a lampwick into small bottles (fig. 39).

1) Geiger's Magazin f. Pharm., 11, pp. 174 & 291; Buchner's Repert., 88, p. 436.
2) Magiae naturalis, lib. decimus, p. 367.
3) Philippe und Ludwig, Geschichte der Apotheker, p. 513.
4) Trommsdorff's Journ. d. Pharm., 11II, p. 242.
5) Buchner's Repert., 14III, p. 481.

Later the Florentine flask shown in figure 40 was also used. The flask used at the present time in the large factories is not only larger, but contains in the upper part in the level of the oily layer a glass stopcock through which the oil can be drawn off from time to time, or an overflow tube through which the oil when it reaches a certain level runs into a receiver (fig. 41).

In the course of time a number of differently constructed receivers for the separation of the volatile oils have been proposed, without however, causing the Florentine flask to be discarded. The first of these was proposed by Amblard of Paris[1]) in 1825. It consisted of a glass tube, open at both ends and drawn out to a taper. This tube was suspended by means of a cork ring at its upper end in a high glass

Fig. 39. Fig. 40. Fig. 41.

mixing cylinder. This cylinder was provided at the top with an overflow tube. The oil collects in the glass tube and can be removed from this after closing the small lower opening, by pouring out as often as desired.

The more salable volatile oils, which were used in larger quantities in the perfume industry that had developed in France in the course of the eighteenth century were still prepared during the first quarter of the present century in the traditional primitive distilling vessels and improved by rectification. In Germany the apparatus shown in figs. 37 and 38 were principally used. While the oils of lavender, rosemary, orange flower, etc., were manufactured in France and rose-oil in Turkey, Germany and Hungary supplied the market with the oils of caraway, fennel, anise, coriander, calamus, peppermint, spearmint,

[1]) Bull. des trav. de la Soc. de Pharm., 1825, p. 247.

valerian, chamomile, and others used in medicine and in the fine arts. In southern France, especially on the sunny slopes of the Alps near the Mediterranean coast, the industry of the aromatic oils developed considerable proportions in the early part of this century. The oils principally used for medicinal purposes, however, were still prepared in apothecary laboratories. In the course of time individual apothecaries and druggists, having made a beginning on a small scale, erected much larger establishments for the preparation of volatile oils. This was done especially in regions suited to the culture of medicinal plants, for instance, in Thuringia on the Saale and the Elbe, in Saxony, Bohemia, Franconia and also in Hungary. Only a few of these, however, remained in existence for any appreciable length of time. As in chemical and other branches of industry, these originally small distilling operations were replaced everywhere, by a larger, more rational and efficient industry. This new industry has worked hand in hand with science and technology. Whereas on the one hand it utilized the results of science, on the other it not only stimulated science but gave direct assistance.

Beginning about the middle of the present century the earlier, simple apparatus took on a different shape in the factories of this larger industry of the volatile oils. The original small distilling apparatus were replaced by larger and more rationally constructed ones which not only effected a better exhaustion of the vegetable matter and thus increased the yield, but also produced oils of a better and purer quality. The apparatus commonly used in the factories about the middle of the present century, were the stills arranged for the so-called water-distillation, and others for the so-called dry steam-distillation.

The first type of stills (fig. 42) is a simply constructed apparatus for the distillation of plant material in water, as well as for the rectification of crude oils by steam. The still is heated by means of steam admitted under pressure through the holes of a ring at the bottom or by allowing the steam to escape directly into the lower double walled jacket *B*. The advantage possessed by the receiving flask *E* over the ordinary Florentine flask consists in the flowing back of the distillation water saturated with oil, through the tube *F*, into the still.

In the distillation with dry steam (fig. 43) the still is filled with the plant material without the addition of water and distilled with steam passing through the material from the bottom upwards. These or similarly constructed steam distilling apparatus are employed even at the present time for the distillation of some of the oils, only in place of the spiral a tube condenser is used.

With the introduction of these apparatus during the middle of this century the volatile oil industry had taken its position as a branch of the rapidly developing chemical industry at large with southern France and central Germany as the principal centers of production. Since then this branch of chemical industry has witnessed great technical improvements which have reacted favorably on its growth and permanent development.

Fig. 42.

Owing to the remarkable development of the entire perfume industry during the second half of this century, the consumption and commerce of the volatile oils assumed entirely unanticipated proportions and importance. Scientific and technical attainment, commercial interests and business competition brought about numerous changes in rapid succession. More rational methods of distillation were devised, large apparatus for the distillation of enormous quantities were constructed, the quality of the product was improved while the price of its production was diminished. Some of the largest stills used in the manufacture of volatile oils have a capacity of 30,000 to 60,000 liters (= 7,926 to 15,852 gallons). Commensurate with their size is their construction and operation, the arrangement for charging and discharging, for the condensation of the vapors in as quick and efficient a manner as possible, and above all their productiveness.

Considerable historical interest is attached not only to the theory and practice of the present art of distillation in its application to the preparation of the volatile oils but to the gradual development of the distilling vessels as well, which are employed for this purpose.

Fig. 48.

On the following pages will be found a number of illustrations of some of the modern stills used in German and French factories. Looking backward it might seem as though no relation existed between the modern giant stills and their prototypes. Yet every one of them is but a link in the long chain of development of the art of distillation. That the process of evolution has been exceedingly rapid during the past ten years does not affect the truth of this statement. Almost every one of these pieces of apparatus has been newly created out of the ruins of its immediate predecessor.

In modern chemical industry Germany unquestionably ranks first. Of the various branches of this industry that of the manufacture of volatile oils and synthetic aromatics has acquired an importance previously unsuspected and with it a correspondingly influential position.

Fig. 44.

Modern Distilling and Rectifying Apparatus.

Fig. 45.

Distilling and Rectifying Apparatus.

Fig. 46.

Apparatus for Distillation with Water.

Fig. 47.

Distilling Apparatus of 30,000 Liter Capacity.

Fig. 48.

Distilling Apparatus for Spices.

Fig. 49.

Great Still of 60,000 Liter Capacity.

Fig. 50.
Distillation of Rose Oil.

Fig. 51.

Distilling Apparatus.

GENERAL PART.

1. THEORETICAL BASIS FOR OBTAINING VOLATILE OILS BY STEAM DISTILLATION.

From the foregoing chapters it becomes apparent that the manufacture of volatile oils was formerly conducted on a small scale, largely after the style of the small mechanic. But a single process for obtaining the oils was used, even though here and there it was conducted in a scientific manner, i. e., corresponding to the status of science at the time. All crude materials and all oils were distilled in the same manner from a distilling apparatus consisting of still, head, and condenser. When the still was heated directly by fire. only one kind of distillation took place, namely water distillation. in which case the material to be distilled was either in the water, or it was suspended dry above the water. If the distillation was conducted with steam. there was a choice between water distillation and dry steam distillation. The introduction of steam as the immediate source of heat was unquestionably in itself an advance. A new impetus, however, was first given to the industry of volatile oils by the use of steam under pressure. That this change should at first have given rise to many difficulties was inevitable.

Out of the large number of small factories and laboratories where oils were distilled on a small scale, a few have risen during the last few decades, and have developed to such an extent as to impart to the entire industry a different aspect. It is, however, not merely the matter of size which separates the past from the present. The rapid development of chemical science is even more responsible for this difference. Through it the preparation of essential oils has developed into a real industry, a branch of chemical technology. The successful investigations of Wallach have not only led the way in chemical science, but they have also done pioneer service in practice.

This development of the volatile oil industry has taken place in a remarkably short period of time. Indeed the unusually rapid growth of chemical knowledge made it difficult for progressive factories to keep abreast with the theoretical advancement, as did also the fact that the simple apparatus and methods no longer sufficed for the new requirements. Some apparatus could be borrowed from other technical branches but much had to be specially designed and made. In addition to the simple stills formerly used, new rather complicated apparatus were constructed. Although these required greater ability and more attention on the part of the operator, they gave in every respect more satisfactory results.

The evolution from the simple to the more highly perfected technique can best be compared to that of the alcohol manufacture from the whisky distillation on a small scale. The simple stills of a past time, which are now only used here and there for the preparation of whisky on a small scale, will scarcely be recognized as the prototypes of the column apparatus with continuous supply of wort and continuous or periodic removal of the wash, as they may now be seen in their most perfect construction and operation in modern alcohol distilleries. Whereas the former whisky distiller first distilled off an intermediate product, and from this by rectification obtained a whisky of only 30 to 40% alcohol, the present column apparatus yields with less attention, time and fuel, a spirit of over 90% alcohol in a single operation.

In the alcohol distillation the conditions are, however, much simpler than in the preparation of the volatile oils. There a few raw materials: potatoes, corn or maize, and only one end product: alcohol. The knowledge of the apparatus is common property. It can be constructed by any reliable coppersmith; and its manipulation is shown by the builder. Here on the other hand, there are numerous crude materials as well as numerous different products, which according to the use to which they are to be put require different methods for production and purification. The apparatus necessary for these various operations must be constructed and perfected by the manufacturer himself, if he does not wish to remain behind in the competition brought about by changes in the methods of production.

The advantages of the present over the former industry are a saving in time, labor and steam, an increase in the yield and especially an improvement in the quality of the oils. Coupled with these are, however, a considerable increase in the cost of the whole plant, and resulting from this, the necessity for production on a large scale.

Briefly described, the manufacture takes place in the following manner. The prepared raw material is filled into the distilling apparatus, which is heated by steam. The vapors saturated with oil particles are condensed in the cooler. The distillate, consisting of water and oil, collects and separates in the receiver, and the crude oil thus obtained is purified when necessary, in order to yield the finished product. The manufacture consists therefore of the following operations:

1. Preparation of the oil-containing material for distillation;
2. Distillation;
3. Cooling and condensing the vapors;
4. Purification of the crude oil.

Preparation of the oil-containing material. The proper preparation of the raw material is one of the most important requirements for the distillation. The essential oils belong to the secretion products in the life process of the plant cell and are found in the cytoplasm of the cells, in the inter-cellular spaces, in resin ducts, (in the conifers) etc. If the cell walls are very tender, the steam may gradually rupture them and bring the volatile oil to boiling. But when they are more or less woody and thick, the raw material must be comminuted before distillation, so that the volatile oil may be removed gradually from the fragments.

The comminution apparatus necessary for this operation are of various designs. Herbs, leaves, and fresh roots are cut; barks, dry roots and dry fruits are ground; woods rasped; seeds crushed. Each raw material requires a method of comminution depending on its properties, which must be further modified according to the method of distillation to be employed. The process of distillation depends wholly upon the proper preparation of the crude material. An insufficient or unsuitable comminution not only raises the cost of the necessary steam, but also decreases the yield of oil. With oils not readily volatile, the material still retains some oil, in spite of an increase in the time of distillation. On the other hand many oils require rapid distillation, as they decompose when in contact with the moist steam, yielding useless decomposition products which are wholly or partly soluble in water, or because they become resinified.

Distillation. The distillation of the essential oils with steam depends upon the physical fact, that the boiling point of a mixture of two liquids which are not miscible must always be lower than that of the most volatile liquid.

The boiling of a liquid consists in the change from the liquid to the gaseous state. A liquid begins to boil when, in the course of heating, its vapor pressure has increased to such an extent that it is able to overcome the outside pressure on the liquid, i. e. when it overcomes the pressure of the atmosphere by heating in an open vessel. The vapor pressure of the heated liquid has then risen to the height of the outer pressure. If the liquid consists of two immiscible liquids, for instance water and a volatile oil, then on heating the mixture, not only does the water give rise to a vapor pressure, but also the volatile oil, and the mixture will therefore be brought to boiling, when the sum of the two vapor pressures is equal to the outer pressure.

The distillation of water and oil of turpentine may serve as illustration. Let the outer pressure be the normal atmospheric pressure of 760 mm. Under this pressure the water, if heated alone would boil at 100° C. and the oil of turpentine, for the greater part, would not boil until about 158°. If, however, a mixture of water and oil of turpentine is heated, the vapor pressure of the water and also that of the oil of turpentine, take part in overcoming the outer pressure of 760 mm. The mixture of water and oil of turpentine will, therefore, boil when the sum of the two vapor pressures is equal to 760 mm. and this takes place at 95.6°, for at this temperature the water has a vapor pressure of 647 mm. and the oil of turpentine 113 mm. (both according to Regnault), or both together 760 mm.

This explains why a mixture of water with a volatile oil, although the boiling point of the volatile oil when distilled alone is usually higher than that of the water, will boil lower than the boiling point of the water and form a mixture of oil and water vapors.

Every liquid or solid body that shows a vapor pressure at the temperature of the water vapor employed can be distilled.[1])

By lowering the boiling point of the volatile oils by distilling them with water or water vapor, it is possible to obtain certain chemical bodies, which when heated by themselves would volatilize only with decomposition at their high boiling points. On the other hand many bodies are but slightly volatile at the low temperature at which the distillation with water or water vapor is conducted and require comparatively large quantities of water vapor for volatilization, so that the production of these difficultly volatile bodies with water vapor is excluded.

---

[1]) Slightly under 100°, when saturated water vapor of any pressure is conducted into the distilling apparatus which is under atmospheric pressure.

The water vapor is the source of the heat for the distillation of the volatile oils. A thorough study of the amount of service obtainable from this heating material is therefore of great importance from an economic standpoint.

Distinction should be made between two ways of heating, the direct and the indirect heating by steam. According to requirements they are employed either separate or together. In the first the steam is conducted into the oil-containing material which is either dry or suspended in water. In the latter the steam does not come into immediate contact with the oil-containing material. The material lies in the water, and the water is heated to boiling by means of steam in the outer jacket of the still, or passing through a coiled tube.

If the indirect heating is accomplished with saturated water vapor of 150°, consequently under a pressure of nearly five atmospheres, the steam is condensed to water of 100°, provided the water separator is working well and removes only water and no steam. To change 1 k. of water from 0° into saturated vapor of 150° there are required 652.8 calories[1]) of heat. If we deduct from this the heat necessary to raise the water from 0° to 100°, namely 100.5 cal., we have for the heat of vaporization of 1 k. of water of 100° into saturated vapor of 150° under pressure, 652.8 — 100.5 = 551.8 cal. This same amount of heat is again liberated when the vapor of 150° is condensed to water of 100°. According to a similar calculation, 1 k. of saturated compressed steam of 180°, used for indirect heating, when changed to water of 100° liberates 661.4 — 100.5 = 560.9 cal., i. e., only about 9 cal. more than saturated steam of 150°.

In the direct heating process, the steam passing through the still will, at the expense of a part of the steam which is condensed, carry with it some of the oil as vapor. In the most favorable case, the ratio of oil vapor and water vapor will be that of their vapor tensions at the temperature at which the distillation is conducted.

For the sake of simplicity, the calculation of the cost of steam for a distillation, may be made as though all the steam was condensed to water of 100° and that the mixture of vapors of water and oil going over into the condenser was formed from liquids of 100°. According to this the amount of service obtainable from the steam is the same with the direct as with the indirect heating process. In practice, however, the two ways of heating are not equally efficient. With the indirect heating

[1]) One calory, abbreviated 1 cal., is the unit quantity of heat. It is the amount of heat required to raise 1 k. of water of about 15° one degree C.)

process all of the heat contained in the steam cannot be utilized, as loss of heat cannot be prevented. On the other hand in the direct heating process, if the steam is conducted into the oil-containing material suspended in water, or into the volatile oil itself, in case of rectification or fractionation a loss of heat is only possible through careless conduction of the steam. If the oil-containing material is dry in the still, the loss of heat in direct heating will be usually large, and with careless manipulation may become so large, that only a small part of the direct steam is utilized, while the larger part goes unused into the condenser. The theoretical heating effect of the steam may also be obtained when the oil-containing material is dry, but this can be done only under certain suppositions.

In order to calculate from the above heat values of the steam the cost of the steam for a distillation, the product of the distillation, i. e., the distillate, must first be more closely examined. The distillate consists of water and volatile oil. The question therefore arises, how much water and how much volatile oil is a certain amount of steam capable of distilling? Evidently the answer to this question depends first of all on the relative weights of the water and the oil in the distillate. This relation, of course, differs with the character of the volatile oils, and is furthermore dependent for each and every oil upon the distilling temperature employed, and the changeable composition of the oil; on the other hand the ratio is fixed, so that 1) with a surplus of steam the water content in the mixture of vapors going into the condenser may indeed be increased; 2) in spite of an economic supply of steam and a generous supply of oil the amount of the former cannot be decreased. The amount of oil which is volatilized with the steam, is determined by the vapor pressures of both at the distilling temperature.

The quantity of oil in the distillate can be determined experimentally if the distillation be so conducted that the vapor contains the greatest possible quantity of oil. It can also be calculated if the vapor pressure of the essential oil at the distilling temperature be known, for the parts by volume in the mixture of the vapors of two immiscible liquids, are to each other as the vapor pressures of these liquids at the boiling temperature of the mixture; or the parts by weight are to each other as the vapor pressures multiplied by the vapor densities or the molecular weights. If

m_1 = the molecular weight of one of the liquids,
p_1 = the vapor pressure at the distilling temperature,
g_1 = the part by weight in the vapor,

and further m_2, p_2, g_2, be the corresponding values of the other liquid then

$$g_1 : g_2 = m_1 p_1 : m_2 p_2.$$

As an example of such a calculation, the distillation of caraway oil with water vapor under atmospheric pressure may be taken. Caraway oil consists of a mixture of limonene and carvone in nearly equal parts by weight. Its boiling temperature by distillation with steam under atmospheric pressure, namely 760 mm., is a few degrees below 100°; at the beginning of the distillation somewhat lower, toward the last nearly 100°. The vapor pressure of the limonene (molecular weight 136) at 57.5° amounts to 12 mm., that of the water (molecular weight 18) at the same temperature is 132 mm. At the distilling temperature of 57.5°, there will, therefore, be in the distillate for every 100 k. of water, 70 k. of limonene. At 176°, the boiling point of the limonene under 760 mm. pressure, the water exerts a vapor pressure of 6962 mm. of mercury; therefore at 176°, with 100 k. of water there distill over 83 k. of limonene. According to this, at a distilling temperature of 96°, for 100 k. of water in the distillate there would be about 75 k. of limonene. An experiment with limonene of sp. gr. 0.850 and still containing some carvone, gave for 100 k. of water in the distillate 60 k. of limonene. Calculation and experiment agree, therefore, fairly well. For carvone (molecular weight 150) the calculation is simpler, because its vapor pressure has been determined at about 100°, namely 104.1°. At this temperature it amounts to 12 mm., that of water at 104.1° being 880 mm. For every 100 k. of water in the distillate there will be therefore 11 k. of carvone. By experiment 9 k. were obtained. Thus are obtained the quantities of oil, which theoretically are distilled from caraway seed with every 100 k. of water under atmospheric pressure; namely, 75 k. of oil at the beginning of the distillation, which amount drops down gradually to 11 k. of oil when pure carvone distills over. But as even the very first parts of the distillate contain some carvone, they will not contain 75 k. of oil for every 100 k. of water, but much less. According to a carefully performed experiment of a caraway oil distillation from a simple still, there were obtained in the beginning of the distillation 37 parts by weight of oil, and past the middle of the distillation 9 parts by weight of oil to every 100 parts by weight of water.

After the heat effect of the steam for the distillation and the relation of water and oil in the distillate have been determined, the heat necessary for obtaining a certain amount of distillate remains to be ascertained.

Aside from the heat consumed in warming the distilling apparatus and its contents to the distilling temperature, the distillation requires a heat supply for the vaporization of liquid bodies at about 100° temperature at atmospheric pressure. This heat of vaporization for 1 k. of water amounts to 536.5 cal. The heat of vaporization of caraway oil or rather of its constituents has as yet not been determined, but it can be fairly accurately calculated by F. Trouton's law,[1] according to which the molecular heat of vaporization is proportional to the absolute temperature of the boiling point. For limonene the calculation gives 65.5 cal., for carvone 66 cal. for the unit of mass. To be sure, these are the heats of vaporization at the boiling temperatures of limonene and carvone, namely at 175° and 230° respectively. The difference from the heats of vaporization at 100° is, however, probably small, as the specific heats of organic bodies in the liquid and in the gaseous state are nearly the same.

As has already been shown, when the steam is utilized to the greatest possible extent, the distillate from caraway oil will contain at the beginning of the distillation 37 k. of oil, beyond the middle of the distillation 9 k. of oil for every 100 k. of water. Consequently there will be in 100 k. of distillate from 73 k. of water and 27 k. of oil, to 92 k. of water and 8 k. of oil, the heat required for 1 k. of distillate therefore is from 409 to 499 cal. In other words 1 k. of saturated steam of 150° may, with its available heat of 551.7 cal., yield in the most favorable case of a caraway oil distillation 1.3 k. to 1.1 k. of distillate. Or when the heat required is based only on the volatile oil, 1 k. of caraway oil costs in the course of the distillation from 1552 cal. to 6275 cal.; or from 2.81 k. to 11.38 k. of saturated steam of 150°; or from 2.77 k. to 11.19 k. of saturated steam of 180°.

The heats of vaporization of the volatile oils are, as already mentioned, small; still smaller are their differences. They cannot, therefore, give rise to a high cost of steam for the distillation. With the different volatility of the essential oils, it is, however, otherwise. The caraway oil distillation of which the distillate contains at the beginning 27% of oil, and later at least 8%, is an example of an easily volatile oil. Other oils, for instance sandelwood oil, require more than the twentyfold amount of steam for distillation, others still more. It is therefore an important requirement of an economic management, to decrease the loss of heat as much as possible, i. e. to conduct the

---

[1]) W. Ostwald, Allgemeine Chemie, II. Ed., vol. 1, p. 354.

distillation in such a manner, that as far as possible, not more than the necessary minimum amount of steam be used.

Incidental to an investigation of the influence of the addition of salt to the water in obtaining the heavy essential oils, the French chemist E. Soubeiran[1]) in 1837 closed his observations with the words: "My experiments definitely show that the preparation of the essential oils, which is considered as a well known operation, is on the contrary worthy of an entirely new investigation." The study of the history of a science is not only useful for a better understanding of the science itself, but is also psychologically interesting and instructive. The mistake of considering studies in any part of a branch of science as concluded, as completely investigated, only because through a series of years no new promising investigations have been recorded, is made quite frequently. The preparation of the volatile oils from the plants was considered to be a well known operation not only 60 years ago, but also a little more than a decade ago. Now, that this industry has so enormously developed through successful investigation of the volatile oils and with the aid of better mechanical appliances, when out of the former "Destillirkunst" there has grown a scientifically conducted industry, which, combined with commercial experience has reached unthought of results, we know, that this development is by no means completed.

The following may serve as the most important characterizations of the different processes of distillation for oil-containing material, or for the rectification or fractionation of the oils themselves:

a) Dry steam distillation with saturated steam under pressure.
b) The same, but with superheated steam.
c) Water distillation.
d) Distillation under diminished pressure.
e) Continuous distillation.

Through modifications in the conducting of a certain process of distillation in an apparatus with the requisite construction, further through suitable combination of these main processes, according to the distilling material, or according to the desired composition of the oil, important advantages may be gained.

In the larger factories the distillation of oil-containing materials by direct heating of the apparatus over a free fire is no longer used.

---

[1]) Liebig's Annalen, 25, p. 245.

There is no limit to the size of the apparatus. If the enormous masses of distilling material are at hand, which such an apparatus requires for its economic operation, there is nothing in the way of erecting as large an apparatus as may be desired.

Cooling and condensing of the vapors. Cold water is used in general for cooling and condensing the vapors from the distilling apparatus. The vapors are conducted into the condenser through a tube which connects the still with the condenser. In the smaller factories, Liebig condensers, spiral condensers, and jacket condensers according to Mitscherlich are used, but in the larger ones they are used only exceptionally for certain purposes. Even moderately large distilling operations require that the cooling effect of the water be used to the best advantage. To condense several cubic meters of distillate in an hour, is an impossibility with these condensers. Tube condensers serve for this purpose in which the vapors are conducted through a bundle of tubes, which are surrounded by flowing water. Vapor and water flow in opposite directions. The capacity for work of these condensers is such that the distillate leaves the condenser at a temperature of 20°, and the condensing water up to 80°.

It is very simple to calculate, if desired, from the steam used from the boiler or for a distilling apparatus, the quantity of water required to condense the vapors. 1 k. of saturated steam of 150° under pressure has taken for its generation from water of 10° 642 cal. of the heat of combustion of the coal. If the distillate is at 20°, then the steam gives back 632 cal. in forming the distillate. If the water flows through the condenser with a beginning temperature of 10° and with an average end temperature of 60°, then 1 k. of water is capable of taking up 50 cal. of heat, according to which 12 to 13 k. of water are required to condense 1 k. of saturated steam under pressure. With water of a beginning temperature of 20° about 16 k. are required for the condensation.

The collection of the distillate in receivers and the separation of the oil from the water according to the different specific gravities of the two, offers nothing of particular interest.

Purification of the crude oil. The odor, taste, and color of the oil, are of the utmost importance in the manufacture. In steam distillation decomposition of the plant material occurs to a greater or less extent. When the distillation is a prolonged one or is carelessly conducted, decomposition of the albumins and albuminous constituents, of ammonia derivatives, of fats and other fatty acid compounds, even of the carbohydrates takes place. Volatile decomposition products are formed, such for instance as sulphuretted hydrogen, ammonia, carbonic acid, trimethylamine, lower and higher fatty acids, acrolein, furfurol, acetaldehyde, phenols of creosote like odor and numerous other bodies, which with few exceptions possess a disagreeable odor, that imparts itself to the volatile oil. In order to understand the character of these decomposition products, it must be remembered, that the process of distillation takes place in the absence of air, for the steam expells the air from the still, so that the reactions of decompositions are largely hydration processes. In addition to this decomposition of the crude material, the volatile oil itself may suffer changes. These may be brought about by the heat, the water and also by the decomposition products of the plant material. They consist of reduction, oxidation, polymerization, condensation, saponification of esters, etc. These changes in the composition of the oil may be slight, but sometimes they are very important. Finally there is still a third source of contamination in the volatile oil, namely that of volatile substances which occur along with the useful volatile oil in the plant. It is true that for the purposes of obtaining the volatile oil, plant material is classified as that containing oil and that which is free from oil. This is correctly expressed if we understand by the term volatile oil, not all volatile substances, but only those which have a commercial value, for there are volatile substances, other than water, in all plants. Volatility is a very common property of organic bodies of low chemical composition, and bodies of this nature, intermediate products of the constructive and destructive metabolism in plant life, are constituents of all plants. The common herbaceous odor of the green parts of plants, the herbaceous and hay-like odor of dried leaves, the woody odor, or the pungent fetid odor of the stems of many blossoms are very apt to cover the odor of a mild smelling volatile oil which is present in only small quantities. This may take place to such an extent, as to make the product wholly worthless.

From this it will be seen, how numerous the causes are for the deterioration of the volatile oils. The foreign odor which is thereby given to the oil, was formerly and probably is even now called "Blasengeruch." This is an old, very inappropriate term, which arose at a time when the various causes of this odor were still unknown. The socalled "Blasengeruch" was formerly the distinguishing property of every crude oil. If it was slight, attempts were made to remove it by long exposure of the oil in open vessels. The more easily volatile impurities gradually evaporate during this treatment, such as sulphuretted hydrogen, acetaldehyde, and ammonia, while the rest are held back by the oil. The more difficultly volatile substances, such as the phenols with a sharp and penetrating odor remain nearly all in the oil. A part of these impurities form new compounds with the constituents of the oils. Finally, the exposure in open vessels is also favorable to the resinification of the oils. On the whole this method for the purification of the oils can only be considered as an inadequate apology.

When the accompanying odor was stronger, or when the oil was colored, it was again distilled, i. e. rectified, and when this was not sufficient, the operation was repeated, hence the terms "twice, thrice rectified oil." At present there are known many methods of purification, entirely different, and each suited for a particular case. In some cases a single rectification by distillation is a useful means for purification, but only when the presence of volatile impurities is small. If properly conducted a single rectification should be sufficient.

2. THE MORE COMMONLY OCCURRING CONSTITUENTS OF VOLATILE OILS.

The volatile oils are widely distributed in the vegetable kingdom, more particularly, however, in the phanerogams. From the cryptogams, a volatile oil has so far been obtained only from the male fern. In most cases the oils exists preformed in the various organs of the plant, the leaves, flowers, fruits, stems and roots. They occur secreted either in glands or in canal-like, intercellular receptacles. Only a few are formed by hydrolysis during the process of preparation from other substances in the plant, e. g. bitter almond oil from amygdalin, mustard oil from sinigrin. Nothing definite is known with regard to their function or their relation to the other constituents of the plant organism. It appears, however, to be well established that they are excretions formed during the life process of the plant, which are of no further importance in the processes of metabolism. It does not follow from this that they are therefore useless, for they act as a means of attracting insects that bring about fertilization, they also protect plants against enemies. These oils are not definite chemical units, but mixtures of substances belonging to many series and classes of compounds. The oil from one and the same plant does not even always possess the same composition. Thus e. g. the distillates from various organs of a plant show differences in odor and physical properties. These differences may become marked even more strongly if plants in different stages of development are distilled. If volatile oils are nevertheless grouped under one generic name, it is because they have several physical properties in common and because almost all of them are prepared in like manner. They have generally been regarded as being closely related to those organic compounds commonly designated as aromatic, but not being of uniform composition they can naturally not be classed with them.

Among the various constituents of an oil, one frequently attracts attention as the principal bearer of the odor. In many instances this constituent is, therefore, the most valuable. The desire to obtain it in a concentrated and pure form, may have given the first impetus to the scientific investigation of essential oils. So long as organic chemistry was in its infancy, these investigations could not be crowned with great success. Only after the methods of investigation had been improved hand in hand with the progress of science, an insight into the nature of volatile oils was obtainable. This insight revealed the fact that the oils are as a rule mixtures of a number of complex substances. How slow this progress was formerly, and how rapid during the past two decades, has been shown in the historical introduction. This chapter will present the results of investigations in so far as they pertain to substances commonly found in volatile oils. For reasons readily understood, only these could be included within the scope of this work. For the same reason a detailed chemical account had to be excluded. Details of chemical and physical properties will be found in the larger chemical handbooks of Beilstein and Ladenburg, and more particularly in the excellent monograph on terpene derivatives by Heusler.

The analysis of a volatile oil is difficult because most of the constituents are liquid, and can, therefore, be separated only by fractional distillation. This operation, imperfect at best, is often rendered more unsatisfactory by the fact that certain constituents are not volatile without decomposition. For this and other reasons it is best to subject an oil to a preliminary examination, the results of which often suggest modifications which simplify the examination considerably. The preliminary examination consists primarily in the determination of the physical properties of the oil and of its elementary composition; also in a study of the behavior of the oil toward certain group reagents, whereby the presence or absence of certain classes of chemical compounds can be ascertained.

Of the physical constants, the specific gravity, the optical properties and the behavior of an oil toward heat and cold allow conclusions to be drawn as to its composition. A specific gravity, e. g. of less than 0.90 indicates the presence of a large amount of terpenes or of compounds of the fatty series. Heracleum oil ($d_{15°}=0.80-0.88$) and oil of rue ($d_{15°}=0.833-8.40$) belong to those with a very low specific gravity. Both consist of derivatives of the fatty series. Oil of orange, sp. gr. 0.848—0.852, consists largely of the terpene limonene; turpentine oil, with a specific gravity of 0.850—0.875, almost completely of hydrocarbons $C_{10}H_{16}$ of the terpene series. A specific gravity higher than 0.90, as is the case with most oils, indicates a mixture of terpenes and possibly their oxygen derivatives; whereas a specific gravity of more than 1.0 indicates the presence of compounds of the aromatic series, or if the compound contains sulphur or nitrogen, of sulphides, nitriles, or isosulphocyanides.

The optical properties, rotation and refraction, are of less importance unless pure chemical compounds are to be examined or adulterations to be looked for. The optical activity of an oil indicates the presence of a compound or compounds with one or more asymmetric carbon atoms; a high index of refraction the presence of a substance or substances with double bonds.

When exposed to low temperatures, a number of oils deposit one or more of their constituents in crystalline form. Some oils, like rose oil, contain crystals even at ordinary temperature, others, like orris oil are butyraceous in consistency. These substances have been designated stearoptenes or camphors and are paraffins, higher members of the series of fatty acids, such as lauric, myristic and palmitic acids, and derivatives of aromatic and hydroaromatic hydrocarbons.

Most oils that can be distilled under ordinary pressure without decomposition begin to boil above 150°. Exceptions to this rule are e. g. those containing sulphur, and the oil of the Digger's pine, which contains normal heptane. In the absence of oxygenated constituents, a boiling point below 200° indicates the presence of terpenes, between 250 and 280° sesquiterpenes, above 300° polyterpenes.

In their elementary composition the volatile oils do not manifest great variety. All contain carbon and hydrogen; most of them also contain oxygen in larger or smaller quantity; few contain nitrogen or sulphur or both. The presence or absence of oxygen can be determined by elementary analysis only. The presence of only a small amount of this element, up to 5 p. c., indicates a high hydrocarbon content. The presence of sulphur, which can be oxidized to sulphuric acid by means of concentrated nitric acid in sealed tubes, indicates sulphides or polysulphides. Nitrogenous compounds are converted into cyanides by heating with metallic sodium or potassium and recognized by means of the Prussian blue reaction. The nitrogen content of an oil is mostly due to nitriles. If sulphur is also found, mustard oils are present which, as a rule, betray their presence by their characteristic odor.

After the elementary composition of an oil has been ascertained, a few group reagents can be applied to learn whether special attention should be given to one class or another of chemical compounds. If an oil shows an acid reaction, it contains acids or phenols. Small amounts of fatty acids occur occasionally as decomposition products of esters present. Larger amounts reveal their presence by the diminution of volume when the oil is shaken with an aqueous solution of caustic or carbonated alkali. The presence of an ester or lactone can be ascertained when an oil is heated with alcoholic potassa of known strength and titrated back with standard acid. This test presupposes the absence of free acids and aldehydes. Alcohols can be converted into acetic esters by heating the oil with acetic acid anhydride. Subsequent saponification will then reveal the presence or absence of an alcohol in the oil. Aldehydes and ketones can be recognized by their addition products with alkali bisulphites, or by their condensation products with hydroxylamine. The latter contain nitrogen and as a rule are difficultly volatile with water vapor. Ethers, which are sometimes present as phenol ethers, can be recognized by means of Zeisel's method.

After these preliminary tests have revealed the presence of a substance belonging to one of the above mentioned classes, it is sometimes possible to separate it without fractional distillation; provided, however, that this method of separation does not change the other constituents of the oil. It should also be noted that these methods of separation never effect a perfect isolation because the other constituents of the oil prevent a part of the substance to be isolated from reacting. When, therefore, the non-reacting portion of the oil is fractionated small amounts of this substance should not be overlooked. Thus it is possible with these group reagents to separate aldehydes and many ketones with acid sulphite solution—a reaction that can at times be facilitated by the addition of alcohol. The crystalline addition product is washed with alcohol and ether and the aldehyde or ketone regenerated by the addition of alkali or dilute acid. Free acids and phenols can be shaken out with aqueous alkali; indifferent substances are then removed from the aqueous solution by shaking it with ether, and then the acid or phenol is set free with dilute mineral acid. If acids and phenols are both present, the former are separated with carbonate solution. Lactones yield salts of the corresponding oxyacids when heated with alcoholic potassa. They are precipitated as lactones or oxyacids by the addition of a mineral acid. Esters that may be present are saponified by this treatment with alkali, aldehydes and ketones, however, are modified thereby and at times destroyed.

If none of these short cuts is possible the oil is fractionated either under ordinary or diminished pressure. If esters are present, the oil is first saponified. It may be taken for granted that the apparatus to be used and the methods to be employed are known. A good guide for distillation under diminished pressure will be found in a small monograph on this subject by Anschütz.[1] The various fractions obtained upon a careful fractionation are examined for compounds, in part according to the methods already mentioned. A trained sense of smell will prove an important additional factor. Suspected compounds are, if possible, converted into crystalline derivatives and thus purified and identified.

Ketones that will not combine with acid sulphites, such as menthone, camphor, fenchone, carvone, are converted into oximes or semicarbazones. Inasmuch as the oximes suffer rearrangement with acids, the semicarbazones are at times to be preferred because the ketones can in most cases be regenerated.

Alcohols are characterized by their capacity to form esters, also phenylurethanes with phenylisocyanate (carbanil). They can be purified by means of difficultly volatile esters of monobasic acids, such as benzoic acid; or by means of acid esters of dibasic acids, such as succinic acid, phthalic acid, etc. Some of the primary alcohols can also be purified by converting them into calcium chloride addition products from which they are easily regenerated by means of water.

From the hydrocarbons traces of oxygenated compounds can be removed by repeated treatment with metallic sodium. If low boiling hydrocarbons are to be distilled under diminished pressure a liquid alloy of potassium and sodium is preferable.

As to the compounds themselves that are found in volatile oils, they belong in part to the aliphatic, in part to the aromatic and hydroaromatic series and are distributed over a large number of classes. The hydrocarbons, especially those of the formula $C_{10}H_{16}$, are widely distributed. Of greater importance, however, are the oxygenated substances, because they are mostly the bearers of the characteristic odor of the oil in which they are contained. In addition to the hydrocarbons there have been found alcohols, aldehydes, acids, esters, ketones, phenols, phenolethers, lactones and oxides, further sulphides, nitriles and isothiocyanates.

---

1) Die Destillation unter vermindertem Drucke.

Hydrocarbons.

a. Aliphatic.

The lowest hydrocarbon of the paraffin series which has been found in a volatile oil is the normal heptane, C_7H_{16} (b. p. 98°). It constitutes the bulk of the oil obtained from the oleoresin of the Digger's pine. *Pinus sabiniana.*

The higher members of the paraffin, and probably of the olefine series also, appear to be quite widely distributed in the vegetable kingdom. They constitute the waxlike coating and secretions on leaves, flowers, fruits etc. In volatile oils, however, they are not met with commonly because of their sparing volatility. Sometimes they separate in crystalline form when the oil is exposed to a low temperature, or they remain behind upon fractional distillation. In the oils of rose and chamomile, however, the amount of paraffin is so large, that the oil congeals even at middle temperature. Apparently these hydrocarbons seldom occur alone, but as mixtures of homologues as has been shown in the case of rose oil. Their melting points seldom if ever agree with those of known members of the series. With the exception of the heptane referred to, they are obtained principally as white, colorless, laminar-crystalline masses which are with difficulty soluble in cold alcohol, but readily soluble in hot alcohol and other organic solvents. They are remarkable on account of their stability toward concentrated acids and oxidizing agents at ordinary temperatures.

The rose oil stearoptene melts at 35° and when distilled in vacuum can be resolved into two fractions melting at 22° and 40—41° respectively. In addition to this solid mixture, paraffin (or olefine) hydrocarbons have been found in the following volatile oils:

Arnica flower oil	M. P.	63°	Sassafras leaf oil	M. P.	58°
Chamomile oil	"	53—54°	Oil of gaultheria	"	65.5°
Dill oil	"	64°	Oil of sweet birch	"	65.5°
Caraway oil (from herb)	"	64°	Oil of wild bergamot	"	62°
Neroli oil (Aurade)	"	55°			

Diolefinic hydrocarbons have so far not been found in volatile oils. The hydrocarbon isoprene, C_5H_8, which belongs here and which is closely related to the terpenes, has been found only as a decomposition product of caoutchouc and of turpentine oil. For its preparation consult Mokiewsky's recent work.[1]

1) Chem. Centrbl., 70¹, p. 589.

However, chain hydrocarbons of the formula of saturation C_nH_{2n-4} with three double bonds have been found. In composition they agree with the terpenes but differ in having a lower specific gravity and in their index of refraction. These hydrocarbons which have been termed "olefinic terpenes" by Semmler have not yet been well characterized. They show a great tendency to resinify, especially when distilled under ordinary pressure.

The first representative of this class was found in oil of bay by Power and Kleber[1] in 1895 and termed myrcene by them. It boils at 167° with partial resinification, under 20 mm. pressure at 67—68°; $d_{15°} = 0.8023$; $n_D = 1.4673$.

The only known reaction by which, in the absence of crystalline addition products, it can be identified is by means of its hydration with glacial acetic and sulphuric acids at 40° according to Bertram's method.[2] Myrcene is thereby converted into an acetate of lavender-like odor, which upon saponification yields linalool. Permanganate oxidizes myrcene to succinic acid.

Similar hydrocarbons have been found by Chapman[3] in oil of hops in 1894; by Gildemeister[4] in Smyrna origanum oil in 1895; and finally by Kleber[5] in sassafras leaf oil in 1896.

b. Aromatic and hydroaromatic.

A hydrocarbon of the aromatic series of the composition $C_{10}H_{18}$ was isolated by Lunge and Steinkauler[6] in 1880 from the oil of the needles of the mammoth fir, *Sequoia gigantea* and was termed sequoiene by them. It consists of laminar crystals which melt at 105° and boil at 290—300° (uncorr.). It is not identical with fluorene and other hydrocarbons of like composition.

Styrene.

The simplest aromatic hydrocarbon with an unsaturated side chain that is found in volatile oils is styrene, $C_6H_5.CH=CH_2$. It occurs in storax oil and recently has been found in xanthorroea resin oil. It is probably formed by the decomposition of cinnamic acid.

Styrene is a colorless, highly refractory liquid of a pleasant odor, which polymerizes to a transparent, glasslike and odorless mass, metastyrene $(C_8H_8)_n$, by being kept for some time, and more rapidly by

---

1) Pharm. Rundsch., 13, p. 61.
2) G. I. P. 80711.
3) Journ. Chem. Soc., 67, p. 54.
4) Arch. d. Pharm., 233, p. 184.
5) Bericht S. & Co., April 1896, p. 71.
6) Berichte, 13, p. 1656; 14, p. 2202.

heating or when in contact with acids. The boiling point of pure styrene is 144—144.5°.

As to its physical properties, the following statements have been made:

B. p. 140° at 760 mm.; $d_{20°}=0.9074$; $n_{\alpha}=1.54030$ (Brühl,[1] 1886).
B. p. 146.2°; $d_{0°} = 0.9251$ (Weger,[2] 1883); $d_{4°}^{17°} = 0.90595$; $n_D = 1.54344$ (Nasini & Bernheimer,[3] 1885).

Pure styrene is optically inactive. Dilute nitric acid or chromic acid mixture oxidize it to benzoic acid.

For identifying styrene the well crystallized styrene dibromide $C_6H_5.CHBr.CH_2Br$ is used, and is obtained by allowing bromine (17 parts) to drop into a solution of the hydrocarbon (10 parts) in twice its volume of ether (Zincke,[4] 1883). On evaporation, the bromide separates in crystals, which after being recrystallized from 80 percent alcohol have the melting point 74—74.5°.

Cymene.

Of the hydrocarbons designated as cymenes only the meta and para compounds are of importance to the chemistry of the volatile oils. While m-cymene has been observed only as a decomposition or "Abbau" product (Kelbe, dry distillation of colophony; Wallach, dehydration of fenchone; Baeyer, splitting off hydrogen from sylvestrene), p-cymene is a frequent constituent of volatile oils. Up to the present it has been found in the volatile oils from *Thymus vulgaris, Th. serpyllum, Th. capitatus, Satureja hortensis, S. thymbra, Monarda punctata, Ptychotis ajowan*, in origanum oil from Trieste and Smyrna, in Roman caraway oil from *Cuminum cyminum*, in the oil from the seeds of water-hemlock, *Cicuta virosa*, and in the oil from *Eucalyptus haemastoma*. Like m-cymene it has also been frequently obtained as a transformation product. Formerly it was thought that all terpenes were related to this hydrocarbon; that they were its hydroderivatives. According to recent investigations, however, this conception is not wholly correct. It is worthy of mention, that several compounds of the formula $C_{10}H_{16}O$ can be changed to p-cymene by the abstraction of water, for instance, camphor and citral.

---

1) Liebig's Annalen, 235, p. 13.
2) Liebig's Annalen, 221, p. 69.
3) Gazz. chim., 15, p. 59; Jahresber. f. Chem., 1885, p. 314.
4) Liebig's Annalen, 216, p. 288.

Cymene is a colorless, pleasant smelling liquid, which possesses the peculiarity of becoming turbid, with separation of water on standing for some time; the reason for this behavior is not known with certainty. perhaps it is due to a gradual oxidation. For common cymene was found:

B. p. 175—176°; $d_{15°}=0.8602$ (Widman,[1] 1891).

B. p. 175.2—175.9° (at 752 mm.); $d\frac{20°}{4°}=0.8551$; $n_D=1.48465$ (Brühl.[2] 1892).

B. p. 173.5—174.5° (at 763 mm.); $d_{15°}=0.8595$; $d_{20°}=0.8588$; $n_D=$ 1.479 (Wolpian,[3] 1896).

The pure hydrocarbon is optically inactive. Dilute nitric acid and chromic acid mixture oxidize it to p-toluic acid and finally to terephthalic acid. Potassium permanganate acts on it only with difficulty and changes it, particularly with heat, into p-oxyisopropylbenzoic acid (m. p. 155—156°) which with dilute hydrochloric acid yields p-isopropenylbenzoic acid (m. p. 255—260°) by splitting off water (R. Meyer & Rosicki,[4] 1883). Oxyisopropylbenzoic acid is characteristic for p-cymene and is used for its identification. For its preparation Wallach in 1891 gave the following directions:[5]

2 g. at a time of the hydrocarbon, prepared as pure as possible, are heated with a solution of 12 g. of potassium permanganate in 330 g. of water on a waterbath with reflux condenser, the mixture being frequently agitated. When the oxidation is complete, the filtrate from the oxides of manganese is evaporated to dryness and the saline residue boiled with alcohol. The potassium salt which is soluble in the alcohol, is decomposed in aqueous solution with dilute sulphuric acid and the precipitated acid recrystallized from alcohol.

The barium salt of the sulphonic acid produced by treating the hydrocarbon with concentrated sulphuric acid is also characteristic for cymene. It $(C_{10}H_{13}.SO_3)_2Ba$ crystallizes in shining, difficultly soluble laminae and contains three molecules of water of crystallization which can be completely driven off at 100°. The sulphone amide which can be prepared from the chloride of this sulphonic acid melts at 115—116°. That cymene sulphonic acid may also result from the action of sulphuric acid on different terpenes must, however, be remembered.

Terpenes.

The principal hydrocarbons occurring in volatile oils have the composition $C_{10}H_{16}$, are of a cyclic nature and belong to the class of

1) Berichte, 24, p. 452.
2) Berichte, 25, p. 172.
3) Pharm. Zeitschr. f. Russl., 35, p. 115.
4) Liebig's Annalen, 219, p. 282.
5) Liebig's Annalen, 264, p. 10.

"terpenes proper." As to their formation in the plant organism nothing definite is known. They possibly are genetically related to oxygenated chain compounds from which hydrocarbons $C_{10}H_{16}$ can be prepared artificially by splitting off water. It is remarkable, however, that the terpene content of an oil is greater the less developed the plant was at the time of distillation.

The majority of the known hydrocarbons of the terpene group are found ready formed in nature: viz., pinene, camphene, limonene, dipentene, phellandrene, sylvestrene, terpinene. With exception of the racemic dipentene and the inactive terpinene, these hydrocarbons occur in both optical modifications, though in different volatile oils and with a varying degree of rotation. It might appear doubtful, therefore, whether dipentene and terpinene are natural hydrocarbons or decomposition products of readily changeable terpenes.

To isolate a terpene in pure form from a volatile oil by means of fractional distillation is mostly impossible. Neither is it necessary, for it suffices to remove oxygenated compounds from the fraction 150—180° by repeated distillation over sodium. The physical constants can then be determined and a characteristic crystalline derivative prepared. These will be described under the individual hydrocarbons.

Pinene.

Pinene is the principal constituent of the distillates from the resinous excretions of different species of *Pinus*, which occur in commerce under the name of turpentine oils. In smaller amounts, usually together with other terpenes, it has been found in a large number of volatile oils. Both optically active modifications occur; American turpentine oil, German and Swedish pine needle oil from *Pinus silvestris*, the tar oil (*Kienöl*) from the same species and the distillate from the needles of *Pinus cembra* consist mainly of d-pinene, while French turpentine oil contains mostly l-pinene.

d-Pinene (australene) has been shown to be present in cypress oil, star anise oil, camphor oil, oil of laurel leaves and berries, mace oil, fennel oil, galbanum oil, coriander oil, niaouli oil, myrtle oil, oil of cheken leaves, eucalyptus oil (*E. globulus*), French basilicum oil, spike oil, and tansy oil. l-Pinene (terebentene) occurs in the following coniferous oils: in the oil from the leaves and cones of *Abies alba*, in the pine needle oils, in the oil of *P. montana* Miller, in English pine needle oil from *P. silvestris*, hemlock oil, in the oils of *Abies sibirica* and *A. canadensis*, further in canella oil, olibanum oil, cajeput oil, valerian and kesso-root oils, in thyme oil, spearmint oil, in the oil from the fruit of *Petroselinum sativum* (parsley seed oil), and the oil of *Asarum europaeum*. In small amounts pinene has also been found in elderberry oil, thuja oil, nutmeg oil, massoy bark oil, sassafras leaf oil, rosemary oil, in peppermint oil, sage oil, lavender oil, and the oils of *Thymus capitatus*, *Satureja thymbra* and *Asafoetida*.

Pinene is one of the few terpenes which can be obtained in a comparatively pure state. It is obtained, although optically inactive, by treating the solid pinene nitrosochloride with aniline in alcoholic solution (Wallach,[1] 1889 and 1890). The pinene thus obtained shows the following properties:

B. p. 155—156°; $d_{20°}=0.858$; $n_{D21°}=1.46553$ (Wallach,[2] 1890).

For the preparation of the active modification of pinene, American or French turpentine oil is used, the fraction boiling below 160° being purified by fractional distillation until the hydrocarbons obtained coincide in boiling point and in the other properties with i-pinene. The highest rotations so far observed are:

for d-pinene from the oil of the needles of the Siberian cedar, *Pinus cembra* (Flawitzky,[3] 1892)

$[\alpha]_{D18°}=+45.04°$ (B. p. 156° corr. at 758 mm B.; $d\frac{20°}{4°}=0.8585$);[3]

for l-pinene from French turpentine oil (Flawitzky,[4] 1879)

$[\alpha]_{D20°}=-48.4°$ (B. p. 155°; $d_{20°}=0.8587$).[4]

If it is necessary to employ solvents in determining the direction and magnitude of the rotation, the influence which these exert must be considered (Landolt,[5] 1877; Rimbach,[6] 1892).

Pinene is a colorless, mobile liquid which, like all terpenes, takes up oxygen from the air on standing and becomes partly resinified. It is readily converted into other terpenes, thus it is changed by a higher temperature (250—270°) or by moist hydrochloric acid into dipentene and its dihydrochloride respectively; by alcoholic sulphuric acid into terpinolene and terpinene, probably with dipentene as an intermediate step. By treating pinene monohydrochloride with sodium acetate or aniline, camphene is formed, pinene nitrosochloride, however, yields under the same conditions pure pinene. In contact with dilute mineral acids, pinene is converted in time into terpin hydrate $C_{10}H_{18}(OH)_2 . H_2O$

1) Liebig's Annalen, 252, p. 132; and 258, p. 343.

2) Liebig's Annalen, 258, p. 344.

3) Journ. f. pr. Chem., II. 45, p. 115.

4) Berichte, 12, p. 2357.

5) Liebig's Annalen, 189, p. 311—317.

6) Zeitschr. phys. Chem., 9, p. 701.

(m. p. 116—117°), and by hydration with sulphuric acid and glacial acetic acid it is converted into terpineol. Oxidizing agents act differently on pinene. While concentrated nitric acid produces so violent a reaction that ignition often takes place, dilute nitric acid, like chromic acid mixture yields besides lower fatty acids and other products, terephthalic acid $C_8H_6O_4$ and terebic acid $C_7H_{10}O_4$. Entirely different results are obtained with potassium permanganate; very dilute permanganate solution induces, as the investigations of Wagner[1] (1894 and 1896) have shown, chiefly the formation of neutral oxidation products, such as pinene glycol, etc. By employing a concentrated solution a monobasic keto acid $C_{10}H_{16}O_3$, pinonic acid (m. p. 103—105°) is formed (Tiemann and Semmler[2] 1895—1896; Baeyer[3] 1896). This on the one hand, yields finally terebic acid, on the other hand the same "Abbau" products as those formed by the oxidation of the derivatives of camphor.

Pinene is an unsaturated hydrocarbon with one double bond which may be removed by addition. By conducting dry hydrochloric acid gas or hydrobromic acid gas into perfectly dry and cooled pinene, the monohydrohalogen derivatives of pinene are formed, of which the hydrochloride, $C_{10}H_{16}HCl$, called "artificial camphor" on account of its camphorlike odor, melts at 125°, the hydrobromide at 90°. Both are converted into camphene on splitting off the hydrohalogen. If bromine in a dry solvent is allowed to act on pinene, one molecule is readily taken up with decoloration. On further addition the absorption takes place only very slowly and is accompanied by evolution of hydrobromic acid. From the product formed by the addition of one molecule of bromine to pinene a dibromide melting at 169—170° may be obtained by distillation with water vapor[4] (Wallach, 1891). A larger yield of this dibromide is obtained by treating pinene with hypobromous acid[5] (Wagner and Ginsberg, 1896). By splitting off hydrobromic acid by means of aniline the dibromide is changed to cymene, while when treated with zinc dust in alcoholic solution it yields a new terpene of the melting point 65—66°, tricyclene[6] (Godlewski and Wagner, 1897).

The compounds which are best suited for the characterization of pinene are pinene nitrosochloride, formed by the addition of nitrosylchloride to pinene, and the nitrolamines which may be prepared from it

---

[1]) Berichte, 27, p. 2270; 29, p. 881.
[2]) Berichte, 28, p. 1345; 29, pp. 529, 3027.
[3]) Berichte, 29, pp. 22, 326, 1907, 1923, 2775.
[4]) Liebig's Annalen, 264, p. 8.
[5]) Berichte, 29, p. 890.
[6]) Chem. Zeitung, 21, p. 98.

by treatment with organic bases. For the preparation of the nitrosochloride Wallach (1888 and 1889) has given the following directions:[1]

A mixture of 50 g. each of turpentine oil (immaterial whether laevo- or dextrogyrate) glacial acetic acid and ethyl nitrite[2] are well cooled in a freezing mixture and 15 cc. of crude (33 percent) hydrochloric acid are gradually added. The nitrosochloride soon separates in a crystalline form, and is obtained in a fairly pure state when it is filtered off with a suction pump and well washed with alcohol. From the filtrate some more nitrosochloride separates on standing in the cold. It is profitable in regard to yield to work with small quantities, as only then can the low temperature be maintained which is necessary for the satisfactory conduct of the reaction; large quantities of pinol $C_{10}H_{16}O$ are formed as a by-product.

The nitrosochloride is a white crystalline powder, which is readily soluble in chloroform and may be again separated from this solution by methyl alcohol. The melting point of the recrystallized compound is 103°. Like its derivatives it is optically inactive. According to observations by Baeyer[3] pinene nitrosochloride is a bisnitroso compound $(C_{10}H_{16}Cl)_2N_2O_2$, which in ethereal solution is changed by hydrochloric acid into hydrochlorcarvoxime.[4] By splitting off hydrochloric acid with alcoholic potassa it is converted into nitrosopinene (m. p. 132°)[5] which has been recognized as the oxime of an unknown ketone $C_{10}H_{14}O$.[6] Aromatic bases, such as aniline and toluidine split off nitrosylchloride, regenerating pinene with the formation of amidoazo-compounds. Quite different is the behavior of the nitrosochloride to bases of the fatty series and those which possess their characteristics, as for instance, benzylamine and piperidine. With primary bases as well as with piperidine nitrolamines result; the secondary bases, like diethylamine, however, produce a splitting off of hydrochloric acid and formation of nitrosopinene.

As the nitrosochlorides of different terpenes show very similar melting points and besides decompose very readily, they are less suitable for characterization than the nitrolamines, which are very stable and crystallize readily; these have been prepared in large numbers, although the compounds produced by the reaction with benzylamine and piperidine

---

1) Liebig's Annalen, 245, p. 251; 253, p. 251.

2) This is easily obtained by allowing a mixture of 200 g. of concentrated sulphuric acid, 1.5 liters of water and 100 g. of alcohol to flow into a solution of 250 g. of sodium nitrite in 1 liter of water and 100 g. of alcohol. The ethyl nitrite which forms at once must be condensed in well cooled receivers. (Wallach.)

3) Berichte, 28, p. 648.

4) Berichte, 29, p. 12.

5) Wallach & Lorentz, Liebig's Annalen, 268 (1891), p. 198.

6) Urban & Kremers, Amer. Chem. Journ., 16 (1894), p. 404, Baeyer, Berichte, 28 (1895), p. 646; Mead & Kremers, Amer. Chem. Journ., 17 (1895), p. 607.

are preferably used. For their preparation[1] the nitrosochloride is treated in excess with the base dissolved in alcohol and heated on a water bath; the nitrolamine formed is separated by the addition of water. The melting point of pinene nitrolpiperidine is 118—119°, that of the nitrolbenzylamine 122—123°.

CAMPHENE.

Camphene is the only hydrocarbon $C_{10}H_{16}$ known in the solid form. In spite of this apparent advantage it has, with the exception of a single case, not been possible to separate it as a crystalline body from a volatile oil. Camphene is found in the vegetable kingdom in both active modifications: as d-camphene (austracamphene of Berthelot) it is contained in ginger oil and spike oil; as l-camphene (terecamphene[2] of Berthelot) in citronella, valerian and kesso oils. It has also been obtained from rosemary oil, camphor oil, French and American turpentine oils, as well as from the oil of *Pinus sibirica*, but only from the last in the solid form. Artificially camphene is obtained in various ways, generally by splitting off hydrochloric acid from pinene monohydrochloride or bromide, or bornylchloride (borneo camphene). It is most readily prepared from isoborneol by the abstraction of water with zinc chloride.

It is a white, crumbling, crystalline mass with a faint camphorlike odor, and shows a great tendency to sublime, but otherwise is much more stable than the other terpenes toward air and light. As it is obtained in a solid form and may be freed from adhering liquid bodies by dissolving in alcohol and carefully adding water, it is one of the few terpenes which can be prepared in a pure state. The following constants are given:

for borneocamphene:

M. p. 48—49°; b. p. 160—161° (Wallach,[3] 1885); $d_{48°}=0.850$; $n_{C48°}=1.4555$ (Wallach,[4] 1888);

M. p. 53.5—54°; $d\frac{58.6°}{4°}=0.83808$; $n_{D58.6°}=1.45314$ (Brühl,[5] 1892);

for camphene from pinene chlorhydrate:

M. p. 51—52°; b. p. 158.5—159.5°; $d\frac{54°}{4°}=0.84224$; $n_{D54°}=1.45514$ (Brühl,[6] 1892);

1) Wallach, Liebig's Annalen, 245 (1888), p. 258; 252 (1889), p. 130.

2) Camphene prepared from pinene hydrochloride is also sometimes designated as terecamphene.

3) Liebig's Annalen, 20, p. 284.

4) Liebig's Annalen, 245, p. 210.

5) Berichte, 25, p. 164.

6) Berichte, 25, p. 162.

for camphene from isoborneol:

M. p. 50°; b. p. 159—160° (Bertram & Walbaum,[1] 1894); 56° at 15 mm.

The rotation of the hydrocarbon artificially prepared from pinene hydrochloride or bornylchloride varies according to the extent of the rotation of the material used and the height and length of time of the temperature employed during the reaction; thus Bouchardat and Lafont[2] (1887) obtained hydrocarbons whose rotation $[\alpha]_D$ was between —80° 37′ and —30° 30′ by treating l-pinene monohydrochloride ($[\alpha]_D$ = —28° 30′) with potassium acetate in alcoholic solution at 150 to 170°.

For a d-camphene obtained from bornyl chloride Kachler[3] observed $[\alpha]_{D85°}$ = +20° (100.3 mm).

Camphene is not as readily inverted into isomers as the other terpenes. Although it is changed by heating for a long time to a high temperature or by treatment with dehydrating agents such as zinc chloride, phosphoric acid anhydride or concentrated sulphuric acid, the resulting products do not have the formula $C_{10}H_{16}$.

As a terpene with one double bond camphene yields addition products with halogens (dibromide m. p. 90°) as well as with hydrohalogens, but not with nitrosylchloride. Whether the hydrohalogen derivatives are identical with the corresponding compounds of borneol or isoborneol has not been definitely settled.[4] The hydrochloride is obtained by conducting gaseous hydrochloric acid into an alcoholic solution of camphene. According to Reychler[4] (1896) it melts at 150—152°; Kachler and Spitzer[5] (1880) report 156—157°.

Oxidizing agents, such as chromic acid mixture, permanganate and nitric acid, do not act in like manner on the hydrocarbon. Thus dilute permanganate solution in the cold yields first camphene glycol $C_{10}H_{16}(OH)_2$, melting at 192°. On further oxidation it produces camphene-camphoric acid $C_{10}H_{16}O_4$, small quantities of a ketone, camphenilone $C_9H_{14}O$ (m. p. 36—38°), homologous with camphor, and principally an acid $C_{10}H_{16}O_3$ of the melting point 171—172° (Wagner's camphenylic acid, Bredt's oxycamphenilanic acid). Nitric acid oxidizes camphene principally to the tribasic camphoylic acid (Marsh and Gardner) $C_{10}H_{14}O_6$ (Bredt's carboxyl-apocamphoric acid). Accompany-

---

1) Journ. f. prakt. Chem. II., 49, p. 8.
2) Compt. rend., 104, p. 694; Bull. Soc. chim. II., 47, p. 488.
3) Liebig's Annalen, 197, p. 97.
4) Jünger & Klages, 1896, Berichte, 29, p. 544; Reychler, ibid. p. 696.
5) Liebig's Annalen, 200, p. 343.

ing this are small quantities of the ketone $C_9H_{14}O$ already mentioned. Chromic acid mixture yields for the main part camphor with a little camphoric acid and other products.

All of these derivatives of camphene, however, do not admit of being used for its characterization. If fairly pure camphene fractions are under consideration, the hydrocarbon may be separated in the form of its chlorhydrate. Camphene may, however, be better identified by converting it into isoborneol. Only when larger quantities of pinene are present at the same time with the camphene the detection is unsatisfactory even with this method, because the inactive terpineol which is formed at the same time with the isoborneol retains the latter in solution and a separation of the mixture can be only partially accomplished.

According to Bertram and Walbaum (1894) camphene can be converted into isoborneol according to the following method:[1]

100 parts of the camphene fraction are heated with 250 parts of glacial acetic acid and 10 parts of 50 percent sulphuric acid for 2—3 hours to 50—60° with frequent shaking; the mixture which at first separates into two layers becomes finally homogeneous and has a slightly reddish color. When the reaction is ended the acetate formed is separated by water, washed repeatedly and then saponified by heating with a solution of 50 g. of potassium hydrate in 250 g. of alcohol. After removing the alcohol the isoborneol is precipitated as a crumbly mass by the addition of water and is purified by recrystallization from petroleum ether. The melting point of isoborneol is about 212°; the determination must, however, be made in a sealed capillary tube on account of its great tendency to sublime. For the further characterization of the isoborneol its bromal compound, m. p. 71—72°[2] may be used.

Fenchene.

A strict proof that fenchene exists in volatile oils has not been furnished up to the present. By the action of different acids Bouchardat and Lafont[3] (1891) have obtained from French turpentine oil and Bouchardat and Tardy[4] (1895) from fractions of the oil of *Eucalyptus globulus* an alcohol of the melting point 42—43°, the identity of which with fenchylalcohol has been shown[5] (1898). It appears, therefore, that this hydrocarbon is to be classed with those terpenes which occur in nature, and for this reason it may be briefly considered.

Up to the present fenchene has been prepared artificially only by the abstraction of water from fenchyl alcohol or by splitting off hydro-

---

1) Journ. f. pr. Chem. II, 49, p. 1.
2) Journ. f. prakt. Chem. II, 49, p. 6.
3) Compt. rend., 113, p. 553; 125, p. 112.
4) Compt. rend., 120, p. 1418.
5) Compt. rend., 126, p. 755.

chloric acid from fenchyl chloride[1] (Wallach 1891). It is a liquid terpene, recalling camphene in odor, and is known in both optically active modifications. As constants have been found:

B. p. 154—155°; $d_{18°} = 0.8660$; $n_{D18°} = 1.4693$.

B. p. 155—156°; $d_{18°} = 0.8670$; $n_{D18°} = 1.47047$;[2] $\alpha_D = \pm 21°$ (W.,[3] 1898).

Repeated experiments in the laboratory of Schimmel & Co. gave the following constants:

B. p. 154—156° (765 mm.); $d_{15°} = 0.8660—0.8665$;
$n_{D16°} = 1.46733—1.46832$.

Fenchene yields no characteristic addition products with halogens, hydrohalogens, or nitrosylchloride which might be used for characterization. Like camphene it can, however, be hydrated with glacial acetic acid and sulphuric acid to an alcohol $C_{10}H_{18}O$, isofenchyl alcohol (m. p. 61.5—62°) which with phenyl isocyanate yields a phenylurethane melting at 106—107°.[4]

Limonene.

Limonene is a very widely distributed terpene. Besides the dextro and laevogyrate forms it also occurs in the inactive modification, called dipentene. As some of the dipentene derivatives are different in properties from the corresponding limonene derivatives it is usually considered as a separate hydrocarbon. d-Limonene (citrene) is contained in the citrus oils, in larger quantities in the oils of orange peel, lemon, and bergamot, in mandarin oil, in Italian limette oil, in neroli and petitgrain oils, further in caraway, dill and Macedonian fennel oil, in celery, erigeron and kuromoji oils. l-Limonene does not occur as frequently. It has been found in the oil from the needles and cones of *Abies alba*, in Russian and American spearmint oil and American peppermint oil.

Since the optically active limonenes could not be regenerated from solid derivatives, recourse had to be taken for their preparation to the fractional distillation of orange peel or caraway oils, and in the case of l-limonene, of oil of *Abies alba*. The most carefully purified hydrocarbon possesses a pleasant lemon like odor, which in the distillates obtained from caraway oil or oil of *Abies alba* changes after keeping for a short time, the odor then reminding of the oils used. This is probably due

1) Liebig's Annalen, 263, p. 148.
2) Liebig's Annalen, 300, p. 313.
3) Liebig's Annalen, 302, p. 376—377.
4) Bericht von S. & Co., October 1898, p. 56.

to the presence of small quantities of foreign substances. The physical constants obtained from such fractions are as follows:

B. p. 175—176°; $d_{15°} = 0.8464$; $n_{D17°} = 1.47568$.

For l-limonene from pine needle oil Wallach found (1888):

B. p. 175—176°; $d_{20°} = 0.846$; $n_{D20°} = 1.47459$.[1]

The rotatory power is not constant. The greatest deviation was found in the laboratory of Schimmel & Co. for a d-limonene fractionated in vacuum from caraway oil $[\alpha]_D = +123°40'$ (no solvent), consequently a still greater value than that found by Kremers[2] (1895). For l-limonene from pine needle oil Wallach[3] (1888 and 1889) has reported $[\alpha]_D = -105°$ (in alcoholic or chloroformic solution), and Tilden and Williamson[4] in 1893 found a similar value $[\alpha]_D = -106°$.

Quite recently Godlewsky[5] has regenerated limonene from the tetrabromide by reduction with zinc dust. The limonene resulting had the following properties:

B. p. 177.5° at 759 mm.; $d_{0°}^{0°} = 0.8585$; $d_{4°}^{0°} = 0.8584$; $d_{4°}^{20°} = 0.8425$; $[\alpha]_D = +125° 36'$ at 20°.

The regenerated limonene again added bromine to form the tetrabromide melting at 104°.

The two limonenes are completely alike in their chemical behavior: both give the same derivatives, differing only in their optical rotation. By mixing equal amounts of d- and l-limonene dipentene results, which hydrocarbon is also formed when the optically active limonenes are heated to a high temperature or treated with acids. In a completely dry state limonene absorbs one molecule of hydrohalogen with the formation of liquid, optically active compounds[6] (Wallach). By replacing the halogen atom by the hydroxyl group this yields optically active terpineol[7] (Semmler, 1895). Only in the presence of moisture does an addition of two molecules of hydrohalogen take place, with the formation of derivatives of dipentene.

Limonene takes up four atoms of bromine and forms the optically active limonene tetrabromide (see p. 115) melting at 104—105°. By the addition of nitrosylchloride there result two (α- and β-) nitrosochlorides[8] (Wallach, 1889) which are prepared in a similar manner to pinene nitrosochloride and are to be considered as physical isomers.

1) Liebig's Annalen, 246, p. 222.

2) Amer. Chem. Journ., 17, p. 692.

3) Liebig's Annalen, 246, p. 222; Wallach & Conrady, ibidem 252, p. 145.

4) Journ. Chem. Soc., 63, p. 293.

5) Chem. Centrbl., 70ı, p. 1241.

6) Annalen, 270, p. 188.

7) Berichte, 28, p. 2190.

8) Liebig's Annalen, 252, p. 109.

The nitrosochlorides are perfectly alike in chemical behavior, both are changed to carvoxime melting at 72° by the abstraction of hydrochloric acid with alcoholic alkali, and yield when treated with bases the same (two) nitrolamines[1] (Wallach, 1892).

Limonene is converted by dilute permanganate solution into the saturated tetratomic alcohol limonetrite (m. p. 191.5—192°)[2] (Wagner, 1890).

A characteristic derivative of limonene, which is often used for identification, the tetrabromide, is prepared according to Wallach's directions[3] (1887):

The fraction is purified as much as possible and diluted with about four times its volume of glacial acetic acid. Bromine is added drop by drop to the well cooled solution as long as this is taken up with decoloration. The crystals separating after standing for some time are collected and recrystallized from acetic ether. The melting point of the pure tetrabromide is 104.5°.

In the preparation of the tetrabromide it is to be observed that the use of absolutely anhydrous reagents is not necessary. In fact these give rise to the formation of an uncrystallizable tetrabromide. Nevertheless the crystallized product is to be considered as the normal bromide.[4] Wallach does not approve of using alcohol and ether as diluents, for they induce the formation of liquid by-products. The same difficulty is met with when the terpene fractions are impure.

Small changes in the method of preparation have been recommended by other investigators. Baeyer and Villiger[5] (1894) dilute the fraction to be investigated with an equal volume of amyl alcohol, and after the addition of twice the volume of ether cool and add the bromine drop by drop. As the ether evaporates the tetrabromide separates out.

Power & Kleber[6] (1894) allow the fraction to be tested for limonene to drop into a cooled mixture of glacial acetic acid and bromine until only a slight excess of bromine is present, then decolorize with an aqueous solution of sulphurous acid and precipitate with water; in this manner they want to avoid the formation of hydrobromic acid always noticed in the usual bromination method, and to prevent the formation of non-crystallizable bromides of isomeric terpenes.

Finally, a combination of both methods has been recommended by Godlewsky[7] (1898), who directs, to allow the solution of the terpene

---

1) Liebig's Annalen, 270, p. 172.

2) Berichte, 28, p. 2815.

3) Liebig's Annalen, 289, p. 8.

4) Liebig's Annalen, 264, p. 14.

5) Berichte, 27, p. 448.

6) Pharm. Rundschau, 12, p. 160; Archiv d. Pharm., 232, p. 646.

7) Chem. Zeitung, 22, p. 827.

in a mixture of equal parts by weight of amyl alcohol and ether to drop into an ethereal solution of bromine, which is to be cooled with ice water during the reaction.

Dipentene.

Dipentene, the optically inactive modification of limonene has often been found occuring naturally, for instance in pine needle oil, citronella and palmarosa oils, in the oils of cubeb, pepper, cardamom, camphor, nutmeg, kuromoji, massoybark, bergamot, limette leaf, in fennel, golden rod, myrtle and kesso root, also in the oils from *Satureja thymbra* and *Thymus capitatus*, as well as in olibanum and elemi oils. In addition to its formation from equal parts of d- and l-limonene it is obtained by the polymerization of the unsaturated aliphatic hydrocarbon isoprene C_5H_8; with terpinene by the abstraction of water from the aliphatic tertiary alcohol linalool $C_{10}H_{18}O$. It is also formed by isomerization of other hydrocarbons $C_{10}H_{16}$, for instance, from pinene, limonene, phellandrene; and results from oxygenated compounds by different transformations, for instance from cineol, terpineol and terpin hydrate.

A relatively pure dipentene is obtained by the dry distillation of caoutchouc. After separating the first fraction containing the isoprene, the fraction boiling at 172—178° is carefully subjected to repeated fractional distillation from sodium. A less pure preparation is obtained from dipentene dihydrochloride by splitting off hydrochloric acid with aniline or sodium acetate and glacial acetic acid[1] (Wallach, 1887) or from crystallized terpineol by the abstraction of water with potassium bisulphate[2] (Wallach, 1893).

Dipentene is distinguished from limonene in its physical properties only by its optical activity; boiling point,[3] specific gravity, and index of refraction are identical with the data determined for limonene. For dipentene from caoutchouc the following data have been found in the laboratory of Schimmel & Co.:

B. p. 175—176°; $d_{20^\circ} = 0.844$; $n_{D20^\circ} = 1.47194$.

Dipentene is relatively stable, for upon heating it is not converted into an isomeric hydrocarbon $C_{10}H_{16}$, but yields polymers. Alcoholic sulphuric acid, however, on heating changes it to terpinene. Its deriva-

1) Liebig's Annalen, 239, p. 3; 245, p. 196. Comp. also Tilden and Williamson. Journ. Chem. Soc., 63, p. 294.

2) Liebig's Annalen, 275, p. 109.

3) Liebig's Annalen, 286, p. 138.

tives are inactive and may be obtained from dipentene itself, as well as by combining equivalent amounts of the corresponding optically active compounds of limonene. They show a few small differences from those of the active limonenes, namely in respect to their melting points, on account of which it appears justifiable to treat dipentene as a separate hydrocarbon. Dipentene behaves like the active limonenes toward hydrohalogen, bromine and nitrosylchloride; the solid addition products formed with two molecules of hydrohalogen exist in two different forms, cis- and trans-modifications, of which the lower melting and more readily soluble is designated as the cis-form (Baeyer,[1] 1893). The higher melting trans-form is the more stable. It always forms, when the reaction takes place with heat. In the cold both forms usually result together. As the trans-form is usually obtained, the following data refer only to this modification. The dipentene dihydrohalides are obtained from limonene as well as from dipentene, when the cooled solutions of the hydrocarbons in ether, glacial acetic acid, etc., are saturated with the respective hydrohalogen. Upon evaporation of the solvent or upon dilution with water the compounds separate as oils which soon crystallize. The dihydrochloride melts at 50° and can be obtained in a crystalline form from its alcoholic solution by the careful addition of water; the dihydrobromide forms rhombic tablets of satin-like lustre and has the melting point 64°: the dihydroiodide crystallizes in different forms and melts at 77—81°. From all of these compounds dipentene may be regenerated by splitting off hydrohalogen.

The nitrosylchloride compound is known in two modifications (α and β) both of which yield inactive carvoxime by splitting off hydrochloric acid with alcoholic potash. When acted upon by bases there are formed from each of the nitrosochlorides two nitrolamines.

For the detection of dipentene the compound usually used is the tetrabromide, the formation of which results in the same manner as was described for limonene tetrabromide; it also forms when concentrated solutions of equal parts by weight of d- and l-limonene tetrabromide are mixed. The crystals are distinguished by their habit, difficult solubility, and higher melting point, 124—125°, from those of the limonene compounds[2] (Wallach, 1888).

For identification the nitrosochloride, which can be changed to inactive carvoxime by heating with alcoholic potash, is also suitable.

---

1) Berichte, 26, p. 2861.

2) Liebig's Annalen, 246, p. 226.

Sylvestrene.

Sylvestrene is one of the rarely occuring terpenes. Up to the present time it has been found in German and Swedish pine needle oil from *Pinus silvestris*, in the oil from the leaves of *Pinus montana*, in pine tar oil and Finnish turpentine oil. In all of these it occurs in the dextrogyrate modification.

It may be obtained in a comparatively pure state by preparing the dihydrochloride from fractions rich in sylvestrene and decomposing this by boiling with aniline[1] (Wallach, 1885), or sodium acetate and glacial acetic acid[2] (Wallach, 1887). The hydrocarbon thus obtained resembles limonene almost completely in its physical and chemical properties. Like this it possesses a pleasant odor, reminding of lemon and bergamot oil, and its specific gravity and boiling point are almost identical with those of limonene.

Atterberg, the discoverer of sylvestrene, ascertained the following physical constants:

B. p. 173—175°; $d_{16°} = 0.8612$; $[\alpha]_{D16°} = +19.5°$.[3]

Wallach reports the following constants:

B. p. 176—177°; $d_{16°} = 0.851$; $n_D = 1.47799$.[4]

B. p. 175—176°; $d_{20°} = 0.848$; $n_D = 1.47573$; $[\alpha]_D = +66.32°$ (in chloroform).[5]

When heated to 250° sylvestrene is polymerized, but is not changed by this physical agent nor by the action of alcoholic sulphuric acid into isomeric terpenes. This hydrocarbon is, therefore, one of the most stable of the terpene group. Like limonene, sylvestrene has two double bonds which may be wholly or partially broken by the addition of hydrohalogen, bromine or nitrosyl chloride. A peculiar behavior is shown by the dihydrochloride, which in opposition to the inactive dipentene dihydrochloride is optically active and by splitting off hydrochloric acid yields active sylvestrene. The tetrabromide, prepared in the same manner as that of limonene, melts at 135—136°, but is obtained with difficulty in the solid form when, as is usually the case in terpene fractions, other hydrocarbons are present. Pure sylvestrene yields, when treated with amyl nitrite and hydrochloric acid, a nitrosochloride melting at 106—107°. With benzylamine this forms a nitrolamine base of the melting point 71—72°.

---

1) Liebig's Annalen, 230, p. 243.

2) Liebig's Annalen, 239, p. 25; 245, p. 197.

3) Berichte, 10, p. 1206.

4) Liebig's Annalen, 245, p. 197.

5) Liebig's Annalen, 252, p. 149.

For the separation of sylvestrene from mixtures, as well as for identification, the dihydrochloride is best suited and is prepared as follows:

The fraction diluted with an equal volume of ether is strongly cooled and saturated with hydrochloric acid gas. After standing for about two days the ether is distilled off and the residue induced to crystallize by strongly cooling. The crystalline mass is freed from oily by-products on a porous plate, and the hydrochloride, first recrystallized from an equal weight of alcohol, is purified by fractional crystallization from ether: the melting point of the pure dihydrochloride is 72° (W.,[1] '85). It must be observed, that in the presence of dipentene or such terpenes as are changed to dipentene dihydrochloride, mixtures of dihydrochlorides are obtained, the melting point of which is the lower, the greater the amount of dipentene dihydrochloride present.

If sylvestrene be dissolved in acetic anhydride and a few drops of concentrated sulphuric acid added to the solution, a beautiful blue coloration is produced. This reaction may be used for the easy detection of sylvestrene, but it is only successful when the fractions to be tested are rich in this hydrocarbon.

Terpinene.

This hydrocarbon is very similar to dipentene, yet sharply differentiated from it. According to Weber[2] (1887) it is contained in cardamom oil and recently it has also been found in marjoram oil by Biltz.[3] It may, however, appear doubtful, whether this terpene is really found in nature, or whether it is formed by the influence of the heat during the distillation from other compounds contained in the respective oils.

Artificially, this hydrocarbon, which distinguishes itself by its stability toward dilute mineral acids, is obtained by the action of boiling alcoholic sulphuric acid on terpenes, such as dipentene and phellandrene, or on oxygenated compounds, such as terpin hydrate, terpineol, dihydrocarveol and cineol. Besides dipentene it is also formed by the action of formic acid on linalool and geraniol. It is further formed, besides isomeric hydrocarbons $C_{10}H_{16}$ and cymene, by the inversion of pinene with alcoholic sulphuric acid or when turpentine oil is shaken with small amounts of concentrated sulphuric acid, avoiding too violent a reaction[4] (Wallach, 1887). The last named method is used when it is desired to obtain fractions containing terpinene. As terpinene has up

1) Liebig's Annalen, 230, p. 241; 239, p. 25.
2) Liebig's Annalen, 238, p. 107.
3) Ueber das ätherische Oel aus *Origanum majorana*. Inaug.-Dissert. Greifswald, 1898.
4) Liebig's Annalen, 239, p. 35.

to the present not been regenerated from a solid derivative, the statement of the physical properties of this hydrocarbon refers not to the pure substance but only to terpinene-containing fractions.

For a terpinene obtained from terpin hydrate and dilute sulphuric acid Wallach reports:

B. p. 179—182°; d = 0.855[1]

and for a purer preparation from dihydrocarveol and sulphuric acid:

B. p. 178—180°; d = 0.847; n_D = 1.48458.[2]

Terpinene has an odor reminding of cymene and resinifies quite rapidly on standing; it has not yet been possible to convert it into isomers. It is therefore very similar to dipentene, but the two hydrocarbons differ in their behavior toward bromine and the hydrohalogens, with which terpinene yields liquid addition products only. A nitrosochloride of this terpene is not known, but nitrolamines can be prepared from the nitrosite mentioned later.

By chromic acid mixture (prepared according to Beckmann) terpinene is readily attacked and completely destroyed (Baeyer,[3] 1894), even in the cold. This behavior may be made use of when terpinene is to be removed from mixtures of pinene, camphene, limonene, terpineols, cineol or pinol, as these compounds are fairly stable toward the oxidizing agent in the cold.

The derivative of terpinene which may serve for the detection of this hydrocarbon is that formed by the addition of nitrous acid, terpinene nitrosite $C_{10}H_{16}N_2O_3$. It is obtained by the action of nascent nitrous acid on the terpene diluted with a solvent. In order to determine rapidly whether this hydrocarbon is present in a fraction of the boiling point of terpinene, Wallach (1887) directs[4] to proceed as follows:

A mixture of 2—3 g. of the fraction with an equal volume of petroleum ether is poured on top of an aqueous solution of 2—3 g. of sodium nitrite and the amount of acid necessary to decompose the latter is added in small portions; when all acid has been added the vessel is immersed for a moment in a hot waterbath and then allowed to stand in the cold. When terpinene is present, the insoluble nitrosite will separate in a few hours, or at latest in the course of two days. This is freed from accompanying oily products by spreading on porous plates, dissolved in glacial acetic acid, again precipitated by water and finally recrystallized from hot alcohol. The purified compound melts at 155°.

---

1) Liebig's Annalen, 230, p. 258; 239, p. 33.
2) Berichte, 24, p. 3991.
3) Berichte, 27, p. 815.
4) Liebig's Annalen, 239, p. 36.

The nitrosite reacts with bases, such as piperidine and benzylamine to form nitrolamines. The nitrolpiperidine base melts at 153—154°, the nitrolbenzylamine base at 137°[1] (Wallach).

PHELLANDRENE.

Phellandrene occurs in nature in both optically active modifications. As d-phellandrene it has been found in bitter- and water-fennel oils, as well as in Macedonian fennel oil, in elemi and schinus oils and in the distillate of the wood of *Caesalpinia sappan*. As l-phellandrene in Australian eucalyptus oil from *Eucalyptus amagdalina*, in pine needle oil. the oil from the needles of *Pinus montana*, in star-anise and schinus oils. Besides these, it is contained in a large number of volatile oils, although in small quantities, for instance in the oils of ginger, curcuma, pepper, camphor, sassafras leaves, Ceylon cinnamon, angelica root and seed, in the oils of German and English dill, in that of dog fennel, wormwood. golden rod, lemon, bay and peppermint, as well as in the oils of *Eucalyptus risdonia* and *Andropogon laniger*.

It is one of the most unstable terpenes, and for this reason phellandrene containing oils must never be redistilled under atmospheric pressure, but are best fractionated in a vacuum when it is desired to detect and isolate this hydrocarbon. The preparation of pure phellandrene has not yet been accomplished; the only compound apparently suited for this purpose, the crystallized nitrosite, decomposes indeed, by treatment with alkali[2] (Wallach, 1895), but no hydrocarbon $C_{10}H_{16}$ is formed by the reaction. Fractions as rich as possible in phellandrene, viz. the portion boiling between 170—172° are therefore taken as representing fairly pure hydrocarbon.

Pesci (1886), who found this hydrocarbon in water fennel oil and named it after its source, reports the following properties:

B. p. 171—172° (760 mm.); $d_{10} = 0.8558$; $[\alpha]_{D10^\circ} = +17° 64'$;[3] $n_D = 1.484$.[4]

For a preparation from Australian eucalyptus oil Wallach (1895) found:[5]

B. p. 65° at 12 mm.; $d_{19^\circ} = 0.8465$; $n_{D19^\circ} = 1.488$.

The optical rotation for d-phellandrene has been found considerably higher than that given by Pesci; for a preparation from schinus oil α_D was found to be + 60° 21'[6] (Gildemeister and Stephan, 1897).

---

1) Liebig's Annalen, 241, p. 315.
2) Liebig's Annalen, 287, p. 374.
3) Gazz chim. ital. 16, p. 225.
4) Jahresber. f. Chem., 1888, p. 1424.
5) Liebig's Annalen, 287, p. 383.
6) Archiv d. Pharm., 235, p. 591.

Phellandrene is very readily changed; even by heating to its boiling temperature it polymerizes. Still more readily is it changed into inactive isomers by the action of acids, thus by hydrohalogen into dipentene, by alcoholic sulphuric acid into terpinene. Well characterized addition products with halogens or hydrohalogens cannot be prepared, the only known solid derivative is phellandrene nitrosite, which is formed by the addition of nitrous acid and which is therefore used for identification.

A mixture of 5 cc. of oil and 10 cc. of petroleum ether is poured on a solution of 5 g. of sodium nitrite in 8 g. of water and the amount of glacial acetic acid (5 cc.) necessary to liberate the nitrous acid is added gradually with frequent shaking; the voluminous crystalline mass produced is filtered off with the aid of an air pump, washed first with water and then with methyl alcohol, and purified by repeatedly dissolving in chloroform and precipitating with methyl alcohol[1] (Wallach and Gildemeister, 1888). The brilliant white crystals thus obtained allow of being recrystallized from acetic ether without decomposition and melt at 105°[2] (Wallach, 1895). The crude compound changes easily and is freed from the oily by-products only with difficulty; by dissolving in a little acetic ether and precipitating with 60 percent alcohol the nitrite can be obtained at once in the form of white crystals[3] (Bertram and Walbaum, 1893).

Both phellandrenes give nitrosites which are identical as to external appearance and in their melting point, but which differ in this respect, that the compound obtained from d-phellandrene is strongly laevogyrate, that from l-phellandrene on the other hand, strongly dextrogyrate. By mixing equal parts by weight of d- and l-phellandrene nitrosite in solution the inactive compound is formed, which in its properties is identical with the optically active modifications. Phellandrene nitrosite, in opposition to terpinene nitrosite, cannot be changed to nitrolamines with bases. If the liquid reaction product formed by treatment with sodium ethylate is reduced there result, as well as by the direct reduction of the nitrosite, derivatives of the hydrocarvone series[4] (Wallach, 1895).

The presence of hydrocarbons of the formulas $C_{10}H_{18}$ and $C_{10}H_{20}$ in volatile oils has not yet been definitely established. The statement made by Andres and Andreef[5] (1892) that a hydrocarbon $C_{10}H_{18}$, presumably the menthene derivable from menthol, occurs in Russian peppermint oil has not been verified.

1) Liebig's Annalen, 246, p. 282.
2) Liebig's Annalen, 287, p. 374.
3) Archiv d. Pharm., 281, p. 298, footnote.
4) Liebig's Annalen, 287, p. 380.
5) Berichte, 25, p. 616.

Whereas menthene is a cyclic hydrocarbon with one double bond, Wassileef[1] has recently prepared a chain hydrocarbon $C_{10}H_{18}$, an olefinic dihydroterpene as it were from active amyl alcohol.

In addition to the terpenes $C_{10}H_{16}$ many volatile oils contain sesquiterpenes, $C_{15}H_{24}$. Like the terpenes these hydrocarbons may be regarded as polymerization products of hemiterpene, C_5H_8. With few exceptions they have not yet been carefully examined. It seems probable, therefore, that some of them will prove identical. Most sesquiterpenes boil between 250 and 280°, have a specific gravity of over 0.90, as separated by fractionation are frequently slightly colored, are less mobile liquids than the terpenes, resinify as readily as do the terpenes and are difficultly soluble in alcohol. As unsaturated hydrocarbons they add halogens, hydrohalogens, nitrosylchloride and the oxides of nitrogen. They frequently occur in oils accompanied by alcohol-like compounds $C_{15}H_{26}O$ (possibly also $C_{15}H_{24}O$) which stand in the same relation to the sesquiterpenes as do the terpineols to the terpenes. However, by treating the sesquiterpenes with the glacial acetic acid-sulphuric acid mixture, only one of them has so far been hydrated. Whether these sesquiterpene alcohols yield upon dehydration the same hydrocarbons $C_{15}H_{24}$ with which they are found in the oils has not yet been determined.

Attention should also be called to the apparent existence of sesquiterpenes corresponding to the olefinic terpenes. Semmler,[2] in 1889 obtained from asarum oil a hydrocarbon $C_{15}H_{24}$ boiling at about 255°, the specific gravity of which (0.873) is much lower than that of most sesquiterpenes. Thuja oils likewise contain fractions of the boiling point of the sesquiterpenes that are characterized by a very low specific gravity (0.85).

Cadinene.

The best known representative of the sesquiterpenes is the widely distributed cadinene. It has so far been found in cade oil (*Oleum cadinum*) from which source it has derived its name, further in the pine needle oils and of *Pinus montana*, the German and Swedish oils of *Pinus silvestris*, the oils of elderberry, savin, cedarwood, cubeb, betel, camphor, paracoto bark, asafetida, galbanum, olibanum, wormwood, golden rod, patchouli, peppermint, ylang-ylang and angosturo bark. As far as examined, cadinene has been found only in the laevogyrate modification.

---

1) Chem. Centrbl., 70¹, p. 775.

2) Chem. Zeitung, 18, p. 1158.

Cadinene can be prepared in a comparatively pure state, as it yields a solid well crystallized dihydrochloride, from which the hydrocarbon may be regenerated by splitting off hydrochloric acid by heating with aniline or sodium acetate and glacial acetic acid[1] (Wallach, 1887).

For the hydrocarbon purified in this manner Wallach found the following constants:

B. p. 272° (uncorr.);[2] 274—275°; $d_{20^\circ} = 0.918$; $n_D = 1.50647$; $[\alpha]_D = -98.56^\circ$.[3]

Upon continued heating with dilute sulphuric acid cadinene is changed, while when treated with hydrohalogens no appreciable influence is noticed, as the optical activity remains unchanged. Especially characteristic are the crystallized addition products formed with two molecules of hydrohalogen, of which the dihydrochloride and the dihydrobromide are used for identification.

For the preparation of the dihydrochloride the fraction boiling between 260—280° is diluted with twice the volume of ether and saturated with hydrochloric acid gas in the cold. After standing for some time the ether is partially removed by distillation. With the complete evaporation of the solvent, crystals of the dihydrochloride separate from the residue. These are freed from oily by-products by spreading on porous plates. They are then washed with alcohol, and recrystallized from acetic ether in which they are readily soluble while warm. The melting point of the pure compound is 117—118°. It is optically active.

The dihydrochloride may also be prepared by using glacial acetic acid saturated with hydrochoric acid in the cold. This method of preparation—treating a glacial acetic acid solution of the sesquiterpene with a glacial acetic acid solution of the respective hydrohalogen—is suitable for the preparation of the dihydrobromide (m. p. 124—125°) and the dihydroiodide (m. p. 105—106°).

The nitrosochloride and nitrosate have been prepared within the last year by Schreiner and Kremers.[4] Of the latter the unusually large yield of over forty per cent was obtained. The nitrosochloride melts at 93—94°, the nitrosate at 105—110°, both with decomposition.

Cadinene is characterized by a color reaction, which under some circumstances may serve to show its presence in oils. The reaction is prettiest when the hydrocarbon, slightly changed by standing for some time, is dissolved in glacial acetic acid and the solution treated with a little concentrated sulphuric acid. The greenish coloration first produced soon changes through blue to red.

---

1) Liebig's Annalen, 238, p. 84.
2) Liebig's Annalen, 271, p. 308.
3) Liebig's Annalen, 252, p. 150; 271, p. 297.
4) Pharm. Arch., 2, p. 249.

CARYOPHYLLENE.

The second well characterized sesquiterpene is caryophyllene. So far it has been found in the oils of cloves and clove stems, the oil of copaiba balsam and that of *Canella alba.* It has not yet been prepared absolutely pure. The physical constants recorded pertain to distillates from clove oil and the oil of clove stems. Caryophyllene prepared from clove oil merely by fractionation contains some benzoyl eugenol[1] which can be removed by saponification. In the sesquiterpene from the oil of clove stems this impurity is not found. The physical constants have been ascertained by several observers, viz.:

Wallach[2] (1892): b. p. 258—260°; $d_{15°} = 0.9085$; $n_D = 1.50094$.
Erdmann[3] (1897): b. p. 119—120° at 9 mm.; 123—124° at 13 mm.; 258—259° at 752 mm.; $d_{24°} = 0.9038$.
Kremers[4] (1898): $d_{20°} = 0.9032$, $n_{D20°}$ 1.50019; $[\alpha]_{D20°} = -8.74°$.
Schreiner and Kremers[5] (1899): b. p. 136—137° at 20 mm.; $[\alpha]_{D20°} = -8.96$; $n_{D20°} = 1.49976$; $d_{20°} = 0.9030$.

The laevogyrate hydrocarbon yields optically active as well as inactive derivatives, the nitrosite e. g. being strongly dextrogyrate. With hydrogen chloride it yields a crystalline dihydrochloride[6] (m. p. 69—70°). It forms a crystalline nitrosochloride, apparently two nitrosites (one blue, the other white), and a nitrosate. From several of these the benzylamine and piperidine bases have been prepared. From both the dihydrochloride and the nitrosochloride a hydrocarbon can evidently be regenerated (Schreiner & Kremers,[7] 1899).

Caryophyllene hydrate which so far has been principally used for the identification of this sesquiterpene is prepared as follows:

To a solution of 1000 g. glacial acetic acid, 20 g. conc. sulphuric acid and 40 g. of water, 25 g. of caryophyllene or so much hydrocarbon as the mixture will retain in solution are added and the mixture heated for some time on a waterbath. When the reaction is completed the mixture is subjected to steam distillation. At first acetic acid and a mobile oil pass over, later a less volatile oil which congeals upon cooling. This is freed from oil by suction and further purified by crystallization from alcohol. Caryophyllene alcohol or hydrate melts at 94—96°. Its identification can be verified by the phenyl urethane derivative which melts at 136—137°.

For the purpose of identification the nitrosite and the benzylamine base obtained from it are especially adapted:

---

1) Journ. f. prakt. Chem. II, 56, p. 146.
2) Liebig's Annalen, 271, p. 298.
3) Journ. f. prakt. Chem. II, 56, p. 146.
4) Pharm. Arch., 1, p. 211.
5) Ibidem, 2, p. 282.
6) Pharm. Archives, 2, p. 296.
7) Pharm. Arch., 2, pp. 295, 296.

To a solution of 5 cc. of caryophyllene in 5 cc. of petroleum ether 5 cc. of glacial acetic acid and then 5 cc. of a saturated solution of sodium nitrite are added. The mixture while still warm from the reaction is rotated and then cooled in ice water with violent shaking. Crystallization usually takes place or may be induced with a fragment of nitrosite. The blue crystals are washed with water and then with alcohol. When recrystallized from hot alcohol, avoiding direct sunlight, the crystals melt at 113°. The benzylamine base from the nitrosite melts at 167°.

Noteworthy is the fact that upon dehydration of caryophyllene alcohol, caryophyllene is not regenerated, but an isomeric hydrocarbon.

	Source.	Boiling point	d	n_D	α_D
Humulene[1]	from hop oil	263—266° 166—171° (60 mm)	0,9001 (20°)	—	—
Cedrene[2] & [3]	from cedar oil	261—262° 131—132° (10 mm)	0,9359 (15°)	1.5015 —	—60° —47°54′
Cedrene[4]	from cedrol & a) P_2O_5 b) formic acid	237° (?) 262—263°	— —	— —	— —80°
Cubebene[5]	from cubebcamphor	255—260°	—	—	—
Guajene[6]	from guajol & $ZnCl_2$	124—128° (13 mm)	0,910 (20°)	1.50114	—
Ledene[7]	from ledumcamphor & dil. sulphuric acid	255°	—	—	—
Patchoulene[8]	from patchoulicamphor and $KHSO_4$	254—256°	0,939 (23°)	1.50094	—
Santalene[9]	from santalol and P_2O_5	260°	—	—	—

1) Chapman, Journ. chem. Soc., 67 (1895), p. 54 & 780.
2) Chapman & Burgess, Chem. News, 74 (1896), p. 95.
3) Rousset, Bull. Soc. chim, III, 17 (1897), p. 485.
4) Walter, Liebig's Annalen, 39 (1841), p. 247; 48 (1848), p. 35; Bericht von Schimmel & Co., October 1897, p. 12, footnote.
5) Schmidt, Berichte, 10 (1877), p. 189.
6) Wallach & Tuttle, Liebig's Annalen, 279 (1894), p. 396.
7) Hjelt, Berichte, 28 (1895), p. 3087.
8) Wallach & Tuttle, Liebig's Annalen, 279 (1894), p. 394.
9) Chapoteaut, Bull. Soc. chim. II, 37 (1882), p. 303.

clovene, is formed. This differs materially from caryophellene and has not yet been found in volatile oils.

Besides these two sesquiterpenes may be mentioned the fairly well characterized humulene. It derives its name from hop oil from which it was prepared by Chapman[1] who first prepared its nitrosochloride, its nitrosite (existing in a blue and a white variety) and nitrosate, also the nitrol benzylamine (m. p. 136°) and the nitrol piperidine (m. p. 153°) bases. It is characterized by the nitrosite, the blue variety of which (m. p. 120°) changes upon repeated recrystallization from alcohol to the white variety (m. p. 165—168°). Humulene also occurs in oil of poplar buds. Fichter and Katz[2] identified it by means of the above mentioned derivatives.

There remain to be mentioned cedrene occuring with cedrol in cedar oil, as well as other hydrocarbons $C_{15}H_{24}$ produced by splitting off water from sesquiterpene alcohols. Their properties are, so far as known, given in the table on the preceding page.

Polyterpenes, namely di- and triterpenes, occur in vegetable balsams and resins. They are possibly formed by polymerization of hydrocarbons C_5H_8 and $C_{10}H_{16}$ and may therefore be found in the higher boiling fractions of volatile oils. The diterpenes are viscid liquids boiling above 300°, the triterpenes resinous masses. Hence they do not invite examination and their properties have been even less investigated than those of the sesquiterpenes.

Alcohols.

Of the monatomic alcohols of the paraffin series only those with less than eight carbon atoms have so far been found in volatile oils. They occur seldom as free alcohols and are mostly combined with fatty acids as esters. The occurrence of fatty alcohols and acids is mostly due to saponification of corresponding esters. Free alcohols may occur when the plant material containing carbohydrates has undergone partial fermentation before being distilled. Thus e. g. ethyl alcohol has been found in the distillate of roses that had been stored for a short time.

Mazé,[3] however, has recently shown that alcohol is formed not only in the process of germination of certain seeds, but also when they are soaked in water. The formation of alcohol in these cases is attributed to a diastatic process similar to that induced by yeast.

---

1) Journ. Chem. Soc., 67, pp. 54 & 780.

2) Berichte, 32, p. 3183.

3) Compt. rend., 128, p. 1608.

As esters and ethers methyl alcohol is widely distributed: as methyl salicylate in numerous plants, as methyl ether e. g. in eugenol which is likewise widely distributed. According to Goeppert (1859), ethyl alcohol as acetate is the odoriferous principle of *Magnolia fuscata* Andrews.[1] This statement has not yet been verified. It has been found as ethyl ester of caproic, caprylic, capric, lauric, palmitic and oleic acids in the volatile oil of saw palmetto by Sherman and Briggs.[2] As butyrate ethyl alcohol is found in the oil of *H. sphondilium.* Isobutyl alcohol (isopropyl carbinol) occurs as ester of isobutyric and angelic acids in Roman chamomile oil; amyl alcohol (of fermentation) as caprinate in cognac oil and as ester of angelic and tiglinic acids also in Roman chamomile oil. Normal hexyl alcohol is found as acetate and butyrate in the oil of *Heracleum giganteum,* also as acetate in the oil of *H. sphondylium.* Optically active hexyl alcohol (methyl ethyl propyl alcohol) has been found combined with angelic and tiglinic acids in Roman chamomile oil. Octyl alcohol is found as acetate in the oils of *H. sphondylium* and *H. giganteum*; as butyrate it is the principal constituent of the oil of *Pastinaca sativa.* Besides hexyl alcohol it is also reported to occur in the oil of male fern.

Of greater interest than the saturated fatty alcohols are the olefinic alcohol citronellol, $C_{10}H_{20}O$ and the two diolefinic alcohols $C_{10}H_{18}O$. linalool and geraniol.

Linalool.

Linalool ("Licareol" of Barbier) is quite widely distributed. It is optically active and occurs in both modifications. So far d-linalool has been found in but one oil, viz. coriander oil and is, therefore, sometimes designated coriandrol; l-linalool, alone or mixed with a little d-linalool, partly free, partly as ester, forms a constituent of the oils of lignaloe, bergamot, neroli, petitgrain, the Italian limette oil, Palermo lemon oil, of the oils of spike, lavender, muscatel sage, thyme, of Russian spearmint oil, German and French basilicum oil, Cretian origanum oil, ylang-ylang oil and of sassafras leaf oil.

To regenerate linalool from a crystallized derivative has so far been unsuccessful. Fractional distillation of the oils, which have previously been saponified, is used for its isolation. The constants given for linalool refer, therefore, always to the products obtained in this manner. If the alcohol obtained is to be freed from indifferent compounds, for instance terpinene, it may be converted into the sodium salt of the acid phthalic

1) Liebig's Annalen, 111, p. 127.

2) Pharm. Archives, 2, p. 110.

ester according to Tiemann[1] (1898). This is soluble in water and can be saponified by alcoholic potassa. The regenerated linalool must be removed from the alcoholic alkaline solution with ether, since it suffers changes when it is distilled from the alkaline solution by steam as the decrease in the rotatory power shows.

According to the material employed and the method of preparation, products are obtained which show slight differences in their properties; in judging the purity of a preparation the following data may serve as a basis:

for l-linalool: b. p. 197—199°; 85—87° at 10 mm.; $d_{15^\circ} = 0.870—0.875$; $n_{D20^\circ} = 1.4630—1.4690$ (Stephan,[2] 1898).

B. p. 86—87° at 14 mm.; $d_{20^\circ} = 0.8622$; $n_D = 1.46108$ (Tiemann,[3] 1898);

for d-linalool: b. p. 85—86° at 12 mm.; $d_{17.5^\circ} = 0.8726$; $n_D = 1.46455$ (Gildemeister,[4] 1895).

The rotatory power is not constant, the greatest deviation so far observed for l-linalool is $[\alpha]_D = -20° 7'$,[5] for d-linalool (coriandrol) $[\alpha]_D = +15° 1'$ (Barbier,[6] 1898). Tiemann reports for d-linalool $[\alpha]_D = + 13° 19'$.

Artificially, linalool is obtained, although only in the inactive state, when geraniol is heated with water to 200° in an autoclave, or when the different chlorides produced by the action of hydrochloric acid on geraniol, are treated with alcoholic potassa (Tiemann[7], 1898). A third method of converting geraniol into linalool consists in passing steam through an aqueous solution of geranyl phthalate of an alkali which is either neutral or rendered slightly alkaline by the addition of an alkali carbonate (Stephan,[8] 1899).

As unsaturated alcohol with two double bonds linalool also shows a capacity for addition. It combines with two molecules of bromine, as well as with the hydrohalogens. With the latter it forms compounds such as $C_{10}H_{18}O . HCl$, $C_{10}H_{18}Cl_2$ which, however, being liquid, are not suited for characterization.

The unsaturated tertiary nature of this alcohol shows itself in its behavior toward reagents. Whereas alkalies scarcely act on it in the cold,[9] organic acids change it either into geraniol or, especially in the presence of small amounts of sulphuric acid, into terpineol. Mineral

1) Berichte, 31, p. 837.
2) Jour. f. prakt. Chem., II, 58, p. 110.
3) Berichte, 31, p. 834.
4) Archiv d. Pharm., 233, p. 179.
5) Compt. rend., 116, p. 1459.
6) Bericht von S. & Co., April 1898, p. 25.
7) Berichte, 31, p. 832.
8) Journ. f. prakt. Chem., II, 60, p. 252.
9) Bull. Soc. Chim., 21, p. 549.

acids, by the abstraction or addition of water, produce compounds with a cyclic structure. Thus terpin hydrate is produced by shaking with 5 percent sulphuric acid (Tiemann & Schmidt,[1] 1895). By heating with glacial acetic acid and acetic acid anhydride there results besides geranyl acetate the acetate of solid terpineol, the rotation of which is opposite in direction of that of the linalool used. Formic acid, at the average temperature of 20°, also converts it into the esters of linalool and of the solid terpineol with opposite rotation (Stephan,[2] 1898). With moderate heat, however, (60—70°), water is split off with formation of the hydrocarbons dipentene and terpinene (Bertram & Walbaum,[3] 1892).

Toward various oxidizing agents linalool behaves differently. With very dilute permanganate solution polyatomic alcohols are probably first formed with simultaneous addition of water. These cannot be isolated in a pure state, and are split up by further oxidation with permanganate or chromic acid mixture into acetone and laevulinic acid (Tiemann & Semmler,[4] 1893). In accordance with this result and in consideration of the fact that linalool is optically active and, therefore, must contain an asymmetric carbon atom, the formula of a dimethyl-2,6-octadiëne-2,7-ol-6

$$CH_3.C(CH_3):CH.CH_2.CH_2.C(CH_3)(OH).CH:CH_2$$

has been suggested for this alcohol.[5] If linalool is oxidized with chromic acid mixture only, it first suffers a rearrangement owing to the acidity of the oxidizing agent and is then changed to the aldehyde of geraniol, citral (Bertram & Walbaum,[6] 1892). The oxidation, however, usually goes further and "Abbau" products of citral are also obtained, namely methyl heptenone, laevulinic acid, etc. The crystallized derivative observed by Bertram & Walbaum when oxidized with peroxide of hydrogen has been shown to be terpin hydrate, the formation of which is in all probability, primarily due to the presence of mineral acid in the peroxide of hydrogen.

Linalool does not add hydrogen when heated with reducing agents, but readily loses its oxygen with the production of the unsaturated hydrocarbon linaloolene $C_{10}H_{18}$. This is formed, when linalool is converted into its sodium compound, or treated with metallic sodium in alcoholic solution, or when heated with zinc dust to 220—230° (Semmler,[7] 1894).

1) Berichte, 28, p. 2137.
2) Journ. f. prakt. Chem., II, 58, p. 109.
3) Journ. f. prakt. Chem., II, 45, p. 601.
4) Berichte, 28, p. 2130.
5) Ibidem, p. 2131.
6) Journ. f. prakt. Chem., II, 45, p. 599.
7) Berichte, 27, p. 2520.

As already mentioned, linalool is found not only in the free state, but also as ester of fatty acids in volatile oils. Particularly important is the acetate, which possesses the characteristic odor of bergamot and constitutes the principal constituent of bergamot oil. In larger or smaller amounts linalyl acetate has been further found in French and English lavender oil, in Italian limette oil, in muscatel sage oil and sassafras leaf oil. Of the other fatty acid esters, the butyrate and probably also the propionate and valerianate are contained in lavender oil. The valerianate also occurs in sassafras leaf oil.

The esters of linalool that occur in volatile oils, are liquids of a more or less strong and pleasant odor. They cannot be distilled under atmospheric pressure without decomposition. Their synthetic preparation meets with difficulty in so far as linalool is fairly sensitive to acids and suffers rearrangement so that, while the products obtained by the heating of linalool with acid anhydrides or by the method of the German Imperial-Patent 80,711 consist for the main part of esters of linalool, they contain also those of geraniol and terpineol.

For the acetic ester isolated from bergamot and lavender oils the following properties are recorded by Tiemann and Semmler[1] (1892):

B. p. 99—105° at 15 mm.; $d_{20^\circ} = 0.8951$; b. p. 97—105° at 15 mm.; $d_{20^\circ} = 0.8972$.

For a preparation containing about 83 percent of acetate distilled from limette oil, Gildemeister[2] (1895) found:

B. p. 101—103° at 13 mm.; $d_{15^\circ} = 0.898$; $\alpha_{D15} = -9°52'$.

Bertram and Walbaum[3] (1892), however, determined the following constants for the acetate prepared by boiling linalool with acetic acid anhydride, which probably also contained the acetic esters of geraniol and terpineol:

B. p. 105—112° at 11 mm.; $d_{15^\circ} = 0.912$.

While the ester isolated from volatile oils rotates polarized light more or less strongly to the left, the synthetic preparation is either almost inactive or dextrogyrate. The dextrorotation increases with the quantity of linalool that has been changed to terpineol.

As linalool yields no solid derivatives, which are suited for identification, it is necessary for this purpose to convert it into citral by oxidation, and to characterize the latter by the citryl-β-naphtocinchoninic acid discovered by Doebner (see citral).

---

1) Berichte, 25, pp. 1184 & 1187.

2) Archiv d. Pharm., 233, p. 181.

3) Jour. f. prakt. Chem., II, 45, p. 598.

Geraniol.

The diolefinic alcohol $C_{10}H_{18}O$ geraniol ("Lemonol" Barbier and Bouveault; "Rhodinol" Erdmann and Huth, Poleck) is isomeric with linalool, but differs from it by its optical inactivity, higher boiling point and higher specific gravity. It is also found in the free state as well as in the form of esters and occurs rather frequently in volatile oils. While it constitutes the main part of palmarosa oil as well as of German and Turkish rose oil, and is contained in appreciable quantities in geranium, citronella and lemon grass oils, it is found in many oils only in small quantities, as in neroli and petitgrain oils, lavender oil and oil of spike, in lignaloe oil, ylang-ylang oil and sassafras leaf oil.

As primary alcohol geraniol produces a crystalline double compound with anhydrous calcium chloride (Jacobsen,[1] 1871), which is insoluble in solvents like ether, ligroin, benzene, chloroform, and is decomposed by water into calcium chloride and geraniol. This property makes the preparation of chemically pure geraniol possible in a very simple manner (see below). For the isolation of this alcohol from mixtures with hydrocarbons, etc., several other methods have been reported. All of them involve the preparation of an acid ester of geraniol either by the action of phthalic acid anhydride on the sodium compound of the crude geraniol (Tiemann and Krueger,[2] 1896), or by heating geraniol with phthalic acid anhydride in a water bath (H. & E. Erdmann,[3] 1897), or in benzene solution (Flatau & Labbé,[4] 1898). This acid ester is saponified with alcoholic potassa either as such or its sodium salt which can be obtained from the silver salt purified by crystallization. These methods, however, possess no advantages over the calcium chloride method. On the contrary they are more complicated and yield no purer product than does the simpler method of Jacobsen.

Geraniol prepared by one or the other of the above methods has a roselike odor and is a colorless, somewhat oily liquid which on longer standing in contact with the air changes through the absorption of oxygen. Its properties are reported as follows:

B. p. 110—111° at 10 mm.; 121° at 18 mm.; 230° under atmospheric pressure (Bertram and Gildemeister,[5] 1897).

B. p. 120.5—122.5° at 17 mm.; $d_{20°} = 0.8894$ (!); $n_{D20°} = 1.4766$, (Tiemann and Semmler,[6] 1893).

---

1) Liebig's Annalen, 157, p. 282.

2) Berichte, 29, p. 901.

3) Journ. f. prakt. Chem., II, 56, p. 17.

4) Compt. rend., 126, p. 1725; Bull. Soc. chim., III, 19, p. 688.

5) Journ. f. prakt. Chem., II, 56, p. 508.

6) Berichte, 26, p. 2711.

B. p. 110.5—111° (corr.) at 10 mm.; $d_{\frac{16°}{4°}} = 0.8812$, (H. and E. Erdmann,[1] 1897).

$d_{15°} = 0.880—0.883$; $n_{D17°} = 1.4766—1.4786$, (Stephan,[2] 1898).

As primary alcohol geraniol is converted by oxidation into the corresponding aldehyde citral, and can be again obtained from this by reduction (Tiemann,[3] 1898). Since citral can be prepared synthetically, geraniol must therefore be also classed with the compounds which can be obtained synthetically. Geraniol (besides terpineol) or its acetate are produced by isomerization from linalool, when this is heated for some time with acetic acid anhydride (Bouchardat,[4] 1893; Stephan,[5] 1898). Obversely, geraniol can be changed to linalool by heating with water in an autoclave to 200°.[6] At a higher temperature hydrocarbons and their polymerization products are formed. The same result is obtained when the chlorides produced by the action of hydrochloric acid on geraniol are treated with alcoholic potassa (Tiemann,[7] 1898). A third method consists in decomposing the phthalate mentioned under linalool (Stephan,[8] 1899).

In general, geraniol is not acted upon by acids to the same extent as linalool. Thus it is quantitatively changed into the acetate by boiling with acetic acid anhydride, but not isomerized. On shaking with dilute sulphuric acid it is changed like linalool, although with greater difficulty, to terpin hydrate (Tiemann and Schmidt,[9] 1895).

Concentrated formic acid exerts, like potassium bisulphate or phosphoric acid anhydride, a dehydrating action on geraniol. While with potassium bisulphate a chain hydrocarbon is said to be formed (Semmler,[10] 1891), the other reagents produce terpenes. Formic acid produces dipentene and terpinene (Bertram and Gildemeister,[11] 1894 and 1896). According to Stephan,[12] formic acid, also a mixture of glacial acetic acid and sulphuric acid, convert geraniol as well as linalool into solid terpineol melting at 35°. Alkalies scarcely act on the alcohol in the cold. If, however, heated to 150° with a concentrated alcoholic alkali solution, a tertiary alcohol $C_9H_{18}O$ is said to be produced (Barbier,[13] 1898) by the splitting off of carbon dioxide.

1) Journ. f. prakt. Chem., II, 56, p. 3; Berichte, 31, p. 359, note 1.
2) Journ. f. prakt. Chem., II, 58, p. 110.
3) Berichte, 31, p. 828.
4) Compt. rend., 116, p. 1253.
5) Journ. f. prakt. Chem., II, 58, p. 111.
6) Bericht von S. & Co., April, 1898, p. 25.
7) Berichte, 31, p. 832.
8) Journ. f. prakt. Chem., II., 60, p. 252.
9) Berichte, 28, p. 2138.
10) Berichte, 24, p. 683.
11) Journ. f. prakt. Chem., II, 49, p. 185; 53, p. 236.
12) Journ. f. prakt. Chem., II, 60, p. 244.
13) Compt. rend., 126, p. 1423.

This statement rests, however, on a mistake, as the alcohol produced is methylheptenol $C_8H_{16}O$ (Schimmel & Co.,[1] 1898; Tiemann,[2] 1898).

The compounds formed by the addition of bromine and the action of hydrohalogen on geraniol are all liquid and quite unstable. That the isomeric chlorides, produced by an excess of hydrochloric acid, yield linalool as well as geraniol when treated with alcoholic potassa, has already been mentioned.

Like all primary alcohols, geraniol on oxidation with chromic acid mixture yields primarily an aldehyde, viz. citral $C_{10}H_{16}O$ (Semmler,[3] 1891). Accompanying it are found further "Abbau" products of this aldehyde, so that the reaction does not at all take place quantitatively. On shaking with very dilute permanganate solution, polyatomic alcohols are propably first formed, which are oxidized by chromic acid mixture to acetone, laevulinic acid and oxalic acid (Tiemann and Semmler,[4] 1895). Since geraniol is a primary alcohol and optically inactive, the following formula has been assigned to it,[5] dimethyl-2,6-octadiëne-2,6-ol-8

$$CH_3.C(CH_3):CH.CH_2.CH_2.C(CH_3):CH.CH_2OH.$$

Like linalool geraniol occurs in volatile oils in the free state and also in the form of esters. As acetate it is contained (beside the normal capronate) in palmarosa oil, further in lemon oil, petitgrain oil and sassafras leaf oil; in the last also as valerianate. Combined with tiglic acid it occurs in geranium oil. As geraniol is fairly stable toward acids, its esters can be prepared artificially from acid anhydrides and geraniol, or also from acid chlorides and geraniol with the addition of pyridine (H. & E. Erdmann,[6] 1897 and 1898). The esters of the fatty acids are all liquid and the weaker in odor, the larger the molecule of the acid radical contained in it. For the acetate prepared by boiling geraniol with acetanhydride Bertram and Gildemeister,[7] (1894) have determined the following constants: b. p. 127.8—129.2°; $d_{15°}=0.9174$; $n_{D15°}=1.4628$. When distilled under ordinary pressure geranyl acetate decomposes, acetic acid being split off.

Of the other esters of geraniol, diphenyl-carbaminic acid ester and the acid phthalic acid ester are worthy of mention, because both crystallize. The former is also a very suitable derivative for the identification of geraniol (see below), while the phthalic acid ester (m. p. 47°) (Flatau & Labbé,[8] 1898) can be used for the preparation of pure geraniol.

---

1) Bericht von S. & Co., October 1898, p. 68.
2) Berichte, 31, p. 2991.
3) Berichte, 24, p. 203.
4) Berichte, 28, p. 2130.
5) Ibid., p. 2132.
6) Journ. f. prakt. Chem., II, 56, p. 14; Berichte, 31, p. 356.
7) Journ. f. prakt. Chem., II, 49, p. 189.
8) Compt. rend., 126, p. 1725.

If it is desired to separate the geraniol as such from a geraniol-containing oil, the following method of Bertram and Gildemeister[1] (1896 and 1897) is used:

Equal parts of oil and very finely powdered calcium chloride are carefully triturated. The mixture, which as a result of the ensuing reaction acquires a temperature of up to 30—40°, is allowed to stand in a cool place in a desiccator. The resulting solid mass is then powdered, triturated with anhydrous ether, benzene or low boiling petroleum ether, transferred to a filter and freed by means of a pump from the compounds not combined with the calcium chloride by repeated washing with ether, etc. The mixture of geraniol calcium chloride and excess of calcium chloride thus obtained is decomposed by water, the separated oil washed several times with warm water and finally distilled with water vapor.

The separation of the geraniol from mixtures by this method is not quantitative. Moreover, the oil to be worked with must consist at least one fourth of geraniol. The separation of geraniol from citronellol is described by Flatau and Labbé,[2] that of geraniol from tertiary alcohols, e. g. terpineol, by Stephan.[3] When only small amounts of material are available, it is better to use for characterization the diphenyl urethane of geraniol $(C_6H_5)_2NCOOC_{10}H_{17}$, first recommended by Erdmann and Huth[4] (1896) for this purpose. For the preparation of this compound the authors cited give the following directions:[5]

1.0 g. of oil, 1.5 g. of diphenylcarbamine chloride and 1.35 g. of pyridine are heated for two hours in a boiling water bath. The product of the reaction is treated with water vapor and the residue solidifying on cooling recrystallized from alcohol. If much citronellol is present at the same time with the geraniol, it is difficult to obtain a pure preparation as citronellol also yields a diphenylurethane, which remains liquid. In this case urethanes are first obtained of a lower melting point (40—50°), which yield the pure diphenyl urethane of geraniol of the melting point 82.2° only after being recrystallized several times from alcohol.

If geraniol is to be still further characterized, it may be converted into citral by oxidation and this into citryl-β-naphthocinchoninic acid (see citral). For this purpose the alcohol must, however, be already fairly pure, and must contain no linalool, as this also yields citral on oxidation with chromic acid mixture.

Citronellol.

Citronellol ("Reuniol" Hesse, Naschold; "Rhodinol" Barbier and Bouveault) $C_{10}H_{20}O$, differs from linalool and geraniol by two atoms

---

1) Journ. f. prakt. Chem., II, 53, p. 233; 56, p. 507.
2) Chem. Centrbl., 70¹, p. 1094.
3) Journ. f. prakt. Chem., II, 60, p. 254.
4) Journ. f. prakt. Chem., II, 53, p. 45.
5) Journ. f. prakt. Chem., II, 56, p. 28.

of hydrogen more in the molecule. Although it appears to stand in close relation to the alcohols cited and especially to geraniol, both being primary, the reduction of geraniol to citronellol has so far been unsuccessful. Geranic acid, however, has been reduced to citronellic acid (Tiemann,[1] 1898).

Citronellol has only recently been obtained in a pure state and characterized as an individual chemical compound. Mixtures of it with geraniol have been formerly described under the name of "Rhodinol" (Eckart,[2] 1891; Barbier and Bouveault,[3] 1894) or "Reuniol" (Hesse,[4] 1894). The "Roseol" of Markownikow and Reformatsky,[5] (1893) which was said to be the main constituent of rose oil, has also shown itself to be a mixture of geraniol and citronellol. These mixtures, considered as individual compounds, were partly obtained from rose oil, partly from the true geranium oils. Now it is known, that there occurs in these oils, besides geraniol a second alcohol $C_{10}H_{20}O$ in larger or smaller quantities, which on account of its identity with the alcohol $C_{10}H_{20}O$ prepared by Dodge by the reduction of citronellal has been called citronellol. In the geranium oils this alcohol occurs as a mixture of both optically active modifications, the laevogyrate predominating. Rose oil, however, contains only l-citronellol. In addition to the free alcohol, its esters of the fatty acids have also been found.

As geraniol and citronellol cannot be separated by fractional distillation, either as such or after conversion into esters, it was difficult to obtain pure citronellol, especially since the calcium chloride method does not effect a quantitative separation. Pure citronellol was first prepared by Wallach, who noticed that geraniol when heated with water in an autoclave to 250° is completely decomposed with the formation of hydrocarbons, whereas citronellol remains unchanged.[6] A method of separation given by Tiemann and Schmidt[7] (1896) depends on the fact that geraniol in etherial solution is changed by phosphorus trichloride partly to hydrocarbons, partly to geranyl chloride; citronellol, however, is converted into a chlorine-containing acid phosphoric acid ester, which is soluble in alkalies and therefore easily separated from the other compounds. The crude citronellol obtained by saponification of the ester is purified by distillation with water vapor. Separation is

1) Berichte, 31, p. 2899.

2) Archiv d. Pharm., 229, p. 355.

3) Compt. rend., 119, pp. 281, 334.

4) Journ. f prakt. Chem., II, 50, p. 472.

5) Journ. f. prakt. Chem., II, 48, p. 293; Berichte. 27, Ref. p. 625.

6) Nachrichten der Kgl. Ges. d. Wiss. zu Göttingen, 1896, I, p. 64.

7) Berichte, 29, p. 921.

also successfully accomplished, when a mixture of both alcohols is heated to 200° with phthalic acid anhydride. Geraniol is destroyed by this treatment, while citronellol is converted into the acid phthalic acid ester, the sodium salt of which is soluble in water and can be saponified by alcoholic potassa. Another method of separating citronellol and geraniol is described by Flatau and Labbé,[1] who make use of the phthalic acid esters.

Artificially, citronellol can be obtained from the corresponding aldehyde, citronellal, by reduction with sodium amalgam and glacial acetic acid (Dodge,[2] 1889; and Tiemann and Schmidt,[3] 1896). The dextrogyrate modification, however, is produced by this treatment. It may also be mentioned here that Wallach[4] (1894 and 1897) has obtained an alcohol $C_{10}H_{20}O$ by splitting the cycle of menthonitrile and inversion of the resulting aliphatic compound. This possesses great similarity to citronellol and is possibly identical with it.

Pure citronellol is a colorless liquid of a pleasant rose-like odor, finer than geraniol. According to the manner of preparation, it shows slight variations in the physical properties. Wallach[5] determined for a citronellol (réuniol) prepared by his method the following constants:

B. p. 114—115° at 12—13 mm.; $d_{22°} = 0.856$; $n_{D22°} = 1.45609$;
$[\alpha]_D = -1° 40'$.

According to Tiemann and Schmidt,[6] (1896), d-citronellol formed by the reduction of citronellal possesses the following properties:

B. p. 117—118° at 17 mm.; $d_{17.5°} = 0.8565$; $n_{D17.5°} = 1.45659$;
$[\alpha]_{D17.5°} = +4°$.

l-Citronellol, prepared from rose oil by the phosphorus trichloride method, boils under a pressure of 15 mm. at 113—114°, has $d_{20°} = 0.8612$, $n_D = 1.45789$ and turns the plane of polarized light 4° 20′ to the left.[7]

For the alcohol obtained from geranium oil from the island of Réunion by a method similar to that of Wallach, Schimmel & Co.[8] found (1898):

B. p. 225—226° at 764.5 mm.; $d_{15°} = 0.862$; $n_{D22°} = 1.45611$;
$[\alpha]_D = -1° 40'$.

---

1) Chem. Centrbl., 70 I, p. 1094.
2) Amer. Chem. Journ., 11, p. 468.
3) Berichte 29, p. 906.
4) Liebig's Annalen, 278, p. 316; 296, p. 129.
5) Nachrichten d. Kgl. Ges. d. Wiss. zu Göttingen, 1896, I, p. 64; Naschold, Beiträge zur Kenntniss aliphatischer Terpenverbindungen. Inaug.-Diss. Göttingen, 1896, p. 56.
6) Berichte, 29, p. 906.
7) Ibid., p. 928.
8) Bericht von S. & Co., April 1898, p. 62.

Citronellol is much more stable than geraniol and is not affected by heating with alkali. When shaken with 10 percent sulphuric acid it is converted by the addition of water into a diatomic alcohol, from which dehydrating agents regenerate citronellol (Tiemann and Schmidt,[1] 1896). To obtain from it a cyclic hydrocarbon $C_{10}H_{18}$ by splitting off water has so far been impossible. As mentioned above several methods for its separation from other alcohols have been based on its stability toward phosphorus trichloride in the cold and the combined action of heat and pressure in the presence of water.

On oxidation, citronellol as primary alcohol is first converted into the aldehyde citronellal $C_{10}H_{18}O$, which can be again changed to the alcohol by reduction with sodium amalgam. The change to the aldehyde, however, is no more quantitative than that of geraniol to citral. Further oxidation products, like citronellic acid etc., are usually obtained (Tiemann and Schmidt,[2] 1897). If citronellol is first hydroxylized by cold dilute permanganate solution and the resulting polyatomic alcohol further oxidized with chromic acid mixture, acetone and β-methyladipinic acid are obtained as decomposition products. The latter, according to the material used, is more or less optically active or inactive with the corresponding variation in melting point from 82—96°. Since citronellol can be converted into a cyclic alcohol, isopulegol, and from this into pulegone, citronellol has been regarded by Tiemann and Schmidt[3] (1896) as dimethyl-2,6-octene-2-ol-8, $CH_3.C(CH_3):CH.CH_2.CH_2.CH(CH_3)CH_2.CH_2OH$.

The esters of citronellol, of which the acetate occurs in volatile oils, are readily obtained by treatment of the alcohol with the respective acid anhydrides. The acetate is a colorless liquid of a pleasant odor, faintly suggesting bergamot oil. According to Naschold[4] it boils under 15 mm. pressure at 121.5°; according to Tiemann and Schmidt[5] (1896) at 119—121° and has $d_{17.5^\circ} = 0.8928$; $n_{D17.5^\circ} = 1.4456$; $[\alpha]_D = +2° 37'$.

The acid phthalic acid ester formed by heating citronellol with phthalic acid-anhydride is distinguished from that of geraniol by being liquid, but gives a well crystallized silver salt, from which pure citronellol can be regenerated (Erdmann and Huth,[6] 1897).

The characterization of citronellol is effected by its oxidation to citronellal (see this) which is easily identified by conversion into

1) Berichte, 29, p. 907.
2) Berichte, 30, p. 34.
3) Berichte, 29, p. 908.
4) Diss., p. 49.
5) Berichte, 29, p. 907.
6) Journ. f. prakt. Chem., II, 56, p. 41.

citronellyl-β-naphtocinchoninic acid or by the semicarbazone melting at 82.5° (Tiemann and Schmidt,[1] 1897).

Of the aromatic alcohols but few have been found in volatile oils. Benzyl alcohol seems to occur as benzoate and cinnamate in Peru and tolu balsam, the cinnamate has also been found in storax. Small amounts of the free alcohol are reported to occur in Peru balsam and the oil of the cherry laurel. Cinnamic alcohol and its reduction product, phenylpropyl alcohol, occur as acetates in cassia oil; the former occurs also as cinnamate (styracin) in storax.

More frequent is the occurrence of hydroaromatic alcohols, both as esters and in the free state. The following representatives of this group have been found: sabinol, $C_{10}H_{16}O$; thujyl alcohol, terpineol, borneol, all of the composition $C_{10}H_{18}O$; and menthol, $C_{10}H_{20}O$. Whereas sabinol, thujyl alcohol and menthol have been found in but one oil each, terpineol and borneol have been found more frequently in volatile oils.

Terpineol.

By the action of dilute sulphuric acid on terpin hydrate the liquid terpineol of commerce is formed, which is not an individual substance, but a mixture of isomeric compounds $C_{10}H_{18}O$. From the results of the oxidation experiments carried out by Tiemann and Schmidt[2] in 1895 the conclusion may be drawn that the solid inactive terpineol of the melting point 35° is contained in it in considerable quantities. In nature, however, only the individual, solid, either optically inactive or active, terpineol appears to be found, the constitution of which was cleared up by the researches of Wallach[3] (1893—1896), and of Semmler and Tiemann[4] (1895—1896). Although liquid terpineols have been isolated from cardamom, kesso, and kuromoji oils and recently also from marjoram oil, it seems probable, however, that these may also be obtained in the solid state, especially since it has been found that the terpineol from cardamom oil can be obtained in a crystalline form.[5]

Solid inactive terpineol occurs in cajuput oil; the laevogyrate modification is contained in niaouli oil, the dextrogyrate in lovage oil, cardamom and marjoram oil, in the latter, however, in the liquid form.

---

1) Berichte, 30, p. 34.
2) Berichte, 28, p. 1783.
3) Liebig's Annalen, 275, pp. 103, 150; 277, p. 110; 291, p. 342; Berichte, 28, p. 1773.
4) Berichte, 28, pp. 1778, 2189; 29, p. 2616.
5) Bericht von S. & Co., Oct. 1897, p. 9.

Besides these, liquid terpineol, in regard to the rotatory power of which no statements are made, has been found in camphor oil, kuromoji oil, kesso and erigeron oil.

Artificially, inactive terpineol is formed according to a report by Bouchardat and Voiry[1] (1887) by the action of very dilute sulphuric acid on terpin, but it is not easy to separate the solid from the liquid terpineol obtained. More successful is the preparation of the solid optically active modifications from d- or l-limonene monohydrobromide by boiling with silver oxide or lead oxide according to the method given by Semmler[2] (1895). In the simplest manner they can be obtained as acetates by the simultaneous action of glacial acetic acid and sulphuric acid (see under camphene, p. 112) on the limonenes, or glacial acetic acid and zinc chloride on the pinenes (Ertschikowsky,[3] 1896). Further, the formation of optically active solid terpineols from linalool by acetic acid anhydride or formic acid (Stephan,[4] 1898), as well as of solid inactive terpineol from geraniol and formic acid[5] are worthy of mention.

The solid terpineol, which possesses the characteristic lilac odor of the liquid compound only to a small extent and is readily soluble in organic solvents, has the following properties:

B. p. 217—218° at 760 mm.; 104—105° at 10 mm.; $d_{15^\circ} = 0.935$—0.940; $n_{D20^\circ} = 1.48084$;[6] m. p. 35°,[7] (Wallach, 1893).

The optical activity varies; the highest deviations observed are, for d-terpineol from cardamom oil $[\alpha]_D = +83° 31'$ (Schimmel & Co.),[8] for l-terpineol from niaouli oil $[\alpha]_D = -2° 10'$ (Bertrand,[9] 1893). With artificially prepared l-terpineol $[\alpha]_D$ has been observed as high as $-117.5°$ (Ertschikowsky[10]).

In their chemical behavior the inactive and active modifications are completely alike; for a few derivatives, however, slight differences in the melting points have been observed. Terpineol is a tertiary unsaturated alcohol, which yields with bromine and nitrosyl chloride addition products, of which the latter as well as the nitrolamines produced from it by treatment with bases, are very well suited for characterization

1) Compt. rend., 104, p. 997.
2) Berichte, 28, p. 2189.
3) Journ. d. russ. phys.-chem. Ges., 28, p. 132; Abs. Bull. Soc. chim., III, 16, p. 1584.
4) Journ. f. prakt. Chem., II, 58, p. 109; II, 60, p. 242.
5) Observations in the laboratory of S. & Co.
6) See p. 110.
7) Liebig's Annalen, 275, p. 104.
8) Bericht von S. & Co., Oct. 1897, p. 9.
9) Bull. Soc. chim., III, 9, p. 436; Comp. rend., 116, p. 1072.
10) Loc. cit. Comp. also Godlewsky, Chem. Centrbl., 70¹, p. 1271.

(see p. 142). By the action of hydrohalogen acids the corresponding dipentene dihydrohalides are produced, of which the dihydriodide $C_{10}H_{18}I_2$ (m. p. 77—78°) formed by shaking with concentrated hydriodic acid may be used for the rapid detection of terpineol (Wallach,[1] 1885). Toward mineral acids and also some of the organic acids the alcohol is rather unstable. Whereas by shaking with dilute sulphuric acid it is converted into terpin hydrate (Tiemann and Schmidt,[2] 1895), by boiling with this reagent water is abstracted with the formation of terpinene, besides a little dipentene and cineol. Similar in action is potassium bisulphate, which yields mostly dipentene; further, phosphoric acid which, besides small amounts of terpinene and cineol, yields mostly terpinolene; and oxalic acid which likewise yields terpinolene (Wallach,[3] 1893; Baeyer,[4] 1894). Acetic acid anhydride, especially on heating, also acts as dehydrating agent with the formation of dipentene, so that it is impossible to convert terpineol quantitatively into its ester by boiling with this reagent (Ginzberg,[5] 1897).

By oxidation with dilute permanganate solution terpineol is first converted into a polyatomic alcohol $C_{10}H_{20}O_3$ (m. p. 122°), from which by oxidation with chromic acid mixture a ketolactone $C_{10}H_{16}O_3$ (m. p. 64°) is produced. The study of this compound has been very useful in the determination of the constitution of terpineol, which accordingly is regarded as Δ^1-terpene-8-ol.[6]

```
      CH3   CH3
        \   /
         C.OH
         |
         CH
        /  \
   CH2 /    \ CH2
      |      |
      |      |
   CH2 \     CH
        \   //
         C
         |
         CH3
```

By the energetic oxidation of the terpineol, also of the ketolactone with chromic acid mixture or nitric acid terpenylic and terebic acids are formed (Tiemann and Mahla,[7] 1896; Godlewsky,[8] 1899).

1) Liebig's Annalen, 230, p. 264.

2) Berichte, 28, p. 1781.

3) Liebig's Annalen, 275, pp. 104—108.

4) Berichte, 27, p. 447.

5) Chem. Centralbl., p. 417; Bericht von S. & Co., Oct. 1897, p. 69.

6) For the literature see p. 189, footnotes 3 & 4.

7) Berichte, 29, p. 2621.

8) Chem. Centrbl., 70I, p. 1241.

Terpineol is of especial interest in so far as it can be converted into derivatives of carvone through the tribromide and nitrosochloride (Wallach,[1] 1894—96).

The alcoholic hydroxyl group of terpineol reacts with phenyl isocyanate, when the two compounds are mixed and allowed to stand some time at room temperature. Sometimes crystals of diphenyl urea first separate, from which the remaining liquid mixture can be removed by treating it with cold anhydrous ether. On careful evaporation of the solvent the urethane separates out in fine needles. Recrystallized from alcohol it possesses the melting point 113° (Wallach,[2] 1893). The compound obtained from optically active terpineol is likewise optically active. The urethane can be used for identification.

For the purpose of identification, the nitrosochloride is especially adapted. According to Wallach[3] (1893) it is prepared in the following manner:

15 g. of terpineol are dissolved in 15 cc. of glacial acetic acid and 11 cc. of ethyl nitrite and after strongly cooling in a freezing mixture a solution of 6 cc. of hydrochloric acid in an equal volume of glacial acetic acid is added drop by drop and with constant shaking; when the reaction is completed the nitrosochloride formed is precipitated by ice water. It separates as an oil, but soon becomes crystalline. The solid product can be purified by recrystallization from hot acetic ether or methyl alcohol and then melts at 112—113°.

By treating the nitrosochloride with piperidine in alcoholic solution terpineol nitrolpiperidine $C_{10}H_{17}(OH)NONC_5H_{10}$ is obtained, which is difficultly soluble in ether and crystallizes from hot methyl alcohol in needles of the melting point 159—160°. This statement by Wallach refers to a preparation obtained from optically inactive terpineol; the nitrolpiperidine obtained from the optically active terpineol melts several degrees lower, namely at 151—152°.[4] With aniline the nitrosochloride yields the terpineol nitrolaniline, m. p. 155—156°.

In connection with terpineol terpin hydrate may be briefly considered. This stands in close relation to terpineol and is formed from it by hydration in the presence of dilute acids. According to older statements it is said to occur in cardamom and basilicum oils. Recent observations, however, corroborating this statement have not been recorded. In all probability terpin hydrate was not originally contained in the oil, but was formed only after standing for some time.

---

1) Liebig's Annalen, 281, p. 140; 291, p. 346.

2) Liebig's Annalen, 275, p. 104.

3) Liebig's Annalen, 277, p. 120.

4) Bericht von S. & Co., April 1897, p. 9.

Borneol.

Borneol occurs in the free state in both optically active modifications, in the form of ester mostly in the laevogyrate modification. The borneo-camphor from ***Dryobalanops camphora*** consists of d-borneol, while the Ngai-camphor (or ***Ngai-fĕn***)[1] from ***Blumea balsamifera*** is the laevogyrate modification of borneol found in nature. d-Borneol has also been found in spike and rosemary oils, as well as in Siam cardamom oil; l-borneol occurs as esters and also in the free state in citronella oil and the oil of ***Matricaria parthenium*** as well as in valerian and kesso oils. In the two latter oils it occurs as acetate and isovalerianate. Further, borneol has been shown to be present in golden rod oil, in the oils from sage and thyme, as well as in the oils from ***Aristolochia serpentaria*** and *A.* ***reticulata***, but no statements as to the direction of rotation of the borneol found are recorded. The acetate of this alcohol possesses the characteristic fir odor and forms a large constituent of many coniferous oils.

Artificially, borneol is most readily obtained from d- or l-camphor by reduction with sodium in alcoholic solution (Wallach,[2] 1885) or in indifferent solvents (Beckmann,[3] 1888); the borneol thus prepared is, however, never pure, but is a mixture of borneol and isoborneol. In alcoholic solution less isoborneol is formed than in indifferent solvents. In the latter case there is obtained as by-product about 5 percent of camphor pinacone (Beckmann,[4] 1897). From the mixture of the two borneols pure borneol can, however, be separated by acetylization, and saponification of the bornyl acetate which crystallizes out.

Pure borneol, recrystallized from ligroin, forms brilliant laminae or plates, which belong to the hexagonal system (Traube,[5] 1894). It possesses an odor somewhat similar to that of camphor, and reminding of amber. It melts at 203—204° (with preparations containing isoborneol at 206—208°); the boiling point lies at 212°. Like camphor, borneol is also somewhat volatile at ordinary temperature, but not to the same extent. The specific gravity is reported by Plowman[6] (1874) to be 1.011 for d-borneol, 1.02 for l-borneol.

Beckmann[7] ('89 & '97) found the rotatory power of d-borneol to be +37.44°. In agreement with this are the statements of Haller[8] (1889)

---

1) Bericht S. & Co., April 1895., p. 74.
2) Liebig's Annalen, 230, p. 225.
3) Berichte, 21. Ref. p. 321.
4) Journ. f. prakt. Chem., II, 55, p. 31.
5) Journ. f. prakt. Chem., II, 49, p. 3.
6) Pharm. Journ., III. 4, p. 710.
7) Liebig's Annalen, 250, p. 353; Journ. f. prakt. Chem., II, 55, p. 31.
8) Compt. rend., 109, 30; see also Haller, Compt. rend., 112 (1891), p. 143, on the influence of the solvent on optical activity.

who determined for the alcohol regenerated from the crystallized acetate, $[\alpha]_D$ to be $+37.63°$. Natural l-borneol possesses according to Beckmann,[1] $[\alpha]_D = -37.74°$, according to Haller,[2] (1889) $[\alpha]_D = -37.77°$. A slightly higher rotatory power, namely $[\alpha]_D = -39° 25'$ was observed for the l-borneol occuring under the name of *Ngai fén* (Schimmel & Co.[3]).

The dextro- and the laevogyrate modifications of borneol are completely alike in their chemical behavior. Although borneol is a saturated alcohol, it forms loose addition products with bromine and hydrohalogens (Wallach,[4] 1885), which, however, are not suited for characterization. As secondary alcohol it is first converted by oxidation into the corresponding ketone $C_{10}H_{16}O$, camphor, a reversal in rotation not taking place. By using nitric acid as oxidizing agent, oxidation products of camphor, like camphoric acid etc., also occur. Toward dehydrating agents, like zinc chloride and dilute sulphuric acid, borneol is very stable (Bertram and Walbaum,[5] 1894). In this respect it differs greatly from the isomeric isoborneol. Phosphorus pentachloride converts it into bornyl chloride. When this is boiled with aniline, hydrogen chloride is split off, and camphene is produced.

As already mentioned, borneol also occurs in combination with fatty acids in volatile oils; more commonly as the acetate of l-borneol, which has so far been shown to be present in the oil of *Abies alba*, hemlock oil, oil of *Picea nigra*, in oil of *Pinus ledebourii* as well as in the oils of *Abies balsamea* and *A. canadensis*, further in valerian and kesso oils. In addition to these it has been shown to be present in the oils of *Pinus montana*, golden rod, *Satureja thymbra* and *Thymus capitatus*, but it is not known in which modification.

Bornyl acetate is the only fatty acid ester of borneol which will crystallize. From petroleum ether it is obtained in rhombic hemihedral crystals, which melt at 29° (Bertram and Walbaum,[6] 1893). Since borneol is fairly stable toward acids, its esters can be prepared either from borneol and acid chlorides or anhydrides. They all possess an odor similar to the acetate, the intensity of which diminishes with increase in the molecular weight of the acid. Tschugaeff[7] (1898) reports on the properties of the fatty acid esters of l-borneol as follows:

---

1) Liebig's Annalen, 250, p. 353; Journ. f. prakt. Chem., II, 55, p. 31.

2) Compt. rend., 108, p. 456; 109, p. 456.

3) Bericht von S. & Co., April 1895, p. 74.

4) Liebig's Annalen, 230, p. 226.

5) Journ. f. prakt. Chem., II, 49, p. 8.

6) Archiv d. Pharm., 231, p. 303.

7) Berichte, 31, p. 1775.

	B. p. (15 mm)	$d\frac{20°}{4°}$	$[\alpha]_D$
l-Borneol	—	—	— 39.00°
Formate	97°	1.0058	— 40.46°
Acetate	107°	0.9855	— 44.40°[1]
Propionate	118°	0.9717	— 42.06°
n-Butyrate	128°	0.9611	— 39.15°
n-Valerianate	139°	0.9533	— 37.08°

Similar statements were made by Bertram and Walbaum[2] in 1893.

For the characterization of borneol the bornyl phenylurethane produced by the action of carbanil is used. It melts at 138—139°[3] and is optically active in the same direction as the borneol from which it is prepared. The addition products formed by the combination of borneol with chloral and bromal, of which that of the chloral melts at 56° (Haller,[4] 1891), that of the bromal at 105—109° (Minguin,[5] 1893), can also be used for the detection of borneol. Borneol may also be converted into camphor by oxidation with Beckmann's chromic acid mixture and this identified by its oxime. When acted on by formic aldehyde in the presence of sulphuric or hydrochloric acid, borneol is converted into diborneol formal $(C_{10}H_{17}.O)_2CH_2$ (Brochet,[6] 1899).

Occasionally it is necessary to separate a mixture of borneol and camphor. The separation can be effected according to Haller's[7] (1889) method by heating the mixture with succinic acid anhydride. Borneol is thereby converted into the acid succinic acid ester, the sodium salt of which is soluble in water and, therefore, easily separated from the camphor. Instead of succinic acid anhydride the corresponding derivative of phthalic acid may be used. The esters of borneol formed by heating with benzoic or stearic acid anhydride are difficultly volatile and can be separated from camphor by distillation with water vapor. The camphor may also be converted into its oxime and this removed from the mixture by shaking with about 25 percent sulphuric acid.

Menthol.

Menthol is found only in the laevogyrate modification as the principal constituent of the peppermint oils, from which it separates in

1) The same value was found by Haller in 1889 for the acetates of the borneol from valerian oil and *Dryobalanops* prepared by boiling with acetanhydride. Compt. rend., 109, pp. 29 & 30.

2) Archiv. d. Pharm., 231, p. 305.

3) Journ. f. prakt. Chem., II, 49, p. 5.

4) Compt. rend., 112, p. 145.

5) Compt. rend., 116, p. 889; Bertram & Walbaum report 98—99° loc. cit.

6) Compt. rend., 128, p. 612.

7) Compt. rend., 108, p. 1308.

crystals on cooling. Artificially, it is obtained by reduction of menthone and pulegone (Beckmann and Pleissner,[1] 1891). With an excess of nascent hydrogen only menthol results from menthone. When solvents are used which do not themselves liberate hydrogen with sodium, some mentho-pinacone also is formed. l- and d-Menthone yield by both methods, also with change in temperature, a strongly laevogyrate menthol mixture. From this l-menthol, melting at 43°, and slightly dextrogyrate isomenthol, $[\alpha]_D = +2°$, melting at 78—81°, can be separated (Beckmann,[2] 1897).

Menthol crystallizes in colorless needles or prisms belonging to the hexagonal system. It is characterized by a strong peppermint odor and cooling taste. Its physical properties are reported as follows:

M. p. 42°; b. p. 211.5° under 735 mm. pressure (Arth,[3] 1886):

M. p. 42.3°; b. p. 212.5° under 742 mm. pressure (corr.): $d_{4°}^{20°} = 0.890$ for solid, $d_{4°}^{44.6°} = 0.8810$ for liquid menthol; $[\alpha]_{D46°} = -49.86°$ for molten menthol (Long,[4] 1892);

B. p. 215.5° under 758 mm. pressure; $\alpha_{D80°} = -43°45'$ for menthol in an overcooled molten condition (Power and Kleber,[5] 1894).

M. p. 43°; $[\alpha]_D = -49.3°$ (in 20 p. c. alcoholic solution); $-57.7°$ in 10 p. c. alcoholic solution (Beckmann,[6] 1889); $n_{C48°} = 1.4479$ (Brühl,[7] 1888).

Menthol is a saturated secondary alcohol, which by the abstraction of water with potassium bisulphate, zinc chloride, etc., is converted into the hydrocarbon menthene $C_{10}H_{18}$. By reduction with hydriodic acid and phosphorus, hexahydrocymene $C_{10}H_{20}$ is produced (Berkenheim,[8] 1892). Upon oxidation with chromic acid mixture the corresponding laevogyrate ketone $C_{10}H_{18}O$, menthone, results (Beckmann,[9] 1889). With potassium permanganate as oxidizing agent compounds are obtained which also result upon the oxidation of menthone, viz. ketomenthylic acid and β-methyladipinic acid, m. p. 88—89°. In accordance with these results the accompanying formula

---

1) Liebig's Annalen, 262, pp. 30 & 32.
2) Journ. f. prakt. Chem., II, 55, pp. 19 & 30.
3) Annales de Chim. et de Phys., VI, 7, p. 488.
4) Chem. Centralbl., 68 II, p. 525.
5) Pharm. Rundschau, 12, p. 162; Archiv d. Pharm., 232, pp. 647 & 658.
6) Liebig's Annalen, 250, p. 327; Journ. f. prakt. Chem., II, 55, p. 15.
7) Berichte, 21, p. 457.
8) Berichte, 25, p. 688.
9) Liebig's Annalen, 250, p. 325.

```
        CH3CH3
          \/
          CH
          |
          CH
         /  \
   CH2 /      \ CHOH
       |      |
       |      |
   CH2 \      / CH2
         \  /
          CH
          |
          CH3
```

has been given to menthol. Jünger and Klages[1] corroborated this formula in 1896 by the conversion of menthone into 3-chlorcymene. That menthol yields cymene when heated with anhydrous copper sulphate to 250—280°, had previously been found by Brühl[2] (1891).

Of the fatty acid esters of menthol, which can be obtained in the same manner as those of borneol, the acetate and isovalerianate have been found in peppermint oil. They have been investigated by Tschugaeff[3] (1898) with special consideration of their rotatory power. According to this investigation, the lower esters possess the following properties:

	B. p. (15 mm)	$d\frac{20°}{4°}$	$[\alpha]_D$
Formate[4]	98°	0.9359	— 79.52°
Acetate	108°	0.9185	— 79.42°
Propionate	118°	0.9184	— 75.51°
n-Butyrate	129°	0.9114	— 69.52°
n-Valerianate	141°	0.9074	— 65.55°

Power and Kleber[5] (1894) report that menthyl acetate boils at 228° under 762 mm. pressure and that $\alpha_D = -72° 15'$. With formic aldehyde menthol forms dimenthol formal $CH_2(OC_{10}H_{19})_2$ melting at 56.5° (Brochet[6]) The methyl ester of menthylxanthogenic acid (m. p. 39°) and menthyldixanthogenate (yellow crystals) are of interest because they can be converted into menthene of a high rotatory power (Tschugaeff,[7] 1899).

With its peculiar physical properties the identification of menthol will hardly offer any difficulty. If it does, however, the menthylphenyl

1) Berichte, 29, p. 304.
2) Berichte, 24, p. 3374.
3) Berichte, 31, p. 364.
4) Solidifies at a low temperature and melts at + 9°.
5) Pharm. Rundschau, 12, p. 162: Archiv d. Pharm., 232, p. 658.
6) Compt. rend., 128, p. 612.
7) Berichte, 32, p. 3332.

urethane produced by the action of phenyl isocyanate may be used for this purpose. This compound, first prepared by Leuckart, melts at 111—112°, and is optically active in the same direction as the menthol from which it is prepared. When heated with sodium ethylate menthol is regenerated, but also inactivated (Beckmann,[1] 1897).

A further derivative, by means of which menthol is easily characterized, is the menthyl benzoate produced by heating it with benzoic acid anhydride. The ester is difficultly volatile with water vapor and melts at 54.5° (Beckmann,[2] 1891 and 1897).

The separation of a mixture of menthol and menthone may be accomplished in the same manner as that given for borneol (Beckmann,[3] 1897).

As previously stated, the sesquiterpenes are frequently accompanied by so-called sesquiterpene alcohols $C_{15}H_{26}O$, which upon dehydration yield hydrocarbons $C_{15}H_{24}$. They are colorless, and when pure mostly odorless. With the exception of the liquid santalol they have a marked tendency to crystallize. The following table contains a compilation of their physical properties together with those of several other alcohols of different composition.

	Formula	M. p.	B. p.	Optical behavior
Maticocamphor	$C_{12}H_{20}O$	94°[4]	—	$[\alpha]_D = -28°$[5]
Kessylalcohol[6]	$C_{14}H_{24}O_2$	85°	300—302° 155—156° (11 mm)	laevogyrate
Santalcamphor[7]	$C_{15}H_{24}O_2$	104—105°	—	—
Cedrol	$C_{15}H_{26}O$	85—86°[8]	282°[9]	—
Cubebcamphor	$C_{15}H_{26}O$	65—67°[10]	248°[11]	laevogyrate
Guaiol[12]	$C_{15}H_{26}O$	91°	288°	—
Ledumcamphor[13]	$C_{15}H_{26}O$	104—105°	148° (10 mm) 282—283°	$[\alpha]_j = +7.98°$
Patchoulialcohol	$C_{15}H_{26}O$	56°[14]	—	$[\alpha]_D = -118°$[15]
Sesquiterpenhydrate from pepper[16]	$C_{10}H_{16}.2H_2O$?	164°	—	—
Santalol[17]	$C_{15}H_{26}O$	liquid ($d_{15°} = 0.980$)	167—169° (10—12 mm)	$[\alpha]_D$ to −31°
Junipercamphor[18]	—	165—166°	—	—
Ylang-ylang-camphor[19]	—	138°	—	—

1) Journ. f. prakt. Chem., II, 55, p. 29.
2) Liebig's Annalen, 262, p. 81; Journ. f. prakt. Chem., II, 55, p. 16.
3) Journ. f. prakt. Chem., II, 55, p. 17.
4) Kügler, Berichte, 16, p. 2841.
5) Traube, Zeitschr. f. Krystallogr., 22, p. 47.
6) Bertram & Gildemeister, Archiv d. Pharm., 228, p. 488.
7—19) See next page.

Aldehydes.

Among the oxygenated constituents of volatile oils characterized by a strong odor are several aldehydes. Of the fatty aldehydes, acetaldehyde is found in the distillate of almost all seeds. Isovalerianic aldehyde and acetaldehyde have also been observed in the distillate of various eucalyptus oils, of cajeput oil and peppermint oil. The aldehyde of oleic acid has been found in orris oil.

Of much greater importance, however, are the so-called aliphatic terpene aldehydes, citral, $C_{10}H_{16}O$ and citronellal, $C_{10}H_{18}O$; and the aromatic aldehydes, benzaldehyde and cinnamic aldehyde. The former are also of special interest on account of their relation to important alcohols occuring in oils, the latter because of the large percentage present in some oils.

Citral.

Citral is the only aldehyde corresponding to the formula $C_{10}H_{16}O$, which has so far been isolated from volatile oils. On account of its close relation to geranial, being its first oxidation product, it is also called geranial. It occurs quite frequently in nature. It was first[20] found in the oil of *Backhousia citriodora* and as it proved to be identical with the constituent to which lemon oil owes its odor, it was called citral. In larger quantity (70—80 percent) it is contained in lemongrass oil and occurs further in orange oil, mandarin and cedron oils, in West Indian limette oil, in the oils of verbena, balm, bay, pimenta, Japanese pepper, sassafras leaves and of *Eucalyptus staigeriana*.

From all these oils citral can be isolated by means of the crystalline bisulphite double compound (see later), which, after previous purification by washing with alcohol and ether, yields citral in a pure state by decomposing it with alkali carbonate.

---

7) Bericht von S. & Co., Oct. 1891, p. 34; Berkenheim, Chem. Centralbl., 64¹, p. 986.

8) Bericht von S. & Co., Oct. 1897, p. 12. Footnote.

9) Walter, Liebig's Annalen, 39, p. 247; 48, p. 35.

10) Schaer & Wyss, Archiv d. Pharm., 206, p. 316; Winckler, Liebig's Annalen, 8, p. 230.

11) Schmidt, Berichte, 10, p. 189.

12) Bericht von S. & Co., April 1892, p. 42; Wallach, Liebig's Annalen, 279, p. 396.

13) Hjelt, Berichte 28, p. 3087.

14) Wallach, Liebig's Annalen, 279, p. 394.

15) Montgolfier, Compt. rend., 84, p. 89.

16) Peinemann, Archiv d. Pharm., 234, p. 241.

17) Observations made in the laboratory of Schimmel & Co.

18) Bericht von S. & Co., Oct. 1895 p. 46.

19) Ibidem, April 1896, p. 62.

20) For the history of citral see Tiemann, Berichte, 31, p. 3278.

Artificially, citral is obtained with a yield of 20—40 percent by the oxidation of geraniol with chromic acid mixture (Tiemann,[1] 1898). The tertiary alcohol linalool also yields the same oxidation product, since a transformation of the linalool to geraniol first takes place by the action of the acid oxidizing agent. In a purely synthetical way citral has been obtained by the distillation of a mixture of calcium gereniate and calcium formate (Tiemann,[2] 1898).

Citral is a mobile, slightly yellowish, optically inactive oil of a penetrating lemon odor, which boils at 228—229° under atmospheric pressure, not entirely without decomposition. Its properties were given by Tiemann and Semmler[3] in 1893 as follows:

B. p. 110—112° at 12 mm.; 117—119° at 20 mm.; 120—122° at 23 mm.; $d_{15°} = 0.8972$; $d_{22°} = 0.8844$; $n_{D15°} = 1.4931$; $n_{D22°} = 1.48611$.

Observations made in the laboratory of Schimmel & Co. on an aldehyde carefully purified by means of the bisulphite compound gave the following results:

B. p. 110—111° at 12 mm.; $d_{15°} = 0.893$; $n_{D17°} = 1.49015$.

As diolefinic aldehyde, citral takes up 2 molecules of bromine, but yields with it no solid compound. Toward acids and acid agents it is very sensitive and is greatly changed by them. Dilute sulphuric acid and potassium bisulphate act very energetically with abstraction of water and formation of cymene. Alkalies also act on citral. When boiled with potassium carbonate solution it is split up into acetaldehyde and methylheptenone $C_8H_{14}O$ (Verley,[4] 1897; Tiemann,[5] 1899). This ketone also accompanies citral, e. g. in lemongrass oil, and is also formed from it by gentle oxidation.

Citral shows all the properties of an aldehyde; thus it is changed by reduction to geraniol (Tiemann,[6] 1898) and reacts with the well-known aldehyde reagents. It shows a peculiar behavior toward sodium bisulphite solution (Tiemann and Semmler,[7] 1893; Tiemann,[8] 1898). If the solution does not contain too great an amount of free sulphurous acid, there separates upon shaking at a low temperature the difficultly soluble, normal crystallized double compound $C_9H_{15}.CH(OH)\text{-}SO_3Na$, which cannot be quantitatively split up by sodium carbonate or caustic soda. If the crystallized compound be allowed to stand with an excess of bisulphite solution at a moderate heat, it will again be dis-

---

1) Berichte, 31, p. 3311.
2) Berichte, 31, p. 827.
3) Berichte, 26, p. 2709.
4) Bull. Soc. chim., III, 17, p. 175
5) Berichte, 32, p. 107.
6) Berichte, 31, p. 828.
7) Berichte, 26, p. 2710.
8) Berichte, 31, p. 3310.

solved with the formation of a dihydrosulphonic acid derivative of citral, $C_9H_{17}.(SO_3Na)_2.CHO$, which no longer regenerates citral with alkali carbonate, but does so with caustic alkali. If the temperature rises too high during the solution of the crystallized compound, caustic alkali also fails to separate citral from the liquid, a sodium salt of a dihydrosulphonic acid derivative which cannot be split up, having been formed. This compound is likewise formed, citral being simultaneously regenerated in part, when the normal bisulphite compound is suspended in water and treated for some time with water vapor until it is dissolved. If the solution of the citral-dihydrosulphonate of sodium is shaken with citral, this is taken up and the disulphonate is changed to citral-hydroxymonosulphonate of sodium which is readily decomposed by alkalies.

The compounds of citral with hydroxylamine, phenyl hydrazine and ammonia, are all liquid and therefore cannot be used for the characterization of citral. The oxime is changed to the nitrile of geranic acid by splitting off water with acetic acid anhydride. With semicarbazide, several well crystallizable semicarbazones (Wallach,[1] 1895; Tiemann and Semmler,[2] 1895; Tiemann,[3] 1898) are produced, which under certain conditions can be resolved into compounds of a definite melting point 164° and 171°, and may therefore be used for the identification of citral (Tiemann[4] 1898).

When oxidized with mild oxidizing agents, e. g. silver oxide in ammoniacal solution, liquid geranic acid is formed (Semmler,[5] 1890–'91), which has an odor similar to that of the higher fatty acids. Upon energetic oxidation with chromic acid mixture methylheptenone results (Tiemann and Semmler,[6] 1893), which on further oxidation with potassium permanganate and chromic acid mixture is decomposed to acetone and laevulinic acid.[7] In harmony with these results citral is regarded as dimethyl-2,6-octadiëne-2,6-al-8,

$$CH_3.C(CH_3):CH.CH_2.CH_2.C(CH_3):CH.CHO.$$

This formula corresponds with that of geraniol and agrees well with the reactions of citral.

It is worth mentioning, that by condensation with acetone, citral yields a ketone $C_{13}H_{20}O$, pseudoionone, which when heated with dilute sulphuric acid is converted into ionone, a compound isomeric with the

1) Berichte, 28, p. 1957.
2) Berichte, 28, p. 2133.
3) Berichte, 31, pp. 821, 2815.
4) Berichte, 31, p. 3331.
5) Berichte, 23, p. 3556; 24, p. 203.
6) Berichte, 26, p. 2718.
7) Ibidem, p. 2128.

irone of orris oil, and which like the latter has the odor of violets (Tiemann and Krueger,[1] 1893).

Inasmuch as citral possesses a penetrating odor, this in itself attracts attention to its presence in volatile oils. For more positive proof the separation of the aldehyde by its bisulphite compound is tried. The regenerated citral is converted by condensation with pyrotartaric acid and β-naphthylamine into the α-citryl-β-naphthocinchoninic acid discovered by Doebner[2] in 1894 who directs this characteristic compound to be prepared as follows:

20 g. of pyrotartaric acid and 20 g. of citral (or the oil in question) are dissolved in absolute alcohol, to this solution 20 g. of β-naphthylamine, likewise dissolved in absolute alcohol, are added and the mixture boiled with a reflux condenser for about 3 hours on a water bath. After cooling, the citrylnaphthocinchoninic acid which has separated in a crystalline state is filtered off and purified by washing with ether. If the acid is too impure it is dissolved in ammonia and separated from the filtered solution by neutralizing with acetic acid. The pure compound thus obtained crystallizes from alcohol in yellow leaflets. According to Doebner[3] it melts at 197°, but the melting point will generally be found at 200° or slightly above.

In the preparation of the naphthocinchoninic acid it must be remembered that with a small citral content α-methyl-β-naphthocinchoninic acid is formed, due to the decomposition of a part of the pyrotartaric acid used. This compound does not melt until 310° and is more difficultly soluble in alcohol than the citrylnaphthocinchoninic acid. It therefore remains in the residue when the crude naphthocinchoninic acid is exhausted with hot alcohol.

Further it must be remembered that when other aldehydes are present with citral the naphthocinchoninic acids resulting from these are formed at the same time. Thus Doebner found in fractions of lemon oil besides the citryl-compound also citronellyl-β-naphthocinchoninic acid (m. p. 225°).

Citral-semicarbazone, m. p. 164°, is prepared in the following manner:

To a solution of 4 parts of semicarbazide chlorhydrate in a little water a solution of 5 parts of citral (or of the fraction to be tested) in 30 parts of glacial acetic acid is added. After a short time considerable quantities of a semicarbazone separate in needles, which after two or three recrystallizations from methyl alcohol have the constant melting point of 164°. From the mother liquid filtered from this semicarbazone the compound melting at 171° can be obtained (Tiemann,[4] 1898—'99).

1) Berichte, 26, p. 2675.

2) Berichte, 27. pp. 354, 2026.

3) Loc. cit.; Berichte, 31. p. 1888; comp. also Berichte, 31, pp. 3197 & 3327.

4) Berichte. 31, p. 3331; and 32, p. 115.

A well crystallized derivative of citral, which is equally well suited for detection as for the approximate quantitative determination, is the citralidenecyanacetic acid $C_9H_{15}.CH:C(CN).COOH$ melting at 122°, and produced by the condensation of citral with cyanacetic acid in the following manner:

Two mol. of sodium hydroxide (as 30 percent soda solution) and one mol. of citral are added to a solution of one mol. of cyanacetic acid in about three times as much water. When pure, the citral dissolves completely on shaking. From the clear solution—purified if necessary by extracting with ether—acids separate the citralidencyanacetic acid in a crystalline form or as an oil which soon solidifies. By dissolving in benzol and precipitating with ligroin it can be obtained in compact yellow crystals (Tiemann.[1] 1898).

Citral likewise forms a solid condensation product with acetyl acetone.

It is obtained when 15.2 g. of citral are condensed with 20 g. of acetylacetone at room temperature with the aid of 10 drops of piperidine. After standing for three days the entire mixture congeals to a solid mass of crystals, from which, by recrystallization from a mixture of alcohol, ether and ligroin, the citralidenebisacetylacetone is obtained in light yellow warts melting at 46—48° (K. Wedemeyer,[2] 1897).

Tiemann[3] recommended to determine the presence of citral by condensing it with acetone to pseudoionone and identifying this by its semicarbazone. This method, however, is more troublesome than the preparation of the citrylnaphtocinchoninic acid, so that the latter will undoubtedly be preferred as a means of identification.

The ionone controversy has caused an unusual activity in the chemical study of citral during the year 1899. The results are of such a nature that it is impracticable to incorporate them into a work of this kind at least at present. Not only do different investigators interpret the same results differently, but different investigators obtained different results in what is supposed to be the same experiment. Furthermore, one and the same investigator has found it necessary to correct his own statements within a few months. It is necessary, therefore, not only to question the arguments made in the excitement of a controversy, but at times the experiments as well. For the benefit of those who desire to go into the details of the subject the following bibliography for 1899 is herewith appended.

BOUVEAULT: Isomeric aldehydes from lemon grass oils.[4]

Nature of isomerism of the two lemonals.[5]

BARBIER: Lemonal from the essential oil of *Lippia citriodora*.[6]

CORIK: Citral in pure oil of lemon.[7]

---

1) Berichte. 31, p. 3829.

2) Ueber Condensationen mittelst aromatischer Basen u. s. w. Inaug. Diss. Heidelberg, p. 24.

3) Berichte, 31, p. 822.

4) Bull. Soc. Chim., [3], 21, p. 419.

5) Ibidem, p. 923.

6) Bull. Soc. Chim., [3], 21, p. 635.

7) C. & D., 54, p. 650.

FLATAU: Essential oils of lemon-grass and citronella.[1]
FLATAU & LABBÉ: Method for separation of citronellal and citral.[2]
LABBÉ: On lemongrass oil.[3]
A polymeride of citral.[4]
On barium acid sulphite addition products of citral and citronellal.[5]
STIEHL: Three lemongrass aldehydes.[6]
TIEMANN: On the action of alkaline and acid reagents on citral.[7]
On the behavior of citral purified according to different methods toward semicarbazide.[8]
On the hydroxysulphonates of cinnamic aldehyde, citronellal and citral.[9]
On the separation of citral from citronellal and methyl heptenone.[10]
On the three lemongrass aldehydes of Mr. W. Stiehl.[11]
On natural citral and the composition of oil of lemon grass.[12]

A number of these articles by the late Professor Tiemann have been published in book form.

VERLEY: Action of acids on citral.[13]
Condensation of citral with cyanacetic acid.[14]
Condensation of citral with malonic acid.[15]

The report of Professor v. Baeyer as expert in the ionone proceedings has been published by A. W. Schade of Berlin.

CITRONELLAL.

The second aliphatic aldehyde with ten carbon atoms found in volatile oils is citronellal $C_{10}H_{18}O$, which occasionally occurs accompanying citral, the dihydroderivative of which it appears to be. Citronellal is distinguished from citral by being optically active, but has up to the present been found only in the dextrogyrate modification. It is probable that citronellal with a low rotatory power is a mixture of both optically active modifications. d-Citronellal is found in citronella oil and in the oil of *Eucalyptus maculata* var. *citriodora*. Its presence has also been shown in the oil of *Eucalyptus dealbata*, in balm oil, and besides citral, in lemon oil. Whether it is contained in mandarin oil has not been definitely determined.

The isolation of this aldehyde from the oils rich in citronellal (citronella oil, oil of *Eucalyptus maculata*) offers no difficulty, since citronellal can be readily separated in the form of its crystalline bisulphite compound. As citronellal is very sensitive to acids, as well

1) Bull. Soc. Chim. [3], 21, p. 158.
2) Chem. Centrbl., 70I, p. 1094.
3) Bull. Soc. Chim. [3], 21, p. 77.
4) Ibidem, p. 407.
5) Ibidem, p. 1026.
6) Journ. pr. Chem., 167, p. 497.
7) Berichte, 32, p. 107.
8) Ibidem, p. 250.
9) Bull. Soc. Chim. [3], 21, p. 196.
10) Berichte, 32, p. 812.
11) Ibidem, p. 827.
12) Ibidem, p. 830.
13) Bull. Soc. Chim. [3], 21, p. 408.
14) Ibidem, p. 413.
15) Ibidem, p. 414.

as to alkalies, alkali carbonate is used for the decomposition of the bisulphite compound. Artificially, citronellal can be obtained by the oxidation of the primary alcohol citronellol $C_{10}H_{20}O$, but the yield in this case is still smaller than when geraniol is oxidized to citral. In this manner the laevogyrate modification which has so far not been found in nature has been prepared from the l-citronellol of rose oil.

By splitting the cycle of menthoxime Wallach[1] (1894 and 1897) has obtained an aldehyde $C_{10}H_{18}O$, designated by him as "Menthocitronellal," which possesses great similarity to the naturally occurring citronellal and is probably identical with it.

Citronellal boils under atmospheric pressure at 205—208°, under 15 mm. pressure at 103—105°; its specific gravity was found at 17.5° to be 0.8538; the index of refraction $n_D = 1.4481$ (Tiemann and Schmidt,[2] 1896). The optical rotation $[\alpha]_D$ was found by Kremers[3] (1892) to be + 8.18°. Recently, however, Tiemann and Schmidt[2] have found $[\alpha]_D = +12° 30'$ for a preparation regenerated from the bisulphite compound.

Citronellal has but one double bond. When reduced with sodium amalgam in alcoholic solution, being constantly kept slightly acid by the addition of acetic acid, it is converted into the primary alcohol citronellol $C_{10}H_{20}O$ (Tiemann and Schmidt[4]). Upon careful oxidation with silver oxide the corresponding oily citronellic acid $C_{10}H_{18}O_2$ is produced (Semmler,[5] 1891 and 1893). Like citral, citronellal is very sensitive toward alkalies and acids; but while citral, when treated with alkali, is split up into acetaldehyde and methylheptenone, citronellal is completely resinified. In contact with acids cymene is formed from citral, by splitting off water. Citronellal, however, is converted into the isomeric isopulegol (Tiemann and Schmidt,[6] 1896—'97). This is isomeric with the alcohol pulegol $C_{10}H_{18}O$, resulting upon reduction of pulegone, and yields when oxidized a ketone $C_{10}H_{16}O$, which can be inverted to natural pulegone. For corrected physical constants of pulegone and isopulegone the more recent work of Harries and Roeder[7] should be consulted. Whether isopulegol, which has been found in commercial citronellal by Tiemann[8] occurs in the natural oils from which citronellal is obtained or whether it is formed upon standing has not yet been determined. Labbé[9] is of the opinion that whereas

---

1) Liebig's Annalen, 278, p. 317; 296, p. 131.
2) Berichte, 29, p. 905.
3) Amer. Chem. Journ., 14, p. 203.
4) Loc. cit., p. 906.
5) Berichte, 24, p. 208; 26, p. 2256.
6) Berichte, 29, p. 913; 30, p. 22.
7) Berichte, 32, p. 3357.
8) Berichte, 32, p. 825.
9) Bull. Soc. Chim. [3], 21, p. 1023.

citronellal when mixed with terpenes and terpene alcohols is relatively stable, when pure it is easily converted into its isomeric cyclic compounds.

Toward sodium bisulphite citronellal behaves much as does citral. Besides the crystallized normal addition product with one molecule of $NaHSO_3$, in which the bisulphite has added itself to the aldehyde group, it also yields hydrosulphonic acid derivatives with one or two molecules of $NaHSO_3$. The addition may take place at the double bond only, or at both the double bond and the carbonyl group (Tiemann,[1] 1898). With hydroxylamine citronellal forms a liquid oxime, which by the abstraction of water is converted into the nitrile of citronellic acid (Semmler,[2] 1893). The semicarbazone resulting with semicarbazide is, so far as present observations show, a single substance, and is well suited for the identification of citronellal. It separates quantitatively when an alcoholic solution containing the aldehyde is shaken with a solution of semicarbazide chlorhydrate and sodium acetate. The crude compound, by recrystallization from chloroform and ligroin is obtained in white leaflets, melting at 84° (Tiemann and Schmidt,[3] 1897).

By energetic oxidation citronellal yields the same products as citronellol, i. e., acetone and β-methyladipinic acid (Tiemann and Schmidt,[4] 1896). Corresponding to citronellol the aldehyde is to be regarded as dimethyl-2,6-octene-2-al-8 $CH_3 . C(CH_3) : CH . CH_2 . CH_2 . CH(CH_3) . CH_2 . CHO$ (T. & S[5]).

Like citral, citronellal reacts with pyrotartaric acid and β-naphthylamine yielding α-citronellyl-β-naphthocinchoninic acid (Doebner,[6] 1894) which can be used for its detection. It is prepared like the citral compound. The crude naphthocinchoninic acid is recrystallized from alcohol containing hydrochloric acid, the hydrochloride obtained is dissolved in ammonia and the ammonium salt decomposed by acetic acid. The compound so purified crystallizes from dilute alcohol in colorless needles, and melts at 225°. By heating it above its melting point it splits off carbon dioxide and is converted into citronellyl-β-naphthochinoline, a base crystallizing from dilute alcohol or ligroin in needles of satin-like lustre melting at 53°.

Citronellal can be identified by the semicarbazone (see above) more rapidly than by the naphthocinchoninic acid.

---

1) Berichte, 31, p. 3305.
2) Berichte, 26, p. 2255.
3) Berichte, 30, p. 34; 31, p. 3307.
4) Berichte, 29, p. 908.
5) Ibidem, p. 918.
6) Berichte, 27, p. 2025.

Of aldehydes with cyclic structure

FURFUROL C_4H_4O

should first be mentioned, the occurrence of which in oil of cloves and oil of clove stems has been observed. Its formation is probably due to the decomposition of a carbohydrate-like compound. Inasmuch as it possesses the property of gradually becoming darker in color it is surmised that the darker color acquired by some oils upon standing is due to the presence of this substance. Furfurol has recently been also found in the oil from musk seed[1] and that of caraway seed.[2] It is identified by means of its hydrazone and by the color reactions it produces with aniline and p-toluidine.

The following carbocyclic aldehydes have been found:

BENZALDEHYDE $C_6H_5.CHO$

in the oils of bitter almond, cherry laurel, wild cherry bark and of *Indigofera galegoides*. It possibly also occurs in the oil of cinnamon leaves.

SALICYLIC ALDEHYDE $C_6H_4.OH^{[2]}.CHO^{[1]}$

(o-oxybenzaldehyde) in spirea oil from *Spiraea ulmaria*, *Sp. filipendula*, *Sp. digitata*, *Sp. lobata*, and in the oil of *Crepis foetida*.

ANISIC ALDEHYDE $C_6H_4.OCH_3^{[4]}.CHO^{[1]}$

(p-oxybenzaldehyde-methylether) in old anise and fennel oils, in which it is formed by the oxydation of anethol.

CUMINIC ALDEHYDE $C_6H_4.C_3H_7^{[4]}.CHO^{[1]}$

(p-isopropylbenzaldehyde) in cumin oil, in the oil of *Cicuta virosa*, also in various eucalyptus oils (*E. haemastoma*, *odorata*, *oleosa*, *populifera*).

VANILLIN

(m-methoxy-p-oxybenzaldehyde) $C_6H_3.OH^{[4]}.OCH_3^{[3]}.CHO^{[1]}$ and

HELIOTROPIN

(protocatechuic aldehyde methylene ether) $C_6H_3 . OOCH_2^{[4,3]} . CHO^{[1]}$ in spirea oil.

CINNAMIC ALDEHYDE $C_6H_5.CH:CH.CHO$

in cassia oil, Ceylon cinnamon oil, the oil of cinnamon leaves and roots, also in the oil of *Cinnamomum loureirii*.

O-CUMARIC ALDEHYDE METHYLETHER

(o-methoxycinnamic aldehyde) $C_6H_4 . OCH_3^{[2]} . CH : CH . CHO^{[1]}$ in cassia oil.

---

1) Bericht S. & Co., Oct. 1899, p. 86. 2) Ibidem, p. 82.

Ketones.

The aliphatic ketones are represented in volatile oils by a few members only. Of the saturated ketones only acetone $CH_3.CO.CH_3$, methyl amyl ketone $CH_3.CO.C_5H_{11}$ and methyl nonyl ketone $CH_3.CO.C_9H_{19}$; of the unsaturated only methyl heptenone $C_8H_{14}O$ has been found. In the distillate, especially of leaves, acetone has frequently been observed together with methyl alcohol, e. g. in that of patchouli, coca, and tea leaves. To what reaction this ketone owes its origin is not known. Methyl amyl ketone is contained in the low-boiling fractions of clove oil and imparts to it a pleasant odor. Methyl nonyl ketone is the principal constituent of oil of rue and separates from this upon cooling (m. p. 15°), and can also be readily separated by shaking with bisulphite solution.

Methyl heptenone.

Of greater interest than the saturated ketones named is the unsaturated methyl heptenone $C_8H_{14}O$, which occurs as a constituent of some volatile oils and is also obtained as a decomposition product of related compounds. Accompanying the bodies closely related to it, linalool, geraniol and citral, it occurs in lignaloe, citronella and lemongrass oils. It evidently owes its existence to the decomposition of these compounds, which can also be brought about artificially by mild oxidation. It can be easily isolated by means of its bisulphite addition product from the fraction 160—180° of the oils named.

As a product of decomposition it was first observed by Wallach (1890) in the dry distillation of cineolic acid anhydride.[1] It was also obtained as a saponification product of geranic acid nitrile (Tiemann & Semmler,[2] 1893) and as oxidation product of citral[3] and finally by the splitting up of the latter with alkalies (Verley,[4] 1897), Synthetically, it has been prepared from amylene bromide and acetyl acetone (Barbier & Bouveault,[5] 1896) also from the iodide of acetopropyl alcohol, acetone and zinc dust (Verley,[6] 1897).

It is a colorless. mobile, optically inactive liquid. with a penetrating amyl acetate-like odor. The physical constants. as given, do not wholly agree. Wallach[7] (1890) found for the ketone formed from cineolic acid anhydride:

---

1) Liebig's Annalen, 258, p. 328.
2) Berichte, 26, p. 2721.
3) Ibidem, p. 2719.
4) Bull. Soc. chim. III, 17, p. 175.
5) Compt. rend., 122, p. 1422.
6) Bull. Soc. chim. III, 17, p. 191.
7) Liebig's Annalen, 258, p. 325.

B. p. 173—174°; $d_{20°} = 0.853$; $n_D = 1.44003$.

Tiemann and Krüger[1] (1895) determined for natural methylheptenone:

B. p. 170—171°; $d_{20°} = 0.8499$; $n_D = 1.4380$;

and Verley[2] (1897) reports for a preparation obtained by splitting up citral:

B. p. 168°; 84° at 56 mm.; $d_{14°} = 0.910$ (!); $n_{D31°} = 1.437$.

According to observations made in the laboratory of Schimmel & Co. methylheptenone. which had been isolated from lemongrass oil and regenerated from the bisulphite compound shows:

B. p. 170—171° (758 mm); $d_{15°} = 0.858$; $n_{D15°} = 1.44388$

and a preparation made from citral by boiling with potassium carbonate solution:

B. p. 173—174°; $d_{15°} = 0.8656$.

Upon reduction with sodium in alcoholic solution methyl heptenone is reduced to the secondary alcohol, methyl heptenol (Wallach,[3] 1893), which is obtained as decomposition product of geraniol, also by the saponification of geranic acid nitrile. It combines with bisulphites to form crystalline addition products; with hydroxylamine and phenylhydrazine to form liquid condensation products and with semicarbazide to form a crystallized semicarbazone, which can be used for identification. Upon oxidation, methylheptenone decomposes in correspondence with the formula

$$CH_3 . C(CH_3) : CH . CH_2 . CH_2 . CO . CH_3$$

(methyl-2-heptene-2-one-6) into acetone and laevulinic acid $C_5H_8O_3$ (Tiemann and Semmler,[4] 1895). Dehydrating agents, like zinc chloride, etc., convert it into dihydro-m-xylene C_8H_{12} (Wallach,[5] 1890).

Methylheptenone is easily recognized by its characteristic amyl acetate-like odor. For purposes of identification it is converted into the semicarbazone, which, like the corresponding citral compound appears to be a mixture of isomers, but is nevertheless obtained of a constant melting point when it is prepared according to the directions given by Tiemann and Krüger[6] (1895):

A solution of 12 g. of semicarbazide hydrochloride and 15 g. of sodium acetate in 30 cc. of water is added to a mixture of 12 g. of methylheptenone and 20 cc. of glacial acetic acid and allowed to stand for some time (½ hour). Upon the addition of water the semicarbazone separates as an oil which soon solidifies and after recrystallization from dilute alcohol melts at 136—138°.

---

1) Berichte, 28, p. 2123.
2) Loc. cit., p. 176.
3) Liebig's Annalen, 275, p. 171.
4) Berichte, 28, p. 2128.
5) Liebig's Annalen, 258, p. 326.
6) Berichte, 28, p. 2124.

A derivative which is also well suited for identification is obtained when methyl heptenone is treated with bromine in the presence of soda solution (Tiemann and Semmler,[1] 1893). Hypobromous acid is added and bromine is substituted with the formation of the well crystallized compound $C_8H_{12}Br_3O.OH$.

3 g. of ketone are shaken with a solution of 3 g. of sodium hydroxide and 12 g. of bromine in 100 to 120 cc. of water. The compound separating as a heavy oil which soon solidifies is taken up with ether, the ethereal solution shaken with dilute soda solution and the residue remaining after evaporation of the ether is recrystallized from ligroin, with the addition of animal charcoal. The melting point of the pure white compound, gradually decomposing when kept, is 98—99°.

With the exception of anisic ketone $C_6H_4.OCH_3^{[4]}.CH_2.CO.CH_3^{[1]}$, which is reported to occur in anise and fennel oils, aromatic ketones have not been found in volatile oils. Hydroaromatic ketones, however, are important constituents of these oils: viz. carvone $C_{10}H_{14}O$, camphor, fenchone, thujone (tanacetone), pulegone, all having the formula $C_{10}H_{16}O$, menthone $C_{10}H_{18}O$, and irone $C_{13}H_{20}O$. All of these possess a characteristic odor. Whereas irone has been found in orris oil only, to which it imparts the odor of violets, the other ketones occur more frequently, the camphor being of most common occurence.

Carvone.

Carvone occurs principally in its optically active variety in volatile oils. The earlier supposition of Wallach, that inactive carvone is contained in the high boiling fractions of thuja oil, has been corrected by this investigator, who has more recently shown that it is not carvone, but a hydrocarvone $C_{10}H_{16}O$, which has probably resulted upon rearrangement from thujone. The dextrogyrate modification of carvone is found with d-limonene in caraway and dill oils, the laevogyrate in spearmint and kuromoji oils.

In order to separate the ketone in a pure state from the respective oil, its property to combine with hydrogen sulphide to form a crystallized compound $C_{10}H_{14}O.H_2S$ is made use of.

A mixture of 20 parts of the carvone fraction, 5 parts of alcohol and 1 part of ammonia (sp. gr. 0.96) is saturated with hydrogen sulphide when carvone hydrosulphide separates out. This is collected and recrystallized from methyl alcohol and then decomposed by treating with alcoholic potassa. The regenerated carvone is purified by steam distillation.

[1]) Berichte, 26, p. 2723.

As an improvement Wallach[1] suggests the following details for the preparation of l-carvone:

50 cc. of the carvone fraction of spearmint oil are diluted with 20 cc. of alcohol and the solution, kept cold with ice, is saturated with hydrogen sulphide. Then sufficient alcoholic ammonia, saturated at 0°, is added until the liquid smells strongly of ammonia. Upon passing more hydrogen sulphide into the solution it congeals shortly to a crystalline mass. The same process is repeated with the mother liquid. The carvone hydrosulphide, when recrystallized from a mixture of three parts of chloroform and one part of alcohol, melts at 190°. For the regeneration of the carvone, 50 g. of the hydrosulphide are digested with a solution of 80 g. of potassa in 400 g. of water. The oil is separated and dried with fused potassa either directly or after distillation with water vapor and finally rectified.

Artificially, derivatives of carvone have been obtained in various ways. Thus, e. g. carvoxime has been obtained from limonene or dipentene nitrosochloride respectively (Goldschmidt and Zürrer,[2] 1885); an oxydihydrocarvoxime from terpineol nitrosochloride (Wallach,[3] 1896); carveolmethylether from limonene tetrabromide and terpineol tribromide respectively and methyl alcoholic potassa (Wallach,[4] 1894). Pinene nitrosochloride is converted by hydrochloric acid in etherial solution into hydrochlorcarvoxime (Baeyer,[5] 1896); and by the decomposition of phellandrene nitrite, hydroderivatives of the carvone series result (Wallach,[6] 1895).

Carvone is a colorless liquid, smelling strongly of caraway and solidifying when strongly cooled (Wallach,[7] 1889). Its physical properties are given as follows:

for d-carvone: b. p. 224°; $d_{4°}^{20°} = 0.9598$; $[\alpha]_{D20°} = +62.07°$ (Beyer,[8] 1883);
$d_{20°} = 0.9538$; $n_{H\alpha} = 1.4751$ (Kanonnikoff,[9] 1881);
$d_{11°} = 0.9667$; $n_D = 1.5020$ (Gladstone,[10] 1886);
$d_{20°} = 0.9606$; $[\alpha]_{D20°} = +62.65$ (Schreiner,[11] 1896);
for l-carvone: b. p. 223—224°; $d_{4°}^{20°} = 0.9593$; $[\alpha]_{D20°} = -62.41°$.[8]

According to observations made in the laboratory of Schimmel & Co., carvone boils at 230° (mercury completely in the vapor) and has the specific gravity 0.964 at 15°.

1) Liebig's Annalen, 305, p. 224.
2) Berichte, 18, p. 1732.
3) Liebig's Annalen, 291, p. 346.
4) Liebig's Annalen, 281, pp. 127 & 140.
5) Berichte, 29, p. 12.
6) Liebig's Annalen, 287, p. 380.
7) Liebig's Annalen, 252, p. 129, Footnote.
8) Archiv d. Pharm., 221, p. 287.
9) Berichte, 14, p. 1699.
10) Jahresb. f. Chemie, 1886, p. 298.
11) Pharm. Review, 14, p. 78.

Carvone is an unsaturated ketone, which forms with hydrochloric acid a liquid addition product, with hydrobromic acid a solid compound, melting at 32° (Goldschmidt & Kisser,[1] 1887). By splitting off hydrobromic acid from the latter, a ketone $C_{10}H_{14}O$ results, isomeric with carvone which has been termed eucarvone by Baeyer[2] (1894). The bodies resulting by the action of bromine, viz. the tribromide $C_{10}H_{14}O.HBr.Br_2$, the tetrabromide $C_{10}H_{14}OBr_4$ and the pentabromide $C_{10}H_{18}OBr_5$, have been investigated by Wallach[3] (1895).

Carvone does not add bisulphite, but with hydroxylamine yields a well crystallized oxime (Goldschmidt,[4] 1884) which, when prepared from the optically active ketone, melts at 72°. By the union of equal quantities of d- and l-carvoxime, inactive carvoxime, melting at 93° results. Artificially, these oximes are obtained by splitting off hydrochloric acid from limonene and dipentene nitrosochlorides (Goldschmidt & Zürrer, 1885; Wallach,[5] 1888). With phenyl hydrazine carvone yields a phenyl hydrazone melting at 109—110° (Baeyer,[6] 1894); semicarbazide combines with d- and l-carvone to form carbazones melting at 162—163° (Baeyer,[7] 1894). The semicarbazone of the inactive carvone melts lower than the active compounds, namely at 154—156° (Baeyer,[8] 1895), differing in this respect from i-carvoxime.

On reduction carvone behaves differently from other ketones. Sodium in alcoholic solution does not convert it, as might be expected, into the corresponding alcohol carveol $C_{10}H_{16}O$, but 4 atoms of hydrogen are taken up at once with the formation of dihydrocarveol (Wallach,[9] 1893). If the reduction is effected with zinc dust in alcoholic alkaline solution, but two atoms of hydrogen are added, yet the resulting compound is no alcohol, but a ketone $C_{10}H_{16}O$, dihydrocarvone (Wallach & Schrader,[10] 1894). Carvone is oxidized by potassium permanganate to oxyterpenylic acid $C_8H_{12}O_5$ (m. p. 190—192°) (Best,[11] 1894; Wallach,[12] 1894). From oxidation results which Tiemann and Semmler (1895) obtained with dihydrocarveol and dihydrocarvone,[13] they have deduced the following formula which had previously been suggested by Wagner, as the most probable one:

---

1) Berichte, 20, pp. 487 & 2071.
2) Berichte, 27, p. 811.
3) Liebig's Annalen, 286, p. 119.
4) Berichte, 17, p. 1578.
5) Berichte, 18, p. 2220; Liebig's Annalen, 245, pp. 256, 268; 246, p. 226.
6) Berichte, 27, p. 811; see also Goldschmidt, Berichte 17, p. 1578.
7) Berichte, 27, p. 1923.
8) Berichte, 28, p. 640.
9) Liebig's Annalen, 275, p. 110.
10) Liebig's Annalen, 279, p. 377.
11) Berichte, 27, p. 1218.
12) Berichte, 27, p. 1495.
13) Berichte, 28, p. 2148.

```
       CH3  CH2
          \ //
           C
           |
           CH
          /  \
      CH2/    \CH2
        |      |
        |      |
      CH \    /CO
           \ /
            C
            |
           CH3
```

With phosphorus pentachloride carvone yields the dichloride $C_{10}H_{14}Cl_2$ which, when heated with quinoline is converted into 2-chlor-cymene (Klages and Kraith,[1] 1899).

Carvone suffers a remarkable rearrangement when it is heated with sulphuric acid, phosphoric acid, phosphorus oxychloride, zinc chloride or alkalies; it is thereby converted into carvacrol, a derivative of benzene: $C_6H_3 . CH_3{}^{[1]} . OH^{[2]} . C_3H_7{}^{[4]}$. With hydrogen chloride the change is almost quantitative. Anhydrous formic acid also causes a like change (Klages[2]). A similar molecular rearrangement takes place with the oxime of carvone when it is boiled with alcoholic sulphuric acid or heated to 230—240° with a strong solution of alkali, being changed to carvacrylamine (Wallach,[3] 1893). By adding carvoxime to concentrated sulphuric acid it suffers rearrangement to p-amidothymol (Wallach,[4] 1894).

When carvone is heated with ammonium formate dihydrocarvylamine is formed (Leuckart & Bach,[5] 1887; Wallach,[6] 1891); the same compound results when carvoxime is reduced with sodium in alcoholic solution (Wallach,[7] 1893). If, however, the reduction is carried on in alcoholic solution with zinc dust and glacial acetic acid, carvylamine $C_{10}H_{14}NH_2$ results in two isomeric forms (Goldschmidt & Kisser,[8] 1887; G. & Weiss,[9] 1893; G. & Fisher,[10] 1897).

Carvone has such a characteristic odor, that it can be recognized without difficulty. If it is to be separated in a pure state from fractions of a volatile oil it is converted into its hydrogen sulphide addition product (see above). It may be mentioned that this compound does

1) Berichte, 32, p. 2550.
2) Berichte, 32, p. 1516.
3) Liebig's Annalen, 275, p. 118; 279, p. 374.
4) Liebig's Annalen, 279, p. 369.
5) Berichte, 20, p. 118.
6) Berichte, 24, p. 3984.
7) Liebig's Annalen, 275, p. 119.
8) Berichte, 20, p. 486.
9) Berichte, 26, p. 2084.
10) Berichte, 30, p. 2069.

not melt at 187° as reported by Beyer,[1] but at 210 to 211° (Claus & Fahrion,[2] 1889).

In order to prove the presence of carvone, the simplest way is to prepare the oxime, taking care, however, that not too great an excess of hydroxylamine be employed, as in this case the oxime may add a second molecule of hydroxylamine forming a compound $C_{10}H_{14}NOH$.-NH_2OH (m. p. 174—175°) (Wallach & Schrader,[3] 1894; Harries,[4] 1898). If the freshly prepared oxime does not crystallize at once, it can sometimes be brought to crystallize by distilling it over with steam.

Camphor.

In order to distinguish d-camphor from Borneo camphor (d-borneol), it is also called Japan or laurus camphor. Besides camphor oil, it is obtained in large quantities by steam distillation from the wood of *Cinnamomum camphora*. In smaller amounts it is contained in the oils of camphor leaves, sassafras leaves, cinnamon root, spike, rosemary, and in the oil from basilicum root from Réunion. l-Camphor occurs only in the oils of feverfew and tansy. Like d-camphor it results upon oxidation of the corresponding borneol with nitric acid.

Synthetically, camphor has been obtained by the dry distillation of the lead (Haller,[5] 1879 and 1896) or calcium (Bredt & Rosenberg,[6] 1896) salt of homocamphoric acid (Haller's hydroxycamphocarbonic acid). This synthesis is, however, only a partial one, since up to the present homocamphoric acid has been prepared only from a derivative of camphor.

Camphor forms a granular, crystalline, colorless, transparent mass, with a great tendency to sublime. It is readily soluble in organic solvents and has a characteristic odor. Its properties, according to different observers are as follows: $d_{18°} = 0.9853$ (determined on l-camphor by Chautard[7]).

M p. 176.3—176.5°; b. p. 209.1° at 759 mm. (mercury thread completely in the vapor) (Förster,[8] 1890);

M. p. 178.4°; $[\alpha]_D = +41.44°$ and $-42.76°$ (Haller,[9] 1887).—M. p. 175°; b. p. 204° (Landolt,[10] 1877).

---

1) Archiv d. Pharm., 221, p. 285.

2) Journ. f. prakt. Chem., II, 39, p. 365; Bericht von S. & Co., April 1898, p. 29.

3) Liebig's Annalen, 279, p. 368.

4) Berichte, 31, pp. 1384, 1810.

5) Contrib. à l'étude du camphre. Thèse. Nancy 1879, p. 84; Bull. Soc. chim., III, 15, p. 323.

6) Liebig's Annalen, 289, p. 5.

7) Jahresbericht f. Chem., 1863, p. 555.

8) Berichte, 23, p. 2988.

9) Compt. rend., 105, p. 229.

10) Liebig's Annalen, 189, p. 333.

M. p. 175°; b. p. 204°; $[\alpha]_D = \pm 44.22°$ in 20 percent alcoholic solution (Beckmann,[1] 1889).

Camphor has always been of interest to chemists and the experimental material on it has grown to fairly large proportions. In harmony with the purpose of this work, only those compounds especially suited for its characterization will be considered.

According to its chemical nature camphor $C_{10}H_{16}O$ is a ketone, which does not combine with bisulphites. Hydroxylamine reacts upon it with the formation of an oxime, from which, however, the pure ketone cannot be regenerated, for when treated with acids water is split off and the nitrile of campholenic acid $C_9H_{15}.CN$ is formed.

If camphor oxime is reduced with sodium in ethyl or better amyl alcoholic solution, two isomeric bornylamines (m. p. 163° and 180°) are formed (Forster,[2] 1898). A similar base, melting at 159—160° is obtained by heating camphor with ammonium formate to 220—230° (Leuckart & Bach,[3] 1887; Wallach & Griepenkerl,[4] 1892).

Upon reduction camphor is converted into the alcohol $C_{10}H_{18}O$, borneol. In indifferent solvents there is also produced some isoborneol besides camphor pinacone (Beckmann,[5] 1894 and '96), while in alcoholic solution a mixture of borneol and isoborneol is principally obtained (Beckmann,[6] 1897).

The oxidation with nitric acid leads to the dibasic camphoric acid $C_{10}H_{16}O_4$ (m. p. 187°) and further to the tribasic camphoronic acid $C_9H_{14}O_6$ (m. p. 139°). From the constitution of the decomposition products of these acids many conclusions have been drawn as to the constitution of camphor. Of the many formulas suggested for this ketone, that of Bredt[7] (1893)

```
CH2 — CH — CH2
|      |      |
|CH3 — C — CH3|
|      |      |
CH2 — C  —  CO
       |
      CH3
```

is at present considered most favorably.

---

1) Liebig's Annalen, 250, pp. 352—358.— On the influence of the concentration and the nature of the solvent on the rotatory power see Landolt loc. cit. and Rimbach, Zeitschr. f. phys. Chem., 9 (1892), p. 701.

2) Journ. Chem. Soc., 73, p. 386.

3) Berichte, 20, p. 104.

4) Liebig's Annalen, 269, p. 347.

5) Berichte, 27, p. 2348; Liebig's Annalen, 292, p. 1.

6) Journ. f. prakt. Chem., II, 55, p. 35.

7) Berichte, 26, p. 3049.

Dehydrating agents act very energetically on camphor. Thus phosphoric acid anhydride produces p-cymene; sulphuric acid and zinc chloride yield other substances besides this hydrocarbon. By the action of iodine carvacrol is obtained.

For the identification of camphor the oxime is used. This compound, discovered by Nägeli[1] in 1883 is preferably prepared according to the method of Auwers,[2] (1889):

To a solution of 10 parts of camphor in 10—20 times the amount of 90 percent alcohol a solution of 7—10 parts of hydroxylamine hydrochloride and 12 to 17 parts of soda solution are added. The mixture is heated in a boiling water bath until the compound first precipitated by water dissolves completely in the soda solution. The oxime precipitated by water is recrystallized from alcohol or ligroin. It melts at 118—119° (Bertram & Walbaum,[3] 1894; Bredt & Rosenberg,[4] 1896), and, when prepared from d-camphor is laevogyrate, whereas the oxime of l-camphor is dextrogyrate (Beckmann,[5] 1889).

For identification the semicarbazone of camphor melting at 236—238° may also be employed, further the compound produced with p-bromphenylhydrazine melting at 101° (Tiemann,[6] 1895).

Fenchone.

Fenchone $C_{10}H_{16}O$ is a compound very similar to camphor, but liquid. It occurs in both optically active modifications in volatile oils. d-Fenchone is contained in fennel oil, while l-fenchone is a constituent of thuja oil. For its preparation in a pure state the fenchone-containing fractions, boiling at about 190—195° are freed from impurities by oxidation with concentrated nitric acid or permanganate solution (Wallach,[7] 1891). Fenchone is very stable toward oxidizing agents and is only slightly acted upon by this treatment. If the fenchone has been fairly well purified by this treatment it will solidify in the cold and can then be completely purified by crystallization and removal of the liquid parts. Artificially, fenchone can be obtained by the oxidation of fenchyl alcohol.

Absolutely pure fenchone is a limpid, somewhat oily liquid, which possesses an intense camphor-like odor and a bitter taste. Its physical properties are given by Wallach[8] (1891—1893) as follows:

M. p. + 5 to 6°; $d_{19°} = 0.9465$, $d_{23°} = 0.943$; $n_{D19} = 1.46306$;
$[\alpha]_D = +71.97°$ and 66.94° resp. (in alcoholic solution).

1) Berichte, 16, p. 497.
2) Berichte, 22, p. 605
3) Journ. f. prakt. Chem., II, 49, p. 10.
4) Liebig's Annalen, 289, p. 6.
5) Liebig's Annalen, 250, p. 354.
6) Berichte, 28, p. 2191.
7) Liebig's Annalen, 263, p. 130.
8) Liebig's Annalen, 263, p. 131; 272, p. 102.

With the exception of the character of rotation, the optically active modifications correspond completely both physically and chemically.

Fenchone does not combine with bisulphites any more than does camphor, but is also indifferent toward phenyl hydrazine; with hydroxylamine, however, it yields an oxime melting at 164—165°, which can be used for characterization (Wallach[1]). The fenchone oxime behaves like camphor oxime in so far as by splitting off water it is converted into the nitrile of fencholenic acid $C_{10}H_{15}N$, isomeric with campholenic acid. Cockburn[8] obtained two fencholenic acids.

Upon reduction fenchone yields fenchyl alcohol $C_{10}H_{18}O$ (Wallach[2]), which melts at 45°. By this treatment a change in rotatory power takes place, so that d-fenchone yields l-fenchyl alcohol and the reverse. Upon oxidation with permanganate fenchone yields besides acetic and oxalic acids dimethylmalonic acid $(CH_3)_2.C(COOH)_2$ (Wallach[4]). When concentrated nitric acid is used isocamphoronic and dimethyl tricarballylic acids are also found. (Gardner & Cockburn,[5] 1898).

Whereas camphor, when dehydrated with phosphoric acid anhydride, yields p-cymene; fenchone, upon like treatment yields m-cymene.

Fenchone therefore appears to be the m-compound corresponding to camphor as the p-compound (Wallach[6]). Concentrated sulphuric acid acts on fenchone with the formation of acetyl xylene $C_6H_3.CH_3^{[1]}.CH_3^{[2]}.CH_3CO^{[4]}$ (Marsh[7]).

For the identification of fenchone the oxime is used, which is best prepared according to the directions given by Wallach.[8]

To a solution of 5 g. of fenchone in 80 cc. of absolute alcohol a solution of 11 g. of hydroxylamine hydrochloride in 11 g. of hot water and 6 g. of powdered potash are added. After standing for some time when some of the alcohol has evaporated the oxime crystallizes out, which should be purified by recrystallization from alcohol, acetic ether, or ether.

Thujone.

Thujone $C_{10}H_{16}O$ (tanacetone Semmler) together with l-fenchone is a constituent of thuja oil, but has also been found in the oils of tansy, wormwood, and sage as well as in the oil of *Artemisia barrelieri*. In all of them it occurs in the dextrogyrate modification only. From oils rich in thujone like tansy oil, and the oil of *Artemisia barrelieri*, the ketone is conveniently separated in the form of its bisulphite compound, which yields pure thujone by decomposition with soda.

---

1) Liebig's Annalen, 263, p. 131; 272, p. 102.

3) Journ. Chem. Soc., 75, p. 501.

2) Ibidem, p. 146.

4) Liebig's Annalen, 263, p. 134.

5) Journ. Chem. Soc., 73, p. 708.

6) Liebig's Annalen, 275, p. 157; 284, p. 324.

7) Proc. Chem. Soc., 15, p. 196.

8) Liebig's Annalen, 272, p. 104.

Thujone is a colorless, pleasant smelling, somewhat oily liquid, for which Semmler reports the following properties:

B. p. 84.5° at 13 mm.; $d_{20°} = 0.9126$; $n_D = 1.4495$.[1]
B. p. 203°; α_D = about +68°.[2]

For the ketone regenerated from the bisulphite compound and the semicarbazone Wallach found:[3]

$d_{19°} = 0.9175$; $n_D = 1.45109$;
$d_{20°} = 0.916$; $n_D = 1.4507$.

As is seen from the physical constants thujone is a saturated ketone, yet it takes up bromine with decoloration and is readily affected by potassium permanganate.

Differing from camphor and fenchone thujone combines with sodium bisulphite. With hydroxylamine it yields an oxime (Wallach,[4] 1893; Semmler,[5] 1892) melting at 54—55°, which upon heating with dilute sulphuric acid is changed to cymidine $C_6H_3 . C_3H_7{}^{[1]} . NH_2{}^{[2]} . CH_3{}^{[4]}$ which yields carvacrol (Semmler[6]). Semicarbazide reacts on the ketone with the formation of a semicarbazone meltihg at 171—172° (Baeyer,[7] 1894).

Upon reduction thujone yields the corresponding secondary alcohol $C_{10}H_{18}O$, thujyl alcohol (Semmler[5]), which had been found with thujone in wormwood oil.

When oxidized with bromine and soda solution the monobasic tanacetogenic acid $C_9H_{14}O_2$ results (Semmler[8]); with permanganate, however, a mixture of two isomeric keto acids $C_{10}H_{16}O_3$, is obtained (Wallach,[9] 1893 & 1897; Semmler,[10] 1892). When subjected to dry distillation, this yields a homologous ketone $C_9H_{16}O$ possessing an odor similar to that of methylheptenone (Wallach,[11] 1893). By building down the keto acids to δ-(ω-) dimethyl laevulinic acid the constitution of thujone has been fairly well cleared up (Tiemann & Semmler,[12] 1897 and 1898).

When thujone is heated to 280° in closed tubes for some time, it is changed to an unsaturated ketone of the same composition $C_{10}H_{16}O$, carvotanacetone (Semmler,[13] 1889), which has an odor resembling that of carvone. The same rearrangement seems to take place even by

1) Berichte, 25, p. 3343.
2) Ibidem, 27, pp. 895 & 897.
3) Berichte, 28, p. 1965.
4) Liebig's Annalen, 277, p. 159.
5) Berichte, 25, p. 3344.
6) Ibidem, 3352.
7) Berichte, 27, p. 1928.
8) Berichte, 25, p. 3346.
9) Liebig's Annalen, 272, p. 111; Berichte, 30, p. 428.
10) Berichte, 25, p. 3347.
11) Liebig's Annalen, 272, p. 116; 275, p. 164.
12) Berichte, 30, p. 429; 31, p. 2311.
13) Berichte, 27, p. 895.

prolonged boiling of the thujone, as may be concluded from the decrease in rotatory power. Another isomeric ketone, isothujone, is formed by heating thujone with dilute sulphuric acid (Wallach,[1] 1895).

For the characterization of thujone the tribromide is well suited, and is best prepared according to Wallach[2] (1893):

5 g. of thujone are dissolved in a large beaker in 30 cc. of petroleum ether and to this solution 5 cc. of bromine are added at once. After several seconds a rather violent reaction takes place with liberation of considerable hydrobromic acid. When this is ended, the solvent is allowed to evaporate and the tribromide separates slowly as a crystalline mass, which is to be freed from adhering oil by washing with cold alcohol and recrystallized from hot acetic ether; the melting point of the pure compound is 121—122°.

Pulegone.

The oil of European pennyroyal contains about 80 percent of a ketone $C_{10}H_{16}O$, pulegone, which has likewise been found in the pennyroyal oil from *Hedeoma pulegioides* and in the oil of *Pycnanthemum lanceolatum*, in all of which it occurs in the dextrogyrate modification.

As European pennyroyal oil consists principally of pulegone, this ketone can be obtained in a fairly pure state by fractional distillation. It can, however, be obtained much purer from the bisulphite compound which separates by shaking the oil diluted with some alcohol (¼ vol.) with sodium bisulphite solution (Baeyer,[3] 1895). The semicarbazone, decomposable by acids, can also be used for the purification of pulegone.

Synthetically, pulegone has been obtained in a round-about way from citronellal (Tiemann & Schmidt,[4] 1896). An isomeric ketone, but not identical with natural pulegone has been obtained by Wallach[5] (1896 & '98) by the condensation of methyl hexanone with acetone.

Pulegone is at first a colorless liquid which on standing for some time becomes faintly yellow, and has a sweet peppermint-like odor.

For the ketone purified only by distillation Beckmann and Pleissner (1891) found:[6]

B. p. 130—131° at 60 mm.; $d_{20°} = 0.9323$; $n_D = 1.47018$; $[\alpha]_D = +22.89°$.

Barbier reports (1892):[7]

B. p. 222—223°; $d_{23°} = 0.9293$; $[\alpha]_D = +25° 15'$.

1) Liebig's Annalen, 286, p. 101; Berichte, 28, p. 1958; 30, p. 426.
2) Liebig's Annalen, 275, p. 179.
3) Berichte, 28, p. 652.
4) Berichte, 29, p. 918; 30, p. 22.
5) Berichte, 29, pp. 1597, 2955; Liebig's Annalen, 300, p. 267.
6) Liebig's Annalen, 262, pp. 3, 4, 20.
7) Compt. rend., 114, p. 126.

For the pulegone prepared from the bisulphite compound Baeyer and Henrich[1] (1895) determined:

B. p. 100—101° at 15 mm.; $[\alpha]_D = +22.94°$

and Wallach[2] (1895):

B. p. 221—222°; d = 0.936; $n_D = 1.4846$.

As an unsaturated compound pulegone yields with bromine a liquid dibromide, with hydrochloric and hydrobromic acids crystallized addition products, which regenerate pulegone when treated with alcoholic potassa (Beckmann & Pleissner,[3] 1891; Baeyer & Henrich,[4] 1895). It also shows the character of a ketone. Upon careful reduction with sodium in alcoholic solution the alcohol $C_{10}H_{18}O$, pulegol, which is difficult to obtain in a pure state, is first formed (Tiemann & Schmidt[5]). Upon further action of hydrogen l-menthol $C_{10}H_{20}O$ results (Beckmann & Pleissner[6]).

Pulegone combines with hydroxylamine, two oximes being known, of which the one has the normal composition $C_{10}H_{16}NOH$ and melts at 118—119° (Wallach,[7]), while the other (m. p. 147°, Tiemann & Schmidt,[8]) described by Beckmann & Pleissner,[9] contains an additional molecule of water. As has lately been determined the latter is produced by the addition of hydroxylamine to the double bond in the pulegone, so that the compound cannot be considered as an oxime (Harries & Roeder,[10] 1898). The semicarbazone formed by the action of semicarbazide melts at 170° (Baeyer & Henrich,[11] 1895).

When pulegone is boiled with anhydrous formic acid, or heated with water to 250° in an autoclave, it takes up water and is resolved into acetone and methyl hexanone $C_7H_{12}O$ (Wallach,[12] 1896). The same decomposition is also brought about by concentrated sulphuric acid (Zelinsky,[13] 1897); also by the addition of bromine and splitting off hydrogen bromide with the aid of heat (Klages[14]). In the latter case m-cresol is also formed.

Upon oxidation with potassium permanganate pulegone splits off acetone and yields β-methyladipinic acid $C_7H_{10}O_4$. In harmony with this result the formula

1) Berichte, 28, p. 653.
2) Berichte, 28, p. 1965.
3) Liebig's Annalen, 262, p. 21.
4) Berichte, 28, p. 653.
5) Berichte, 29, p. 914.
6) Liebig's Annalen, 262, p. 30.
7) Liebig's Annalen, 277, p. 160; 289, p. 347.
8) Berichte, 30, p. 26.
9) Liebig's Annalen, 262, p. 6.
10) Berichte, 31, p. 1809.
11) Berichte, 28, p. 653.
12) Liebig's Annalen, 289, p. 338.
13) Berichte, 30, p. 1532.
14) Berichte, 32, p. 2564.

```
       CH3CH3
         \/
         C
         ||
         C
        / \
   H2C/     \CO
     |       |
     |       |
   H2C\     /CH2
        \ /
        CH
         |
        CH3
```

has been assigned to pulegone (Semmler,[1] 1892) which also explains the ready decomposition into acetone and methyl hexanone.

Pulegone has an odor very similar to that of menthone. Both ketones can, however, be readily distinguished by their derivatives. For the characterization of pulegone the semicarbazone prepared in the usual manner, or the normal oxime may be used. For the preparation of the latter Wallach,[2] 1896, has given the following directions:

To a solution of 10 g. of pulegone in three times its volume of absolute alcohol a solution of 10 g. of potassium hydrate in 5 g. of water is added. To the liquid warmed to about 80° are added quite rapidly with constant agitation, 10 g. of hydroxylamine hydrochloride dissolved in 10 g. of water, and the mixture kept for 10 minutes more at 80°. If, after cooling, the mixture is poured into water, the oxime usually separates at once in the solid form. After recrystallization from ether or petroleum ether it melts at 118—119°.

According to Baeyer & Henrich[3] (1895) the identification can be accomplished more rapidly with the characteristic bisnitrosopulegone.

A solution of 2 cc. of pulegone or oil containing pulegone in 2 cc. of ligroin and 1 cc. of amyl nitrite is well cooled in a freezing mixture and treated with a very small amount of hydrochloric acid. In a very short time the bisnitroso compound separates in the form of fine needles, which are obtained pure by spreading on a porous plate and washing with petroleum ether.

Menthone.

l-Menthone occurs together with menthol in peppermint oil and probably also in the oil of buchu leaves. Recently it has been detected in geranium oil from Réunion.

As it does not combine with bisulphite and cannot be freed from accompanying compounds by fractional distillation alone, it can be

1) Berichte, 25, p. 3515.

2) Liebig's Annalen, 289, p. 347.

3) Berichte, 28, p. 654; compare Baeyer & Prentice, Berichte, 29, p. 1078.

isolated only by means of its oxime or semicarbazone. It should be remembered, however, that in the decomposition of these compounds, which is usually effected by means of dilute sulphuric acid, the rotatory power of the ketone is changed.

By the oxidation of menthol with the chromic acid mixture recommended by Beckmann it is easy to obtain a l-menthone with normal rotatory power (Beckmann,[1] 1889).

Pure menthone is a mobile, limpid liquid, which possesses a peppermint odor and a cooling, bitter taste. For the ketone obtained by oxidation from menthol Beckmann found:

B. p. 207°; $d_{20^\circ} = 0.8960$; $n_{D12^\circ} = 1.4525$; $[\alpha]_D = -28.18^\circ$.[2]

Binz records (1893):

$d_{24^\circ} = 0.8934$; $[\alpha]_D = -27.67^\circ$.[3]

According to Wallach (1895) menthone regenerated from the semicarbazone (m. p. 184°) possesses the following properties:

B. p. 208°: $d = 0.894$; $n_D = 1.4496$.[4]

If l-menthone is treated at a low temperature with concentrated sulphuric acid, it is converted into the dextrogyrate isomer (Beckmann,[5] which has so far not been found in volatile oils.

Menthone belongs to those ketones which do not yield addition products with bisulphites. Upon reduction with sodium in alcoholic solution it is converted into the corresponding secondary alcohol l-menthol, $C_{10}H_{20}O$ There are also formed small quantities of a slightly dextrogyrate isomenthol; by using indifferent solvents menthopinacone also results (Beckmann,[6] 1897).

With hydroxylamine in alcoholic-aqueous solution it combines readily to form l-menthoxime melting at 59° (Beckmann,[7] 1889; Wallach,[8] 1893). When this is treated with dehydrating agents it is converted into an aliphatic nitrile $C_9H_{17}CN$, which by further changes yields compounds which show great similarity to the members of the citronellal series (Wallach,[9] 1894 & '97). Semicarbazide reacts on menthone with the formation of the semicarbazone, which crystallizes in needles and melts at 184° (Wallach,[10] 1895; Beckmann,[11] 1896).

1) Liebig's Annalen, 250, p. 325.
2) Liebig's Annalen, 250, p. 327.
3) Zeitschr. f. phys. Chem., 12, p. 727.
4) Berichte, 28, p. 1963.
5) Liebig's Annalen, p. 884.
6) Journ. f. prakt. Chem., II, 55, pp. 18 & 30.
7) Liebig's Annalen, 250, p. 330.
8) Liebig's Annalen, 277, p. 157.
9) Liebig's Annalen, 278, p. 308; 296, p. 120.
10) Berichte, 28, p. 1963.
11) Liebig's Annalen, 289, p. 366.

If menthone is oxidized with a solution of chromic acid in glacial acetic acid, a liquid keto acid $C_{10}H_{18}O_3$ (keto- or oxymenthylic acid) is first produced (Beckmann & Mehrländer,[1] 1896), which, on further oxidation with potassium permanganate or chromic acid mixture is converted into the dibasic β-methyladipinic acid (Arth's[2] β-pimelinic acid) consequently into the same "Abbauproduct" which results from pulegone and citronellal (Beckmann and Mehrländer;[3] Manasse and Rupe,[4] 1894). When oxidized with Caro's reagent (persulphate with concentrated sulphuric acid) it yields the corresponding ε-lactone (Baeyer,[5] 1899).

When isoamyl nitrite and hydrochloric acid are allowed to act on menthone in the cold, there is formed besides bisnitrosomenthone $(C_{10}H_{17}O.NO)_2$, the oxime of keto-menthylic acid, the first oxidation product of menthone (Baeyer & Manasse,[6] 1894; B. & Oehler,[7] 1896).

When bromine (2 mol.) acts on menthone (1 mol.) in chloroformic solution a crystallized dibrommenthone, $C_{10}H_{16}Br_2O$ (m. p. 79—80°) results, which can be converted into thymol by splitting off hydrobromic acid by means of chinoline (Beckmann & Eichelberg,[8] 1896).

Those rearangements, as well as the conversion of menthone into 3-chorcymene effected by Jünger and Klages[9] (1896) agree with the formula

```
        CH3CH3
          \/
          CH
          |
          CH
         /  \
   H2C/      \CO
     |        |
     |        |
   H2C\      /CH2
        \  /
         CH
         |
         CH3
```

assigned to it. This also expresses the relation of menthone to pulegone.

In order to identify menthone, the semicarbazone or the oxime is employed, the preparation of which is accomplished in the usual manner. For further characterization the ketone may be reduced to menthol and this converted into the benzoic acid ester (see menthol).

---

1) Liebig's Annalen, 289, p. 368.
2) Annales de Chim. et Phys., VI, 7, p. 488.
3) Liebig's Annalen, 289, p. 378.
4) Berichte, 27, p. 1818.
5) Berichte, 32, p. 3625.
6) Berichte, 27, pp. 1918 and 1914.
7) Berichte, 29, p. 27.
8) Berichte, 29, p. 418.
9) Berichte, 29, p. 315.

A diketone, viz. diacetyl $CH_3.CO.CO.CH_3$ remains to be mentioned. It has been isolated from the distillate from caraway seeds[1] and in all probability occurs in the distillate from cloves and probably in other oils. It can be identified by its dioxime melting at 234.5°.

Acids, Esters, Lactones, Oxides.

The aqueous distillate obtained by the distillation of volatile oils with steam occasionally contains free fatty acids, viz. acetic, propionic, butyric or valerianic acid. These acids, like methyl and ethyl alcohol, are probably decomposition products of esters contained in the parts of plants subjected to distillation.

Those acids contained in oils in the form of esters are obtained as salts when the oils are saponified. Formic acid is supposed to be contained in valerian oil as bornyl formate. Acetic acid occurs frequently in the form of esters as linalyl, geranyl, and bornyl acetates, which are characterized by their peculiar odor. Of the other fatty acids, propionic and butyric acids have been found in lavender oil; butyric acid as ethyl ester in the oil of *Heracleum sphondylium*, as octyl ester in the oil of *Pastinaca saviva;* valerianic acid and its esters in the oils of valerian, citronella and angelica root; methyl ethyl acetic acid in the oil of angelica root and seed; n-capronic acid in palmarosa oil and the oil of *H. sphondylium;* caprinic acid as ethyl and amyl ester in cognac oil; laurinic acid free in the oil of arnica flowers; myristic acid in orris oil and oil of nutmeg; and palmitic acid in the oils of ambrette seeds, vetiver, wormwood, celery and arnica flowers.

Of the unsaturated acids, angelic and tiglinic acids occur as esters in cumin oil; tiglinic acid in geranium oil from Réunion; oleic acid in orris oil.

Of the aromatic acids, benzoic acid has been found in ylang-ylang oil and in the oil of tolu balsam; cinnamic acid in storax oil, and as methyl ester in the oil of *Alpinia malaccensis*. Much more widely distributed is salicylic acid which occurs as methyl ester in the oils of species of *Gaultheria* and *Betula*, also those of spiraea, senega root and tea. Methyl salicylate occurs in small amounts but is widely distributed in leguminous plants and in other plant families.[2] In oil of cloves salicylic acid probably occurs as acetsalicylic ester of eugenol.

---

1) Bericht S. & Co., Oct. 1899, p. 32.

2) Comp. Bericht von Schimmel & Co., Oct. 1898, p. 51, also Jahresbericht des botanischen Gartens zu Buitenzorg, 1897, p. 37.

Methyl salicylate is unquestionably one of the most widely distributed plant constituents. Its presence in a plant is not restricted to any one part. Probably the ester does not as a rule exist as such in plants, but in form of a glucoside, for frequently the distillate from fresh vegetable matter yields no reaction for methyl salicylate.

Oxyacids and their anhydrides, the lactones, occur in volatile oils; oxymyristic and oxypentadecylic acid have been found in the oils of angelica root and of angelica seed respectively.

Of the lactones, sedanolid from celery oil and cumarin and hydrocumarin are characterized by their powerful odor. On the other hand, alantolactone (helenin) from alant oil and the lactone $C_{10}H_{16}O_2$ from peppermint oil have but a faint odor.

The oxides are represented by but one compound, cineol or eucalyptol $C_{10}H_{18}O$.

Cineol.

Cineol (eucalyptol, terpane of Bouchardat and Voiry) like methylsalicylate is widely distributed. It is the principal constituent of cajeput oil, the oils of *Artemisia cina* and *Eucalyptus globulus* and has been found in larger or smaller quantities in the following oils: galangal oil, zedoary oil, different cardamom oils, in the oil from the seeds of *Amomum melegueta*, kaempferia oil, camphor oil, laurel leaf and berry oils, myrtle oil, in the oils of *Melaleuca leucadendron var. lancifolia*, *M. acuminata*, *M. decussata* and *M. uncinata*, in cheken leaf oil, canella oil, in the oils of *Eucalyptus odorata*, *cneorifolia*, *oleosa*, *dumosa*, *amygdalina*, *rostrata*, *populifera*, *corymbosa*, *resinifera*, *baileyana*, *microcorys*, *risdonia*, *leucoxylon*, *macrorrhyncha*, *capitellata*, *eugenioides*, *obliqua*, *punctata*, *loxophleba*, *crebra* and *hemiphloia*, in oil of *Artemisia vulgaris*, milfoil oil, iva oil, in sage and basilicum oil, in the oil of *Lavandula dentata*, *L. stoechas* and *L. vera*, in rosemary oil, as well as in peppermint and Russian spearmint oils.

The preparation of this body from oils rich in cineol, like that of *Eucalyptus globulus*, is not difficult, insofar as the cineol, purified as far as possible by fractional distillation possesses the property of crystallizing when cooled. If the object is to detect and to isolate small quantities of cineol, the hydrobromic acid compound is used, which by decomposition with water yields cineol.

Cineol has been observed as inversion product in the preparation of terpineol from terpin hydrate and dilute sulphuric acid; it results further by boiling terpineol with dilute sulphuric acid or oxalic acid.

Pure cineol is a colorless liquid of camphor-like odor, which crystallizes in the cold and is optically inactive. For the cineol prepared from its hydrochloric acid addition product Wallach found:

B. p. 176°; $d_{20°}=0.9267$; $n_D=1.45839$.[1]

According to observations made in the laboratory of Schimmel & Co. cineol purified by crystallization melts at —1°; it boils at 177° (764 mm.), and $d_{15°}=0.930$, $n_{D17°}=1.45961$.

Cineol yields characteristic loose addition products with bromine, iodine, hydrochloric and hydrobromic acids,[2] phosphoric acid, α- and β-naphthol, and iodol, which can be used in part for its isolation and characterization. By the action of dehydrating agents it is converted into dipentene (Wallach & Brass,[3] 1884). It can also be converted directly into dipentene derivatives by proper treatment; thus for instance, dipentene hydriodide results by merely conducting dry hydriodic acid into cineol. The oxygen in cineol is combined as oxide oxygen and, therefore, this compound reacts neither with hydroxylamine nor with phenyl hydrazine, nor does sodium in alcoholic solution act upon it.

When oxidized with warm solution of potassium permanganate dibasic cineolic acid, $C_{10}H_{16}O_5$ (m. p. 196 to 197°) is formed (Wallach & Gildemeister,[4] 1888). The cineolic acid anhydride produced by the action of acetanhydride on this acid yields, when subjected to dry distillation, methyl heptenone, $C_8H_{14}O$, which also occurs naturally.

Cineol possesses a characteristic odor, which often directs the attention to it. For rapid detection, the reaction with iodol given by Hirschsohn[5] (1893) is especially suited. In a few drops of the oil to be tested a little iodol is dissolved by moderately heating; if cineol is present, the crystalline addition product consisting of equal molecules of the components soon separates. The melting point of the compound, recrystallized from alcohol or benzene, is about 112° (Bertram & Walbaum,[6] 1897).

---

1) Liebig's Annalen, 289, p. 22; 245, p. 195.
2) Liebig's Annalen, 225, pp. 300, 303; 230, p. 227; 246, p. 280.
3) Liebig's Annalen, 225, p. 310.
4) Liebig's Annalen, 246, p. 268.
5) Pharm. Zeitschr. f. Russl., 32, pp. 49 & 67.
6) Archiv d. Pharm., 235, p. 178.

If the cineol is to be isolated as such, the cineol fraction diluted with about an equal volume of petroleum ether is well cooled and saturated with dry hydrobromic acid gas. The white crystalline precipitate which soon separates is filtered with a force-pump and washed with petroleum ether. The fairly stable hydrobromide thus obtained melts at 56—57° and is readily decomposed by water into cineol and hydrobromic acid.

For the further characterization the cineolic acid produced by oxidation with warm potassium permanganate solution may be used.

Phenols and Phenol ethers.

Several of the phenols and their ethers have acquired considerable commercial importance and are, therefore, prepared on a large scale. Thus thymol is largely used on account of its antiseptic properties, anethol is used in the manufacture of liquors, from eugenol and safrol the valuable vanillin and heliotropin are prepared.

Of the simpler representatives, p-cresol methyl ether is found in ylang-ylang oil (cananga oil), phlorol (m-ethyl phenol?) is found as isobutyric ester and as methyl ether in the oil of arnica root, the ethyl ether of hydroquinone occurs in star anise oil, and the methyl ether of thymohydroquinone in the oil of arnica root.

Of the higher homologues of phenol, the two isomeric phenols $C_{10}H_{14}O$, thymol and carvacrol are of special interest on account of their relationship to terpene derivatives.

Carvacrol.

Carvacrol (isopropyl-o-cresol), $C_6H_3 . CH_3^{[1]} . OH^{[2]} . C_3H_7^{[4]}$, is the principal constituent of the Trieste and Smyrna origanum oil. It has also been found in the oils of *Monarda fistulosa*, of thyme and wild thyme, of *Satureja hortensis*, *S. montana* and of *Pycnanthemum lanceolatum*, also in spearmint oil and the oil of *Schinus molle.*

Artificially it can be obtained from the isomeric carvone by treatment with potassa, hydrochloric, sulphuric, phosphoric, and formic acids (comp. p. 163); also from camphor by heating it with iodine.

When freshly distilled it is a colorless, thick oil, which solidifies in the cold. Its constants are reported as follows:

B. p. 119° at 16 mm.; $d_{20°} = 0.9782$; $n_D = 1.5228$ (Semmler,[1] 1892).

---

1) Berichte, 25, p. 3353.

For carvacrol from origanum oil, Gildemeister[1] (1895) found:

M. p. + 0.5°; b. p. 235.5—236.2° (742 mm.); $d_{15°}=0.980$, $d_{20°}=0.976$; $n_{D20°}=1.52338$;

and for a preparation prepared from carvone:

M. p. + 0.5°; b. p. 236—236.5° (742 mm.); $d_{15°}=0.983$, $d_{20°}=0.979$; $n_{D20°}=1.52295$.

For the detection of carvacrol the phenyl carbaminic acid ester, first prepared by Goldschmidt[2] (1893) can be used. According to Gildemeister[1] it melts at 140°. For this purpose the nitroso derivative which can readily be prepared according to Klage's[3] (1899) method should also prove serviceable.

Thymol.

Thymol (isopropyl-m-cresol) $C_6H_3 . CH_3^{[1]} . OH^{[3]} . C_3H_7^{[4]}$ is found in considerable quantity besides p-cymene in ajowan oil, but also occurs in the oils of *Monarda punctata*, *Morula japonica* and *Cunila mariana*, in oil of wild thyme, and in thyme oil.

Its artificial formation from dibrom menthone by splitting off hydrobromic acid by means of quinoline is interesting.

Thymol forms colorless, transparent crystals, having the odor of thyme which melt at 50—51° and distill at 232° (231.8°) (Pinette,[4] 1886). Eykmann[5] (1893) determined $d_{9.6°}=0.9816$, Nasini and Bernheimer[6] (1885) report $d_{4°}^{24°}=0.96895$ and $n_D=1.51893$.

Chemically it can be readily characterized by its nitroso derivative (Klages,[7] 1899).

The most phenols and phenol ethers as well as the most important ones belong to the olefinic phenols, i. e. phenols with olefinic side-chains. They mostly contain the group C_3H_5, either the allyl or propenyl group; myristicin alone apparently makes an exception, having a butenyl group. When heated with alcoholic potassa or sodium ethylate those substances containing an allyl group are converted into their isomers with a propenyl group.

---

1) Arch. d. Pharm., 233, p. 188.
2) Berichte, 26, p. 2086.
3) Berichte, 32, p. 1518.
4) Liebig's Annalen, 243, p. 46.
5) Recueil des trav. chim. des P.-B., 12, p. 177.
6) Gazz. chim. ital., 15, p. 59. Jahresb. f. Chem., 1885, p. 314.
7) Berichte, 32, p. 1518.

CHAVICOL.

Chavicol (p-allylphenol), $C_6H_4 . C_3H_5^{[1]} . OH^{[4]}$ occurs in betel leaf oil from Java and in bay oil.

It is a colorless liquid, boiling at about 237° and according to Eykmann[1] has $d_{18°} = 1.033$ and $n_D = 1.5441$.

METHYL CHAVICOL.

Methyl chavicol (estragol) was first found in anise bark oil and later also in estragon oil.[2] Accompanying its isomeric propenyl compound anethol, it occurs in the oils of anise, star anise, and fennel, and has been shown to be present in bay oil, in German and French basilicum oils and in the oil of *Persea gratissima*.

Methyl chavicol is a colorless liquid with a faint anise-like odor, which boils at 215—216° (corr.) (Grimaux,[3] 1893). At 11.5° $d = 0.979$ and $n_D = 1.5244$ (Eykman,[1] 1890). The ether isolated from estragon oil has, according to the observations made in the laboratory of Schimmel & Co., the following properties:

B. p. 97—97.5° at 12 mm., 86° at 7 mm.; $d_{15°} = 0.9714$—0.972; $n_{D16°} = 1.52355$—1.52380.

Methyl chavicol is characterized by changing it into the solid anethol by boiling with alcoholic potassa or by converting it into homoanisic (p-methoxyphenylacetic) acid melting at 86°.

ANETHOL.

Anethol (p-propenylanisol), $C_6H_4 . C_3H_5^{[1]} . OCH_3^{[4]}$, is the main constituent of anise and star anise oils and is also contained in considerable quantity in fennel oil. It has also been shown to be present in the oil of *Osmorrhiza longistylis*.

It is a white crystalline mass, with a strong anise-like odor, the properties of which with slight disagreements are given as follows:

M. p. 21.1°; b. p. 233—233.5° (751 mm.); $d\frac{21.6°}{4°} = 0.9855$ (Moreau & Chauvet,[4] 1897);

M. p. 21°; $d_{25°} = 0.986$; $n_{D18°} = 1.56149$ (Stohmann,[5] 1892); $d_{11.5°} = 0.999$; $n_D = 1.5624$ (Eykman[1]).

---

1) Berichte, 23, p. 862.

2) Bericht von Schimmel & Co., April 1892, pp. 17 & 41; April, 1894, p. 28.

3) Compt. rend., 117, p. 1089.

4) Bull. Soc. chim., III, 17, p. 411.

5) Sitzungsber. d. Akad. d. Wiss. Leipzig, 1892, p. 307.

When oxidized with chromic acid it is converted into anisic acid, with potassium permanganate into p-methoxyphenylglyoxylic acid $C_6H_4.OCH_3{}^{[4]}.CO.COOH^{[1]}$ (m. p. 89°). With bromine the well crystallized monobrom anethol dibromide $C_6H_3Br(OCH_3).C_3H_5Br_2$ (m. p. 107—108°) results (Hell & Gärttner,[1] 1895).

Eugenol, betelphenol and safrol are derivatives of allyl- and propenyl-dioxybenzene respectively, which are not known as such.

Eugenol.

Eugenol (allylguaiacol), $C_6H_3.C_3H_5{}^{[1]}.OCH_3{}^{[3]}.OH^{[4]}$, is contained in large amounts in the oil of cloves and clove stems. It is also found in the oils of cinnamon leaves, Ceylon cinnamon, sassafras, massoy bark, pimenta, bay, and canella, also in camphor oil and culilawan oil.

It is a faintly yellow liquid, with an intense clove-like odor and burning taste with the following properties:

B. p. 252° at 749 mm.; 123° at 12—13 mm. (Erdmann,[2] 1897); $d_{14.5°} = 1.072$; $n_D = 1.5439$ (Eykman,[3] 1890).

With ferric chloride in alcoholic solution eugenol produces a blue coloration. Upon oxidation (best as acetic acid ester) it yields vanillin and vanillic acid, besides small quantities of homovanillic acid.

For its characterization the benzoic acid ester, m. p. 69—70°, which can be prepared with benzoyl chloride is well adapted.

Besides eugenol some aceteugenol is also found in clove oil, but not in clove stem oil. It is frequently accompanied by its methyl ether (allyl veratrol) $C_6H_3.C_3H_5{}^{[1]}.OCH_3{}^{[3]}.OCH_3{}^{[4]}$, which occurs in bay and culilawan oils, in paracoto bark oil, in oil of *Asarum europaeum*, in citronella oil and probably in matico oil. The methyl ether possesses an odor reminding of eugenol but much weaker; it boils at 248—249° (128—129° at 11 mm.), $d_{11°} = 1.041$, $n_D = 1.5373$ (Bertram & Gildemeister,[4] 1889; Eykmann[3]).

With bromine methyl eugenol yields the tribrom methyl eugenol $C_6H_2Br.(OCH_3)_2.C_3H_5Br_2$, crystallizing in pretty needles melting at 78° (Wassermann;[5] B. & G.[4]). By oxidation with potassium permanganate it is converted into dimethyl dioxybenzoic acid (veratric acid) melting at 179—180° (Comp. Wallach & Rheindorff,[6] 1892). Both compounds are well suited for the detection of methyl eugenol.

---

1) Journ. f. prakt. Chem. II, 51, p. 424.
2) Journ. f. prakt. Chem. II, 56, p. 146.
3) Berichte, 23, p. 862.
4) Journ. f. prakt. Chem. II, 39, p. 354.
5) Jahresber. f. Chem., 1879, 520.
6) Liebig's Annalen, 271, p. 306.

Safrol.

Safrol (shikimol of Eykman), the principal constituent of sassafras oil, is contained in considerable quantity in camphor oil and has been found in cinnamon leaf oil, star anise oil and massoy bark oil. It is the methylene ether of an allyl pyrocatechin $C_6H_3 . C_3H_5^{[1]} . OOCH_2^{[3,4]}$.

It is a colorless or faintly yellow liquid which has the following properties, according to observations made in the laboratory of Schimmel & Co.:

B. p. 233° (759 mm.); $d_{15°} = 1.108$; $n_{D17°} = 1.53836$.

On cooling it solidifies to a mass of crystals which do not melt until + 11°. Subjected to careful oxidation with potassium permanganate a glycol is first formed, which by further oxidation is converted into homopiperonylic acid $C_6H_3 . OOCH_2 . CH_2 . COOH$ (m. p. 127°); when the oxidation is effected with chromic acid mixture piperonal (heliotropin) and piperonylic acid $C_6H_3 . OOCH_2 . COOH$ (m. p. 228°) are obtained.

Asarone and myristicin are representatives of olefinic trioxybenzenes occurring in volatile oils.

Asarone.

Asarone (propenyl trimethoxybenzene) $C_6H_2 . C_3H_5^{[1]} . (OCH_3)_3^{[2,4,5?]}$, is contained in the oil of *Asarum europaeum* and has recently been found in matico oil. It melts at 62°, yields with bromine a well crystallized dibromide (m. p. 85—86°) and is converted by oxidation with potassium permanganate into asarylic acid (trimethoxybenzoic acid, $C_6H_2(OCH_3)_3 . COOH$, m. p. 144°). Its synthesis from asarilic aldehyde has been accomplished by Gattermann and Eggers[1] (1899).

As derivatives of a tetratomic, unsaturated phenol the isomeric apiols $C_6H . C_3H_5 . O_2CH_2 . (OCH_3)_2$ should be mentioned. The common (or parsley) apiol (m. p. 29.5—30°) is contained in the oil of parsley seed and in that of the Venezuelan camphor wood. The isomeric liquid (or dill apiol) is obtained from the high boiling fractions of the East Indian dill oil. The latter differs from the former by yielding isoapiol, m. p. 44°, when heated with sodium ethylate.

1) Berichte, 32, p. 290.

Compounds Containing Nitrogen and Sulphur.

(Nitriles, Sulphides, Mustard Oils.)

Compounds containing nitrogen and sulphur occur frequently when parts of plants rich in albuminous (protoplasm) and similar substances are distilled, e. g. fresh herbs or seeds. The more volatile compounds, such as ammonia, trimethylamine, sulphuretted hydrogen (given off in considerable quantity in the distillation of caraway), and hydrocyanic acid escape mostly in the course of the distillation, and are dissolved in small quantity in the aqueous distillate. Occasionally they combine with constituents of the oil.

Hydrocyanic acid, the nitrile of formic acid, is formed during the distillation of a large number of plants.[1] Together with benzaldehyde it occurs in *Plectronia dicocca* and *Indigofera galegoides*, in the oils of bitter almonds, cherry laurel and wild cherry bark.

The nitrile of phenyl acetic acid constitutes the principal constituent of the oils of *Tropaeolum majus* and of *Lepidium sativum*; the nitrile of phenyl propionic acid in the oil of *Nasturtium officinale*. Gadamer,[2] however, has recently shown that several of the cruciferous oils containing nitriles in place of isothiocyanates are products of decomposition rather than products of hydrolysis of the underlying glucosides. When nitriles are obtained upon distillation the ferment has not had an opportunity to act.

Of carbon disulphide small amounts have been found in the oil of black mustard. Dimethyl sulphide has been isolated from the low boiling portions of American peppermint oil. Sulphides and polysulphides have been found besides other sulphur derivatives in the oils of garlic and onion. Vinyl sulphide and its polysulphides constitute the principal constituent of *Allium ursinum*.

As a separate group should be mentioned the esters of isothiocyanic acid, which are characterized by their pungent odor, and which are commonly designated as mustard oils. Their typical representative is the common (allyl) mustard oil from *Brassica nigra* and *B. juncea*. It has also been found in the following plants: *Alliaria officinalis*, *Capsella bursa pastoris*, *Cardamine sp.*, *Sisymbrium sp.*, *Cochlearia armoracia* and *Thlaspi arvense*. Secondary butyl mustard oil constitutes the principal constituent of the oil of spoonwort; p-oxybenzyl mustard oil (sinalbin mustard oil) is obtained from the seeds of *Sinapis alba*; phenyl ethyl mustard oil has been found in the oil of mignonette root.

[1] Jahresbericht des botanischen Gartens zu Buitenzorg, 1897, p. 87.

[2] Archiv f. Phar., 237, p. 111.

3. THE EXAMINATION OF VOLATILE OILS.

The special methods of testing the more important volatile oils are described in detail in the following chapters. In order to avoid repetition, however, it will be expedient to discuss briefly in a special chapter the general methods used in the investigation and the most common adulterants met with.

The practice of adulteration of the volatile oils, which is probably as old as the manufacture itself, had in the beginning a certain justification, as with the incomplete technical equipment of the early times the addition of fatty oils, turpentine oil, or alcohol was often necessary in order to extract from the plants their odorous principle. Later, when the preparation of the pure oils was already known, the practice of making these additions was still retained.

Even thirty years ago, it was customary to distill coriander with the addition of orange oil and to put the distillate on the market as coriander oil. Since pure coriander oil can now be prepared without difficulty, the product obtained by using orange oil, as it is found now and then even at the present time, must be considered as adulterated and if the foreign ingredient is not made known, its sale is a fraud.

The adulteration need not always be by the addition of a less valuable body, it sometimes consists in that the more valuable constituent of the oil has been partially removed. The effect is the same, whether from a caraway oil of the specific gravity 0.910 so much carvone be removed that an oil of the specific gravity 0.890 remains behind, or whether the same result is attained by the addition of limonene to the same oil.

Although the adulterations themselves mostly find a sufficient explanation in the profitableness and the pecuniary advantage to the adulterator, it cannot, however, be denied, that often the ignorance of the consumer, and above all the desire to buy as cheaply as possible, is the cause of the spurious composition of many an oil. More than once the producer may have been induced to adulterate, because he found no buyers for his pure products at a reasonable price, while his adulterating competitor was able to do a lucrative business at lower prices.

The main reason for the extensive adulteration to which volatile oils have been subjected at times, is to be sought in the fact that the detection of adulterants was very difficult and often entirely impossible.

Owing to the development of the chemistry of the terpenes and their derivatives, great progress has been made during the last ten or fifteen years in the detection of adulterations. Knowing the composition of not a small number of volatile oils, it has become possible not only to distinguish between a pure and an adulterated oil, but also to judge the quality of these oils. This is effected by estimating the amount of the principal, or the most important constituent. In lavender oil, bergamot oil, petitgrain oil and others, the amount of esters present are therefore determined: in thyme oil, clove oil, bay oil, and Cretian origanum oil the amount of phenols are estimated; in cassia oil and lemongrass oil the amount of aldehyde. The assay of santalwood oil shows how much santalol, that of palmarosa oil, how much geraniol is contained in the oil. The quality of the oils named finds numerical expression in the percentage strength of the active constituents such as esters, phenols, aldehydes and alcohols.

In a second class of oils, whose composition is likewise sufficiently known, an assay is not yet possible. The reason for this is twofold: first the value of the oil depends not upon a single constituent but upon the blending of the properties of several; and, secondly the chemical methods of investigation are not sufficiently developed.

With these oils, the examination is restricted as a rule to the determination of the normal composition of the oil and the absence of commonly used adulterants. Such oils are lemon oil, orange oil, rosemary oil and spike oil, which should be tested particularly for turpentine oil.

The incomplete knowledge of the composition and the defectiveness of the methods of testing most of the oils, do not at present allow of an investigation resting on a rational chemical basis. With this class of oils the entire examination consists in determining the physical constants. As the average and limit values of specific gravity, optical rotation, solubility, etc., of the more common oils are well known through observations extending over many years, variations from these values call the attention of the investigator to adulterants.

Indeed, the physical behavior of an oil is in general very well suited to indicate rapidly the addition of foreign substances: the investigation of volatile oils should therefore begin with the determination of the physical properties, no matter whether the investigation be for practical or for scientific purposes. After this, the special methods are to be used, such as saponification, acetylization, aldehyde and phenol determinations, and finally, if it appears necessary, the tests for turpentine oil, fatty oil, alcohol or petroleum should be applied.

It is of course evident that when the practical value of an oil is to be considered the examination of the odor[1] and the taste must accompany the chemical investigation, for these are the properties on account of which the volatile oils are used in the perfume and soap industries, in the manufacture of candies and liquors, and partly at least in medicine.

It is necessary, or at least greatly to be desired, to have for comparison a sample of a genuine, faultlessly distilled oil. A few drops each, of the genuine oil and of the oil to be tested, are put on a strip of filter paper, and compared by smelling alternately of both. This odor test is repeated after the larger part of the oil has volatilized, and in this manner easily volatile as well as more difficultly volatile foreign substances may be recognized.

It is possible, however, to give only a very imperfect expression in language of odor and taste perceptions; moreover, odor and taste are purely subjective and are also quite differently developed faculties in each individual. The perceptions made with the sense of odor and of taste do not admit of expression and comparison by means of figures like other observations. An adulteration may, therefore, be subjectively recognized, but cannot be objectively proven. A good sense of smell is in spite of this limitation of great value in the examination, as it often directs the investigation into the proper channel in the shortest time.

Poorly distilled oils, possessing an empyreumatic odor or "Blasengeruch," or oils carelessly kept but otherwise unadulterated are almost altogether recognized by the sense of smell, rarely by any other means of investigation.

Determination of the Physical Properties.

Specific gravity. On account of its ready determination the specific gravity is the most frequently taken and, therefore, the best

[1]) Attention may here be called to the interesting work by H. Zwaardemaker, Die Physiologie des Geruchs, Leipzig, 1895.

known property of the volatile oils. Even concerning the seldom and little investigated oils statements of their density are to be found. Further, as the maximum and minimum values of the more commonly used oils are fixed, the determination of the specific gravity belongs to the most important and also easiest means of investigation. The specific gravity of an oil is changeable within certain limits, and dependent, outside of age, on the manner of the distillation and also upon the source and the state of ripeness of the plant material used. The extent of this variation is so different in the individual oils that no general rules can be formulated. With normal bergamot oil, for instance, the specific gravity lies between 0.883 and 0.886. The difference between the highest and the lowest density, therefore, amounts in this case to only three places in the third decimal. As a rule, however, the limit values lie much further apart.

Most oils are lighter, some, however, heavier than water, especially those which contain larger amounts of oxygenated constituents of the aromatic series (e. g. wintergreen oil, clove oil, sassafras oil). The lowest specific gravity of all volatile oils is that of heracleum oil with 0.800 and of rue oil with 0.833; the highest is that of wintergreen oil with 1.187.

The determination is practically performed with an hydrostatic balance according to Mohr or Westphal,[1] as the accuracy attained with this instrument, if rightly handled, is sufficient. When only small quantities of an oil are at disposal a pycnometer is used.

In the determination the temperature is of course considered, and usually +15° C.[2] is selected for reasons of convenience. Only with those oils which are viscous at +15°, or wholly or partly solidified, the density is determined at a correspondingly higher temperature.

Optical rotation. The optical rotation is such a characteristic property of most of the volatile oils that its determination should never be omitted in an examination. Especially adapted for this purpose is the half-shadow polariscope according to Laurent.[3] If the dark color of the oil does not allow of making the observation in a 100 mm. tube, which is that usually employed, one of 50 or even 20 mm. may be

1) The hydrostatic balance with steel axes by F. Sartorius, Göttingen, may be especially recommended.

2) In this book, when not otherwise stated, the statements of the specific gravity are for a temperature of + 15° C.

3) With regard to the manipulation of the polariscope the reader is referred to the well-known work by H. Landolt, "Das optische Drehungsvermögen organischer Substanzen", II. Edit., Braunschweig, 1898.

used. In the following, α_D is the observed angle of rotation in a 100 mm. tube with sodium light, and $[\alpha]_D$ is the specific rotation as calculated by the formula $[\alpha]_D = \frac{\alpha}{l.d}$ where α is the observed angle of rotation, l the length of the tube in millimeters and d the specific gravity of the liquid. When no special mention is made of the temperature, room temperature is to be understood. In general it is not necessary, although desirable, to make the observation at a fixed temperature, as the natural variations in the rotation of an oil are usually greater than the differences due to a variation in temperature of several degrees. Exceptions to this are the oils of lemon and orange, the rotation of which is relatively strongly influenced by even small changes in temperature. It is necessary, in order to get comparable figures, to determine the rotation of these two oils at $+20°$ or else to reduce the result to this temperature by calculation. The details of this will be found in the description of these oils in the special part.

Refraction. The determination of the index of refraction n_D has been recommended by several investigators for the examination of the volatile oils. Between chemical constitution and refraction there exist, as is well known, certain relations, and in many cases conclusions as to the position and number of double bonds may be drawn from the molecular refraction. Chemical individuals carefully purified are, however, necessary in order to obtain useful results. As the refractive coefficients of the constituents of the volatile oils are on the whole only slightly different from each other, they are not so well suited for the detection of adulterations as are other methods of examination. The addition of turpentine oil, for instance, influences the refraction of lemon oil only slightly, but changes the rotation to a marked degree.[1] As the refraction changes rapidly with the temperature, this must always be considered in making observations of the refractive coefficients of volatile oils.

Congealing point. With certain oils, especially anise, star anise, fennel and rue oils, the congealing point gives a good basis for judging the quality. With the first three oils a high congealing point shows a large content of anethol, with rue oil one of methyl nonyl ketone.

The determination of the congealing point can be very well performed with Beckmann's well-known apparatus for the determination of the molecular weight by the lowering of the freezing point. A few slight changes make it especially suited for this purpose.

---

1) Bericht von S. & Co., October 1893, p. 50.

They consist principally in doing away with the cork connections which hinder the free inspection of the mercury thread of the thermometer. The laboratory of Schimmel & Co.[1] has the form shown in fig. 52. The battery jar A serves as the receptacle for the cooling liquid or freezing mixture. The glass tube B hanging in the metal cover serves as an air jacket for the freezing tube C and prevents the premature congealing of the oil to be tested. The freezing tube C is wider at the top and becomes narrower at the place where it rests on the edge of the tube B. In order to retain C in a fixed position three glass projections are fastened on the inside of the tube B, about 5 cm. below its upper edge. The thermometer, which is graduated into half degrees is held in position in a metal plate by three springs which allow of sliding the thermometer up or down.

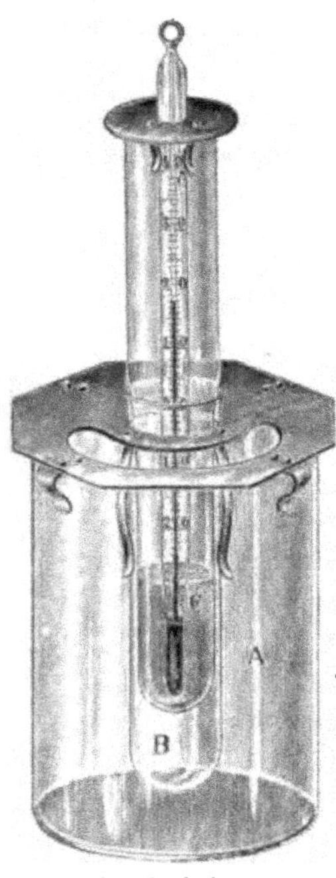

½ actual size.

Fig. 52.

In conducting the determination with anise and star anise oil the battery jar is filled with cold water and pieces of ice, but with fennel oil a freezing mixture of ice and salt is used. Then pour into the freezing tube so much of the oil to be tested that it stands at a height of about 5 cm. in the tube and bring the thermometer, which must not touch the sides of the tube at any place, into the liquid. During the cooling the overcooled oil is to be protected from disturbances, which would produce a premature congealation.[2] When the thermometer has sunk about 10° below the congealing point, (with anise and star anise oil to about 6 to 8°) crystallization is to be induced by rubbing and scratching the thermometer against the sides of the tube. If this proceeding should not prove successful a small crystal of congealed oil or some solid anethol is brought into the liquid, when congealation will take place with liberation of heat. The solidification is hastened by continued stirring with the thermometer, the mercury thread of which rapidly rises and finally reaches a maximum which is called the congealing point of the oil.

Behavior on boiling and fractional distillation. Inasmuch as the volatile oils are mixtures of substances with different boiling points it is improper to speak, as is so often done, of the boiling point of a volatile oil. It is more correct to speak of a boiling temperature, by which is meant the temperature interval within the limits of which the oil distills over in a single distillation from an ordinary distilling

---

1) Bericht von S. & Co., October 1898, p. 49.

2) A premature congealation often takes place when the oil has not been filtered clear, as suspended dust particles may give rise to congealation.

flask (fig. 53) without the application of a fractionating arrangement. A thermometer with a shortened scale and with the entire mercury thread in the vapor is used.

The observations recorded by different observers of the amounts of the same oil which distill over between certain limits of temperature, seldom agree, because the results are not only greatly influenced by the form of the distilling flask, but also by the rapidity of the distillation and the height of the barometer. It is therefore necessary, in the examination of certain fractions of individual oils, to use flasks of fixed dimensions and to observe a certain rapidity of distillation. For testing

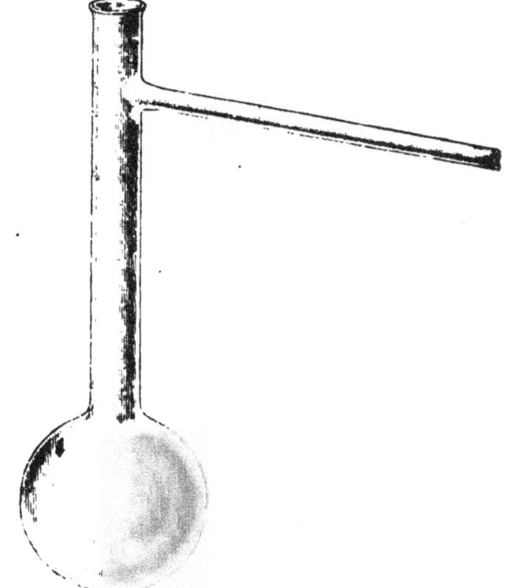

Fig. 53.

lemon oil, rosemary oil and spike oil, Schimmel & Co. use distilling flasks according to Ladenburg of the dimensions given in fig. 54.[1] From 50 cc. of the oils mentioned, 5 cc. are distilled over in such a manner that about 1 drop falls in a second, and the distillate tested in the polariscope, as will be described more in detail under the individual oils.

When the different constituents are to be isolated from an oil, the fractional distillation must be repeated many times and preferably by employing one of the well known distilling columns. It is best, in order

---

1) Bericht von S. & Co., October 1898, p. 46.

to avoid decomposition, to distill the fractions boiling above 200° in vacuum. Oils containing esters must first be saponified, as the acids, which are easily split off by the boiling, disturb the fractionation and may act upon the other constituents of the oils.

Solubility. The volatile oils are readily soluble in the ordinary solvents, such as alcohol, ether, chloroform, benzene, acetic ether, carbon disulphide, petroleum ether, paraffin oil etc. Mention of this general property is not made in the description of the individual oils. A phenomenon which is sometimes considered as an incomplete solubility may be mentioned here. The turbidity noticed by mixing certain oils

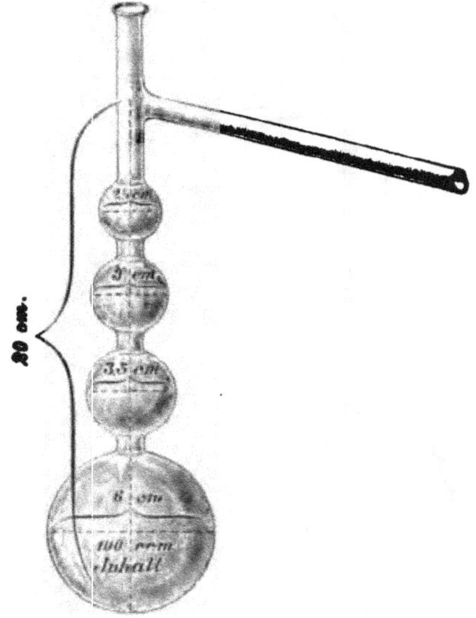

Fig. 54.

with petroleum ether, paraffin oil or carbon disulphide is caused by the small amount of water which the oils have retained from their preparation. The richer an oil is in oxygen, the more water it is capable of dissolving and the more cloudy does it become with petroleum ether.[1] The turbidity does not take place when the oil has been thoroughly dried with anhydrous sodium sulphate.

---

1) When an oil rich in oxygen, such as bergamot oil, is mixed with one rich in terpenes, as turpentine oil or orange oil, the mixture becomes turbid by the separation of water.

Although all oils are readily soluble in strong alcohol, only some of them are entirely soluble in dilute alcohol. For the last class this property becomes a practical and rapid means of examination. The presence of the difficultly soluble turpentine oil can, for instance, be readily shown in this manner in the oils soluble in 70 percent[1] alcohol. The solubility determination is very simple.

Bring into a small graduated cylinder (fig. 55) ½—1 cc. of the oil to be tested and add small portions of the alcohol at a time until with vigorous shaking solution is effected.

Fig. 55.

If an oil which is soluble under normal conditions does not dissolve, it is sometimes possible to draw conclusions as to the adulterant from the character of the turbidity and the separation of the insoluble part. Petroleum floats on the 70 percent alcohol, whereas fatty oil settles in drops at the bottom.

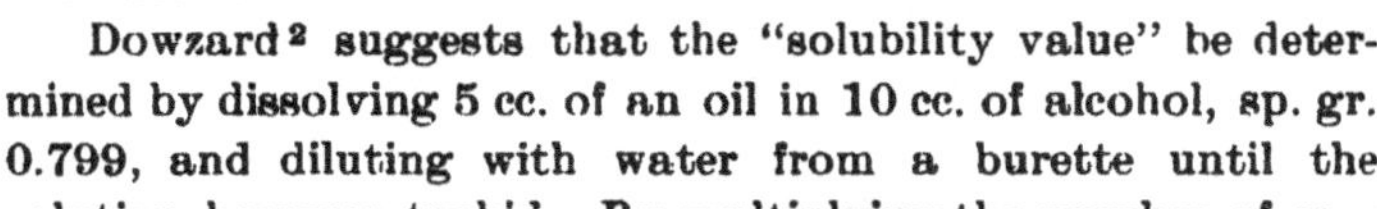

Dowzard[2] suggests that the "solubility value" be determined by dissolving 5 cc. of an oil in 10 cc. of alcohol, sp. gr. 0.799, and diluting with water from a burette until the solution becomes turbid. By multiplying the number of cc. of water by 100 the "solubility value" is obtained.

Duyk[3] recommends the use of a solution of sodium salicylate in water 1:1. The alcohols, aldehydes, and ketones, are much more soluble than the sesquiterpene hydrates, phenol ethers and esters. The hydrocarbons are practically insoluble. In special cases this solution may prove valuable.

Chemical Methods of Testing.

The rational examination of a volatile oil in a chemical manner is only then possible, when its composition or at least its main constituents are known. The chemical investigation must be directed as much as possible toward the isolation and quantitative estimation of the constituents recognized as being the most valuable. The methods of testing must, therefore, conform to the analysis of the oil. If this really self-evident supposition had earlier been generally recognized, those methods of investigation, which are designated as quantitative reactions, as for instance the iodine absorption, or Maumené's sulphuric acid test, which

---

[1]) The statements in this book always refer to volume percent. 90 percent alcohol corresponds to the *Spiritus*, and 70 percent to the *Spiritus dilutus* of the German Pharmacopoeia.

[2]) Chemist & Druggist, 58, p. 749.

[3]) Bull. de l'Acad. Roy. de Méd. de Belgique, 1899, p. 508; Ann. de Pharm., 5, p. 348.

had given good results with the fatty oils, would not have been applied offhand to the volatile oils.

The fatty oils are a group of chemically closely related bodies; they are glycerides of the fatty and oleic acid series. The constituents of the volatile oils, however, recruit themselves from the greatest variety of classes of bodies. Among them may be found terpenes, sesquiterpenes, paraffins, alcohols, aldehydes, ketones, phenols, ethers and esters. It should be no matter for surprise, therefore, that the methods of testing which are useful with the fatty oils, fail completely with the volatile oils. Nor is there any sense in subjecting the fatty and the volatile oils to the same reactions, just because they both bear the same designation "oils."

The application of Hübl's iodine addition method to volatile oils has been recommended by Barenthin[1] (1886), Kremel[2] (1888), Williams[3] (1889), Davies[4] (1889) and Snow[5] (1889). By a direct comparison of the results of these separate observers it could not have been difficult for Cripps[6] (1889) to show the utter uselessness of this method.

The use of bromine in place of iodine was first suggested by Levallois[7] (1884) and later by Klimont[8] (1894).

In Maumené's test the fatty oil to be investigated is mixed in a certain proportion with concentrated sulphuric acid and the rise in temperature which takes place is observed. Its application to volatile oils was recommended by Williams[9] (1890) as well as by Duyk[10] (1897) but it has found just as little favor in practice as the others named above.

With these methods are to be classed also the much recommended color reactions. They mostly consist in bringing together a volatile oil and e. g. sulphuric acid or nitric acid, whereby some coloration is produced, which only in rare cases can be ascribed to a definite chemical change. As the shades of color produced are difficult to describe, and often change from one to the other and may therefore easily give rise to mistakes, the color reactions in general are to be designated as useless. This does not exclude, however, the occasional use of a color reaction in the detection of adulterants. It is, however, never to be considered as conclusive in itself.

---

1) Archiv d. Pharm., 224, p. 848.
2) Pharm. Post, 21, pp. 789, 821.
3) Chem. News, 60, p. 175.
4) Pharm. Journ., III, 19, p. 821.
5) Pharm. Journ., III, 20, p. 4.
6) Chem. News, 60, p. 286.
7) Compt. rend., 99, p. 977.
8) Chemiker Zeitung, 18, p. 641.
9) Chem. News, 61, p. 64.
10) Bull. de l'Acad. roy. de med. de Belgique.

Besides the methods of testing already enumerated, many others have been suggested in the course of time, which however have acquired as little practical importance as these. Only such methods are here to be discussed as have really proven satisfactory in the investigation of volatile oils.

Saponification. Through scientific investigation it has been established that many volatile oils contain ester-like compounds, the components of which are alcohols, usually of the composition $C_{10}H_{18}O$ or $C_{10}H_{20}O$ on the one hand, and acid radicals of the fatty series, on the other.

The esters, which are nearly without exception of a pleasant odor, are often to be considered as the most important constituents of the oils. Thus, linalyl acetate is the carrier of the odor in bergamot oil; the same ester is found in lavender oil and also occurs along with other compounds in petitgrain oil. The esters of borneol, found in the different pine-needle oils, play an important part in the formation of the pine aroma. Menthyl acetate is found in the peppermint oils and the geranyl ester of tiglic acid in the different geranium oils. The quantitative estimation of the ester is always valuable for judging the oils, even when the esters are of little consequence to the odor. But much more important, and really the only rational method as a test of quality is the determination in all cases where the esters are the carriers of the characteristic odor, as with bergamot oil and with lavender oil. The estimation is made according to the method of quantitative saponification, as it has long been used in the analysis of the fats. Its application to the volatile oils was first made by A. Kremel[1] (1888). This suggestion did not reach a practical importance until through scientific investigation the nature of the saponifiable compounds was learned.

Kremel distinguishes acid number (S. Z.), ester number (E. Z.), and saponification number (V. Z.). The acid number expresses how many mg. KOH are necessary to neutralize the amount of free acid contained in 1 g. of oil. The ester number gives the amount of potassa in milligrams used in the saponification of the ester contained in 1 g. of oil. The saponification number is the sum of acid number and ester number. As the volatile oils usually contain only a small amount of free acid, this may in general be neglected. Only old, partly spoilt oils, tend to show somewhat higher acid numbers.

It must be mentioned that in all oils which contain aldehydes, the

1) Pharm. Post, 21, pp. 789, 821.

ester determination by saponification cannot be made, as a consumption of alkali takes place which increases with the length of the reaction, due to the decomposition of the aldehyde, but which gives no information as to the amount of aldehyde decomposed.

No. 56.

The saponification is conducted in a wide necked flask of potash glass of 100 cc. capacity (fig. 56). A glass tube about 1 m. in length and passing through a stopper serves as a reflux condenser. About 2 g. of oil are weighed accurately to 1 cg. into such a flask and 10 to 20 cc.[1] of an hemi normal alcoholic potassa solution are added. Previous to this the oil should be tested for free acid, an alcoholic solution of phenolphthaleïn being used as indicator. The flask provided with the condensing tube is heated for half an hour[2] on a steam bath, and after cooling, the contents are diluted with about 50 cc. of water and the excess of alkali titrated back with half normal sulphuric acid.

From the saponification number found, the percentage content of esters of the formula $C_{10}H_{17}OCOCH_3$ and $C_{10}H_{19}OCOCH_3$ may be read directly in the tables to be found on page 204 at the end of this chapter.

Acetylization. Many volatile oils contain as important constituents alcohols of the formula $C_{10}H_{18}O$ and $C_{10}H_{20}O$, for instance, borneol, geraniol, terpineol, linalool, thujylalcohol, menthol and citronellol. For the quantitative estimation their behavior toward acetic acid anhydride, with which they form acetic esters when heated, may be used. The reaction takes place according to the equation:

$$C_{10}H_{18}O + (CH_3CO)_2O = C_{10}H_{17}OCOCH_3 + CH_3COOH.$$

The reaction is quantitative with borneol, geraniol (Bertram and

1) In most cases 10 cc. are sufficient, only with unknown oils 20 cc of alkali solution are taken.

2) With bergamot oil the saponification is complete within 10 minutes. In order to be certain in all cases, the heating can be continued somewhat longer, without detrimental effect. Compare Bericht von Schimmel & Co., October 1895, p. 16. The method used by Helbing (Helbing's Pharmacological Record, No. 30, p. 4) of saponifying in a closed vessel, consequently under pressure, gave, as the experiments by Schimmel & Co. show, for bergamot oil from 1—2 percent higher results than did the saponification in an open flask. The reason for this is, that when saponified under pressure the linalool is attacked by the potassa, which is thereby used up, while by boiling with potassa in an open flask the alcohol is not affected. Later (1895), Helbing and Passmore (Chem. & Drug., 47, p. 585) coroborated that the saponification in an open vessel with a reflux condenser was preferable to saponification in an autoclave. Neither does the so-called cold saponification appear to be applicable to volatile oils. Moreover, it has the disadvantage of requiring a longer time. For, according to Henriques, (Zeitschrift für angewandte Chemie, 1897, p. 399) linalool after 12 hours action of the alkali gave the saponification number 4.2, and geraniol by the same treatment one of 2.8.

Gildemeister,[1] 1894) and menthol (Power and Kleber,[2] 1894) and makes an accurate determination of these bodies possible.[3] Less fortunate are the conditions with linalool and terpineol as these on boiling with acetic acid anhydride are partly decomposed by splitting off water with the formation of terpenes. Comparative figures can, however, be also obtained with these alcohols, if the same amount of acetic acid anhydride is always used and the boiling continued for the same length of time. With linalool there was found as a favorable result after 2 hours of boiling, a deficiency of 15 percent of alcohol.[4]

Terpineol behaves toward acetic acid anhydride as follows:

Time of boiling:	Terpinyl acetate formed:
10 minutes	51.2 percent
30 "	75.5 "
45 "	84.4 "
2 hours	77.9 "

With terpipeol, therefore, heating longer than 45 minutes has a detrimental effect.

Acetic acid anhydride acts differently on aldehydes. While citronellal is converted quantitatively into isopulegol acetate, with citral there are formed indefinite amounts of saponifiable products[5] as yet unknown.

Fig. 57.

For the quantitative acetylization[6] 10 to 20 cc. of the oil, mixed with an equal volume of acetic acid anhydride and 1—2 g. of dry sodium acetate, are boiled uniformly from 1—2 hours in a small flask provided with a condensing tube which is ground into the neck of the flask (fig. 57). After cooling, some water is added to the contents of the flask and then heated from ¼ to ½ hour on a water bath, to decompose the excess of the acetic acid anhydride. The oil is then separated in a separating funnel, and washed with soda solution and water until the reaction is neutral. Of the acetylized oil dried with anhydrous sodium sulphate 2 g. are saponified according to the method described on

1) Journ. f. prakt. Chem., II, 49, p. 189.

2) Pharm. Rundsch., 12, p. 162; Archiv d. Pharm., 232, p. 658.

3) Even with these alcohols an indiscriminate application of this otherwise excellent method to all oils containing them should be avoided. It is known that acetic acid combines with terpenes to form acetates of corresponding alcohols. In fact preliminary experiments show that pinene and limonene may assay in part as alcohol when this method is used. *E. K.*

4) Bericht von S. & Co., April 1893, p. 38.

5) Bericht von S. & Co., Oct. 1896, p. 34.

6) Bericht von S. & Co., Oct. 1894, p. 63.

page 194. The amount of alcohol, based on the original unacetylized oil, corresponding to the saponification number, can be found in the tables on page 204 at the end of this chapter.

Aldehyde determination. The quantitative determination of cinnamic aldehyde depends on its property to form a water soluble addition product with an excess of sodium bisulphite. When a known amount of an oil containing cinnamic aldehyde is shaken sufficiently long with a hot concentrated bisulphite solution, the aldehyde content of the oil is about equal to the decrease in volume. The manipulation of the aldehyde determination and the necessary apparatus are described in detail under cassia oil (see this). Citral behaves similar to cinnamic aldehyde, and is able to form a water soluble bisulphite addition product. The same method is, therefore, applicable to oils rich in citral, as for instance, lemongrass oil.

For the determination of the aldehydes in lemon oil this method is, however, not suited, on account of the formation of the citronellal addition product, which is insoluble in water and floats partly in the aqueous, partly in the oily layer and makes an accurate reading impossible. Besides, the bisulphite method is hardly accurate enough for lemon oil, as this contains only 6—8 percent of aldehydes.

Another method, to determine the carbonyl oxygen and thus the aldehyde and ketone content in volatile oils has been worked out by Benedikt and Strache[1] (1893). The oil to be investigated is heated with a weighed amount of phenyl hydrazine, after some time the hydrozone which has formed is separated by filtration, and the unchanged phenyl hydrazine in the filtrate oxidized with boiling Fehling's solution. By this treatment all the nitrogen of the phenyl hydrazine which has not taken part in the reaction is liberated as gas. From the volume of the collected nitrogen the amount of the unused phenyl hydrazine can be calculated. From this the amount which has gone into combination is known and consequently the amount of the ketone or aldehyde present. The amount of carbonyl oxygen, expressed in 1/10 percents is designated as carbonyl number.

With bitter almond oil (benzaldehyde), cumin oil (cuminic aldehyde), and rue oil (methyl nonyl ketone) this method yields fairly good results. With cassia oil, caraway oil, fennel oil and lemon oil, however, the determinations fall far too low.[2]

1) Monatsh. f. Chem., 14, p. 270.
2) Bericht von S. & Co., Oct. 1893, p. 48.

The problem of a general method for determining the aldehyde and ketone content of volatile oils, is therefore not yet satisfactorily solved. Perhaps it is possible to reach better results by suitable changes in the method described.

A method devised by Kremers and Schreiner specially for the estimation of carvone in carvone-containing oils will be discussed under caraway oil.

Phenol determination. For an approximately accurate phenol determination in volatile oils, the property of the phenols to combine with alkalies to form compounds which are soluble in water, is used. If a measured quantity of an oil is shaken with a solution of alkali, the content of phenols can be approximately determined from the diminution in volume. The soda solution is to be five percent. A greater concentration is not allowable, as stronger phenol-alkali solutions take up relatively large amounts of the remaining constituents of the oil, and thus give results which are too high. The manipulation of such a determination is described under thyme oil.

Thymol and carvacrol are determined by Kremers and Schreiner[1] (1896) quantitatively in volatile oils, by precipitating the phenols from the alkaline solution as their iodine compounds and titrating back the iodine added in excess. The details of this method can be found under thyme oil.

Schryver[2] has based a new method for the determination of phenols on the fact that hydroxy compounds liberate ammonia from sodium amide. The determination of thymol in oil of thyme and of eugenol in clove oil is reported to give good results.

According to Thoms[3] (1891) eugenol is quantitatively changed to benzoyl eugenol and weighed as such. This method is especially to be recommended with clove oil and the oil of cinnamon leaves, as shaking out with alkali gives results which are too high. The manipulation for Thoms' method for estimating eugenol is given under clove oil.

Methyl number. A number of volatile oils contain as important constituents methyl and ethyl ethers of phenols and acids, the alkyl groups of which can be determined according to Zeisel's method[4] (1885). Benedikt and Grüssner[5] (1889) have recommended the quantitative methoxyl determination for the practical and scientific investigation of

---

1) Pharm. Review, 14, p. 22.
2) J. Soc. Ch. Ind., 18, p. 553.
3) Berichte der pharm. Gesellsch., 1, p. 278.
4) Monatsh. f. Chemie, 6, p. 989.
5) Chem. Zeitung, 13, pp. 872, 1087.

volatile oils, and shown its usefulness by a series of illustrations. They designate as methyl number that number which tells how many mg. of methyl 1 g. of substance splits off when boiled with hydriodic acid. Ethyl or propyl and isopropyl are here considered as having been replaced by the equivalent amount of methyl. The weighed amount of silver iodide is therefore in all cases calculated as methyl.

The vapors of methyl iodide formed by boiling from 0.2 to 0.3 g. of the oil to be investigated with hydriodic acid (sp. gr. 1.70, to which, according to Herzig[1] (1888) 8 percent of acetic acid is added) are first passed through some warm water in which some phosphorus is suspended, so as to retain any iodine vapors which may have been carried over. After the methyl iodide has passed through this apparatus it is passed into an alcoholic solution of silver nitrate and the separated silver iodide weighed.

A very convenient apparatus has been designed by L. Ehmann[2] (1890) for carrying out this determination.

Gregor[3] (1898) has recently suggested to replace the phosphorus suspended in water by a solution of one part each of potassium bicarbonate and arsenous acid in 10 parts of water, by which not only the iodine vapors but also any hydriodic acid carried over is retained. For collecting the methyl iodide Gregor uses a 1/10 normal silver nitrate solution which has been acidified with nitric acid and titrates back the silver not used for the precipitation of the silver iodide with 1/10 normal potassium sulphocyanate solution according to Volhard.

Of the oils investigated by Benedikt and Grüssner the following gave no methyl numbers: the oils of wormwood, bitter almond, angelica, bergamot, caraway, lemon, copaiba-balsam, coriander, cubeb, elemi, eucalyptus, geranium, juniper, cherry laurel, lavender, spearmint, peppermint, olibanum, oil of *Pinus montana*, savin, East Indian and West Indian sandalwood, turpentine and valerian.

High methyl numbers were given by anise oil, star anise oil and fennel oil on account of their content of anethol and methyl chavicol, by clove oil, and oil of cinnamon leaves on account of their content of eugenol. With wintergreen oil the high methyl number is due to the methyl salicylate, with parsley oil to the apiol.

The determination is only applicable with oils absolutely free from alcohol, as ethyl alcohol itself gives a methyl number, from which it follows, that this method may also be used for the quantitative

---

1) Monatsh. f. Chemie, 9, p. 544.

2) Chem. Zeitung, 14, p. 1767.

3) Monatsh. f. Chemie, 19, p. 166.

estimation of alcohol in those oils which in their pure condition do not contain any methoxyl groups.[1]

The Detection of Some of the More Common Adulterants.

Turpentine oil. Turpentine oil may be considered as the adulterant most often used. Very often it can be recognized by its characteristic odor, especially in those oils which contain no pinene, as this is the main constituent of turpentine oil. In general, its presence causes changes in the specific gravity, solubility, boiling temperature and optical rotation. It must here be remembered that there are dextro- as well as laevogyrate turpentine oils.

The positive proof of the presence of turpentine oil in oils, which in their pure state contain no pinene, is furnished by the isolation of and characteristic derivatives of pinene. The constituents boiling in the neighborhood of 160°, are fractionated out and, according to the method described on page 109, the pinene nitrosochloride, as well as pinene nitrolbenzylamine or pinene nitrolpiperidine are prepared.

If pinene is a normal constituent of the oil, the addition of turpentine oil can be recognized by a comparison of the physical properties of the lowest boiling fractions of the adulterated oil with the corresponding portions of the pure oil. The detection of turpentine oil in rosemary oil furnishes an example (see this).

Cedar wood, copaiba and gurjun balsam oils. On account of their cheapness and faint odor, these three oils belong to the most favored and the most dangerous of adulterants. They may, however, be detected without difficulty in most cases by means of their physical properties, which differ from those of many of the volatile oils, namely

---

1) The methyl numbers of some of the oils investigated by Benedikt and Grüssner can only be explained by the presence of alcohol. It is greatly to be regretted, that the physical properties of the investigated oils are not given, and their purity cannot therefore be judged. With the Ceylon cinnamon oil designated as No. 22 in the article cited, Benedikt and Grüssner calculate from the methyl number found, 25.7, a eugenol content of 28.1 percent. As pure Ceylon cinnamon oil contains only 4—8 percent of eugenol, it follows from the determination (providing that the methyl number is due only to eugenol), that the oil was adulterated with the oil of cinnamon leaves, which is rich in eugenol. From this illustration it may be seen that the reported methyl numbers are to be used with care. It would therefore be a valuable piece of investigation to repeat these determinations on oils whose purity could be previously determined in another manner, in order to give to this really fine method the basis necessary for practical purposes. For the scientific investigation of the volatile oils the methoxyl determination is extremely valuable as it gives information of the presence or absence of phenol ethers or acid esters, which contain a methyl, ethyl, or propyl group.

their difficult solubility in 70 to 90 percent or even stronger alcohol, their high specific gravity (above 0.900), their high boiling temperature, above 250°, and finally their rotatory power.

All three oils turn the plane of polarized light more or less to the left. With copaiba balsam oil the angle of rotation α_D lies between -7 and $-35°$,[1] with cedar wood oil between -30 and $-40°$, and with gurjun balsam oil between -35 and $-130°$(!).

At present only copaiba balsam oil can be detected in a chemical way, in that the fraction boiling about 260° is converted into caryophyllene hydrate melting at 94—96°, or into caryophyllene nitrolpiperidine (see p. 125) melting at 141—143°, or into caryophyllene nitrosite and the corresponding nitrolbenzylamine base (see p. 126). As yet no characteristic derivatives of the sesquiterpenes of cedar wood and gurjun balsam oils are known.

Alcohol. The addition of alcohol to a volatile oil always results in a lowering of the specific gravity. When an oil containing alcohol is dropped into water, the drops do not remain clear and transparent, as is the case with pure oils, but become opaque and milky.

For the definite identification of alcohol the suspected oil is heated until it just begins to boil,[2] and the first few drops that come over are collected in a test tube and filtered, to remove any oil globules which may also have come over, through a filter moistened with water. The filtrate is made strongly alkaline with dilute potassa solution, and treated, after heating to 50—60°, with a solution of iodine in potassium iodide, until the solution remains slightly yellow. If alcohol is present small crystals of iodoform will separate in a short time at the bottom of the liquid. It must here be remembered that other bodies, such as aldehydes, acetone and acetic ether also yield iodoform under the given conditions.

Larger amounts of alcohol may be removed from volatile oils by shaking with water, from which the alcohol may be again removed by distillation and identified by the iodoform reaction. If the shaking out is done in a graduated cylinder, the increase of the watery layer corresponds approximately to the amount of the alcohol.

According to Hager it is better to use glycerin, because with this the two layers separate better, and a more accurate reading is possible.

The alcohol content may also be approximately calculated if the specific gravity is determined before and after shaking with water.

---

[1]) Umney in 1895 (Pharm. Journ., III, 24, p. 215), found the oil of a socalled copaiba balsam from west Africa to be dextrogyrate, $\alpha_D = +20° 42'$. Dextrogyrate gurjun balsam oils are also said to exist.

[2]) All alcohol is not removed by heating on a water-bath.

As already mentioned on page 198 the amount of the alcohol in an oil which in itself does not give a methyl number, may be determined quantitatively by Zeizel's methoxyl method.

Fatty oil. The volatile oils adulterated with fatty oil leave a permanent fatty stain when evaporated on paper. With high boiling and difficultly volatile oils, however, similar residues are left, which may give rise to mistakes. Fatty oil is insoluble in 90 percent alcohol.[1] For the separation of the fatty oil from the volatile oil the latter is distilled off by steam distillation or removed by evaporation in an open dish on the waterbath, where it must be remembered that some volatile oils, as bergamot, lemon, orange, anise and star-anise oil leave a residue of several percent even when they are not adulterated.

The presence of the fat may be shown qualitatively in the residue by heating with potassium bisulphate in a test tube. Penetrating odors of acrolein show its presence. By igniting the residue on a platinum foil the characteristic odor of burning fat is noticed.

As the fatty oils give saponification numbers which lie between 180 and 200, the amount of the fat added can be determined quantitatively either directly in the volatile oil itself or in the distillation residue, by saponification.

Oils that have been adulterated with cocoa nut oil solidify wholly or in part in a freezing mixture. Cocoa nut oil has been found in cananga oil, citronella oil and palmarosa oil and detected in this manner.

Mineral oil, petroleum. Mineral oil, paraffin oil, kerosene, petroleum and petroleum fractions are insoluble in alcohol and can therefore be detected without difficulty in volatile oils; besides, they are often easily recognized by their low specific gravity. Palmarosa oil to which some mineral oil has been added is only partly soluble in 70 percent alcohol. If the insoluble portion is successively treated with 90 percent and absolute alcohol, an' oil remains, which in the beginning it is true is colored brown by sulphuric or nitric acids, but in general resists the action of these acids and also of alkalies, and on saponification with alcoholic potassa gives no saponification number.

The mineral oils have different boiling points. The hydrocarbons of illuminating oil boil at about the same temperature as the terpenes. Lower boiling fractions are said to be sometimes used for adulterating turpentine oil. A higher boiling mineral oil, about 250°, has been

1) Only castor oil is soluble in 90 percent, but is insoluble in 70 percent alcohol.

found in citronella and in gingergrass oils. The petroleum fractions of a lower boiling point are easily volatile with water vapor, the higher fractions not at all, or at least to only a small degree.

A method for the quantitative estimation of mineral oils consists in weighing the residue left after having removed the volatile oil by oxidation with fuming nitric acid, as described under turpentine oil (see this). Attention should be called to the fact that some volatile oils, such as rose oil, chamomile oil, orange flower oil and others, contain larger or smaller amounts of paraffins as natural constituents.

Chloroform. This body, which has been found occasionally in volatile oils (for instance, in cognac oil and sandalwood oil) can be isolated by distillation on a water bath and identified by the isonitrile reaction. This consists in that a small amount of the suspected distillate is treated with a few drops of aniline and alcoholic caustic soda solution, and gently heated. In the presence of chloroform the noxious and exceedingly disagreeable smelling vapors of phenylisonitrile are formed.

Table

for Calculating the Amount of Alcohols of the Formula $C_{10}H_{18}O$ and $C_{10}H_{20}O$,

as well as acetic esters of these alcohols, from the saponification numbers, found before and after acetylizing.

In order to save time-robbing calculations the following table (p. 204 et seq.) has been worked out in the laboratory of Schimmel & Co.,[1] from which the amounts of the alcohols $C_{10}H_{18}O$ (geraniol, linalool, borneol, isopulegol, etc.) and $C_{10}H_{20}O$ (menthol, citronellol) as well as their acetic esters, corresponding to the individual saponification numbers, can be read off. From this compilation it is further possible to see how much alcohol in the original oil corresponds to the saponification number of the acetylated oil.

A few examples may serve to illustrate the above statements.

The saponification number 109 has been found by saponifying a bergamot oil. In the left half of the table headed $C_{10}H_{18}O$ look up the number 109, and there will be found in the column which bears the heading "Acetate", the number 38.15, which gives the percentage content

[1] Bericht von Schimmel & Co., Oct. 1897, p. 82.

of linalyl acetate in the bergamot oil. This 38.15 p. c. of acetate corresponds, as may be seen in the next column headed "Alcohol", to 29.98 p. c. of linalool.

By the saponification of a peppermint oil in the original unacetylated state, the saponification number 20 is found, and by the saponification after boiling with acetic acid anhydride, the saponification number 179. It is found that in the right half of the table the saponification number 20 = 7.07 p. c. of acetate $C_{10}H_{19}OCOCH_3$, in this case menthyl acetate, and corresponds to 5.57 p. c. of menthol. The saponification number 179 shows that 57.60 p. c. of total menthol are contained in this peppermint oil. If the total amount of menthol is 57.60 p. c., and that of menthol as ester 5.57, the peppermint oil contains 57.60 minus 5.57 or 52.03 p. c. of free menthol.

In the calculation of this table the following formulas were applied:

For the alcohol $C_{10}H_{18}O$:

1. $\frac{196 \times \text{Sap. No.}}{56}$ = percent of ester
2. $\frac{154 \times \text{Sap. No.}}{56}$ = percent of alcohol
3. $\frac{a \times 15.4}{s - (a \times 0.042)}$ = percent of alcohol in the original oil

For the alcohol $C_{10}H_{20}O$:

1. $\frac{198 \times \text{Sap. No.}}{56}$ = percent of ester
2. $\frac{156 \times \text{Sap. No.}}{56}$ = percent of alcohol
3. $\frac{a \times 15.6}{s - (a \times 0.042)}$ = percent of alcohol in the original oil.

In these formulas:

a = the number of cubic centimeters of normal potassa solution used.
s = the number of grams of the acetylated oil used for the saponification.

	$C_{10}H_{18}O$			$C_{10}H_{20}O$			
Sap. No.	Acetate	Alcohol	Alcohol in the orig. oil	Acetate	Alcohol	Alcohol in the orig. oil	Sap. No.
1	0,35	0,28	0,27	0,35	0,28	0,28	1
2	0,70	0,55	0,55	0,71	0,56	0,56	2
3	1,05	0,83	0,83	1,06	0,84	0,84	3
4	1,40	1,10	1,10	1,41	1,11	1,12	4
5	1,75	1,38	1.38	1,77	1,39	1,40	5
6	2,10	1,65	1,66	2,12	1,67	1,68	6
7	2,45	1.93	1,94	2,47	1,95	1,96	7
8	2,80	2,20	2,21	2,83	2,23	2,24	8
9	3,15	2,48	2,49	3,18	2,51	2,52	9
10	3.50	2,75	2,77	3,54	2,79	2,81	10
11	3,85	3,03	3.05	3,89	3,06	3,09	11
12	4,20	3,30	3,33	4,24	3,34	3,37	12
13	4,55	3,58	3,61	4,60	3,62	3,66	13
14	4,90	3,85	3,89	4,95	3,90	3,94	14
15	5.25	4,13	4,17	5,30	4,18	4,23	15
16	5,60	4,40	4,45	5,66	4,46	4,51	16
17	5,95	4,68	4,74	6,01	4,74	4,80	17
18	6,30	4,95	5,02	6,36	5,01	5,08	18
19	6,65	5,23	5,30	6,72	5,29	5,37	19
20	7,00	5,50	5,58	7,07	5,57	5,66	20
21	7,35	5,78	5,87	7,42	5.85	5,94	21
22	7,70	6.05	6.15	7,78	6,13	6,23	22
23	8,05	6,33	6,44	8,13	6.41	6,52	23
24	8,40	6,60	6,72	8.49	6,69	6,81	24
25	8,75	6,88	7,01	8,84	6.96	7.10	25
26	9,10	7.15	7,29	9.19	7,24	7,39	26
27	9,45	7.43	7.58	9,55	7,52	7,68	27
28	9,80	7,70	7,87	9,90	7,80	7,97	28
29	10,15	7.98	8,15	10,25	8,08	8,26	29
30	10,50	8,25	8,44	10,61	8,36	8.55	30
31	10,85	8,53	8.73	10,96	8,64	8,84	31
32	11,20	8,80	9,02	11.31	8,91	9.13	32
33	11,55	9,08	9,31	11.67	9,19	9.43	33
34	11,90	9,35	9.59	12,02	9,47	9.72	34
35	12.25	9,63	9,88	12.37	9,75	10.01	35
36	12,60	9,90	10,17	12 73	10,03	10.31	36
37	12,95	10,18	10,47	13,08	10,31	10,60	37
38	13,30	10,45	10,76	13,44	10,59	10.90	38
39	13,65	10,73	11,05	13,79	10,86	11,19	39
40	14,00	11,00	11,34	14,14	11,14	11,49	40

	$C_{10}H_{18}O$			$C_{10}H_{20}O$			
Sap. No.	Acetate	Alcohol	Alcohol in the orig. oil	Acetate	Alcohol	Alcohol in the orig. oil	Sap. No.
41	14,35	11,28	11,63	14,50	11.42	11,78	41
42	14,70	11,55	11,93	14,85	11,70	12,08	42
43	15,05	11,83	12,22	15,20	11,98	12,38	43
44	15,40	12,10	12,51	15,56	12,26	12,68	44
45	15,75	12,38	12,81	15,91	12,54	12,97	45
46	16,10	12,65	13,10	16,26	12,81	13,27	46
47	16,45	12,93	13,40	16,62	13.09	13.57	47
48	16,80	13,20	13,69	16,97	13.37	13,87	48
49	17,15	13,48	13,99	17,32	13,65	14,17	49
50	17,50	13,75	14,29	17,68	13,93	14,47	50
51	17,85	14,03	14,58	18,03	14,21	14,77	51
52	18,20	14,30	14,88	18,39	14,49	15,07	52
53	18,55	14,58	15,18	18,74	14,76	15,38	53
54	18,90	14,85	15,48	19,09	15,04	15,68	54
55	19,25	15,13	15,77	19,45	15,32	15,98	55
56	19,60	15,40	16,07	19,80	15,60	16.28	56
57	19,95	15,68	16,38	20,15	15,88	16,59	57
58	20,30	15,95	16,68	20,51	16.16	16,89	58
59	20,65	16,23	16,98	20,86	16,44	17,20	59
60	21,00	16,50	17,28	21,21	16,71	17,50	60
61	21,35	16,78	17,58	21,57	16,99	17,81	61
62	21,70	17,05	17,88	21,92	17,27	18,11	62
63	22,05	17,33	18,18	22,27	17.55	18,42	63
64	22,40	17,60	18,49	22,63	17,83	18,73	64
65	22,75	17,88	18,79	22,98	18,11	19,04	65
66	23,10	18,15	19,10	23,34	18 39	19,34	66
67	23.45	18,43	19,40	23,69	18,66	19,65	67
68	23.80	18,70	19.70	24,04	18,94	19,96	68
69	24,15	18,98	20,01	24,40	19,22	20,27	69
70	24,50	19.25	20,32	24,75	19,50	20,58	70
71	24,85	19,53	20,62	25,10	19.78	20,89	71
72	25,20	19,80	20,93	25,46	20,06	21,20	72
73	25,55	20,08	21,24	25,81	20,34	21,51	73
74	25,90	20,35	21,55	26,16	20,61	21,83	74
75	26,25	20,63	21,85	26,52	20.89	22,14	75
76	26,60	20,90	22,16	26,87	21,17	22,45	76
77	26,95	21,18	22,47	27,22	21,45	22,77	77
78	27,30	21,45	22,78	27,58	21,73	23,08	78
79	27,65	21,73	23,09	27,93	22,01	23,39	79
80	28,00	22,00	23,40	28,29	22,29	23,71	80

	$C_{10}H_{18}O$			$C_{10}H_{20}O$			
Sap. No.	Acetate	Alcohol	Alcohol in the orig. oil	Acetate	Alcohol	Alcohol in the orig. oil	Sap. No.
81	28,35	22 28	23.72	28,64	22,56	24,02	81
82	28,70	22,55	24,03	28,99	22,84	24,34	82
83	29,05	22,83	24,34	29,35	23,12	24,66	83
84	29,40	23,10	24,65	29,70	23,40	24,97	84
85	29,75	23,38	24,97	30,05	23,68	25,29	85
86	30,10	23,65	25,28	30,41	23.96	25,61	86
87	30,45	23,93	25,60	30,76	24,24	25,93	87
88	30,80	24,20	25,91	31,11	24.51	26,25	88
89	31,15	24,48	26,23	31.47	24,79	26,57	89
90	31,50	24,75	26,54	31,82	25,07	26,89	90
91	31,85	25,03	26,86	32,17	25,35	27.21	91
92	32,20	25,30	27,18	32,53	25,63	27,53	92
93	32,55	25,58	27,49	32,88	25,91	27,85	93
94	32,90	25,85	27,81	33,24	26,19	28,17	94
95	33,25	26,13	28,13	33,59	26,46	28,49	95
96	33,60	26,40	28,45	33,94	26,74	28,82	96
97	33,95	26,68	28,77	34,30	27,02	29,14	97
98	34,30	26,95	29,09	34,65	27,30	29,47	98
99	34,65	27,23	29,41	35,00	27,58	29,79	99
100	35,00	27,50	29,73	35,36	27,86	30,11	100
101	35.35	27,78	30,05	35,71	28,14	30,44	101
102	35,70	28,05	30,37	36,06	28,41	30,77	102
103	36,05	28,33	30,70	36,42	28,69	31,09	103
104	36,40	28,60	31,02	36,77	28,97	31,42	104
105	36,75	28.88	31,34	37,12	29,25	31.75	105
106	37,10	29,15	31,67	37,48	29,53	32,08	106
107	37.45	29,43	31,99	37,83	29,81	32,41	107
108	37,80	29,70	32,32	38,19	30,09	32,74	108
109	38,15	29,98	32.64	38,54	30,36	33.07	109
110	38,50	30,25	32,97	38,89	30,64	33,40	110
111	38,85	30,53	33,30	39,25	30,92	33.73	111
112	39,20	30,80	33,62	39,60	31,20	34,06	112
113	39,55	31,08	33,95	39,95	31,48	34,39	113
114	39,90	31.35	34,28	40,31	31,76	34,73	114
115	40,25	31,63	34,61	40,66	32,04	35,06	115
116	40,60	31,90	34,94	41,01	32,31	35,39	116
117	40,95	32,18	35,27	41,37	32,59	35,73	117
118	41,30	32,45	35,60	41,72	32,87	36,06	118
119	41,65	32,73	35.93	42,07	33,15	36.40	119
120	42,00	33,00	36,26	42,43	33,43	36,73	120

	$C_{10}H_{18}O$			$C_{10}H_{20}O$			
Sap. No.	Acetate	Alcohol	Alcohol in the orig. oil	Acetate	Alcohol	Alcohol in the orig. oil	Sap. No.
121	42,35	33,28	36,60	42,78	33,71	37,07	121
122	42,70	33,55	36,93	43,14	33,99	37,41	122
123	43,05	33,83	37,26	43,49	34,26	37,75	123
124	43,40	34,10	37,60	43,84	34,54	38,08	124
125	43,75	34,38	37,93	44,20	34,82	38,42	125
126	44,10	34,65	38,27	44,55	35,10	38,76	126
127	44,45	34,93	38,60	44,90	35,38	39,10	127
128	44,80	35,20	38,94	45,26	35,66	39,44	128
129	45,15	35,48	39,27	45,61	35,94	39,78	129
130	45,50	35,75	39,61	45,96	36,21	40,13	130
131	45,85	36,03	39,95	46,32	36.49	40,47	131
132	46,20	36,30	40,29	46,67	36,77	40,81	132
133	46,55	36,58	40,63	47,02	37,05	41,16	133
134	46,90	36,85	40,97	47,38	37,33	41,50	134
135	47,25	37,13	41,31	47,73	37,61	41,84	135
136	47,60	37,40	41,65	48,09	37,89	42,19	136
137	47,95	37,68	41,99	48,44	38,16	42,53	137
138	48,30	37,95	42,33	48,79	38,44	42,88	138
139	48.65	38,23	42,67	49,15	38 72	43,23	139
140	49,00	38,50	43,02	49,50	39,00	43,58	140
141	49,35	38,78	43,36	49,85	39,28	43,92	141
142	49,70	39,05	43,71	50,21	39,56	44.27	142
143	50,05	39,33	44,05	50,56	39,84	44,62	143
144	50,40	39,60	44,39	50,91	40,11	44,97	144
145	50,75	39,88	44,74	51,27	40,39	45,32	145
146	51,10	40,15	45,09	51,62	40,67	45,67	146
147	51,45	40,43	45,44	51,97	40,95	46,02	147
148	51,80	40,70	45,78	52,33	41,23	46,38	148
149	52,15	40,98	46,13	52,68	41,51	46,73	149
150	52,50	41,25	46,48	53,04	41,79	47,08	150
151	52,85	41,53	46,83	53,39	42,06	47,44	151
152	53,20	41,80	47,18	53,74	42,34	47,79	152
153	53,55	42,08	47,53	54,10	42,62	48,15	153
154	53,90	42,35	47,88	54,45	42,90	48,50	154
155	54,25	42,63	48.23	54.80	43,18	48,86	155
156	54,60	42,90	48,58	55,16	43,46	49,21	156
157	54,95	43,18	48,94	55,51	43,74	49,57	157
158	55,30	43,45	49,29	55,86	44,01	49,93	158
159	55,65	43,73	49,65	56,22	44,29	50,29	159
160	56,00	44,00	50,00	56,57	44,57	50,65	160

	$C_{10}H_{18}O$			$C_{10}H_{20}O$			
Sap. No.	Acetate	Alcohol	Alcohol in the orig. oil	Acetate	Alcohol	Alcohol in the orig. oil	Sap. No.
161	56,35	44,28	50,36	56,92	44,85	51,01	161
162	56,70	44,55	50,71	57,28	45,13	51,37	162
163	57,05	44,83	51,07	57,63	45,41	51,73	163
164	57,40	45,10	51,42	57,99	45,69	52,09	164
165	57,75	45,38	51,78	58,34	45,96	52,46	165
166	58,10	45,65	52,14	58,69	46,24	52,82	166
167	58,45	45,93	52,50	59,05	46,52	53,18	167
168	58,80	46,20	52,86	59,40	46,80	53,55	168
169	59,15	46,48	53,22	59,75	47,08	53,91	169
170	59,50	46,75	53,58	60,11	47,36	54,28	170
171	59,85	47,03	53,94	60,46	47,64	54,64	171
172	60,20	47,30	54,31	60,81	47,91	55,01	172
173	60,55	47,58	54,67	61,17	48,19	55,38	173
174	60,90	47,85	55,03	61,52	48,47	55,75	174
175	61,25	48,13	55,40	61,87	48,75	56,12	175
176	61,60	48,40	55,76	62,23	49,03	56,48	176
177	61,95	48,68	56,13	62,58	49,31	56,85	177
178	62,30	48,95	56,49	62,94	49,59	57,23	178
179	62,65	49,23	56,86	63,29	49,86	57,60	179
180	63,00	49,50	57,22	63,64	50,14	57,97	180
181	63,35	49,78	57,59	64,00	50,42	58,34	181
182	63,70	50,05	57,96	64,35	50,70	58,71	182
183	64,05	50,33	58,33	64,70	50,98	59,09	183
184	64,40	50,60	58,70	65,06	51,26	59,46	184
185	64,75	50,88	59,07	65,41	51,54	59,84	185
186	65,10	51,15	59,44	65,76	51,81	60,21	186
187	65,45	51,43	59,81	66,12	52,09	60,59	187
188	65,80	51,70	60,19	66,47	52,37	60,97	188
189	66,15	51,98	60,56	66,82	52,65	61,35	189
190	66,50	52,25	60,93	67,18	52,93	61,72	190
191	66,85	52,53	61,31	67,53	53,21	62,10	191
192	67,20	52,80	61,68	67,89	53,49	62,48	192
193	67,55	53,08	62,06	68,24	53,76	62,86	193
194	67,90	53,35	62,43	68,59	54,04	63,24	194
195	68,25	53,63	62,81	68,95	54,32	63,63	195
196	68,60	53,90	63,19	69,30	54,60	64,01	196
197	68,95	54,18	63,57	69,65	54,88	64,39	197
198	69,30	54,45	63,95	70,01	55,16	64,78	198
199	69,65	54,73	64,33	70,36	55,44	65,16	199
200	70,00	55,00	64,71	70,71	55,71	65,55	200

	$C_{10}H_{18}O$			$C_{10}H_{20}O$			
Sap. No.	Acetate	Alcohol	Alcohol in the orig. oil	Acetate	Alcohol	Alcohol in the orig. oil	Sap. No.
201	70,35	55,28	65,09	71,07	55,99	65,93	201
202	70,70	55,55	65,47	71,42	56,27	66,32	202
203	71,05	55,83	65,85	71,77	56,55	66,71	203
204	71,40	56,10	66,23	72,13	56,83	67,09	204
205	71,75	56,38	66,62	72,48	57,11	67,48	205
206	72,10	56,65	67,00	72,84	57,39	67,87	206
207	72,45	56,93	67,39	73,19	57,66	68,26	207
208	72,80	57,20	67,77	73,54	57,94	68,65	208
209	73.15	57,48	68.16	73,90	58,22	69,04	209
210	73,50	57,75	68,55	74,25	58,50	69,44	210
211	73,85	58,03	68,93	74,60	58,78	69,83	211
212	74,20	58,30	69,32	74,96	59,06	70,22	212
213	74,55	58,58	69,71	75,31	59,34	70,62	213
214	74,90	58,85	70,10	75,66	59,61	71,01	214
215	75,25	59,13	70,49	76,02	59,89	71,41	215
216	75,60	59,40	70,88	76,37	60,17	71,80	216
217	75,95	59,68	71,28	76,72	60,45	72,20	217
218	76,30	59,95	71,67	77,08	60,73	72,60	218
219	76,65	60,23	72,06	77,43	61,01	73,00	219
220	77,00	60,50	72,45	77,79	61,29	73,40	220
221	77,35	60,78	72,85	78,14	61,56	73,80	221
222	77,70	61,05	73,25	78,49	61,84	74,20	222
223	78,05	61,33	73,64	78,85	62,12	74,60	223
224	78,40	61.60	74,04	79,20	62,40	75,00	224
225	78,75	61,88	74,44	79,55	62,68	75,40	225
226	79,10	62,15	74,84	79,91	62,96	75,81	226
227	79,45	62,43	75.23	80,26	63,24	76,21	227
228	79,80	62,70	75,63	80,61	63,51	76,62	228
229	80,15	62,98	76,03	80,97	63,79	77,02	229
230	80,50	63,25	76,44	81,32	64,07	77,43	230
231	80,85	63,53	76,84	81,67	64,35	77,83	231
232	81,20	63,80	77,24	82,03	64,63	78,24	232
233	81,55	64,08	77,64	82,38	64,91	78,65	233
234	81,90	64,35	78,05	82,74	65,19	79,06	234
235	82,25	64,63	78,45	83,09	65,46	79,47	235
236	82,60	64,90	78,86	83,44	65.74	79,88	236
237	82,95	65,18	79,27	83,80	66,02	80,29	237
238	83,30	65,45	79,67	84,15	66,30	80,71	238
239	83,65	65,73	80,08	84,50	66,58	81.12	239
240	84,00	66,00	80,49	84,86	66,86	81,53	240

	$C_{10}H_{18}O$			$C_{10}H_{20}O$			
Sap. No.	Acetate	Alcohol	Alcohol in the orig. oil	Acetate	Alcohol	Alcohol in the orig. oil	Sap. No.
241	84,35	66,28	80,90	85,21	67,14	81,95	241
242	84,70	66,55	81,31	85,56	67,41	82,36	242
243	85,05	66,83	81,72	85,92	67,69	82,78	243
244	85,40	67,10	82,13	86,27	67,97	83,20	244
245	85,75	67,38	82,54	86,62	68,25	83,61	245
246	86,10	67,65	82,96	86,98	68,53	84,03	246
247	86,45	67,93	83,37	87,33	68,81	84,45	247
248	86,80	68,20	83,78	87,69	69,09	84,87	248
249	87,15	68,48	84,20	88,04	69,36	85,29	249
250	87,50	68,75	84,62	88,39	69,64	85,71	250
251	87,85	69,03	85,03	88,75	69,92	86,14	251
252	88,20	69,30	85,45	89,10	70,20	86,56	252
253	88,55	69,58	85,87	89,45	70,48	86,98	253
254	88,90	69,85	86,29	89,81	70,76	87,41	254
255	89,25	70,13	86,71	90,16	71,04	87,83	255
256	89,60	70,40	87,13	90,51	71,31	88,26	256
257	89,95	70,68	87,55	90,87	71,59	88,69	257
258	90,30	70,95	87.97	91.22	71,87	89,11	258
259	90,65	71,23	88,40	91,57	72,15	89,54	259
260	91.00	71,50	88,82	91,93	72,43	89,97	260
261	91,35	71,78	89,25	92,28	72,71	90,40	261
262	91,70	72,05	89,67	92,64	72,99	90,83	262
263	92,05	72,33	90,10	92,99	73,26	91,27	263
264	92,40	72,60	90,52	93,34	73,54	91,70	264
265	92,75	72,88	90,95	93,70	73,82	92,13	265
266	93,10	73,15	91,38	94,05	74,10	92,57	266
267	93.45	73,43	91,81	94,40	74,38	93,00	267
268	93,80	73,70	92,24	94,76	74,66	93,44	268
269	94,15	73,98	92,67	95,11	74.94	93,87	269
270	94 50	74,25	93,10	95,46	75,21	94,31	270
271	94.85	74,53	93,54	95,82	75,49	94,75	271
272	95,20	74,80	93,97	96,17	75,77	95,19	272
273	95,55	75,08	94,40	96,52	76,05	95,63	273
274	95,90	75,35	94,84	96,88	76,33	96,07	274
275	96.25	75,63	95,28	97,23	76,61	96,51	275
276	96,60	75,90	95,71	97,59	76,89	96,96	276
277	96,95	76,18	96.15	97,94	77,16	97,40	277
278	97,30	76,45	96,59	98,29	77,44	97,84	278
279	97,65	76,73	97,03	98,65	77.72	98,29	279
280	98,00	77,00	97,47	99,00	78,00	98,73	280

	$C_{10}H_{18}O$			$C_{10}H_{20}O$			
Sap. No.	**Acetate**	**Alcohol**	**Alcohol in the orig. oil**	**Acetate**	**Alcohol**	**Alcohol in the orig. oil**	**Sap. No.**
281	**98,35**	**77,28**	**97,91**	**99,35**	**78,28**	**99.18**	**281**
282	**98.70**	**77,55**	**98.35**	**99,71**	**78,56**	**99.63**	**282**
283	**99.05**	**77,83**	**98.80**	**100,06**	**78,84**	**100.08**	**283**
284	**99,40**	**78,10**	**99.24**	—	—	—	—
285	**99,75**	**78.38**	**99,68**	—	—	—	—
286	**100,10**	**78,65**	**100,13**	—	—	—	—

4. LIST OF PLANTS, ARRANGED ACCORDING TO FAMILIES,[1] FROM WHICH VOLATILE OILS ARE OBTAINED.

Polypodiaceae.

Aspidium filix mas—Oil of male fern.

Pinaceae.

Larix decidua—Oil of larch turpentine, larch needle oil.
Pinus australis, P. taeda, P. cubensis, P. palustris—American turpentine oil.
Pinus pinaster—French turpentine oil.
Pinus laricio—Austrian turpentine oil.
Pinus silvestris—Pine tar oil, and pine needle oil.
Pinus ledebourii—Pine tar oil, Russian turpentine oil, and Siberian pine needle oil.
Pinus khasya, P. merkusii—Burma turpentine oil.
Pinus montana—Oil from needles.
Pinus cembra—Oil from needles.
Pinus sabiniana—Californian turpentine oil.
Picea excelsa—Oil from oleoresin, oil from needles.
Picea alba—Oil of needles.
Picea nigra—Oil of needles.
Abies alba—Pine (white fir) needle oil, oil of cones of white fir, oil from oleoresin.
Abies balsamea—Needle oil.
Abies canadensis—Needle oil. } Oil from Canada balsam.
Abies fraseri—Needle oil.
Abies reginae amaliae—Oil from cones.
Cedrus libani—Lebanon cedar oil.
Sequoia gigantea—Sequoia oil.
Callitris quadrivalis—Sandarac oil.
Thuja occidentalis—Thuja oil.
Thuja orientalis—Oil from the roots.
Cupressus sempervirens—Cypress oil.
Chamaecyparis obtusa—Hinoki oil.
Juniperus communis—Oil of juniper berries.

---

1) According to Engler's Syllabus. Second edition. Berlin, 1898.

Juniperus oxycedrus—Oil of berries.
Juniperus phoenicea—Oil of berries.
Juniperus sabina—Oil of savin.
Juniperus virginiana—Oil of cedar wood (cedar oil), oil of cedar leaves.

PANDANACEAE.

Pandanus odoratissimus—Oil of flowers.

GRAMINEAE.

Andropogon schoenanthus—Palmarosa oil, and gingergrass oil.
Andropogon citratus—Lemongrass oil.
Andropogon muricatus—Vetiver oil.
Andropogon nardus—Citronella oil.
Andropogon odoratus—Oil of herb.
Andropogon laniger—Camel grass oil.

PALMAE.

Serenoa serrulata—Saw palmetta oil.

ARACEAE.

Acorus calamus—Calamus oil, oil from leaves, Japanese calamus oil.

LILIACEAE.

Sabadilla officinalis—Sabadilla oil.
Aloë vulgaris—Aloë oil.
Xanthorrhoea hastile—Xanthorrhoea oil.
Allium sativum—Garlic oil.
Allium cepa—Onion oil.
Allium ursinum—Oil from entire plant.

IRIDACEAE.

Crocus sativus—Saffron oil.
Iris florentina, I. pallida, I. germanica—Orris oil.

ZINGIBERACEAE.

Curcuma longa—Curcuma oil.
Curcuma zedoaria—Zedoary oil.
Kaempferia rotunda—Oil of root.
Hedychium coronarium—Oil of flowers.
Alpinia galanga—Galangal oil.
Alpinia malaccensis—Oil of roots.
Alpinia nutans—Oil of roots.
Zingiber officinale—Ginger oil.
Elettaria cardamomum—Oil of cardamom (Ceylon), Malabar (Madras) cardamom oil.

Amomum cardamomum Siam cardamom oil.
Amomum melegueta—Oil of paradise grains.
Amomum aromaticum—Bengal cardamom oil.
Amomum spec.?—Kamerun cardamom oil.
Amomum angustifolium—Korarima cardamom oil.

Piperaceae.

Piper nigrum—Oil of black pepper.
Piper longum—Oil of long pepper.
Piper ovatum—Oil of leaves.
Piper lowong—Oil of fruit.
Piper clusii—Aschanti pepper oil.
Piper cubeba—Oil of cubebs.
Piper angustifolium—Matico oil.
Piper betle—Oil of betle leaves.
Potomorphe umbellata—Oil of leaves.
Artanthe geniculata—Oil of leaves.
Ottonia anisum—Oil of roots.

Salicaceae.

Populus nigra—Oil of buds.

Myricaceae.

Myrica gale—Oil of Dutch myrtle.
Myrica cerifera—Oil of bay-berry.
Myrica asplenifolia—Oil of sweet fern.

Juglandaceae.

Juglans regia—Walnut leaf oil.

Betulaceae.

Betula lenta—Oil of sweet birch (wintergreen oil).

Moraceae.

Humulus lupulus—Oil of hops.
Cannabis sativa—Hemp oil.

Santalaceae.

Santalum album—Oil of sandalwood (East Indian).
Santalum preissianum—South Australian sandalwood oil.
Santalum cygnorum—West Australian sandalwood oil.
Santalum yasi—Fiji sandalwood oil.
Unknown species—African sandalwood oil.

Aristolochiaceae.

Asarum europaeum—Oil of root.
Asarum canadense—Oil of Canada snake root (wild ginger).

Aristolochia serpentaria—Oil of Virginia snake root.
Aristolochia clematitis—Oil of roots.

CHENOPODIACEAE.

Chenopodium ambrosioides* var. *anthelminticum—Oil of American wormseed.

RANUNCULACEAE.

Paeonia moutan—Oil of roots.
Nigella sativa—Oil of seeds.
Nigella damascena—Oil of seeds.

MAGNOLIACEAE.

Michelia champaca—Champaca oil.
Michelia longifolia—Oil of flowers.
Illicium verum—Star anise oil.
Illicium anisatum—Japanese star anise oil.
Drimys winteri—Oil of Winter's bark.

ANONACEAE.

Cananga odorata—Ylang ylang oil and cananga oil.

MYRISTICACEAE.

Myristica fragrans—Oil of mace, oil of nutmeg.

MONIMIACEAE.

Peumus boldus—Oil of boldo leaves.
Atherosperma moschata—Atherosperma oil.
Citriosma oligandra—Oil of leaves and bark.
Citriosma cujabana—Oil of leaves and bark.
Citriosma apiosyce—Oil of leaves and bark.
Unknown species—Oil of para coto bark.

LAURACEAE.

Cinnamomum camphora—Camphor oil.
Cinnamomum zeylanicum—Oil of cinnamon (Ceylon), oil of cinnamon leaves, oil of roots.
Cinnamomum cassia—Cassia oil.
Cinnamomum loureirii—Japanese cinnamon oil.
Cinnamomum kiamis—Oil of bark.
Cinnamomum culilawan—Culilawan bark oil.
Cinnamomum wightii—Oil of bark.
Cinnamomum oliveri—Oil of bark.
Persea gratissima—Oil of leaves.
Persea caryophyllata—Oil of clove bark.
Nectandra puchury—Pichurim oil.

Nectandra caparrapi—Caparrapi oil.
Ocotea caudata—Guayana lignaloe oil.
Ocotea species?—Ocotea oil.
Nectandra myriantha.
Nectandra or *Ocotea* species?—Venezuela camphor wood oil.
Sassafras officinale—Sassafras oil (bark), oil of sassafras leaves.
Cryptocaria moschata—Cryptocaria oil.
Cryptocaria pretiosa—Oil of bark.
Laurus nobilis—Oil of laurel leaves, oil of fruit.
Lindera sericea—Kuromoji oil.
Oreodaphne californica—Oil of mountain laurel.
Benzoin odoriferum—Spicewood oil.
Tetranthera citrata—Tetranthera oil.
Unknown species—Massoy bark oil.

Cruciferae.

Lepidium sativum—Pepper-grass oil.
Thlaspi arvense—Oil of herb.
Cochlearia officinalis—Oil of spoonwort.
Cochlearia armoracia—Oil of horseradish.
Alliaria officinalis—Oil of hedge garlic.
Brassica nigra, *B. juncea*—Oil of mustard.
Sinapis alba—Oil of white mustard.
Nasturtium officinale—Oil of watercress.
Raphanus sativus and *R. niger*—Oil of common radish.

Resedaceae.

Reseda odorata—Mignonette oil, oil of root.

Hamamelidaceae.

Liquidambar orientale—Oil of liquid storax.
Liquidambar styracifluum—Oil of sweet gum, oil of leaves.
Altingia excelsa—Rasamala oil (wood).

Rosaceae.

Spiraea ulmaria—Spiraea oil.
Rosa damascena—Rose oil.
Rosa centifolia—Rose oil.
Prunus amygdalus, *P. armeniaca*, *P. persica*—Bitter almond oil.
Prunus laurocerasus—Cherry laurel oil.
Prunus virginiana—Oil of wild cherry bark.

Leguminosae.

Copaifera officinalis C. guajanensis, *C. coriacea*, *C. langsdorffia*, *C. confertiflora*, *C. oblongifolia*, *C. rigida*—Copaiba balsam oil.
Myrocarpus fastigiatus—Cabriuva wood oil.

Genista tridentata—Carqueja oil.
Indigofera galegoides—Oil of leaves.
Caesalpinia sappan—Sappan oil.
Toluifera balsamum—Tolu balsam oil.
Myroxylon peruiferum—Oil of leaves.
Glycyrrhiza glabra—Oil from root.
Glycyrrhiza glandulifera—Oil from root.

Geraniaceae.

Pelargonium odoratissimum, *P. capitatum*, *P. roseum*—Oil of rose geranium.

Tropaeolaceae.

Tropaeolum majus—Oil of nasturtium.

Erythroxylaceae.

Erythroxylon coca—Oil of coca leaves.

Zygophyllaceae.

Bulnesia sarmienti—Guaiac wood oil.

Rutaceae.

Xanthoxylum piperitum—Japanese pepper oil.
Xanthoxylum hamiltonianum—Oil of seeds.
Ruta graveolens—Oil of rue.
Boronia polygalifolia—Boronia oil.
Barosma betulina, *B. serratifolia*—Oil of buchu leaves.
Empleurum serrulatum—Oil of leaves.
Pilocarpus jaborandi—Jaborandi oil.
Cusparia trifoliata—Oil of angustura bark.
Toddalia aculeata—Toddalia oil.
Citrus limonum—Oil of lemon.
Citrus aurantium—Oil of sweet orange, oil of nerolf (Portugal).
Citrus bigaradia—Oil of bitter orange, oil of neroli, oil of petit grain.
Citrus bergamia—Oil of bergamot.
Citrus medica—Cedrus oil.
Citrus medica var. *acida*—West Indian limette oil.
Citrus limetta—Italian limette oil, oil of leaves.
Citrus nobilis—Oil of mandarins.
Citrus decumana—Grape fruit oil.
Schimmelia oleifera—West Indian sandalwood oil.
Dacryodes hexandra—Oil from oleoresin.

Burseraceae.

Commiphora abyssinica, *C. schimperi*—Oil of myrrh
Balsamodendron kafal—Opoponax oil.
Boswellia carteri—Oil of frankincense.

Canarium spec.?—Elemi oil.
Icica heptaphylla—Conima resin oil.
Bursera aloexylon—(Mexican) lignaloe oil.

Meliaceae.

Cedrela odorata and other species—Cedrela wood oil.

Polygalaceae.

Polygala senega—Oil of senega root.
Polygala variabilis, P. oleifera, P. calcarea, P. depressa, P. nemorivaga—Methyl salicylate.

Euphorbiaceae.

Croton eluteria—Cascarilla oil.
Stillingia silvatica—Stillingia oil.

Anacardiaceae.

Pistacia lentiscus—Mastic oil.
Pistacia terebinthus—Chios turpentine oil.
Schinus molle—Schinus oil.

Vitaceae.

Vitis vinifera—Cognac oil.

Tiliaceae.

Tilia ulmifolia, T. platyphyllos—Oil of linden flowers.

Malvaceae.

Hibiscus abelmoschus—Oil of ambrette seeds.

Theaceae.

Thea chinensis—Tea oil.

Dipterocarpaceae.

Dryobalanops camphora—Borneo camphor oil.
Dipterocarpus turbinatus and others—Gurjun balsam oil.

Cistaceae.

Cistus creticus, C. ladaniferus—Labdanum oil.

Canellaceae.

Canella alba—Canella oil.

Turneraceae.

Turnera aphrodisiaca, T. diffusa—Damiana leaf oil.

Lythraceae.

Lawsonia inermis—Henna oil.

Myrtaceae.

Myrtus communis—Oil of myrtle.
Myrtus cheken—Cheken leaf oil.
Pimenta officinalis—Oil of pimenta.
Pimenta acris—Oil of bay.
Eugenia caryophyllata—Oil of cloves, oil of clove stems.
Melaleuca leucadendron—Cajeput oil.
Melaleuca viridiflora—Niaouli oil.
Melaleuca leucadendron var. *lancifolia*, *M. acuminata*, *M. decussata*, *M. ericifolia*, *M. genistifolia*, *M. linariifolia*, *M. squarrosa*, *M. uncinata*, *M. wilsonii*—contain in the leaves cajeput-like oils.
Eucalyptus globulus, *E. odorata*, *E. cneorifolia*, *E. oleosa*, *E. dumosa*, *E. amygdalina*, *E. rostrata*, *E. populifera*, *E. corymbosa*, *E. resinifera*, *E. baileyana*, *E. microcorys*, *E. risdonia*, *E. leucoxylon*, *E. hemiphloia*, *E. crebra*, *E. macrorrhyncha*, *E. capitellata*, *E. eugenioides*, *E. obliqua*, *E. punctata*, *E. loxophleba*, *E. dextropinea*, *E. laevopinea*, *E. maculata*, *E. citriodora*, *E. dealbata*, *E. planchoniana*, *E. staigeriana*, *E. haemastoma*, *E. piperita*, *E. diversicolor*, *E. fissilis*, *E. goniocalyx*, *E. gracilis*, *E. lehmanni*, *E. longifolia*, *E. occidentalis*, *E. pauciflora*, *E. stuartiana*, *E. terreticornis*, *E. tessellaris*—yield the various eucalyptus oils.
Backhousia citriodora—Backhousia oil.

Araliaceae.

Aralia nudicaulis—Oil of rhizome.

Umbelliferae.

Coriandrum sativum—Coriander oil.
Cuminum cyminum—Cumin oil.
Apium graveolens—Oil of celery seed, oil of herb.
Petroselinum sativum—Parsley oil, oil of root, oil of herb
Cicuta virosa—Cicuta oil.
Cicuta maculata—Oil of fruit.
Carum carvi—Caraway oil.
Carum ajowan—Ajowan oil.
Pimpinella anisum—Anise oil.
Pimpinella saxifraga—Oil of root.
Pimpinella nigra—Oil of root.
Foeniculum vulgare—Fennel oil.
Meum athamanticum—Oil of root.
Silaus pratensis—Oil of fruit.
Oenanthe aquatica—Oil of water fennel.
Levisticum officinale—Oil of lovage (root), oil of seed, oil of herb.
Archangelica officinalis—Oil of angelica (root), oil of seed, oil of herb.

Angelica refracta (anomala?)—Japanese angelica root oil.
Ferula asa foetida—Oil of asafetida.
Ferula rubricaulis—Galbanum oil.
Ferula sumbul—Oil of sumbul.
Dorema ammoniacum—Ammoniac oil.
Peucedanum oreoselinum—Oil of fresh herb.
Peucedanum ostruthium—Oil of root.
Peucedanum graveolens—Dill oil.
Anethum sowa—East Indian dill oil.
Peucedanum sativum—Pastinaca oil.
Peucedanum grande—Oil of fruit.
Peucedanum officinale—Oil of root.
Heracleum sphondylium—Oil of fruit.
Heracleum giganteum—Oil of fruit.
Daucus carota—Oil of carrots.
Osmorrhiza longistylis—Oil of root.

PIROLACEAE.

Monotropa hypopitys—Oil of stem.

ERICACEAE.

Ledum palustre—Oil of Labrador tea.
Gaultheria procumbens—Oil of wintergreen.
Gaultheria punctata—Oil of leaves.
Gaultheria leucocarpa—Oil of leaves.

PRIMULACEAE.

Primula veris—Oil of root.

CONVOLVULACEAE.

Convolvulus scoparia, C. floridus—Oil of rhodium.

VERBENACEAE.

Verbena triphylla—Verbena oil.
Lantana camara—Oil of herb.
Vitex trifolia—Oil of leaves.

LABIATAE.

Rosmarinus officinalis—Oil of rosemary.
Lavandula vera—Oil of lavender.
Lavandula spica—Oil of spike.
Lavandula stoechas—Oil of herb.
Lavandula dentata—Oil of herb.
Lavandula pedunculata—Oil of herb.
Nepeta cataria—Oil of catnep.
Nepeta glechoma—Oil of gill-over-the-ground.
Salvia officinalis—Oil of sage.

Salvia sclarea—Oil of herb.
Monarda punctata—Horsemint oil.
Monarda didyma—Oswego tea oil.
Monarda fistulosa—Wild bergamot oil.
Melissa officinalis—Balm oil.
Hedeoma pulegioides—Pennyroyal oil.
Hyssopus officinalis—Hyssop oil.
Satureja hortensis—Oil of summer savory.
Satureja montana—Oil of winter savory.
Satureja thymbra—Oil of herb.
Origanum vulgare—Oil of wild marjoram.
Origanum majorana—Oil of sweet marjoram.
Origanum hirtum, *O. smyrnaeum* and others—Oil of Cretian origanum.
Thymus vulgaris—Oil of thyme.
Thymus serpyllum—Oil of wild thyme.
Thymus capitatus—Oil of herb.
Lycopus virginicus—Oil of bugle weed.
Mentha piperita, *M. arvensis* var. *piperascens*—Oil of peppermint.
Mentha silvestris var. *crispa*, *M. viridis*—Oil of spearmint.
Mentha aquatica—Oil of herb.
Mentha arvensis—Oil of herb.
Mentha canadensis—Oil of Canada mint.
Mentha pulegium—Oil of European pennyroyal.
Pogostemon patchouly—Patchouly oil.
Pogostemon comosus—Oil of leaves.
Ocimum basilicum—Oil of sweet basil.
Morula japonica—Oil of herb.
Cunila mariana—Oil of herb.
Lophantus anisatus—Oil of herb.
Pycnanthemum lanceolatum—Oil of mountain mint.
Pycnanthemum incanum—Oil of herb.

SOLANACEAE.

Fabiana imbricata—Oil from leaves.

CAPRIFOLIACEAE.

Sambucus nigra—Oil of flowers.

VALERIANACEAE.

Valeriana officinalis—Oil of valerian.
Valeriana officinalis var. *angustifolia*—Kesso oil.
Valeriana celtica—Oil of root.
Nardostachys jatamansi—Oil of root.

COMPOSITAE.

Eupatorium foeniculaceum—Oil of entire plant.
Ageratum conyzoides—Oil of herb.

Solidago odora—Golden rod oil.
Solidago rugosa—Oil of herb.
Erigeron canadensis—Oil of flea bane.
Blumea balsamifera, Bl. lacera—Blumea camphor.
Helichrysum stoechas—Oil of herb.
Inula helenium—Oil of elecampane.
Osmitopas asteriscoides—Oil of herb.
Ambrosia artemisiaefolia—Oil of rag weed.
Anthemis nobilis—Oil of Roman chamomile.
Anthemis cotula—Oil of mayweed or dog fennel.
Achillea millefolium—Oil of milfoil.
Achillea nobilis—Oil of herb.
Achillea moschata—Oil of flowering herb.
Achillea coronopifolia—Oil of herb.
Achillea ageratum—Oil of herb.
Matricaria chamomilla—Oil of German chamomile.
Matricaria parthenium—Oil of flowering plant.
Tanacetum vulgare—Oil of tansy.
Tanacetum balsamita—Oil of herb.
Pyrethrum indicum—Kiku oil.
Artemisia vulgaris—Oil of mug-wort.
Artemisia dracunculus—Esdragon oil.
Artemisia cina—Wormseed oil.
Artemisia absinthium—Oil of wormwood.
Artemisia gallica—Oil of herb.
Artemisia barrellieri—Oil of herb.
Artemisia glacialis—Oil of herb.
Erechthites hieracifolia—Oil of fire weed.
Arnica montana—Oil of arnica flowers, oil of arnica root.
Saussurea lappa—Oil of root.
Carlina acaulis—Oil of root.
Spaeranthus indicus—Oil of root.

SPECIAL PART.

1. Oil of Male Fern.

ORIGIN. The anthelmintic properties of the rhizome of male fern, *Aspidium filix mas* Sw., are said to be due in part to small quantities of a volatile oil, which was first prepared by Bock[1] in 1851 by distillation with water vapor. The yield varies according to the season in which the rhizome is collected. Ehrenberg[2] in 1893 obtained from air-dried rhizome freshly collected in June 0.025 p. c.; from rhizome collected in the months from September to November he obtained 0.04 to 0.045 p. c. of volatile oil respectively.

PROPERTIES. The volatile oil of male fern is a light yellow liquid, with an intense male fern odor, and an aromatic, afterwards burning taste. It is readily soluble in ether and absolute alcohol. Its specific gravity lies between 0.85 and 0.86. Most of the oil boils between 140 and 250°. Above this temperature decomposition takes place and dark colored products distil over.

COMPOSITION.[2] Besides free fatty acids (principally butyric acid) oil of male fern contains the hexyl and octyl esters of the fatty acids from butyric acid upwards, probably up to pelargonic acid. The substances to which the peculiar physiological action is due have not yet been isolated.

2. Oil of Angiopteris Evecta.

This fern, *Angiopteris evecta* Hoffm., yields an aromatic oil, said to be used in the South Sea Islands for perfuming cocoanut oil (Maiden[3]).

3. Oil of Polypodium Phymatodes.

Polypodium phymatodes L. (*Pleopeltis phymatodes* G. Moore) like the above fern yields an oil also said to be used in the South Sea Islands for perfuming cocoanut oil.[3]

---

1) Arch. d. Pharm., 115, p. 262.
2) Arch. d. Pharm., 281, p. 845.
3) Useful native plants of Austr., p. 258.

OILS OF THE ABIETINEÆ.

Turpentine Oil, Pine Tar Oil ("Kienöl"), and the Pine Needle Oils ("Fichtennadelöl").

The term turpentine oil is applied in a more restricted sense to the product obtained by distillation of turpentine with water vapor. "Kienöl" is the product obtained by the dry distillation of coniferous roots rich in resin. The name turpentine oil, in a more general sense, is sometimes applied to these oils also, but this use of the term should be avoided. The oils obtained by the distillation with water vapor of the leaves or one year old cones of various conifers are grouped under the collective term of pine needle oils.

As to chemical composition, the oily distillates of the turpentine, the wood, and the leaves have in common one characteristic constituent, viz. pinene. The turpentine oils proper consist almost exclusively, the pine tar oils very largely of this hydrocarbon; whereas in the pine needle oils the place of the pinene is frequently taken by limonene or oxygenated substances, notably bornyl acetate.

Turpentine Oils Proper.

HISTORY. The oils obtained by the distillation of the oleoresins of various *Abietineæ* were known to the ancients as cedar oil (πισσέλαιον)[1] and later became known as turpentine oil. The oil as well as the resin, the colophonium, were used by seafaring people. Taking into consideration the perfection early reached by the Chinese and Japanese in the preparation of varnishes and lacquers, it may be supposed that coniferous oils were distilled and used by them. However this may be, the oils of the *Abietineæ* obtained in a crude manner have evidently been the first volatile oils that found commercial use and technical application.

The name turpentine oil seems to have been introduced during the period of Greek civilization. Like the older synonyms (cedar oil, etc.) it apparently was used as a collective term. It is of Persian origin,[2] and may have been derived from the Cyprian species *Pistacia terebinthus* L.

As far as is known to history, the preparation of turpentine oil probably had its origin in the Caucausus and its south-western

---

[1]) Herodoti, Historiae, Lib. II, p. 85.—Dioscorides, De mat. med., Lib. I. p. 84. Editio Kühn-Sprengel, 1829, 1, p. 98. — Plinius, Naturalis historiae libri., Lib. 15, cap. 6—7; and Lib. 16, cap. 22.

[2]) Flückiger, Pharmacognosie, p. 77.

spurs. In central Europe it became known during the middle ages, somewhat later also in northern Europe. The North American industry had its origin in the dense and extensive pine forests of the South Atlantic States and developed in the beginning of the eighteenth century especially in Virginia and Carolina.[1]

Inasmuch as the crude turpentine oil found little or no use in either household economy or in religious rites, it is but seldom mentioned in early literature. Attention has already been called on pp. 5, 20, 21, 22 and 51 to mentionings by the older writers. Since its introduction into medicine, the mediaeval works on distillation and materia medica make mention of the oil. In addition to the references by Villanovus and Lullus, who lived in the thirteenth century, to which attention has been called on pp. 20 and 21, mention is made of oil of turpentine by the following writers of the fifteenth century: Saladinus of Asculo[2] and the canon Johann of Santo Amando of Doornyk;[3] during the sixteenth century by Walter Ryff,[4] Conrad Gesner,[5] Joh. Baptista Porta,[6] Valerius Cordus[7] and Adolphus Occo.[8]

Attention has already been called to the synonymous usage during the seventeenth century of the designations of alcohol and turpentine as *aqua ardens* and *spiritus*. The name *Spiritus terebinthinae* has maintained itself as a popular term up to this time. As *huile aetherée* it seems to have been first designated in the year 1700.

The early observations made in connection with oil of turpentine concerned its behavior at low temperatures. As early as 1794 Margueron[9] claims to have observed that the oil, when reduced to a temperature of —22° R. solidifies to a crystalline mass. Crystals had already been observed by Geoffroy in 1727 in the neck of the retort while distilling the oil. In conformity with the practice of designating as camphor all solid substances separating from volatile oils, these needle-like crystals, presumably pinol hydrate, were called turpentine camphor.

---

1) Peter Kalm's Reise, etc., vol. 2, pp. 418, 556; vol. 3, pp. 298, 305, 528.
Schöpf, Reise, etc., vol. 2, pp. 220, 228, 278.
Michaux, Histoire des arbres, etc.
2) Saladini Asculani, Compendium, etc.
3) Expositio Janis de Santo Amando.
4) Ryff, Neu gross Destillirbuch, etc., fol. 180.
5) Ein kostlicher theurer Schatz Euonymi Philiatri, vol. 1, p. 288.
6) Gio. Batt. Portae, Magiae naturalis.
7) Dispensatorium Noricum. 1546.
8) Pharmacopoea pro Republica Augustana. 1564.
9) Journ. de Chim. et de Phys., 2, p. 178; Crell's Chem. Annal., II, pp. 195, 310 and 430.

While making the socalled *Liquor antarthriticus Pottii,* in the preparation of which hydrogen chloride is passed into turpentine oil, the apothecary Kind,[1] of Eutin, in 1803 obtained a solid crystalline mass[2] which he considered to be artificially prepared camphor. This compound was examined by Gehlen[3] in 1819 and by Dumas[4] in 1833. The first elementary analysis of the oil was made by Houton-Labilliardière[5] in 1817. In the same year this oil served as the first volatile oil of which the angle of rotation was ascertained.

Origin. The various species of *Pinus, Picea* and *Abies,* of the family *Pinaceae,* growing in the temperate zones, contain in special ducts under the bark an oleoresin, which in some species collects in cavities. If the cambium under the bark is wounded, or if the cavities are tapped, the oleoresin exudes, sometimes turbid, sometimes clear. This oleoresin, which since antiquity has been known as turpentine, is a solution of resin in volatile oil. This solution is mostly rendered turbid by the intimate admixture of aqueous sap. Even when exposed to the air it mostly remains turbid and frequently congeals to a granular or crystalline, honey-like mass.

This turpentine, when distilled either by itself or with water vapor, yields the turpentine oil; and there remains a residue which, when remolten and strained, constitutes rosin or colophony.

In former centuries, oil of turpentine—most likely in part also cedar oil—was distilled in the eastern countries bordering on the Mediterranean, especially in Asia Minor. When, with the development of the arts and industries as well as with that of navigation, it became more important and found wider application, greater advantage was taken of the coniferous forests in the more northern European countries. The principal producers were, and are in part to-day, the central European mountainous sections, rich in conifers, from Hungary and Galicia to Spain and Portugal; also the large coniferous forests of Russia, of eastern Germany, and of the southern provinces of the Scandinavian countries. In the course of the previous century, North America has been added with its apparently inexhaustible coniferous forests, and with its turpentine industry that surpasses that of all other countries. In spite of the economic waste of two centuries, America still is and, no doubt will continue to be the principal producer of these articles of commerce.

---

1) Trommsdorff's Journ. der Pharm., 11ii, p. 132.

2) Pinene monohydrochloride, $C_{10}H_{17}Cl$.

3) Gehlen's allgem. Journ. f. d. Chem., 6, p. 462.

4) Ann. de Chim. et Phys., II, 52, p. 400; Liebig's Annalen, 9, p. 56.

5) Journ. de Pharmacie, II, 4, p. 5.

With reference to geographical origin, the American and French oils only come into consideration in the world's commerce. The other commercial brands have only local significance.

GENERAL PROPERTIES. Freshly distilled turpentine oil is a colorless, limpid liquid of a peculiar odor which varies somewhat according to its source. The French oil, e. g. reminds somewhat of juniper and has a finer and milder odor than the American oil with its decided terebinthinate odor. The sharp odor of old oil is said to be due to an aldehyde,[1] $C_{10}H_{16}O_3$, which is supposed to be formed by the exposure of the oil to the oxygen of the air.

Turpentine oil is volatile at ordinary temperature. During evaporation, a part of the oil resinifies, taking up oxygen. The residue, which is sticky at first, gradually takes on the consistency of rosin.

Owing to the presence of traces of free formic and acetic acids, crude oil of turpentine has a faint acid reaction. For this reason it has to be rectified with milk of lime for certain purposes. When air is not sufficiently excluded, oxidation products with an acid reaction are soon reformed.

Of physiological interest is the fact that turpentine oil, when taken internally, or when the vapors are but inhaled, imparts to the urine a peculiar violet odor. This peculiarity is shared by all pinene-containing oils. Other terpenes do not possess this property. The inhalation of the vapors of turpentine oil for a longer period produces an unpleasant affection of the kidneys known as painter's disease.

SOLUBILITY. Turpentine oil is difficultly soluble in alcohol, especially in dilute alcohol. Owing to this property, it can often be readily recognized when added to other oils as an adulterant. In the course of time, however, the solubility of turpentine oil in alcohol undergoes considerable change. Whereas the solubility of most oils diminishes with age, that of turpentine oil increases.[2] To effect a clear solution of freshly distilled oil, more alcohol is necessary, than for an old oil which has been exposed to the air. The plausible explanation for this observation lies in the formation of more readily soluble oxygenated compounds.

---

[1]) If, according to Schiff (1896), the oil from which the aldehyde has been removed by shaking with sodium acid sulphite solution, be distilled in an atmosphere of carbon dioxide, an almost odorless product is obtained. The odor, however, is again rapidly developed when the oil is exposed to the air. Chem. Ztg., 20, p. 361.

[2]) A normal American turpentine oil which originally was soluble in not less than 6 p. of 90 p. c. alcohol, after standing seven weeks gave a clear solution with 3 p.; French oil, which had been standing for four years in a bottle that was not completely filled, gave a clear solution with but 1 p. of 90 p. c. alcohol.

On account of this changeability, little stress is to be laid on the solubility as a test. In general, a good turpentine oil will be soluble in from 5 to 12 parts of 90 p. c. alcohol.

Solubility of turpentine oil in alcohol of different strength according to Ledermann and Godeffroy.[1]

Variety of Oil.	Strength of alcohol in percentage by volume				
	70	80	85	90	95
	parts requisite to effect solution:				
1. French, crude	66	18	14	7	2
2. French, rectified	80	16—17	12	6.5—7	2—2.4
3. American, crude	56	20	12	5	2
4. American, rectified	60—64	17—19	12—14	5—6	2.2
5. Austrian, crude	—	—	—	6	—
6. Austrian, rectified	—	—	13	8	3
7. Polish ("Kienöl")	—	—	—	5	—
8. Russian ("Kienöl")	49	16	11	5.6	2

Turpentine oil is soluble in almost any proportion in ether, chloroform, carbon disulphide, benzene, petroleum ether, glacial acetic acid, also in the fatty oils. When mixed with other volatile oils, turbidity occasionally results. Turpentine oil is an excellent solvent for fats, resins and most varieties of caoutchouc.

Boiling Point. By far the greater portion of turpentine oil, namely from 75—80 p. c., boils between 155 and 162°. Above 162° the mercury rises rapidly and there remains in the flask a soft, colophony-like resin. Concerning the quantities that distill over at the various temperatures, also the specific gravity and rotatory power of these fractions of rectified American turpentine oil, Kremers[2] has recently made a detailed report.

On account of the oxidation and polymerization products formed, the boiling point of a resinified oil lies considerably higher. Of the pine tar oils also a larger portion boils above 162°, because of the presence of dipentene and sylvestrene.

Rotatory Power. An important distinctive feature of the several oil varieties is their different rotatory power. French oil is always strongly laevogyrate, American oil mostly faintly dextrogyrate and but seldom laevogyrate. Austrian oils have been found to be dextro- as well as laevogyrate. The angle of rotation is given in connection with the respective oils.

---

1) Zeitschr. d. allg. öst. Apt.-Ver., 15, p. 381; Jahresber. f. Pharm., 1877. p. 394.
2) Pharm. Review, 15, p. 7.

INFLUENCE OF AIR AND LIGHT. It is a well-known fact that turpentine oil when allowed to stand in open vessels undergoes rapid change, especially if water is present. The oil becomes viscid, the specific gravity increases and the boiling point rises, the solubility in 90 p. c. alcohol increases, the originally neutral oil becomes acid, resinifies, it becomes "rancid" as this change is technically designated.[1] Formerly such an oil was termed ozonized, because it acted like a strong oxidizing agent.

Most of these changes are referable to a slow oxidation caused by the oxygen of the air. Schoenbein[2] had assumed that in this process the oil was charged with ozone, the oxygen of the air being activated by the turpentine oil. It was shown later, however, by Kingzett,[3] Bardsky,[4] and Papasogli[5] that the turpentine oil contained no ozone[6] but hydrogen peroxide.

---

1) A normal American turpentine oil, sp. gr. 0.867, which had been kept in a stoppered but partly filled bottle, after standing seven weeks had a sp. gr. of 0.897. It was now soluble in 3.5 p. by volume of 90 p. c. alcohol, whereas the original oil had required 6 p. to effect a clear solution.

Another sample of American oil, after having stood for some time had acquired a sp. gr. of 0.913 and gave a clear solution with but 3 p. of 90 p. c. alcohol.

A French oil, which had been kept for four years in a partly filled but well stoppered bottle, had undergone the following changes:

	Original normal oil.	The same after 4 years.
Specific gravity	0.871	1.009
Optical Rotation	—29° 55′	—19° 18′

Whereas the original oil required 20 p. by volume of 80 p. c. alcohol to effect a clear solution, the oxidized oil was soluble in 1 p. of the same alcohol and was miscible in all proportions in 90 p. c. alcohol.

2) Liebig's Annalen, 102, p. 133.

3) Journ. Chem. Soc., 27, p. 511; Pharm. Journ., III, 5, p. 84; ibidem, 6, p. 325; ibidem, 7, p. 261; ibidem, 9, pp. 772 and 811; ibidem, 20, p. 868; Ch. N., 69, p. 148; comp. also Robbins, Pharm. Journ., III, 9, pp. 748, 792, 872. According to Kingzett, the ultimate products of the slow oxidation of turpentine oil, camphoric acid and hydrogen peroxide, do not result directly Peroxide of camphor, $C_{10}H_{14}O_4$, is first formed which, in the presence of water, breaks up into camphoric acid and hydrogen peroxide.

$C_{10}H_{14}O_4$	+	$2H_2O$	=	H_2O_2	+	$C_{10}H_{16}O_4$
Camphor peroxide		Water		Hydrogen peroxide		Camphoric acid

However, Kingzett did not succeed in isolating the hypothetical camphor peroxide. Later (1864) he tried to explain the change in a different way. Chem. News, 69, p. 148.

4) Bardsky (Chem. Centralbl. 1882, p. 808) found hydrogen peroxide and, as he believes, nitrous acid in the water with which oxidized turpentine oil had been shaken.

5) According to Papasogli (Chem. Centralbl. 1888, p. 1548) the water, which has stood in contact with turpentine for a longer period, contains hydrogen peroxide, camphoric acid (m. p. 176°), formic acid, acetic acid and an acid $C_{10}H_{18}O_3$, isomeric with campholenic acid. — The oxydized oil itself is said to contain oxysilvinic acid.

6) The accuracy of these investigations have been questioned by no one, and the presence of hydrogen peroxide in oxidized turpentine oil can be definitely assumed. Nevertheless, many text-books contain the statement that old turpentine oil, in fact oils in general, contain ozone. Inasmuch, however, as ozone and hydrogen peroxide decompose each other according to the following equation

$$O_3 + H_2O_2 = H_2O + 2O_2$$

(Schoene, Liebig's Annalen, 196, p. 239), the presence of ozone is necessarily excluded.

As Loew[1] has first shown, turpentine oil which has been oxidized in the presence of moisture contains other substances besides hydrogen peroxide. Oxidized turpentine oil liberates iodine from potassium iodide, a behavior not attributable to hydrogen peroxide. This action, as was already assumed by Kingzett,[2] must evidently be attributed to organic peroxides which decompose with water so far as ultimately to yield hydrogen peroxide. Presumably peroxide hydrates are formed as intermediate products. The explanation of this process has recently been given in a contribution by Engler and Weissberg.[3]

These investigators demonstrate that, when absolutely dry turpentine oil is activated, neither hydrogen peroxide nor ozone results. They further show that turpentine oil activates oxygen most readily at a temperature of 100°; that above 100° no activated oxygen is formed, but that it is then used for the oxidation (destruction) of the oil. 1 cc. of turpentine oil can activate 100 cc. of oxygen at 100°.

Turpentine oil, whether moist or dry, when charged with oxygen has a capacity to convey the oxygen to such substances as are not directly oxydizable with atmospheric oxygen. Thus, as has already been stated, iodine is liberated from potassium iodide, indigo solution is bleached, arsenous acid is oxidized to arsenic acid. Activated turpentine oil retains its properties for years if kept in the dark. But little is as yet known concerning the oxidation products resulting.

The presence of the following substances has been definitely ascertained: formic acid, acetic acid and camphoric acid, $C_{10}H_{16}O_4$.[4] Besides these, a small amount of an aldehyde, corresponding in its composition with camphoric aldehyde, $C_{10}H_{16}O_3$, has been found.[5] It possesses a narcotic odor and is probably the cause of the peculiar odor of the old "rancid" turpentine oil.

These oxidation processes, accompanied by the same changes of the oil, are accelerated when warm air saturated with water vapor is passed through the oil.[6]

---

1) Zeitschrift für Chemie, II, 6, p. 609; Chem. Centralbl., 1870, p. 821.

2) Loc. cit.

3) Berichte, 31, p. 3046.

4) Chem. Centralbl., 1888, p. 1548.

5) Schiff, Chemiker Zeitung, 20, p. 361. The aldehyde seems to form most readily when turpentine oil is kept in partly filled, not well stoppered bottles in diffused light. The amount of the unstable aldehyde does not exceed 1 p. c. As has already been stated, it can be removed by shaking the oil with sodium acid sulphite solution. When exposed to the air in a watch glass, the aldehyde resinifies, loses its narcotic odor and no longer reacts with rosaniline sulphate.

6) An oil, sp. gr. 0.864, after having been treated in this manner for 44 hours had a sp. gr. of 0.949. Kingzett also observed a considerable rise in the boiling point.

By the action of direct sunlight on moist turpentine oil in the presence of air, or better still, of oxygen, pinol hydrate, $C_{10}H_{18}O_2$, results. According to the solvent used, pinol hydrate crystallizes in laminae or needles. Its inactive modification melts at 131°.[1]

COMPOSITION. The first elementary analysis, referred to on p. 41, revealed the fact that turpentine oil consists of hydrocarbons $C_{10}H_{16}$. This was verified by later investigations, which also showed that different oils reveal physical differences in so far as some of them turn the plane of polarized light to the left, others to the right. Berthelot[2] designated the laevogyrate hydrocarbon terebenthene, the dextrogyrate one australene. Inasmuch as the two modifications are merely physical isomers, but chemicaly identical, Wallach[3] introduced the terms l- and d-pinene.

Since pinene is one of the most labile terpenes, it is not surprising that some of its decomposition products are formed during the production of turpentine oil, and thus get into the oil. As already stated, the oil contains traces of formic and acetic acids, also resin acids. At higher temperatures, these act on the pinene so that dipentene and presumably polymeric terpenes result. At least both are found as constant companions of the pinene in the turpentine oil.

Certain observations and indications render it probable that camphene and fenchene belong to the normal constituents of turpentine oil. Since the boiling point of both of these lies very close to that of pinene, a direct proof to which no objection can be raised is impossible. Nevertheless, it may be assumed that the indirect proof for the presence of camphene has been furnished.

Armstrong and Tilden[4] found camphene in the so-called terebene, the product resulting by the action of concentrated sulphuric acid on turpentine oil. They assumed that the camphene resulted according to a reaction analogous to the one by which it can be obtained from pinene hydrochloride, a possibility not to be denied off-hand. Power and Kleber,[5] however, consider it more probable that the camphene is contained originally in the oil, and that its presence becomes noticeable only after the destruction and removal of the pinene.

The results of the investigation by Bouchardat and Lafont[6] also seem to permit a similar conclusion. They heated French turpentine oil

1) Liebig's Annalen, 80, p. 106; ibidem, 259, p. 318; Journ. Chem. Soc., 59, p. 315.

2) Compt. rend., 55, pp. 496 and 544; Liebig's Annalen, Suppl. II, p. 226.

3) Liebig's Annalen, 227, p. 300.

4) Berichte, 12, p. 1753.

5) Pharm. Rundschau, 12, p. 16.

6) Compt. rend., 113, p. 551.

with benzoic acid anhydride for fifty hours. From the product of reaction they isolated camphene, also esters of isoborneol and fenchyl-alcohol, whose formation is probably due to the presence of camphene and fenchene.

The same investigators[1] obtained, by the action of sulphuric acid on French turpentine oil and the subsequent treatment with alcoholic potassa, two potassium salts of the composition $C_{10}H_{17}OSO_2OK$, which upon hydrolysis with acids yielded borneol and potassium hydrogen sulphate on the one hand, and l-fenchyl alcohol and the same potassium salt on the other. The formation of these salts likewise indicates the presence of camphene and fenchene.

The presence of l-camphene in American turpentine oil was shown by Schimmel & Co.[2] in the following manner. By treating fraction 160—161° of American turpentine oil (sp. gr. 0.869; $[\alpha]_D = +1°16'$) with glacial acetic acid and sulphuric acid according to Bertram's method,[3] isobornyl acetate resulted which upon saponification yielded isoborneol (m. p. of the phenyl urethane derivative 138°). Inasmuch as pure pinene, regenerated from the nitroso chloride does not yield isoborneol when treated in the same manner, the formation of isoborneol is explained by assuming the presence of camphene in the American oil.

Adulterations. On account of its cheapness, turpentine oil has been used largely as an adulterant. It itself has been subject to adulteration with petroleum and resin oils.

Petroleum. Common illuminating oil, also the lighter as well as the heavier fractions of petroleum have been used. Illuminating oil and the lighter petroleum fractions are recognized by the lowering of the specific gravity; also by the lower flashing point, which lies at 33—34° for pure turpentine oil. The heavier fractions of petroleum are not volatile with water vapor and, therefore, remain behind, when the oil is rectified with water vapor, as a fluorescent residue which is not affected by concentrated nitric and sulphuric acids.

Socalled *patent turpentine oils* are frequently mixtures of petroleum products with turpentine oil or camphor oil. Sometimes they are nothing but petroleum hydrocarbons brought upon the market under various fanciful names. Such designations are:[4] Canadian turpentine oil, patent turpentine, turpentyne, turpenteen, larixolin, paint oil, and others. According to Dunwody,[5] mixtures of turpentine oil and

1) Compt. rend., 125, p. 111.
2) Bericht von S. & Co., Oct. 1897, p. 68.
3) Journ. f. prakt. Chem., II, 49, p. 1.
4) Pharm. Centralh., 38, p. 131.
5) Amer. Journ. Pharm., 62, p. 288.

petroleum differ from pure turpentine oil by their different solubility in 99 p. c. acetic acid. Absolute acetic acid (99.5 to 100 p. c.) is said to be miscible in all proportions with petroleum as well as with turpentine oil. An acetic acid prepared by mixing 99 cc. of glacial acetic acid and 1 cc. of water makes a clear solution with turpentine oil in the ratio of 1:1, but not with petroleum. For the complete solution of turpentine oil-petroleum mixtures the following quantities of this acetic acid (99 + 1) are required.

Petroleum	1	2	3	4	5	7	8	cc.
Turpentine oil	9	8	7	6	5	3	2	"
Solution effected with acetic ac. (99+1)	40	60	80	110	150	230	270	cc.

The Prussian Revenue Commission has recently[1] directed the use of a test based on the different solubility of turpentine oil and mineral oils in aniline for the detection of patent turpentine oils.[2]

"In a cylinder of 50 cc. capacity and calibrated in cc., also provided with a glass stopper, 10 cc. of the oil in question are poured and 10 cc. aniline are added. The cylinder is stoppered and the mixture thoroughly shaken. If after five minutes the solution is not uniform, but has separated into two layers, the oil is a patent turpentine oil."

Conradson[3] detects petroleum in turpentine oil in the following manner:

50 cc. of oil are evaporated on a waterbath down to from 1 to 2 cc. If the oil was free from petroleum, the residue should be soluble in from 5—10 cc. of glacial acetic acid. If 10 p. c. or more of petroleum was present, the mixture is turbid and separates into two layers upon standing.

For the quantitative estimation of the petroleum, the turpentine oil is oxidized with fuming nitric acid and the residue, not acted upon by the acid, is weighed as mineral oil.

According to Burton[4] 100 cc. of oil are carefully cooled in a spacious flask connected with a reflux condenser and 300 cc. of fuming nitric acid are slowly added from a dropping funnel. The residual oil is washed with hot water and weighed. Control experiments with known mixtures showed that the determination is more accurate for the higher boiling fractions of petroleum than for the lower ones, for the latter are somewhat attacked by nitric acid. A petroleum product boiling at 250° yielded 34.1 p. c. in place of 35 p. c.;

---

1) According to the former official directions for testing for purposes of revenue, an oil was to be regarded as pure if an increase in temperature of 25° C. resulted when the oil was shaken with an equal volume of fuming hydrochloric acid, or with one volume of hydrochloric acid (1.12) and ⅔ vol. of English sulphuric acid. As was to be expected, this process frequently yielded unreliable results and, as a consequence, was frequently criticised.

2) Chem. Zeitung, 22, p. 884.

3) Chem. Centralbl., 1897, II, p. 449.

4) Amer. Chem. Journ., 12, p. 102.

a petroleum product boiling at 75° yielded 17.9 p. c. in place of 20 p. c.; and 28 p. c. in place of 30 p. c. A similar method is described by Allen[1] who employs 400 cc. of fuming nitric acid instead of 300 cc. for the same quantity of oil.

The older method of Armstrong[2] is based on the fact that turpentine oil is polymerized with concentrated sulphuric acid and thus largely converted into products not volatile with water vapor. Inasmuch as paraffin-like hydrocarbons as well as cymene, which are stable toward sulphuric acid, result, the process with nitric acid is to be preferred.

Resin oil. A second adulterant of turpentine oil is the resin oil, a product of the destructive distillation of rosin. The adulteration is said to be practiced within moderate limits, since the addition of more than 5 p. c. renders the turpentine oil sticky and imparts to it an unpleasant odor. According to Baudin,[3] the adulteration can be detected by the fatty stain, which the adulterated oil leaves on paper upon evaporation.

Aignan[4] believes in the possibility of detecting resin oil both qualitatively and quantitatively in French turpentine oil by means of the higher rotatory power of the former. The formula which he suggests for the calculation of the amount of the adulterant leads one to suppose that he assumed that resin oil as well as turpentine oil always has the same rotatory power. That this is not true at least for turpentine oil is shown by the variations in the angle of rotation recorded by various observers and enumerated under the description of the French oil. This, naturally, disposes of Aignan's formula.

Aignan distinguishes between three kinds of resin oil:

1) *Huile blanche de choix rectifiée* $[\alpha]_D$[5] $= -72°$.
2) *Huile blanche fine rectifiée*......... $[\alpha]_D = -32°$.
3) *Huile blanche rectifiée*................ $[\alpha]_D = -21°$.

It is assumed that only No. 1 of these oils is used for the purpose of adulteration. This varies somewhat in its rotatory power from that of French turpentine oil, but could easily be given a normal rotation by a slight admixture of No. 2.

An apparently better method for the detection of resin oil is given later in a contribution by the same author:[6] The oil is distilled

[1]) Chem. Centralbl., 1890 II, p. 125.
[2]) Journ. Soc. Chem. Industry, Dec. 1882; Pharm. Journ., III, 13, p. 584.
[3]) Chem. Centralbl., 1891, I, p. 813.
[4]) Comp., rend., 109, p. 944.
[5]) As $[\alpha]_D$. Aignan apparently does not consider the specific rotatory power, but as becomes apparent from the rotatory power of turpentine mentioned later, the angle of rotation observed in a 250 mm. tube!
[6]) Compt. rend., 124, p. 1367.

under 60 mm. pressure and the residue remaining above 100° is examined with the polariscope. The residue from pure French turpentine oil is laevogyrate; that of an oil adulterated with more than 5 p. c. resin oil is dextrogyrate.

Zune[1] employes the refractometer as a means of detecting resin oil. The oil is submitted to fractional distillation, three-fourths being distilled over: the index of refraction of the first fourth and of the residue being determined. The difference in the two indices for a pure oil amounts to from 0.0035 to 0.004; in the presence of resin oil, it is larger and amounts to 0.006 for 1 p. c. of adulterant.

PRODUCTION AND COMMERCE. With reference to importance and value of production of turpentine oil, the United States of North America occupy the first position as is shown by the total arrivals at the principal staple ports:

From April 1, 1896, to March 31, 1897: 465,578 barrels;
From April 1, 1897, to March 31, 1898: 439,304 barrels.

The barrel contains 50 gallons or about 150 k. If an average production of 450,000 barrels per year is assumed, the grand total[2] will be 67,500,000 bbls., with an approximate value of $7,500,000.00.

The consumption of turpentine oil in the United States amounted to:

From April 1, 1896, to March 31, 1897: 111,560 barrels;
From April 1, 1897, to March 31, 1898: 130,722 barrels.

Total export from the United States:

1894:	12,618,407 gallons	=	abt. 252,000	barrels;
1895:	14,652,738 "	=	" 293,000	"
1896:	17,431,566 "	=	" 348,000	"
1897:	16,820,000 "	=	" 336,000	"

Thirty years ago, North and South Carolina only came into consideration in the production and export of turpentine oil. The forests in these states having been largely exhausted, the industry has gone farther south. Thus, e. g. Charleston, S. C., formerly the important port has ceased to export turpentine oil. The business has been transferred to the ports Savannah and Brunswick, both in Georgia. Savannah is the most important port in the world's market for turpentine oil. It has exported more than one-half of the total American output.

From New York turpentine oil is exported principally in cases to non-European countries.

1) Compt. rend., 114, p. 490.

2) For 150 years.

During the last three years the turpentine oil industry has largely developed in Florida and Alabama. It is feared that Pensacola will become a strong competitor of the present ports.

The most important European markets for American turpentine oil are:

London.

Import 1896: 24,940 tons = abt. 165,000 bbls.
" 1897: 25,140 " = " 168,000 "

Hamburg.

Import 1896: 7,974,700 kilo = abt. 58,200 bbls.
" 1897: 8,608,100 " = " 57,400 ".

Antwerp.

Import 1896: 7,170,000 kilo = abt. 47,800 bbls.
" 1897: 6,144,500 " = " 40,963 "

The market price of American turpentine oil is not only affected by supply and demand, but is frequently influenced by speculations and, therefore, subject to considerable variation. In January, 1897, the price per 100 k. f. o. b. Hamburg was M. 43.50. It rose to M. 47.00 in the beginning of June, and dropped again to M. 42.50 by the middle of July. It again rose to M. 54.00 by the beginning of October, but dropped to M. 50.50 by the end of September. Still greater were the variations in price in 1898. Beginning with M. 52.00 in January, the price gradually rose to M. 72.00.

Of much lesser importance is the production of turpentine oil in France. Nevertheless, it is sufficient to supply the demand of that country for the importation of American turpentine oil has been made prohibitive by a duty of Fr. 24.00 per 100 k. The industry is confined to the Dep. des Landes. The principal commercial centers are Mont de Moreau, Bayonne and Bordeaux.

Under these conditions but small quantities can be exported. According to official statistics, the export of turpentine oil from France amounted to:

1896: 1,937,990 Kilo;
1897: 1,411,500 "

Of these amounts Hamburg received:

1896: 430,100 Kilo;
1897: 992,500 "

Antwerp:

1896: 970 Colli;
1897: 530 "

Qualitatively the French oil holds the first position among all commercial brands and in the arts is frequently preferred to the American. Its market value is, as a rule, from $2.00 to $2.50 higher per 100 kilo than the American.

Among the commercial varieties of turpentine oil, the Russian[1] comes next in importance.

According to the records published in "Die Productionskräfte Russlands".(1898), a work issued by the Imperial Russian Ministry of Finance the value of the oil of turpentine annually produced in Russia is Rb.[2] 1,500,000. The greatest part of this is used by the country itself.

The industry has its seat in the gouvernements to the north of the Volga and in Poland. It is conducted on a small scale and is in the hands of the peasants.

The export amounted to:

1885:	160,000	Pud,[2]	value	Rb.	498,000.
1890:	295,000	"	"	"	811,000.
1892:	365,000	"	"	"	826,000.
1893:	300,000	"	"	"	853,000.
1894:	258,000	"	"	"	717,000.

Corresponding to the quality, the commercial value of the Russian turpentine oil is much less than that of the American and French. It is indeed the poorest of all the commercial varieties.

The Russian import duty on turpentine oil of 60 kopeken[2] gold per pud, or of nearly $3.00 per 100 k. has the object of a duty for revenue only.

In the Antwerp lists of imports there appears regularly for several years a Spanish turpentine oil. The importation amounted to:

1896:	2,644	Colli:
1897:	7,520	"

4. American Oil (Spirit) of Turpentine.

Oleum Terebinthinae Americanum.—Amerikanisches Terpentinöl.—Essence de Térébenthine Américaine.

History and Origin. The enormous areas which were at one time and in part are to-day densely covered with pine forests acquire their maximum luxury and broadest expanse in the South Atlantic States,

---

[1]) The figures enumerated, in all probability, refer primarily to the pine tar oil obtained by destructive distillation. Whether turpentine oil proper, obtained from turpentine, is at all made on a commercial scale in Russia is not known.

[2]) Rubel = 100 Kopeken = abt. 80¢; Pud = 16.38 k.

from Virginia to the Gulf States, in the eastern part of Canada, in British North America and on the western slopes of the Pacific States.

The first mentioned area, which is also the largest, comprises the states North and South Carolina, Georgia, and Alabama and within it there has developed the principal turpentine industry. Up to the middle of the last century the products of this industry were tar and pitch which were used principally in ship building and as naval supplies and hence were termed "naval stores."[1] The preparation of turpentine oil seems to have begun as late as the middle of the eighteenth century in North Carolina and Virginia. Professor Kalm, the Swedish traveler, who is known as a careful observer, and who explored the Atlantic provinces of the then British colonies from Quebec to Virginia during the years 1749 and 1750, reports concerning the preparation of tar and pitch only.[2] Later travelers and reports first make mention of the preparation of turpentine, turpentine oil and colophony in Carolina. Among these are Dr. Johann David Schoepf, who traversed the Atlantic States from Canada to Florida[3] in 1783 and 1784; also François André Michaux, who about twenty years after the longer stay of his father, the well known botanist André Michaux, traveled in North America at the beginning of this century.[4]

Up to the year 1820 the consumption of turpentine, turpentine oil and colophony was restricted to the limited demands of the home industries. The exportation of oil and rosin to England was unimportant. Up to 1830 the manufacture of turpentine was restricted to the coast: between the Tar river in the north, and Cape Fair river in the south; while the ports New Bern, Wilmington and Washington in North Carolina served as collective points. The distillation of turpentine was conducted in cast iron stills.

At the beginning of the thirties the application of turpentine in the industries experienced considerable extension. This was caused primarily by the increased use of paints accompanying the increase in wealth; by the development of the varnish, lacquer and caoutchouc industries; and,

---

1) The oldest mention concerning tar and pitch, and the preparation of turpentine in Virginia is to be found in vol. 1 of the "Calendar of State Papers. Colonial Series" for the year 1574 to 1660 in the Public Record Office in London. Of the year 1610 this volume contains "Instructions for suche thinges as are to be sente from Virginia," also a printed pamphlet: "The Booke of the Commodities of Virginia." Both mention pitch, tar, rosin and turpentine among the products of Virginia. The former also contains brief directions for the method of preparation of turpentine which is still in vogue. (Dan. Hanbury, in Proc. A. Ph. A., 19, p. 491.)

2) Kalm, Reise, etc., vol. 2, pp. 418, 474; vol. 3, pp. 305, 528.

3) Schöpf, Reise, etc., vol. 2, pp. 141, 247.

4) Michaux, Histoire des arbres, tom. 1, p. 78.

finally, by the use of a mixture of turpentine oil and alcohol as an illuminating agent, which was introduced since 1839 as camphine and under other fanciful names. Up to the introduction of petroleum products (kerosene) about 1860, this was the cheapest illuminating material. The improvements, made in the course of the thirties, by Comstock, Hancock, Macintosh, Chaffee and especially by Luedersdorff in the processes employed in the caoutchouc industry did much to bring about a larger consumption of turpentine oil.

This increase in the consumption brought about an increase in the turpentine industry in 1834, and caused the introduction of better distilling apparatus which insured a larger yield (Ashe[1]). The exportation of American turpentine oil and colophony to England and other countries assumed large proportions only after the removal of the import duty in England in 1846. It was interrupted, however, during the years 1861 to 1865 of the civil war, the period of industrial and commercial stagnation.

Up to the year 1837, the opinion prevailed in Carolina that the pine forests to the south were not adapted to the production of turpentine on account of differences in climate and soil. In the year mentioned, experiments conducted on a large scale showed this opinion to be erroneous. As a result of the great demand and of increasing speculation, the industry spread rapidly to South Carolina and Georgia, and later to Alabama and Mississippi (Mohr[2]). With the introduction of the more readily transportable copper stills, the distillation was more and more conducted at the place of production, so that the turpentine farms began to supply the finished though crude products of distillation, oil and rosin, in place of turpentine, to the ports along the coast. One other circumstance contributed to the transfer of the distillation to the immediate source of production of the crude material. This was the fact that the increased demand for turpentine caused a corresponding overproduction of rosin. This could not be disposed of and consequently suffered a corresponding depreciation in value.

This disparity was equalized toward the end of the sixties. The opening up of new territories of production brought about an overproduction in turpentine oil as well, which was felt all the more because petroleum products took its place as an illuminating agent and also superseded it in various branches of the arts and industries. On the other hand, colophony found new and large application. The

1) The forests, forest lands and forest products, etc.

2) The timber pines of the Southern U. S., p. 69; also Pharm. Rundschau, 2, p. 187.

turpentine industry in the Southern states thus developed unimpeded. With the establishment of cheap means of transportation by rail and by water, all the conditions were given for a prosperous growth to the enormous dimensions on which this industry is being conducted at present.

At best, the larger pine trunks can be made to yield turpentine for from 15 to 20 years before they are exhausted. Of late years, the exhausted trees are no longer abandoned to wind and fire, but are taken to the mills to be sawed into logs and boards; to paper pulp factories; or are converted into charcoal.

It is true that waste, the clearing of forests and temporary destruction by fire has driven the turpentine industry away from the coast into the interior. A diminution or exhaustion of the American turpentine industry, however, need scarcely be feared. Not only does there still exist an enormous wealth of pine forests, but the removed trunks are replaced by new shoots from the roots and also by more rational methods of cultivation. Former methods of waste have not been without their lessons. Manufacturers as well as the general population have become alive to the necessity of displacing former irrational methods and waste by more rational ones which insure a continuation of the industry.

The pines which principally constitute the largest forests of the Southern states and which are used in the manufacture of turpentine are four in number: *Pinus palustris* Miller (*P. australis* Michaux), long-leaved or Southern pitch pine, which is the most important in this industry; *P. taeda* L., Loblolly or Rosemary pine; *P. heterophylla* Elliot (*P. cubensis* Griesebach), Cuban pine, Swamp pine, Slash pine: and *P. echinata* Miller (*P. mitis* Micheaux), short-leaved yellow pine.

PREPARATION. The method of production is in general the same everywhere. On account of the extent and density of the pine forests, the size and diameter of the trees, the larger yield due to a favorable climate, the lack of forest cultivation and the absence of control, the methods employed were those that insured the largest yield and quickest returns and had but little heed of the permanence of the forests.

The operation, mostly by negroes, of the turpentine farms, begins in the first dry days of spring, as a rule in April. The trees are "boxed" about 1 to 1½ foot above the ground (diagonally across the trunk of the tree). The length of such a box is about 14 inches, its greatest depth 6 to 7 inches, so that it has the capacity of more than a quart. If the trunk is very thick, a similar "box" is sometimes cut on the other side of the tree.

As soon as the sap begins to flow in spring, a strip of bark 2 inches wide and 8 inches long is removed from the trunk at either end of the box ("cornering"). Then the surface above the box between the two bare strips is deprived of its bark down to the splint ("hacking," "chipping"). The exudation of the sap soon begins and varies with the temperature and the formation and secretion of the sap ("bleeding"), the exposed surfaces above the boxes are increased every one or two

Fig. 58.
BEGINNING TO WORK IN A VIRGIN FOREST: BOXING

weeks. In this manner the chipped surface above the box is lengthened or widened or both (fig. 58) as long as warm weather lasts, mostly to the end of October.

In the beginning and during very warm weather, the boxes are filled every 2 to 4 weeks and are emptied by means of a flat ladle into wooden pails (fig. 59) and from these into barrels. With the advent of the cooler season the exudation ceases. The exposed surfaces and boxes are then freed from adhering and dried oleoresin ("scrape") and allowed to heal or they are again cut during the next spring. If allowed to heal, the operation of bleeding is performed on another part of the trunk during the following or the third spring.

The turpentine farms are usually laid out so as to supply one copper still with a capacity of 800 gallons or 20 bbls. of turpentine. For the operation of this, 4000 acres of forest with good growth are necessary. This is divided into 20 lots, of which each contains about 10,000 boxes yielding olecresin. Inasmuch as many trunks according to their

Fig. 59.
Scraping and Dipping in a Turpentine Orchard.
Tree to the right shows scarified surface.

diameter contain two, three, sometimes even four boxes, the above number is distributed over about 4—5000 trees usually occupying an area of two acres.

With a rational treatment the yield from 10,000 boxes for each collection amounts to about 40 to 50 bbls. of turpentine of 280 lbs. each.

During the first years, and before the capacity of the trees has been reduced by excessive bleeding, the average yield of a turpentine farm is about 270 bbls. of 280 lbs. each of dip turpentine and about 70 bbls. of scrapings at the end of the season. Each barrel of dip turpentine yields about 7 gal. of oil upon distillation, the scrapings about 3 gals. The total yield of a farm, therefore, amounts to about 2,200 gals., or 50 bbls. @ 45 gals. of oil, and about 260 bbls. of rosin.

During the first two or three months the collected turpentine is almost white and constitutes the best quality, "virgin dip." This brand is brought into the market as "water white" and "window glass." Inferior, more or less yellow brands are marked N. or M. K. Minor shades are indicated by the various letters of the alphabet, so that A indicates the poorest, and N the best grades of the varieties next to the W. G.

During the second year of the operations, the yield of the dip is about 10 bbls. less than during the first, whereas the scrapings increase to about 120 bbls. The quantity of turpentine oil during the second year amounts to about 40 bbls. of 45 gals. each, and about 200 bbls. rosin. The quality of the oil is the same as that of the first; the rosin, however is mostly darker. During the third year the yield of turpentine diminishes.

The continued use of the same cups and exposures, and the annual increase of the latter bring about a deterioration of the oleoresin, due to evaporation and action of the air: a smaller yield of oil is obtained and the quality of the rosin is poorer. It is more advantageous, therefore, to allow the trees to rest for several years and then to tap them at a different place.

As has already been stated, the distillation of the turpentine has been conducted for years in copper stills on the larger turpentine farms. In the course of the distillation, and after the still has been heated, a thin but continous stream of water from the higher placed condenser is allowed to flow into the body of the still until the distillation is ended. The liquid rosin is allowed to flow off through an exit tube at the bottom of the still. It is strained through wire sieves and filled into barrels for shipment.

The total average output of a farm of the described size, which is operated mostly for four years only, is about 120,000 gals. of turpentine oil, 52,000 bbls. of rosin of first quality, 4,000 bbls. of rosin of second grade, and 2,400 bbls. of poor quality (Mohr[1]).

---

1) Pharm. Rundschau, 2, pp. 168, 187; 12, p. 211.

With the wasteful methods still largely in vogue, and in order to obtain the largest possible immediate returns from such a venture, the timber, after four years' production of oleoresin, is cut either entirely or in part, provided a saw mill is located in the neighborhood or the wood can be disposed of to advantage in some other way. Each acre yields about 25,000—30,000 ft. of lumber or 50—60 cords of fuel wood.

According to the Forestry Division of the U. S. Department of Agriculture, the area involved in the turpentine industries of the Southern states during 1890 amounted to 2,300,000 acres of pine forests.

Of late years, the distillation of the enormous wastes of the saw mills has increased largely. For this purpose the saw dust and other waste, especially of the older, dying trees are used. While gradually drying, the lower portion of these trees accumulate a larger amount of oleoresin. This wood, permeated by dried oleoresin, is known as "light wood" and is used for underground constructions, such as railroad ties and as fuel for boilers and locomotives. Upon distillation, it yields about 2 to 2½ p. c. of pine tar oil, also tar, pyroligneous acid, pitch and charcoal. The yield from 600 lbs. of fairly dried wood rich in resin from *Pinus australis* is about 21 to 22 lbs. of tar oil, 95 lbs. of pyroligneous acid, 150 lbs. of tar, and 127 lbs. of charcoal (Mohr[1]).

Properties. The specific gravity of crude American turpentine oil lies as a rule between 0.865 and 0.870. However, lighter (sp. gr. 0.858) as well as heavier (up to 0.877, Kremers[2]) oils occur. Freshly distilled or rectified oil is mostly lighter than the crude or old oil.

When fractionally distilled, about 85 p. c. distill over between 155 and 163°. Inasmuch as the American species of pine yield dextrogyrate[3] as well as laevogyrate[4] oils, the rotatory power of the commercial oils necessarily varies. According to the prevalence of the one or other species at the locality of production it may be more or less strongly dextrogyrate, or even laevogyrate. Armstrong[5] observed in

---

1) Pharm. Rundschau, 2, p. 199.

2) Pharm. Review, 15, p. 8.

3) $[\alpha]_D$ of the oleoresin from *P. palustris* = — 18.665°,
$[\alpha]_D$ of the oil from *P. palustris* = + 28.98°,
$[\alpha]_D$ of the oleoresin of *P. cubensis* = — 32.428°,
$[\alpha]_D$ of the oil from *P. cubensis* = + 9.6°.
E. Kremers, Pharm. Rundschau, 13, pp. 135. 136.

4) According to Long (Journ. Amer. Chem. Soc., 16, p. 844; Abstr. Chem. Centralbl., 1895, I, p. 156) the oil of *Pinus glabra* is strongly laevogyrate, $\alpha_D = -31.5°$ to $-35°$. According to the same author (Chem. Centralbl., 1898, I, p. 885) the oil of *Pinus palustris* is also laevogyrate. This is contradictory to the observations made by Kremers (see footnote 3).

5) Pharm. Journ., III. 18, p. 584.

1883 that the rotatory power of 28 samples from Wilmington varied from + 13° 33′ to + 14° 17′; that of samples from Savannah from + 9° 30′ to + 12° 4′. Two samples from Savannah recently examined were slightly laevogyrate: $\alpha_D = -0°40'$ to $-2°5'$.

5. French Oil of Turpentine.

Oleum Terebinthinae Gallicum. — Französisches Terpentinöl. — Essence de Térébenthine Française.

Origin and Preparation. The larger part of the turpentine oil prepared in Europe is obtained from the turpentine from the *Pin maritime, Pinus pinaster* Solander (*Pinus maritima* Poiret) and is obtained by steam distillation. This tree, which attains a height of 40 m., thrives best along the western half of the southern shore of the Mediterranean as far as the Bay of Biscay. The production takes place principally in the western French dune districts (*Landes*) of the Départments de la Gironde and des Landes. The principal commercial centers are Bordeaux and Bayonne.

In order to obtain the turpentine, one or two narrow notches (*carre*) down to the splint are cut near the base of the trunk in spring when the sap begins to rise. These wounds are made 30 to 40 cm. long and 10 cm. wide. Below the notch a tin or earthenware vessel is fastened so that the oleoresin as it exudes flows into it. According to temperature and duration of the flow, the incisions are extended upward from time to time (*pinage*), so that, in the course of 4 to 5 years they constitute parallel stripes, 4 m. in length, along the trunks of the trees. Only after the abandoned incision (*carre*) has healed, is another made on the opposite side in the following spring. The vessels in which the turpentine collects are emptied whenever necessary. The soft turpentine (*gemme*) is mostly distilled in the neighborhood of the place of collection. That which has dried along the *carres* is removed in fall and brought as galipot (*barras*) into the market.

As a rule the trees are tapped (*gemmage*) only after they have reached an age of 20 to 30 years. Under rational treatment, a normal tree can then be used for the production of turpentine for almost 100 years. Trees that show signs of exhaustion, after a final tapping made in such a manner as to produce the largest yield possible (*gemmage à morte*) are cut down at the end of the season for the wood. The trees are reproduced by shoots from the roots.

100 Trunks of the *Pin maritime* in western France, on an average, yield 360 k. of turpentine, which upon distillation yields 15 to 18 p. c.

of turpentine oil.[1] When the turpentine is to be distilled, it is first sufficiently heated in casks so that it can be strained through a wire sieve, which removes fragments of bark and leaves, into the copper stills of about 300 liters capacity. The distillation is conducted over direct fire while a current of steam is passed into the still. The residual resin (*brai*) is transferred to barrels and brought as yellow colophonium into the market.

PROPERTIES. The specific gravity of French turpentine oil[2] varies from 0.859 to 0.872, as a rule from 0.865 to 0.876. Formerly higher values were occasionally given. The optical rotation was first ascertained by Biot in 1818. Later the following angles were observed by different investigators: $\alpha_D = -20°$ to $-40°$;[3] $-33°$ to $38°$ (Pereira,[4] 1845); $-30°$ to $-30°30'$ (Armstrong,[5] 1883); and $[\alpha]_D = -39°50'$ (Lafont,[6] 1888). In its other properties as well as in its composition, the French turpentine oil agrees with the American oil.

6. Austrian Turpentine Oil.

Oesterreichisches (Neustädter) Terpentinöl.

ORIGIN AND PREPARATION. The turpentine industry in Austria is confined principally to the "Wiener Wald." In the vicinity of Moedling, Voeslau and Neukirchen, *Pinus laricio* Poiret serves for the production of turpentine. The method pursued is the same as that in America, but trees at least 50 years old only are tapped. Near the base, a cavity is cut into the trunk, into which the exuding oleoresin flows. Above this cavity (*Grandel*) the bark and splint are gradually removed by means of a bent hammer up to a height of about 15 inches. From time to time, when necessary, the turpentine is removed. During the following spring 15 inches more of bark are removed above the old wound. In order that the oleoresin may not flow too slowly and get dry, a gutter is made from the newly exposed surface to the cavity below.

1) Comp. under Bibliography: Petzholdt, p. 88; Curie; Mathieu, pp. 537—540; Desnoyers.

2) The Spanish turpentine oil mentioned on p. 289 can not be distinguished from the French. A sample of this oil examined in the laboratory of Schimmel & Co. possessed the following properties: sp. gr. 0.8478; $[\alpha]_D = -28°4'$; soluble in 7 parts of 90 p. c. alcohol. Upon distillation the following fractions were obtained: 157—159°, 50 p. c.; 159—161°, 28 p. c.; 161—166°, 10 p. c.; 166—170°, 2 p. c.; 170—190°, 5 p. c.; residue 5 p. c.

3) Bericht von S. & Co., April 1897, p. 47.

4) Pharm. Journ., I. 5, p. 70.

5) Ibid., III, 18, p. 584.

6) Compt. rend., 106, p. 140.

The present annual production of turpentine oil in the "Wiener Wald" is said to be about 4,000 cwt.[1] The principal commercial center of this industry is Vienna (Neustadt).

Properties. In the laboratory of Schimmel & Co. a single sample of oil of unquestionable source was examined. Sp. gr. 0.866; angle of rotation +3°46′; it was soluble in 6—8 parts of 90 p. c. alcohol. Upon distillation the following fractions were obtained:

From	159—159°,	21 p. c.	α_D = +6°18′
"	159—160°,	56 "	" = +4°25′
"	160—167°,	18 "	" = +0°18′
	Residue,	5 "	"

Ledermann and Godeffroy[2] observed laevo-rotation in two Austrian turpentine oils. Taking into consideration the above, it would seem doubtful as to whether they had Austrian oil or not.

7. Galician Turpentine Oil.

Of this oil, of which the botanical source is unknown, but one sample appears to have been examined. It had a sp. gr. of 0.863; α_D = +17°18′; and was soluble in 4.5 parts of 90 p. c. alcohol.

Upon distillation in a Ladenburg flask the following fractions were obtained:

	Up to 162°,	7 p. c.,	α_D = +17°15′
from	162—165°,	18 "	" = +19°4′
"	165—169°,	38 "	" = +18°7′
"	169—173°,	13 "	" = +17°28′
"	173—177°,	10 "	" = +15°37′
"	177—210°,	12.5 "	" = +12°2′
	Residue,	1.5 "	

Judging from the relatively large amount of high boiling fractions, the conclusion seems justified that this oil must have a different composition from those previously enumerated.

8. Turpentine Oil from Venetian (Larch) Turpentine.

History. Larch turpentine was known to the Romans. It is mentioned in the writings of Vitruvius,[3] a contemporary of Caesar, also in those of Dioscorides, Pliny,[4] and Galen. During the middle ages, larch turpentine was one of the most highly prized balsams. The name

---

1) According to a communication by A. Kremel of Vienna.
2) Jahresber. d. Pharm., 1877, p. 394.
3) Vitruvius, De architectura, Vol. 2, p. 9; Dioscorides, De mat. med., vol. 1, p. 95.
4) Plinius, Historiae de plantis, p. 575.

Venetian turpentine was applied to it during the fifteenth century,[1] because it was brought from Venice, at that time the center of the drug commerce, into the market.

The first mention of larch oil (*Oleum laricis*) in medical treatises is found in the works of Matthiolus[2] and Conrad Gesner.[3]

ORIGIN AND PREPARATION. Venetian turpentine is principally collected in southern Tyrol in the neighborhood of Meran, Mals, Bozen and Trient, also in Steiermark from *Larix decidua* Miller (*L. europaea* D. C., *Pinus larix*, L.), a larch that flourishes in the central European mountainous districts. According to experience, and as has been demonstrated botanically by H. von Mohl[4] in 1859, the heart wood only contains numerous resin ducts, although they occur in limited numbers in all parts of the outer wood and bark. The production of larch turpentine, therefore, is different from that described heretofore. By means of a wide auger, one or several holes are bored in spring to the center of the tree. These are closed with a wooden stopper and opened in fall in order to remove the oleoresin which has collected, with an iron spoon. The hole is again closed for the further accumulation of oleoresin during the next summer.

If only one or two holes are drilled into the tree, the yield amounts to only several hundred grams, but remains the same for many years. If a larger number of holes are drilled into the tree, and the balsam is allowed to flow out, several pounds of balsam can be obtained in a single summer. In this case, however, the tree is exhausted after a number of years and the wood suffers considerable reduction in value.

A more scientific management of the forests, for the production of turpentine is, therefore, more remunerative (Wesseley,[5] 1853).

In the French Dauphiné and about Briançon the oleoresin is reported to be obtained in a similar manner. The holes, however, are drilled in a straight up and down line and each hole is provided with a short tin or wooden tube. After the exudation ceases, the openings are closed and again opened after 2 to 3 weeks. After the first reopening the flow is said to be greater than when the holes are first drilled. The tapping goes on from March to September. Large trees yield about 3—4 k. of turpentine in a year. After 40—50 years the trees are exhausted.[6]

---

1) Flückiger, Pharmacognosie, p. 80.
2) Matthioli, Opera, I, p. 108.
3) Philiatri, Ein köstlicher Schatz, p. 289.
4) Bot. Zeitung, 17, pp. 329 and 377.
5) Die österreichischen Alpenländer und ihre Forste, p. 369.
6) G. Planchon et E. Collin, Les Drogues simples d'origine végétale, Tom 1, p. 70.

Inasmuch as there is no real demand for Venetian turpentine oil, the distillation is not conducted on a large scale. The yield is about 13.5 to 15 p. c.

Properties and Composition.—The oil distilled from Venetian turpentine has a sp. gr. of 0.878 and a rotatory power $\alpha_D = -11°$.[1] The turpentine itself, according to Flückiger, is dextrogyrate $\alpha_D = +9.5°$.[2] Upon fractional distillation the oil comes over between 155—190°, the larger portion at 157°. If the lowest fraction is saturated with hydrogen chloride, crystallized chlorhydrate, $C_{10}H_{16}.HCl$ results, thus showing the presence of pinene.

9. Turpentine Oil from Canada Balsam.

History. Canada balsam, which was probably long known to the American Indians and employed by them, is first mentioned in European reports of travel by Marc Lescarbot[3] who traveled in Canada during the years 1606 and 1607. He declares the balsam equal in value to the Venetian. In the European market, however, it is not found before the eighteenth century (Flückiger[4]).

Origin and Preparation. Canada balsam, also known as balsam of fir, balsam of Gilead, is used externally as well as in the preparation of elastic collodium, for the mounting of microscopic preparations, etc. Its source is the balsam fir, *Abies balsamea* Miller (*Pinus balsamea* L.), which grows in British America, the northern and northwestern United States. It is collected principally in the Laurentine hills in the province of Quebec. Somewhat farther south, in the northern Alleghanies, the double balsam fir, *Abies fraseri* Purch, and the hemlock spruce, *Abies canadensis* Michaux, *Tsuga canadensis* Carr. are also utilized.[5]

These conifers secrete the oleoresin in resin ducts between the bark and the splint. The tedious collection is done mostly by descendants of the Indians, who camp in the forests during the summer. For the tapping of the pustules and the collection of the balsam they use small iron cans with a pointed lip. With this they puncture the bark covering the pustules readily recognizable on the trunk and the larger branches.

---

1) Bericht von S. & Co., April 1897, table on p. 46.

2) Neues Jahrb. f. Pharm., 31 (1869), p. 78; and Jahresber. d. Pharm., 1869, p. 37; Pereira (1845) states that the turpentine itself is laevogyrate (Pharm. Journ., I, 5, p. 71).

3) Hist. de la Nouvelle-France, pp. 805, 811, 820.

4) Documente zur Gesch. d. Pharm., p. 92.

5) The latter constitutes large forests along the lower St. Lawrence, in Nova Scotia, New Brunswick and westward as far as Minnesota.

The oleoresin exudes slowly into the cans, which are emptied daily and then thrust into a fresh pustule. One man collects little more than ½ gal., or 2¼ k., daily; with the aid of children he may collect double the amount. After every season the trees must be allowed to rest for from one to two years, because of the insufficiency of the accumulation of oleoresin. The principal place of export for Canada balsam is Quebec. The annual production has been given at greatly varying figures. It probably does not exceed 20,000 k. during the best years. (Stearns,[1] 1859; Saunders,[2] 1877.)

Upon distillation the balsam yields 16—24 p. c. of oil.

Properties and Composition. The oil obtained from the dextrogyrate balsam is laevogyrate. Its odor corresponds to that of the ordinary oil of turpentine. It begins to distill at 160°, the major portion coming over at 167°, a smaller amount even above 170°.

When saturated with hydrogen chloride the oil does not yield a crystalline hydrochloride directly. A compound of the formula $C_{10}H_{16}.HCl$ was obtained by Flückiger[3] only after treatment of the product of reaction with fuming nitric acid. The presence of pinene, rendered probable by this experiment, was proven by Emmerich[4] (1895) by means of the nitrosochloride, and the nitrolbenzylamine base melting at 122°. The pinene of Canada balsam is l-pinene.

10. Turpentine Oil from Strassburg Turpentine.

History. Strassburg turpentine, which was known and used at the time of the Romans, has long disappeared from the market being displaced by cheaper varieties of turpentine. The oil has probably never been distilled on any other than a small scale and principally for scientific purposes.

Origin and Preparation. The oleoresin of *Abies alba* Miller (*Abies pectinata* D. C., *Pinus picea* L.) which is widely distributed in central Europe, is found in pustules like that of *Abies balsamea* Miller. Upon raising the bark, the pustules are readily recognizable and are opened. The exuding balsam is collected in small tin cans. The yield is small and the collection of larger quantities tedious. Up to the middle of the seventies, Strassburg turpentine was still collected near Mutzig and Barr in the Vosges mountains. For local trade and use, the collection may still take place: in general, however, this aromatic turpentine possesses historic interest only.

---

1) Americ. Journ. Pharm., 31, p. 29.

2) Proc. Amer. Pharm. Assn., 25, p. 387.

3) Jahresber. f. Pharm., 1869, p. 37; Pharmacographia, p. 618.

4) Amer. Journ. Pharm., 67, p. 135.

PROPERTIES AND COMPOSITION. The laevogyrate turpentine, according to its age, yields up to 24 p. c. of a laevogyrate oil which boils at 163° and has a sp. gr. of 0.861. Although no crystalline hydrochloride could be obtained (Flückiger,[1] 1869) there can be little or no doubt but that this oil contains pinene.

11. Burma Turpentine Oil.

The distillate of the turpentines of two pines indigenous to and widely distributed in Burma, viz. the *Pinus khasya* and *Pinus merkusii* have been examined by Armstrong[2] (1891 and 1896). Aside from the rotatory power, both are like the French oil of turpentine. Armstrong is of the opinion that India might supply its entire demand for turpentine oil from Burma if a turpentine industry could be established.

The turpentine from *Pinus khasya* yielded upon distillation 13 p. c. of oil of the sp. gr. 0.8627, and a specific rotatory power $[\alpha]_D = +36° 28'$.

The turpentine from *Pinus merkusii* yielded almost 19 p. c. of oil, the sp. gr. of which was 0.8610 and the specific rotatory power $[\alpha]_D = +31° 45'$.

12. Russian Turpentine Oil.[3]

As shown in the communication of Schkatelow[4] (1898), turpentine is collected from *Pinus silvestris* L. in the gouvernements Archangelsk and Wologda. Whether or not the oil is distilled from the oleoresin is not mentioned.

Golubeff[5] (1888) examined an oil distilled from *Pinus sibirica*[6] with superheated steam. From fraction 162°, cooled to 0°, a solid, optically active camphene, $C_{10}H_{16}$, separated, which melted at 30° and boiled at 159.° From fraction 230° a solid substance was also obtained, which, however, was not examined farther.

---

1) Jahresber. f. Pharm., 1869, p. 38; Pharmacographia, p. 615.

2) Pharm. Journ., III, 21, p. 1151; 56, p. 870.

3) In commerce the pine tar oil distilled in Russian Poland is commonly designated as Russian turpentine oil.

4) Journ. d. russ. phys.-chem. Ges., 20, p. 477; Jahresber. f. Pharm., 1888, p. 10; Chem. Centralbl., 1889, I, p. 106.

5) Journ. d. russ. phys.-chem. Ges., 20, p. 585; Chem. Centralbl., 1888, II, p. 1622.

6) Whether the oil was obtained from the oleoresin does not become apparent from the abstract available. This, however, may be assumed, since the oil from the needles of *Pinus sibirica* (*Abies sibirica*) does not contain camphene.

13. Turpentine Oil from Picea Excelsa.

Rothtannenterpentinöl.

The turpentine from *Picea excelsa* Lk. (*Pinus picea* Duroi) collected in the neighborhood of Naples, being distilled for experimental purposes, yielded 18.3 p. c. of an oil of the sp. gr. 0.866. The optical rotation α_D was $+3° 5'$ at 18° and the saponification number = 0. The oil had the agreeable odor of pine needles.[1]

14. Turpentine Oil from Pinus Sabiniana.

The turpentine of the nut or digger pine, *Pinus sabiniana*, which is indigenous to California and the western slopes of the Sierra Nevada, yields upon distillation an oil that differs totally in properties and composition from the other turpentine oils.

According to Wenzell[2] (1872) the crude oil boils principally between 101—105° and consists almost entirely of a hydrocarbon, abietene, the sp. gr. of which is 0.694 at 16.5°. At the beginning of the seventies it was sold in the San Francisco market under the names abietene, erasine, aurantine and theoline as a substitute for petroleum benzin for the removal of stains.

Abietene is not affected by hydrochloric, sulphuric, and nitric acids in the cold and, according to Thorpe[3] is identical with heptane. The examination by Schorlemmer[4] (1883), Venable[5] (1880), render it probable that this hydrocarbon is normal heptane, identical with that from petroleum.

The source of this oil has been a question of dispute, it having been referred to *P. ponderosa* and to *P. jeffreyi* as well; the oil has even been suspected of consisting merely of a petroleum fraction. The classification and synonomy of these pines is in an unsatisfactory condition and may well give rise to misunderstandings.[6]

Pine Tar Oils.

Kienöle.

Besides pyroligneous acid, pine tar oil is obtained in several districts of Europe rich in pine forests, as a by-product in the manufacture of charcoal and wood tar by the dry distillation of the resiniferous roots of the common pine, *Pinus silvestris* L. In Scotland *Pinus lede-*

1) Bericht von S. & Co., Oct. 1896, p. 76.
2) Americ. Journ. Pharm., 44, p. 97: Pharm. Journ., III, 2, p. 789.
3) Liebig's Annalen, 198, p. 364.
4) Annalen, 217, p. 149.
5) Berichte, 18, p. 1649.
6) Pharm. Review. 18, p. 165.

bourii Endl. is also used (Tilden,[1] 1878). In continental Europe, most of the pine tar oil is made in eastern Germany, in Poland, in Finnland and in other parts of northern Russia, also in Sweden.

The crude pine tar oil contains tarry, empyreumatic admixtures which are removed by rectification with milk of lime. Nevertheless, the oil retains an empyreumatic odor. Though resembling turpentine oil, this odor renders it inferior.

In a general way it can be said that all pine tar oils have like própertics and like composition. Their components are dextro-pinene, dextro-sylvestrene, and dipentene.[2] From turpentine oil they differ principally by their sylvestrene content.[3]

15. German Pine Tar Oil.

Dentsches Kienöl.

The distillation of tar is conducted only in a few places in the vicinity of Torgau (Province of Saxony); the production of charcoal, tar and pitch have become the prime object. While the trunks of the pines are converted into lumber or fuel, the roots rich in resin are submitted to dry distillation after the splint has been removed to be used as fuel. With the present rational method of production[4] the distillation of tar is conducted in a sugarloaf-like oven, built of firebrick, which is heated by a wood fire circulating outside of these walls. The floor, which inclines toward the center, is provided with a large exit tube through which all of the liquid products of distillation escape into a pit of masonry. Through a lateral branch of the exit tube, the more volatile aqueous and ethereal products of distillation pass through a condenser and are collected in large Florentine flasks. In this manner the lighter oils are separated from the heavier pyroligneous acid.

The crude pine tar oil is rectified over milk of lime and charcoal. It is used in the preparation of iron paints, for the cleaning of type and electros, and in the preparation or dilution of the cheaper grades of paints.

---

1) Pharm. Journ., III, 8, p. 589.

2) In order to ascertain whether sylvestrene and dipentene are to be regarded as original components or as decomposition products due to destructive distillation, Aschan and Hjelt (Chem. Zeit., 18 [1894], p. 1566) distilled the trunk wood of the Scotch pine (fir) with water vapor. They found pinene and sylvestrene, but no dipentene. It would appear, therefore, that sylvestrene is an original constituent of the pine oil, but that dipentene must be regarded as a product of inversion of the pinene due to heat.

3) So far, sylvestrene has been found in the roots, the wood and the needles of *Pinus silvestris* L., also in the needles of *Pinus montana* Duroi. Apparently it is found in pinus species only and not in other genera of the *Abietineae*.

4) According to personal observations made at a large tar distillery in the neighborhood of Torgau.

Properties. German pine tar oil has a terebinthinate, empyreumatic odor, a light yellow color, sp. gr. 0.865 to 0.870, and a rotatory power, α_D, of $+18$ to $+22°$. Upon distillation of a normal pine oil the following fractions were obtained:

From 160—165°,	21.6 p. c.		From 175—180°,	6.4 p. c.
" 165—170°,	50.4 "		Residue,	6.4 "
" 170—175°,	15.2 "			

Composition. German pine tar oil contains dextro-pinene, dextro-sylvestrene, and dipentene and does not differ in composition from either the Polish or Swedish oils.

16. Polish or Russian Pine Tar Oil.

Polnisches oder Russisches Kienöl.

(Also known as Russian turpentine oil.)

In Russia also, pine tar oil is but a by-product in the manufacture of charcoal in the tar distilleries. These are very common in Russia and to be found wherever there are large pine forests. The principal seat of the charcoal and tar industry is in Archangelsk, Wologda and the neighboring gouvernements,[1] also in Poland.[2] It has almost invariably the character of an home industry and is conducted in a primitive manner in ovens made of clay.

The crude pine tar oil of the smaller manufacturers is principally used for home consumption. The larger distilleries which also purchase the crude oil from the smaller plants, purify the oil by rectification over milk of lime and charcoal. The pyroligneous acid is mostly converted into calcium acetate.

The sp. gr. of the Russian pine tar oil is 0.862—0.872, the angle of rotation $\alpha_D = +15°25'$ to $+24°$ (Armstrong,[3] 1883). It boils between 155—180° (Wallach,[4] 1885; Tilden,[5] 1877). Upon distillation of a normal Russian pine tar oil, Tilden obtained between 160—171°, 10 p. c.; between 171—172°, 63 p. c.; and between 172—185°, 24 p. c.

Polish pine tar oil was examined in 1877 by Tilden and in 1887 by Flawitzky.[6] Dextro-pinene was found present; a terpene boiling between

[1]) Kowalewski, Die Produktivkräfte Russlands. German translation. Leipzig 1898, pp. 254. 255.

[2]) According to a private communication (1896) from Prof. G. Wagner of Warschau there are in Russian Poland, particularly in the gouvernements Lublin, Lomsa and Suwalki nearly 100 tar distilleries, each one of which produces on an average about 1,500—2,500 kilo of pine tar oil.

[3]) Pharm. Journ., III, 13, p. 586.

[4]) Liebig's Annalen, 230, p. 245.

[5]) Pharm. Journ., III, 8, p. 447; Journ. Chem. Soc., 33, p. 80.

[6]) Berichte, 20, p. 1956.

171—172° which Tilden supposed to be sylvestrene, though he did not succeed in preparing the chlorhydrate, m. p. 72°; and cymene, identified by means of bromine and sulphuric acid.[1] Wallach in 1885 substantiated the presence of pinene.[2] He also identified the sylvestrene in fraction 170—180° by means of the crystalline hydrochloride, m. p. 72°. In the fractions boiling about 180°, he also identified dipentene by means of its tetrabromide, and a terpene which yielded a liquid addition product with bromine (terpinene?).

17. Swedish Pine Tar Oil.

Schwedisches Kienöl.

Swedish pine tar oil has a sp. gr. 0.871, and a rotatory power $\alpha_D = +14° 48'$. According to the examination of Atterberg[3] (1877) it contains dextro-pinene, b. p. 156.5—157.5° (pinene chlorhydrate, m. p. 131°); and a hitherto unknown terpene, b. p. 173—175°, which Atterberg characterized by a dichlorhydrate melting at 72—73° and which he termed sylvestrene. Wallach[4] (1885) also examined this oil.

18. Pine Tar Oil from Finnland.

Finländisches Kienöl.

Since closed ovens are used in the production of wood tar in Finnland, a large quantity of pine tar oil is obtained as by-product from the trunks of the pines,[5] spruces and firs. Two varieties of pine tar oil from Finnland have been examined by Aschan and Hjelt[6] (1894).

1. From southern Finnland. After five repeated fractionations the following fractions were obtained; 1) 155—160°, 7.1 p. c.; 2) 160—165°, 30.2 p. c.; 3) 165—170°, 22.6 p. c.; 4) 170—175°, 20.1 p. c.

The first fraction consisted of pinene (hydrochloride, m. p. 123—124°; nitrosochloride, nitrosopinene). In fraction 170—174° sylvestrene (dihydrochloride, m. p. 72°) and dipentene (dihydrochloride, m. p. 49—50°) were found.

2. An oil from northern Finnland differed from the former in the relatively larger amount of higher boiling fractions. Fractions 160—165°, also 165—170° were insignificant, whereas fraction 170—174 represented 32.2 p. c., and fraction 174—178°, 21.2 p. c. In the lower fractions

1) In the presence of terpenes this reaction cannot be considered a positive proof.
2) Liebig's Annalen, 230, p. 245.
3) Berichte. 10, p. 1202.
4) Liebig's Annalen, 230, p. 240.
5) The forests of Finnland consist approximately of 77 p. c. pines and 12 p. c. spruce.
6) Chem. Zeitung, 18, pp. 1566, 1699, 1800.

pinene was found, whereas the higher fractions consisted chiefly of dipentene. Sylvestrene, though probably present in small quantities, could not be identified.

Mention has already been made (on p. 255, footnote 2) of the distillate with water vapor which was prepared in order to ascertain whether sylvestrene and dipentene are normal constituents of the original oil.

Pine Needle Oils.

Fichtennadelöle.

Distillates from the needles and cones of the Abietineae.

The collective term pine needle oil comprises the fragrant oils from the fresh needles and young twigs, also the one year old cones of pines, spruces, larches and firs.

If the above designation is not infrequently a misnomer, the names applied to specific oils in price lists are often no more correct, so that frequently no conclusion can be drawn from the name as to the source of the oil. Current names, though wrong, are difficult to displace. However, in the following presentation the correct names only will be used.

This state of affairs is partly due to the confusion existing in the Latin as well as in the German (we might also add English) nomenclature of the conifers. For this reason, the older literary references to coniferous oils must be cautiously accepted.

On account of their balsamic and refreshing odor, these oils are being used in the preparation of various pine needle essences for sprays in living rooms and sick chambers, in the preparation of aromatic baths, and in finer perfumery and the soap industry. They have, therefore, become current articles of commerce.

The cheaper pine needle oils are frequently adulterated with turpentine oil. Inasmuch as pinene is a normal constituent of these oils, its mere presence does not signify adulteration. The addition of a large percentage of turpentine oil, however, can readily be demonstrated by comparing quantitatively the fractions of a suspected oil with those of a genuine oil. In the adulterated oil the fractions distilling about 160° or below 170° are much larger than in the pine needle oil. As a basis for comparison, quantitative fractionations are recorded in connection with the more important oils. Variations in specific gravity are not great and, therefore, of less importance than the examination of the rotatory power of the oils. Besides, the saponification number of the oils adulterated with turpentine oil is much smaller than that of the genuine oils and, therefore, may serve in a general way as a good test for the purity of the pine needle oils.

19. Pine-Needle Oil from Abies Alba.

Edeltannennadelöl.

This oil is distilled principally in Switzerland and Tyrol (in the Puster valley) from the needles and twigs of *Abies alba* Miller (*Abies pectinata* D. C., *Abies excelsa* Lk.) (Germ. *Edeltanne, Weisstanne* or *Silbertanne*).

PROPERTIES.[1] This oil is a colorless liquid possessing a balsamic odor; sp. gr. 0.869—0.875; $\alpha_D = -20°$ to $-59°$. It is soluble in 5 parts of 90 p. c. alcohol. It contains 4.5 to 10.9 p. c. of ester (bornyl acetate). Upon distillation, 8 p. c. come over below 170°, and 55 p. c. between 170—185°. Above this temperature partial decomposition results, the bornyl acetate splitting off acetic acid.

COMPOSITION. Bertram and Walbaum[1] ascertained the presence of the following constituents: l-pinene (pinene nitrol benzylamine, m. p. 122—123°), l-limonene (tetrabromide, m. p. 104°), l-bornyl acetate,[2] and a sesquiterpene not yet identified.

20. Oil from Cones of Abies Alba.

Edeltannenzapfenöl. Templinöl.

This oil is obtained in several sections of Switzerland and the Thuringian forest by distilling the one year old cones of *Abies alba* Miller, collected in August and September, with water vapor.

PROPERTIES. This oil is colorless and possesses an agreeable balsamic odor reminding somewhat of lemon and orange. Its sp. gr. is 0.853 to 0.870; its $\alpha_D = -60°$ to 76°. The amount of ester (calculated as bornyl acetate) is 0.5 to 4 p. c. With 6 parts of 90 p. c. alcohol, the oil gives a clear solution. It is characterized by a large percentage of l-limonene. Inasmuch as this strongly optically active terpene is its most valuable constituent, the rotatory power of the oil is taken as a criterion of its value. The stronger the laevorotation and the lower the sp. gr., the more limonene the oil contains.

---

1) Bericht von S. & Co., Oct. 1892, p. 21; April 1893, p. 29; Archiv d. Pharm., 231, p. 291.

2) Hirschsohn, Pharm. Zeitschr. f. Russland, 31, p. 593.

Upon distillation 11 p. c. came over between 150—170°, and 37 p. c. between 170—185°. Above this temperature partial decomposition accompanied by splitting off of acetic acid takes place.

Composition. Of the older investigations of Templin oil those of Flückiger[1] (1856) and of Berthelot[2] (1856) should be mentioned. These investigators studied the action of strong acids on the oil and obtained terpin hydrate, a terpene monohydrochloride and a terpene dihydrochloride. The formation of these compounds is to be attributed to the presence of pinene and limonene as shown later.

In 1885 Wallach[3] examined a "templin oil" designated "Fichtennadelöl" and found its principal constituents to be pinene and l-limonene. Bertram and Walbaum[4] (1893) showed that the pinene also is laevogyrate and that "Templinöl" consists principally of l-pinene and l-limonene. They also found a trace of an ester, the nature of which could not be ascertained, but which in all probability is bornyl acetate, also found in the oil from the needles. On account of the large percentage of l-limonene, this oil is the best source for this hydrocarbon.

21. Pine Needle Oil from Picea Excelsa.

Fichten- or Rothtannennadelöl.

The "Fichtennadelöl" proper is obtained by steam distillation from the fresh needles and twigs of the Norway spruce, the *Picea excelsa* Lk., (*Picea vulgaris* Lk.). The yield is about 0.15 p. c. (Bertram and Walbaum.[5]) As far as is known, it is not a commercial article.

The odor of the colorless oil is as pleasantly aromatic as that of the oil from the needles and cones of *Abies alba* Miller. Its sp. gr. is 0.880 to 0.888, and its $\alpha_D = -21°41'$ to $-37°$. Upon fractional distillation Bertram and Walbaum[6] obtained 20 p. c. between 160—170°, and 50 p. c. between 170—185°. Above this temperature decomposition took place. Umney[7] (1895) obtained 41 p. c. between 163—173°, 16 p. c. betw. 173—176°, 13 p. c. betw. 176—185°, 14 p. c. betw. 185—220°, and 16 p. c. residue. The amount of ester (calculated as bornyl acetate) is 8.3 to 9.8 p. c.

Composition. Fraction 160—170° of "Fichtennadelöl" contains l-pinene (nitrolbenzylamine, m. p. 122—123°; nitrosopinene, m. p. 132°). Fraction 170—175° consists of a mixture of l-phellandrene (nitrite, m. p.

---

1) Vierteljahrsschr. f. prakt. Pharm., 5, p. 1; Jahresber. f. Chem., 1855, p. 642.

2) Journ. d. Pharm. et Chim., III, 29, p. 38.

3) Wallach, Liebig's Annalen, 227, p. 287.

4) Archiv d. Pharm., 231, p. 293.

5) Ibidem, p. 295.

6) Ibidem, p. 296.

7) Pharm. Journ., 55, p. 162.

101°) and dipentene (dihydrochloride, m. p. 50°). The higher fractions contain l-bornyl acetate and cadinene (dihydrochloride, m. p. 118°).[1]

22. Pine Needle Oil from Pinus Montana.

Latchenkiefer- oder Krummholzöl.

It is obtained by steam distillation from the needles and young twigs of *Pinus montana* Miller (*Pinus pumilio* Haenke, *Pinus mughus* Scop., Ger. *Latschen-* or *Zwergkiefer, Legföhre*) principally in the Austrian Alps, more especially in Tyrol (Puster valley), also in Hungary and Siebenbürgen.

Of the yield obtained at these places, no reliable data are on hand. An experimental distillation in Leipzig with fresh twigs obtained from Siebenbürgen and Hungary yielded 0.26 p. c.[2] and 0.68—0.71 p. c.[3] of oil respectively. The young wood without the needles yielded 0.27 p. c. of oil.

PROPERTIES. "Latschenkiefer" oil has a pleasant balsamic odor. It is colorless and has a sp. gr. of from 0.865 to 0.875. The oils, alluded to above, which had been distilled in Leipzig from material that had partly dried during the long transportation, could, therefore, not be considered as normal oils. Their sp. gr. was as high as 0.892.[3] The optical rotation of the normal oil varies from —4° 30′ to —9°; the percentage of ester, calculated as bornyl acetate, varies from 5—7 p. c. Upon fractional distillation, nothing came over between 160—170°, but 70 p. c. were obtained between 170—180° (Bertram & Walbaum[4]). An oil distilled by Umney[5] (1895) yielded 2 p. c. betw. 155—165°, 59 p. c. betw. 165—180°, 21 p. c. betw. 180—200° and 18 p. c. above 200°.[6]

COMPOSITION. According to Bertram & Walbaum[7] "Latschenkieferöl" contains but little l-pinene[8] in the lower fractions (pinene nitrolbenzylamine, m. p. 122—123°). In the following fractions l-phellandrene (nitrite, m. p. 102°), sylvestrene[9] (dihydrochloride, m. p. 72°) and bornyl acetate; in the highest fractions cadinene (dihydrochloride, m. p. 118°) have been found.

---

1) Archiv d. Pharm., 231, p. 296.
2) Bericht von S. & Co., Oct. 1898, p. 19.
3) Ibidem, Oct. 1896, p. 76.
4) Loc. cit., p. 297.
5) Pharm. Journ., 55, p. 162.
6) This seems to show conclusively that the requirements of the seventh edition of the Austrian Pharmacopoeia, viz. sp. gr. 0.85, and b. p. 170°, are inaccurate.
7) Loc. cit., p. 297.—Comp. also Liebig's Annalen, 116, p. 328.
8) First found by Atterberg (1881) and described by him as terebentene. Berichte, 14, p. 2531.
9) Atterberg suspected the presence of sylvestrene in this oil.

23. Pine Needle Oil from Pinus Silvestris.

Kiefernadelöl.

From the needles of *Pinus silvestris* L., Germ. *Kiefer* or *Föhre.*

German "Kiefernadelöl". The needles from *P. silvestris,* which grows in all parts of Germany, have been repeatedly distilled for experimental purposes. Although but slightly inferior to other pine needle oils in odor, the oil has not found its way into perfumery and the soap industry and consequently is not an article of commerce.

Fresh leaves distilled in July yielded 0.55 p. c.,[1] leaves distilled in December 0.45 p. c. of volatile oil (Bertram & Walbaum[2]).

PROPERTIES.[2] The odor of the oil resembles that of the pine needle oils from *Abies alba* Miller and *Pinus montana* Miller but is less delicate. Its sp. gr. is 0.884—0.886 and its $\alpha_D = +7° 3'$ to 10°. Upon fractional distillation 10 p. c. came over between 160—170°, 46 p. c. betw. 170—185°. With 10 p. of 90 p. c. alcohol the oil gives a clear solution. The ester content, calculated as bornyl acetate, is 3.2—3.5 p. c.

COMPOSITION. The German oil, like the pine needle oils previously mentioned, contains pinene, but the pinene differs from that found in these oils in being dextrogyrate[1] (nitrolbenzylamine, m. p. 122—123°). It also contains d-sylvestrene. Its dihydrochloride first melted below 50° and only after repeated recrystallization acquired a constant melting point of 72°. Inasmuch as, according to Wallach, dipentene dihydrochloride causes a material depression of the melting point of sylvestrene dihydrochloride, the assumption that dipentene is present may be justified.

Upon saponification, acetic acid was found which was evidently combined with an alcohol (probably borneol or terpineol) not yet identified. In the highest fractions cadinene (dihydrochloride, m. p. 118°) was found.

Swedish "Kiefernadelöl" is distilled in Sweden, more particularly in the district Jönköping, and is brought into the market as *Schwedisches Fichtennadelöl.* It is used for hygienic and medical purposes, in inhalations for affections of the lungs, as admixture to baths, and in sprays for sick chambers.

PROPERTIES. The Swedish oil corresponds in its general properties with those of the German oil. Its sp. gr. is 0.872, its rotatory power $+10° 40'$, and upon distillation yields 44 p. c. distillate at 160—170°

---

[1]) Archiv d. Pharm., 231, p. 300.

[2]) Ibidem; and Ber. von S. & Co., Oct. 1896, p. 76.

and 40 p. c. at 170—185°. It contains d-pinene (nitrolbenzylamine, m. p. 122—123°), d-sylvestrene (dihydrochloride, m. p. 72°), and small amounts of an ester (3.5 p. c. calculated as bornyl ester), the nature of which has not yet been definitely ascertained but, judging from the odor, appears to be bornyl acetate (Bertram and Walbaum,[1] 1893).

English "Kiefernadelöl." From the German and Swedish oils, the English is distinguished by its laevorotation.

Umney[2] (1895) distilled the needles of the Scotch fir, *Pinus silvestris* L. at different times of the year and obtained 0.5 p. c. of oil in June, and 0.133 p. c. in December. Its sp. gr. varied from 0.885—0.889, and its rotatory power, $\alpha_D = -7.75°$ to $-19°$. The ester content, calculated as bornyl acetate, amounted to 2.9—3.5 p. c. Upon fractionation of the two oils, the following results were obtained:

	Oil distilled, In June.	In December.
157—167°	8 p. c.	13 p. c.
167—177°	27 "	24 "
177—187°	20 "	9 "
187—197°	3 "	6 "
197—240°	7 "	7 "
240—252°	6 "	4 "
Residue	29 "	37 "

COMPOSITION. The lowest fraction deviated the ray of polarized light 13° to the left (in a 100 mm. tube) and possessed all the properties of l-pinene. Fraction 171—175° was slightly dextro rotatory (+0.75°), and corresponds in its properties to dipentene[3] and gave with glacial acetic and sulphuric acids the characteristic violet sylvestrene reaction.

With the exception of the opposite rotation, it may be assumed that the English oil has the same composition as the German and Swedish oils.

24. Hemlock or Spruce Needle Oil.

Hemlock- or Spruce-Tannennadelöl.

The needles and young twigs used in the distillation of this oil seem to be contributed by three different species: 1) *Abies canadensis* Michx., (*Tsuga canadensis* Carriere), the hemlock (Ger. *Spruce- Hemlock-* oder *Schierlings-Tanne*) which occurs throughout North America from Canada to Alabama and westward as far as the Pacific; 2) *Picea alba* Lk. or

1) Archiv d. Pharm., 281, p. 299.

2) Pharm. Journ., 55, pp. 161, 542.

3) Derivatives of this hydrocarbon were evidently not prepared. Neither did Umney succeed in obtaining the dihydrochloride of sylvestrene.

white spruce (Ger. *weisse Tanne*); and 3) *Picea nigra* Lk., the black spruce (Ger. *schwarze Tanne*). The two latter are equally widely distributed with the first. In the collection of the leaves and twigs it seems highly probable that no distinction is made between these three species, so that a commercial oil may contain variable amounts of the oils from all three. In fact the oils of these three species, being regarded as identical, are brought into the market under the common name of hemlock or spruce oil. Inasmuch as the oils are alike in properties and composition, quantitatively, the confusion in this case may be regarded as being of little or no consequence.

PROPERTIES AND COMPOSITION. Hemlock or spruce oil is colorless, of an agreeable balsamic odor,[1] sp. gr. 0.907—0.913, $\alpha_D = -20° 54'$ (Bertram and Walbaum[2]) to $-23° 55'$ (Power[3]). Fractional distillation yielded 11 p. c. between 150—170°, and 37 p. c. betw. 170—185°. Above this temperature decomposition sets in with the formation of acetic acid.

Hemlock oil contains l-pinene (nitrolbenzylamine, m. p. 122—123°), 36 p. c. l-bornyl acetate and sesquiterpenes not identified.

Hunkel[4] examined an oil distilled by himself from the fresh twigs of *Abies canadensis* Michx. Its sp. gr. at 20° was 0.9288, $[\alpha]_D = -18.399°$. It contained 51.5—52 p. c. of l-bornyl acetate. l-Pinene was identified by means of the nitrolbenzylamine base, m. p. 122°.

An oil distilled in September consisted principally of about equal parts of l-bornyl acetate and l-pinene.

25. Needle Oil from Picea Nigra.

Schwarzfichtennadelöl.

The oil from *Picea nigra* Lk. already mentioned under the previous heading is almost identical with hemlock oil. Its sp. gr. is 0.922 at 20°, $\alpha_D = -36.367°$ at 20°, $[\alpha]_D = -39.45°$. The oil distills between 160—230°, the principal fraction between 212—230°. It contains 48.85 p. c. bornyl acetate (Kremers,[5] 1895).

26. Needle Oil from Balsam Fir.

Balsamtannennadelöl.

The fresh twigs and young cones of the North American balm of Gilead fir yield upon distillation with water vapor an oil similar to the preceding ones.

1) The oil with which Hunkel worked was by no means very agreeable as to odor. Kleber, however, regards to the odor of the commercial oil as very pleasant. *E. K.*

2) Archiv. d. Pharm., 231, p. 294.

3) Descr. Catalogue of ess. oils, p. 58.

4) Pharm. Review, 14, p. 85.

5) Pharm. Rundschau, 13, p. 135.

Properties and Composition. The sp. gr. of the oil is 0.8881 at 20°; $\alpha_D = -28.91°$ at 20°; $[\alpha]_D = -32.55°$. Upon distillation the following fractions were obtained: up to 160°, 1.5 p. c.; 160—170°, 47.7 p. c.; 170—185°, 29.2 p. c.; 185—210°, 16.2 p. c.; residue 5.4 p. c.

This oil contains 17.6 p. c. of bornyl acetate. Fraction 160—165°, which is laevogyrate, yielded a nitrosochloride melting at 101°. Although the yield was too small to make any of its derivatives it may be assumed that the fraction in question contains l-pinene (Hunkel,[1] 1895).

27. Siberian Pine Needle Oil.

Sibirisches Fichtennadelöl.

This oil is distilled from the needles and young twigs of *Abies sibirica*[2] (Ger. *sibirische Fichte*); on a large scale in the gouvernement Wjätka, in northern Russia. On account of its strong, balsamic odor, the oil is used in perfuming "Fichtennadel" soaps and the cheaper "Tannenduft" preparations.

Properties and Composition. The oil is colorless, has a sp. gr. of 0.905—0.920; $\alpha_D = -40$ to $-42°$. The ester content (bornyl acetate) varies from 29—36 p. c. With equal parts of a 90 p. c. alcohol it makes a clear solution. Fraction 160—163° contains l-pinene (nitrolbenzylamine, m. p. 122—123°). Camphene could not be identified. In the higher fractions l-borneol was contained as acetic ester (Hirschsohn,[3] 1892). Besides bornyl acetate, it contains the ester of another terpene alcohol, possibly terpineol.[4]

28. Needle Oil from Pinus Cembra.

Zirbelkiefernadelöl.

The needles (without the branches) of the Siberian cedar, *Pinus cembra* L. (Ger. *sibirische Ceder, Arve*) yield upon distillation with water vapor 0.88 p. c. of oil. Its rotatory power, $\alpha_D = +29.1°$. It consists chiefly of d-pinene (monohydrochloride, m. p. 125°). Fraction 156° has a very high specific rotatory power $[\alpha]_D = +45.04°$.

On account of this high rotatory power, the oil is very serviceable in the preparation of a strongly dextrogyrate pinene (Flawitzky,[5] 1892).

---

1) Amer. Journ. Pharm., 67, p. 9.

2) According to a statement of the firm of R. Koehler & Co., of Moscau; comp. also Bericht von Schimmel & Co., April 1886, p. 15. According to Prof. Menthin of Warsau, however, this oil is obtained from the needles of *Larix sibirica* Ledebour (*Pinus ledebourii* Endl.).

3) Pharm. Zeitschr. f. Russl., 30, p. 593.

4) Ber. v. S. & Co., Oct. 1896, pp. 42 & 76.

5) Journ. f. prakt. Chem., II, 45, p. 115.

29. Oil from the Cones of Abies Reginae Amaliae.

The fruit of *Abies reginae amaliae* Heldr., which grows in the forests of Arcadia, contains so large an amount of volatile oil that it exudes when the fruit is compressed. Upon distillation of the crushed fruit more than 16 p. c. of volatile oil has been obtained (Buchner and Thiel,[1] 1864).

Properties and Composition. The sp. gr. of the oil is 0.868, its rotatory power — 20°. It begins to boil at 156°, the boiling point remains constant for some time at 170° and finally rises to 192°.

As shown by elementary analysis, the oil consists principally of hydrocarbons $C_{10}H_{16}$. A solid hydrochloride was not obtained. This is evidently due to the fact that besides pinene the oil probably contains limonene and dipentene. A mixture of hydrochlorides is thus obtained from which it is difficult to separate the individual compounds.

30. Larch Needle Oil.

Lärchennadelöl.

The needles of the larch, *Larix decidua* Mill. upon distillation yield but 0.22 p. c. of oil of the sp. gr. 0.878; $\alpha_D = + 0° 22'$.[2] It forms a clear solution with 5 and more parts of 90 p. c. alcohol. Saponification number 28.3, S. Z. after acetylization 46.

If it is assumed that the ester of this oil is bornyl acetate, as is the case in most coniferous oils, and that the alcohol is borneol, the above figures indicate a percentage of 8.1 p. c. of bornyl acetate in the original oil and 16.1 p. c. in the acetylized oil. With reference to the alcohol, the same figures indicate 6.53 p. c. of borneol as ester, and 6.14 p. c. of free borneol, a total of 12.67 p. c.

Upon distillation the following fractions were obtained: 160—165°, 30 p. c. ($\alpha_D = + 4° 15'$); 165—170°, 24 p. c.; 170—180°, 16 p. c.; 180—190°, 8 p. c.; 190—200°, 4 p. c.; 200—230°, 9 p. c.; residue 9 p. c.

Larch needle oil has a pleasant, refreshing pine needle odor. The small yield and high price, as well as the difficulty of obtaining larger quantities prevent its practical application.

31. Sequoia Oil.

Upon distillation of the needles of the Californian giant, *Sequoia gigantea* Torrey (*Wellingtonia gigantea* Lindl.) which had been cultivated in Zürich, Lunge and Steinkauler[3] (1880) obtained a volatile oil which partly solidified at ordinary temperature.

---

1) Journ. f. prakt. Chem., 92, p. 109.

2) Bericht von S. & Co., Oct. 1897. p. 66.

3) Berichte, 13, p. 1656; 14, p. 2202.

The greater portion of the distillate consisted of a hydrocarbon $C_{10}H_{16}$, boiling at 155°, which possessed a pleasant terebinthinate odor, had a sp. gr. of 0.8522 and a rotatory power $[\alpha]_j = + 28.8°$. With hydrogen chloride, a white, camphor-like hydrochloride crystallizing in needles was obtained. Although these facts indicate pinene, Lunge and Steinkauler regard it as a new terpene on account of its high rotatory power.[1]

Fraction 227—230° had a sp. gr. 1.045, angle of rotation + 6°, and its odor reminded of peppermint oil. From the elementary analysis the formula $C_{18}H_{20}O_{8}$ was calculated. Between 280 and 290° a small amount of a heavy yellow oil of empyreumatic odor was obtained.

Further there is contained in the oil a colorless hydrocarbon, which crystallizes in laminae melting at 105° and boiling between 290—300° (uncorr.). It is called sequojene and probably has the formula $C_{13}H_{10}$ being isomeric with fluorene.

32. Sandarac Resin Oil.

According to Balzer[2] (1896) African sandarac from *Callitris quadrivalvis* Vent. contains about 1 p. c. of volatile oil which can be obtained by steam distillation. The oil has a brownish color, a pleasant, strongly aromatic odor reminding of the odor of pines. In the cold it becomes viscid and apparently separates a stearoptene-like substance.

33. Thuja Oil.

Oleum Thujae.—Thujaöl.—Essence de Thuya.

Origin and Production. Thuja oil obtained by distillation with water vapor of the leaves and twigs of the *arbor vitae, Thuja occidentalis* L. (Ger. *Lebensbaum*). The yield varies according to the season from 0.4 to 0.65 p. c. It is largest in spring (March) and decreases toward summer (June, Jahns,[3] 1883). The small amount necessary to supply the demand, is distilled principally in North America.

Properties. Thuja oil is a limpid, colorless liquid or of a yellowish or greenish-yellow color. It has a characteristic strong, camphor-like odor of tansy, and a bitter taste. Sp. gr. 0.915—0.935; $\alpha_D = -5$ to $-14°$. With 3—4 parts of 70 p. c. alcohol it forms a clear solution. It boils between 160 and 250°. The first runnings contain an acid liquid consisting principally of acetic acid with a little formic acid (Jahns[3]).

1) Since then Flawitzky (1892) (Journ. f. prakt. Chem., II, 45, p. 115) has isolated a pinene with much higher dextro-rotation. See needle oil from *Pinus cembra* on p. 265.

2) Archiv d. Pharm., 234, p. 311.

3) Archiv d. Pharm., 221, p. 748.

COMPOSITION. Thuja oil was first examined by Schweizer[1] in 1843 without definite results. Jahns succeeded in separating three substances by fractional distillation. He obtained a dextrogyrate terpene fraction between 156—161°, a dextrogyrate oxygenated fraction between 195—197°, and a laevogyrate oxygenated fraction between 197—199°. The two latter fractions had the formula $C_{10}H_{16}O$ and inasmuch as they appeared to be optical isomers, he named them d- and l-thujol.

A few years later (1892) Wallach's[2] investigation clearly revealed the composition of thuja oil. By means of the nitrosochloride the identity of fraction 160° with d-pinene was established. The fractions up to 190° also contain a substance that is acted upon by potassa with the formation of potassium acetate and probably contains an acetic acid ester. Fraction 190—200° contains two chemically different ketones of the formula $C_{10}H_{16}O$, l-fenchone, and d-thujone. The purification of l-fenchone, Wallach accomplished by acting on fraction 190—195° with potassium permanganate and nitric acid, thereby completely destroying the thujone whereas the very stable fenchone remained almost unaffected. The fenchone thus obtained corresponds in all its properties, with the exception of opposite optical rotation, with the ketone obtained from fennel oil (comp. p. 166). To the second ketone $C_{10}H_{16}O$, which boils slightly higher, Wallach applied the name thujone. Properties and derivatives of this substance are described on p. 167.

From fraction 220° of the oil, Wallach[3] (1894) obtained an inactive oxime melting at 93—94° which corresponded with optically inactive carvoxime. Upon hydrolysis with sulphuric acid it yielded an oil volatile with water vapor, which had the odor of carvone and which boiled between 220—230°. The properties of the hydro-sulphide addition product, however, did not agree with those of the corresponding compound obtained with inactive synthetic carvone. Later on it was shown by analysis that the oxime is not a derivative of i-carvone, as Wallach first supposed, but of hydrocarvone, $C_{10}H_{16}O$[4] (1894). Inasmuch as Semmler[5] has shown (1894) that thujone, when heated to higher temperatures, is converted into carvotanacetone = hydrocarvone, it seems probable that hydrocarvone is not contained in the original oil but is formed by fractional distillation from thujone.

1) Journ. f. prakt. Chem., 30, p. 376; Liebig's Annalen, 52, p. 398.
2) Liebig's Annalen, 272, p. 99.
3) Liebig's Annalen, 275, p. 182.
4) Liebig's Annalen, 279, p. 384.
5) Berichte, 27, p. 895.

34. Thuja Root Oil.

The oil from the root of *Thuja occidentalis* L. has a deep-brown color, an odor reminding of that of thymoquinone, and a sp. gr. 0.979. The yield is about 2.75 p. c.[1]

35. Oil of Cypress.

Oleum Cupressi. — Cypressenöl. — Essence de Cyprès.

Origin and History. Cypress oil which was recommended by Bravo in 1892 against whooping-cough, was introduced into commerce by Schimmel & Co. in 1894.[2] It is distilled from the leaves and young branches of *Cupressus sempervirens* L. According to the season and the freshness of the material the yield varies from 0.6 to 1.2 p. c.

Properties. Cypress oil is a yellowish liquid, the odor of which is pleasant and reminds of cypress, but which after evaporation leaves an odor that reminds distinctly of labdanum and is ambra-like. Sp. gr. 0,88—0.89, $\alpha_D = +4$ to $+14°$. The oil is soluble in 4—5 parts of 90 p. c. alcohol.

Composition. Cypress oil consists chiefly of terpenes, principally of d-pinene (nitrolbenzylamine, m. p. 124°). Sylvestrene and sesquiterpenes appear also to be present. From the last runnings a crystalline substance occasionally separates, which crystallizes from alcohol in fine needles, from petroleum ether in compact crystalline masses. In its behavior, "cypress camphor" resembles cedar camphor or cedrol, from which it differs, however, by being optically inactive. Probably it is the optically inactive modification of cedrol. In addition, cypress oil contains small amounts of esters.

36. Hinoki Oil.

The white wood of the hinoki tree, *Chamaecyparis obtusa* Endl. (*Retinospora obtusa* Sieb. et Zucc.), which is cultivated in Japan, is used in the construction of the Shintô temples and the manufacture of lacquered goods.[3] The odor of the oil distilled from the leaves resembles that of the oil from the savin or thuja. It has a remarkably low boiling point. About half of the oil distills between 110 and 160°, the other half between 160 and 210°.[4]

---

1) Bericht von S. & Co., Apr. 1892, p. 43.
2) Ibidem, Oct. 1894, p. 70; and Apr. 1895, p. 22.
3) Rein, Japan, Leipzig, 1886, vol. 2, p. 277.
4) Bericht von S. & Co., April 1889, p. 44.

37. Oil of Juniper Berries.

Oleum Juniperi. — Wacholderbeeröl. — Essence de Genièvre.

Origin and History. The several species of the genus *Juniperus* of the family *Cupressineae* are evergreen trees or shrubs. Of these, *Juniperus communis* L. is distributed over the entire northern hemisphere, whereas the very similar *Juniperus oxycedrus* L. is found wild only in the Mediterranean countries east as far as the Caucasus and west as far as Madeira. The aromatic odor of all parts of the juniper, especially of its fruits, when burned has in all probability attracted early attention. It is also probable that the juniper tar oils, obtained by destructive distillation, were known and used previously to those obtained by water distillation from the wood and the fruits. Of the latter two that obtained from the wood seems to have been first known and used.[1]

Apparently, juniper was first used as a domestic remedy by the Greeks and Romans. The fruits of both species, which grow together in the Mediterranean countries, were merely distinguished as the larger and smaller juniper berries.

The empyreumatic oil obtained by the dry distillation of the juniper wood was already known to the Romans, and was used for medicinal purposes during the middle ages. The distilled juniper wood oil is repeatedly mentioned in the mediaeval treatises on distillation (*Destillirbücher*), and is enumerated together with the oil from the fruits in medical treatises and price ordinances (*Taxen*) of the sixteenth century. The yield of the oil from the fruits was determined by Cartheuser[2] (1738), and Spielmann.[3]

Production. The distillation of the berries and the preparation of the extract go hand in hand. After the mashed berries have been distilled with water, the residue in the still is extracted with hot water. The aqueous liquid is evaporated in vacuum pans to the consistency of a soft extract and is brought into the market as *Succus*[4] or *Roob Juniperi*. The yield of both oil and extract varies considerably in different years. The quality of the juice likewise varies considerably. In some years it is so rich in sugar that it congeals to a fairly solid mass.

---

1) See p. 18.

2) Fundamenta materia medicae, vol. 2, p. 346.

3) Ibid., vol. 2, p. 272.

4) The extract obtained in this manner does not correspond to the *Succus Juniperi* of the German Pharmacopoeia since according to the method of preparation of the latter a part of the oil remains in the extract.

On an average Italian berries yield 1—1.5 p. c., Bavarian 1—1.2 p. c., Hungarian 0.8—1 p. c. of oil. A smaller yield of oil, 0.6—0.9 p. c., is obtained from East Prussian, Polish, Thuringian and Frankish juniper berries. A distillation of Swedish berries yielded but 0.5 p. c. of oil.

These data contradict the opinion expressed by Maier[1] (1867) that the oil content of juniper berries grown in northern countries is greater than that of berries grown in southern countries. The above data indicate the opposite to be true. However, the collected data are not sufficient to settle this question finally.

Fairly considerable quantities of oil of juniper berries are brought into the market from the Hungarian "Komitat" Trencsin. Judging from experience, this Hungarian oil is of minor quality and not normal in composition. It is probably obtained as a by-product in the manufacture of gin.

Oil of juniper berries is used principally in the preparation of gin (*Steinhager, Genièvre*) and liquors, also to a limited extent medicinally, particularly in veterinary practice.

Properties. Juniper berry oil is a limpid liquid, colorless or slightly greenish. Old oil is more viscid, reacts acid and smells more or less rancid. Fresh oil possesses a peculiar terebinthinate odor, and a balsamic, burning, somewhat bitter taste. In regard to its physical constants, it shows considerable variation according to origin and method of preparation. The sp. gr. lies between 0.865—0.882, and in a normal oil between the narrower limits of 0.867—0.875. The oil is mostly laevogyrate up to $-11°$, seldom inactive, now and then slightly dextrogyrate. In alcohol, particularly in dilute alcohol, it is but slightly soluble. To effect solution 8—10 parts of 90 p. c. alcohol are requisite. With some oils a clear solution cannot be effected with 90 p. c. alcohol in any proportion. The solubility further diminishes with increasing age. With chloroform, carbon disulphide, benzene and amyl alcohol, juniper berry oil is miscible in all proportions.

The previously mentioned Hungarian oil has a sp. gr. 0.862—0.868, and is optically active up to $-18° 48'$. In its solubility in alcohol it behaves like the German. So-called extra strong juniper oil, which is obtained either by fractional distillation or by shaking out with alcohol of varying strength, loses its original greater solubility within a short period.

---

1) Die ätherischen Oele, p. 102.

Composition. Fraction 155—162° contains pinene[1] (Wallach, 1885): fraction 260—275° cadinene, as shown by the preparation of the dihydrochloride melting at 118°.[2] The fractions between 162 and 260° have not yet been specially examined, although they probably contain the characteristic constituents of the oil. The supposition that the peculiar odor of the oil is due to esters[2] had to be dropped since the oil retains its odor after saponification. The amount of saponifiable substances is small, for the saponification figures are very low: e. g. 3.3, 3.4, 3.5, 3.7. Kremel[3] (1888), however, has found figures as high as 7.4 and 16.4.

Alcoholic constituents which can be converted into esters by boiling with acetic acid anhydride, are present in small quantity only. In three normal oils the following saponification numbers were found after esterification: 18.3, 22.5, 22.9. Substances containing a methoxyl group are likewise not present. A further constituent is a substance obtained from the last runnings of an oil that had been standing in a cool place for a long time. It separated in the form of fine needles. After several recrystallizations from alcohol, fine needle-shaped crystals were obtained which melted at 165—166°.[4] Similar crystallizations have been formerly observed and have been described in older literature[5] as juniper camphor, juniper berry stearoptene, juniper berry hydrate.[6]

Adulterations of juniper berry oil have seldom been definitely proven, because a moderate addition of turpentine oil cannot be readily established, since pinene is a regular constituent of the oil and since the addition of turpentine oil but slightly changes the physical constants of the oil. The addition of alcohol is ascertained according to the method described on p. 200.

38. Oil from Juniper Wood.

The so-called juniper wood oil is nothing but turpentine oil distilled from juniper wood or branches, or turpentine oil to which some juniper berry oil has been added. The juniper wood oils of commerce correspond in their properties with such products. They are used externally as domestic remedies or in veterinary practice.

1) Liebig's Annalen, 227, p. 288.
2) Bericht von S. & Co., April, 1890, p. 48.
3) Pharm. Post, 21, p. 828.
4) Bericht von S. & Co., Oct. 1895, p. 46.
5) Of the older contributions on juniper berry oil the following may here be mentioned: Blanchet (1833), Liebig's Annalen, 7, p. 167; Dumas (1835), Liebig's Annalen, 15, p. 159; Soubeiran and Capitaine (1840), Liebig's Annalen, 34, p. 324.
6) Blanchet, loc. cit.; Buchner (1825), Repert. d. Pharm., 22, p. 425. From the insufficient data of the older authors it does not become apparent whether they had the same substance as Schimmel & Co. or possibly terpin hydrate.

39. Oil from the Berries of Juniperus Phoenicea.

A commercial variety of red juniper berries obtained from Smyrna and probably the fruit of *Juniperus phoenicea* L., yielded upon distillation 1 p. c. of oil. Its sp. gr. was 0.859, $\alpha_D = -4° 55'$ at 16°. The oil corresponds in all of its properties with those of the oil from *Juniperus communis*.[1]

40. Juniper Oil from Juniperus Oxycedrus.

The reddish-brown berries of *Juniperus oxycedrus* L., growing in Dalmatia and Istria, yield about 1.3[2] to 1.5[1] p. c. of oil which is quite terebinthinate in odor, reminding but faintly of juniper. The sp. gr. of the oil is 0.851[2] to 0.854;[1] $\alpha_D = -4° 40'$[1] to $-8° 30'$.[2] It does not produce a clear solution with 10 p. of 95 p. c. alcohol.

The odor of the oil distilled from the fresh twigs in Spain reminds of that of the finer pine needle oils.[3]

41. Oil of Savin.

Oleum Sabinae. — Sadebaumöl. — Essence de Sabine.

Origin and History. Savin, *Juniperus sabina* L. is indigenous to the temperate zones of the old world, but is not widely distributed.

Savin was used medicinally and in veterinary practice by the Romans. It seems probable that the name *Sabina* has been derived from the mountainous country of the Sabines lying to the north-east of Rome. Dioscorides and Pliny mention the plant among those being used medicinally, but distinguish between two varieties. The differences, however, were probably those of source only and slight morphological variations. Charlemagne in the ninth century mentioned it in his *Capitulare* and thus caused its cultivation in the northern Alps. The abbess Hildegard of Bingen mentions savin as a remedy in her writings; it is also one of the 77 remedies praised by Otto of Meudon (Macer Floridus). In England the tree seems to have been cultivated and used before the Norman conquest.

---

1) Bericht von S. & Co., Oct. 1895, p. 45.

2) Observation in the factory of Schimmel & Co.

3) Bericht von S. & Co., October 1889, p. 54.

Cade oil, which is obtained in southern France by the dry distillation of the branches and the wood of *Juniperus oxycedrus*, and which is used medicinally, is of special interest because it contains a high percentage of cadinene. This sesquiterpene, which was named by Wallach (Liebig's Annalen, 238, p. 82; 271, p. 297) is widely distributed in volatile oils.

During the period in which distilled waters were in general use, *Aqua sabinae* was also officinal and is enumerated in the treatises on distillation mentioned on pp. 23 *et seq.*

The distilled oil is first mentioned in the price ordinance of Frankfurt-on-the-Main for 1587 and was described by Begninus at the close of the seventeenth century.[1] Concerning the yield of the oil Hoffmann seems to have made the first experiments about 1715.[2] Wedel examined the oil in 1707 according to the methods in vogue at his time.[3] The first chemical examination was made by Dumas in 1835.[4]

PREPARATION. Oil of savin is prepared by distillation with steam of the leaves and twigs (*Summitates sabinae*). The yield varies between 4 and 5 p. c. according to the time of collection and the freshness of the material. The leaves used for this purpose come principally from Tyrol. In southern France the oil is also distilled, but apparently the French oil of the market is invariably adulterated with large quantities of turpentine oil.

PROPERTIES. Oil of savin is a colorless or yellowish liquid of an unpleasant narcotic odor and a bitter, pungent, camphor-like taste. Sp. gr. 0.910—0.930; $\alpha_D = +42$ to $+60°$; saponification number 115—125. It is soluble in ½ part or more of 90 p. c. alcohol. Of 80 p. c. alcohol, 15—20 volumes are requisite, but a perfectly clear solution is not always effected. The oil distills between 175 and 250° leaving considerable residue. Below 200° at most 25 p. c. distill over.[5]

COMPOSITION. The principal constituent is an alcohol, "sabinol", which occurs partly free, partly combined with acetic acid as ester.[6] Sabinol, which can be obtained by fractionation of the saponified oil, boils at 210—213° (under 20 mm. pressure between 105—107°). Its odor reminds of thujone; the odor of the acetate reminds of savin. Inasmuch as oil of savin yields a higher saponification number after acetylization than before, a part of the alcohol possibly exists in the free state. The acid combined with sabinol was shown to be acetic acid by the analysis of the silver salt.

If the formula assumed for sabinol, $C_{10}H_{18}O$, is correct, the saponification numbers 115—125 correspond to a content of 40—44 p. c. of acetate. The amount of free alcohol in an oil examined was about 10 p. c.

---

1) Tyrocinium chymicum, vol. III, p. 27.
2) Opera omnia, Liber 65. Observatio 1.
3) Dissertatio de Sabina. Jenae 1707.
4) Liebig's Annalen, 15, p. 159.
5) Pharm. Journ., III, 25 (1895), p. 1045.
6) Bericht von S. & Co., Oct. 1895, p. 89.

According to a more recent investigation by Fromm,[1] (1898) the formula of the alcohol is not $C_{10}H_{18}O$, but $C_{10}H_{16}O$. When pure it boils at 208—209°, its acetic ester $C_{10}H_{15}OCOCH_3$ at 222—224°. In small part the sabinol of the savin oil is combined with an acid that boils at 247°. Oxidized with potassium permanganate, sabinol yields quantitatively tanacetogen dicarbonic acid, $C_9H_{14}O_4$, which melts at 140°.

In the highest fractions of savin oil, Wallach[2] (1887) found cadinene, $C_{15}H_{24}$. In addition. it is probable that terpenes are present, but nothing reliable as to their nature is known. From the results published by Dumas[3] in 1835 the conclusion that either pinene or camphene is present might seem warranted. Dumas isolated a fraction 155—161° which had the composition $C_{10}H_{16}$. Gruenling[4] (1878) obtained terephthalic and terebinic acids upon oxidation of fraction 161°. This also indicates pinene. Inasmuch, however, as savin oil does not contain any considerable quantities boiling below 175°, and since it is frequently adulterated with turpentine oil, these statements should be accepted cautiously.

According to Umney[5] oil of savin is said to contain a considerable amount of polyterpenes of the b. p. 226°. This, however, does not seem very probable since the boiling point of the polyterpenes lies in the neighborhood of about 300°.

The chemists of Schimmel & Co.[6] have isolated an aldehyde or ketone from fraction 220—250° which, when regenerated from the sodium acid sulphite addition product, possesses an odor which faintly reminds of that of cumin aldehyde. The unstable hydrazone melts at 40.45°; the more stable oxime at 85°.

Examination. The principal adulterant of oil of savin is turpentine oil. The oil distilled in southern France seems to enter the market only after the addition of a large amount of turpentine oil.

The addition of French turpentine oil can be recognized by the lowering of the sp. gr., the reduction or inversion of the angle of rotation, the lowering of the saponification number and the diminution of the solubility in alcohol. Of an oil, adulterated with turpentine oil. more than 25 p. c. distills over below 200°.

1) Berichte, 31, p. 2025.

2) Liebig's Annalen, 238, p. 82.

3) Liebig's Annalen, 15, p. 159.

4) Beiträge zur Kenntniss der Terpene. Inaug. Diss., Strassburg, p. 27. —See also Levy (1885), Berichte, 18, p. 3206.

5) Loc. cit.

6) Bericht von S. & Co., Apr. 1900, p. 40.

42. Oil of Red Cedar Wood.

Oleum Ligni Cedri.—Cedernholzöl.—Essence de Bois de Cèdre.

ORIGIN AND PREPARATION. The Virginia or red cedar, *Juniperus virginiana* L., a shrub or tree growing as high as 15 m., is distributed throughout the United States of North America. Its wood is used in the manufacture of cigar boxes, lead pencils and small ornaments. It is adapted to this purpose on account of its uniform structure, its mild sandal-wood-like odor and because it is not attacked by insects.

For the distillation of the oil, the waste from the lead pencil manufactory is used, yielding from 2.5 to 4.5 p. c. The exhausted chips are then utilized by the furriers in the preparation of skins.

A very inferior oil is obtained in this country as a by-product from the drying chambers of the lead pencil factories. These chambers are so constructed that the escaping vapors from the cedar wood can be condensed. In this case, however, the high boiling constituents of the wood remain behind and only the more volatile constituents are obtained. As a result the oil thus obtained is more mobile, its odor is both less fine and less permanent than that of the normal, and not serviceable for perfumery.

PROPERTIES. Oil of cedar wood is almost colorless, somewhat viscid, and sometimes contains crystals of cedar camphor. It has a peculiar mild but persistent odor. The inhalation of the vapors causes the urine to acquire a violet odor. Sp. gr. 0.945—0.960; $a_D = -30$ to $-40°$; index of refraction $n_D = 1.505$ at 17°. In alcohol it is rather difficultly soluble, for 1 p. of oil requires 10—20 p. of 90 p. c. alcohol to effect solution.

COMPOSITION. The most interesting constituent of cedar wood oil, the so-called cedar camphor, was first examined by Walter[1] (1841). He describes the oil as a soft, slightly reddish mass, permeated with crystals. The camphor, purified by expression and by recrystallization from alcohol, melted at 74° and boiled at 282°. By the action of phosphoric acid anhydride, Walter obtained from it the hydrocarbon cedrene, which boiled at 237°. The formula, $C_{16}H_{28}O$, assigned to cedar camphor by Walter, was declared improbable by Gerhardt,[2] who suggested $C_{15}H_{26}O$, the correctness of which was verified by more recent investigations.

---

1) Liebig's Annalen, 39, p. 247. 2) Lehrbuch d. org. Chemie, vol. 4, p. 378.

The liquid constituents were examined by Chapman and Burgess.[1] By fractional distillation of the oil, they prepared cedrene (b. p. 261—262°; sp. gr. 0.9359; $\alpha_D = -60°$) and compared it with the hydrocarbon obtained by the dehydration of fraction 301—306° of sandal wood oil, which according to the opinion of Chapoteaut[2] (1882) was identical with the former. Chapman and Burgess, however, came to the conclusion that both hydrocarbons were very similar but not identical.

In a recently conducted investigation of cedarwood oil, Rousset[3] (1897) obtained the following results: cedrene (obtained by fractional distillation of the volatile oil) is a sesquiterpene, $C_{15}H_{24}$, and boils under 10 mm. pressure at 131—132°; $[\alpha]_D = -47° 54'$. Attempts to establish the number of double bonds in the molecule, by the addition of hydrogen chloride and hydrogen bromide, failed on account of the lack of stability of the addition products. Upon oxidation of cedrene with chromic acid in glacial acetic acid, a liquid ketone, cedrone, results which boils at 147—151° under 7.5 mm. pressure. It has the composition $C_{15}H_{24}O$ and upon reduction is converted into the alcohol isocedrol, isomeric with the cedar camphor or cedrol (m. p. 84°). When heated with acetic acid anhydride in a sealed tube, only a part of the cedrol is converted into the acetic ester, another part is dehydrated yielding a sesquiterpene. Treated with benzoyl chloride, no ester is obtained but only the hydrocarbon $C_{15}H_{24}$. Inasmuch as neither aldehyde nor ketone result upon oxidation, cedrol is to be regarded as a tertiary alcohol.

On account of the readiness with which cedrol loses a molecule of water, its quantitative estimation by means of acetylization is impossible. A stronger dehydrating agent than acetic acid is formic acid, with the aid of which cedrol can be quantitatively converted into the hydrocarbon in the cold.[4] The latter boils at 262—263° and is strongly optically active, $\alpha_D = -80°$. It has, therefore, almost the same boiling point as the natural cedrene from the oil and is probably identical with it.

Cedrol is not always contained in cedar wood oil. After having looked for this substance for years in vain in the laboratory of Schimmel & Co., it has recently been observed more frequently. The cause for this peculiar phenomenon has not yet been ascertained.

---

1) Proc. Chem. Soc. 1896, p. 140.

2) Bull. Soc. chim., II., 37, p. 303.

3) Bull. Soc. chim., III., 17, p. 485.

4) Observation made in the Laboratory of Schimmel & Co.

Examination. Adulterations of cedarwood oil have not yet been observed. This cheap oil, however, is frequently used for the adulteration of other oils, for which purpose it is well adapted on account of its feeble odor. It can be recognized by its high sp. gr., its high boiling point, its strong laevorotation, and by its slight solubility in alcohol.

43. Oil of Cedar Leaves.

Oleum Foliorum Cedri. — Cedernblätteröl. — Essence des Feuilles de Cèdre.

According to the observations made in the laboratory of Fritzsche Bros.,[1] the oil of cedar leaves of American commerce is never what it ought to be, namely the oil from the leaves of *Juniperus virginiana* L. This is partly due to the fact that the name cedar is applied in this country to two totally different trees, viz. *Juniperus virginiana* and *Thuja occidentalis*. It is true, a distinction between them as red and white cedar is generally made, but the distillers of the oil evidently pay little or no attention to it. They not only use the leaves of both species indiscriminately, but also those of other conifers. It is not surprising, therefore, that the cedar leaf oils of commerce vary considerably in their properties.

The specific gravities of a number of commercial oils varied from 0.868 to 0.920, the optical rotation from $-3°40'$ to $-24°10'$. Some of these oils were soluble in 4 or 5 vols. of 70 p. c. alcohol, others not. All of these oils had a more or less thuja-like odor.

Genuine cedar leaf oil seems to have been distilled but once.[2] The yield was 0.2 p. c. This oil had the following properties:[3] sp. gr. 0.887; $\alpha_D = +59°25'$; insoluble in 10 parts of 80 p. c. alcohol. The odor was pleasant, somewhat sweetish. Upon fractional distillation most of the oil distilled below 180°, of this the greater part between 173°—176°. The sp. gr. of this fraction was 0.847; optical rotation $+89°$ (at 20° in a 100 mm. tube). Upon bromination it yielded a tetrabromide melting at 104—105°. This fraction, therefore, consisted almost entirely of d-limonene. Of the lower fractions but small quantities supposed to contain pinene were obtained. Although a nitrosochloride was successfully prepared, no crystalline piperidine and benzylamine bases could be obtained. Possibly, as seems to be the case with lemon oil, there are here terpenes, closely related to those already known, but which have not yet been definitely characterized.

1) Bericht von S. & Co., April 1898, p. 13.
2) In the factory of Fritzsche Bros. Comp. Bericht von S. & Co., Apr. 1894, p. 56.
3) Bericht von S. & Co., Apr. 1898, p. 18.

The fractions boiling higher than limonene were treated with alcoholic potassa because they contained small amounts of saponifiable substances (saponification number of the original oil 10.9, of the acetylized 39.1). The acids thus obtained yielded a silver salt corresponding with silver valerianate upon analysis. The acid-free oil then yielded upon distillation under diminished pressure a fraction which under ordinary pressure boiled at 210—215°. Upon cooling, this congealed to a crystalline mass which upon expression and sublimation revealed itself as borneol melting at 203—204°. In the highest fractions cadinene could be readily identified by means of its characteristic hydrochloride.

Genuine cedar leaf oil, therefore, consists principally of limonene with cadinene, and some borneol, also small amounts of bornyl esters.

44. Lebanon Cedar Oil.

Origin and History. Whether cedar wood was distilled in antiquity or not does not become apparent from the statements by Dioscorides and other writers. Dioscorides[1] mentions as medicinal plants a number of pinus species, among them apparently also the cedar of Lebanon. The *Cedrus libani* Barr. (*Pinus cedrus* L., *Abies cedrus* Poir. *Larix cedrus* Mill.) was considered during antiquity as one of the noblest trees. As such, and on account of its durable wood, it is often mentioned in the books of the Old Testament.

Upon distillation, the comminuted wood yields 2.9 p. c. of oil.[2]

Properties. The yellowish-brown oil has a very pleasant cedar odor and might find practical application if a sufficient quantity of the wood were to be had cheap. Sp. gr. 0.985; $\alpha_D = -10° 48'$.[2]

45. Pandanus Oil.

In India, Arabia and Persia, the flowers of *Pandanus odoratissimus* L. (Fam. *Pandanaceae*) are highly esteemed on account of their fragrance and their supposed medical virtue.[3] The Mohammedan physicians use the aqueous decoction of the bruised stems against various diseases. Among the Hindoos, the aqueous distillate of the flowers is used as a preventive against small-pox. If the distilled water is to be used as a perfume, it is sometimes prepared with rose water and the addition of sandal wood oil. An odoriferous fatty oil, made by maceration of

---

1) De mat. med. Editio Kühn-Sprengel, Vol. 1, pp. 93, 140, 880.

2) Bericht von S. & Co., Apr. 1892, p. 41.

3) Dymock, Warden and Hooper, Pharmacographia Indica. Part. VI, p. 585.

the flowers with sesame oil, is also much used. According to Holmes[1] (1880) the volatile oil has a very pleasant, decidedly honey-like odor. Concerning the other properties of the oil nothing is known.

Andropogon Oils.

Origin and History. The genus *Andropogon* of the family *Gramineae* is distributed well-nigh over all parts of the globe. A number of species, which are indigenous particularly to the East Indies, the islands of the Indian Archipelago and northern Africa, are odoriferous on account of the volatile oil they contain. The oils distilled within recent periods from these grasses, some of which are being cultivated, are: palmarosa or rusa oil, also known as Indian geranium oil; citronella oil; lemon-grass or Indian verbena oil; and vetiver oil.

These aromatic grasses have been used on account of their fragrance for various purposes during antiquity: for the aromatization of wine; also of earthenware wine cups, the so-called Rhodian cups; in the preparation of ointments and oils; an incense in religious rites; and as couches during festivities. In Sanskrit writings, in the Old Testament, and in other documents of antiquity, these grasses are referred to under various names. The spices and annointing oils mentioned in the biblical translations and other ancient writings as *narde, stakte, schönus*, etc., apparently have been used synonomously for the fragrant andropogon grasses and their roots. Of these, it may be supposed that *Andropogon laniger* Desf. was the best known and most used during antiquity, inasmuch as it was more widely distributed throughout northern India, Thibet, Persia and Arabia as far as Egypt, Nubia and Ethiopia than the other species. Originally, however, and again in more modern times, the term narde was applied only to the aromatic root of the valerianaceous *Nardostachys jatamansi* D. C., indigenous to the Himalayas of northern India, perhaps also to *Valeriana celtica* L. indigenous to the European Alps.

The Greek and Roman writers possibly referred to the same aromatic andropogon species when they used the words *ξοῖνος* or *σχοῖνος*, also *juncus*. In the occident they apparently have never been cultivated nor introduced in the dried condition.

The first mention of andropogon grasses by European travelers is to be found in the works of Garcia da Orta, van Rheede tot Draakenstein,—who was governor of the Dutch East-India company on the

---

1) Pharm. Journ., III. 10, p. 685.

Malabar coast about the middle of the seventeenth century,—and of G. E. Rumpf (Rumphius, also Plinius indicus), Dutch governor in Amboyna during the second half of the seventeenth century. The first sample of a distilled andropogon oil, a lemon-grass oil, is said to have been brought to Europe from the Moluccas in 1717. However, the distillation of these oils on a large scale and their introduction into the commerce of the world and into industry apparently first began in 1820. In this year the botanist William Roxburgh, who was Director of the Botanical Garden at Calcutta for a long time, mentions lemon-grass oil as coming from the Moluccas.[1] In 1832 the first large assignment was received in London. Since then, it, as well as the palmarosa oil and somewhat later citronella oil, have been finding increased application in perfumery and especially in the soap industry. As a result of the ever increasing demand, the cultivation of these aromatic grasses, especially of the citronella grasses in Ceylon, has increased considerably so that these oils are now exported in large quantities.

Besides Asia and Africa, several aromatic andropogon species thrive in Central and South America, on the West Indies and in Australia. With the increasing consumption of these oils in the soap industry, these countries as well as Italy may become producers in the course of time. Already samples of lemon-grass oil from southern Brazil have been sent to Europe.[2]

46. Palmarosa (East Indian Geranium) Oil.

Oleum Palmarosae seu Geranii Indicum.—Palmarosaöl.—Essence de Géranium des Indes.

Origin and Production. Palmarosa oil, also known as Indian grass oil, rusa oil, Indian or Turkish geranium oil, is the oil from the leaves of *Andropogon schoenanthus* L. (Family *Gramineae*). This plant is widely distributed in India proper and is also frequently found in tropical West Africa.[3] The designation Turkish geranium oil, which, on account of its incorrectness, has now quite generally been dropped, has come down to us from a time when the oil entered the European market via Constantinople. From Bombay it was shipped to the ports of the Red Sea, and thence was conveyed over the land route through Arabia to Constantinople. After having been treated in a special manner, it was here used on a large scale for the adulteration of rose oil.

1) Flora indica, vol. 1, p. 280.

2) Bericht von S. & Co., Apr. 1896, p. 68.

3) Berichte d. pharm. Ges., 7, p. 501.

The preparation of the oil is described in the Pharmacographia indica of Dymock, Warden and Hooper[1] in the following manner: "The oil distillers in Khandesh[2] call the grass *Motiya*, when the inflorescence is young and of a bluish-white color; after it has ripened and become red, it is called *Sonfiya*. The oil obtained from it in the first condition has a more delicate odor than that obtained from the ripened grass. The Motiya oil is usually mixed with the second kind, which by itself would not fetch a good price in the European market. The grass grows freely, though not very widely, on open hillsides in West Khandesh, especially in Akráni. The makers are Mussulmans, who, at the close of the rains, about September, when the grass is ripening, buy it from the Bhils, stack it, and set furnaces at the sides of brooks where wood and water are plentiful. A large pit, four feet long by two wide and two and a half deep, is dug, and a furnace (*chula*) prepared. Over this furnace is placed a copper or iron caldron, large enough to hold from 30 to 50 pots of water. After pouring in some water, the caldron is filled to the brim with chopped grass, and a little more water is added. The mouth of the caldron is carefully closed with an iron or copper plate, made fast with wheat dough. From a hole in this lid, a bamboo tube, wrapped in a piece of cloth, plastered with the flour of *Udid* (*phaseolus mungo*, Linn., black var.), and bound with ropes, passes into a second closed caldron, sunk to the neck in running water. The steam from the grass is condensed in the second caldron, which, when full, begins to shake. The tube is then skillfully removed, and the contents of the caldron poured into a third similar vessel and stirred. Then the oil begins to appear on the surface, and is slowly skimmed off. The distillate is returned with fresh grass to the still." The yield amounts to about 0.3—0.4 p. c.

COMPOSITION. The older statements concerning the botanical origin and physical properties of palmarosa or Indian grass oil are so contradictory that it seems doubtful as to whether the authors really had palmarosa oil.[3]

---

1) Part. VI, p. 558; comp. Archiv d. Pharm., 234, p. 321.

2) Bombay presidency.

3) Stenhouse (Liebig's Annalen, 50, p. 157) in 1844 reports on an investigation of East Indian grass oil from *Andropogon Ivarancusa*, the odor of which resembled that of rose oil, the taste that of oil of lemon. Upon distillation it yielded a hydrocarbon, $C_{10}H_{16}$, boiling at 170°. It may be assumed with some degree of certainty that this grass oil was not palmarosa oil, but citronella oil from *Andropogon nardus* L., sometimes also designated *Andropogon Ivarancusa* Roxb.; for citronella oil contains a terpene boiling at 160°, viz. camphene, whereas palmarosa oil does not contain such low boiling constituents.

Gladstone (Journ. Chem. Soc., 17, p. 1; Jahresb. f. Chem., 1863, p. 548) describes

The first examination with unobjectionable material was carried out by Jacobsen[1] in 1871. He established the fact that the principal constituent of the oil is an alcohol, $C_{10}H_{18}O$, boiling at 232—233°, to which he gave the name geraniol. He further discovered the calcium chloride compound of geraniol, which is readily decomposed into its components by water, and which has become of such great importance in the purification of this alcohol. Semmler[2] later (1890) verified the correctness of the formula $C_{10}H_{18}O$ and recognized geraniol as belonging to the unsaturated chain compounds. Geraniol thus became the first representative of a new and important class of constituents of volatile oils designated as aliphatic terpene alcohols.

The amount of geraniol in palmarosa oil varies from 76—93 p. c. Of this the greater portion exists in the free state, and about 5.5—11 p. c. are present as ester. The acids of the ester content of the oil are, as was shown by Gildemeister and Stephan[3] (1896), approximately like parts of acetic and normal capronic acids. Acetic acid was identified by means of its silver salt (calculated: 64.69 p. c. Ag.: found: 64.54 p. c. and 64.37 p. c. Ag.). The capronic acid isolated from palmarosa oil boils at 205—206°, has a sp. gr. of 0.935 at 15°, is optically inactive and, therefore, identical with normal capronic acid. (Analysis of silver salt: calculated: 48.43 p. c. Ag.; found: 48.24—48.55 p. c. Ag.)

Of terpenes but a small amount is present in palmarosa oil, viz. about 1 p. c. of dipentene (tetrabromide, m. p. 125°; nitrolbenzylamine, m. p. 109—110°). Judging from the odor of the oil, traces of methyl heptenone are also present.

Besides geraniol, Flatau and Labbé[4] claim to have found a second alcohol, viz. citronellol. Their method of separation of these two alcohols is as follows:

The oil is first saponified and the saponified product distilled under diminished pressure. Fraction 120—140°, under 30 mm. pressure, is boiled with an equal weight of phthalic acid anhydride and an equal volume of benzene for an hour in a flask connected with a reflux condenser. Soda is then added, the product of reaction dissolved in water, and the solution extracted with ether. Hydrochloric acid is finally added whereby the acid phthalic acid esters of geraniol and citronellol are liberated.

---

an Indian geranium oil, sp. gr. 0.948 at 21°, which he considers identical with the East Indian grass oil from *Andropogon Ivarancusa*. Which oil it was in reality does not become apparent from the statements of the author.

1) Liebig's Annalen, 157, p. 282.

2) Berichte, 23, p. 1098.

3) Archiv der Phar., 234, p. 321.

4) Compt. rend., 126, p. 1725; Bull. Soc. chim., III, 19, p. 688.

These esters are now dissolved in ligroin and the solution cooled down to —5°. At this temperature it is said that the geranyl ester crystallizes out quantitatively whereas the citronellyl ester remains in solution. The isolated acid esters are saponified with alcoholic potassa and the free alcohols purified by rectification.

In this manner Flatau and Labbé claim to have demonstrated the presence of 63 p. c. of geraniol and 17 p. c. of citronellol in palmarosa oil. A repetition of this experiment in the Laboratory of Schimmel & Co.[1] has revealed the fact that the liquid ester regarded as the acid citronellyl phthalic acid ester was also fairly pure geranyl ester as was shown by the calcium chloride compound of the alcohol obtained by saponification. If citronellol is contained in palmarosa oil, the proof for which may be considered to be still wanting, it can be present in traces only.

A second statement by the same chemists appears to be based on a similar weak foundation. Flatau and Labbé[2] have claimed that palmarosa oil contains a saturated fatty acid $C_{14}H_{28}O_2$, m. p. 28°. According to the investigations of Schimmel & Co.,[3] pure palmarosa oil does not contain such an acid. The observation of the French chemists may possibly be referred to the presence of cocoa nut oil or some other fatty oil which had been added as adulterant.

Properties. Palmarosa oil is colorless or light yellow, and has a pleasant odor reminding of roses. The optical rotation varies, some oils being slightly dextrogyrate, others slightly laevogyrate, still others inactive. Angles of rotation varying from +1° 41′ to −1° 55′ have been observed.

It is soluble in 3 or more parts of 70 p. c. alcohol. The saponification number lies between 20 and 40. After acetylization the saponification number lies between 230 and 270. Jeancard and Satic[4] have recently observed that the amount of free acid increases if the oil is kept in partly filled containers.

Examination. Palmarosa oil is frequently adulterated. The following adulterants have been found: gurjun balsam oil, cedar wood oil, turpentine oil, petroleum (kerosene, paraffin oil) and cocoa nut oil.[5] They are all indicated by their solubility in 70 p. c. alcohol. Oils adulterated with cocoa nut oil solidify when exposed to the temperature of a freezing mixture. Petroleum and turpentine oil lower the sp. gr., whereas fatty

---

1) Bericht von S. & Co., Oct. 1898, p. 67.
2) Compt. rend., 126, p. 1726.
3) Bericht von S. & Co., Oct. 1898, p. 29.
4) Bull. Soc. Chim., III, 23, p. 37. Comp. also Bericht von S. & Co., Apr. 1900, p. 26.
5) Bericht von S. & Co., Apr. 1888, p. 22; Apr. 1889, p. 20; Oct. 1890, p. 28.

oils raise it. In doubtful cases it is well to resort to an acetylization. Oils with less than 75 p. c. total geraniol content are to be rejected.

PRODUCTION AND COMMERCE. For the last 20 years Bombay has been the principal center and port of exportation, whereas formerly almost the entire output was taken by caravans over land from East India to Cairo and Constantinople. From these places it was distributed to the brokers.

The oil is distilled in the Khandesh district northeast of Bombay. Pimpalner, Akrani, Nandurbár, Sháháda and Taloda are the principal places of production. In skins it is conveyed on the bullock's back over the Kundaibári pass to Surat, and by Dhulia and Manmad to Bombay. Here it is transferred to large tinned copper vessels of 100 to 200 lbs. capacity. These are not boxed, but surrounded by a network of ropes for the purpose of more easy handling.

In 1879 the total production was estimated at 3,600 k. Since then it has risen enormously and may amount to about 20,000 k. at present.

47. Ginger-grass Oil.

Ginger-grass oil is an inferior quality of palmarosa oil, or a mixture of the latter with much (up to 90 p. c.) turpentine oil or mineral oils.

Occasionally other grasses are also used in the distillation (*Andropogon laniger?*), for some ginger-grass oils possess a phellandrene-like odor which is entirely wanting in the palmarosa oil. Such an oil had the sp. gr. 0.897, $\alpha_D = -2° 8'$ and was soluble in 70 p. c. alcohol. Its petroleum ether solution gave a faint phellandrene reaction with sodium nitrite and glacial acetic acid (Gildemeister & Stephan,[1] 1896).

48. Lemon-grass Oil.

Oleum Andropogonis Citrati.—Lemongrasöl.—Essence de Lemongrass. Essence de Verveine des Indes.

ORIGIN AND PREPARATION. Lemon-grass, *Andropogon citratus* D. C., which is cultivated all over India, is used, in the form of an infusion, both internally and externally against all possible diseases. The oil, which is officinal in the Pharmacopoeia of India, is especially esteemed by the Hindoos as a remedy against cholera. For the production of the oil, the grass is cultivated on a large scale only on the Malabar coast in Travancore on the western slope of the mountains, north of Ajengo. The distillation is conducted in a primitive manner in plain stills.[2]

---

1) Archiv d. Pharm., 284, p. 326. 2) Pharmacographia Indica, Part VI. p. 564.

Lemon-grass oil is also distilled near Singapore and in Ceylon. *Andropogon citratus* is also cultivated in S. Thomé, where oil has been distilled experimentally[1] (1897). In Brazil (Porto Alegre) an attempt to produce the oil has also been made.[2]

Composition. Nothing about the chemical composition of lemon-grass oil was known until 1888 when citral, an aldehyde $C_{10}H_{16}O$ possessing a strong lemon-like odor, was recognized as the principal constituent in the laboratory of Schimmel & Co.[3] Two years later Dodge[4] described this substance as citriodoric aldehyde. He, however, did not have the aldehyde in a pure state inasmuch as he describes it as slightly dextrogyrate, boiling at 225°. Pure citral is inactive and boils between 228° and 229°.

Inasmuch as Barbier and Bouveault[5] obtained three semicarbazones[6] by the action of semicarbazide on lemon-grass oil, melting resp. at 171°, 160° and 135°, they propounded the hypothesis that the citral from lemon-grass consists of three isomeric aldehydes. Since semicarbazones of the same substance frequently occur in different modifications, which are distinguished by their melting points, solubility, and behavior toward acids, it must be considered in this case—as was, in fact, proven by Tiemann[7]—that there are no isomeric aldehydes but isomeric semicarbazones. Flatau[8] arrived at a like conclusion.

W. Stiehl[9] also assumes the presence of three isomeric aldehydes $C_{10}H_{16}O$ in lemon-grass oil: of citral (geranial), 10 p. c.; citriodoric aldehyde, 40—50 p. c.; and of allolemonal 25—30 p. c. Judging from the method of preparation this citriodoric aldehyde, according to Schimmel & Co.,[10] corresponds to the citral of commerce. Inasmuch as Semmler[11] (1891) has shown that this is identical with the geranial from geraniol, citral and the citriodoric aldehyde of Stiehl may be regarded as identical.

---

1) Ber. d. pharm. Ges., 7, pp. 358, 501; 8, p. 28.

2) Bericht von S. & Co., Apr. 1896, p. 68.

3) Bericht von S. & Co., Oct. 1888, p. 17.

4) Amer. Chem. Journ., 12, p. 553.

5) Compt. rend., 121 (1895), p. 1159.

6) The formation of different semicarbazones from citral was first observed by Wallach (1895). Berichte, 28, p. 1957; comp. also Tiemann & Semmler, Berichte, 28, p. 2188 and Tiemann, Berichte, 31, p. 821.

7) Berichte, 32, p. 115.

8) Bull. Soc. Chim., III, 21, p. 158.

9) Journ. f. prakt. Chem., II, 58, p. 51.

10) Bericht von S. & Co., Oct. 1898, p. 66.

11) Berichte, 24, p. 203; 31, p. 3001.

Neither has Stiehl's third aldehyde stood the test of criticism. Doebner[1] (1898) found that Stiehl's allolemonal consists of about equal parts of citral and a non-aldehyde substance.

From the exhaustive investigations of Tiemann[2] (1898) which can merely be referred to here, it becomes well nigh certain that the bulk of lemon-grass oil consists of citral. Besides citral, it contains traces of citronellal and possibly also of an isomer of citral (Doebner,[3] 1898).

In the first fraction of lemon-grass oil Barbier and Bouveault[4] (1894) found methyl heptenone. Inasmuch as they could not obtain hydro-m-xylene by means of zinc chloride, nor a solid bromide upon the addition of bromine, they declared this methyl heptenone different from the one obtained by Wallach[5] from cineolic acid.

However, according to Schimmel & Co.[6] the methyl heptenone obtained from lemon-grass oil is identical with that obtained by Wallach. The methyl heptenone regenerated from the bisulphite compound from lemon-grass oil boils at 173—174°, sp. gr. 0.853 at 20°, index of refraction, $n_D = 1.43996$. Wallach records the following physical constants for methyl heptenone from cineolic acid: b. p. 173—174°, sp. gr. 0.853 at 20°, n_D 1.44003 at 20°. If the ketone from lemon-grass oil is heated with zinc chloride, hydro-m-xylene, C_8H_{12} (b. p. 132—135°) results in considerable quantity. Upon nitration this yields dinitro-m-xylene (m. p. 92°) and the very characteristic trinitro-m-xylene which melts at 181—182° and is difficultly soluble in alcohol. The same substances have been obtained from the methyl heptenone from cineolic acid so that a doubt as to the identity of the methyl heptenone from lemon-grass oil and of the compounds of like composition obtained by Wallach and by Tiemann and Semmler[7] can no longer exist.

Geraniol,[8] the alcohol corresponding to the aldehyde citral, is contained in the highest fractions of the oil and has been isolated by means of its calcium chloride compound. The presence of linalool[8] in fraction 198—200° is also rendered probable. To furnish direct proof that the alcohol in this fraction is linalool is a difficult matter in the presence of citral and geraniol. Labbé[9] has found capronic and caprinic acids, which he supposes are present in form of the geraniol ester.

---

1) Berichte, 31, p. 3195.

2) Berichte, 31, pp. 3278, 3297, 3324; 32, pp. 107, 115.

3) Berichte. 31, p. 1891.

4) Compt. rend., 118, p. 983.

5) Liebig's Annalen, 258, p. 319.

6) Bericht von S. & Co. Oct. 1894, p. 32. Comp. also Tiemann & Semmler, Berichte, 28, p. 2126, footnote.

7) Berichte, 26, p. 2721.

8) Bericht von S. & Co., Oct. 1894, p. 32.

9) Bull. Soc. Chim., III, 21, p. 159.

The presence of a terpene, b. p. 175°, $\alpha_D = -5° 48'$, has been observed by Barbier and Bouveault in several oils. It yielded a liquid bromide from which small amounts of solid substance separated which melted at 85°. In other oils this terpene could not be found and possibly must be regarded as an adulteration. According to Stiehl[1] lemon-grass oil contains dipentene and possibly limonene.

PROPERTIES. As the name indicates, lemon-grass oil has an intense lemon-like odor and taste. It is a reddish-yellow or brownish-red, mobile liquid of sp. gr. 0.899—0.903. On account of the dark color, the rotatory power mostly cannot be determined. In a few instances in which it was possible, an angle of from + 1° 25′ to — 3° 5′ was observed. The oil is readily soluble in alcohol, even in dilute alcohol, for it gives a clear solution with 2 and more parts of 70 p. c. alcohol.[2] The citral content amounts to 70—75 p. c.

EXAMINATION. Adulterations with fatty oil or petroleum are recognized by the incomplete solubility in 70 p. c. alcohol. The citral assay, which should not yield less than 70 p. c. aldehyde, supplies a further clue as to the quality of the oil. The method of assay is the same as the assay of cinnamic aldehyde in cassia oil:

10 cc. of oil are shaken with bisulphite solution, the number of cc. of oil going into solution being regarded as citral. The necessary apparatus and the details of procedure are described under cassia oil. Inasmuch as methyl heptenone likewise reacts with bisulphite, one might suspect that the aqueous solution not only contains the citral, but also the methyl heptenone so that the apparent citral content was in reality that of citral + methyl heptenone. However, since the bisulphite compound of methyl heptenone is broken up into its components when heated in the waterbath, this ketone is contained in the oily layer, citral in the aqueous solution. The assay, therefore, shows the correct citral content.[3]

PRODUCTION AND COMMERCE. Lemon-grass oil is principally produced in East India, more particularly in the province Travancore south of Cochin. The principal commercial center is Trivandrum. The total exports from the Malabar coast were:

---

[1]) Loc. cit.

[2]) An oil distilled in Porto Alegre, Brazil, showed a peculiar behavior in regard to solubility since it did not give a clear solution even with 98 p. c. alcohol. Its sp. gr. was 0.895, angle of rotation —0° 8′, citral content 77 p. c. The oil, which has not yet become an article of commerce, is obtained from cultivated plants. In rainy years, four crops of this grass can be harvested; in dry years only three. The yield of oil from the fresh grass varies from 0.24 to 0.4 p. c. according to the season (Bericht von S. & Co., April 1896, p. 68.)

[3]) Comp. Tiemann, Berichte, 31, p. 3324.

Season 1891—92	1,450 Cases	Season 1894—95	2,370 Cases
" 1892—93	1,863 "	" 1895—96	3,070 "
" 1893—94	2,332 "	" 1896—97	3,000 "

Each case contains 12 wine bottles with a total net content of about 7½ k.

Some oil is also distilled in Ceylon, but only here and there and then only in small quantities. Statistics of production are wanting.

In the Straits Settlements the distillation is no longer conducted on as large a scale as formerly. The output is at most 2,000—3,000 lbs., for the total export of volatile oils from Singapore to England in 1896 amounted to only 14,165 lbs. Engl., of which at least two-thirds was citronella oil.

49. Oil of Vetiver.

Oleum Andropogonis Muricati.—Vetiveröl.—Essence de Vétiver.

Origin and Preparation. *Andropogon muricatus* Retz., known in India as Vetiver or Cus-Cus, is a perennial grass the leaves of which are odorless, but the long fibrous roots of which possess a peculiar strong odor reminding somewhat of myrrh. The plant is found all along the Coromandel coast, in Mysore, in Bengal and in Burma in moist, heavy soil particularly along the shores of rivers. It grows also in Réunion, Mauritius and the Philippines, further in Porto Rico and Jamaica[1] and in Brazil (Peckolt[2]).

In India the root is used in the manufacture of artistic mats and wicker baskets. Inasmuch as the distillation of the oil is difficult on account of its sparing volatility and viscid consistency, it is usually conducted with sandal wood or sandal wood oil. This oil is seldom exported, whereas the root constitutes a regular article of commerce. Tuticorin (south Coromandel coast) is the principal place of export. The root is of a reddish color and often contaminated with red sand. A half distilled root is frequently found in commerce and can be recognized by its light color. The yield obtained in Europe from the Indian root varies from 0.4 to 0.9 p. c. In Réunion larger quantities of oil are distilled, but the root is not exported.

Properties. Vetiver oil is the most viscid of all volatile oils. Its color is a dark yellow to dark brown. It has an intensive and very persistent odor which is not pleasant to some persons. On account

---

1) Pharmacographia indica, part VI, p. 571; Odorographia, vol. 1, p. 309.

2) Catalogue of the National Exposition in Rio 1866, p. 22 and 48.—Pharm. Rundschau, 12, p. 110.

of its sparing volatility it is used in finer perfumery primarily for fixing the more volatile odors. At 15° vetiver oil is heavier than water. Oils distilled in Germany had a sp. gr. of 1.015—1.030 at this temperature. On account of the tough consistency of the oil no claim for great accuracy can be made for these figures. At higher temperatures the oil is lighter than water. One sample had a sp. gr. of 0.994 at 44°.

On account of the dark color, the angle of rotation cannot be ascertained in most instances. In one instance $\alpha_D = +27° 40'$ was observed, in another +25 to 26°.[1] Vetiver oil forms a clear solution with 1½ to 2 p. of 80 p. c. alcohol. Upon the further addition of alcohol the solution becomes turbid.

A vetiver oil distilled in Germany boiled between 144—200° under 23 mm. pressure. Between 144—164°, 8 p. c. were obtained; from 164—170°, 10 p. c.; 170—180°, 24 p. c.; 180—185°, 30 p. c.; 185—200°, 20 p. c.; residue, 8 p. c. The saponification number lies between 60 and 80.

The oils distilled in Réunion are less viscid and specifically lighter. In a number of samples the sp. gr. varied from 0.982—0.998 at 30°. Being light in color the angle of rotation could be readily determined: $\alpha_D = +29°$ and $+36°$ being observed in two cases. These oils possessed the same solubility as those described above. Their faint odor and the difference in physical properties would lead one to suspect adulteration, but, owing to the want of knowledge of chemical composition, this cannot be decided. It is by no means excluded, however, that these differences are attributable to differences in the material distilled, for it seems but reasonable to assume that the Réunion oils are distilled from fresh roots.

Peckolt[2] found the sp. gr. 0.996 at 15° and 0.972 at 13° respectively for two oils distilled from fresh Brazilian root. It should be noted, however, that only 10 kilo of root were used and that it would be difficult to obtain the less volatile constituents from such small amounts.

Examination. The examination is restricted to the determination of the sp. gr., solubility and, if possible, of the rotatory power. Under all circumstances it should yield a clear solution with two parts of 80 p. c. alcohol. This permits of the recognition of fatty oils which are frequently used as adulterants.[3]

The admixture of sandal wood oil would not be recognized by the solubility test, but by the lower sp. gr., also by the change in the

1) Bericht von S. & Co., Oct. 1897, p. 62.

2) Footnote 2, p. 289.

3) Bericht von S. & Co., Apr. 1893, p. 59.

rotatory power. Inasmuch as sandal wood oil is laevogyrate, a vetiver oil adulterated with it will possess a lower rotatory power than a pure oil.

50. Citronella Oil.

Oleum Citronellae. — Citronellöl. — Essence de Citronelle.

Origin. Citronella oil is distilled from the grass *Andropogon nardus* L. This plant grows principally in Ceylon, the Malaccan peninsula, also in India proper, and occurs frequently in tropical East Africa.[1] According to Winter,[2] one of the largest distillers of citronella oil in Ceylon, two varieties of *Andropogon nardus* are used in the preparation of the oil. In Ceylon, which produces most of the oil, *Lana Batu* is the most widely distributed variety. It is said to have first been found in Matara, a city in the district of like name in southern Ceylon (comp. map). As will be shown later, this variety yields an oil relatively poor in geraniol, and contains methyl eugenol, has a high specific gravity and constitutes the bulk of the oil of commerce. The second grass variety, which is cultivated in Ceylon only in the neighborhood of Baddagama, comes from Malacca and grows on good soil only. Apparently this variety is generally cultivated in the Straits Settlements, also in Java[3] and is known as *Maha pangiri*. The oil from this variety has a lower specific gravity. At least the oil from Java has a lighter color and may in general be pronounced the finer oil.[4]

Preparation.[5] The cultivation of citronella grass in Ceylon is confined to the Southern Province, mainly between the Gin Ganga in the northwest and the Walawi Ganga in the east. It is found growing on the slopes of the hills. The individual tufts of grass grow at small, irregular intervals, attaining a height of 1 meter. According to the statement of competent dealers, as much as 40,000 to 50,000 acres are at present under cultivation.

The plants require little or no care, provided the formation of seeds is prevented by regular harvests. Otherwise the tufts become too dense, become yellow within, and spoil. In general a distinction is made between two harvesting periods. The first and principal season is in July and August, the second lasts from December until February. The yield averages from 16—20 bottles (of 22 oz. each) per acre for the summer,

1) Ber. d. pharm. Ges., 7, p. 501.
2) Chemist and Druggist, 52, p. 646.
3) Bericht v. S. & Co., Oct. 1899, p. 18.
4) Bericht v. S. & Co., Apr. 1900, p. 11.
5) According to the Bericht von S. & Co., Oct. 1898, p. 11.

and from 5—10 bottles for the winter season. Exact data cannot be given, for the yield naturally varies with the weather, and the age and location of the plantation. Thus e. g. a plantation yields less and less oil with increasing age, though climate and soil conditions are favorable. When the plantation has reached the age of 15 years new plants have to be raised if the industry is to be profitable.

The distilleries are mostly located at the base of ridges where cool water is to be had in sufficient quantity.

The construction of the distilleries is not at all primitive, as is shown in the following sketch. The success of the natives, who constitute the majority of the producers, is really remarkable. The machinery is housed under a long roof which serves as a protection against the sun. The

Elevation.

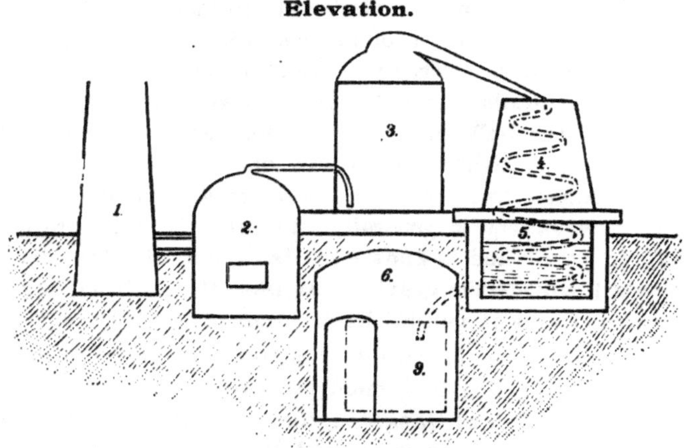

Fig. 60.

1. Chimney. 2. Boiler. 3. Stills. 4. Condenser. 5. Condensing basin. 6. Locked cellar for storage of the distillate. 7. Roof (fig. 61). 8. Water supply pipe (fig. 61). 9. Basin for storage of distillate.

regulation steam boiler, provided with safety valve and water gauge, rests on a solid foundation. The cylindrical iron stills, mostly 6—7 ft. high and 3—4 ft. in diameter, rest on a base and are provided with a common, interchangeable alembic. This connects with a spiral cooler in a large wooden barrel resting on a lowered water basin. The condensation products of the cooler empty into a basin still lower than the one constituting part of the cooler, and in a locked compartment. The complete arrangement is shown by the accompanying sketches. (Fig. 60; and fig. 61, p. 293.)

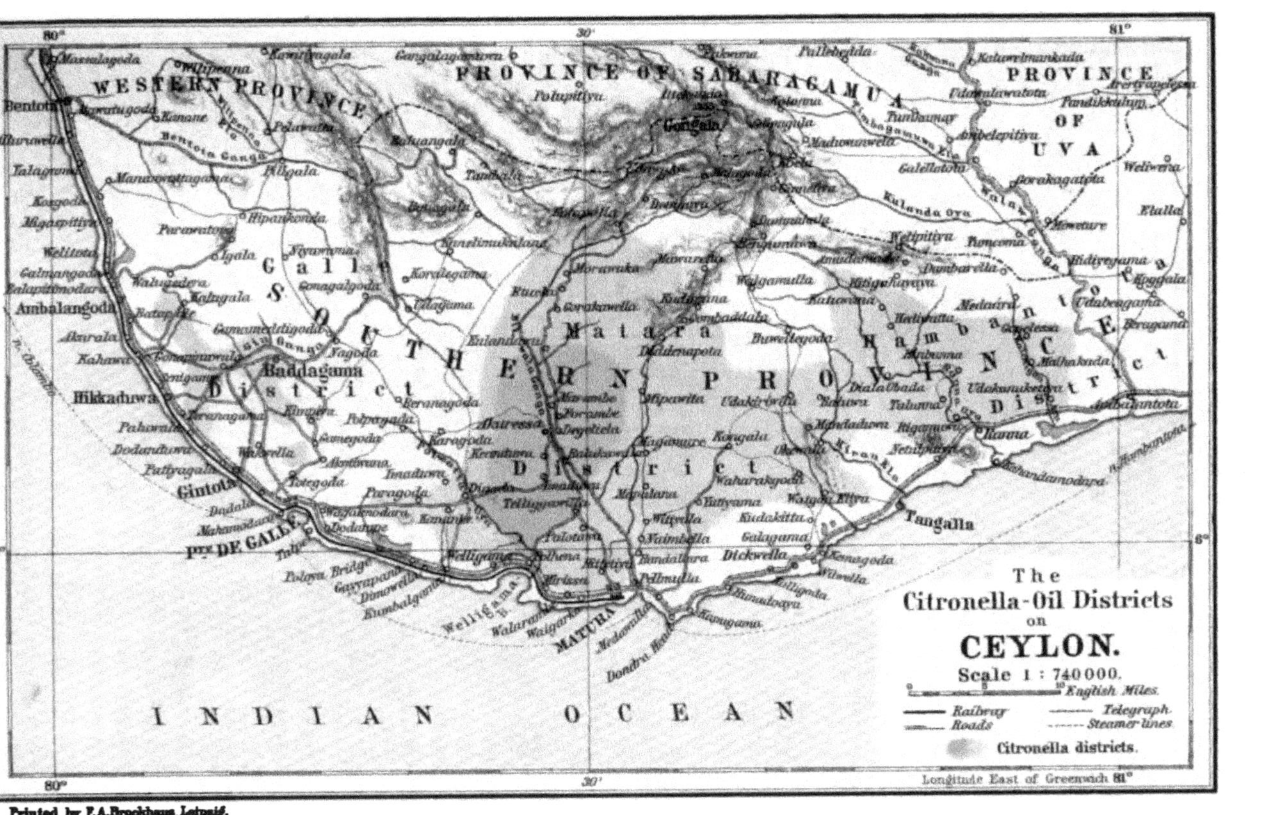

Printed by F.A.Brockhaus, Leipzig.

The distillation is conducted with direct steam without the addition of water to the grass. The water in the barrel condenser, when warm, is used for feeding the boiler; whereas the water in the lower basin serves to effect complete condensation. The distillate is stored without effecting a separation of the oil. After a definite period, the proprietor visits his distilleries to remove the oil. The aqueous distillate is poured away.

A charge of dry grass, for only such is used, is distilled in about 6 hours. The exhausted grass, after being dried in the sun, is used as fuel. The Southern Province is extremely poor in wood. For this reason the distillation must be discontinued during the rainy season when the exhausted grass can no longer be dried.

Ground-plan.

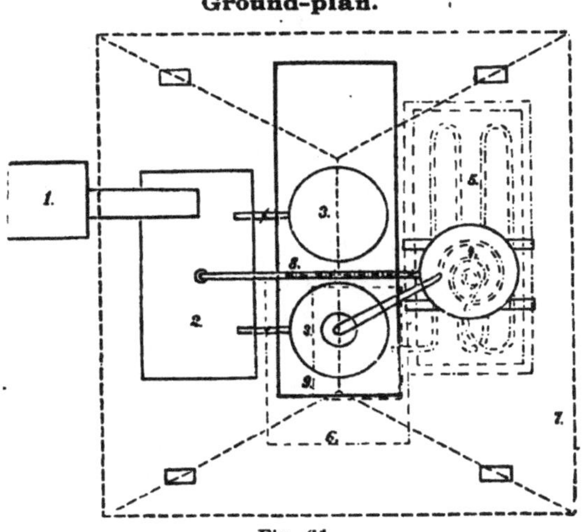

Fig. 61.

A still 7 ft. high and 4½ ft. in diameter yields about 16—20 bottles of 22 oz. each per day, or about 360—440 oz. The work never being controlled no accurate figures as to yield are obtainable.

Besides the apparatus described above, such with direct fire are said to be in use in some districts. However, most of the citronella oil is to-day obtained by steam distillation. If the distillation is conducted over direct fire, water is added to the grass. It should be stated the grass is not dried before distillation. Nevertheless it is not moist when put into the apparatus, for as a rule, several hours elapse from the time when it is cut to the time when it is distilled. With a temperature of about 65—70° in the sun, this suffices to remove a large portion of the moisture.

The districts of production are more clearly indicated by the accompanying map. The number of stills in operation in Ceylon is estimated at 600, with an annual output of 1,000,000 lbs. of citronella oil.

COMPOSITION. The principal bearer of the citronella odor and, therefore, the characteristic constituent of the oil is citronellal. It is an aldehyde of the formula $C_{10}H_{18}O$. Although present to the extent of from 10—20 p. c. only, it first attracted the attention of investigators. Owing to the instability of the aldehyde, its name as well as opinions regarding its composition have undergone repeated changes.

Gladstone[1] in 1872 found the boiling point of "citronellol," as he designated the substance, at 199—205° and assigned to it the formula $C_{10}H_{16}O$. Wright[2] in 1874 states that it boils at 210°. His analyses agree with the formula $C_{10}H_{18}O$. By shaking the oil with alkali bisulphite and regeneration with acid, Kremers[3] obtained an aldehyde which decomposed when distilled. One of the fractions upon analysis yielded results corresponding with an heptoic aldehyde $C_7H_{14}O$. Schimmel & Co.[4] designated the compound, regenerated from the bisulphite derivative with soda and boiling between 205—210°, "citronellon," without deciding as to its ketone or aldehyde nature. Dodge[5] then pronounced it an aldehyde without, however, bringing any proof as to the aldehyde nature of the "citronella aldehyde." This proof was supplied by Semmler[6] (1891) who oxidized the "citronellon" to an acid with the same number of carbon atoms, the citronellic acid, $C_{10}H_{18}O_2$. For this reason the substance is now designated as citronellal, a more rational name designating its chemical character. For the properties and derivatives of

---

1) Journ. Chem. Soc., 25, p. 1; Pharm. Journ., III, 2, p. 746; Jahresb. f. Chemie, 1872, p. 815.

2) Pharm. Journ., III, 5, p. 233. Gladstone as well as Wright erroneously mention *Andropogon schoenanthus* as the plant from which citronella oil is derived. Their work, however, shows with sufficient certainty that the oils examined by them were true citronella oil from *Andropogon nardus* L. and not the oil from *A. schoenanthus*. On the other hand, the "Untersuchung des Oels von *Andropogon Ivarancusa*" by Stenhouse in 1844 (Liebig's Annalen, 50, p. 157), which he calls East Indian grass oil, does not pertain to palmarosa oil, but in all probability to citronella oil. Comp. "Ueber Palmarosaöl" by G. E. Gildemeister and K. Stephan. Archiv der Pharm., 234, p. 323.

3) Proc. Am. Pharm. Assoc., 35, p. 571; Chem. Centralbl., 1888, p. 898.

4) Bericht von S. & Co., Oct. 1888, p. 17.

5) Dodge assigned to it the correct formula $C_{10}H_{18}O$. Amer. Chem. Journ., 11, p. 456; Chem. Centralbl., 1890, I, p. 127.

6) Berichte, 24, p. 210.

citronellal see p. 154. A detailed chemical examination of a *Lana Batu* oil has recently been made by Schimmel & Co.[1]

Citronella oil contains but 10—15 p. c. of terpenes. In fraction 157—164° camphene has been found (Bertram & Walbaum,[2] 1894). By passing hydrogen chloride into an ethereal solution of this fraction there is obtained, in addition to liquid products, a solid chloride, which, when heated under pressure with water at 100°, yields quantitatively camphene. By treating the same fraction with glacial acetic acid and sulphuric acid, isoborneol results. Furthermore there is present a second terpene of like boiling point which has not yet been identified.[2] Fraction 172—177° contains dipentene (tetrabromide, m. p. 124°).[2] Limonene[3] also is present and has been identified as tetrabromide melting at 105°.

Of oxygenated constituents of an alcoholic nature the presence of borneol and geraniol has been determined with certainty. On account of the small amount present (1—2 p. c.), the former can be isolated only with difficulty. The melting point was found at 203—204°, its specific rotatory power $[\alpha]_D = -31.82°$, hence laevo borneol.[4] Geraniol[5] constitutes about one-half of the citronella oil. The alcohol can be obtained pure by treating the corresponding fraction with calcium chloride and decomposing the calcium chloride geraniol with water.[6]

Basing his conclusion on somewhat unsatisfactory analytical data, Dodge[7] assumes the presence of an alcohol $C_{10}H_{20}O$, b. p. 222°, identical with citronellol (citronellyl alcohol) obtained by the reduction of citronellal. This claim still requires verification.[8] Methyl heptenone,[9] acetic and valerianic acids[10] as esters and linalool[9] have also been found.

Another constituent, methyl eugenol, has been found by Schimmel & Co.[11] This probably occurs only in the *Lana Batu* oil, and to it the higher specific gravity of this variety possibly is due. From the highest boiling portion of the oil a fraction was obtained which possessed the odor of methyl eugenol and which upon oxidation with

1) Bericht von S. & Co., Oct. 1899, pp. 18—22.

2) Journ. f. prakt. Chem., II, 49, p. 16.

3) Bericht von S. & Co., Oct. 1899, p. 15.

4) Bericht von S. & Co., Apr. 1894, p. 15.

5) Ibid.; Oct. 1898, p. 11.

6) The preparation of geraniol from citronella oil is patented. G.I.P. 76,485. Chem. Zeitung, 18, p. 1856.

7) Footnote 5, p. 294.

8) Flatau and Labbé claim to have found 6 p. c. of citronellol in citronella oil. As shown on p. 284 under palmarosa oil, the method which they employed to effect the separation of citronellol and geraniol is altogether unreliable.

9) Bericht von S. & Co., April 1895, 21.

10) Kremers, footnote 3, p. 294.

11) Bericht von S. & Co., Oct. 1898, p. 17.

potassium permanganate yielded veratric acid melting at 179°. According to a more recent investigation 8 p. c. of methyl eugenol was found present.[1]

If this occurrence explains the higher sp. gr. of the *Lana Batu* variety, the cause for the higher optical rotation is to be sought in the larger content of strongly laevogyrate camphene. The lowest fractions of the oil in which the methyl eugenol had been found, had an angle of rotation $\alpha_D = -55° 0'$.[2]

According to the recent investigation of Schimmel & Co. already referred to, the composition of the *Lana Batu* oil may be summarized as follows: a liquid camphene, dipentene, limonene, citronellal (28.2 p. c.), geraniol (32.9 p. c.), linalool, borneol, terpineol (?), methyl eugenol (8 p. c.), a light sesquiterpene (sp. gr. 0.8643), a heavy sesquiterpene (sp. gr. 0.912).[3]

So far as examined the oil from the *Mana pangiri* variety has the same constituents, but the two oils differ decidedly in percentage composition, the *Mana pangiri* variety containing up to 91 p. c. of acetylizable constituents:[4] 50.45—55.34 p. c. citronellal, 38.15—31.87 p. c. geraniol, and only 0.78—0.84 p. c. of methyl eugenol. The low percentage of the last mentioned substance accounts for the low specific gravity.

Properties. Citronella oil is a yellow or yellowish-brown liquid, sometimes green due to the presence of copper. Brownish oils frequently become green when exposed to the air. This change does not take place if the copper is removed by shaking the oil with acid.[5] The odor of citronella oil is pleasant and lasting. Frequently the odor is described as resembling that of balm. This comparison is not pertinent inasmuch as the odor of balm resembles citral, and not citronellal, the bearer of the citronella odor.

With regard to physical properties, the oils of the two varieties of grass previously described differ decidedly.

The first kind, which may be designated as Singapore oil and which is regarded as the better quality, has a sp. gr. 0.886—0.900, rotatory power $\alpha_D = -0° 34'$ to $-3°$, and a geraniol content of from 80—91 p. c.

The oil distilled from *Lana Batu* constitutes the bulk of the commercial oil. Its sp. gr. is 0.900—0.920, $\alpha_D = -5°$ to $-21°$, the geraniol content varies from 50 to at most 70 p. c.

---

1) Bericht von S. & Co., Oct. 1899, p. 22.

2) Bericht von S. & Co., Oct. 1898, p. 17.

3) No mention is made of the citral found by Flatau, who claims to have obtained 25—30 p. c. of citronellal and 2—5 p. c. of citral. (Bull. Soc. Chim., III, 21, p. 158.

4) Bericht von S. & Co., Apr 1900, p. 11.

5) Pharm. Journ., III, 21, p. 922.

The solubility of both oils is about the same. A good citronella oil yields a clear solution with 1—2 parts of an 80 p. c. alcohol. This solution, as a rule, remains clear if the quantity of alcohol is increased to 10 vols. Sometimes, however, the increase of the alcohol to 5—10 vols. causes a slight turbidity, but even upon longer standing in a closed vessel no oily drops separate.

Citronella oil does not yield a definite saponification number. On account of the slow decomposition of the citronellal by means of alkali the results will vary with the length of boiling.

EXAMINATION. In testing the oil, the main attention is to be directed against the most common adulterants, fatty oils and petroleum. The latter reduces the sp. gr. of the oil decidedly, whereas the fatty oils scarcely produce a change in this respect. Both adulterants can be detected by the behavior of the oil toward 80 p. c. alcohol. The following table[1] shows the solubility of pure and intentionally adulterated citronella oils.

	Spec. gravity at 15°	Proportion of oil to 80 p. c. of alcohol			
		1 : 1	3 : 4	1 : 5	1 : 10
Pure commercial oils	0.897	soluble	soluble	soluble	soluble
	0.908	"	"	"	"
	0.900	"	"	"	"
	0.905	"	"	"	"
German oil[2]	0.895	"	"	"	"
The same with					
10 percent petroleum	0.887	"	"	slightly turbid	milky
20 percent petroleum	0.877	"	"	milky	"
30 percent petroleum	0.868	"	milky	"	"
Commercial oil adulterated with petroleum	0.883	milky	"	"	"
Commercial oils adulterated with fatty oils	0.893	turbid	turbid	turbid	turbid
	0.895	"	"	"	"
	0.898	"	"	"	"

Conclusions as to the kind and quantity of the adulterant can be drawn from the appearance of the turbidity, also from the behavior of the substance separating. Petroleum produces a milky white turbidity, whereas fatty oils produce a turbid, but not a milky mixture. After prolonged standing, fatty oil collects in the form of drops at the bottom of the liquid, while petroleum rises to the surface.

1) Bericht von S. & Co., Oct. 1889. p. 21.

2) Distilled from dry citronella grass in Leipzig.

Citronella oil adulterated with fatty oils is soluble neither in 1—2 parts nor in 10 parts of 80 p. c. alcohol. An oil adulterated with petroleum gives a clear solution with 1—2 parts, but becomes turbid upon the addition of more solvent and separates oily drops at the surface only after prolonged standing.

Strictly speaking, adulteration can be established only after the separation of oily drops, for a mere turbidity is frequently produced by pure commercial oils. The exact proof for adulteration with fatty oils and petroleum is described on p. 201.

Acetylization will be found useful in passing judgment on an oil: the higher the geraniol content, the more valuable the oil. However, the geraniol content found upon acetylization does not correspond to the real alcohol content. Inasmuch as in the process of acetylization, citronellal is quantitatively converted into the acetic ester of isopulegol $C_{10}H_{18}O$ (Tiemann and Schmidt[1]), the citronellal content is determined with that of the geraniol. The term "geraniol content," therefore, is improper, but for the sake of simplicity is retained in practice. A method for the determination of geraniol in the presence of citronellal has not yet been worked out.

Production and Commerce. Ceylon and the Straits Settlements on the Malaccan peninsula, are the countries which produce citronella oil. The accompanying map shows the producing districts in Ceylon. An area of 40,000—50,000 acres is said to be under cultivation in the island, with 600 stills in operation, which have an annual capacity of more than a million pounds of oil.

At the time of publication of the German edition of this work the largest quantity of oil had been exported in 1898: viz. 1,365,917 lbs., valued at about $350,000.00. This record was broken in 1899 with a total of 1,478,756 lbs. The countries to which this oil was shipped, with the amounts, are:

	For 1898.	For 1899.
England	696,869 lbs.	766,594[2] lbs.
America	618,999 "	667,332 "
Germany	22,883 "	1,335[2] "
India	10,100 "	7,537 "
Australia	10,633 "	25,865 "
France	3,440 "	1,467 "
China	2.249 "	8,626 " (China and Singapore)
Singapore	504 "	
Africa	250 "	
Total	1,365,917 lbs.	1,478,756 lbs.

[1]) Berichte, 29. p. 918.

[2]) These two figures are misleading inasmuch as the German importations amounting to about 250,000 lbs. are included in the English imports.

The enormous increase in production and consumption is strikingly shown by the following statistics of exportation:

Export of citronella oil from Ceylon.

1887	551,706 lbs.
1888	659,967 "
1889	641,465 "
1890	909,942 "
1891	603,974 "
1892	844,502 "
1893	668,520 "
1894	908,471 "
1895	1,182,255 "
1896	1,132,141 "
1897	1,182,867 "
1898	1,365,917 "
1899	1,478,756 "

In comparison with these numbers the production of the Straits Settlements near Singapore is insignificant. The total area of the citronella estates in the peninsula is estimated at 2000 acres at the highest. The annual production of oil may not exceed 30,000 lbs. Qualitatively, however, this oil is preferable.

51. Oil of Andropogon Odoratus.

The grass *Andropogon odoratus* Lisboa, which serves the natives on the west coast of India proper as a domestic remedy, yields upon distillation as much volatile oil as *Andropogon schoenanthus*.

Properties. According to Dymock,[1] the oil is of a dark red color; sp. gr. 0.931 at 31°; $\alpha_D = -22.75°$; $[\alpha]_D = -24.43°$.

An oil distilled from fresh grass possessed an odor resembling that of pine needle oil; sp. gr. 0.915; $\alpha_D = -23° 10'$.[2]

52. Camel-grass Oil.

Origin. From Dioscorides up to the middle of the last century the herb of *Andropogon laniger* Desf. has been carried in apothecary shops as *Herba schoenanthi* or *Squinanthi*, as *Juncus odoratus* or as *Foenum camelorum*.[3] The plant is widely distributed throughout northern Africa and Arabia, also in northern India as far as Thibet. In the desert it constitutes the principal food of the camels. Upon distillation of the dried grass, as offered for sale in the Indian bazaars, Dymock[4] obtained 1 p. c. of oil.

1) Pharmacographia indica, part VI, p. 571.
2) Bericht von S. & Co., Apr. 1892, p. 44.
3) Pharmacographia, II. edit., p. 728.
4) Pharmacographia indica, part VI, p. 564.

Properties and Composition. Dymock observed the sp. gr. 0.905 at 29.5° and the angle of rotation $\alpha_D = -4°$. Schimmel & Co.[1] found the sp. gr. 0.915 at 15° and the angle of rotation $\alpha_D = +34° 38'$. The odor of camel-grass oil reminds of that of elemi oil, which resemblance appears to be due to its phellandrene content.[1] The oil distills between 170—250°.

53. Oil of Saw Palmetto.

Origin and History. The oil of saw palmetto is obtained from the berries of *Serenoa serrulata* Michx. Hook, f., (Family *Palmae*), indigenous to the southern United States, especially to Florida. The oil is apparently first mentioned in 1894 by Sherrard,[2] who claims to have obtained it from the chloroformic extract of the berries. Coblentz[3] in 1895 obtained a small quantity of a volatile oil by distillation. The oil had, however, been distilled in much larger quantities as early as 1890 by J. U. Lloyd.[4]

Preparation. The oil is obtained by distillation from the fresh berries. The yield of oil is about 1.2 p. c. The dried berries do not yield the oil.[5] An oil has also been obtained by expressing the juice from the berries and collecting the oily layer (Sherman and Briggs[6]), and by siphoning off the oily layer which separates from the fluid extract (Lloyd[5]).

Properties. The distilled oil of saw palmetto has a pleasant fruity odor, accompanied by a heavier, almost disagreeable odor. When distilled it is of a green to brownish color, which, however, disappears when distilled in a vacuum. The oil distilled in 1890 by J. U. Lloyd had in 1900 a sp. gr. of 0.8682 at 20°. The oil, when distilled under diminished pressure, was colorless and had the following properties:

B. p. 60—170° at 18 mm.; $d_{20°}^{20°} = 0.8679$; $n_{D20°} = 1.41233$.

The oil is optically inactive (Schreiner[5]).

Two samples of oil obtained from the fluid extract had the sp. gr. 0.8651 and 0.8775. They yielded by distillation with steam from 4 to 5 p. c. of a green to brownish oil of the sp. gr. 0.8650 and 0.8653 respectively. An oil obtained by Sherman and Briggs[6] by expression from the berries preserved in alcohol boiled from 70—270° under a pressure of 16 mm.

1) Bericht von S. & Co., Apr. 1892, p. 44.
2) Proc. Am. Pharm. Assn., 42, p. 312.
3) Proc. New Jersey Pharm. Assn., 1895, p. 68.
4) Personal communication from J. U. Lloyd.
5) Pharm. Rev., May, 1900.
6) Pharm. Archives, 2, p. 101.

COMPOSITION. Sherman and Briggs[1] have made a thorough chemical study of the oil of saw palmetto obtained by expression from the berries preserved in alcohol. The oil was subsequently fractionated under diminished pressure. They found this oil to consist to the extent of about 63 p. c. of free fatty acids, caproic, caprylic, capric, lauric, palmetic, and oleic acids, and about 37 p. c. of ethyl esters of these acids. No glycerides were found in the oil obtained from the pulp of the berries, although their presence was demonstrated in the fixed oil from the seeds, in which the same acids, with the addition of stearic acid, were found as in the oil from pulp. The fruity odor of the oil is due to the ethyl esters. The fact that the oil consists so largely of free fatty acids and of ethyl esters of these acids, and inasmuch as the berries are kept in alcohol, has led to the suspicion that the esters may be formed by the action of the alcohol on the free acids.[2] This suspicion is supported by the fact that no volatile oil is obtained from the dry berries.

54. Calamus Oil.

Oleum Calami. — Calamusöl. — Essence de Calamus.

ORIGIN AND HISTORY. Calamus, *Acorus calamus* L., Family *Araceae*, though frequently found but in isolated places, occurs in the moderate zones of the entire northern hemisphere. It also occurs in Japan (Thunberg) China, Cochin China (Loureiro), and Burma, the East Indies (Roxburgh), the Philippines and some of the islands of the Indian Archipelago, and has, furthermore, been found on Tana lake in Abyssinia (Heuglin). Inasmuch as the plant does not multiply by means of seeds (Kerner) in moderate and colder climates, but by the branching of the rhizome, it is assumed that the wide, but ofttimes isolated distribution of calamus may have been brought about by transplanting.

Calamus is reported to have been cultivated in the thirteenth century in Poland; in Germany first during the sixteenth century; whence it became more widely distributed. Calamus is also indigenous to North America, being found from Nova Scotia south to Florida and westward to Minnesota, Iowa and Kansas. Of botanists Schöpf first observed it in 1783 in Pennsylvania and New Jersey.

The distilled oil of calamus is first mentioned in the price ordinance of Frankfurt of 1582 and in the Dispensatorium Noricum of 1589. The yield of oil obtainable upon distillation of the rhizome was determined

---

1) Pharm. Archives, 2, p. 101.

2) Pharm. Rev., May 1900.

at the beginning of the eighteenth century by Hoffmann and Neumann; and about the middle of the eighteenth century by Cartheuser. The first investigations of the oil appear to have been made by Wedel in 1718 and by Trommsdorff in 1808. Later examinations are by Martius in 1832, Schnedermann in 1842, Gladstone in 1863, and Kurbatow in 1873.

PREPARATION. The rhizome, which contains oil in all its parts, is collected late in summer and in fall. Upon distillation the fresh root which contains 70—75 p. c. of water yields about 0.8 p. c., the dried unpeeled rhizome about 1.5—3.5 p. c. of oil. The root bark as well as the peeled roots when dried and distilled by themselves produce a smaller yield than the unpeeled rhizome. A plausible explanation seems to be that when peeled, the rhizome is cut into thin strips whereby oil is lost by both volatilization and resinification. A calamus oil of inferior quality, with properties deviating from those of the ordinary oil, is distilled in Galicia.

On account of its aroma, calamus oil is used in the manufacture of liquors and of snuff, and finds less application than formerly in general medicine and the preparation of household remedies.

COMPOSITION. Since 1832 calamus oil has been examined repeatedly.[1] The results obtained so far, however, do not afford a satisfactory insight into its chemical nature. It is not even known what substance constitutes the bulk of the oil, much less to what its peculiar properties are due. The investigation of Kurbatow[2] (1874) revealed the presence of 5 p. c. of a terpene ($C_{10}H_{16}$) boiling at 158—159°, which yielded a solid derivative with dry hydrogen chloride melting at 63°. Though this melting point is extremely low—pinene hydrochloride melts at 125° —it is probable that this terpene is identical with pinene.

Fraction 255—258°, sp. gr. 0.932 at 14° and constituting 2.5 p. c. of the oil, contains a sesquiterpene that does not combine to form a crystalline derivative with hydrogen chloride. At a still higher temperature a blue oil distills over. The highest boiling fractions contain a subtance (phenol?) which, in alcoholic solution, produces a greenish-brown color with ferric chloride (Flückiger[3]).

---

1) Martius (1832), Liebig's Annalen, 4, pp. 264 & 266. — Schnedermann (1842), ibidem, 41, p. 374. — Gladstone (1864), Journ. Chem. Soc., 17, p. 1; Jahrb. f. Chem., 1863, pp. 546, 547.

2) Liebig's Annalen, 173, p. 4. The older statements are to be accepted with caution, for some of them, like those of Gladstone, are evidently based on adulterated oils.

3) Pharmacognosie, p. 352.

Calamus oil also contains saponifiable substances and a small amount of an alcoholic substance as becomes apparent from the saponification number (40—50) after acetylization.

Properties. Calamus oil is somewhat viscid in consistency, of a yellow to brownish-yellow color, possesses a camphor-like, aromatic odor and a corresponding bitter, burning, spicy taste. Its sp. gr. is 0.960 to 0.970; angle of rotation $\alpha_D = +10$[1] to $+31°$;[2] saponification number 16—20, after acetylization 40—50. Calamus oil is soluble in almost all proportions in 90 p. c. alcohol, but rather difficultly soluble in more dilute alcohol. Of a 50 p. c. alcohol about 1,000 parts are necessary to effect a clear solution. (The solubility of the Japanese oil described below is very different.) Upon distillation of a normal calamus oil nothing came over below 170°; between 170—275° 32 p. c. were obtained; from 275—300°, 60 p. c.; leaving a residue of 8 p. c.

Examination. Owing to the lack of knowledge of its chemical composition, the examination of calamus oil is confined principally to the determination of its optical rotation, sp. gr., and solubility in 90 p. c. alcohol.

Adulterations that cannot be readily recognized by their odor are turpentine oil, cedar wood oil and gurjun balsam oil. The first mentioned can be recognized by a lower sp. gr. and angle of rotation, and, if present in large quantity, by the diminution of solubility in 90 p. c. alcohol, and by fractional distillation.

Cedar wood oil and gurjun balsam oil may influence the sp. gr. but little, but reduce the dextro-rotation materially or change it to laevo-rotation. Their presence may further be indicated by their sparing solubility in 90 p. c. alcohol.[3] Inasmuch as the boiling points of both oils lie within the limits of those of calamus oil, they cannot be recognized by fractional distillation.

55. Oil from Calamus Herb.

The fresh green parts of calamus, *Acorus calamus* L., yield upon distillation with water vapor an oil which closely resembles that from the rhizome. Its sp. gr. is 0.964; $\alpha_D = +20°44'$.[4]

---

1) Bericht von S. & Co., Apr. 1895, p. 15.
2) The angle of rotation of the Galician oil is frequently lower than + 10°.
3) Bericht von S. & Co., Apr. 1895, p. 16.
4) Bericht von S. & Co., Apr. 1897, p. 8 of table.

56. Japanese Calamus Oil.

Japanese calamus root, which does not differ morphologically from the ordinary rhizome, is probably derived from the same plant as the latter. It is possible, however, that it is derived from *Acorus spuriosus* Schott, which is common in Japan and the rhizome of which is said to scarcely differ from that of *A. calamus* L. (Holmes,[1] 1879).

The Japanese root is much richer in oil than the common, containing 5 p. c. The oil has a sp. gr. 0.992 and boils from 210—290°. Collected in two fractions, the lower one possessed the characteristic calamus odor, whereas the higher boiling one had the odor of sesquiterpenes.[2]

Japanese oil is more readily soluble in alcohol than the German. 1 p. of the Japanese oil dissolves in 500 p. of 50 p. c. alcohol, the German being soluble in not less than 1,000 parts.[2]

57. Cevadilla Seed Oil.

Sabadillsamenöl.

This oil is obtained by distillation of the comminuted seeds of *Sabadilla officinalis* (*Liliaceae*), or of the fat obtained by benzin extraction in the preparation of veratrine (Opitz,[3] 1891). The yield from the fresh seed is about 0.32 p. c., from old seed much less. The sp. gr. lies between 0.902 and 0.928. Upon distillation the bulk passes over between 190—250°, the esters present being largely decomposed. After saponification the oil distills principally between 220—250°, $[\alpha]_j = -9° 10'$.

From the alkaline saponification liquid, oxymyristic acid $C_{14}H_{28}O_3$, (m. p. 51°), and veratric acid, $C_9H_{10}O_4$ (m. p. 179—180°), have been obtained. These acids seem to be present in the original oil as methyl and ethyl esters. Lower aliphatic aldehydes have also been found.

58. Aloe Oil.

The faint but characteristic odor of aloes is due to a minute quantity of volatile oil. Upon distillation of 500 lbs. of Barbadoes aloes from *Aloë vulgaris* Lam. var. *Aloë barbadensis*, T. and H. Smith & Co. of London in 1880 obtained 2 fl. dr. of oil. It was a light yellow, mobile liquid, sp. gr. 0.863, b. p. 266—271°.[4]

1) Pharm. Journ., III, 10, p. 102.
2) Bericht von S & Co., Apr. 1889, p. 7.
3) Archiv d. Pharm., 229, p. 265.
4) Pharm. Journ., III, 10, p. 618.

59. Xanthorrhoea Resin Oil.

Upon distillation of the Australian yellow xanthorrhoea resin (acaroid resin, yellow grass tree gum) from *Xanthorhoea hastile* R. Br. Schimmel & Co.[1] obtained 0.37 p. c. of a yellow oil, storax-like in odor. Sp. gr. 0.937, $\alpha_D - 3°14'$, saponification number 74.3, acid number 4.9, ester number 69.4. The free acid was isolated by shaking with dilute sodium hydroxide solution and recognized by means of its melting point 133°, as cinnamic acid. From the saponification lye cinnamic acid was separated in considerable quantity, 200 g. of oil yielding about 40 g. of cinnamic acid recrystallized from water.

The saponified oil boiled between 145—240°. From the low boiling portions a fraction 145—150° possessing the properties of styrene was isolated. Upon the bromination in a cold ethereal solution styrene dibromide, m. p. 74—75°, was obtained in fine needles.

60. Garlic Oil.

Upon distillation of the entire plant, *Allium sativum* L., 0.005—0.009 p. c. of an oil[2] are obtained, which is yellow in color and possesses a most disagreeable garlic-like odor. Sp. gr. 1.046—1.057; optically inactive.

The oil was first examined chemically in 1844. Based upon this examination, Wertheim[3] arrived at the conclusion that garlic oil consisted principally of allyl sulphide $(C_3H_5)_2S$. This view was held for almost 50 years and has passed into all text books without being put to a test in a single instance.

When Semmler[4] reexamined the oil in 1892 the fact was revealed that garlic oil does not contain a trace of allyl sulphide.[5]

Analysis revealed the presence of sulphur besides carbon and hydrogen, also the absence of oxygen and nitrogen. Inasmuch as the oil decomposed when distilled under ordinary pressure it was fractionated under reduced pressure. Under 16 mm. pressure it distilled between 65—125°. Semmler isolated the following substances:

1. A disulphide $C_6H_{12}S_2$ (abt. 6 p. c.), b. p. 66—69° under 16 mm. It is probably allyl propyl disulphide, $C_3H_5S-SC_3H_7$.

---

1) Bericht von S. & Co., Oct. 1897, p. 66.

2) Bericht von S. & Co., Oct. 1889, p. 52; and Oct. 1890, p. 25.

3) Liebig's Annalen, 51, p. 289.

4) Archiv d. Pharm., 230, p. 434.

5) This naturally disposes of the statements concerning other oils supposed to be identical with garlic oil, such as those of *Thlaspi arvense* L. and *Alliaria officinalis* L.

2. A disulphide $C_6H_{10}S_2$ (abt. 60 p. c.) constitutes the bulk of the oil and is the bearer of the pure garlic odor. Sp. gr. 1.0237 at 14.8°; b. p. 79—81° under 18 mm. Its constitution probably is $C_3H_5S-SC_3H_5$.
3. A substance $C_6H_{10}S_3$ (abt. 20 p. c.). Sp. gr. 1.0845 at 15°; b. p. 112—122° under 16 mm. Probable constitutional formula: $C_3H_5S-S-SC_3H_5$.

The residue shows a still higher sulphur content and possibly has the composition $C_6H_{10}S_4$.

Inasmuch as allyl sulphide boils between 36—38° under 15.5 mm. pressure, and the first fraction of garlic oil between 60—65°, the presence of allyl sulphide in garlic is excluded. Neither could the sesquiterpene of Beckett and Wright,[1] b. p. 253.9° be found in the oil examined by Semmler.

61. Onion Oil.

The pungent, persistent odor of the garden onion, *Allium cepa* L., is due to a volatile oil, of which a yield of 0.046 p. c. is obtained when the entire plant is distilled.[2]

Onion oil is a dark brown, mobile liquid, sp. gr. 1.0410 at 8.7°[3] or 1.036 at 19°;[2] $\alpha_D = -5°$.

Inasmuch as the oil is decomposed by distillation under ordinary pressure, this operation must be conducted in vacuum. Under 10 mm. pressure the oil distills almost completely between 64—125°. According to Semmler[3] the principal constituent is a disulphide $C_6H_{12}S_2$ (b. p. 75—83° under 10 mm., sp. gr. 1.0234 at 12°) which upon reduction with zinc dust is converted into a substance $C_6H_{12}S$ (b. p. 130°). Nascent hydrogen reduces $C_6H_{12}S_2$ to $C_6H_{14}S_2$ (b. p. 68—69° under 10 mm). The oil also contains a higher sulphide which zinc dust reduces to the compound $C_6H_{12}S$. Finally, another sulphur compound was found in onion oil which is possibly identical with one of the high boiling compounds of asafetida oil. Neither allyl sulphide nor terpenes are contained in onion oil any more than in garlic oil.

62. Oil from Allium Ursinum.

Bärlauchöl.

All parts of *Allium ursinum* L., leaves, flowers and corms, possess an extremely penetrating garlic odor. Upon distillation of the entire

---

1) Journ. Chem. Soc., 1, p. 1.
2) Bericht von S. & Co., Apr. 1889, p. 44.
3) Archiv d. Pharm., 230, p. 443.

plant 0.007 p. c. of an oil is obtained which is highly refractive, and of a dark brown color and burning taste. Its odor, though garlic-like, is quite distinct. Sp. gr. 1.015 at 13°. It boils almost entirely between 95 and 106°. According to Semmler[1] it consists principally of vinyl sulphide, $CH_2:CH-S-CH:CH_2$ (b. p. 101°, sp. gr. 0.9125). In addition this oil contains vinyl polysulphides, also traces of a mercaptan, and an aldehyde not yet characterized.

63. Saffron Oil.

Origin and History. The spicy stigmas of the saffron, *Crocus sativus* L., N. O. *Iridaceae* were used medicinally, and for their color and flavor during antiquity, and up to the middle ages were regarded as one of the more valued spices.

The distilled oil of saffron is first mentioned by Ryff[2] and Gesner,[3] and is enumerated in the municipal price ordinance of Nürnberg of 1613. The older medical treatises do not mention it. The yield was apparently first determined in 1670.[4] Saffron and its constituents were further examined by Lagrange and Vogel in 1810.[5] In 1821 Henry examined the coloring matter of saffron and arrived at the conclusion that the yield of oil was almost doubled if for every ounce of saffron 8 oz. of common salt and 4 oz. of potassa are added to the aqueous distillate.[6]

Preparation, Properties and Composition. Upon distillation of saffron with water in a current of carbon dioxide a small amount of an oil is obtained which is but slightly yellow in color, and possesses an intense saffron odor. It readily absorbs oxygen, becoming viscid and brownish in color.

An elementary analysis gave figures agreeing with the formula $C_{10}H_{16}$. The bulk of the oil, therefore, seems to consist of a terpene.[7] This also results when the aqueous solution of picrocrocin, the bitter principle of saffron, is heated. The reaction is supposed to be expressed by the following equation:

$$\underset{\text{Picrocrocin}}{C_{38}H_{66}O_{17}} + \underset{\text{Water}}{H_2O} = \underset{\text{Crocose}}{3C_6H_{12}O_6} + \underset{\text{Terpene.}}{2C_{10}H_{16}}$$

1) Liebig's Annalen, 241, p. 90.

2) Neu gross Destillirbuch, fol. 188.

3) Ein köstlicher Schatz, fol. 222.

4) Hertodt, Crocologia. Dissertatio, Jenae 1671.

5) Ann. de Chim., 80, p. 185.—Trommsdorff's Journ. d. Pharm., 21, I, p. 206.

6) Trommsdorff's Neues Journ. d. Pharm., 6, p. 65.—Berl. Jahrb. f. Pharm., 24, I, p. 160.

7) Inasmuch as the terpenes possess but a slight odor it seems probable that the characteristic saffron odor is due to the presence of a small amount of an oxygenated substance.

64. Orris Oil.

Oleum Iridis. — Irisöl. — Essence d'Iris concrète. Beurre de Violettes.

Origin and Preparation. The steam distillation of the rhizome of white flag for the production of orris oil is a practice of modern times. Orris root itself, however, has been used since antiquity on account of its mild fragrance and taste.

Iris germanica L., *I. pallida* L. and *I. florentina* are the three species of *Iris*, N. O. *Iridaceae*, which are indigenous to the Mediterranean countries. In these as well as in central European countries they are cultivated and widely distributed as decorative plants.

White flag, especially the first two species, is cultivated for commercial purposes principally in the province of Florence, the districts of Greve, Dicomano, Pelago, Regello, Bagno a Ripoli, Pontassieve, Galluzzo, S. Casciano in Val di Pesa and Montespertoli being the principal centres. The best root is being cultivated in the communities S. Polo and Castellina belonging to the Greve district.

The cultivation has gradually extended to the neighboring provinces of Florence which produce an equally good product: in Arezzo, Castelfranco di Sopra and Lore Ciuffenna in the province of Arezzo; in Grosseto in the province of like name; in Faenza in the province of Ravenna; and in Terni, in the province of Perugia.

Orris root is cultivated on hills and the slopes of mountains—never in the valleys—mostly in large sunny clearings or in rows among the grape vines, and but seldom in extended fields. It flourishes in stony, dry soil. Usually the rhizome is harvested after three years, but if the price is high it is cut when two years old. The freshly cut rhizomes are laid in water to facilitate the peeling and then spread on terraces to dry, a process that requires about 14 days.

When dry, the rhizome is turned into the form in which they are used for teething or into rosaries. The powder is used in the preparation of sachets. The less well-shaped roots, fragments and wastes from the peeling and cutting of the dried rhizome are sold for the distillation of orris oil.

The principal places of export of Italian or Florentine orris root are Livorno, Verona and Triest.

For the purpose of distillation the Florentine root is almost exclusively employed. The Veronese root, derived from *Iris germanica*, is inferior and seldom used for distillation. Still poorer is the Morocco or Mogadore orris root, also derived from *I. germanica*. It is darker in color and has but a faint odor. In order to render it lighter in color it is sometimes bleached with sulphur dioxide, thus rendering it unsuitable for distillation.

Occasionally Indian root[1] has appeared in the London market. It was, however, of such poor quality—due probably to poor methods of collection and treatment—as to be unfit for distillation.

Upon steam distillation[2] orris root yields but 0.1 to 0.2 p. c. of volatile oil. The distillation is rendered difficult and tedious by the frothing, due principally to the high starch content of the rhizome. To facilitate distillation, the addition of sulphuric acid has been suggested, whereby the starch is partly converted into sugar. This process, however, did not stand the test of experience, inasmuch as it affected the odor of the oil.

On account of the small yield and high price, the use of the oil is restricted principally to finer perfumery. In recent years, orris oil is used together with ionone in the preparation of artificial violet perfume.

PROPERTIES. Orris oil is a yellowish-white to yellow mass, of rather firm consistency at ordinary temperature, and an intensive odor reminding of the dried orris root (*Veilchenwurzel*); it melts at about 44—50° to a yellow or yellowish-brown liquid.[3] Orris oil is slightly dextrogyrate. The acid number, about 213—222, corresponds to 85—90 p. c. of myristic acid. The saponification number is 2—6.

COMPOSITION. The bulk of orris oil, about 85 p. c., is composed of the completely odorless myristic acid.[4] The bearer of the violet odor is irone, a ketone of the formula $C_{13}H_{20}O$ (Tiemann and Krüger,[5] 1893). Besides these two constituents, the oil contains small amounts of the methyl ester of myristic acid, oleic acid and its esters, also oleic aldehyde.

For the preparation of the irone, the rhizome is exhausted with ether, and the ethereal residue is distilled with water vapor. In the

1) Bericht von Schimmel & Co., Oct. 1896, p. 45.

2) Prof. H. Hirzel in his "Toiletten-Chemie," 4. ed. (1892), p. 215, states that orris oil cannot be the normal distillate of orris root, because myristic acid is not volatile with water vapor. The fact, however, is that the oil is obtained by steam distillation and since myristic acid is its principal constituent, there can be no doubt that the acid is volatile with water vapor under the conditions under which the distillation is conducted.

3) In order to avoid detrimental over heating, the oil is melted by placing the bottle in warm water having a temperature of about 55—60°.

4) Flückiger (1876), Archiv d Pharm., 208, p. 481. Of older references comp. Vogel, Journ. de Pharm., II, 1, p. 483; Trommsdorff's Journ. d. Pharm., 24, II, p. 64; and Dumas, Journ. de Pharm., II, 21, p. 191; and Liebig's Annalen, 15, p. 158.

5) Berichte, 26, p. 2675.

residue there remains a part of the myristic acid, iregenin, iridic acid, and esters of myristic and oleic acid, whereas part of the myristic acid and its methyl ester, oleic acid and one of its esters, and oleic aldehyde distill over with the irone. Upon repeated steam distillation of the first distillate, irone passes over first and can thus be separated somewhat from the other constituents.

The irone is farther purified by converting it into its phenyl hydrazone and regenerating it by means of sulphuric acid. Perfectly pure irone is obtained by converting it into the crystallizable ironoxime and regenerating it by means of dilute sulphuric acid.

In the preparation of irone from orris oil, the myristic acid is first removed by shaking with dilute potassa solution. This solution is shaken with ether, the ether extract distilled fractionally with water vapor and the irone purified as stated above.

Irone has a sp. gr. of 0.939 at 20°, boils at 144° under 16 mm. pressure, and in a 100 mm. tube turns the plane of polarized light about 40° to the right. In water it is almost insoluble, but readily soluble in alcohol, ether, chloroform, benzene and ligroin.

The odor of pure irone is pungent and in a concentrated state seems to differ entirely from that of the violet. The latter, however, becomes apparent when irone is dissolved in a large amount of alcohol and the solution evaporated.

ADULTERATION. A liquid or semi-liquid oil is sometimes found in commerce under the name of orris oil, which has been obtained from orris root by distillation with cedar oil or other oils, or which is merely a mixture of such oils with some orris oil. Several years ago a mixture of 97.5 p. of acetanilid and 2.5 p. of orris oil was brought into commerce as "irisol" and sold at an enormous price.

65. Curcuma Oil.

ORIGIN AND PREPARATION. The curcuma plant, *Curcuma longa* L., N. O. *Zingiberaceae*, is indigenous to southern Asia, and is cultivated on account of the yellow dyestuff which it contains, in India and in southern and eastern China. Upon distillation with water vapor, curcuma root yields 5—5.5 p. c. of volatile oil.

PROPERTIES. Curcuma oil is an orange-yellow, slightly fluorescent liquid, which has a slight odor of curcuma and a sp. gr. of 0.942. With ½ to 1 vol. of 90 p. c. alcohol it produces a clear solution which is rendered milky upon the addition of more alcohol.

Composition. According to Bolley, Suida and Daube[1] (1868) curcuma oil begins to boil at about 220° and is in complete ebullition at 250°. Above this temperature decomposition takes place. Upon the addition of ammonium sulphide to fraction 230—250° they obtained crystals which were regarded as carvone hydrosulphide. Flückiger[2] (1876), however, could not obtain a hydrogen sulphide addition product with any of the fractions, thus indicating the absence of carvone.

Jackson and Menke[3] (1882) analysed the fraction distilling between 193—198° under 60 mm. pressure, assigned to it the formula $C_{19}H_{28}O$, and called it turmerol. It is an alcohol which yields a chloride, turmeryl chloride, with strong hydrochloric acid. Its sp. gr. is 0.9016 at 17°, $[\alpha]_D = +33.52°$. Under ordinary pressure turmerol boils at 285—290° with decomposition. With an excess of hot permanganate solution it is oxidized to terephthalic acid. Chromic acid yields different oxidation products. Ivanow-Gajevsky[4] obtained valerianic and capronic acids from fraction 280—290°. The lowest boiling portion of curcuma oil consists of phellandrene.[5]

66. Oil of Zedoary.

Oleum Zedoriae.—Zitwerwurzelöl.—Essence de Zédoaire.

Origin and History. The rhizome of the zingiberaceous plant *Curcuma zedoaria* Roscoe (*C. zerumbet* Roxburgh) is at present brought into commerce from Ceylon via Bombay. The plant has been cultivated in the island for a long period because some of the inhabitants are fond of it as a vegetable (Flückiger;[6] Dymock[7]).

The distilled oil of zedoary is first mentioned in the price ordinance of Berlin of 1574,[8] in those of Worms and Frankfurt-on-the Main of 1582, also in the Dispensatorium Noricum of 1589. Examinations as to yield and properties of the oil were later made by Neumann[9] (1785), Dehne[10] (1779) and Geoffroy[11] (1757.)

---

1) Journ. f. pr. Chem., 108, p. 474. Jahresb. f. Pharm., 1868, p. 47.
2) Berichte, 9, p. 470.
3) Amer. Chem. Journ., 4, p. 368.
4) Berichte, 5, p. 1102.
5) Bericht von S. & Co., Oct. 1890, p. 17.
6) Pharmacognosie, p. 369.
7) Materia medica of Western India, p. 772.
8) Estimatio materiae medicae . . . in gratiam et usum publicum civitatum Marchiae Brandenburgensis. Autore Matthaeo Flacco. Berolini 1574.
9) Goettling's Almanach für Scheidekünstler, 1785, p. 118.
10) Crell's Chemisches Journal, 3, p. 20.
11) Tractatus de materia medica, vol. 3, p. 265.

PROPERTIES AND COMPOSITION. Oil of zedoary is a somewhat viscid, oily liquid. In thin layer its color is somewhat greenish, in thicker layer strikingly greenish black; in transmitted light its color is reddish. The odor reminds of that of ginger, but differs from the latter by a camphor-like odor due to the presence of cineol. Its sp. gr. is 0.990—1.01. The rotatory power has not yet been observed on account of the dark color of the oil. It it soluble in 1½—2 vol. of 80 p. c. alcohol.

Upon distillation only a small amount of oil comes over below 240°. In this fraction cineol[1] was shown to be present by means of the hydrobromide. The bulk of the oil distills between 240—300°, a small amount above 300°.

67. Kaempheria Oil..

Kaempferia rotunda L., N. O. *Zingiberaceae*, formerly supplied the *Rhizoma zedoariae rotundae* of the apothecary shops. Upon distillation the root yields 0.2 p. c. of a yellow colored oil.[2] Its pleasant odor, at first camphor-like, is afterward decidedly estragon-like. The sp. gr. of the fresh oil varies from 0.886—0.894[2] at 26°. Another sample, evidently older, had a sp. gr. of 0.945[3] at 15°, $\alpha_D = +13° 4'$ at 14°. Upon distillation half of the oil distilled below 200°, and the greater portion of the other half about 240°.[2] The oil contains cineol.[3]

68. Hedychium Oil.

The oil of the flowers of *Hedychium coronarium* L., (*Zingiberaceae*), which is cultivated in Java, has a pleasant, delicate, but faint odor. Sp. gr, 0.869, $\alpha_D = -0° 28'$.[4]

From the flowers of the Brazilian plant, Peckolt[5] obtained 0.026—0.029 p. c. of oil, sp. gr. 0.869.

69. Oil of Galangal.

Oleum Galangae. — Galgantöl. — Essence de Galanga.

ORIGIN AND HISTORY. Galangal, *Alpinia officinarum* Hance, (*Zingiberaceae*), is originally indigenous to the island Hai-nan. It is still cultivated in the same and in the opposite peninsula Lei-tschou and the neighboring coasts, also in Siam. The spicy rhizome seems to have

1) Bericht von S. & Co., Oct. 1890, p. 58.
2) S'Lands Plantentuin Buitenzorg, 1893, p. 55.
3) Bericht von S. & Co., Apr. 1894, p. 57.
4) Bericht von S. & Co., Apr. 1894, p. 58.
5) Pharm. Rundschau, 11, p. 287.

been used by the Chinese during antiquity. It is mentioned in the *Ayur-vedas* Susrutas, by Plutarch and by the medical writers of the middle ages. In German literature it occurs since the eighth century.

The rhizome is mentioned in the *Dispensatorium Noricum*, but the oil seems to have been distilled later. The first mention of the oil occurs in the price ordinance of Frankfurt-o.-M. of 1587.

PRODUCTION. Upon distillation of the comminuted root with water vapor 0.5—1 p. c. of oil is obtained.

PROPERTIES. Galangal oil constitutes a greenish-yellow, slightly viscid liquid of a camphor-like odor, the taste of which is at first slightly bitter then somewhat cooling. The sp. gr. lies between 0.915 and 0.925, the angle of rotation between —1° 30′ and —3° 30′. It is miscible with ½ and more parts of 90 p. c. alcohol; of 80 p. c. alcohol 10—20 parts are necessary to effect a clear solution. It boils between 170 and 275°.

COMPOSITION. The only known constituent of galangal oil is cineol [1] identified by means of the hydrogen bromide addition product. It is this constituent to which the camphor-like odor of the oil is due.

70. Oil from Alpinia Malaccensis.

The fresh rhizome of *Alpinia malaccensis* Roscoe (*Ladja goah*), which grows wild in Java, yields upon distillation ¼ p. c. of an oil with a pleasant odor.[2] Its sp. gr. varies between 1.039 and 1.047 at 27°. In a 200 mm. tube it deviates the plane of polarized light from 0.25 to 1.5° to the right. Upon cooling, the oil becomes almost completely solid, handsome long needles of methyl cinnamate being formed. The leaves of the same plant also contain methyl cinnamate.

71. Oil from Alpinia Nutans.

Alpinia nutans, Roscoe, also contains in its rhizome a volatile oil,[3] the sp. gr. of which is 0.95 at 29°. A large portion of the oil distills below 230°. Fraction 255—265° yielded upon saponification with methyl alcoholic potassa an acid melting at 134°—probably cinnamic acid.

72. Ginger Oil.

Oleum Zingiberis.—Ingweröl.—Essence de Gingembre.

ORIGIN AND HISTORY. The rhizome of *Zingiber officinale* Roscoe (*Amomum zingiber* L.), which is originally indigenous to southern Asia

1) Bericht von S. & Co., Apr. 1890, p. 21.
2) Kon. Akad. v. Wetensch. te Amsterdam 1898, p. 550.
3) S'Lands Plant. te Buitenzorg, 1897, p. 86.

and which is cultivated there to-day as well as in the south Asiatic archipelago and in other countries, seems to have been used as a spice by the ancient Chinese and Indians. The Greeks and Romans with whom it was a favorite spice obtained it via the Red Sea route and, therefore, considered it an Arabian product. In the third century, however, it was already enumerated among the Indian products brought to Europe via the Red Sea and Alexandria. Ginger was introduced into the West Indies by the Spanish about the middle of the sixteenth century. As early as 1547 ginger was shipped from Jamaica to Spain. from San Domingo in 1585 and from Barbadoes in 1654.

The first mention of distilled ginger oil is found in the spice ordinance of Copenhagen of 1672. The yield was determined in the course of the eighteenth century by Neumann,[1] Gesner,[2] Geoffroy,[3] and Cartheuser.[4]

The rhizome yields about 2—3 p. c. of volatile oil. At present African ginger is mostly used for the distillation of the oil. Of interest is the observation that ammonia is formed in the process. Its formation has not yet been satisfactorily explained.

Properties. Ginger oil possesses the aromatic, not very strong. but persistent odor of the rhizome, but not the pungent taste of the latter. It is viscid and of a greenish-yellow color.

The sp. gr., which lies between 0.875 and 0.885, seems very low if the amount of high boiling constituents is taken into consideration. The angle of rotation varies between —25 and —45°. Ginger oil is one of the least soluble in alcohol, 50—100 parts of a 96 p. c. alcohol being requisite to effect a clear solution. When distilled over a direct flame. the oil passes over between 155 and 300°, but a considerable residue of decomposition products remains in the flask.

The distillate from Japanese ginger differed from the oil from African or Jamaica ginger. Inasmuch, however, as but a few kilos of the rhizome were used in the experiment, the oil may not have been normal. Its sp. gr. was as high as 0.894; its angle of rotation was dextrogyrate, viz. $\alpha_D = 9° 40'$; and it was soluble in 2 p. of 90 p. c. alcohol. No positive test for phellandrene was obtained.

Moreover, an oil distilled in Japan,[5] did not differ in its properties from the ordinary oils. Its sp. gr. was 0.883, $\alpha_D = -26° 32'$. With sodium nitrite and glacial acetic acid it yielded a positive test for phellandrene.

---

1) Chymia medica, vol. 2, p. 688.

2) Dissertatio de Zingibere. Altdorf, 1728, p. 18.

3) Tractatus de mat. med., vol. 2, p. 265.

4) Elementa Chymiae, vol. 2, p. 62.

5) Bericht von S. & Co., Oct. 1893, p. 46.

COMPOSITION. Of ginger oil only a few of the minor constituents are known;[1] of the substances to which the characteristic odor is due nothing is known. In the lowest boiling portions terpenes are found. The angle of rotation of fraction 155—165° is + 63° 13′ (opposite in direction to that of the oil). This fraction contains dextro-camphene (Bertram & Walbaum.[2]) When treated with glacial acetic acid and sulphuric acid it yields an acetate which upon saponification yields isoborneol. m. p. 212°, the bromal derivative of which melts at 71°. Fraction 170° contains phellandrene[2] and yields a nitrite melting at 102°.

The bulk of ginger oil passes over between 256—266° and consists principally of a sesquiterpene (Thresh,[3] 1881) $C_{15}H_{24}$, not yet defined, which does not yield a solid compound with hydrogen chloride.

73. Ceylon Cardamom Oil.

Oleum Cardamomi.—Ceylon Cardamomenöl.—Essence de Cardamome.

HISTORY. Cardamoms, the capsular fruits of different species of *Elettaria* and *Amomum*, family *Zingiberaceae*, appear to have been widely used during antiquity, being known by the names *Ela* (*Ayurvedas*), *Amomis*, *Amomum* and *Card-amomum*, though some doubt exists as to the synonomy of these terms.

Distilled oil of cardamom was known about 1540; having been distilled by Valerius Cordus.[4] The yield was determined by Neumann,[5] Martius,[6] Cartheuser and Spielmann.[7]

ORIGIN AND PREPARATION. The cardamom oil of commerce is not distilled from the officinal Malabar cardamom from *Elettaria cardamomum* White et Matton, but from the long Ceylon cardamom. This variety was designated *E. c.* var. *β* by Flückiger.[8] Formerly it was considered a separate species, the *E. major* Smith.

The Ceylon cardamoms grow wild in the forests of the inner and southern provinces of the island. The ground fruits are distilled yielding 4—6 p. c. of oil. In an experiment in which the seeds and the empty capsules were distilled separately, the former yielded 4 p. c., the latter 0.2 p. c. of oil.

---

1) The first investigation of ginger oil by Papousek in 1852 threw no light on its composition. Sitzungsberichte der Akademie der Wissenschaften zu Wien, 9, p. 315.—Liebig's Annalen, 84, p. 352.

2) Journ. f. prakt. Chemie, II, 49, p. 18.

3) Pharm. Journ., III, 12, p. 243.

4) De artificiosis extractionibus, fol. 226,

5) Chemia medica, vol. 1, p. 828.

6) Schweigger's Journ. f. Chem. und Phys., 3, p. 311.

7) Cardamomi historia et vindiciae.

8) Pharmacographia, 2nd ed., p. 644.

Properties. The Ceylon cardamom oil is light yellow, somewhat viscid. It possesses the strong aromatic odor of the cardamoms, and a pleasant, cooling taste. Sp. gr. 0.895—0.905; $\alpha_D = +12$ to $+15°$. The oil is clearly soluble in 1—2 and more parts of 80 p. c. alcohol. With 70 p. c. alcohol it yields a turbid solution. The saponification number lies between 30 and 70.

The above mentioned oil from the seed has the following properties: sp. gr. 0.908, $\alpha_D = +13°14'$; that of the empty capsules: sp. gr. 0.908, $\alpha_D = +9°48'$.

Composition. After repeated fractionation of Ceylon oil, Weber[1] in 1887 obtained the following four principal fractions: 1.) 170—178°; 2.) 178—182°; 3.) 182—190°; 4.) 205—220°. By passing hydrogen chloride into fraction 1, a dihydrochloride, $C_{10}H_{16}2HCl$, melting at 52° was obtained. With bromine no crystalline bromide could be obtained. On account of the somewhat high melting point (2° higher than that of dipentene dihydrochloride), Weber leaves it undecided whether the dihydrochloride obtained is due to the presence of dipentene or not.

Fraction 2) contains terpinene, a terpene which is characterized by the nitrosite melting at 155° and which had not been found previously in any volatile oil.

Fraction 4) consists of terpineol. Weber obtained with hydrogen chloride a chloride melting at 52°, and with hydrogen iodide an iodide melting at 76°.

Although neither dipentene tetrabromide, obtained by Wallach[2] from terpineol, nor phenyl terpinyl urethane[3] could be obtained, the negative results of Weber must be attributed to impurities which act as disturbing factors in the test.

In the course of the distillation water was split off, formic and acetic acid were also formed. The latter indicate the presence of esters, the amount of which according to the saponification number (comp. properties) varies from about 10—20 p.c. From the residue of the fractionation a solid substance separated which, upon recrystallization from alcohol, constitutes light laminae of a silvery lustre melting at 60—61°.

74. Malabar Cardamom Oil.

Origin. On account of their high price, the officinal Malabar (and Madras) cardamoms from *Elettaria cardamomum* White et Matton are

---

1) Liebig's Annalen, 238, p. 98.
2) Liebig's Annalen, 230, p. 266.
3) Ibid., p. 267.

seldom used in the manufacture of the volatile oil. The yield varies between 2 and 8 p. c.

PROPERTIES. The odor of Malabar cardamom oil differs but slightly from that of the Ceylon oil. Sp. gr. 0.933 (Haensel[1]) to 0.943 (Schimmel & Co.[2]); $\alpha_D = +26°$ [1] to $+34° 52'$.[2] It is soluble in 4 and more parts of 70 p. c. alcohol. Saponification number 132.

COMPOSITION. In an old specimen of Malabar cardamom oil, Dumas and Péligot[3] (1834) found prismatic crystals of terpin hydrate, $C_{10}H_{16} . 3H_2O$. Their presence was undoubtedly due to their formation from terpineol shown to be present later. According to Schimmel & Co.,[4] the high saponification number is due to terpinyl acetate. The silver salt from the saponification liquid corresponds with silver acetate. (Found 64.8 and 65 p. c., Ag., calculated 64.69 p. c.) From the saponified oil crystallized d-terpineol was obtained by fractional distillation under diminished pressure (150—164° under 14 mm.). M. p. 35—37°; $\alpha_D = +81° 37'$ (in the molten condition). It was chemically identified by means of dipentene dihydriodide (m. p. 78—79°); terpinyl phenyl urethane (m. p. 112—113°); (also optically active $[\alpha]_D = 33° 58'$ at 20° in 10 p. c. alcoholic solution); and terpineol nitrosochloride. The piperidide from the last melted at 151—152°, eight degrees lower than that of the inactive isomer. The oil also contains cineol.

In a recent investigation Parry[5] has also shown the presence of limonene.

75. Siam Cardamom Oil.

ORIGIN. Under the name of "Camphor seeds," so called on account of their camphor-like odor, the seeds of the Siam cardamom from *Amomum cardamomum* L. occasionally find their way into the London market. Upon distillation, Schimmel & Co.[4] obtained 2.4 p. c. of oil.

PROPERTIES. At ordinary temperature, this oil constitutes a semi-solid mass having the odor of camphor and borneol. In order to redissolve the crystals, the oil had to be heated to 42°, at which temperature the oil had a sp. gr. of 0.905 and an angle of rotation of $+38° 4'$. The saponification number was 18.8, after acetylization 77.2, corresponding to 22.5 p. c. of borneol in the original oil. The oil is soluble in 1.2 vol. of 80 p. c. alcohol.

---

1) Südd. Apoth.-Zeitung, 1896, p. 688.
2) Bericht von S. & Co., Oct. 1897, p. 8.
3) Ann. d. Chim. et Phys., II, 57, p. 335.
4) Bericht von S. & Co., Oct. 1897, p. 9.
5) Pharm. Journ., 63, p. 105.

Composition. In order to separate the stearoptene, the oil was cooled in ice and placed in a centrifuge. From 800 g. of oil, 100 g. of crystals were obtained. From the petroleum ether solution, about 40 g. of almost pure borneol separated upon cooling. When purified by means of the benzoyl ester, it melted at 204°, and in a 10 p. c. alcoholic solution had a specific rotatory power $[\alpha]_D = +42° 55'$ at 20°.

The petroleum ether mother liquid, upon evaporation, left a granular mass which, after recrystallization from 80 p. c. alcohol, melted at 176—178° and possessed all the properties of camphor. The oxime melted at 118°. The rotatory power determined in alcoholic solution at 20° was $[\alpha]_D = +45° 17'$.

Thus the crystals which separate from Siam cardamom oil[1] consist of d-borneol and d-camphor, present in about equal parts.

76. Oil from Grains of Paradise.

Origin and History. The seeds of the zingiberaceous *Amomum melegueta* Roscoe, which is indigenous to the coast of tropical West Africa, were formerly much used as spice and were known in the apothecary shops as *Grana paradisi*, *Semina cardamomi majoris* or *Piper melegueta.* The plant grows from the Congo to the Sierra Leone and a part of this coast district is known by the name of the drug as the Pepper- or Melegueta coast. Upon distillation on a small scale of the grains of paradise, 0.3 p. c.[2] were obtained, upon a larger scale 0.75 p. c. of oil.[3] This oil was distilled by Porta[4] at the beginning of the seventeenth century and used medicinally.

Properties. The oil from the grains of paradise is a yellowish liquid, of a spicy but scarcely characteristic odor. Sp. gr. 0.894;[5] $\alpha_D = -3° 58'$. It begins to boil at 236°, the principal portion passing over between 257—258°. The results of an elementary analysis of this fraction corresponded with the formula $C_{20}H_{32}O$.

77. Bengal Cardamom Oil.

Origin. Bengal cardamoms, from *Amomum aromaticum* Roxb.,[6] upon distillation yield 1.12 p. c. of volatile oil.

1) The camphor mentioned by Flückiger in Pharmacographia, 2nd ed., p. 647, should probably be referred to the oil from Siam cardamoms and not to that from Ceylon cardamoms.

2) Pharmacographia, 2nd ed., p. 658.

3) Bericht von S. & Co., Oct. 1897, p. 10.

4) De Destillatione, lib. IV, C. 4.

5) Flückiger (l. c.) gives the sp. gr. as 0.825, which may be due to a printer's error.

6) According to E. M. Holmes, Bengal cardamoms are derived from *Amomum aromaticum* Roxb., and not as Flückiger states in Pharmacographia from *Amomum subulatum* Roxb.

PROPERTIES. This oil[1] has a light yellow color, and possesses a decided odor of cineol; sp. gr. 0.920 at 15°; $a_D = -12° 41'$. It forms a clear solution with one and more parts of 80 p. c. alcohol. Upon distillation the bulk of the oil passes over below 220°, leaving, however, considerable residue in the flask.

COMPOSITION. The only known constituent of the oil is cineol. It was identified by means of the hydrobromide, regeneration of the pure cineol boiling at 175—176°, sp. gr. 0.924, and conversion into cineolic acid (m. p. 197°).

Inasmuch as the characteristic cardamom odor is wanting, the Bengal oil cannot take the place of the Ceylon oil and is, therefore, without practical value.

78. Cameroon Cardamom Oil.

ORIGIN. The plant from which the so-called Cameroon cardamom is obtained, is not yet known. Upon distillation of the fruit Schimmel & Co. obtained 2.33 p. c. of oil.

PROPERTIES. The oil smells strongly of cineol and cannot, therefore, be taken as a substitute for the Ceylon oil. Sp. gr. 0.907[2] to 0.9071 (Haensel[3]); $a_D = -20° 34'$[2] to $-20° 30'$.[3] The oil forms a clear solution with 7—8 p. of 80 p. c. alcohol.

COMPOSITION. The only known constituent is cineol, identified by means of the cineol-iodol reaction.[2]

79. Korarima Cardamom Oil.

Korarima cardamoms, formerly known as *Cardamomum majus*, have the form and size of a small fig. They come from the countries south of Abyssinia, but are seldom found in the European markets. The mother plant of this species, the *Amomum angustifolium* Sonnerat attains a height of 4—5 m. and has orange-yellow blossoms and scarlet fruits. The latter are eaten by the natives. The plant is common along the rivers in British Central Africa, and occurs also in the islands of Mauritius and Madagascar (Hanbury[4]). The roots have a slight ginger-like taste, the leaves are more aromatic.[5] The oil from the fruits was distilled by Schimmel & Co.[6] in 1877, a yield of 2.13 p. c. being obtained.

---

1) Bericht von S. & Co., Apr. 1897, p. 48.

2) Bericht von S. & Co., Oct. 1897, p. 10.

3) Südd. Apoth. Zeitung, 36, p. 688.

4) Science Papers, p. 112.

5) Bull. Royal Gardens Kew., No. 142, p. 288.—Ref. Apt.-Zt., 13, p. 872.

6) Bericht von S. & Co., Jan. 1878, p. 7.

80. Oil of Black Pepper.

Oleum Piperis.—Pfefferöl.—Essence de Poivre.

Origin and History. The unripe, dried berries from *Piper nigrum* L., a climber of the family *Piperaceae*, which was originally indigenous to southern India, are obtained from plants cultivated in different parts of southern India, in numerous islands of the Indian Archipelago, the Philippines and in the West Indies.

The ripe berries, after the removal of the peri- and mesocarp, constitute the white pepper of commerce.

Pepper has been highly prized since antiquity. Like gold it was used as a medium of exchange and as an article of tribute. Pepper was used as a symbol of the spice trade. Even in Rome the dealers in spices were known as *Piperarii*, later in France as *Pebriers*, and in England as pepperers.

Inasmuch as the more common spices were frequently distilled, the distilled oil of pepper may have been known during the middle ages. It is first mentioned by Saladin and described by Valerius Cordus, later by Giov. Batt. Porta. Winther of Andernach, who distilled the oils of pepper, cinnamon, cloves etc., about 1750, first described the process of distillation. In medical treatises, it is first mentioned in the 1589 edition of the Dispensatorium Noricum, in the "Apothekertaxe" of Berlin for 1574 and that of Frankfurt for 1582. Rheede also described oil of pepper in 1688.

The first examinations of the constituents of pepper were made by Neumann[1] and by Gaubius,[2] later by Willert,[3] Oerstedt[4] and Pelletier.[5]

Preparation. The comminuted black pepper, the unripe fruit of *Piper nigrum* L., upon distillation with water vapor yields 1 to 2.3 p. c. of volatile oil. Worthy of notice is the formation of ammonia, which is also observed in the distillation of several other oils, viz. the oils of ginger, pimenta and cubebs.

White pepper, which is obtained from the ripe berries, also the pericarp removed in its preparation, contains volatile oil.[6] Whether this agrees with the oil from black pepper is not known.

1) Gründliche mit Experimenten erwiesene Chymie. Editio C. H. Kessel. 1789. Vol. 2, Part 4, p. 9.

2) Gaubii, Adversariorum varii argumenti liber unus. 1771. Chap. 5, p. 55.

3) Trommsdorff's Journ. der Pharmacie, 20, II, p. 44.

4) Schweigger's Journ. für Chemie und Physik, 29, p. 80.

5) Trommsdorff's Neues Journ. d. Pharm., 6, II, p. 288.

6) Lucä (Trommsdorff's Taschenbuch für Chemiker und Pharmaceuten, 1822, p. 81) obtained 1.61 p. c. of oil from white pepper.

PROPERTIES. A colorless or yellowish-green liquid of more or less distinct phellandrene-like odor, and a mild, by no means sharp taste. Sp. gr. 0.87 to 0.90. It deviates the ray of polarized light either to the right or left. The degree of deviation as observed on rather meagre material varies from $-5°2'$ to $+2°27'$. In alcohol it is rather difficultly soluble, mostly requiring about 15 p. of 90 p. c. alcohol to effect a clear solution. On account of the large phellandrene content, the phellandrene test with sodium nitrite and acetic acid, is as a rule, though not invariably, successful directly without fractionation.

COMPOSITION. As shown by the first analyses[1] (1835) oil of pepper is almost free from oxygen. This fact was verified by the examination of Eberhardt[2] (1887) who obtained 87.26 p. c. C. and 10.81 p. c. H. as the results of an elementary analysis. The highest fractions of the oil boiling between 170 to 310° were colored green. Fraction 169.5 to 171° had the composition of a terpene; fraction 176° yielded terpin hydrate when treated with alcohol and acid.

Although fraction 176 to 180° upon bromination yielded dipentene tetrabromide melting at 122 to 123°, Eberhardt regarded the terpene as not identical with any of the known terpenes. As was shown later by Schimmel & Co.,[3] oil of pepper contains phellandrene, which is evidently laevogyrate, the fraction from which it was obtained turning the plane of polarized light 10° to the left.

The formation of the tetrabromide by Eberhardt from fraction 176 to 180° reveals the presence of dipentene. It is still a question, however, whether dipentene is contained in the original oil or whether it was formed from the phellandrene by the repeated distillation of the oil.[4]

It also remains undecided to which terpene the terpin hydrate owes its origin. That dipentene can be converted into terpin hydrate is known. Whether phellandrene also will undergo this change has not yet been ascertained.

---

[1]) An analysis of the oil by Dumas (Liebig's Annalen, 15, p. 159) yielded results agreeing with the formula $C_{10}H_{16}$; comp. also Soubeiran & Capitaine (1840), Liebig's Annalen, 34, p. 326.

[2]) Archiv d. Pharm., 225, p. 515.

[3]) Bericht von S. & Co., Oct. 1890, p. 39.

[4]) Wallach states (Liebig's Annalen, 287, p. 372): that oils containing phellandrene should not be distilled, at least not repeatedly, under ordinary pressure, since phellandrene undergoes changes when subjected to such treatment.

81. Pepper Oil from Long Pepper.

Origin and History. The spikes with the compact berries of two species of *Piper* or *Chavica*, viz. *P. officinarum* D. C. (*C. officinarum* Miqu.) and *P. longum* L. (*C. roxburghii* Miqu.) are known as long pepper. Both species grow in the islands of the Indian Archipelago, the latter also in the Philippines, in southern India, Bengal, Malabar and Ceylon. The fruit is collected while immature and dried.

The distilled oil from long pepper is mentioned in the drug price ordinances of 1589 and was admitted to the Dispensatorium Noricum of 1589. The first examination of long pepper was made by Winckler in 1827.[1]

Preparation and Properties. Upon distillation, long pepper yields 1 p. c. of a viscid, yellowish-green oil of sp. gr. 0.861. It distills between 250 and 300°, has a mild taste like pepper oil and an odor reminding of ginger.[2]

82. Oil from Piper Ovatum.

Besides other constituents, the leaves of *Piper ovatum* (?) contain a terpene (Dunstan and Garnett,[3] 1895).

83. Oil from Ashanti Pepper.

According to Herlant[4] the fruits of *Piper clusii* D. C, Ashanti pepper, contain 11.5 p. c. of volatile oil.

84. Oil of Cubebs.

Oleum Cubebarum.—Cubebenöl.—Essence de Cubèbe.

Origin and History. The berries from the piperaceous climber *Piper cubeba* L. (*Cubeba officinalis* Miquel) are mostly brought into commerce via Batavia and Singapore. The shrub is indigenous to the large Sunda islands and is there cultivated, also in Ceylon and other tropical islands.

Cubebs were first examined by Wedel[5] in 1704, soon after by Neumann,[6] in 1810 by Trommsdorff,[7] and in 1821 by Vauquelin.[8] The

1) Archiv d. Pharm., 26, p. 89.

2) Bericht von S. & Co., Apr. 1890, p. 48. Comp. also Dulong (1825), Journ. de Pharm. II, 11, p. 59.—Trommsdorff's Neues Journ. d. Pharm., 11, I, p. 104.

3) Chem. News, 71, p. 88.

4) Acad. Roy. de Méd. de Belgique, 1894, p. 115; Jahresb. d. Pharm., 1895, p. 142.

5) De cubebis. Dissertatio. Jenae 1705.

6) Lectiones chymici.

7) Trommsdorff's Journ. der Pharm., 20, I, p. 69.

8) Trommsdorff's Taschenb. für Chem. und Pharm., 1822, p. 195.

volatile oil was known to Valerius Cordus before 1540;[1] in price ordinances it first receives mention in that of Frankfurt-on-the-Main in 1582.

PREPARATION. Comminuted cubebs upon distillation with water-vapor yield 10—18 p. c. of oil. In the course of the distillation ammonia is given off, the formation of which is no more understood in this case than in the distillation of ginger, pepper, pimenta and other drugs.

PROPERTIES. Cubeb oil is a viscid, light green or bluish-green oil. It is colorless only when the last portions of the distillation, which are blue, have not been added to the product. It possesses the characteristic cubeb odor and a warm, camphor-like taste, which finally becomes grating. Sp. gr. 0.915 to 0.930; $\alpha_D = -25$ to $-40°$.

The solubility in 90 p. c. alcohol varies greatly. Some oils, presumably from old cubebs, are soluble in an equal part, others require 10 p. c. of this alcohol to effect solution.[2]

The bulk of the oil distills[3] between 250—280°. A quantitative fractionation yielded the following result:[4]

1)	from 175—250° =	9.2	p. c.
2)	" 250—260° =	26.8	"
3)	" 260—270° =	47.6	"
4)	" 270—280° =	7.2	"
	above 280° =	9.2	"
		100.0	p. c.

Oils distilled from old cubebs, which contain the cubeb camphor described below, are heavier than the normal and can be recognized by their behavior toward potassium. A piece of potassium or sodium, immersed in such an oil, loses its lustre and is covered with a crust, whereas an oil distilled from fresh cubebs does not attack the metal (Schmidt,[5] 1870).

COMPOSITION. From the behavior toward potassium it becomes apparent that the oil from fresh cubebs is almost devoid of oxygen, consisting only of terpenes and sesquiterpenes. On the other hand the reaction of the oil from old cubebs is to be attributed to cubeb camphor.

---

1) Annotationes, 1561, fol. 226.

2) The requirement of the U. S. P., according to which the oil is to be soluble in an equal volume of 90 p. c. alcohol is, therefore, too exacting.

3) Oils containing cubeb camphor undergo, on distillation, partial decomposition with the elimination of water. Schaer & Wyss (1875), Archiv d. Pharm., 206, p. 322.

4) Pharm. Journ., III, 25, p. 951.

5) Archiv d. Pharm., 191, p. 28.

From the lowest boiling fractions a small amount can be collected between 158—163° by systematic fractionation. It consists of a laevogyrate terpene (Oglialoro,[1] 1875), ($\alpha_D = -35.5°$) presumably pinene or camphene. Furthermore dipentene was identified (dihydrochloride, m. p. 48—49°) in the fraction below 200° (Wallach,[2] 1887).

As already stated, the bulk of the oil distills between 250 and 280° and consists of two laevogyrate sesquiterpenes.[2] The one, b. p. 262—263°, has the lesser rotatory power, does not combine with hydrogen chloride, and has not yet been further examined. The second yields a dihydrochloride,[3] $C_{15}H_{24}.2HCl$, melting at 118° and is identical with cadinene. Although pure cadinene boils at 274—275°, almost the total amount of the hydrocarbon is obtained in lower boiling fractions from which the dihydrochloride separates upon saturation with hydrogen chloride.

Cubeb camphor[4] is a sesquiterpene hydrate, possibly an alcohol of the composition[5] $C_{15}H_{25}OH$. It is laevogyrate, crystallizes in rhombic forms and melts according to different authors at 65°,[6] 67°[7] and 70°.[8] It is rather unstable, decomposing when kept over sulphuric acid into sesquiterpene and water.[6] It boils at 148° giving off part of its water.[9] The dehydration is complete when heated for some time to 200—250°. The resulting sesquiterpene has not yet been examined.

The fact that cubeb camphor is found only in old cubebs, leads to the supposition that it is formed by the hydration of the oil when the berries are exposed to a moist atmosphere.

85. Oil from Piper Lowong.

From the fruit of *Piper lowong* Bl., known as false cubebs, Peinemann[10] in 1896 obtained by distillation with water vapor 12.4 p. c. of an almost colorless oil with a sp. gr. of 0.865.

---

1) Gazz. chim. ital., 5, p. 467; Berichte, 8, p. 1357.

2) Liebig's Annalen, 238, p. 78.

3) This was first obtained by Soubeiran and Capitaine in 1840 (Liebig's Annalen, 34, p. 323), and recognized by them as sesquiterpene dihydrochloride. Later it was examined by Schmidt, Schaer & Wyss, also by Wallach.

4) Cubeb camphor was first observed by Teschemacher at the beginning of this century. It was also examined by Müller in 1832, Liebig's Annalen, 2, p. 90; by Blanchet & Sell in 1833, Liebig's Annalen, 6, p. 294; by Winckler in 1833, Liebig's Annalen, 8, p. 203; by Schmidt in 1870, Archiv d. Pharm., 191, p. 23, and Berichte, 10, p. 188; also by Schaer & Wyss in 1875, Archiv d. Pharm., 206, p. 316.

5) Schmidt, Schaer & Wyss, loc. cit.

6) Schmidt, loc. cit.

7) Schaer & Wyss, loc. cit.

8) Winckler, loc. cit.

9) This low boiling point for a sesquiterpene is rather remarkable and probably due to the partial decomposition.

10) Archiv d. Pharm., 234, p. 238.

Upon distillation it can be resolved into two principal fractions, but partial decomposition takes place whether distilled under ordinary or diminished pressure.

Fraction 1, constituting 40 p. c. of the oil, boils between 165—175°, and is colorless; sp. gr. 0.854 at 20°; $\alpha_D = +22°$.

Fraction 2, constituting 34 p. c. of the oil, boils between 230—255°, and is colored yellow; sp. gr. 0.9218; optically inactive.

At 270° a bluish-green fraction is obtained.

From a fraction 110—148° under 17 mm. small crystals separated upon standing, which after recrystallization from chloroform melted at 164°.[1] An elementary analysis yielded results corresponding with the rather improbable formula $C_{10}H_{16} \cdot 2H_2O$.

86. Matico Oil.

Oleum Foliorum Matico.—Maticoöl.—Essence de Matico.

Origin. *Piper angustifolium* Ruiz et Pavon, indigenous to South America, is stated to be the mother plant of the matico leaves. Inasmuch, however, as the term *Matico* is applied to a number of plants, the leaves of which can hardly be distinguished from the genuine, it is not surprising that mistakes are frequent and that the genuine article cannot be had at times. Recently imported matico leaves differ but little in appearance from those formerly in commerce, but considerably in the amount of oil they contain and in the properties of this oil. Whereas formerly 1—3.5 p. c. of an oil lighter than water was obtained, now 3—6 p. c. of an oil heavier than water results.

Properties. The matico oil of former years had a sp. gr. of 0.93—0.99 and was slightly dextrogyrate. It was a viscid, more or less dark colored liquid the odor of which reminded of cubebs and mint.

The more recent matico[2] oil is yellowish-brown; has a sp. gr. 1.06—1.13; slightly laevogyrate $[\alpha]_D = -0° 25'$, or dextrogyrate up to $+5° 34'$. The odor reminds of that of *Asarum europaeum.* The oil is soluble in 10 p. of 80 p. c. alcohol and in equal parts of 90 p. c. alcohol.

Composition. Matico camphor which was discovered by Flückiger[3] and which has been examined as to its physical properties only, is the only known constituent of the old matico oil. It was obtained by distilling off those portions of oil boiling below 200°. From the residue

1) Possibly this substance is identical with the stearoptene, of like melting point, found in juniper oil.

2) An oil, which evidently was obtained from a third variety of leaves (yield 0.8 p. c.), had a sp. gr. of 0.922 and a rotatory power $\alpha_D = -27° 28'$.

3) Pharmacognosie. III ed., p. 747.

the camphor crystallized out in hexagonal prisms 2 cm. long and 5 mm. thick. Frequently it also crystallized from the oil when exposed to the cold.

Matico camphor, which when pure is odorless and tasteless, is readily soluble in alcohol, ether, chloroform, benzene and petroleum ether. The melting point of the crystals lies at 94° (Hintze[1]). It is optically active (Traube[2]). In chloroformic solution $[\alpha]_D = -28.73°$ at 15°. In a molten condition, calculated for 15°, $[\alpha]_D = -29.17°$. The specific rotatory power of the crystals is about eight times as great, being —240° for a plate 100 mm. thick.

The identity of matico camphor with ethyl camphor, $C_{10}H_{15}(C_2H_5)O$, was regarded probable by Kügler[3] who based his supposition on an elementary analysis. The fact, however, that ethyl camphor[4] is a liquid having the odor of camphor, whereas matico camphor is odorless with a great tendency to crystallize, seems to speak against this assumption.[4]

The new matico oil contains no matico camphor. From an oil distilled by Schimmel & Co.[5] (sp. gr. 1.077; $\alpha_D = -0°25'$) a substance separated which, after repeated crystallization from petroleum ether melted at 62° and which proved to be asarone. The dibromide melted at 85—86°, and the asarylic acid, obtained by oxidation with permanganate, at 144°. The oil possibly also contains methyl eugenol, for upon oxidation with permanganate it yielded an acid melting at 174° which probably was identical with veratric acid.

87. Oil from Artanthe Geniculata.

The leaves of *Artanthe geniculata* Miq., known in Brazil as "false jaborandi," yield small amounts of a light greenish oil, of a spicy, somewhat mint-like odor and pungent, burning taste.[6]

88. Oil of Betel Leaves.

Oleum Foliorum Betle.—Betelöl.—Essence des Feuille des Bétel.

ORIGIN. The ancient custom of betel chewing practiced in the Malayan Archipelago and in southern China consists in chewing a betel leaf (*Sirih*)

---

1) Tschermak's Mineral. Mitth., 1874, p. 227.
2) Zeitschr. f. Kryst., 22, p. 47.
3) Berichte, 16, p. 2841.
4) Compt. rend., 68, p. 222.
5) Bericht von S. & Co., Oct. 1898, p. 87.
6) Peckolt, Pharm. Rundschau, 12, p. 286. — *Artanthe geniculata* Miq. and *Piper angustifolium* Ruiz et Pavon are according to Lürssen, Medic. Pharm. Botanik, vol. 2, p. 514, synonymous terms. This oil, therefore, is probably identical with one of the oils described under matico oil.

from *Piper betle* L. (*Chavica betle* Miq.) with some lime, gambier (from *Uncaria gambir* Roxb.) and a piece of betel nut from *Areca catechu* L.[1] The betel leaves owe their spicy, burning taste to a volatile oil which has been repeatedly prepared and investigated, but which has not yet found any practical application worth mentioning.

History. When and by whom betel oil was first distilled is not known: it is highly probable that the preparation of the oil by Kemp[2] in 1885 was not the first. A superficial examination of Siam betel oil in the laboratory of Schimmel & Co.[3] in 1887 revealed the presence of a phenol in fraction 250—260°, which seemed to correspond with eugenol. Eykman,[4] who in 1888 examined an oil distilled by himself in Java, found no eugenol, but a new phenol which he called chavicol. A second examination in the laboratory of Schimmel & Co. revealed the fact that Siam betel oil contains neither eugenol nor chavicol, but a third phenol, a previously unknown isomer of eugenol, which Bertram and Gildemeister[5] in 1889 termed betel phenol.

Properties. Betel oil is a light yellow to dark brown liquid of aromatic, somewhat creosote-like odor, reminding of tea, and with a pungent taste.

The sp. gr. varies between 0.958 and 1.044; the oil from fresh leaves being lighter both in weight and color than that distilled from the dried material. The rotatory power was observed on three samples of oil from fresh leaves. Of these two were laevogyrate (α_D up to $-1°45'$) and one dextrogyrate ($\alpha_D = +2°45'$).

With ferric chloride the alcoholic solution of betel oil produces a greenish to bluish-green color.

Source, yield, physical properties and composition of the betel oils thus far examined are given in the following table (see next page).

Composition. Of the two phenols found in betel oil. betel phenol and chavicol, only the former has been found in all oils and may, therefore, be regarded as the characteristic constituent of betel oil. Chavicol, which so far has been found only in the Java oil, does not occur in the Manila or Siam oils.

---

1) The details concerning betel chewing and the constituents of the "Betel-Apotheke" owned by almost every Malay family are described in Tschirch's work. "Indische Heil- und Nutzpflanzen" (Berlin 1892) p. 188. See also True, Pharm. Review, 14, pp. 130, 177.

2) Pharmacographia indica, part VI, p. 188.

3) Bericht von S. & Co., Oct. 1887, p. 34.

4) Chemiker-Zeitung, 12, p. 1388.

5) Journ. f. prakt. Chemie, II, 39, p. 349.

	Origin, Material used and Yield	Sp. Gr.	α_D	Constituents
I.	Siam.[1] Leaves dried on hot plates. Yield 0.6 p. c.	1.024 at 15°	—	Betel phenol, Cadinene.
II.	Siam.[1] Leaves dried by exposure to the sun. Yield 0.9 p. c.	1.020 at 15°	—	
III.	Manila.[2] Fresh leaves. Yield 0.27 p. c.	1.044 at 15°	—	Betel phenol.
IV.	Java.[3] Fresh leaves.	0.959 at 27°	−1°45′	Chavicol, Betel phenol, Sesquiterpene.
V.	Java.[4] Probably fresh leaves.	0.958 at 15°	+2°45′	Not examined.
VI.	Bombay.[5] Fresh leaves.	0.9404 at 28°	Slightly laevogyrate	

Betel phenol (chavibetol), $C_{10}H_{12}O_2$, contains the same side chains as its isomer eugenol, but in different positions.

```
      C3H5                C3H5                C3H5
       |                   |                   |
       C                   C                   C
      / \                 / \                 / \
  HC/     \CH         HC/     \CH         HC/     \CH
   |       |           |       |           |       |
   |       |           |       |           |       |
  HC\     /COH        HC\     /COCH3      HC\     /CH
      \ /                 \ /                 \ /
     COCH3                COH                 COH
   Eugenol.           Betel phenol.        Chavicol.
```

Betel phenol[6] is a liquid with marked refractive power, and possesses a peculiar, not unpleasant but persistent betel odor, which differs greatly

1) Bertram & Gildemeister, Journ. f. prakt. Chemie II, 39, p. 349.—Bericht von S. & Co., Apr. 1888, p. 8; Apr. 1889, p. 6; Oct. 1889, p. 6; Apr. 1890, p. 6; Oct. 1891, p. 5.

2) Bericht von S. & Co., Apr. 1891, p. 5, and Oct. 1891, p. 5.

3) Berichte, 22, p. 2786.

4) Bericht von S. & Co., Oct. 1893, p. 45.

5) Pharm. Journ., III, 20, p. 749.

6) Bertram & Gildemeister, loc. cit.

from that of the closely related eugenol. When pure, betel phenol boils without apparent decomposition at 254—255° (at 131—132° under 12—13 mm. pressure) and has a sp. gr. of 1.067 at 15°. In alcoholic but not in aqueous solution it yields an intensely bluish-green color with ferric chloride.

Betel phenol can be identified by means of its benzoyl derivative, which crystallizes in laminae and melts at 49—50°. (Benzoyl eugenol melts at 69—70°). Acetyl betel phenol boils at 275—277° and melts at —5° (acetyl eugenol melts at 30—31°), and upon oxidation with permanganate yields acet-isovanillic acid, m. p. 207°.

Chavicol,[1] also a phenol, has so far been definitely shown to be present only in betel oil from Java. Chavicol or para hydroxy allyl benzene (comp. p. 179) boils at 237° and has a sp. gr. of 1.041 at 13°. Its aqueous solution is colored intensely blue with ferric chloride, the alcoholic solution but faintly blue.

An oil distilled by de Vrij in Java, when examined in the laboratory of Schimmel & Co.[2] was found to contain in addition to betel phenol a second phenol, which is probably identical with chavicol. The benzoyl compound crystallized in long needles and melted at 72—73°.

Another oil distilled in Manila from fresh leaves and examined by Schimmel & Co.[3] contained no other phenol than betel phenol. The phenol separated from the soda solution had a constant boiling point of 128—129° under 11 mm. pressure and yielded exclusively a benzoyl compound crystallizing in laminae and melting at 50°.

Another constituent, possibly to be found in all betel oils, is cadinene, $C_{15}H_{24}$. So far this hydrocarbon has been isolated from the Siam oil[4] only (dihydrochloride, m. p. 118°). Probably the sesquiterpene (b. p. abt. 260°, sp. gr. 0.917) found by Eykman in the betel oil from Java is also cadinene. The Java oils from fresh leaves contain a considerable amount of low boiling constituents. Eykman[5] did not succeed in isolating or identifying a pure terpene of constant boiling point from the fractions between 173—190°. Probably several terpenes are present, but apparently no pinene. Fraction 173—175° (sp. gr. 0.848 at 16°; $\alpha_D = -5°20'$) yielded neither a solid bromide nor a crystalline hydrochloride. Fraction 190—220° contains substances having a minty odor (menthone or menthol?).

---

1) Eykman, loc. cit.
2) Bericht von S. & Co., Apr. 1890, p. 7.
3) Bericht von S. & Co., Oct. 1891, p. 5.
4) Bertram & Gildemeister, loc. cit.
5) Loc. cit.

The oil distilled from dried Siam leaves was devoid of low boiling fractions, only a few drops coming over below 200°. Whereas the absence of terpenes in this oil may be accounted for by the character of the crude material, the absence of chavicol in Siam and Manila oils and the occurrence of this substance in Java oil may be accounted for by differences in climate and soil of the respective countries.

89. Oil from Potomorphe Umbellata.

The fresh leaves of *Potomorphe umbellata* Miq. (Family *Piperaceae*) contain 0.05 p. c. of an oil, the taste and odor of which remind of pepper (Peckolt[1]).

90. Oil from Ottonia Anisum.

From 12 kilo of the air-dried root of wild jaborandi, *Ottonia anisum* Spreng. (*Serronia jaborandi* Guill.) 10.54 g. (=0.088 p. c.) of a viscid oil were obtained. It possesses a peculiar, somewhat pepper-like odor and a burning taste which benumbs the tongue. Sp. gr. 1.035 (Peckolt[2]).

91. Oil from Poplar Buds.

The leaf-buds of the black poplar, *Populus nigra* L., (Family *Salicaceae*), which were formerly found in apothecary shops under the name of *Oculi populi* and used in the preparation of ointments, yield about ½ p. c. of volatile oil upon distillation with water vapor. It is light yellow in color and has a pleasant odor reminding somewhat of chamomile. It is insoluble in 70—90 p. c. alcohol, but yields a clear solution with ½ p. or more of 95 p. c. alcohol. Sp. gr. 0.900—0.905; $\alpha_D = +1° 54'$[3] to $+5° 54'$; saponification number 13.

Poplar bud oil distills between 255—265°[3] and consists principally of a hydrocarbon $(C_5H_8)x$ boiling at 260—261°. Piccard[4] (1873) judges from its vapor density that it is a diterpene, $C_{20}H_{32}$, whereas its boiling point indicates a sesquiterpene.

Fichter and Katz[5] in a recent investigation of this oil, found it to consist principally of a sesquiterpene boiling at 263—269°. The compound was identified as humulene by means of its nitrosochloride, nitrosate, both the blue and the white nitrosite, and the piperidine and benzylamine bases prepared from these. Besides humulene there is pro-

---

1) Pharm. Rundschau, 12, p. 241.
2) Pharm. Rundschau, 12, p. 287.
3) Bericht von S. & Co., Apr. 1887, p. 36.
4) Berichte, 6. p. 890; 7, p. 1486.
5) Berichte, 32, p. 3183.

bably contained in the oil a second sesquiterpene, and about ⅓ p. c. of paraffin, $C_{24}H_{50}$, m. p. 53—68°. The bodies to which the pleasant odor of the oil is due passed over in the lowest fraction.

92. Dutch Myrtle Oil.

Upon distillation of the fresh leaves of *Myrica gale* L. (Family *Myricaceae*), 0.65 p. c. of a brownish-yellow oil (Rabenhorst,[1] 1837), sp. gr. 0.876, are obtained. At 17.5° it congeals partly, at 12.5° completely to a crystalline mass. The odor is peculiar balsamic, pleasant; the taste mild at first, then temporarily burning and permanently astringent. It is said to contain 70 p. c. camphor (?).

93. Bayberry Oil.

The leaves of *Myrica cerifera* L. (Family *Myricaceae*) yield 0.021 p. c. of volatile oil,[2] of greenish color, and very pleasant aromatic, spicy odor. Sp. gr. 0.886; $\alpha_D = -5° 5'$.

94. Sweet Fern Oil.

The dried leaves of *Myrica asplenifolia* Endl. (*Comptonia asplenifolia* Aiton) yield upon distillation about 0.08 p. c. of volatile oil. It has a strong spicy, cinnamon-like odor. Sp. gr. 0.926. When cooled in a freezing mixture the oil congeals.[3]

95. Oil from Walnut Leaves.

The aromatic odor of the fresh leaves of *Juglans regia* L. (Family *Juglandaceae*) is attributable to the volatile oil they contain. Upon distillation of 800 k. of fresh leaves, Schimmel & Co.[4] obtained 235 g. of oil. It is yellowish-green in color, solid at ordinary temperature, and has a pleasant tea-like odor.

96. Sweet Birch Oil (Wintergreen Oil).

Oleum Betulae Lentae. — Birkenrindenöl. — Essence de Betula.

Origin and History. Cherry birch or sweet or black birch, *Betula lenta* L. (Family *Betulaceae*) is a tree 15—20 m. high which grows on good forest soil throughout southern Canada and the northern United

---

1) Repert. Pharm., 60, p. 214. Gmelin, Organ. Chemie, IV Ed., Vol. 7, p. 885.

2) Bericht von S. & Co., Oct. 1894, p. 78. Comp. also Hambright (1868), Am. Journ. Pharm., 35, p. 193.

3) Bericht von S. & Co., Oct. 1890, p. 50.

4) Bericht von S. & Co., Oct. 1890, p. 49.

States, westward as far as Minnesota and Kansas, and to the south as far as Georgia and Alabama. When chewed its reddish-bronze colored bark develops a peculiar fragrance and taste, and on this account has been used by the natives for chewing and in the preparation of refreshing and medicinal beverages. Next to turpentine oil, the oils of sassafras, wintergreen and birch bark are among the first oils obtained by distillation in the United States. The similarity in odor and taste of birch bark oil with true oil of wintergreen from *Gaultheria procumbens* was known before 1818 (Bigelow[1]). The chemical identity of the principal constituent of both, however, was demonstrated by Proctor[2] in 1843. As the demand for wintergreen oil increased, sweet birch bark was distilled indiscriminatly with wintergreen leaves or even distilled alone as substitute[3] so that the commercial natural oil is at present obtained almost exclusively from the bark of *Betula lenta* L.

Preparation. For purposes of distillation, the young trunks and branches were formerly used. These were cut into pieces 1—4 inches in length which were macerated for 12 hours previous to distillation. For the latter operation stills like those described under wintergreen oil were used. More recently the bark of the trunk and the larger branches is peeled off in late summer and either cut, or torn by means of toothed rollers, and freshly distilled with water from copper stills. If wintergreen grows abundantly in the neighborhood it is added to the bark in the still. According to the abundance and the convenience with which the one or the other can be obtained, preference is given to the cheaper material. According to Kennedy, maceration for 12 hours is considered indispensable to a good yield. A ton of 2,240 lbs. of birch bark yields about 5 pounds of oil = 0.23 p. c. A like amount of wintergreen yields about 18 lbs. of oil.[4] Upon rational distillation, however, as much as 0.6 p. c. of oil can be obtained from the bark.

Proctor already recognized in 1843 that the oil did not preexist in the bark, but results upon the interaction of two constituents in the presence of water, similar to the formation of the oils of bitter almonds,

1) American Medical Botany, vol. 2, pp. 28 and 241.

2) Am. Journ. Pharm., 15, p. 241.

3) *Betula lenta* and *Gaultheria procumbens* grow together in the wooded mountainous regions of the North Atlantic states. In a report on the wintergreen oil industry, Kennedy (Am. Journ. Pharm., 52, p. 49) in 1882 called attention to the fact that the cost of collection of birch bark was but ½ of that of wintergreen leaves; also that the yield of oil from the bark was but 0.23 p. c., whereas that from wintergreen was 0.80 p. c. According to these figures the cost of production of the oil from birch bark is but half as great as that of wintergreen leaves.

4) Am. Journ. Pharm., 54, pp. 49—58.

mustard, etc.[1] According to more recent investigations by Schneegans[2] these substances are betulase, a ferment, and the glucoside gaultherin[3] which crystallizes with one molecule of water:

$$C_{14}H_{18}O_8 + H_2O = C_6H_4<^{OH}_{COOCH_3} + C_6H_{12}O_6$$

Gaultherin Water Methyl salicylate Grape sugar.

PROPERTIES. Oil of sweet birch is a colorless or yellowish colored oil sometimes colored red due to the presence of traces of iron. As to odor and taste, it can scarcely be distinguished from pure methyl salicylate, but differs decidedly in both respects from the oil of *Gaultheria procumbens.* Sp. gr. 1.180 to 1.187. It is optically inactive, differing in this respect from the slightly laevogyrate gaultheria oil. It forms a clear solution with 5 p. of 70 p. c. alcohol at ordinary temperature, and with aqueous potassa or soda solution upon gentle heating. Water shaken with a drop of the oil produces a deep violet color with ferric chloride. By distillation over a free flame, sweet birch oil passes over between 218—221°.

COMPOSITION. Inasmuch as no commercial distinction is made between birch oil and gaultheria oil, it remains doubtful whether the discovery of salicylic acid by Cahours[4] (1844) was made in connection with an oil of sweet birch, a true wintergreen oil, or a mixed oil. Proctor possibly distinguishes between the two. It is quite certain that the oil examined by Cahours must have been adulterated for he found 10 p. c. of a terpene boiling at 160° which he terms "gaultherilene" and which presumably was pinene from admixed turpentine oil. Recent examinations of genuine oils have shown the absence of terpenes in both oils.

According to Power and Kleber[5] (1895) sweet birch oil contains as much as 99.8 p. c. of methyl salicylate.[6] If the oil diluted with ether is shaken with an aqueous 7.5 p. c. potassa solution, the methyl salicylate passes into the aqueous solution as potassium methyl salicylate and can readily be separated from the other constituents remaining in ethereal solution.

---

1) Am. Journ. Pharm., 15, p. 241.

2) Journ. d. Pharm. v. Elsass-Lothr., 28, p. 17.

3) Arch. d. Pharm. 282, p. 487.

4) Ann. de Chim. et Phys., III. 10, p. 327.—Liebig's Annalen, 48, p. 60; 52, p. 327.

5) Pharm. Rundschau, 13, p. 228.

6) The statement by Trimble & Schröter (Am. Journ. Pharm., 61, p. 398, and 67, p. 561), that the oils of gaultheria and sweet birch contain benzoic acid, ethyl alcohol and a sesquiterpene was not substantiated by the examination of a large amount of authentic oil. See Power (Pharm. Rundsch., 7, p. 288), also Power & Kleber (Pharm. Rundschau, 13, p. 228). These constituents evidently do not occur in the genuine oil.

The residue remaining upon evaporation of the ether was separated into two parts by steam distillation. The non-volatile residue congealed upon cooling. It was readily soluble in ether, more sparingly in alcohol. and crystallizes in laminae which are not attacked by sulphuric or nitric acids. They consist of paraffin which, judging from an elementary analysis and m. p. determination (m. p. 65.5°) may be regarded as triacontane, $C_{30}H_{62}$. The portion volatile with water vapor is readily soluble in 80 p. c. alcohol, boils between 230—235° (under 25 mm. pressure at abt. 135°), and has the composition $C_{14}H_{24}O_2$. It is an ester and upon saponification is resolved into its compounds; an alcohol $C_8H_{16}O$ and an acid $C_6H_{10}O_2$.

Examination. Inasmuch as the oils of sweet birch and gaultheria are the heaviest of all known oils, the presence of foreign oils or of petroleum causes a lowering of the specific gravity. The solubility in 70 p. c. alcohol is also diminished by the presence of most adulterants. The property of methyl salicylate to form water soluble salts with the alkalies (e. g. $C_6H_4[OK]COOCH_3$) affords a convenient means of detecting adulterations. A suitable method is that directed by the U. S. Pharmacopoeia:

If to 1 cc. of oil contained in a capacious test-tube, 10 cc. of sodium hydrate T. S. be added, and the mixture agitated, a bulky, white, crystalline precipitate will be produced; then if the test tube, loosely corked, be allowed to stand in boiling water for about five minutes, with occasional agitation, the precipitate should dissolve, and form a clear, colorless or faintly yellowish solution, without the separation of any oily drops. either at the surface or at the bottom of the liquid (absence of other volatile oils or of petroleum).

Still more convenient is the use of a 5 p. c. caustic potash solution which dissolves methyl salicylate immediately without the formation of a precipitate, leaving other oils undissolved.

Chloroform can also be detected in this way. Five percent of foreign constituents can be recognized in this way. Inasmuch as the odor of methyl salicylate disappears completely in the alkaline solution, the nature of the adulterant can usually be recognized by its odor.

The amount of methyl salicylate can be determined by isolating and weighing the salicylic acid or volumetrically.

For the gravimetric determination Ewing[1] (1892) recommends the following process:

1.5—2 g. of the oil are saponified in a flask of 50 cc. capacity with a slight excess of soda solution. To the liquid, transferred to a separating funnel,

---

1) Proc. Am. Pharm. Ass., 40, p. 196.

hydrochloric acid is added in excess and the liberated salicylic acid shaken out with ether. In order to remove traces of sodium chloride from the ethereal solution, this is shaken with water. The ethereal solution is evaporated in a tared capsule on a waterbath. The residual salicylic acid is weighed after having been dried to constant weight over sulphuric acid.

This method of assay presupposes a pure oil, for foreign additions such as sassafras oil, would pass into the ethereal solution with the salicylic acid and remain upon evaporation of the solvent.

For the volumetric estimation, E. Kremers and M. James[1] give the following directions:

5 g. of oil to be dissolved in an excess of standard soda solution and the solution to be boiled for five minutes to effect saponification. The excess of soda solution is ascertained with the aid of normal acid and the number of ccs. of normal alkali consumed in the saponification multiplied by 0.152, the methyl salicylate factor.

The assay can also be effected by determining the amount of salicylic acid present according to the method of Messinger and Vortmann.[2] This method is based upon the supposed fact that in the presence of much alkali salicylic acid is converted by iodine into a diiodo salicylic acid iodide. The excess of iodine is ascertained by titration with standard thiosulphate solution. The reaction is supposed to take place according to the following equation:

$$C_6H_4(OH)COOK + 3NaOH + 6I = C_6H_2I_2(OI)COOK + 3KI + 3H_2O$$

The application of this method to wintergreen oil has been suggested by Kremers and James:[1]

A weighed quantity of oil (abt. 3 g.) is saponified with normal alkali, care being taken to employ at least 7 molecules of potassium hydroxide for every molecule of methyl salicylate. The saponified liquid is diluted to either 250 or 500 cc. Of this solution 5 or 10 cc. respectively are heated in a flask to 60°. Decinormal iodine solution is then added until the color is permanently yellow. Upon shaking a deep red precipitate results. Upon cooling, the solution is acidified with dilute sulphuric acid and diluted with water to 250 or 500 cc. In an aliquot part (abt. 100 cc.) of the filtrate the excess of iodine is ascertained by means of N/10 thiosulphate solution.

$$\frac{\text{1 Mol. Methyl salicylate}}{\text{6 Atoms Iodine}} = \frac{151.64}{759.2} = 0.19974314.$$

By multiplying the amount of iodine found (in ccs.) with the factor 0.19974314 the amount of methyl salicylate present is found, from which the percentage of ester in the oil can be calculated.

A good oil should contain at least 98 p. c. of methyl salicylate.

---

1) Pharm. Review, 16, p. 130.

2) Berichte, 22, p. 2321; 23, p. 2755.

97. Oil of Hops.

Oleum Humuli Lupuli.—Hopfenöl.—Essence de Houblon.

ORIGIN AND HISTORY. On account of their peculiar aromatic odor and taste, the fruit of *Humulus lupulus* L. (Family *Moraceae*), which is cultivated in most countries, has been used since the middle ages for the aromatization of barley beer. Hops, therefore, are mentioned in literature since the eighth century.[1] Medicinally hops are being used only comparatively recently. The medical use of lupulin was suggested in 1820 by Ives,[2] a New York physician.

The volatile oil of hops seems to have been first distilled from the glands by Payen and Chevallier,[3] but thus far has found but little application.

PREPARATION. Oil of hops is prepared from the strobiles as well as from the lupulin. The yield from hops varies from 0.3—1 p. c., from lupulin as much as 3 p. c. are obtained. The latter, however, is less pleasant and, therefore, inferior. The aqueous distillate—especially that from lupulin—is strongly acid and contains valerianic acid (Personne) and probably also butyric acid (Ossipow). Unbleached hops must be used for distillation, otherwise the oil has an unpleasant sulphurous odor.

PROPERTIES. Oil of hops is a light yellow to reddish-brown thin liquid which becomes viscid upon prolonged standing. It has an aromatic odor and its taste is not bitter. Sp. gr. 0.855—0.880; $\alpha_D = +0°28'$ to $+0°40'$. In alcohol it is very difficultly soluble, especially older oils do not yield a clear solution with even 95 p. c. alcohol. This may possibly be attributed to polymerization products resulting from so-called olefinic terpenes present in the oil.

COMPOSITION. Although oil of hops has frequently been examined chemically,[4] not a single constituent had been isolated and characterized until recently.

In 1895 Chapman[5] showed that, aside from small amounts of olefinic terpenes, oil of hops consists principally of a sesquiterpene, humulene. The oxygenated constituents, however, to which the oil evidently owes its peculiar aroma, were not investigated by Chapman.

---

1) Pharmacographia, p. 551.

2) Silliman's Journ. of Sc. & Arts, 1820, p. 302.

3) Journ. de Pharmacie, 8, pp. 214 & 583.

4) Payen & Chevallier, loc. cit. — Wagner (1858), Journ. f. prakt. Chem., 58, p. 351. — Personne (1884), Compt. rend., 38, p. 309.—Kühnemann (1877), Berichte, 10, p. 2281. —Ossipow (1888 & '86), Journ. f. prakt. Chem., II, 28, p. 48, and 34, p. 238.

5) Journ. Chem. Soc., 67, pp. 54 & 780.

The fact that the analysis of fraction 166—171 (sp. gr. 0.789) agrees with the formula $C_{10}H_{17}$, seems to render it probable that it consists of a mixture of two hydrocarbons, viz. $C_{10}H_{18}$ and $C_{10}H_{16}$. The latter is possibly identical with one of Semmler's[1] olefinic terpenes. Bromine and hydrochloric acid are absorbed energetically, but do not yield crystalline derivatives. Upon oxidation, oxalic acid alone was obtained.

Almost two-thirds of the oil consist of humulene, a sesquiterpene boiling at 263—266°, sp. gr. 0.9001 at 15°. Isolated by fractionation it is obtained either slightly laevo- or dextrogyrate. When perfectly pure it is probably inactive.

With bromine and hydrogen chloride liquid addition products only are obtained. A characteristic derivative was obtained by passing nitrosyl chloride into the chloroform solution. This nitrosochloride melts at 164—165° and may serve for its characterization. The humulene nitrolpiperidide obtained from it melts at 153°, the nitrolbenzylamine base at 136°.

Caryophyllene from oil of cloves yields similar derivatives which are, however, quite different in melting point. Caryophyllene also readily yields a crystalline hydrate, whereas humulene does not. The latter, therefore, is to be regarded as a new distinct representative of the class of sesquiterpenes.[2]

ADULTERATION. Oil from copaiba balsam is mentioned as an adulterant of hop oil. The higher sp. gr. and larger angle of rotation should, however, readily betray its presence.

According to Chapman,[3] the statement made by Personnes that the oil upon oxidation yields valerianic acid is incorrect. This acid, however, results upon oxidation of the extract of hops with permanganate.

98. Oil of Hemp.

The various statements concerning the properties and composition of the volatile hemp oil differ considerably.

According to Personne[4] (1857) the oil from *Cannabis indica* is lighter than water and congeals at 12—15° to a butyraceous mass. It is said to consist of two hydrocarbons: the liquid canabene, $C_{18}H_{20}$,

---

1) Berichte, 24, p. 682.
2) For a detailed comparison of these two hydrocarbons see Pharm. Arch., 2, p. 278.
3) Chem. Centralbl., 1898, II. p. 860.
4) Journ. de Pharm. et de Chim., III, 31, p. 48.

22

which boils at 235—240°, and cannabene hydride, $C_{12}H_{24}$, which crystallizes from alcohol in scales with a fatty lustre.

Valente[1] (1880) examined an oil that was obtained from Italian *Cannabis indica.* It consisted principally of a sesquiterpene, $C_{15}H_{24}$; b. p. 256—258°; sp. gr. 0.9299 at 0°; $[\alpha]_D = -10.81°$. With hydrogen chloride it yielded a solid hydrochloride. The same sesquiterpene is found in the oil from the male inflorescence of *Cannabis gigantea.*

Upon steam distillation of the female flowering plant, Vignolo[2] (1885) obtained a mobile aromatic oil which boiled between 248 and 268° and did not congeal when cooled to —18°. When distilled over sodium, a stearoptene remained behind which was not further examined, whereas a sesquiterpene (b. p. 256°; sp. gr. 0.897 [?] at 15.3°) passed over. The formula $C_{15}H_{24}$ was established by elementary analysis and vapor density determination. A crystalline hydrochloride was not obtained.

Schimmel & Co.[3] obtained upon distillation of the non-flowering herb of *Cannabis indica* 0.1 p. c. of a mobile oil with narcotic but not unpleasant odor. At 0° it congealed to a butyraceous mass. Sp. gr. 0.932.

Whether the sesquiterpene found in the oil is identical with any of the known compounds does not become apparent from the scant literature on the subject. The cannabene hydride of Personnes is possibly nothing more or less than paraffin which is frequently found in volatile oils.

99. Sandalwood Oil.

Oleum Ligni Santali.—Ostindisches Sandelholzöl.—Essence de Santal.

History. On account of its peculiar, very pleasant odor, as well as on account of its durability, sandalwood from *Santalum album* L. has been used since antiquity. Medicinally and economically it has an important and very interesting history.

In mediaeval writings and in the later treatises on distillation, however, sandalwood is but seldom mentioned, the oil having been used only in recent periods for medicinal and other purposes. Previous to this century the oil has been distilled by Saladin[4] (1488), Gesner[5] (1555), and Hoffmann[6] (1722). Neumann and Dehne[7] (1780) ascertained the yield, and Chapoteaut[8] first examined the oil in 1882.

---

1) Gazz. chim. ital., 10, p. 540, & 11, p. 191.—Berichte, 18, p. 2481, & 14, p. 1717.

2) Gazz. chim. ital., 25, I, p. 110.

3) Bericht von S. & Co., Oct. 1895, p. 57.

4) Compendium aromatariorum, fol. 349.

5) Ein köstlicher Schatz, p. 246.

6) Observatorium physico-chemicarum, p. 69.

7) Crell's Chemisches Journal, 3, p. 18.

8) Bull. Soc. chim., II, 37, p. 303.

In Ceylon, distilled sandalwood oil is reported to have been used for the embalming of the dead bodies of native princes since the ninth century.[1]

ORIGIN AND PREPARATION.[2] *Santalum album* L. (Family *Santalaceae*), a tree 6—10 m. high with dense foliage, is indigenous to the mountains of India and either grows wild or is cultivated in dry open places, rarely in forests.[3] The wood of the trees growing in dry rocky mountainous soil is harder and richer in oil than that of trees cultivated in fertile soil. The territory which produces sandalwood is a strip 250 miles long which extends from the Nilgiri mountains to the north and northwest through Mysore and Coimbatore to Canara. The tree grows from sea-level up to altitudes of 1000 m. The trees are property of the state. In the presidencies of Madras and Mysore the wood and the roots are arranged according to grade into 18 classes[4] and sold at auction. Wood destined for the European market is shipped via Tellichery or Bombay. A considerable amount is used in Asia for ritualistic purposes, and some of it is used in the primitive distillation of the oil at the place of production. Bidie[5] describes the process as follows:

"The body of the still is a large globular clay pot, with a circular mouth, and is about 2½ feet deep by about 6 feet circumference at the bilge. No capital is used, but the mouth of the still, when used, is closed with a clay lid, having a small hole in its centre, through which a bent copper tube about 5½ feet long is passed for the escape of the vapor. The lower end of the tube is conveyed inside a copper receiver, placed in a large porous vessel containing cold water. When preparing the sandalwood for distillation the white or sap wood is rejected, and the heart wood is cut into small chips, of which about 2 maunds or 50 lbs. are put into the still. As much water is then added as will just cover the chips, and the distillation is carried on slowly for ten days and nights, by which time the whole of the oil is extracted. As the water from time to time gets low in the still, fresh supplies are added from the heated contents of the refrigerator."

According to another report[6] the distillation lasts 21 days and yields 2.5 p. c. of oil of a sp. gr. 0.980 and higher. The prolonged distillation is sufficient explanation for the dark color and high specific

---

1) Pharmacographia, II ed., p. 599.

2) J. L. Pigot, conservator of forests in Mysore, India, has written an interesting pamphlet on the sandalwood cultivation in India accompanying the exhibition of the British Colonial Governments at the Paris Exposition 1900.

3) Holmes, Pharm. Journ., III, 16, p. 819.—Petersen, ibid., p. 757.—Kirkby, ibid., p. 857.—Sawer, Odorographia, vol. I, p. 315.

4) Bericht von S. & Co., Oct. 1898, p. 44.

5) Holmes, loc. cit.

6) Chemist and Druggist, May 26, 1894.

gravity[1] of this oil contaminated with decomposition products. Only a small part of it is brought to Europe, most of it being exported to China and Arabia.

Sandalwood is also produced in eastern Java, and in the islands Sumba (Soemba or Tjendana) and Timor. This variety is brought into the market as Macassar sandalwood via Macassar (in the Celebes). It does not contain as much oil as the East Indian wood, but the oil can hardly be said to be inferior.

Owing to a more complete comminution as well as to perfected apparatus for distillation a larger yield of oil is obtained in Europe than in India: from East India sandalwood 3—5 p. c., from Macassar wood 1.6—3 p. c. The European oil is light colored, has a pleasant odor and compares favorably with the Indian oil which is contaminated with empyreumatic products.

Properties. East Indian sandalwood oil is a rather viscid, light yellowish to yellow liquid of peculiar, faint but persistent odor, and of an unpleasant, resinous, harsh taste. The sp. gr. of a normal oil lies between 0.975 and 0.980. The Indian oils, as already stated, have a higher specific gravity and a darker color due to decomposition products. The angle of rotation varies between narrow limits, viz. —17 to —19°. With 5 p. of 70 p. c. alcohol, sandalwood oil yield a clear solution which is not rendered turbid by the further addition of alcohol. Upon age, especially under the influence of light and air, its solubility diminishes and it then yields turbid mixtures in the ratio mentioned.

The santalol content determined by acetylization and calculated for $C_{15}H_{26}O$ is 93—98 p. c. Under 14 mm. pressure about 95 p. c. of the oil passes over between 155—170°,[2] under ordinary pressure between 275—295° (Umney,[3] 1895). The saponification number is 5—15.

Composition. According to the investigation of Chapoteaut[4] (1882) sandalwood oil consists almost exclusively of two substances $C_{15}H_{24}O$ and $C_{15}H_{26}O$. The former boils at 300° and is regarded as an aldehyde, whereas $C_{15}H_{26}O$, with a boiling point of 310° is supposed to be the corresponding alcohol. Phosphoric acid anhydride abstracts water from the oil, two hydrocarbons resulting: to the one boiling at 245°

---

1) Conroy has shown experimentally that the sp. gr. of an oil is increased from 0.989 when heated with water to 50° for ten days (Chemist and Druggist, Aug. 19, 1893).

2) Bericht von S. & Co., Oct. 1893, p. 37.

3) Pharm. Journ., III, 25, p. 1044. The statement of Chapman and Burgess, however, (see under composition) do not fully agree with these data.

4) Bull. Soc. chim., II, 37, p. 303.

the formula $C_{15}H_{22}$ is assigned; and to the other, b. p. 260°, the formula $C_{15}H_{24}$. When the oil is heated with glacial acetic acid to 150° the acetic ester of santalol, $C_{15}H_{25}O.COCH_3$ results, also a substance $C_{30}H_{46}O$ (?).

Inasmuch as the low specific gravity of Chapoteaut's oil (0.945) would seem to indicate that it was adulterated, the above results should be accepted with caution. Furthermore he did not prepare a single crystalline derivative with which to prove the correctness of his formulas. For the aldehyde character of the substance $C_{15}H_{24}O$ he does not supply a single proof.

Chapman and Burgess[1] have more recently published a preliminary report on santalol and the hydrocarbon obtained by its dehydration.

Santalal, which they also regard as an aldehyde, was isolated by fractional distillation from the oil. The substance thus obtained boils at 301—306° and upon oxidation with potassium permanganate yielded santalenic acid, which crystallizes in thin laminae and melts at 76°. The hydrocarbon obtained by the dehydration of santalal with phosphorus pentoxide boils at 140—145° under 25 mm. pressure; sp. gr. 0.9359 at 15°; $[\alpha]_D = +5° 45'$. It is unsaturated and combines directly with the hydrides of chlorine and bromine. Nitroso and nitrosochloride derivatives could not be obtained.

As was first shown by Parry[2] (1895) sandalwood oil contains saponifiable substances. It is, therefore, assumed that part of the santalol occurs in the oil as ester. The acetylization of the oil, described under the next heading, shows that the bulk of the oil consists of an alcohol. Whether the formula used in the calculation, viz. $C_{15}H_{26}O$, which was assigned to santalol by Chapoteaut is correct or not will have to be decided by further investigation. Schimmel & Co.[3] have recently obtained saponification numbers with acetylized santalol which had been regenerated from the phthalic acid compound which do not agree with the formula $C_{15}H_{26}O$. The results agree better with $C_{15}H_{22}O$, or else santalol is a mixture of different alcohols.

Von Soden and Müller[4] found in the non-alcoholic portion of the oil a sesquiterpene boiling from 261—262°; sp. gr. 0.898; $\alpha_D = -21°$. It could not be identified as one of the known sesquiterpenes. It adds hydrochloric and hydrobromic acid and bromine, but yields no crystalline derivative. By hydration with glacial acetic acid and sulphuric acid a

1) Proc. Chem. Soc., 1896, p. 140.
2) Pharm. Journ., 55, p. 118.
3) Bericht von S. & Co., Apr. 1900, p. 44
4) Pharm. Ztg., 44, p. 258.

small amount of an alcohol boiling at 160—165° under 7 mm. pressure was obtained. They also found an acid melting at 154° in the oil.

Guerbet[1] has recently investigated sandalwood oil, but his results must be taken with caution as the oil investigated had a very low sp. gr. and was therefore either adulterated or else an abnormal distillate. He found the following constituents: 1. Two sesquiterpenes: α-santalene, b. p. 252—252.5°, sp. gr. 0.9134 at 16°, $\alpha_D = -13.98°$; β-santalene, b. p. 261—262°, sp. gr. 0.9139 at 0°, $\alpha_D = -28.55°$. The latter is identical with that found by Von Soden and Müller. 2. A mixture of two alcohols, $C_{15}H_{26}O$, of different rotatory power. 3. An aldehyde, $C_{15}H_{24}O$, santalal, b. p. 180° at 40 mm.; it has a strong pepper-like odor. Its semicarbazide melts at 212°. 4. An acid, $C_{15}H_{24}O_2$, santalic acid, a viscid, colorless liquid boiling at 210—212° at 20 mm. 5. An acid, $C_{10}H_{14}O_2$, teresantalic acid, m. p. 157°. Small amounts of strongly smelling substances, to which the odor of the oil is principally due were found in the more volatile portions of the oil.

Examination. Inasmuch as the physical constants of sandalwood oil are subject to slight changes only, additions of almost any nature can usually be detected by means of the specific gravity, optical rotation, and solubility in 70 p. c. alcohol. Cedar wood oil, the principal adulterant, is readily recognized by an increase in the optical rotation, lowering of the sp. gr. and the diminution of solubility. Very similar changes are produced by the oils from copaiba- and gurjun balsam; the former, however, usually diminishes the angle of rotation.

West Indian sandalwood oil, which is sometimes used as adulterant, is dextrogyrate and is very difficultly soluble in alcohol. It is reported that the oil distilled in India is sometimes adulterated with castor oil or with the fatty oil from the seeds of the sandalwood tree, which is used as illuminating oil. Additions of sesame oil, paraffin oil and linseed oil are also reported as not being infrequent. All of these, no doubt, betray their presence by their insolubility in 70 p. c. alcohol and by their high saponification number (naturally with the exception of paraffin oil).

The best means of ascertaining the purity of a sandalwood oil, or the amount of an adulterant, is to determine the santalol content. Good oils mostly contain from 93—98 p. c., never less than 90 p. c. santalol.

Parry[2] first suggested the conversion of santalol, $C_{15}H_{26}O$, into its acetic ester, $C_{15}H_{25}O.COCH_3$, by heating it with glacial acetic acid

[1]) Compt. rend., 130, p. 417; Journ. de Pharm. et Chim., VI, 9, p. 224.
[2]) Pharm. Journ., 55, p. 118.

Fig. 62.

Santalwood Tree.

in a closed vessel to 150°; and to saponify the ester with alcoholic potassa. According to Schimmel & Co.,[1] however, it is more expedient to make the alcohol assay of volatile oils with acetic anhydride.

The process is conducted as follows:[2]

About 20 g. of sandalwood oil are gently boiled for 1½ hour with an equal volume of acetic acid anhydride and a small amount of fused sodium acetate. The product is washed with water and soda solution and the resulting oil dried with anhydrous sodium sulphate. Of the dried oil, 2—5 g. are boiled with an excess of N/1 potassa V. S. and the excess of alkali ascertained by titration with N/1 sulphuric acid V. S.

The amount of santalol is calculated with the aid of the following formula:

$$P = \frac{a \times 22.2}{s-(a \times 0.42)}$$

P = Santalol content of the original oil.
a = Number of cc. N/1 potassa V. S. consumed.
s = Amount of acetylized oil, expressed in grams, used for saponification.

In most instances the determination of the physical constants will suffice to distinguish a pure oil from one adulterated. Positive assurance can, however, be had by making a determination of the santalol content. It is, therefore, as unnecessary as it is unscientific to resort to color reactions which have been suggested.

100. South Australian Sandalwood Oil.

Santalum preissianum Miq. known as "quandong" in Australia bears edible fruits known as native peaches. The wood is dark brown, of very dense and tough texture and unusually hard and heavy. It contains 5 p. c. of a viscid, cherry-red oil, sp. gr. 1.022. The odor is pleasantly balsamic reminding somewhat of roses. Upon standing the oil separates crystals, which by recrystallization are obtained in prisms melting at 104—105°.[3]

COMPOSITION. The crystalline constituent of the oil has been examined by Berkenheim[4] (1892). He found the melting point 101—103° and assigned to it the formula $C_{15}H_{24}O_2$. The substance is an alcohol, the acetic ester of which crystallizes in hexagonal plates melting at 68.5 to 69.5°. With phosphorus trichloride it yields a chloride, $C_{15}H_{23}OCl$, m. p. 119—120.5°; phosphorus pentachloride does not act on the

1) Bericht von S. & Co., Oct. 1895, p. 41.
2) Bericht von S. & Co., Apr. 1897, p. 40.
3) Bericht von S. & Co., Apr. 1891, p. 49; and Oct. 1891, p. 38.
4) Zeitschr. d. Russ. phys. chem. Ges., 24, p. 688; Abstr. Chem. Centralbl., 1893, I, p. 986.

alcohol. The methyl ether obtained by means of the sodium compound of the alcohol is liquid. Potassium permanganate oxidizes it to a liquid acid $C_7H_{14}O_2$.

101. West Australian Sandalwood Oil.

The wood of *Santalum cygnorum* Miq. (*Fusanus spicatus* R. Br.) is exported from Freemantle, W. Austr., and is known in the Singapore market as Swan river sandalwood. In India and China it is used as substitute for the Indian sandalwood from *Santalum album*. The wood contains 2 p. c. of oil having an unpleasant resinous odor; sp. gr. 0.953[1]—0.965 (Parry[2]); $\alpha_D = +5° 20'$.

West Australian sandalwood oil, therefore, has very different properties from those of the East Indian oil and cannot be used as a substitute for the latter.

The oil was distilled as early as 1875 by Schimmel & Co.; recently the distillation of the oil has been taken up in Freemantle.[3]

Parry[2] found the saponification numbers 1.1—1.6. After acetylization he obtained saponification numbers which seemed to indicate an apparent santalol content of 75 p. c. Whether the alcohol of this oil is identical with that of *S. album* has not yet been established.

102. Fiji Sandalwood Oil.

The wood of *Santalum yasi* Seem[4] from the Fiji islands was exhibited at the Colonial Exposition at South Kensington in 1886. Upon distillation it yielded 6¼ p. c. of a volatile oil with a faint but not very delicate odor, thus rendering it unfit for perfumery.[5] Sp. gr. 0.9768; $\alpha_D = -25.5°$ (Mac Ewan,[6] 1888).

103. African Sandalwood Oil.

The botanical origin of this oil is not known. The wood designated sandalwood, from which this oil was distilled, was dark brown in color, very hard and tough and had been brought to Europe from Tamatave (Madagascar) via Zanzibar. Upon distillation it yielded 3 p. c. of a ruby red oil of the consistency of East Indian sandalwood oil, which it also resembled in odor.[7] Sp. gr. 0.969.

1) Bericht von S. & Co., Oct. 1888, p. 86; and April 1891, p. 48.
2) Notes on Santal Wood Oil, p. 9; Chem. & Drug., 58, p. 708.
3) Bericht von S. & Co., Oct. 1898, p. 45.
4) Pharm. Journ., III, 16, p. 757 & 820.
5) Bericht von S. & Co., Apr. 1888, p. 39.
6) Pharm. Journ., III, 18, p. 661.
7) Bericht von S. & Co., Apr. 1891, p. 49.

The wood is possibly identical with *Hasoranto*, a wood occurring in northern Madagascar which is said to possess properties similar to those of sandalwood.[1]

104. Oil of Asarum Europaeum.

Oleum Asari Europaei.—Haselwurzöl.—Essence d'Asaret.

The root of *Asarum europaeum* L. (Family *Aristolochiaceae*), which grows in the shady hard wood forests of Europe, Siberia and the Caucasus, upon distillation yields 1 p. c. of a viscid brown oil heavier than water with an aromatic odor and a peppery, burning taste. Often the oil congeals soon after distillation, sometimes crystals of asarone separate only after prolonged standing. Sp. gr. 1.018—1.068. On account of its dark color, the angle of rotation has not yet been observed.

COMPOSITION. The separation of a solid substance from the oil was first observed by Görz[2] in 1814. Lassaigne and Feneulle[3] (1820) seem to have regarded the stearoptene obtained by the distillation of the root with water vapor, as camphor. Further reports on this substance were made by Gräger[4] in 1830 and by Blanchet and Sell[5] in 1833 who made the first elementary analysis of asarum camphor. Schmidt[6] (1845) studied principally the crystallographic characters of the substance to which he assigned the name asarone which is still in use. Rizza and Butlerow[7] recognized that asarone contains three methoxy groups and assigned to it the formula $C_{12}H_{16}O_3$, which was later shown to be correct. Poleck and Staats,[8] however, found at first the formula $C_8H_{10}O_2$, later $C_{18}H_{17}O_3$ and finally $C_{18}H_{18}O_3$.

The relative position of the three methoxy groups was ascertained by Will[9] in 1888 who showed that asarone is a derivative of oxyhydroquinone. The recently accomplished synthesis of asarone by Gattermann[10] shows it to have the formula $C_6H_2 . C_3H_5 . (OCH_3)_3 = 1:2:4:5$.

1) Odorographia, vol. 1, p. 325.

2) Pfaff, System der Mat. Med., vol. 3, p. 280.

3) Journ. de Pharm., 6, p. 561.—Trommsdorff's Neues Journ. d. Pharm., 5, II, p. 71.

4) Dissertatio de Asaro Europaeo. Götting.

5) Liebig's Annalen, 6, p. 296.

6) Liebig's Annalen, 53, p. 156.

7) Berichte, 17, p. 1159.—Zeitschr. d. Russ. phys. chem. Ges., 19, I, p. 1; Berichte 20, Referate, p. 222.

8) Berichte, 17, p. 1415.—Chem. Zeitung, 9, p. 1464.—Jahresb. f. Pharm., 1885, p. 381.—Tagebl. der 59. Versammlung deutscher Naturforscher, 1886, p. 127.—Jahresb. f. Pharmacie, 1886, p. 288.

9) Berichte, 21, p. 614.

10) Berichte, 32, p. 289.

Eykman[1] made a study of the structure of the C_3H_5-radical in 1889. From the index of refraction and the dispersion he came to the conclusion that asarone was a propenyl and no allyl-derivative. Properties and derivatives of asarone are described on p. 181.

Upon fractionation of the constituents accompanying the asarone, Petersen[2] obtained a laevogyrate fraction boiling at 162—165° which contained l-pinene. Upon direct bromination it yielded a liquid monobromide; after heating to 250° dipentene tetrabromide melting at 122° was obtained.

The higher boiling constituents distilled principally at about 250° yielding a fraction of the composition $C_{11}H_{14}O_2$. With sodium nitrite and acetic acid a nitrite melting at 118° was obtained. When heated with hydrogen iodide, methyl iodide was split off. Upon oxidation with potassium permanganate, veratric acid was formed. Petersen (1888), therefore, regarded the substance boiling at 250° as methyl eugenol. Mittmann[3] (1889), however, is of the opinion, that it is not methyl eugenol, but methyl isoeugenol, basing his conclusion on comparisons made with the synthetic methyl ether prepared from bay oil eugenol; also with the natural ether occurring in bay oil. Inasmuch, however, as the phenol from bay oil, from which Mittmann prepared the methyl ether, must have been a mixture of eugenol and chavicol, as was shown later, the methyl ether of Mittmann can hardly have been pure and thus his comparisons are rendered valueless. It, therefore, still remains an open question whether asarum oil contains the methyl ether of eugenol or of isoeugenol.

The highest boiling fraction is colored green by a substance not yet examined.

105. Oil of Canada Snake-root.

Oleum Asari Canadensis.—Canadisches Schlangenwurzelöl.—Essence de Serpentaire du Canada.

Asarum canadense L. is known in the United States as Canada snake-root, wild ginger, and Canadian asarabacca. The rhizome contains a fragrant volatile oil which is used extensively in perfumery in North America. Upon distillation, the dry rhizome yields 3.5—4.5 p. c. of oil. The rootlets contain somewhat less oil than the rhizome.

1) Berichte, 22, p. 3172.
2) Archiv d. Pharm., 226, p. 89.
3) Archiv d. Pharm., 227, p. 543.

PROPERTIES. The oil has a yellow to yellowish-brown color and a strong but pleasant aromatic odor and taste. It is soluble in 2 p. of 70 p. c. alcohol; sp. gr. 0.93—0.96.

COMPOSITION. According to Power[1] (1880) the oil contains the following subtances:

1. A very small amount of an inactive terpene, $C_{10}H_{16}$, boiling at 163—166° which possibly is identical with pinene.[2]

2. Asarol alcohol $C_{10}H_{18}O$; b. p. 196—199°; sp. gr. 0.874 at 17°; $\alpha_D = +4°$. Its odor reminds of coriander and in the oil it is partly combined with acetic and valerianic acids as ester. It appears to be identical with linalool.

3. An alcohol $C_{10}H_{18}O$, isomeric with the former but boiling betw. 222—226°. Its odor resembles that of geranium oil and possibly consists largely of geraniol.

4. Methyl eugenol. Fraction 254—257° has the composition $C_{11}H_{14}O_2$, and upon oxidation with permanganate yields an acid which Petersen has recognized as veratric acid. Methyl eugenol may, therefore be considered as one of the constituents of the oil of *Asarum canadense* L.

5. Fraction 275—350° is blue, has no constant boiling point, and is indefinite in composition.

106. Oil of Virginia Snake-root.

The rhizome and roots of *Aristolochia serpentaria* L. and of *A. reticulata* Nutt. (Family *Aristolochiaceae*) are official in the U. S. Pharmacopoeia under the title of Serpentaria. As the rhizomes of both plants are similar morphologically and as to their therapeutic properties, so the oils from both are closely related.

Aristolochia serpentaria L. upon distillation yields 1—2 p. c. of oil of valerian-like odor and a sp. gr. of from 0.98—0.99. Spica[3] (1887) has shown the principal constituent to be borneol.

Aristolochia reticulata Nutt., distilled in small quantities by Peacock[4] (1891) yielded but 1 p. c. of a golden-yellow oil of camphor- and valerian-like odor. Sp. gr. 0.974—0.978; $\alpha_D = -4°$. The oil contained a terpene boiling at 157°, probably pinene; also borneol, which is combined with an acid not yet identified.

---

1) On the constituents of the rhizome of *Asarum canadense* L. Dissertation, Strassburg; Proc. Am. Pharm. Assoc., 28, p. 464.

2) Comp. Petersen, Arch. d. Pharm., 226, p. 128; Berichte, 21, p. 1064; also Power, Pharm. Rundschau, 6, p. 101.

3) Gazz. chim. ital., 17, p. 318; Jahresb. f. Pharm., 1887, p. 45.

4) Am. Journ. Pharm., 63, p. 257.

107. Oil from Aristolochia Clematis.

Osterluzeiöl.

The oil from the rhizome of *Aristolochia clematis* L. was prepared by Winckler[1] in 1849 and by Frickhinger[2] in 1851, the former obtaining 0.4 p. c. Walz[3] in 1853 distilled the entire plant and obtained a golden-yellow, viscid oil of acid reaction and a sp. gr. of 0.903.

108. American Wormseed Oil.

Oleum Chenopodii Anthelmintici.—Chenopodiumöl oder amerikanisches Wurmsamenöl.—Essence de Semen Contra d'Amérique.

Origin. In the neighborhood of Baltimore the oil is distilled from the entire plant, *Chenopodium ambrosioides* L. var. *anthelminticum* Gray.[4] Westminster in western Maryland is the center of production. The fruits are said to owe their anthelmintic properties to the oil they contain. Upon distillation, the fruit yields 0.6—1 p. c., the leaves about 0.35 p. c. of volatile oil.

Properties. The odor of the colorless or yellowish oil is very penetrating, offensive, camphor-like; the taste bitter and burning. The sp. gr. of good commercial oils has been found to be 0.97; the rotatory power between —5 and —6°. A clear solution resulted with 10 p. of 70 p. c. alcohol.

Oils with a lower sp. gr. and slight dextro rotation were not soluble in 70 p. c. alcohol and were shown to be adulterated with American turpentine oil. The data obtained from an oil distilled by Schimmel & Co. show that a lower sp. gr. and a lesser solubility alone cannot be regarded as proof of adulteration. In order to justify the conclusion that turpentine oil is present, pinene, which does not appear to occur in the pure oil, should be shown to be present.

An oil distilled by Schimmel & Co. from the seed[5] had a sp. gr. of 0.900; $\alpha_D = -18° 55'$. The oil from the leaves had a sp. gr. of 0.879;[6] and an angle of rotation of $-32° 55'$. Neither of these two oils was clearly soluble in 70 p. c. alcohol.

The seed of *Chenopodium ambrosioides* L., a closely related plant, are used in Brazil as a popular anthelmintic. According to Peckolt[7] they contain a volatile oil with a strong aromatic odor, bitter, burning taste, sp. gr. 0.943.

---

1) Jahrb. f. prakt. Pharm., 19, p. 71.
2) Repert. f. d. Pharm., III., 7, p. 1.
3) Jahrb. f. prakt. Pharm., 26, p. 65.
4) Am. Journ. Pharm., 22, p. 304; 26, p. 503.
5) Bericht von S. & Co., Apr. 1894, p. 56.
6) Ibid., p. 57.
7) Pharm. Rundschau, 13, p. 89.

Fig. 68.
Wormseed Oil Distillation in Maryland.

The leaves of this plant yielded upon distillation 0.25 p. c. of a volatile oil with a repulsive camphor-like, narcotic odor, reminding of trimethylamine, and sp. gr. 0.901.[1]

COMPOSITION. The constituents of American wormseed oil have as yet been but little investigated. According to an examination conducted by Garrigues[2] in 1854 the oil contains a hydrocarbon boiling at 176° (possibly limonene), and a liquid substance $C_{10}H_{16}O$.

109. Oil from Paeonia Moutan.

The rootbark of *Paeonia moutan* Sims. (Family *Ranunculaceae*) is much used as a drug in Japan and China. On the inner surface as well as on surfaces made by fracture, white prismatic crystals of paeonol are found. These can be obtained by distillation with water vapor, or better by extraction with ether. The crude oil (yield 3—4 p. c.)[3] is purified by shaking the ethereal solution with soda solution, which takes up the impurities only; then combining the paeonol with sodium hydrate and regenerating it with sulphuric acid.

Paeonol was first isolated by Martin and Jagi[4] (1878). Basing their conclusions on the elementary analysis of a calcium derivative, these chemists regarded it as a fatty acid closely related to caprinic acid. According to Nagai[5] (1891) paeonol has an aromatic odor, crystallizes in colorless, shining needles melting at 50°, and has the composition $C_9H_{10}O$.[6] It is sparingly soluble in cold water, readily in hot water, alcohol, ether, benzene, chloroform and carbon disulphide. The aqueous as well as the alcoholic solution is colored reddish-violet by ferric chloride. Aqueous solutions of the caustic alkalies dissolve paeonol and form well crystallizable derivatives.

According to Nagai, paeonol is p-methoxy-o-hydroxyphenyl methyl ketone: $C_6H_3.COCH_3.OH.OCH_3=1:2:4$. When fused with potassa, or when boiled with hydriodic acid, paeonol yields resacetophenone, $C_6H_3.COCH_3.OH:OH=1:2:4$, m. p. 142°. Acetyl paeonol, m. p. 46.5° is oxidized with permanganate to p-methoxysalicylic acid, $C_6H_3.COOH.OH.OCH_3=1:2:4$. The oxime of paeonol crystallizes in fine needles, the phenyl hydrazone in light yellow needles melting at 170°.[6] Tahara[7] in 1891 prepared paeonol synthetically by methylation of resacetophenone.

---

1) Bericht von S. & Co., Apr. 1891, p. 49.

2) Am. Journ. Pharm., 26, p. 405.

3) Berichte, 19, p. 1776.

4) Archiv d. Pharm., 218, p. 335.

5) Berichte, 24, p. 2847.

6) Tiemann in 1891 (Berichte, 24, p. 2854,) found the m. p. of pure paeonol at 48°. Further derivatives of paeonol are described: Berichte, 25, pp. 1284, 1806, and 29, p. 1754.

7) Berichte, 24, p. 2459.

110. Oil of Nigella Sativa.

The seeds of *Nigella sativa* L. (Ger. *Schwarzkümmel*) yield 0.46 p. c. of a volatile oil of a yellow color, which does not fluoresce and has an unpleasant odor. Sp. gr. 0.875; $\alpha_D = +1°26'$. It boils between 170—260°.[1]

111. Nigella Oil from Nigella Damascena.

The seeds of *Nigella damascena* L., which are sometimes also called *Schwarzkümmel*, yield upon distillation[2] 0.5 p. c. of an oil with a beautiful blue fluorescence, that possesses the agreeable odor and taste of the wild strawberry. Sp. gr. 0.895 to 0.906; $\alpha_D = +1°4'$. In 90 p. c. alcohol the oil is imperfectly soluble, but miscible with absolute alcohol in all proportions.

The fluorescence of the oil is due to damascenine, $C_{10}H_{15}NO_3$, which boils at 168° and melts at 27°. According to Schneider[3] it belongs to the alkaloids: with acids it forms crystallizable salt, with the chlorides of platinum, gold and mercury it forms double salts. With the general alkaloidal reagents it yields the reactions characteristic of alkaloids: with iodine potassium iodide a purplish-brown precipitate; with potassium bismuth iodide a brown; with potassium mercuric iodide and phospho molybdic acid white precipitates. Solutions of damascenine containing an excess of nitric acid are colored a beautiful violet upon standing (damascenine red). Upon heating of the nitrate, damascenine blue results.[4]

Pommerehne[5] has recently studied damascenine which was isolated by extracting with dilute hydrochloric acid, making alkaline with soda and shaking out with petroleum ether. The blue fluorescent petroleum ether solution is then again shaken with hydrochloric acid and the acid solution evaporated. Pommerehne sets up the formula $C_8H_9(OCH_3)_2NO$, based on an analysis of the platinum double salt and a methyl determination.

---

1) Bericht von S. & Co., Apr. 1895, p. 74.

2) Bericht von S. & Co., Oct. 1894, p. 55.

3) Pharm. Centralh., 31, pp. 173 and 191. Comp. also Inaug. Dissertation of the same author, Erlangen 1890.

4) The earlier reports are frequently contradictory, because the examinations were conducted in part with the seeds of *Nigella damascena* L., in part with those of *Nigella sativa* L.—Greenish (1882), Pharm. Journ., III, 12, p. 681.—Reinsch (1841), Jahrbuch f. prakt. Pharm., II, 4, p. 384.—Flückiger (1854), ibid., III, 2. p. 161.

5) Archiv d. Pharm., 237, p. 475.

112. Champaca Oil.

The oil distilled from the flowers of the champaca tree, *Michelia champaca* L. (Family *Magnoliaceae*), which grows in Java and the Philippine islands, possesses decided fragrance, which strongly resembles that of the flowers of *Acacia farnesiana* Willd., but also reminds of that of violet and ylang-ylang.[1] Inasmuch as the oil is very costly it can be had, if at all, in comparatively small quantities only.

The rather mobile oil is of a light yellow, or reddish-yellow to brownish color. Sp. gr. 0.907—0.935; $\alpha_D = -12°18'$ to $-52°$. Saponification number 77 (one determination). In alcohol it is difficultly soluble, for it does not even give a clear solution with 10 p. of 90 p. c. alcohol.

As to its composition, hardly anything is known. Like ylang-ylang oil it has a rather high ester content, and like this oil contains benzoic acid (m. p. 121°) which sometimes crystallizes from the oil.[2]

113. Oil from Michelia Longifolia.

The flowers of *Michelia longifolia* Bl. (Family *Magnoliaceae*), which grows in Java, yield upon distillation a limpid, very volatile oil, the odor of which reminds of basilicum. Sp. gr. 0.883; $\alpha_D = -12°50'$.[3]

114. Star Anise Oil.

Oleum Anisi Stellati. — Sternanisöl. — Essence de Badiane.[4]

The various species of *Illicium* are indigenous to middle and southern East-Asia, Japan, and the islands of the Chinese and Indian seas. On account of the similarity of the fruits of the different species, the statements concerning the origin of the various commercial varieties have been contradictory up to the present time. Linné first called the tree, which belongs to the *Magnoliaceae*, *Badanifera anisata*,[5] later *Illicium anisatum*.[6] A variety cultivated in Japan, especially in

1) Guaiac oil from *Bulnesia sarmienti* (see this) which has recently been introduced into the market as champaca wood oil has nothing in common with genuine champaca oil.

2) Bericht von S. & Co., Apr. 1882, p. 7; Apr., 1894, p. 58; Oct. 1894, p. 10; Apr. 1897, p. 11.

3) Bericht von S. & Co., Apr. 1894, p. 59.

4) In Europe, star anise was formerly known as Anis de la Chine, de la Sibérie, *Foeniculum sinense*, Badian. The last name, taken from the Arabian *Bádiyán* for fennel, was employed by Pierre Pomet, the author of the Histoire général des drogues (livre 1, p. 48), published in 1694, and remained in use for a long time.

5) Mat. med. e regno vegetabile, lib. 1, p. 180.

6) Species plantarum, p. 664.

Fig. 64.
Star Anise Plantation in the Vicinity of Langson (Tonkin).

the groves of Buddhist temples, was described in 1690 by Kämpfer,[1] in 1781 by Thunberg[2] and again in 1825 by von Siebold,[3] who termed

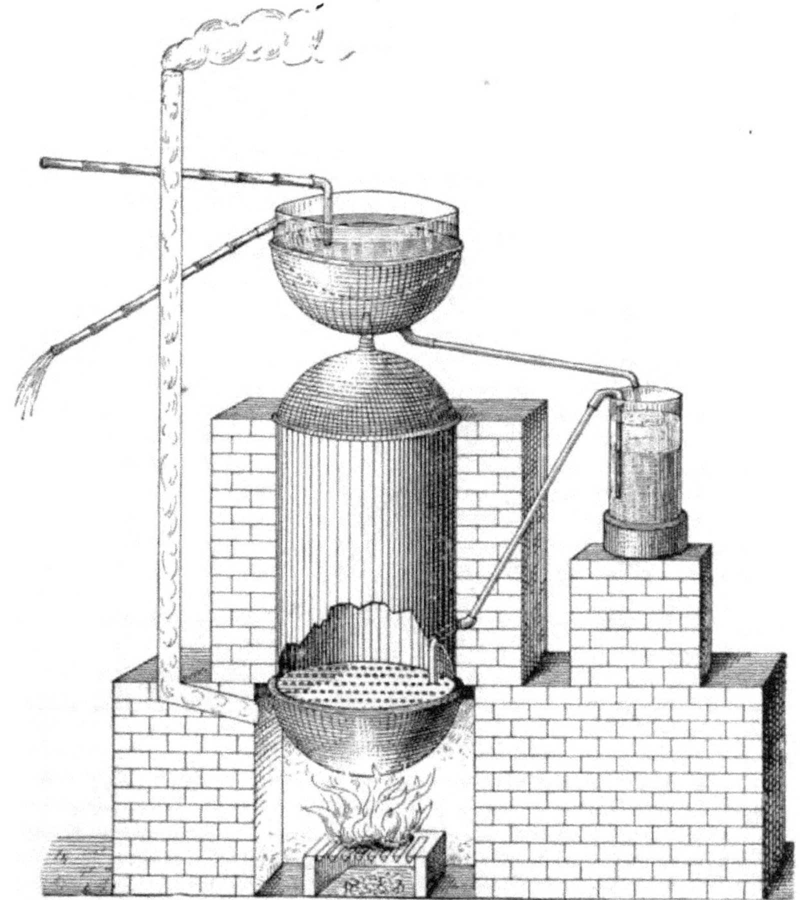

Fig. 65.
Tonkin Distilling Apparatus.

it *Illicium japonicum*, which term he changed to *I. religiosum* in 1837. Jos. Hooker the younger, showed in 1886 that the official and com-

1) Amoenitates exoticae, p. 880.

2) Flora japonica, p. 235.

3) Het gezag van Kämpfer, Thunberg, Linnaeus en anderen, omtrent den botanischen oorsprong van den steranijs des Handels. Leiden, 1837, p. 19.—Rein, Japan, 1886, vol. 2, pp. 160 and 307.

mon star anise is not derived from the species termed *I. anisatum* by Linné, and named the tree *I. verum*.

The volatile oil was distilled in the course of the eighteenth century, but did not find wider application until the nineteenth century. As to its constituents and percentage of volatile oil, star anise was examined by Neumann and Cartheuser,[1] later in 1818 by Meissner.[2]

Preparation. At present, star anise oil is imported principally from the northwestern provinces of China and from the French colony Tonkin. In both districts it is prepared on a large scale.

The distillation is conducted in a primitive manner. According to English[3] and French reports the method in Tonkin is the same as in the Chinese provinces.

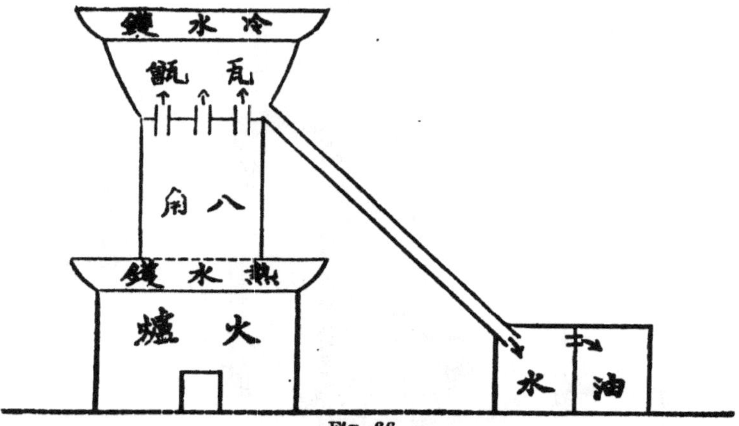

Fig. 66.
Chinese Distilling Apparatus.

The still (fig. 65) consists of a strong, vapor-tight wooden barrel or of an iron cylinder, the bottom of which is frequently perforated. This rests on an hemispherical or flat iron kettle fastened in a hearth of masonry and heated by direct fire. The upper part of the barrel or cylinder is so constructed that the charging and discharging can take place from above, and that the vapors escape through a central opening and tube into the condenser situated above. This consists usually of an earthenware jar covered by a tin kettle. The latter is filled with water, supply and exit tubes consisting of bamboo.

The barrel or cylinder is filled with comminuted star anise. After all joints have been sealed, the water in the kettle below is heated. The vapors

1) Elementa chymiae-medicae, vol. 2, p. 327.

2) Buchholz, Taschenb. f. Scheidekünstler. u. Apoth. auf das Jahr 1818 und 1819, p. 1.

3) Decennial Reports on the trade, etc., in China and Corea. Statistical series No. 6, 1882 to 1891, p. 659.

of water and oil condense on the lower surface of the hemispherical upper wall of the condenser and flow through a tinned tube from the condenser basin into a basin usually made of wood and lined with tin.

The Chinese apparatus is not essentially different. The receiver, however, consists of two chambers and is so arranged that the lighter oil flows into the second chamber, whereas the water is prevented from flowing over by means of a siphon (fig. 66).

In Tonkin a kind of Florentine flask is used and the aqueous distillate is allowed to flow back into the still.

As a rule, the stills contain 3 piculs = 180 k. of star anise. The distillation lasts about 45 hours. The yield averages 5½ k. or about 3 p. c. of oil, which is shipped in lead canisters of 7.5 k.

COMPOSITION. Anethol is the most important and valuable constituent of star anise oil. The congealing point shows whether the oil is relatively rich in anethol; an exact method of assay has not yet been devised. Inasmuch as from 80 to 90 p. c. of anethol can be obtained from good anise oil by repeated freezing, the true content may be regarded as somewhat higher. Anethol was first recognized in star anise oil by Cahours[1] in 1841, who established the identity between the stearoptene of fennel, anise and star anise oils. Shortly afterward, Persoz[2] obtained anisic acid from anethol by oxidation with chromic acid. He called the acid badianic acid. For properties and derivatives of anethol see p. 179.

The remaining 10—20 p. c. of the oil consist of a mixture of at least five different substances. The lowest fraction 157—175° contains two terpenes.[3]

1) d-Pinene: b. p. 157—163°; $\alpha_D = +21°30'$; m. p. of the nitrolbenzylamine 122—123°.

2) l-Phellandrene: b. p. 170—175°; $\alpha_D = -5°40'$; m. p. of the nitrite 102°.

3) Methyl chavicol (para methoxy allyl benzene), the isomer of anethol (para methoxy propenyl benzene), has not yet been isolated in a pure form from star anise oil. Judging from the properties of the corresponding fraction its presence can not well be doubted.[4] If star anise oil is freed as far as possible from anethol by repeated freezing, a mobile oil is obtained which suffers striking changes when boiled with alcoholic potassa (Eykman's method). The boiling point rises, the index of refraction is increased, and upon freezing large amounts of

---

1) Compt. rend., 12, p. 1218; Liebig's Annalen, 35, p. 313.
2) Compt. rend., 13, p. 433; Liebig's Annalen, 44, p. 311.
3) Bericht von S. & Co., Apr. 1893, p. 56.
4) Bericht von S. & Co., Oct. 1895, p. 6.

anethol are again separated. This has evidently been formed by the action of the potassa on the preexisting methyl chavicol.

4) Hydroquinone ethyl ether,[1] $C_6H_4(OH).OC_2H_5$, but mere traces of which are present, can only be obtained by shaking out large amounts of star anise oil with aqueous alkali. When pure it crystallizes in colorless laminae of pearly lustre which melt at 64°.

5) Safrol is considered to be present though a satisfactory proof of its presence has not yet been given. Oswald[2] (1891) attributes the difference in odor of star anise oil from that of anise oil to safrol.

Two substances, which are found not only in star anise oil, but in every oil containing anethol are anisic aldehyde and anisic acid. They are said not to preexist in the oil but are formed by oxidation due to exposure to air. The older the oil, the larger the amount of these two substances present.

PROPERTIES. Star anise oil is a colorless or yellowish liquid of great refractive power. On account of the high percentage of anethol it congeals in the cold. It has an anise-like odor; an intensely sweetish taste; sp. gr. 0.98 to 0.99; rotation slightly laevogyrate, α_D up to $-2°$.[3] The oil forms a clear solution with three or more parts of 90 p. c. alcohol. The congealing point lies between + 14 and + 18°.[4]

Under certain conditions, especially in closed vessels and when slowly cooled, star anise oil can be cooled far below its congealing point without being solidified and may then remain liquid for a long time. The congealing is usually induced by some external impetus, by a particle of dust or by shaking, and takes place the more readily the lower the temperature of the oil. The best way to cause a cooled oil to solidify is to drop into it a particle of anethol or to scratch the inner side of the dish with a glass rod.

It is noteworthy that an old oil which has been kept for a longer period in partly filled containers or which has been exposed to the air by repeated melting, gradually loses its power to congeal on account of the partial oxidation of anethol to anisic aldehyde and anisic acid.

---

1) Bericht von S. & Co., Oct. 1895, p. 6.

2) Archiv d. Pharm., 229, p. 86.

3) In a few instances slight dextro rotation has been observed.

4) Recently oils have found their way into the market that had a lower congealing point but were unadulterated. The properties of five of such oils were as follows: sp. gr. 0.988—0.998; $\alpha_D = +0°11'$ to $+0°32'$; congealing point $+8\frac{3}{4}$ to $13\frac{3}{4}°$; all soluble in $1\frac{1}{2}$ p. of 90 p. c. alcohol. These "flower oils," as they have been designated, are not, however, obtained from the flowers of the star anise tree, but from the unripe fruits which are picked to facilitate the ripening of the other fruits. Though not adulterated, these oils cannot be regarded as the equivalent of the oil from the ripe fruit and should not be sold in its place. (Bericht von Schimmel & Co., Oct. 1898, p. 47.)

In order to distinguish between star anise oil and anise oil a color reaction[1] with alcoholic hydrogen chloride has been recommended, which, however, does not yield reliable results. Star anise oil is said to give a yellowish to a brownish color with this reagent, whereas anise oil is colored blue or red according to the degree of concentration of the acid.

EXAMINATION. In order to detect adulterations and to test for a normal anethol content, the specific gravity, solubility and congealing point should be determined.

Formerly adulterations were never observed and it is only in recent years that it has occurred to the Chinese to add petroleum to the oil. This addition causes a lowering of the specific gravity and of the congealing point and also influences the solubility in 90 p. c. alcohol. Whereas pure oil dissolves without turbidity in three or more parts of 90 p. c. alcohol, oil adulterated with petroleum produces a turbid mixture from which the petroleum separates in the form of drops upon prolonged standing.

In order to isolate the petroleum, the oil is distilled with water vapor and the first distillate collected separately, and treated first with concentrated sulphuric acid and then with concentrated nitric acid. The volatile oil is thus destroyed whereas the petroleum remains almost unchanged and can be recognized by its general properties. High boiling mineral oils are non-volatile with water vapor and will be found in the residne.

The following table[2] shows in how far the properties of star anise oil are modified by the addition of from 5 to 10 p. c. of petroleum.

	Sp. gr.	Congealing point	Solubility in 90 p. c. alcohol
Pure oil	0.986	+ 18°	1:2.2 and more.
The same with 5 p. c. petroleum	0.978	+ 16¼°	not clearly soluble in 10 parts of 90 p. c. alcohol.
The same with 10 p. c. petroleum	0.970	+ 14¾°	

From the above data it becomes apparent that adulteration with petroleum is more readily recognized by the determination of the specific gravity and solubility than by the congealing point. For this reason an oil should be required to stand the tests of specific gravity and solubility even if it conforms with the requirements made with regard to the congealing point.

1) Eykman (1881). Mittheil. d. deutsch. Gesellsch. f. Natur- und Völkerkunde Ostasiens, 28.—Umney, Pharm. Journ., III, 19, p. 647.—Squire, ibid., III, 24, p. 104.—Umney, ibid., III, 25, p. 947.

2) Bericht von S. & Co., April 1897, p. 42.

On the other hand, unadulterated oils that are normal with regard to specific gravity and solubility, may be defective as to anethol content. Inasmuch as the oil is valuable in the ratio of its anethol content, and as the congealing point rises with the latter, the congealing point may be regarded as a direct measure of the quality of the oil.

Determination of the congealing point. When taking a sample of the oil, care should be taken that the contents of the canister are completely melted and well mixed.

200 g. of oil are transferred to a flask and filtered if necessary.[1] An accurate thermometer, indicating half degrees, is dipped into the oil and this cooled by means of chopped ice or ice water. During the process of cooling shaking is to be avoided, neither is the liquid to be stirred with the thermometer in order to avoid premature congealing. After the oil has been cooled to about +6 to 8°, crystallization is induced by the addition of a particle of solid anethol or solid star anise oil, or by scratching the inner wall of the flask with the thermometer. During the solidification, the mixture is continuously stirred in order to facilitate the process of congealing. In the course of this operation the temperature rises rapidly. The highest point which the mercury reaches is the congealing point.

The determination can be made with less material with the aid of Beckmann's apparatus for the determination of molecular weights described on p. 187.

In order to obtain uniform results, care should be exercised to induce crystallization at a temperature of about 6 to 8°. The melting point of the oil would be just as serviceable as the congealing point were it not for the fact that the latter can be ascertained much more accurately. The congealing point of star anise oil should not be below 15°. The oil imported from Tonkin usually has a higher congealing point.

PRODUCTION AND COMMERCE. The production of star anise is restricted to the Chinese districts Lung-chow (province Kwang-si), Po-se (province Kwang-tung), and to the neighborhood of Lang-son in the French colony of Tonkin. The yield is subject to considerable fluctuations, being only 1,922 piculs in 1895, whereas it was 16,138 piculs in 1892. During the last 10 years between 3,000 and 13,000 piculs of star anise, or 180,000 to 780,000 k., taken for export, were shipped exclusively via Pak-hoi to Hong-kong which is the world's market for this drug.

A still larger quantity of star anise is used for the distillation of the oil. The two Chinese districts mentioned above produce about 1,500 to

[1]) Oils that are not perfectly clear can be cooled below the congealing point only with difficulty. The suspended solid particles induce crystallization as soon as the temperature has been reduced to several degrees below the congealing point.

2,000 piculs or 3,000—4,000 cases which are also shipped to Hong-kong via the port Pak-hoi. The Tonkin-Chinese border district has produced about 400 piculs or 800 cases in recent years. This is taken by rail and steamer to the Tonkin-Chinese port Hai-phong whence, favored by special duty regulations, it is shipped exclusively to France. On account of its high anethol content, it has met with special favor in recent years and is preferred to the Chinese oil.

Total export of star anise oil during the last 10 years:

Year	From Pak-hoi and Lappa (Macao)	From Hai-phong (Tonkin)	Total
1888	976 Piculs[1]	150 Piculs	1126 Piculs
1889	810 "	200 "	1010 "
1890	1294 "	230 "	1524 "
1891	507 "	288 "	795 "
1892	1803 "	— "	1803 "
1893	862 "	— "	862 "
1894	1998 "	— "	1998 "
1895	489 "	13 "	502 "
1896	2369 "	435 "	2804 "
1897	1398 "	399 "	1797 "

The price of the oil is also subject to considerable fluctuations. During the last twenty years it has varied between $1.30 and $2.25 per pound. Since several years the congealing point serves as basis for its valuation. The principal Hongkong and Tonkin firms are familiar with the method and have the necessary equipment.

115. Star Anise Leaf Oil.

The natives of the Pé Sé district obtain an oil by distilling the leaves and even the branches of star anise trees. The yield is about 0.75 p. c. (Simon[2]).

In odor the oil differs slightly from that of the fruit, the odor of anisic aldehyde being far more pronounced. It has a higher sp. gr. than the oil from the fruit, 0.9878 at 15.5°; $\alpha_D = +1°$.[3] According to Umney[3] the proportion that boils above 230° is quite different from that of the oil from the fruit.

	Star anise fruit oil.	Star anise leaf oil.
Below 225°	20 p. c.	10 p. c.
226—230°	65 "	60 "
Above 230°	15 "	30 "

[1]) 1 Picul = 2 cases with a net content of 80 kilo. During the years 1897 and 1898 the total export has been valued at $550,000.00 and $875,000.00 respectively.

[2]) Chem. & Druggist, 53, p. 875.

[3]) Chem. & Druggist, 54, p. 323.

The anethol content is small and its congealing point correspondingly low, for which reason it has been designated as liquid star anise oil[1] and appears to be used as an adulterant and a substitute for the oil from the fruit.

116. Japanese Star Anise Oil.

The oil from the leaves of *Illicium religiosum* Sieb. (Japanese *Shikimi*), the Japanese star anise tree, which is ill-famed on account of its poisonous fruits, was examined by Eykman[2] in 1881. Upon distillation of the leaves he obtained 0.44 p. c. of volatile oil, sp. gr. 1.006 at 16.5°. $[\alpha]_D = -8° 6'$. It contains a terpene (b. p. 173—176°; sp. gr. 0.855; $\alpha_D = -22.5°$) and 25 p. c. of a liquid anethol which yields nitroanisic acid melting at 174°.

According to a second contribution by the same author[3] (1885) the shikimi leaf oil consists of eugenol, a terpene "shikimene", (b. p. 170°; sp. gr. 0.865) and "shikimol" or safrol. It is noteworthy that Eykman determined on "shikimol" the constitution of safrol $C_6H_3:O_2CH_2(C_3H_5)$.

The fruits of the Japanese star anise tree also contain a volatile oil[4] of a disagreeable odor, which has nothing in common with that of the true star anise oil. Sp. gr. 0.984; $\alpha_D = -4° 5'$.

117. Winter's Bark Oil.

According to Arata and Canzoneri[5] (1889), the genuine Winter's bark of *Drimys winteri* Forst. contains 0.64 p. c. of a dextrogyrate volatile oil. This consists principally of a hydrocarbon, "winterene" boiling at 260—265°, which the authors regard as a triterpene. Probably, however, it is a sesquiterpene.

118, 119. Ylang-Ylang Oil and Cananga Oil.

Oleum Anonae. Oleum Canangae.—Ylang-Ylangöl. Canangaöl.—Essence de Ylang-Ylang. Essence de Cananga.

Origin and History. Ylang-ylang oil is obtained by the distillation of the flowers of *Cananga odorata*, Hooker et Thomson (Family *Anonaceae*). This tree occurs in the Malay Archipelago and in the Philippines and is cultivated in the latter and throughout southern

---

1) Chem. & Druggist, 58, pp. 840, 875.

2) Mitth. d. deutsch. Ges. für Natur- u. Völkerkunde Ostasiens, 28; Berichte, 14, Ref., p. 1720.

3) Rec. trav. chim. des P. B., 4, p. 32; Berichte, 18, Ref., p. 281.

4) Bericht von S. & Co., Sept. 1885, p. 29; Oct. 1893, p. 46.

5) Jahresb. f. Pharm., 1889, p. 70.

Asia. The first statement in European literature concerning this tree is made by the English botanist John Ray (born 1628, died 1705), who described the tree under the name *Arbur Sagnisen*. Rumpf, a contemporary of Ray described and illustrated it as *Borga Cananga*. Lamarck called the tree *Uvaria* or *Unonia odorata*. Roxburgh became acquainted with the tree, which had been transplanted from Sumatra to the botanical garden in Calcutta, in 1797. The first correct illustrations of flower and fruit were published by Blume in 1829.

Notwithstanding the exceeding fragrance of the blossoms and th. desire of the European colonial powers to find profitable commercial articles in the Malay Archipelago, the manufacture of ylang-ylang oil was first conducted by Germans in the island of Luzon in the early sixties. The oil became known to wider circles through its exhibition at the Paris Industrial Exposition in 1878 by the Manila merchants Oskar Reymann and Adolf Roensch.

Ylang-ylang oil is distilled in the islands of Luzon and Java. The blossoms of the trees growing in dense forests are said to be less fragrant and poorer in oil than those of cultivated trees or trees standing in clearings.

In the process of distillation, the oxygenated constituents and the esters, the substances to which the fragrance of the cananga blossoms is due, pass over first; later the sesquiterpenes predominate. In Luzon, and recently also in Java, the first part of the distillate is kept separate and brought into the market as ylang-ylang oil, whereas the second part of the distillate or the entire oil is sold as cananga oil.

On account of the delicacy of the flavor, the collection and distillation of the blossoms requires special knowledge and care. For this reason but few distillers have succeeded so far in manufacturing a uniformly good article of like qualities.

Composition. Although the ylang-ylang oil consists principally of the lower and cananga oil of the higher fractions, the difference between these two oils is of a quantitative rather than a qualitative nature.

The first examination of ylang-ylang oil was made by Gal[1] in 1873. Upon saponification of the oil with alcoholic potassa he obtained benzoic acid, thus indicating the presence of a benzoic ester. Reychler,[2] who later (1894) made a more careful examination of both oils, also found acetic acid to be present as ester.

---

1) Compt. rend., 76, p. 1482.

2) Bull. Soc. chim., III, 11, pp. 407, 576 and 1045; ibid. 13, p. 140.

Of the alcoholic constituents which are possibly combined with these acids, the following have been identified: l-linalool (b. p. 196—198°; sp. gr. 0.874 at 20°; $\alpha_D = -16° 25'$) and geraniol (b. p. 230°; sp. gr. 0.885 at 20°) which was separated from the corresponding fraction by means of the calcium chloride compound.

In addition to these esters, the characteristic odor of the oil is partly due to the presence of the methyl ether of para cresol, $CH_3 . C_6H_4 . OCH_3$, which boils at 175° and upon oxidation yields anisic acid melting at 178°.

The high boiling fractions, especially of the cananga oil, contain cadinene (dihydrochloride, m. p. 117°). This sesquiterpene is accompanied by a colorless and odorless substance[1] crystallizing in needles that melt at 138°, which evidently belongs to the class of sesquiterpene hydrates. In the lower boiling portions a very small amount of a terpene, presumably pinene,[2] is contained.

Besides these substances, the oil contains a very small amount of a compound that combines with acid sulphite, the chemical nature of which, however, has not yet been determined. The violet color reaction, which the oil produces with ferric chloride, indicates the presence of a phenol (Flückiger,[3] 1881) the isolation of which would demand a large quantity of this costly oil.

Properties of Ylang-Ylang Oil. The ylang-ylang oil from Manila, which alone is of importance commercially, constitutes a light yellow liquid of exceeding fragrance. Sp. gr. between 0.930 and 0.950; $\alpha_D = -38$ to $-45°$. In alcohol the oil is but difficultly soluble. Generally a clear solution is obtained with ½ to 2 volumes of 90 p. c. alcohol but, as a rule, this solution becomes turbid upon the addition of more alcohol. Inasmuch as the odor is largely dependent on the esters present, the determination of the saponification number (between 75 and 120 for good oils) is to be recommended. With ferric chloride the alcoholic solution of the oil produces a violet color reaction.

Ylang-ylang oils of other sources vary considerably in their physical properties from those of the Manila oil. The oil examined by Gal,[4] which is reported to have been obtained from the West Indies, had a sp. gr. of 0.980 at 15° and $\alpha_D = -28°$. An oil distilled in Réunion[5] had a sp. gr. of 0.974.

---

1) Bericht von S. & Co., Apr. 1896, p. 62.
2) Bericht von S. & Co., Apr. 1896, p. 67.
3) Arch. d. Pharm., 218, p. 24; also Reychler, loc. cit.
4) Compt. rend., 76, p. 1482.
5) Bericht von S. & Co., Oct. 1890, p. 47.

The differences existing in the properties of pure oils render difficult the detection of adulterations. To pass judgment on the quality of ylang-ylang oil is, therefore, often a matter to be decided by the perfumer rather than by the chemist.

Properties of Cananga Oil. The odor of Cananga oil is similar to that of ylang-ylang oil, but not as fine. Owing to the different methods of preparation, the physical properties are subject to considerable variations: sp. gr. 0.91—0.94; $\alpha_D = -17$ to $-55°$.

In 90 p. c. alcohol cananga oil is not completely soluble. Of 95 p. c. alcohol 1½ to 2 vols. are mostly necessary, but the solution becomes turbid on the further addition of alcohol. However, no drops separate from the resulting opalescent liquid upon standing.

The saponification number is 10—30. In general an oil with a high saponification number may be regarded as the best, provided this is not due to the presence of cocoa nut oil.

Cananga oil from dried flowers.[1] Dried cananga blossoms from Samoa, known there as *Mosoi* yielded 1 p. c. of oil upon distillation. Its odor was somewhat different from that of the oil from fresh blossoms but had the same general character. Like ordinary cananga oil, it contained benzoic acid and had a sp. gr. of 0.922.

Adulteration. In recent years adulteration with cocoa nut oil has been observed repeatedly in connection with oils coming via Amsterdam.[2] Such addition can readily be recognized. Specific gravity and rotatory power are influenced but little, but the saponification number is considerably increased and the oil ceases to be soluble in 95 p. c. alcohol. From the turbid mixture of oil and alcohol drops of oil separate upon standing.

Cocoa nut oil can also be recognized without difficulty by exposing the suspected oil to the temperature of a freezing mixture of ice and salt. Whereas a pure oil remains liquid, an oil adulterated with larger amounts of cocoa nut oil congeals to a butter-like mass. This test can readily be carried out by anyone and should never be omitted.

Small amounts of fat as well as other oils that do not congeal at low temperatures remain behind upon distillation with water vapor and will be found in the residue. It should be borne in mind, however, that unadulterated oils also leave a residue of about 5 p. c., which amount should be deducted in an approximate quantitative determination.

---

1) Bericht von S. & Co., Oct. 1890, p. 48.

2) Bericht von S. & Co., Oct. 1897, p. 8.

120, 121. Oils of Mace and Nutmeg.

Oleum Macidis. Oleum Nucis Moschati.—Macisöl. Muskatnussöl.—Essence de Macis. Essence de Muscade.

Origin and History. The nutmeg tree, *Myristica fragrans* Houtthuyn (*M. officinalis* L., *M. moschata* Thunberg), which belongs to the family *Myristicaceae* and which attains a height of 20 meters, is indigenous to the Molucca, Banda and Sunda islands. It has been cultivated extensively in these islands and also in other countries.

The distilled oils of nutmeg and mace were well known to the authors of the "Destillirbücher" about the middle of the sixteenth century and later Cordus[1], Ryff[2], Gesner[3], Porta[4], Winther[5] and others repeatedly mention the oils. Both oils are first mentioned in the apothecaries' price ordinances of Berlin of 1574, those of Frankfurt and Worms of 1582, and in the 1589 edition of the Dispensatorium Noricum. These oils were first examined by Neumann,[6] (1749), Valentini[7] (1719), and Bonastre[8] (1824). Upon distillation nutmegs yield 8—15 p. c., mace 4—15 p. c. volatile oil.

Properties of Oil of Mace. It is colorless or yellowish, becoming reddish-yellow with time. It has a pleasant though strong mace odor which with age becomes unpleasant and terebinthinate. The taste is mild at first, then pungent, aromatic. As a rule the specific gravity of oil of mace is somewhat higher than that of oil of nutmeg and varies between 0.890 and 0.930; $\alpha_D = +10$ to $+20°$. The oil is clearly soluble in 3 parts of 90 p. c. alcohol.

Properties of Oil of Nutmeg. This is a mobile, colorless liquid which, owing to the absorption of oxygen, becomes viscid with age, has a characteristic nutmeg odor and a spicy taste. According to the nature of the crude material, the sp. gr. of the distillate varies between 0.865 and 0.920; $\alpha_D = +14$ to $+28°$. The oil is clearly soluble in 3 parts of 90 p. c. alcohol. When distilled, about 60 p. c., pass over below 180°. When evaporated, a small amount of a fatty substance, consisting principally of myristic acid, remains behind.

Oil of nutmeg clearly resembles oil of mace in all of its properties and is scarcely distinguishable from the latter.

1) Annotationes, etc., De artifi. extr. lib., fol., 226.
2) Neu gross Destillirbuch. Ed. Frankf.-a.-M., fol. 181.
3) Ein köstlicher Schatz, etc., p. 215.
4) Magiae naturalis. Lib. de dest., pars 1, p. 378.
5) Guintheri, Andernacei de medicina veteri et nova. Basilae 1571, pp. 680—685.
6) Chymia medica, vol. 2, part 3, p. 437.
7) Macis vulgo sed perperam, Muskatenblume dicta. Dissertatio. Giessen 1719.
8) Trommsdorff's N. Journ. der Pharm., 8, II, p. 231.

COMPOSITION. On account of the great similarity between the two oils, frequently no commercial distinction is made. As a result, the true origin of the various oils that have been examined is not ascertainable. For this reason it will be expedient to discuss the composition of both at the same time. This can be done without hesitation because so far no qualitative difference has been observed.[1] Quantitatively they vary in so far as oil of nutmeg generally contains more terpenes than oil of mace.

Although numerous older examinations are recorded,[2] individual constituents were but recently definitely characterized and determined. However, there are still a number of problems which demand further investigation.

The following constituents have thus far been found:

1. Pinene. This terpene was found by Wallach[3] in the lowest boiling fraction of the oil (pinene nitrolbenzylamine, m. p. 123°). It is present as an almost inactive mixture of dextro- and laevogyrate pinene. As early as 1862 Schacht had isolated pinene, his "macene" and prepared from it the solid hydrochloride.

2. Dipentene.[4] This is present in fraction 175—180° and has been identified as tetrabromide.[5]

3. Myristicol. The contradictory statements concerning this substance make a revision very desirable. According to Gladstone, myristicol has the composition $C_{10}H_{14}O$, boils at 224°, has the sp. gr. 0.9466 and a rotatory power of +31°. Wright gives the boiling point as 212° and considers the formula $C_{10}H_{18}O$ to be correct. This formula at least explains the formation of cymene by the action of zinc chloride or phosphorus pentachloride on myristicol. Distilled under a pressure of 10 mm., myristicol is found in fraction 70—144° (about 15 p. c.) of the oil, sp. gr. 0.913 (Semmler[5]).

---

1) Koller pronounces both oils identical (N. Jahrb. Pharm, 28, p. 186; Jahresb. f. Chem., 1864, p. 586).

2) John (1821), Chemische Schriften, 6, p. 61. Abstr. in Journ. f. Chem. u. Phys. von Schweigger u. Meinecke, 88, p. 250. — Mulder (1889), Journ. f. prakt. Chem., 17, p. 102, and Liebig's Annalen, 31, p. 71. — Schacht (1862) Archiv d. Pharm., 162, p. 106. — Cloëz (1864), Compt. rend., 58, p. 133. — Gladstone (1872), Journ. Chem. Soc., 25, p. 1. Abstr. in Jahresb. f. Chem., 1872, p, 816. — Wright (1878), Journ, Chem. Soc., 26, p. 549. Abstr. in Jahresb. f. Chem., 1878, p. 369. — Pharm. Journ., III, 4, p. 311.

3) Liebig's Annalen, 227, p. 288; 252, p. 105.

4) The so-called myristicene of Gladstone, like the terpene (b. p. 165°) of Cloëz, was a mixture of dipentene and pinene. Wright's statement that oil of mace contains cymene is open to question inasmuch as he used concentrated sulphuric acid in order to identify it. This reagent, however, is not permissible when terpenes are present. Semmler could not find cymene in the oil.

5) Berichte, 28, p. 1808; 24, p. 3818.

4. Myristicin,[1] $C_{12}H_{14}O_3$.[2] After treatment with metallic sodium, the highest boiling fraction (b. p. 142—149° under 10 mm. pressure) contains a substance that melts at 30°, has a sp. gr. of 1.1501 at 25°, an intense odor of mace and has been termed myristicin. According to Semmler it is oxymethyl oxymethylene butylene benzene.

$$C_6H_2\begin{cases}C_4H_7 & [1]\\ -O\diagdown_{CH_2} & [3]\\ -O\diagup & [4]\\ OCH_3 & [5]\end{cases} \quad \text{or} \quad C_6H_2\begin{cases}C_4H_7 & [1]\\ -OCH_3 & [3]\\ -O\diagdown_{CH_2} & [4]\\ O\diagup & [5]\end{cases}$$

Whether the position of the side chains is as indicated in the first or second formula has not yet been ascertained.

According to a private communication by Professor Semmler, more recent observations seem to agree with those made in the laboratories of Schimmel & Co., viz. that the oil originally contains a phenyl ether which undergoes modifications in the processes to which the oil is submitted during its examination.

The fact that the fraction which after the treatment with sodium yields myristicin, in its original state does not solidify when exposed to a freezing mixture,[3] and the coefficient of refraction being much less before than after the treatment, renders it highly probable that the C_4H_7-group changes its position under the influence of sodium. The same change evidently takes place here that eugenol undergoes when converted into isoeugenol, of safrol into isosafrol, of methyl chavicol into anethol. Should this prove to be the case, the solid substance should properly be termed isomyristicin, whereas the name myristicin should be made to apply to the original liquid substance.

5. Myristinic acid. According to the duration of the process of distillation, more or less of this acid is found in the oil and will sometimes crystallize out.

6. Phenol-like substance. The last portions obtained by fractional distillation produce an emerald green color with ferric chloride, thus indicating the presence of a phenol.[4]

122. Oil of Boldo Leaves.

The fragrant, dried leaves of *Peumus boldus* Mol. (*Monimiaceae*), a tree indigenous to Chili, yield upon distillation 2 p. c. of an aromatic

1) Not to be confounded with the stearoptene myristicin, so called by John and Mulder, which sometimes crystallizes from old oils and which, as has been shown by Flückiger in 1874 (Pharm. Journ., III, 5, p. 136), consists of myristic acid.

2) Wright assumed the formula $C_{10}H_{18}O_2$ for myristicin.

3) Observation in the laboratories of Schimmel & Co.

4) Semmler, loc. cit.

oil, the odor of which is cymene-like and reminds slightly of peppermint. Sp. gr. 0.915 to 0.945; slightly laevogyrate, $a_D = -1°40'$; it boils between 175 and 250° and produces a green color with ferric chloride.[1] The oil contains terpenes and oxygenated substances that have not yet been closer examined (Hanausek[2]).

123. Atherosperma Oil.

The bark of *Atherosperma moschata* Labillard. (*Monimiaceae*) yields upon distillation 1 p. c. of volatile oil (Maiden[3]). On account of its sassafras-like odor the tree is also called Victorian sassafras. According to Gladstone,[4] the oil has a yellowish-brown color and a sp. gr. of 1.0386 at 20°. It is slightly dextrogyrate, + 7° in a 10 inch tube. It begins to distill at 221° and passes over almost completely at 224°.

124. Oil of Citriosma Oligandra.

On account of its unpleasant odor, the tree *Citriosma oligandra* Jul., (*Monimiaceae*) is known in Brazil as *Negra Mina*, *Catinga de negra*, or *Catingueira*. The leaves upon distillation yield 0.54 p. c. of volatile oil which is of a light yellow color with greenish fluorescence and an odor that faintly reminds of bergamot. Sp. gr. 0.899 (Peckolt[5]).

125. Oil of Citriosma Cujabana.

Citriosma cujabana Mart. is known in Brazil as *Limoeiro domato*, and *L. bravo* or wild lemon tree. The fresh leaves yield 0.18 p. c., the fresh twigs 0.07 and the fresh bark 0.22 p. c. of volatile oil. The oil is a mobile liquid, sp. gr. 0.894, the odor of which resembles that of a mixture of bergamot and lemon oils (Peckolt[5]).

126. Oil of Citriosma Apiosyce.

Citriosma apiosyce Mart. is known in Brazil as *Limoeiro bravo*, *Cidreira melisse*, or *Café bravo*, wild coffee tree. All parts of the shrub, especially the leaves and unripe fruits, have a strong odor of melissa (lemon balm) or lemon. From the fresh leaves Peckolt obtained 0.14 p. c., from the twigs 0.06 p. c. of volatile oil.[5]

---

[1]) Bericht von S. & Co., Apr. 1888, p. 48. — Claude Verne, Etude sur le Boldo. Thèse présentée et soutenue à l'Ecole Supérieure de Pharmacie de Paris. Paris 1874. According to Pharm. Journ., III, 5, p. 405.

[2]) Jahresb. f. Pharm., 1877, p. 79.

[3]) Useful native plants of Australia, p. 254.

[4]) Journ. Chem. Soc., 1864, 17, p. 5. Ref. Jahresb. f. Chem., 1863, p. 545. — Chem. News, 1871, 24, p. 283.

[5]) Berichte d. deutsch. pharm. Ges., 6, p. 93.

127. Oil of Paracoto Bark.

Paracoto bark is brought into commerce from Bolivia and derived from an unknown species of the *Monimiaceae*.[1] The oil which is obtained as a by-product in the preparation of cotoin, was first examined by Jobst and Hesse[2] in 1879, later (1892) by Wallach and Rheindorff.[3]

PROPERTIES. Paracoto bark oil is a mobile, colorless liquid of a very pleasant odor. Sp. gr. 0.9275; $\alpha_D = -2.12°$.[2]

COMPOSITION. 1. According to Jobst and Hesse, fraction 160° (sp. gr. 0.8727; $[\alpha]_D = +9.34°$) is a hydrocarbon $C_{12}H_{18}$ named α-paracotene by them.

2. Fraction 170—172° (sp. gr. 0.8846; $[\alpha]_D = -0.63°$) has the formula $C_{11}H_{18}$ assigned to it and is designated β-paracotene.

Both fractions, which evidently consisted of slightly impure terpenes, were wanting almost entirely in the oil examined by Wallach and Rheindorff.

3. The principal fraction of the oil consists of laevogyrate cadinene[3] (m. p. of the dihydrobromide 121°, of the dihydrochloride 118°).

4. After the dihydrobromide has been removed there remains an oil which contains methyl eugenol (m. p. of the bromide 78°, of the veratric acid obtained by oxidation 179—180°).

According to Wallach, the α-, β- and γ-paracotol of Jobst and Hesse, to which the formulas $C_{15}H_{24}O$ and $C_{28}H_{40}O_2$ had been assigned, are not chemical individuals but mixtures principally of laevogyrate sesquiterpene and inactive methyl eugenol. As to α-paracotol, the possibility is not excluded that it may be a natural hydrate of cadinene. In this case, however, the formula would have to be $C_{15}H_{26}O$, not $C_{15}H_{24}O$.

128. Oil of Camphor.

HISTORY. In China, camphor has been prepared and used in early antiquity. The first documental reference to it is found in the writings of the Arabian prince Imru-l-Kais of the sixth century; also in the Koran, in which it is mentioned as a refrigerant in connection with the beverages of the blessed in paradise. Aetius of Amida, also in the sixth century, mentions camphor as a remedy. When the palace in Madain of Choroës II, king of the Sassanides was pillaged in 636, much camphor

---

1) Möller (Anatomie der Baumrinden) is of the opinion that the bark is that of a *Monimiacea*, whereas Vogl (Commentar zur österreich. Pharmacopöe) believes that it belongs to a *Lauracea*.

2) Liebig's Annalen, 199, p. 75.

3) Liebig's Annalen, 271, p. 300.

was taken in addition to musk, ambra, sandalwood and other oriental aromatics.[1] Later, when Arabian sailors and merchants sailed to India and through the Indian Archipelago to China, they became acquainted not only with the principal ports from which the camphor was shipped from southeastern Asia to the Mediterranean countries, but also with the sources of supply. Thus the Arabian merchants and physicians became acquainted with the different varieties which entered the world's market in part from China, in part from Sumatra. Marco Polo, in the thirteenth century, also became acquainted with the camphor in Sumatra and Borneo. E. Kaempfer was the first to describe the mode of preparation in Japan. Medicinally camphor was used by the Arabians, also in Italy as early as the eleventh century and during the twelfth century in Germany.

ORIGIN AND PREPARATION OF CAMPHOR OIL. This oil which is a by-product in the Japanese camphor industry, has but recently been introduced into the European market. The camphor laurel, *Cinnamomum camphora* Fr. Nees et Ebermaier (*Laurus camphora* L.) is a stately forest tree found in several provinces of southern China, in the islands Hai-nan, Formosa and several other Japanese islands.

In the older trees camphor is found in a crystalline condition in cavities of the trunk. Principally, however, it is found dissolved in a volatile oil that permeates the entire tree. This oil occurs most abundantly in the roots, less abundantly in the trunk and still less in the branches and leaves. The amount of oil also seems to vary with the height of the trunk, diminishing with increasing height. With the age of the tree and the density of the wood, the amount of oil, however, increases.

The ratio of solid camphor and camphor oil seems to vary with the age of the tree, the season and the temperature. The younger trees yield upon distillation more camphor oil and less camphor. The same is true with increasing temperature, so that the yield of oil is greater in summer than in winter.

In recent times, and more particularly since the acquisition of Formosa, Japan has become the principal producer of camphor and camphor oil. The method of preparation is generally as follows:[2]

On the slope of a hill near a supply of water a sufficient area is leveled. On this a stove is erected with crude stones having a height of about 1 m.

---

1) Weil, Geschichte der Chalifen. Mannheim, 1846, p. 75.

2) D. E. Grassmann, Der Campherbaum. Mittheilungen der Deutschen Gesellschaft für Natur- u. Völkerkunde Ostasiens. Tokio, 1895, 6, pp. 277—328.

and an inner diameter of 0.70 m. The opening for the fuel supply is rather small, 0.40 x 0.30, and covered with a roof. On this the distilled chips (fig. 67 *b*) are dried to be used later as fuel. A flat kettle provided with a strong, perforated, wooden cover is placed on top of the stove (fig. 68 *b*) and over this a barrel or tub. This has the shape of a truncated cone (fig. 67 *c*)

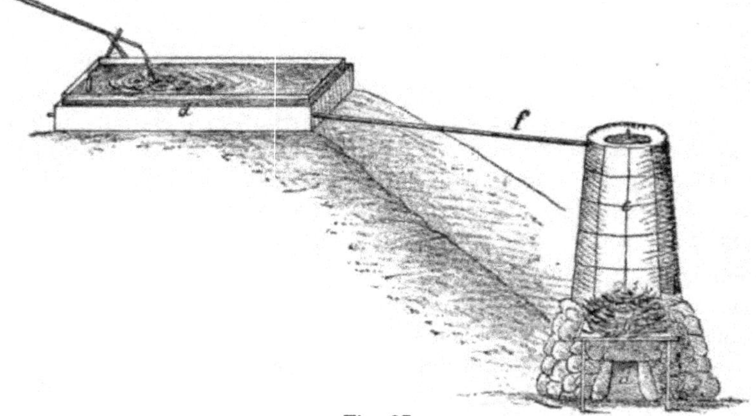

Fig. 67.

and is 1.15 m. high; the upper diameter being 0.30 m. the lower 0.87 m. The cover to the kettle fits as a bottom to the tub. To one side, just over the cover of the kettle there is a rectangular opening in the tub 0.30 m. high and 0.25 m. wide (fig. 68 *e*). The head-piece of the barrel or tub consists of a

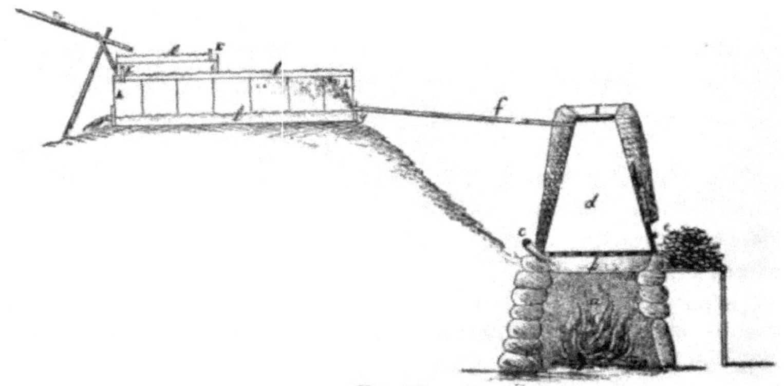

Fig. 68.

removable, well fitting cover provided with an opening that can be closed by a plug. The tub is coated with a layer of clay 0.15 m. thick and held together by a network of bamboo. Near the top a bamboo tube 2 m. long is carefully inserted, which leads to a condenser farther up on the slope. This condenser, in its simplest form (fig. 69), consists of two boxes of which the upper serves for the condensation of the camphor and the lower for the reception of the cooling water. The upper box is 1.60 m. long, 0.90 m. broad and 0.42 m. high. The inverted bottom is covered with water, the sides projecting 10—12 cm. over the bottom. The camphor vapors enter the box over the surface of the water. To pass them through the water has not proved successful. By means of partitions 18.5 cm. apart the condenser is divided into sections. These partitions are each provided with a rectangular opening at the top, one in the right corner, the other in the left corner, etc, so that the camphor vapors must pursue a circuitous path. From the last chamber the vapors can pass out through a bamboo tube slightly plugged with straw. A lateral tube allows the water to flow from the bottom (= cover) of the upper box into the lower. The upper box inverted with its opening downward is placed into the lower box which is somewhat longer and wider but not as high, so that the water in the latter rises to about one-half of the height of the former on all sides. A lateral exit tube allows the excess of water to flow out. In order to avoid a rapid heating of the water in the condenser, this is protected by a roof made of boards.

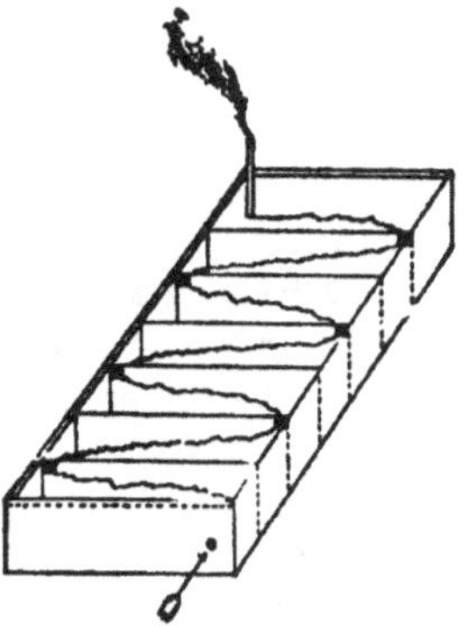

Fig. 69.

Frequently a third smaller box, 0.80 m. long, 0.54 m. broad and 0.25 m. high, is placed over the upper one. This box is likewise open on the lower side, standing in water 10 cm. deep. The sides likewise project over the top so that a layer of water 5 cm. high remains on it. The vapors from the last chamber of the larger condenser pass through a tube (fig. 68 *h*) into this smaller box where more camphor is condensed. This small upper condenser is provided with a small exit tube for the vapors.

The following utensils are used in connection with this apparatus: a wooden shovel of the shape of a spoon-oar for shoveling the distilled chips into the fire place; also an iron poker. In order to protect the stove and tub a roof of straw and reed is constructed, also a wall of straw matting to the windward (valley) to protect against draft and rain.

The production of camphor is conducted as follows. After the kettle has been filled with water, the chips are introduced through the upper opening into the tub, and all cracks and crevices carefully luted so that the vapors cannot escape. Only a moderate fire is maintained.

In the course of the distillation the kettle is frequently refilled through the opening *c* (fig. 68). The vapors pass through the perforation into the tub (fig. 68 *d*) and convey the camphor vapors from the heated chips through the bamboo tube (fig. 68 *f*) into the condenser where they are deposited. In the early part of the distillation, camphor oil only is found in the condenser, later also solid camphor. Most of the camphor condenses behind the 3rd, 4th and 5th chamber of the seven-part condenser. The tub will hold 112.5 k. of chips, a quantity that can be distilled in 24 hours. The exhausted chips are removed from the lateral opening (fig. 68 *e*). Every week the condenser is opened and the camphor and camphor oil contained therein removed.

On the surface of the water in the condenser there collects a granular crystalline mass resembling a mixture of snow and ice, which in the first few chambers is mixed with camphor oil and is colored yellow. On the walls of the condensing chambers the pure, white crystalline camphor is deposited. The yellowish or brownish-black camphor oil floats on the surface where it mixes with the granular camphor.

Properties. Normal camphor oil, i. e. the oil with all constituents as it is obtained by the distillation of the camphor wood, is a semi-solid mass, or a liquid mixed with camphor. Usually, however, the product which remains after the camphor has been removed by filtration and expression is known as camphor oil. This was the oil that came into the market about the middle of the eighties. It still contained large quantities of camphor which could be obtained by fractionation and cooling. At present this operation appears to be carried out in Japan as well, for the oil now found in the market is almost free from camphor.

Distinction is made between two kinds of oil which often vary largely in their composition. The measure of their value is their specific gravity.

1. White Camphor Oil. This consists of the lowest boiling fractions and contains terpenes almost exclusively, also some cineol. Sp. gr. up to about 0.890.

2. Black Camphor Oil. It is that portion of the oil that boils higher than camphor and contains safrol, eugenol and sesquiterpenes. Sp. gr. mostly from 0.900 up to over 1.000.

The normal oil, as well as the individual fractions, is dextrogyrate. All fractions have the odor of camphor. In the so-called black oils, however, this odor becomes subordinate to that of safrol.

The oil from the roots and leaves is like that obtained from the wood. Camphor roots[1] yielded upon distillation 4 p. c. of an oil that was partly solid at ordinary temperature. When completely liquified its sp. gr. at 45° was 0.957. The liquid portion of the oil boiled between 165—270° and possessed the properties of ordinary camphor oil.

An oil obtained from dry leaves[1] (yield 1.8 p. c.) also contained solid camphor. The liquid portion boiled between 170—270° and could not be distinguished from ordinary camphor oil.

Two oils distilled from fresh leaves were reported on by Hooper[2] in 1896. The leaves from which the first oil was distilled were obtained

---

1) Bericht von S. & Co., Oct. 1892, p. 7. 2) Pharm. Journ., 56, p. 21.

from a tree in the government garden at Utakamand (India) and yielded 1 p. c. of oil: sp. gr. 0.9322 at 15°; $\alpha_D = +4° 32'$. It contained about 10—15 p. c. of camphor. The second batch of leaves was from Naduvatam, in the Nilgiri mountains. The oil contained 75 p. c. of camphor, by far more than did the first oil. After the camphor had been removed, the oil had a sp. gr. of 0.9814; $\alpha_D = +27°$.

COMPOSITION. Camphor oil is an exceedingly complex mixture of hydrocarbons and oxygenated substances, which belong to a large number of classes of chemical compounds. Aldehydes are represented by acet-aldehyde; the ketones by camphor; the alcohols by terpineol; the phenols and the phenol ethers by eugenol and safrol; the oxides by cineol; the terpenes by pinene, phellandrene and dipentene; and finally the sesquiterpenes by cadinene.

Arranged according to their boiling points, they may be enumerated in the following order:

1. Acet-aldehyde, CH_3CHO. This substance, which is possibly found in all oils to the extent of mere traces, makes its presence known when large quantities of oils are rectified. In part it is removed with the aqueous distillate, in part it is also found in the lowest terpene fractions.

2. Dextro-pinene,[1] $C_{10}H_{16}$, has been found in the fraction boiling below 160° and was identified by means of its nitrosochloride and conversion into nitrosopinene melting at 130°.

Yoshida[2] (1885) obtained a strongly laevogyrate fraction, $[\alpha]_j = -71.1°$, which was identified as l-pinene (chlorhydrate, nitrosochloride, nitrosopinene). Inasmuch as the lowest boiling fractions have always been found to be dextrogyrate, Yoshida's observation is rather remarkable and arouses the suspicion that the oil examined was not a normal oil (Bertram & Walbaum,[3] 1894).

Whether camphene, the presence of which seems to be indicated, is really present or not has not yet been decided.[4]

3. Phellandrene,[1] $C_{10}H_{16}$. This is present in such small quantity that its identification by means of the nitrite (m. p. 102°) caused difficulties.

---

1) Bericht von S. & Co., April 1889, p. 8.
2) Journ. Chem. Soc., 47, p. 779.
3) Journ. f. pract. Chem., II, 49, p. 19.
4) It is also noteworthy that Yoshida does not mention any fraction boiling above 230°, of which a considerable quantity is contained in camphor oil.

4. Cineol, $C_{10}H_{18}O$, constitutes 5—6 p. c. of the camphor oil. It has been isolated by means of the hydrobromide.[1]

5. Dipentene, $C_{10}H_{16}$, has been found by Wallach[2] and was identified by means of the tetrabromide melting at 123°. An impure dipentene dihydrochloride (m. p. 42° instead of 49°) had been isolated by Lallemand[3] by passing hydrochloric acid into the fraction boiling at 180°.

6. Camphor, $C_{10}H_{16}O$, which is economically the most important constituent, crystallizes in the receiver when the thermometer rises to 200°.

7. Terpineol, $C_{10}H_{18}O$. Inasmuch as it was impossible to obtain this alcohol in such a degree of purity by fractional distillation to enable its identification as such, its presence in fraction 215—220° was determined in several ways. Upon boiling with dilute sulphuric acid, terpinene and dipentene resulted. Upon standing with dilute acids terpin hydrate was formed.[4] When shaken with concentrated hydriodic acid, dipentene dihydriodide[5] melting at 77° was formed.

8. Safrol, $C_{10}H_{10}O_2$, b. p. 232°, was found by J. Bertram in 1885 in Schimmel & Co.'s laboratory. This firm has since then prepared it in large quantities from camphor oil.[6]

9. Eugenol,[7] $C_{10}H_{12}O_2$, is present only in relatively small amounts. It is obtained by shaking the higher fractions with dilute soda solution.

10. Cadinene, $C_{15}H_{24}$, is the principal constituent of the high boiling portions. By passing hydrogen chloride into fraction 260—270° a good yield of cadinene dihydrochloride melting at 117° is obtained.[8]

11. Blue oil. The portions distilling over last, consist of a blue oil which has been frequently observed in volatile oils and of which no characteristic derivative has yet been prepared.

A few words about the so-called "camphorogenol," which is reported to boil at 212—213° and is said to have the formula $C_{10}H_{18}O_2$, remain to be said. According to Yoshida[9] it polymerizes when boiled, but at the same time separates camphor. When acted upon by dilute nitric acid, acetic acid anhydride or benzoic acid, camphor is said to be formed; and when heated with sodium and alcohol, it is converted into borneol.

---

1) Bericht von S. & Co., Oct. 1888, p. 8.
2) Liebig's Annalen, 227, p. 296.
3) Liebig's Annalen, 114, p. 196.
4) Bericht von Schimmel & Co., Apr. 1888, p. 9.
5) Bericht von S. & Co., Apr. 1889, p. 8.
6) Ibidem, Sept. 1885, p. 7.
7) Ibidem, Apr. 1886, p. 5.
8) Ibidem, Apr. 1889, p. 9.
9) Yoshida, loc. cit.

As a matter of fact, such a substance evidently does not exist in camphor oil. "Camphorogenol" apparently is a mixture consisting principally of camphor and terpineol. Such a mixture behaves toward the above mentioned reagents like "camphorogenol."[1] When boiled, the terpineol decomposes into terpenes and water and the camphor, which is less soluble in the terpenes, partly separates out. If the terpineol is oxidized with nitric acid, the more resistant camphor remains behind. Treatment with sodium reduces the camphor to borneol.

USE. The oil from which the camphor has been removed is used by the poorer classes in Japan for illuminating purposes. It is also used as a solvent for resins in the manufacture of lacquer. From the soot, obtained by burning the oil, India ink is made.

In Europe camphor oil is used for the safrol it contains. The waste oil that results in the isolation of safrol is utilized as light and heavy camphor oil.

Light camphor oil, sp. gr. 0.89 to 0.92 consists of the low boiling fractions. Inasmuch as it has the property of hiding the penetrating odor of the cheap fats and tallow, it is used for perfuming ordinary soaps. It is also used as a substitute for turpentine oil over which it has some advantages. Thus e. g. it surpasses other cleansing media in the cleaning of types, plates and parts of machinery. It is further a good solvent for resins[2] and caoutchouc and is used in the manufacture of lacquers and varnishes.

Heavy camphor oil, sp. gr. 0.96 to 0.97, boils between 240 and 270° and is likewise added to cheap soaps and lubricants.

129. Oil of (Ceylon) Cinnamon.

Oleum Cinnamomi Zeylanici.—Ceylon Zimmtöl.—Essence de Canelle de Ceylan.

ORIGIN AND PREPARATION. The genus *Cinnamomum*, Fam. *Lauraceae*, consists of evergreen, aromatic trees and shrubs. Its several species yield volatile oils that are much used. The Ceylon cinnamon, *C. zeylanicum* Breyne contains volatile oil in the underground portions of the shrub as well as in those overground. Whereas the oils obtained from the various parts of the cassia, *C. cassia* Bl., seem to be almost identical, the oils obtained from the bark, leaves and root respectively of the Ceylon cinnamon vary considerably.

---

1) Bericht von S. & Co., Apr. 1888, p. 9.

2) G. Bornemann, Technische Mittheil. f. Malerei, 1892, 9, p. 5.

C. zeylanicum is indigenous to the forests of Ceylon. The bark, which was originally collected from the wild shrub, is at present collected from the cultivated plant only.[1]

For the distillation of the oil, the waste that results in the peeling of the bark, is used. These shavings and fragments have been exported as "chips" since 1867. The preparation of the oil from chips was introduced into Germany in 1872 by the firm of Schimmel & Co. of Leipzig. The yield from the chips is about 0.5 to 1 p. c.

The oil produced in Ceylon is but rarely the pure distillate from the bark. Commonly it contains much oil from the leaves, which are possibly distilled with the bark. It is also possible that the oil from the bark is subsequently adulterated with leaf oil.

History. Cinnamon and cassia are among the oldest spices, and the history of the drug as such is a very interesting one. The history of the oil, however, is of more recent date.

When in the course of the fifteenth century the distillation of aromatic waters for medicinal purposes became a common practice, cinnamon bark was undoubtedly distilled for the preparation of its water. The heavy oil which sinks to the bottom of the distillate was in all probability observed by the apothecary or chemist. The Canon St. Amando of Doormyk seems to have been the first to isolate besides bitter almond oil and oil of rue, a number of other volatile oils, among them the oil of cinnamon. Valerius Cordus had prepared the oil about 1540. At that time it may already have found application in medicine, so that it was included in the first edition of the Dispensatorium Noricum. Lonicer soon afterward distilled the oils from the spices, among them cinnamon oil, in a new peculiar apparatus. In the price ordinances cinnamon oil is first mentioned in that of Berlin of 1574, also in that of Frankfurt-o.-M. of 1582. Winter of Andernach distilled and described the oil in 1570, and J. B. Porta in 1589.

The early observations made on cinnamon oil, like those made on other oils, restricted themselves to the crystalline deposit formed upon prolonged standing. Such crystal formations were observed (among others) by Ludovici about 1670, later by Slare in England, by Boerhaave and Gaubius in Holland. The latter regarded them as camphor; Du Menil and Stockmann supposed them to be benzoic acid, Dumas and Péligot in 1831 recognized them to be cinnamic acid. C. Bertagnini prepared pure cinnamic aldehyde in 1852.

1) The cultivation, preparation and packing of Ceylon cinnamon is described in detail in A. Tschirch's Indische Heil- und Nutzpflanzen, p. 86.

The yield of volatile oil from cassia and cinnamon barks was determined by the following observers: G. W. Wedel in 1707; Fr. Cartheuser, C. Neumann and Ph. F. Gmelin in 1763; J. F. A. Goettling about 1803; by Dehne about the same time; also by Buchholz in 1813.

Properties. The oil is a light yellow liquid which has the agreeable fine delicate odor of the Ceylon cinnamon, and a spicy, sweet, burning taste.

The sp. gr. lies between 1.024 and 1.040; optically it is slightly laevogyrate, α_D = up to —1°. The oil is clearly soluble in 2 parts of 70 p. c. alcohol. It contains between 65—75 p. c. of cinnamic aldehyde.

Composition. Blanchet[1] in 1833 observed that in the distillation of Ceylon cinnamon two oils separate in the receiver, one being lighter than water, the other heavier, whereas Chinese cinnamon yielded only the heavy oil. The older chemists, however, concerned themselves only with the heavy oil. Dumas and Péligot[2] in 1834 ascertained, that like the oil of cassia, the heavy oil of the Ceylon cinnamon consisted principally of cinnamic aldehyde. That the heavy oil of the Ceylon cinnamon contains eugenol as well was observed later. The amount of eugenol in the oil from the bark is but 4—8 p. c., much less than in the leaf oil.

Of the constituents of the light oil but one is known, viz. phellandrene, which has been isolated from the fraction boiling about 175° and has been identified by the nitrite[3] melting at 102°. The substance to which the characteristic odor of the Ceylon cinnamon oil is due is not yet known.

Adulteration and Examination. Ceylon cinnamon oil is frequently adulterated with the much cheaper oil from the leaves. None of the oils from Ceylon appear to be pure bark oils, for all of the oils from this source examined thus far contained considerable quantities of leaf oil. Whether the two oils are mixed, or whether the leaves have been distilled with the bark, naturally cannot be decided.

Inasmuch as the addition of leaf oil increases the eugenol content and diminishes that of the cinnamic aldehyde, its recognition is not difficult. All that is necessary is to determine with approximate quantitativeness the amount of both substances present. After the determination of the specific gravity, which is increased by the addition of larger amounts of leaf oil, the aldehyde content is determined according

1) Liebig's Annalen, 7, p. 163.
2) Ann. de Chim. et Phys., 57, p. 305; Liebig's Annalen, 14, p. 50.
3) Bericht von S. & Co., Oct. 1892, p. 47.

to the method described under cassia oil on p. 390. If the aldehyde content is less than 65 p. c. or more than 75 p. c., the oil is suspicious.

If the oil contains less than 65 p. c. of aldehyde, the amount of eugenol in the non-aldehyde portion is determined according to one of several ways. The most accurate results are obtained according to the somewhat complex method of Thoms,[1] the eugenol being converted into benzoyl eugenol and weighed. (Comp. under oil of cloves.) Umney[2] has suggested to shake the oil with 10 p. c. potassa solution and to determine the eugenol content from the diminution in the volume of the oil. Inasmuch, however, as the strong potassa solution dissolves appreciable amounts of non-phenol constituents, the results are inaccurate and about 10 p. c. higher than those obtained according to Thoms' method. The error can be reduced by using a 5 p. c. potassa solution. For ordinary purposes at least, this method can be considered of sufficient accuracy to ascertain whether an oil is pure or adulterated. Adulteration may be assumed if the eugenol content of the original oil exceeds 10 p. c.

The extent to which the adulteration with leaf oil is carried on becomes apparent from the result of examinations published by Umney[2] and Schimmel & Co.[3] Every one of the seven oils was strongly adulterated with leaf oil, three containing at least 30 p. c., the other four not less than 50 p. c. The four last mentioned oils had the following properties:

	Sp. gr.	a_D	Percentage of cinnamic aldehyde	Percentage of Eugenol (according to Thoms)
1)	1.039	—0° 55′	29 p. c.	41.9 p. c.
2)	1.040	—0° 28′	28 "	39.1 "
3)	1.041	—0° 57′	29 "	47.7 "
4)	1.049	—0° 22′	24 "	45.7 "

The following reaction enables the qualitative distinction between the bark and leaf oils:[2]

A drop of genuine cinnamon oil dissolved in 5 drops of alcohol produces a pale green color with ferric chloride, whereas leaf oil and oil adulterated therewith produce a deep blue color.

The cinnamon oils adulterated with cassia oil have a higher specific gravity and, as a rule, also a higher cinnamic aldehyde content. Oils with more than 75 p. c. aldehyde, if not suspicious, are of inferior value. Whereas the cinnamic aldehyde content is the criterion of value of the cassia oil, this is not the case with the cinnamon oil, for in the latter

---

[1]) Berichte der pharm. Ges., 1, p. 279.
[2]) Pharm. Journ., III, 25, p. 949.
[3]) Bericht von S. & Co., Oct. 1895, p. 48.

the non-aldehyde constituents are the more valuable as is shown by the fact that the cinnamon oil with the smaller aldehyde content commands several times the price of the cassia oil.

130. Oil of Cinnamon Leaves.

Oleum Foliorum Cinnamomi.—Zimmtblätteröl.—Essence de Feuilles de Cannelle de Ceylan.

For a time the oil from the leaves of the genuine cinnamon, *Cinnamomum zeylanicum* Breyne, was designated commercially as cinnamon root oil. After Schimmel & Co. had shown in 1892 that an oil, distilled by themselves from Ceylon cinnamon leaves (yield 1.8 p. c.) corresponded in all its properties with the so-called root oil, the erroneous designation could no longer be retained. Two oils, which corresponded in every respect with the home distillate and which were correctly designated as cinnamon leaf oil, had previously been obtained from the Seychelles and from the botanical garden in Buitenzorg, Java.[1]

The oil formerly exported as cinnamon leaf oil was a thick, viscous oil of the consistency of West Indian sandalwood oil. It has since disappeared from commerce and nothing definite is known as to its origin.

Properties. Cinnamon leaf oil is of a light color, rather mobile, and has the odor of cloves and cinnamon. Sp. gr. 1.044—1.065; $\alpha_D = -0°5'$ to $+1°18'$. With 3 parts of 70 p. c. alcohol it forms a clear solution, which, however, sometimes becomes turbid upon the further addition of alcohol.

Composition. Stenhouse[2] (1855) found large amounts (70—90 p. c.) of eugenol in cinnamon leaf oil. Schaer[3] (1882) later confirmed this report. In the oil distilled by themselves, Schimmel & Co. found cinnamic aldehyde (0.1 p. c.). Weber[4] made a detailed examination of two different oils. The first, obtained from the Seychelles, had a sp. gr. of 1.0552 at 18.5° and contained eugenol (benzoyl eugenol, m. p. 69—70°), cinnamic aldehyde (phenyl hydrazone, m. p. 167°), and terpenes that were not identified. The other, still purchased by Schimmel & Co. under the erroneous name of cinnamon root oil, was also leaf oil. Its sp. gr. was 1.041 at 19°. It varied somewhat in its composition from the preceding. Weber found eugenol, safrol (piperonylic acid, m. p. 226 to 227°), terpenes and benzaldehyde (phenyl hydrazone, m. p. 150—151°), but no cinnamic aldehyde.

---

1) Bericht von S. & Co., Apr. 1892, p. 45; Oct. 1892, p. 47.

2) Liebig's Annalen, 95, p. 103.

3) Archiv d. Pharm., 220, p. 492.

4) Archiv d. Pharm., 230; p. 232.

Stenhouse had found benzoic acid which is probably combined with an alcohol. Later investigators do not mention this constituent. Whether the benzaldehyde found by Weber is an original constituent of the oil or an oxidation product of the cinnamic aldehyde, remains undecided.

131. Oil of Cinnamon Root.

The oil of cinnamon root is an almost colorless liquid with a strong odor of camphor. Even at ordinary temperature a part of the camphor,[1] which is identical with the ordinary laurus camphor, separates out. Other constituents are cinnamic aldehyde and a hydrocarbon (Holmes,[2] 1890). That the roots contain camphor has long been known. It was mentioned by Trommsdorff,[3] also by Dumas and Péligot[4] (1835).

Cinnamomum zeylanicum affords an interesting illustration of a plant, the root, leaves and bark of which afford oils of totally different composition. In the root oil we find camphor as characteristic constituent, in the leaf oil eugenol predominates and in the bark oil cinnamic aldehyde.

132. Oil of Cassia.

Oleum Cinnamomi Cassiae. — Cassiaöl, Chinesisches Zimtöl, Zimtblüthenöl. — Essence de Cannelle de Chine.

Origin and Preparation. Although the plant which yields the oil of cassia, the *Cinnamomum cassia* Bl., has been known for a long time, considerable uncertainty existed as to the part of the plant from which the oil was obtained. It was supposed, by way of illustration, that the unripe fruits, the *Flores cassiae* of commerce, were the source of the oil which for this reason was called "Zimtblüthenöl." As early as 1881,[5] however, it became known that near In-lin, north of Pak-hoi, the leaves of the cassia bush were distilled for oil. The oil was, however, reported as being thicker, darker and less fragrant than the cassia oil. H. Schroeter,[6] in the description of his travels through the cassia district in 1886, also mentions the production of oil from leaves. It was supposed, however, that the leaves yielded but a small amount of an inferior oil. In order to ascertain the properties of the leaf oil, Schimmel & Co. secured, with the aid of Melchers & Co. of Canton, not only the leaves, but all parts of the cassia shrub that might come into consider-

1) Bericht von S. & Co., Oct. 1892, p. 46.
2) Pharm. Journ., III, 20, p. 749.
3) Handbuch der Pharmacie, p. 666.
4) Liebig's Annalen, 14, p. 50.
5) Deutsches Handels-Archiv, Sept. 2, 1881, p. 262.
6) Bericht über eine Reise nach Kwangsi. Im Herbst 1886 unternommen.

ation in the distillation of the oil.[1] Upon distillation, the bark, flowers, flower stalks, twigs and leaves yielded oils that were similar in properties and aldehyde content. Inasmuch as the bark, the flowers and flower-parts are excluded on account of their high price, the leaves and branches alone can come into consideration as far as the production of the commercial oil is concerned. As a matter of fact, the leaves only, with possibly some waste from the bark, are used for distillation.

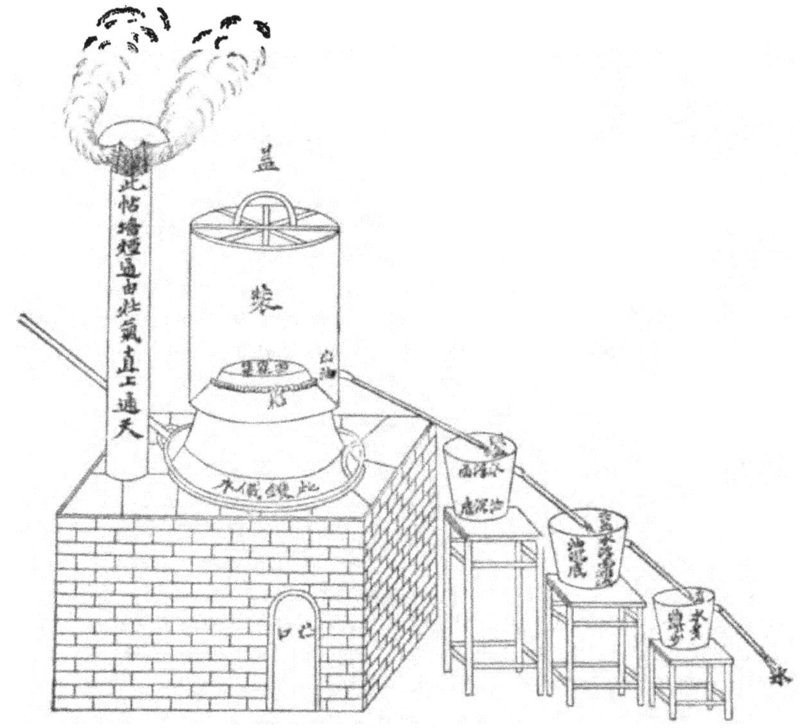

Fig. 70.
Cassia Oil Distillation in China.

This was definitely ascertained by O. Struckmeyer, who made a trip through the Cassia district in 1895 at the request of Siemssen of Hongkong.[2]

The distillation is conducted in the valleys of the provinces Kwang-si and Kwang-tung at points where a sufficient water supply can be had.

1) Bericht von S. & Co., Oct. 1892, p. 12. 2) Ibidem, Oct. 1896, p. 11.

The apparatus in use[1] (fig. 70) consists of a brick oven with an iron pan let into the masonry. On this rests a wooden cylinder lined on the inside with sheet-iron and open above. After the cylinder has been charged with leaves and half filled with water, the opening is closed with a peculiar shaped helmet of sheet-iron. The joint is sealed with wet cloths pressed between. The helmet has a groove on the inner side into which the condensed vapors flow and thence through a tube into the receivers. The receivers are arranged cascade-like and in them is deposited the heavy oil. The water is used for the next distillation.

Cassia oil is shipped in lead canisters with a capacity of 7.5 k., of which 4 are packed in a case. The interstices are filled with the glumes of rice. The form and color of the labels on the canisters have long remained unchanged.

History. The history of cassia oil will be found with that of Ceylon cinnamon on p. 378.

Properties. Chinese cinnamon oil or cassia oil, when pure, is a mobile liquid, yellowish to brownish in color, and strongly refractive. Its odor is cinnamon-like, the taste burning and intensely sweet, without the unpleasant, grating after-taste that is perceptible in oils adulterated with rosin. Sp. gr. 1.055—1.065. Inasmuch as the specific gravity of the cinnamic aldehyde and the other constituents of the oil is almost the same, differences in the aldehyde content are not revealed by changes in the specific gravity.

Cassia oil is optically inactive or has but a slight rotatory power either to the right or to the left. It is readily soluble in 1—2 p. of 80 p. c. alcohol. Its behavior toward 70 p. c. alcohol varies in so far as most oils yield clear solutions with 3 parts of the solvent, whereas some otherwise good oils produce turbid, opalescent solutions. This peculiar behavior is possibly traceable to zinc cinnamate which is frequently present in the oil. The oil boils with partial decomposition between 240—260°; acetic acid is split off, and a viscid residue of 6—8 p. c. remains. For further details see under Examination.

If four drops of cassia oil, well cooled with ice water, are mixed with a like amount of nitric acid, they unite forming a crystalline mass. This reaction, which is mentioned in the German Pharmacopoeia as a reaction of identity, depends on the union of the nitric acid with the cinnamic aldehyde to an unstable addition product, which is readily split up into its component parts by water. The reaction cannot be used as a test of purity for even strongly adulterated oils will produce crystals. In order to carry out the reaction, the oil must be well cooled,

---

1) Bericht von S. & Co., Apr. 1893, p. 11; Oct. 1896, p. 13.

for if the heat of the reaction becomes too great, only oily products result. The principal constituent of the oil, which supplies a criterion for its valuation, is the cinnamic aldehyde. A good oil contains as much as 75—90 p. c.

COMPOSITION. The principal constituent of cassia oil as well as of cinnamon oil is the cinnamic aldehyde. The differences in the two oils due to the accompanying constituents were known to the chemists who investigated the Chinese and Ceylon oils during the first half of this century. Blanchet[1] (1833) e. g. calls attention to the fact that cassia oil has a much more pungent odor than the Ceylon cinnamon oil. Dumas and Péligot[2] (1834) also point out differences in the oils. With oils distilled by themselves they came to the conclusion that the cinnamic acid, obtained by the oxidation of the oil, stood in the same relation to the cinnamon oil or cinnamic aldehyde, as benzoic acid does to bitter almond oil or benz-aldehyde. In other words they recognized that cinnamon oil consists essentially of cinnamic aldehyde. The same investigators discovered the interesting, though unstable substance $C_9H_8O.NO_3H$, which results upon the addition of nitric acid to cinnamic aldehyde in the cold, and which is remarkable on account of its tendency to crystallize.

Of other investigations of the same period, those of Mulder[3] (1840) and of Bertagnini[4] (1853) may be mentioned. Bertagnini investigated the compounds resulting by the action of the acid alkali sulphites upon the aldehyde, the composition of which has only recently been definitely established by Heusler[5] (1891).

Cinnamic aldehyde, $C_6H_5.CH:CH.CHO$, is a light yellow, strongly refractive liquid with an intensely sweet taste. At low temperatures it solidifies and melts at $-7.5°$.[6] Sp. gr. 1.064 at 15° (1.0497 at $\frac{24°}{4°}$).[7] Only under greatly diminished pressure does it boil without decomposition: at 128—130° under 20 mm. pressure.[8] For purposes of identification the phenyl hydrazone, which melts at 168°, is well adapted.[9]

The "cassia stearoptene" of Rochleder[10] is a crystalline deposit only seldom found in old cassia oil. Its constitution was determined by

---

1) Liebig's Annalen, 7, p. 164.

2) Ann. de Chim. et Phys., 57, p. 305; Abstr. in Liebig's Annalen, 12, p. 24; 13, p. 76; 14, p. 50.

3) Liebig's Annalen, 34, p. 147; Journ. f. prakt. Chem., 15, p. 307; 17, p. 303; 18, p. 385.

4) Liebig's Annalen, 85, p. 271.

5) Berichte, 24, p. 1805.

6) Zeitschr. f. phys. Chem., 16, p. 24.

7) Liebig's Annalen, 235, p. 18.

8) Berichte, 17, p. 2110.

9) Berichte, 17, p. 575.

10) Ber. d. Academ. d. Wissensch. zu Wien, mathem. phys. Kl., 1850, p. 1. (Abstr. Chem. Centrbl., 1851, p. 46), and ibid., vol. 12, p. 190. (Abstr. Chem. Centrbl., 1854, p. 701).

Bertram and Kuersten[1] in 1895. This particular specimen crystallized out of the last fraction obtained in the rectification of an oil of cassia. In the pure state it consists of well formed, hexagonal, yellow plates which melt at 45—46°. They have a very persistent, unpleasant odor and are identical with methyl ortho cumaric aldehyde, $C_6H_4(OCH_3)$ CH:CH.CHO, a substance that can be obtained synthetically by the condensation of methyl salicylic aldehyde and acet-aldehyde.

A constituent that has no very favorable influence on the odor and taste of the cassia oil is the cinnamyl acetate which was found in the oil by Schimmel & Co.[2] in 1889. Upon saponification of the non-aldehyde constituents of the oil with alcoholic potassa cinnamic alcohol is obtained on the one hand and acetic acid on the other. The former crystallizes from ether in hard, white crystals (b. p. 137° at 11 mm.). Cinnamyl acetate boils between 135—140° under 11 mm. pressure and has an unpleasant odor and a grating taste. Besides this, small amounts of a second ester seem to be present, viz. the phenyl propyl ester of acetic acid. This conclusion is drawn from the boiling point of an alcohol that accompanies the cinnamic alcohol after saponification.

On account of the readiness with which cinnamic aldehyde oxidizes, cassia oil always contains free cinnamic acid, but the amount is much less than one might expect from the changeability of the pure aldehyde. It is present to the extent of about 1 p. c.[3] The free acid has the undesirable property of dissolving the lead of which the canisters are made. A crystalline sediment observed by Hirschsohn[4] in 1891 consisted of lead cinnamate. Subsequently he examined a number of commercial oils and out of twelve samples eleven were found to contain lead. The test for lead is made by shaking a few drops of the oil with sulphuretted hydrogen water. According to the amount of lead present, the oil is colored red to black. For medicinal or culinary purposes only the rectified oil, free from lead, should be used.

Adulterations and their Detection. Formerly cassia oil was adulterated only with fatty oils, cedar wood oil and gurjun balsam oil.

1) Journ. f. prakt. Chem., II, 51, p. 316.

2) Bericht von S. & Co., Oct. 1889, p. 19.

3) Evidently the cinnamyl acetate largely prevents the oxidation as is shown by the following experiment. A very old oil of cassia containing 77.7 p. c. of aldehyde was set aside for a year in shallow dishes covered with perforated filter paper. Heat, light and air had free access. At the end of this period the cinnamic acid content had increased from 0.7 p. c. to 8.5 p. c. Resinification had been but slight, for the non-volatile residue (comp. tests), was but slightly greater after the experiment than before. Under like conditions pure cinnamic aldehyde would rapidly have been converted into a mass of crystals of cinnamic acid.—(Ber. of S. & Co., Oct. 1890, p. 10.)

4) Pharm. Zeitschr. f. Russland, 30, p. 790.

The detection of these adulterants was very simple, since the specific gravity was lowered and the oil was no longer soluble in 80 p. c. alcohol. The presence of the oils of cedar wood and gurjun balsam could furthermore be readily ascertained by their strong laevo-rotation. Less easy of detection was the adulteration with rosin and petroleum which was much practiced in Macao and Hongkong especially during the late eighties. At first colophonium only seems to have been used, but inasmuch as larger additions of rosin thickened the oil and increased its specific gravity, this discrepancy was equalized by the addition of petroleum. Inasmuch as this method of adulteration did not influence the solubility in 80 p. c. alcohol any more than the specific gravity, it was not discovered for some time. Detection, however, followed when the addition of rosin became too large. The commercial oils of that period had a very unpleasant odor, a dark brown color and the consistency of a thick lacquer. The taste which was but slightly sweetish soon gave way to an unpleasant and persistently grating sensation in the mouth. Upon rectification with water vapor, as much as 40 p. c. of a resinous mass remained in the still. The distillate separated in two layers: one sank to the bottom of the water, the other floated on it. The latter consisted, as was found upon examination, of petroleum.

The demand to be able to detect on a small scale this form of adulteration, led to the distillation test of Schimmel & Co.[1] This test consists in distilling a weighed or measured quantity of the oil and weighing the residual rosin.

Fifty grams of cassia oil are weighed in a tared fractionation flask of 100 cc. capacity (fig. 53, p. 189). A tube 1 m. in length is attached as condenser and the oil distilled with the aid of a direct flame. At first water passes over with "cracking," the thermometer then rises rapidly to 240° and the bulk of the oil distills between 240 and 260°. The end of the distillation is indicated by the appearance of white fumes arising from the decomposition of the residue. The thermometer at the same time rises to 280—290°. Upon cooling the residue in the flask is weighed. The residue of a good oil is viscid and tough and constitutes 6—8, at most 10 p. c. of the oil. The residue of an oil adulterated with rosin is hard and brittle and correspondingly larger.

Instead of weighing the oil, 50 cc. can be measured in a pipette and the distillate be collected in a graduated cylinder. By deducting the number of cc. of distillate from 50, the amount of residue can be ascertained with sufficient accuracy.

The petroleum can be detected in the distillate. The distillate from pure oils forms a clear solution with 70—80 p. c. alcohol. If petroleum is present,

---

1) Bericht von S. & Co., Oct. 1889, p. 15.

no perfect solution results:[1] the liquid is turbid at first, but upon standing the supernatant petroleum can be removed and its behavior toward sulphuric and nitric acids be tested (comp. p. 201.)

In order to detect rosin in cassia oil without material loss, Gilbert[2] suggests that several grams of oil be heated in a watchglass in a drying oven at a temperature of 110—120° until of constant weight. According to Gilbert, the determination of the acid number also yields useful results for the detection of rosin. An oil which upon distillation yielded 6 p. c. of residue had an acid number of 13. After 20 p. c. of rosin (acid number 150) had been dissolved in the oil, the acid number rose to 40. A cassia oil with 28 p. c. of residue yielded an acid number of 47.

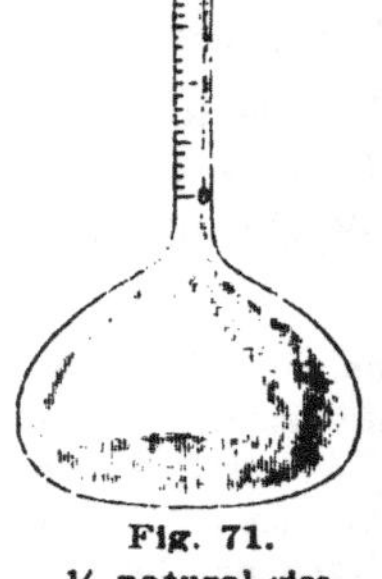

Fig. 71.
⅙ natural size.

According to Hirschsohn[3] (1890), an alcoholic solution of lead acetate, which has the property of producing precipitates with a solution of rosin, can be used for the detection of rosin in cassia oil.

To a solution of 1 p. of cassia oil in 3 p. of 70 p. c. alcohol, a freshly prepared solution of lead acetate in 70 p. c. alcohol, saturated at ordinary temperature, is added drop by drop to the extent of one-half the volume of the oil solution. If a precipitate is produced, the oil was adulterated with rosin. As little as 5 p. c. of rosin can be detected in this way.

Aldehyde Assay. Inasmuch as the value of cassia oil depends on its cinnamic aldehyde content, the aldehyde assay is of the greatest importance in determining the value of the oil. The method worked out by Schimmel & Co.[4] is universally recognized[5] and the oil is sold according to the results obtained by it in the principal markets of the world, such as Hongkong, London, Hamburg and New York.

For the assay a special flask (cassia flask, aldehyde flask, fig. 71) of about 100 cc. capacity is used. The neck of the flask is about 13 cm. long, has an

1) As already stated, cassia oil that contains rosin as well as petroleum, is soluble in 80 p. c. alcohol, whereas otherwise pure cassia oil to which petroleum has been added does not form a clear solution with 80 p. c. alcohol, the petroleum separating in the form of oily drops.

2) Chem. Zeitung, 18, p. 1406.

3) Pharm. Zeitschr. f. Russl., 29, p. 255.

4) Bericht von S. & Co., Oct. 1890, p. 12.

5) In the description of his travels through the cassia districts, O. Struckmeyer relates that he found the implements for the aldehyde determination according to Schimmel & Co. on a plantation near Loting-chow in the interior of China. Bericht von S. & Co., Oct. 1896, p. 12.

inner diameter of 8 mm., and is calibrated into 1/10 cc. The neck has a capacity of about 6 cc. The zero mark is fixed slightly above the point where the flask is narrowed into the neck.

By means of a pipette, 10 cc. of oil are transferred to the flask and an equal volume of 30 p. c. sodium acid sulphite solution is added. The mixture is shaken and the flask placed in a boiling water bath. After the solid mass has become liquid more acid sulphite solution is added, the mixture being constantly heated and occasionally shaken, until fully three-fourths of the flask is filled. The solution is heated until no more solid particles are visible and the odor of cinnamic aldehyde has disappeared. When the clear oil floats upon the salt solution, flask and contents are allowed to cool and sufficient acid sulphite solution is added to raise the lower limit of the oily layer to the zero point of the scale. The number of cc. of oil is read off, and by deducting this number from ten, the aldehyde content is ascertained. Accurately speaking, percentage by volume and not percentage by weight is determined in this manner. Inasmuch, however, as the specific gravity of the aldehyde and the non-aldehyde constituents is almost the same, the two are practically identical.

The chemical reactions involved can be explained by means of the following equations (Heusler,[1] 1891): At first the insoluble aldehyde addition product is formed.

$$C_6H_5 \,.\, CH : CH \,.\, CHO + NaHSO_3 = C_6H_5 \,.\, CH : CH \,.\, CHO \,.\, NaHSO_3$$

Cinnamic aldehyde — Acid sulphite of sodium — Cinnamal hydroxy sulphonate of sodium.

When boiled with water, two molecules of this addition product break up into a molecule of cinnamic aldehyde and one of sulpho cinnamal hydroxy sulphonate of sodium.

$$2C_6H_5 . CH : CH . CHO . NaHSO_3 = C_6H_5 . CH . CH . CHO +$$

Cinnamal hydroxy sulphonate of sodium — Cinnamic aldehyde

$$C_6H_5CH_2 . CH(SO_3Na) . CHO . NaHSO_3$$

Sulpho cinnamal hydroxy sulphonate of sodium.

In order to convert all of the aldehyde into the second compound soluble in water it is necessary to use an excess (2 mol.) of sodium acid sulphite. The salts of the sulpho cinnamal hydroxy sulphonic acid are readily soluble in water, and not decomposed by boiling water. They are decomposed by destructive distillation or when heated with caustic soda or sulphuric acid.

The aldehyde content of a good cassia oil amounts to at least 75 p. c., and only in rare instances is higher than 90 p. c. However, oils have been obtained from China which contained only 35—50 p. c. of aldehyde, yet no adulteration could be detected. According to the Chinese, these oils were distilled from young leaves, an assertion the correctness of which cannot be ascertained.

---

[1]) Berichte, 24, p. 1805; Bericht von S. & Co., April 1890, p. 13.

In order to be able to assay oils with but 35 p. c. of aldehyde according to the method given above, it is necessary to use 5 cc. of oil instead of 10 cc. and to make corresponding changes in the calculation.

Oils obtained from the different parts of the Cassia Shrub.

All parts of the cassia shrub yield oils with a high aldehyde content.[1] The differences in the specific gravity, though they are but slight, indicate that the non-aldehyde constituents vary somewhat as to composition. In the distillation of the commercial oil only the leaves and twigs are commonly used. The bark and the so-called cassia buds command too high a price, whereas the flower stalks cannot be obtained in sufficient quantity. However, it may happen occasionally that all of these parts are distilled.

1. Oil from the bark, the *Cassia lignea* of commerce. Yield 1.5 p. c.; sp. gr. 1.035; aldehyde content 88.9 p. c.

2. Oil from the cassia buds, the *Flores cassiae* of commerce. Yield 1.6 p. c.; sp. gr. 1.026; aldehyde content 80.4 p. c.

3. Oil from the stalks of the cassia buds. Yield 1.7 p. c.; sp. gr. 1.046; aldehyde content 92 p. c.

4. Oil from the leaves. Yield 0.54 p. c.; sp. gr. 1.056; aldehyde content 93 p. c

5. Oil from the twigs. Yield 0.2 p. c.; sp. gr. 1.045; aldehyde content 90 p. c.

6. Oil from a mixture of leaves, petioles and young twigs. Yield 0.77 p. c.; sp. gr. 1.055; aldehyde content 93 p. c.

Inasmuch as the commercial oil is distilled from the materials enumerated under No. 6, its properties should correspond with those found and recorded above. The fact that the commercial oils never equal the above in quality is probably due to the crude method common in China, namely the distillation over a free flame.

Production and Commerce. Cassia, also known commercially as *Cassia lignea* in order to distinguish it from other cinnamon varieties, is produced in a comparatively limited territory, namely in the provinces Kwang-si and Kwang-tung. The cassia district lies between the degrees 110 and 112 eastern longitude. It is bounded on the north by the Si-kiang or West river, and extends to the south as far as the 23° 3′ northern latitude. The principal plantations are in the vicinity of Tai-wo, Yung and Sih-leong on the Sang-kiang, also at Loting-chow on the Lintan river.

---

1) Bericht von S. & Co., Oct. 1892, p. 12.

The Tai-wo product is preferred in the market. The annual output of cassia varies between 50,000 and 80,000 piculs or 3,000,000 and 4,800,000 k. Canton and Hongkong are the principal markets. From the latter point 54,032 piculs were exported during 1896.

The distillation is conducted in the district outlined above. As material for distillation the leaves, petioles and twigs, which result as waste in the production of the *Cassia lignea*, are used. Whereas the cassia bark is transported down the Si-kiang, the natural water route, to Canton, the oil is conveyed in lead containers over the mountains to Pak-hoi and then by ship to Hongkong. This is done to avoid the high tariff levied at the Likin stations on the way to Canton.

The annual production of cassia oil varies between 2,000 and 3,000 piculs, or 4,000 and 6,000 cases of 30 k. each, with a value of $250,000 to $375,000. Hongkong is the principal center. The aldehyde content serves as criterion of the value of the oil. Following the example of Schimmel & Co., the principal firms in Hongkong are equipped to make the aldehyde assay and have acquired a considerable degree of accuracy in making the test.

133. Japanese Cinnamon Oil.

The various parts of *Cinnamomum loureirii* Nees are used in Japan as cinnamon, the root bark *Komaki* being especially esteemed. Upon distillation Shimoyama[1] obtained 1.17 p. c. of a light yellow, strongly refractive oil, sp. gr. 0.982 at 15°. Like the genuine cinnamon oil it contains cinnamic aldehyde. The non-aldehyde oil has the odor of lavender, boils at 175—176° and, as shown by analysis, consists principally of a terpene.

134. Oil from the Bark of Cinnamomum Kiamis.

From the massoy bark of *Cinnamomum kiamis* Nees Bonastre[2] (1829) distilled an oil that separated into a light and a heavy oil and from which a camphor crystallized out. The light oil was almost colorless, mobile, and its odor reminded of sassafras. The heavy oil was thicker, less volatile, had a fainter odor and a strong taste of sassafras. The massoy camphor, which is likewise heavier than water, consists of a white, soft, odorless and tasteless powder which is soluble in hot alcohol and ether.

---

[1]) Mitt. der med. Fak. Tokio, vol. 8, No. 1. Abstr. Apoth.-Zeit., 11, p. 587.

[2]) Journ. de Pharm., 15, p. 204.

135. Culilawan Oil.

The oil from culilawan bark from *Cinnamomum culilawan* Bl. was first prepared by Schloss[1] in 1824 who describes it as an oil having the odor of the oils of clove and cajeput. The yield is 4 p. c.; sp. gr. 1.051. With 3 or more parts of 70 p. c. alcohol it produces a clear solution.

The principal constituent is eugenol (benzoyl eugenol, m. p. 70—71°) of which the oil, assayed according to Thoms' method contains about 62 p. c. (Gildemeister & Stephan,[2] 1897). A fraction 249—252° contained methyl eugenol (monobrom methyl eugenol dibromide, m. p. 78—79°; veratric acid, m. p. 179—180°[5]). Fraction 100—125° under 10 mm. pressure possibly contains terpineol.

136. Oil from Cinnamomum Wightii.

The bark of *Cinnamomum wightii* Meissn. which grows in the mountainous districts of southern India, yielded upon distillation 0.3 p. c. of oil, which boiled from 130—170° and the sp. gr. of which was 1.010 at 15°. The plant that yields this bark was formerly falsely designated *Michelia nilagirica.*[3]

137. Oil from Cinnamomum Oliveri.

From the bark of *Cinnamomum oliveri* Bail., a tree known in Australia as black, brown or white sassafras, R. T. Baker[4] obtained upon distillation with water vapor 0.75—1 p. c. of an oil of a golden-yellow color and a very agreeable odor. Sp. gr. 1.001; $\alpha_D = +22$ to 22.3°. The oil distilled between 213—253°, 54 p. c. passing over between 230—253°.

The lowest fraction gave the iodol reaction for cineol. With alkali a small amount of a phenol, presumably eugenol, was separated, which gave a blue color with ferric chloride. Acid sulphite removed from the oil about 2 p. c. of an aldehyde, having the odor of cinnamon and which presumably is cinnamic aldehyde. Upon cooling the oil to —12°, crystals separated which, however, dissolved again in the oil when removed from the freezing mixture (safrol?).

The leaves of this tree, which was formerly incorrectly designated as *Beilschmiedia obtusifolia*, yield upon distillation (770 oz. to the ton) an oil which has a decided odor of sassafras.[5]

---

1) Trommsdorff's Neues Journ. der Pharm., 8, II, p. 106.

2) Archiv d. Pharm., 235, p. 588.

3) Bericht von S. & Co., Oct., 1887, p. 36; Apr. 1888, p. 46.

4) Proc. Linnean Soc. of N.-S.-Wales, 1897, part 2, p. 275; abstr. Pharm. Ztg., 42, p. 859.

5) Bericht von S. & Co., Apr. 1887, p. 38.

138. Oil from Persea Gratissima.

The leaves of *Persea gratissima* Gaertn. (from the botanical garden in Genoa) yielded upon distillation 0.5 p. c. of a light greenish oil which closely resembles estragon oil in odor and taste.[1] Sp. gr. 0.9607; $\alpha_D = +1° 50'$; n_D at 18.2° = 1.5164. These data indicated the presence of fairly pure methyl chavicol. Anethol was excluded because of the absence of a sweet taste and because no crystals separated when the oil was exposed to a freezing mixture.

An attempt to convert a part of the oil, of which but 10 g. were obtainable, into anethol by means of alcoholic potassa failed on account of an accident. The remaining oil, when oxidized with permanganate, yielded an acid melting at 183° which possessed the properties of anisic acid, thus indicating methyl chavicol.

139. Oil from Clove Bark.

The bark of the Brazilian tree *Persea caryophyllata* Mart. (*Dicypellium caryophyllatum* Nees) was formerly used in the apothecary shops as *Cortex caryophyllati.* It was also known as *Cassia caryophyllata* (Ger. *Nelkenrinde, Nelkenzimt,*[2] *Nelkencassie* and *Nelkenholz*).

Upon distillation of the powdered bark, Trommsdorff[3] (1881) obtained 4 p. c. of a light yellow oil heavier than water which had the odor of cloves. With potassa, soda, and ammonia the oil formed crystalline compounds from which it could be regenerated upon the addition of sulphuric acid. This property as well as odor and density render it probable that this oil, like the oil of cloves, contains eugenol.

140. Oil of Pichurim Beans.

The volatile oil of the cotyledons of *Nectandra pichury major* Nees and *N. p. minor* Nees, the pichurim beans of commerce, was first prepared by Robes[4] in 1799 and then by Bonastre[5] in 1825. It was examined by Mueller[6] in 1853. On account of the large amount of starch present he added sulphuric acid to prevent the formation of starch paste and obtained 0.7 p. c. of an oil of greenish-yellow color and the characteristic pichurim odor.

---

[1]) Bericht von S. & Co., Oct. 1894, p. 71.

[2]) The oil from the leaves of the Ceylon cinnamon was for a time designated as *Nelkenzimtöl.*

[3]) Trommsdorff's Neues Journ. d. Pharm.. 23, I, p. 7.

[4]) Berlin. Jahrb. d. Pharm., 5, p. 60.

[5]) Ibidem, 37, p. 160: Repert. f. d. Pharm., I, 21, p. 201.

[6]) Journ. f. prakt. Chem., 58, p. 463.

The oil distills between 180—270°. After repeated fractionation, Mueller obtained a hydrocarbon (pinene?) with a constant boiling point at 150°. Fraction 165—170° had an orange odor and seems likewise to have been a terpene. Fraction 255—256° had a deep indigo blue color. From the highest fractions crystals of "pichurim Fettsäure" (lauric acid) separated. Inasmuch as the pichurim beans have a distinct safrol odor, this substance is probably a constituent of the oil.

141. Caparrapi Oil.

According to B. F. Tapia[1] (1898) the oil of *Nectandra caparrapi* is long known in Columbia as caparrapi oil. The tree is popularly known as *Canelo,* probably on account of the cinnamon-like odor of the bark. The oil is obtained according to a method similar to that in vogue in this country for the production of turpentine, and is used as a substitute for copaiba in this country.

In commerce the oil is found more or less colored in accordance with the temperature employed in the removal of the water. It contains a monobasic acid, $C_{15}H_{26}O_3$, which melts at 84.5° and which can be obtained in a crystalline form only from the light colored oils. The acid free oil consists in large part of a sesquiterpene alcohol, $C_{15}H_{25}OH$, called caparrapiol, which is converted into caparrapene, $C_{15}H_{24}$, by dehydrating agents. The alcohol as well as the hydrocarbon are readily polymerized, especially by heat, so that in the water distillation about three-fourths of the oil is resinified.

142. Guayana Linaloe Oil.

The oil from *Ocotea caudata* is described together with Mexican linaloe oil, which see.

143. Ocotea or Laurel Oil from Guayana.

An oil designated laurel oil from Demerara (British Guayana) was examined by Stenhouse[2] in 1842. According to Christison, it is obtained from a species of the genus *Ocotea.* Incisions are made near the root, penetrating the oil cavities beneath the bark. The exuding oil is a yellowish liquid which has a pleasant terebinthinate odor, sp. gr. 0.864 at 13.3° and boils between 149—162.5°. Upon standing, a mixture of

---

[1]) Bull. Soc. chim., III, 19, p. 638.

[2]) Philos. Magaz. and Journ. of Science, 20, p. 273; 25, p. 200. Liebig's Annalen, 44, p. 309; 50, p. 155. Comp. also Hancock, Bull. des sciences, math. phys. et chim. Févr. 1825, p. 125. Abstr. Trommsdorff's Neues Journ. d. Pharm., 11, I, p. 171.

oil, alcohol and dilute nitric acid yielded crystals which upon recrystallization were obtained in the form of white, rhombic prisms melting at 150°. The oil, therefore, appears to consist of pinene and the crystals obtained by Stenhouse were probably terpin hydrate although the melting point of this substance lies at 116—117° and not at 150°.[1]

144. Venezuelan Camphorwood Oil.

Through the kindness of Professor H. H. Rusby,[2] Fritzsche Bros. of New York obtained some Venzuelan camphorwood, the botanical origin of which has not been determined, but which probably belongs to either the genus *Nectandra* or *Ocotea* (Family *Lauraceae*). The wood, which was in the form of large billets, was rather soft and easily split, has a silky lustre and possesses a faint odor reminding somewhat of borneol.

Upon distillation of the comminuted wood, 1.15 p. c. of a light yellow oil was obtained that had an unusually high sp. gr., viz. 1.155, and $\alpha_D = +2°40'$.[3] The oil had a faint odor reminding of that from *Asarum canadense* and congeals at ordinary temperature to a crystalline mass. Drained and recrystallized this was obtained in the form of handsome, colorless prisms which melt at 28.5°. Concentrated sulphuric acid dissolves these with a blood-red color. When boiled with alcoholic potassa they are converted into a substance which recrystallizes from alcohol in plates and melts at 55—56°. These properties show that the crystals which separate from the oil, and which constitute about 90 p. c. of it, consist of apiol. A careful comparison showed them to be identical in all respects with the apiol from parsley oil.

145. Sassafras Oil.

Oleum Sassafras.—Sassafrasöl.—Essence de Sassafras.

Origin and History. The sassafras tree, *Sassafras officinalis* Nees (Family *Lauraceae*) is widely distributed in North America from Canada to Florida and Alabama, and westward as far as Kansas[4] and the northern part of Mexico. In the southern states it frequently attains a height of 15 m., the trunk occasionally having a diameter of ½ m.

Numerous older statements to the contrary, the older bark and wood are odorless. The green parts of the tree, when crushed, smell faintly aromatic, but not of safrol. The wood of the roots and especially the root bark are more rich in oil cells.

1) Gmelin, Organ. Chem., IV ed., vol. 7, p. 282, the melting point of the substance obtained by Stenhouse is given as 125°.

2) Comp. Rusby, Botany of Venezuela, Pharm. Journ., 57, p. 292.

3) Bericht von S. & Co., Apr. 1897, p. 52.

4) Proc. Am. Pharm. Assoc., 29, p. 446.

As already mentioned under the oils of sweet birch and wintergreen (p. 331) next to turpentine oil, the oil of sassafras was the first volatile oil distilled in a primitive fashion in North America. On account of the pleasant aroma, the root bark was chewed by the natives as *Pavame.* It was also mixed with smoking tobacco (Rafinesque[1]), and added as aromatic to refreshing beverages and was used as a remedy. On account of its peculiarity, the sassafras tree is said to have attracted the attention of the Spaniards at their first landing in Florida under Ponce de Leon in 1512, also under de Soto in 1538. They are said to have regarded it as a kind of cinnamon tree. Afterward the sassafras tree soon came to be regarded as a valuable medicinal plant and article of luxury of the new world, not only by the Spaniards and French in Florida, but also in Mexico. As late as the first half of this century the bark, leaves and buds were used in the middle and central states as a substitute for Chinese tea (Lloyd[2]).

As early as 1582 sassafras wood and bark became known in Germany as a new American drug and were used under the name of *Lignum pavanum, L. floridum, L. xylomarathrum* (Ger. *Fenchelholz*).[3] In 1610 young plants were brought to England and cultivated.[4] Bark and wood were apparently first distilled by Angelus Sala in 1620, who mentions that the oil is heavier than water.[5] Schroeder's Pharmacopœa medico-chymica, published in Frankfurt-on-the-Main in 1641, is the first pharmacopœia that gives directions for the distillation of the oil, whereas the municipal price ordinance of Frankfurt-on-the-Main of 1587 already enumerates *Oleum ligni sassafras.* Hoffmann in Halle distilled the oil in 1715 and describes it as being colorless and specifically heavy.[6] In 1738 Maud, an Englishman, observed the formation of large crystals of sassafras camphor.[7] Early examinations of the oil were made by Muschenbroeck, by Neumann,[8] and by Dehne,[9] the first thorough one was made by Grimaux and Ruotte[10] in 1869.

A detailed historic account of the sassafras tree as a drug-yielding plant was given by J. U. Lloyd[2] in 1898, who, however, does not enter into the history of the volatile oil. Schoepf,[6] who was a careful observer

1) Med. Flora or Manual of the Med. Bot. of the U. S. of N. Am., vol. 2, p. 285.

2) Historical study of Sassafras. Pharm. Review, 16, p. 450; Pharm. Era, 20, p. 608.

3) Documente zur Geschichte der Pharmacie, p. 30.

4) Pharmacographia, p. 587.

5) Opera physico-medica, p. 84.

6) Observationes physico-chymici. Observatio 1, p. 13.

7) Phil. Trans. of the Roy. Soc., 8, p. 248.

8) Chym. med., vol. 2, pars 3, p. 248.

9) System d. Mat. med., vol. 4, p. 242.

10) Compt. rend., 68, p. 928.

and who traveled through the Atlantic states in 1783 and 1784 repeatedly refers to the sassafras tree, but does not mention the oil. Evidently the distillation of the oil did not become an industry until the close of the last or the early part of this century. Possibly it was fostered by the growth of the industry in medical specialties ("patent medicines"), as was the case with the oils of sweet birch and wintergreen.

The process of distillation seems to have been generally very primitive.[1]

The still consisted of a barrel with a perforated bottom, firmly fitted into a conical iron kettle. The top of the barrel could be removed to charge the still. It is provided with an opening for the reception of a tin tube which passes through a barrel of cold water and serves as condenser.

In recent years the distillation is conducted in a somewhat more rational manner.

The stills, made of 3 in. pine planks, are 4—5 ft. high, about 12 ft. square and strengthened by iron bands. One of the sides is provided with two close fitting doors, an upper one for charging the still, and a lower one for removing the exhausted material. The wood is split, or sawed into thin pieces. The steam, generated in a boiler, enters the still at the bottom and the distillate is cooled in a coil condenser and collected in a large copper flask of 20 gal. capacity. About 2 in. from the bottom this flask is provided with a stopcock through which the oil is drawn off from time to time. The exhausted wood is dried and used as fuel. Such a still has a capacity for 20,000 lbs. of wood and the distillation of this quantity lasts about 48—50 hours. The root bark yields 6—9 p. c. of oil, the woody part of the root less than 1 p. c.

Up to the middle of this century the oil was distilled principally in Pennsylvania, Maryland and Virginia, and Baltimore and Richmond were the principal commercial centers for the oil. At the beginning of the war in 1860, not less than 50,000 lbs. of sassafras oil were sold annually in Baltimore alone (Sharp[2]). Since the sixties, considerable quantities of the oil have also been distilled in New Jersey, New York, Ohio, Indiana, Tennessee and the New England states, until the practical extinction of the tree rendered the industry unprofitable.

Properties. Sassafras oil is a yellowish or reddish-yellow liquid having the odor of safrol and an aromatic taste. Sp. gr. 1.070—1.080; $\alpha_D = +3$ to $+4°$; soluble in all proportions in 90 p. c. alcohol. If, according to the U. S. Pharmacopœia, 5 drops of oil are mixed with 5 drops of nitric acid, a violent reaction results: the oil is first colored red and is later converted into a red resin (Bonastre[3]).

1) Proc. Am. Pharm. Assoc., 14, p. 211: Am. Druggist 1887, p. 45; New Remedies. 1888. p. 224; Oil, Paint & Drug Reporter, Sept. 14, 1891; Pharm. Journ., III, 22, p. 491.

2) Am. Jour. Pharm., 35, p. 18. Comp. also W. Procter, ibid. 33, p, 1; 38, p. 484.

3) Comp. p. 400, footnote 2.

Fig. 73.
Comminuting apparatus and stills in Lexington, Va.

Composition. To its principal constituent, the safrol, $C_{10}H_{10}O_2$, the oil owes its odor as well as its principal properties. This substance was observed by Binder[1] in 1821, it having crystallized out of the oil. The earlier investigators[2] contented themselves by studying the action of chlorine, bromine, etc. on the oil as well as on the safrol. Upon saturating the oil with chlorine, then neutralizing the mixture with lime and distilling with water vapor, Faltin[3] observed another constituent of the oil, the ordinary camphor.

The term safrol was introduced by Grimaux and Ruotte[4] (1869) who ascertained its composition to be represented by the formula $C_{10}H_{10}O_2$. They also isolated a hydrocarbon $C_{10}H_{16}$, b. p. 155—157°, and named it safrene. They further observed the presence of a phenol which had the odor of eugenol and produced a green color with ferric chloride. The identity of this phenol with eugenol was later established by Pomeranz[5] (1890).

A detailed study of the composition of the oil was made by Power and Kleber[6] in 1896. From the air-dried root bark 7.4 p. c. of oil were obtained, sp. gr. 1.075, $\alpha_D = +3°16'$. After most of the safrol had been removed by freezing, the eugenol (benzoyl eugenol, m. p. 69°) was shaken out with caustic alkali solution. The oil thus deprived of safrol and eugenol was fractionated. Fraction 155—160° was found to consist of pinene (pinene nitrolbenzylamine, m. p. 123°); fraction 160—175° contained phellandrene (nitrite). The so-called safrene of Grimaux and Ruotte is, therefore, pinene. From a higher fraction dextro camphor (oxime, m. p. 115°) was isolated. The quantitative reduction of the camphor to borneol and the acetylization of the latter showed that camphor was present to the extent of 6.8 p. c. in the oil. Inasmuch as fraction 260—270° gave the characteristic sesquiterpene reaction with sulphuric and glacial acetic acids, the presence of such a hydrocarbon is assumed, though a solid hydrochloride could not be obtained.

According to the investigation of Power and Kleber, the composition of sassafras is as follows:

---

1) Buchner's Repert. f. d. Pharm., 11, p. 346.

2) Bonastre, 1828, Journ. de Pharm., II, 14, p. 645. Abstr. Trommsdorff's Neues Journal, 19, I, p. 210; Saint Evre, 1844, Ann. de Chim. et Phys., III, 12, p. 107; Compt. rend., 18, p. 705. Abst. Liebig's Annalen, 52, p. 396.

3) Liebig's Annalen, 87, p. 376.

4) Compt. rend., 68, p. 928.

5) Monatsh. f. Chem., 11, p. 101.

6) Pharm. Review, 14, p. 101.

Safrol	80.0 p. c.
Pinene / Phellandrene	10.0 "
d-Camphor	6.8 "
Eugenol	0.5 "
High boiling substances (sesquiterpene and residue)	3.0 "
	100.3 "

ADULTERATION. Sassafras oil is frequently adulterated with camphor oil. A perfectly pure article appears to be of rare occurrence. Inasmuch as camphor oil contains all of the constituents found in sassafras oil, the detection of the former is exceedingly difficult, but its presence may be indicated by strong variations in the specific gravity and other physical properties. Under the name of artificial sassafras oil, fractions of camphor oil having a specific gravity similar to that of true sassafras oil are sold.

146. Oil of Sassafras Leaves.

As far as is known, the oil of sassafras leaves has been prepared but once. Power and Kleber[1] obtained upon the distillation of fresh sassafras leaves 0.028 p. c. of an oil with a very agreeable odor reminding of that of lemons. Sp. gr. 0.872, $\alpha_D = +6°25'$.

The low boiling fractions consist of a mixture of pinene (nitrolbenzylamine) and myrcene. This representative of the olefinic terpenes which have been found e. g. in the oils of bay, hops, origanum and rosemary, was hydrated according to Bertram's method. The resulting alcohol had the odor and boiling point of linalool. This alcohol was further identified by its oxidation product citral and the condensation product of the latter with naphthylamine and pyrotartaric acid. The third terpene identified is phellandrene (nitrite). In addition, the presence of a paraffin melting at 58° and of a sesquiterpene was established. Of oxygenated constituents the oil contains citral (citral-β-naphtho cinchoninic acid) and the two isomeric alcohols linalool and geraniol. These alcohols occur free as well as in the form of acetic and valerianic esters.

147. Cryptocaria Oil.

Cryptocaria moschata Mart. is a tree 10—15 m. high and is known in Brazil by the popular name *Nos moscado do Brasil*, Brazilian

1) Pharm. Review, 14, p. 108.

nutmeg. The ripe fruits are somewhat smaller than nutmegs and have a strong aromatic odor reminding of laurel, sassafras, cajeput and nutmeg.

Ten kilos of the powdered fruit deprived of their pericarp yielded upon distillation 37 g. of volatile oil (Peckolt[1]). This is a mobile liquid, light yellow in color, of a penetrating, aromatic odor and a burning, spicy taste. Sp. gr. 0.917. It is soluble in 90 p. c. alcohol in all proportions.

148. Oil of Cryptocaria Pretiosa.

The bark of *Cryptocaria pretiosa* Mart., (*Mespilodaphne pretiosa* Nees et Mart., Family *Lauraceae*), which is indigenous to northern Brazil on the Rio Negro, has a pleasant aromatic, cinnamon-like odor. Upon distillation the bark yielded 1.16 p. c. of an oil of cinnamon-like odor and a sp. gr. of 1.118. Although the odor seems to indicate the presence of cinnamic aldehyde, no crystalline addition product was obtained with sodium acid sulphite solution.[2]

149. Oil of Laurel Leaves.

Oleum Lauri Foliorum. — Lorbeerblätteröl. — Essence de Laurier.

Origin and History. The laurel, *Laurus nobilis* L. is indigenous to Asia Minor, Syria and the Silician Taurus and has been extensively cultivated. During classical antiquity it acquired great significance as a symbol of victory but apparently found no other application than as a decorative plant. During the middle ages bark and leaves were used medicinally. The laurel oil of commerce is distilled from the leaves. The yield varies from 1—3 p. c. according to the quality of the leaves.

Properties. The oil of the laurel leaves is a light yellow liquid with a pleasant odor, which first reminds of that of cajeput oil, and later becomes sweetish. Sp. gr. 0.920—0.930; $\alpha_D = -15$ to $-18°$; 2—3 p. of 80 p. c. alcohol are requisite to effect solution.

Composition. The oil begins to boil at 158°, the thermometer rising rapidly to 168°. This fraction contains pinene (nitrolpiperidine base, m. p. 118°). Fraction 176° contains cineol as shown by the hydrobrom addition product. The fractions above 180° had a decided odor of anethol.[3] Inasmuch, however, as the oil has no sweetish taste, which ought to be the case if anethol were present, this odor seems to indicate methyl chavicol, the isomer of anethol. The oil also contains small amounts of eugenol (benzoyl eugenol, m. p. 70°).[3]

---

1) Pharm. Review, 14, p. 248.

2) Bericht von S. & Co., Apr. 1898, p. 68.

3) Bericht von S. & Co., Apr. 1899, p. 31.

150. Laurel Oil from Berries.

History. Upon boiling and expression of the laurel berries a green mixture of fatty and volatile oil is obtained (*Oleum laurinum*) which is semi-liquid at ordinary temperature. It was used during antiquity and in older literature is mentioned among the substances employed in the preparation of ointments; also among the aromatics. It was contained in the first edition of the Dispensatorium Noricum of 1543. The volatile oil also seems to have been used medicinally. It is also mentioned in the price ordinances of Frankfurt of 1582 as well as in later ones. At present it seems to find little or no practical application and is only now and then prepared for scientific purposes. The yield from the berries is about 1 p. c.

Properties. The oil from the berries is somewhat more viscid than that from the leaves; its odor is less agreeable. On account of the lauric acid which it contains, some specimens solidify at temperatures above 0°. Sp. gr. 0.915—0.935. The angle of rotation of a single oil was found to be —14° 10′. The same oil was insoluble in 80 p. c. alcohol, but was soluble in ½ and more parts of 90 p. c. alcohol.

Composition.[1] The lower fractions contain the same constituents as the leaf oil (Wallach,[2] 1889), viz. very little pinene (pinene nitrolbenzylamine, m. p. 122—123°) and much cineol (cineol hydrobromide). The supposed laurene of Brühl[3] (1888) has revealed itself as a mixture of the above two constituents. Fraction 250°, sp. gr. 0.925, $\alpha_D = -7.2°$, has the composition $C_{15}H_{24}$ and is therefore a sesquiterpene (Blas,[4] 1865).

Another constituent of the oil, the amount of which will vary according to the duration of the distillation, is lauric acid. It can be removed by shaking the oil with lye and melts when pure at 43°.[5] In addition the oil contains ketones and alcoholic constituents which form solid compounds with sodium. Liberated by means of water they form a viscid oil which distills between 71—184° under 20 mm. pressure.

The statement made by Gladstone[6] that this oil contains eugenol has not been substantiated by the investigations of Blas and Müller.

---

1) The first investigations were carried out by Bonastre, 1824, Journ. de Pharm., 10, p. 36; 11, p. 8; Abstr. Repert. f. d. Pharm., I, 17, p. 190; and by Brandes, 1840, Archiv d. Pharm., 72, p. 160.

2) Liebig's Annalen, 252, p. 97; also Müller, Berichte, 25, p. 547.

3) Berichte, 21, p. 157.

4) Liebig's Annalen, 134, p. 1.

5) Blas and Müller, loc. cit.

6) Journ. Chem. Soc., II, 2, p. 1. Abstr. in Jahresb. f. Chem., 1863, p. 545.

Although Gladstone designates his oil as "Bay oil from the berries of *Laurus nobilis*," it seems rather probable that he examined genuine bay oil which consists largely of eugenol.[1]

151. Kuro-moji Oil.

Oleum Kuromoji.—Kuromojiöl.—Essence de Kuro moji.

Lindera sericea Bl. is a shrub quite generally distributed throughout the mountainous regions of Japan.[2] All parts of the plant contain small amounts of volatile oil, even the wood which is used in the manufacture of tooth-picks. The oil, which was introduced into the market in 1889 by Schimmel & Co., is obtained from the leaves and young twigs. The distillation is conducted on a small scale by farmers who are satisfied to work up the material growing on their farms.

PROPERTIES. Kuro-moji oil is of a dark yellow color and has a fine aromatic, balsamic odor. Sp. gr. 0.890 to 0.905; $\alpha_D = -0°4'$.

COMPOSITION. According to Kwasnik[3] (1892) the oil contains two terpenes and two oxygenated constituents, viz. d-limonene (tetrabromide, m. p. 104°), dipentene (tetrabromide, m. p. 124°), terpineol (phenyl-terpinyl urethane, m. p, 109.5°), and l-carvone (hydrosulphide, m. p. 214).

152. Mountain Laurel Oil.

All parts of the Mountain laurel or California bay tree[4] *Umbellularia californica* Nutt. (*Oreodaphne californica* Nees, *Tetranthera californica* Hook. et Arn., *Californischer Lorbeerbaum*) contain volatile oil, especially the leaves which yield 2.4[5] to 4[6] p. c. The oil is of a light yellow color and has a pungent aromatic odor reminding of nutmeg and cardamom. Strongly inhaled it attacks the mucous membrane and causes the flow of tears. The taste is warm and camphor-like. Sp. gr. 0.936 to 0.940.

According to Heany[6] the oil boils between 175 and 245° and contains a hydrocarbon boiling at 175°, also an oxygenated constituent called oreodaphnol, b. p. 210°, sp. gr. 0.960.

Stillmann[5] found in fraction 167—168° a substance of the composition of the "terpinol" of Wiggers and List,[7] also umbellol, boiling at 215—216° and having the composition $C_8H_{12}O$.

1) Even much more recently no difference has been made in England between bay oil and the oil from laurel or bay berries. Comp. e. g. Ashton in Chemist and Druggist, July 2, 1892.

2) Chemist and Druggist, 47. p. 502.

3) Archiv d. Pharm., 230, p. 265.

4) Ber. d. pharm. Ges., 6, p. 56.

5) Berichte, 18, p, 630.

6) Am. Journ. Pharm., 47, p. 105; Pharm. Journ., III, 5, p. 791.

7) As shown by Wallach, terpinol is no chemical unit, but a mixture of dipentene, terpinene, terpinolene and terpineol (Liebig's Annalen, 230, p. 251).

153. Spicewood Oil.

All parts of the North American spicewood, spice bush or fever bush (*Benzoin odoriferum* Nees, *Laurus benzoin* L., *Benzoëlorbeer-Strauch*), but especially the bark and the berries contain volatile oil.[1] The bark is used as a domestic remedy.

1. The oil of the bark and twigs which was introduced into the market as spicewood oil by Fritzsche Bros. in 1885, has the odor of wintergreen, sp. gr. 0.923 and boils between 170—300°. It consists of hydrocarbons and 9—10 p. c. of methyl salicylate. By treating 200 g. of oil with caustic soda, 16 g. of salicylic acid were obtained.

2. The berries contain 4—5 p. c. of an aromatic, spicy oil with a camphor-like odor; b. p. 160—270°; sp. gr. 0.850—0.855.

3. The leaves contain about 0.3 p. c. of oil with a very agreeable odor. Sp. gr. 0.888.

154. Tetranthera Oil.

The fruits of *Tetranthera citrata* Nees (Family *Lauraceae*), sometimes called citronella fruits, are regarded as a panacea in India. Upon distillation they yield about 5.5 p. c. of an oil which has a strong odor resembling that of verbena oil and for this reason has been designated Japanese verbena oil. Sp. gr. 0.890—0.896. It boils between 180—240°, and contains in addition to a terpene, citral as principal constituent.[2]

155. Oil of Massoy Bark.

The botanical origin of the New Guinea bark, from which Schimmel & Co. have since 1888 distilled their massoy bark oil, is not known. Upon comparison of this bark with that of the genuine bark from *Massoia aromatica* Beccari, Holmes[3] (1888) arrived at the conclusion that the two were not identical, but that the bark of Schimmel & Co. was very similar to the *Cortex culilabani papuanus* of Hanbury's collection, the botanical origin of which is also unknown. Wender[4] (1891), on the other hand is of a different opinion. He states that according to its anatomical structure, the massoy bark of Schimmel & Co. is unquestionally that of a lauraceae, and that in other respects it corresponds best with the bark of *Sassafras goesianum* or *Massoia aromatica*.

1) Bericht von S. & Co., Oct. 1885, p. 27; Oct. 1890, p. 49.

2) Bericht von S. & Co., Oct. 1888, p. 44. The sp. gr. 0.980 mentioned in the report is evidently due to a printer's error and should be 0.890. Two later distillates had the sp. gr. 0.894 and 0.896 respectively.

3) Pharm. Journ., III, 19, pp. 465 and 761.

4) Zeitschr. d. allgem. österr. Apoth. Ver., 29, p. 2.

The massoy bark from German Guinea yields upon distillation 6.5—8 p. c. of a volatile oil of a pleasant clove- and nutmeg-like odor. Sp. gr. 1.04—1.06.

COMPOSITION. According to Schimmel & Co.[1] the oil consists principally of eugenol (75 p. c.) and safrol. In addition to these substances Woy[2] isolated a fraction boiling at 172° in which he supposed he had found a new terpene, "massoyene." Wallach[3] (1890), however, showed that it consisted of a mixture of three terpenes: pinene (nitrolbenzylamine, m. p. 123° and nitrosopinene, m. p. 133°), limonene (nitrolbenzylamine, m. p. 93°), and dipentene (tetrabromide, m. p. 123°).

156. Oil of Garden Cress.

In 1846 Pless[4] demonstrated that the oils of *Lepidium ruderale* L., *L. sativum* L., and *L. campestre* R. Br. are heavier than water and contain sulphur. He states that the oil is ready formed in the herb but not in the seeds, where it results through the action of the water.

The composition of the oil from garden cress, *L. sativum* L. was ascertained by Hofmann[5] in 1874. The herb was collected immediately after flowering and distilled in wooden vats[6] with water vapor. The oil being completely dissolved in the aqueous distillate was extracted by shaking with benzene. The yield was 0.115 p. c. The crude oil was of a light yellow color but became colorless upon rectification. Three-fourths of the oil boiled at 231.5°, the boiling point of benzyl cyanide, $C_6H_5.CH_2.CN$. The identity of this substance was established by its conversion into phenyl acetic acid melting at 77°, and by the analysis of the silver salt of this acid. The lower fractions contained a small amount of sulphurated compounds, the composition of which was not determined. Gadamer[7] has recently investigated this oil and found it to be identical in composition with the oil from *Troepoelum majus*. It consists principally of benzyl isocyanate, $C_6H_5.CH_2.N:CS$, and small amounts of benzyl cyanide. The oil is formed by the action of a ferment on the glucoside glucotropaeolin, $KO.SO_2.O.C(:N.CH_2.C_6H_5).S.C_6H_{11}O_5+2H_2O$.

The fresh herb of *Lepidium latifolium* L. yields upon distillation an oil which is heavier than water and contains sulphur. (Steudel[8]).

---

1) Bericht von S. & Co., Oct. 1888, p. 42.

2) Archiv d. Pharm., 228, pp. 22 and 687.

3) Liebig's Annalen, 258, p. 340; Archiv d. Pharm., 229, p. 116.

4) Liebig's Annalen, 58, p. 39.

5) Berichte, 7, p. 1293.

6) During the distillation of this and other sulphur-containing oils, the use of copper vessels is to be avoided, for the copper will cause partial decomposition of the oils with formation of copper sulphide.

7) Berichte, 32, p. 2326; Archiv d. Pharm., 237, p. 508.

8) Dissert. de Acredine nonnull. Vegetabil. Tübingen 1805.

157. Oil of Thlaspi Arvense.

According to Pless[1] (1846) this oil is obtained by first macerating the herb or seed with water and then distilling. It is colorless, possesses a peculiar penetrating odor and leek-like taste, reminding alike of garlic and mustard.

If the oil is saturated with ammonia and the product distilled with water, there remains a residue of thiosinamine (m. p. 72°), thus indicating the presence of allyl sulphocyanate. This distillate yielded with platinum chloride the same addition product obtained by Wertheim[2] (1844) from garlic oil. Pless, therefore, regarded it as allyl sulphide. Semmler[3] (1892), however, has shown that Wertheim's conclusions are erroneous, that oil of garlic contains a group of different sulphides, but does not contain allyl sulphide. The presence of allyl sulphide in the oil of *Thlaspi arvense* L. is, therefore, highly improbable.

158. Oil of Spoonwort.

Oleum Cochleariae.—Löffelkrautöl.—Essence de Cochléaria.

Origin and History. Spoonwort, *Cochlearia officinalis* L. grows wild in the neighborhood of the coast-line of the northern continents and in several mountain ranges of the central European Alps. It is also largely cultivated. Upon distillation the plant yields an oil possessing a faint mustard odor.

During the middle ages spoonwort was regarded as a remedy against scurvy. The distilled oil of spoonwort seems to have been known and to have been used medicinally since the middle of the sixteenth century. This use, however, does not appear to have been general, for the oil is not mentioned in any edition of the Dispensatorium Noricum, neither in the Pharmacopoea Augustana nor in Schroeder's Pharmacopoea medicochymica of Frankfurt-on-the-Main. It occurs, however, in the municipal ordinance of Frankfurt of 1587. Later it occurs also in inventories of apothecary shops of Braunschweig and Dresden of 1640 and 1683 respectively. At the beginning of the eighteenth century it was distilled by Hoffmann in Halle and described by him.

Preparation. According to Gadamer[4] (1898), the dried herb, which need not be in blossom, is macerated with water to which powdered white mustard has been added and then distilled. The yield is about 0.23 p. c. or more.

1) Liebig's Annalen, 58, p. 36.
2) Liebig's Annalen, 51, p. 298.
3) Archiv d. Pharm., 230, p. 434.
4) Apotheker Zeitung, 13, p. 679; Arch. d. Pharm., 237, p. 92.

PROPERTIES. Spoonwort oil is optically active; Gadamer observed the angle $\alpha_D = +55.27°$. Upon distillation of 42.8 g. the following fractions were obtained: 150—154°, 6.3 g.; 154—156°, 12.2 g.; 156—158°, 10.0 g.; 158—162°, 12.0 g.; residue 2 g. The sp. gr. of these fractions varied from 0.941 to 0.943; the angle of rotation from +51.41° to +62.87°.

COMPOSITION. Simon[1] (1840) gives the boiling point of the oil at 156—159°. He found that it contains sulphur and yields a thiosinamine-like derivative with ammonia. Gieseler[2] regarded the oil as being free from nitrogen but containing oxygen and pronounced it an oxysulphide of allyl. Hofmann[3] (1869) showed the oil to have the composition of secondary butyl isosulphocyanate. From the bruised fresh herb[4] mixed with water he obtained 0.034 p. c. of an oil distilling between 158—165°. After repeated fractionation, fraction 161—163° upon analysis yielded results corresponding with those for butyl mustard oil. Comparison with synthetic secondary butyl isosulphocyanate, $CH_3CH_2.(CH_3).CH.N:C:S$, revealed the identity of the two. It is a colorless liquid, sp. gr. 0.944 at 12°, b. p. 159.5°, and possesses the characteristic odor of the oil of spoonwort. When heated with ammonia to 100° it yields a thio-urea[5] which melts at 133° and is optically active.

According to Moreigne,[6] spoonwort oil contains raphanol. Inasmuch as the commercial, so-called artificial spoonwort oil is not secondary but isobutyl mustard oil it should not be substituted for the natural product in medicinal preparations.

159. Oil from Horse Radish.

The investigations of Gadamer[7] have rendered it highly probable that the formation of the oil of the root of *Cochlearia armoracia* L., to which the pungent taste of horse radish is due, is caused by the presence of sinigrin. When the juicy root is grated, the glucoside is broken up by a ferment present with the formation of mustard oil. When distilled in glass vessels, the root yields about 0.05 p. c. of oil.

The crude oil is light yellow in color and has the consistency of cinnamon oil; the rectified oil is colorless. Sp. gr. 1.01. The odor is

---

1) Poggend. Annalen, 50, p. 377.

2) De Cochlearia officinali ejusque oleo dissertatio. Berol. 1857.

3) Berichte, 2, p. 102; 7, p. 508.

4) According to Simon, dry herb yields an oil upon distillation only when white mustard meal has been added.

5) Gadamer, loc. cit.

6) See under radish oil, p. 417.

7) Archiv d. Pharm., 235, p. 577.

penetrating, causes tears to flow and cannot be distinguished from that of mustard oil. Like the latter, the oil of horse radish causes burning and blistering when brought in contact with the skin.

Oil of horse radish contains sulphur. Based on the results of elementary analyses as well as the formation of thiosinamine, Hubatka[1] (1843) concluded that the oil has the same composition as mustard oil. The statements of Hubatka have later been verified by Sani.[2] When kept in contact with water in a closed vessel for years, the oil disappeared and in its place acicular crystals were formed that had a silvery lustre and the odor first of horse radish, then of peppermint and finally of camphor (Einhof,[3] 1807).

160. Oil of Hedge Garlic.

From the roots of *Alliaria officinalis* Andr. (*Sisymbrium alliaria* Scop., Ger. *Lauchhederich*), Wertheim[4] (1844) obtained upon distillation 0.033 p. c. of an oil that could not be distinguished from mustard oil and which yielded thiosinamine melting at 74°. Judging from the odor the herb contains the same constituents as garlic oil. According to Pless[5] (1846), the oil of the seeds consists of about 9/10 mustard oil and 1/10 garlic oil.

161. Oil of Mustard.

Oleum Sinapis. — Senföl. — Essence de Moutarde.

Origin and History. As black mustard are designated the seeds of several mustard plants belonging to the *Cruciferae*, viz. *Brassica nigra* Koch (*Sinapis nigra* L.) and *B. juncea* Hooker fil. et Thompson (*S. juncea* L.). *B. nigra* belongs to the European-Asiatic flora and is cultivated in most civilized countries, especially in those of central Europe; whereas *B. juncea* is cultivated on a larger scale in southern Russia, India and North America. The largest amount of seeds are used in the manufacture of table mustard, less for medicinal purposes and for the distillation of mustard oil.

The first statement which indicates a knowledge of the fact that mustard oil can be obtained by distillation with water, is found in the writings of Porta; another in the writings of the Parisian apothecary Nic. le Febvre. Boerhaave, however, seems to have been the first to prepare the volatile mustard oil in 1732 and to have called attention

---

1) Liebig's Annalen, 47, p. 153.

2) Accad. Linc., 1892; Abstr. Bericht von S. & Co., Apr. 1894, p. 50.

3) Neues Berlinisches Jahrbuch, 5, p. 365.

4) Liebig's Annalen, 52, p. 52.

5) Liebig's Annalen, 58, p. 38.

to its properties. That it contains sulphur was observed by Thibierge of Paris in 1819. Boerhaave and Murray observed the great density of the oil; Jul. Fontenelle determined the specific gravity in 1824.

Undoubtedly those who prepared mustard oil knew that the volatile oil does not preexist in the seed but is produced by the action of water. Yet the first to call attention to this fact were Glaser[1] in 1825, Boutron and Robiquet[2] in 1831, and independently of these Fauré[3] as well as Guibourt,[4] both in 1831. Shortly after (1833), Dumas and Pelouze[5] made the first elementary analysis of the oil and discovered the thiosinamine, which mustard oil forms with ammonia. That mustard oil is produced by the action of a ferment was noticed by Boutron and Fremy[6] (1840). They isolated myrosin by extracting the seed with alcohol and obtained mustard oil by allowing this substance to act on the aqueous extract of the seed which had previously been extracted with alcohol. Sinigrin or myronate of potassium, was first prepared by Bussy[7] (1840). He termed the underlying acid *acide myronique* and with Robiquet[8] determined its physical constants and its behavior toward reagents. The knowledge of the chemical composition of the oil was advanced by Will[9] (1844) and simultaneously by Wertheim[10] who regarded the mustard oil as allyl sulphocyanate. Ludwig and Lange[11] in 1860 confirmed the existence of sinigrin and its decomposition by ferment action into mustard oil, sugar and potassium acid sulphate. This reaction was made more clear by the detailed studies of Will and Koerner[12] (1863). Artificial mustard oil had been prepared by the action of allyl iodide on potassium sulphocyanate by Zinin[13] in 1855 and by Berthelot and de Luca[14] also in 1855. The natural oil was therefore regarded as an ester of thiocyanic acid. Oeser,[15] however, showed in 1865 that allyl thiocyanate and natural mustard oil possess different properties.

The true constitution of mustard oil as the ester of the isothiocyanic acid was recognized by Billeter[16] and Gerlich[17] in 1875. They showed

1) Repert. f. d. Pharm., I, 22, p. 102.
2) Journ. de Pharm., II, 17, p. 294; Geiger's Magazin f. Pharm. und Exper. Kritik, 36, pp. 64 and 67.
3) Journ. de Pharm., II, 17, p. 299; 21, p. 464.
4) Journ. de Pharm., II, 17, p. 360.
5) Ann. de Chim. et Phys., II, 58, p. 181; Liebig's Annalen, 10, p. 324.
6) Journ. de Pharm., II, 26, pp. 48 and 112; Liebig's Annalen, 34, p. 230.
7) Journ. de Pharm., II, 26, p. 39; Liebig's Annalen, 34, p. 223.
8) Journ. de Pharm., II, 26, p. 110.
9) Liebig's Annalen, 52, p. 1.
10) Liebig's Annalen, 52, p. 54.
11) Zeitschr. f. Chemie und Pharm., 3, pp. 430, 577.
12) Liebig's Annalen, 125, p. 257.
13) Journ. f. prakt. Chem., 64, p. 504; Liebig's Annalen, 95, p. 128.
14) Compt. rend., 41, p. 21.
15) Liebig's Annalen, 134, p. 7.
16) Berichte, 8, pp. 464 and 820.
17) Berichte, 8, p. 650; Liebig's Annalen, 178, p. 89.

that by the interaction of allyl iodide and potassium sulphocyanate, allyl sulphocyanate is first formed and that upon heating this is converted into its isomer, the allyl isosulphocyanate. As an explanation of this form of isomerism, Hofmann[1] in 1868 had suggested that in the true thiocyanates the carbon is directly united with the sulphur, but in the iso compounds with the nitrogen.

Inasmuch as the possibility was not excluded that by the ferment action on sinigrin allyl sulphocyanate may first be formed, Schmidt[2] allowed this action to take place at low temperature. He ascertained that even at 0° allyl isosulphocyanate is formed and only traces of the normal isomer. The last uncertainty in connection with the hydrolysis of sinigrin was removed by Gadamer[3] (1897). He showed that the formula of sinigrin is $C_{10}H_{16}NS_2KO_9$ and not $C_{10}H_{18}NS_2KO_{10}$ as was supposed by Will and Koerner, also that the hydrolysis takes place by the addition of the elements of one molecule of water.

PREPARATION. The mustard which is used in the manufacture of mustard oil is obtained either from *Brassica nigra* K. (*Sinapis nigra* L.) which is cultivated in Holland, Apulia or the levant, or more frequently from *B. juncea* Hook. fil. et Thomson (*S. juncea* L.), a species which is cultivated on a large scale in Sarepta, in the Russian gouvernement Saratow, also in East India.

Mustard oil is not contained as such in the seeds, but is formed by ferment action. The ground mustard seed is deprived of its fatty oil with the aid of hydraulic presses. The press-cakes are mixed with tepid water and allowed to undergo fermentation, and then distilled with water vapor. The yield varies between 0.5—0.75 p. c. of the original seed.[4]

---

1) Berichte. 1, p. 28.

2) Berichte, 10, p. 187.

3) Archiv d. Pharm., 235, p. 44.

4) For the quantitative estimation of mustard oil in the seed, the method of Dieterich (Helfenberger Annalen, 1896, p. 382) can be employed: 5 g. of the bruised seed are allowed to ferment for two hours at a temperature of from 20—25°, when 10 g. of alcohol are added. The flask is connected with a condenser, the tube of which dips into the receiver containing 30 cc. of ammonia. The distillation is continued until 50—60 cc. have passed over. To the ammoniacal liquid an excess of silver nitrate solution is added, the precipitated silver sulphide is collected on a filter, washed, dried and heated to a red heat. The reduced silver is then weighed and its weight multiplied by 0.4938. The product gives the amount of mustard oil in 5 g. of seed.

Gadamer (Arch. d. Pharm., 235, p. 58) has modified the process somewhat. Before the distillation he adds a definite quantity of N/10 silver nitrate V. S. known to be in excess to the ammonia. After the distillation, the distillate is diluted to a definite volume, set aside for 12 hours and filtered through a dry filter into a dry flask. In an aliquot part of the filtrate, acidulated with nitric acid, the excess of silver is titrated back with potassium sulphocyanate.

At a temperature exceeding 70° no fermentation takes place, because the myrosin is coagulated and rendered inactive.

The fermentation is brought about by the albuminoid ferment, myrosin, which acts on the sinigrin in the presence of water with the formation of mustard oil, dextrose and potassium acid sulphate.

$$\underset{\text{Sinigrin}}{C_{10}H_{16}NS_2KO_9} + \underset{\text{Water}}{H_2O} = \underset{\text{Mustard oil}}{CSNC_3H_5} + \underset{\text{Dextrose}}{C_6H_{12}O_6} + \underset{\text{Potassium acid sulphate.}}{KHSO_4}$$

Beside this reaction others take place resulting in the formation of allyl cyanide and carbon disulphide, which are never entirely wanting in the oil. Contact with the copper of the still, or prolonged contact with water cause the separation of sulphur and the formation of allyl cyanide.

$$\underset{\text{Mustard oil}}{C_3H_5NCS} + \underset{\text{Copper}}{Cu} = \underset{\text{Allyl cyanide}}{C_3H_5CN} + \underset{\text{Copper sulphide.}}{CuS}$$

When the distillation has been conducted in a careless manner, the amount of allyl cyanide can be so large that the oil becomes lighter than water (Will,[1] 1863). (Sp. gr. of allyl cyanide is 0.835 at 17.5°).

How the carbon disulphide always found in the oil (in the artificial oil as well) is formed has not yet been clearly demonstrated. It has been shown[2] that when mustard oil is boiled with water for an hour in a flask connected with a reflux condenser no carbon disulphide is formed. This substance, however, is formed in appreciable quantities together with carbon dioxide when mustard oil is heated for several hours with water in sealed tubes under pressure at a temperature of 100—105°. It may be assumed that in the nascent state mustard oil has a greater capacity for reaction and that in this condition water may effect its decomposition in a manner expressed by the following equation.

$$\underset{\text{Mustard oil}}{2C_3H_5NCS} + \underset{\text{Water}}{2H_2O} = \underset{\text{Allyl amine}}{2C_3H_5NH_2} + \underset{\text{Carbon dioxide}}{CO_2} + \underset{\text{Carbon disulphide.}}{CS_2}$$

Carbon disulphide is also formed by prolonged contact of water with mustard oil. For the detection of this substance see under Examination.

As to the time requisite to complete the fermentation, 80 minutes are required for the hydrolysis of pure sinigrin.[3]

Inasmuch as the potassium acid sulphate, which is split off in the course of the reaction, acts on the mustard oil in the process of formation, a better yield can be obtained by neutralizing with alkali (when

---

1) Liebig's Annalen, 125, p. 278.

2) Archiv d. Pharm., 235, p. 58.

3) Gadamer, loc. cit.

pure sinigrin is used). An excess of alkali is to be carefully avoided, because it also diminishes the yield. Calcium carbonate has also been found serviceable. With mustard meal, however, it is quite different. The addition of calcium carbonate not only is of no advantage, but its presence acts detrimentally. The cause for this has not yet been explained. Possibly the base sinapine, which is contained in the mustard enters into the reaction by being liberated by the calcium carbonate, forming a non-volatile thiosinamine-like compound with the mustard oil.[1]

COMPOSITION. Besides small variable amounts of allyl cyanide and carbon disulphide, mustard oil consists almost entirely of allyl mustard oil or allyl isothiocyanate, C_3H_5NCS. Possibly traces of the isomeric allyl thiocyanate, and higher boiling (polymeric?) unknown compounds are present.

The chemical reactions of mustard oil are, therefore, largely those of allyl isosulphocyanate which can be found in any chemical reference work. Here only those will be mentioned that are necessary to understand the tests.

For the quantitative determination of allyl isosulphocyanate in mustard oil, its property of forming a solid, non-volatile compound with ammonia is utilized. If to mustard oil an excess of ammonia and alcohol are added, the odor of both mustard oil and ammonia disappears, gradually in the cold but more rapidly when heated, while crystals of thiosinamine are formed.

$$\underset{\text{Mustard oil}}{C{\begin{matrix}=N.C_3H_5\\=S\end{matrix}}} + \underset{\text{Ammonia}}{\begin{matrix}H\\NH_2\end{matrix}} = \underset{\text{Thiosinamine}}{C{\begin{matrix}/NH.C_3H_5\\-NH_2\\\backslash S\end{matrix}}}$$

Thiosinamine or allyl thio urea crystallizes in rhombic prisms which melt at 74°, possess a faint leek-like odor and taste, and are readily soluble in water, alcohol and ether. When a small amount of mustard oil is mixed with twice its volume of concentrated sulphuric acid a violent reaction sets in with the formation of carbon oxysulphide,[2] sulphur dioxide[3] and allylamine sulphate. The last remains as a faintly colored liquid which sometimes solidifies in the test tube.

Artificial mustard oil is obtained by the interaction of allyl iodide and potassium thiocyanate in alcoholic solution. Allyl thiocyanate is first formed which under the influence of heat rearranges itself to the isothiocyanate.

---

1) Gadamer, loc. cit.

2) Berichte, 1, p. 182.

3) Archiv d. Pharm., 196, p. 214.

C_3H_5I	+	NCSK	=	C_3H_5NCS	+	KI
Allyl iodide		Potassium thiocyanate		Mustard oil		Potassium iodide.

Mustard oil also results by the dry distillation of potassium allyl sulphate and potassium thiocyanate.

$C_3H_5KSO_4$	+	CNSK	=	C_3H_5NCS	+	K_2SO_4
Potassium allyl sulphate		Potassium thiocyanate		Mustard oil		Potassium sulphate.

PROPERTIES. Mustard oil is a mobile, colorless or yellowish, strongly refractive liquid, which is optically inactive and which has a very strong odor causing tears to flow. When brought in contact with the skin it draws blisters. The specific gravity varies in accordance with the method of preparation between 1.016 and 1.022 and sometimes rises up to 1.030. Mustard oil is soluble in 160—300 p. of water, and in 10 p. of 70 p. c. alcohol. With 90 p. c. alcohol, with ether, amyl alcohol, benzene and petroleum ether it is miscible in all proportions. It boils principally between 148—156° (b. p. of pure allyl isothiocyanate is 150.7°). Exposed to light, mustard oil gradually becomes of a reddish-brown color. At the same time a film, consisting of a substance composed of carbon, nitrogen, hydrogen and sulphur, is deposited on the inner wall of the bottle.

EXAMINATION. The German Pharmacopœia contains a series of tests, that have been taken almost unchanged from the U. S. Pharmacopœia of 1890, which generally suffice for the characterization of a good mustard oil.

1. The sp. gr. is to be 1.018—1.029 according to the U. S. P., 1.016—1.022 according to the Ph. Ger. A slight variation either way, however, is not sufficient indication of adulteration, for the reasons mentioned under Preparation.

2. The requirement of both pharmacopœias that the boiling point is to lie between 148—150° cannot be construed literally inasmuch as allyl isothiocyanate boils at 150.7° (mercury in vapor). Furthermore, a slight decomposition invariably takes place during the distillation, so that some of the oil distills above 152°. Inasmuch as allyl cyanide is always present, the first fractions cannot have the same specific gravity as the original oil. Large differences in this respect will be accompanied by a lowering of the boiling point of the first fractions and should not be tolerated.

3. Sulphuric acid test. If to 3 g. of the oil 6 g. of sulphuric acid be gradually added, the liquid being kept cool, the mixture, upon subsequent agitation, will evolve sulphur dioxide, but will remain of a light yellow color, and at first perfectly clear, becoming afterwards thick

and occasionally crystalline, while the pungent odor of the oil will disappear. In the presence of petroleum, petroleum ether, chloroform or undue amounts of carbon disulphide, the mixture of mustard oil and sulphuric acid is rendered turbid and upon standing forms two layers. Other volatile oils would be rendered dark in color by the sulphuric acid.

4. Ferric chloride test. Diluted with 5 p. of alcohol, mustard oil is not changed by ferric chloride. Adulteration with oil of cloves would be recognized by the production of a bluish-green color.

5. The thiosinamine test according to the U. S. P. is to be made as follows:

"If a mixture of 3 g. of the oil and 3 g. of alcohol be shaken in a small flask with 6 g. of ammonia water, it will become clear after standing for some hours, or rapidly when warmed to 50° C. (122° F.) and usually deposit without becoming colored, crystals of thiosinamine [allyl-thio-urea, $CS.N_2H_3(C_3H_5)$].

"To determine the proportion of thiosinamine obtainable from the oil, decant the mother-water from the crystals, and evaporate it gradually in a tared capsule, on a water-bath, adding fresh portions only after the ammoniacal odor of each preceding portion has disappeared. Then add the crystals from the flask to those in the capsule, rinsing them out of the flask with a little alcohol, and heat the capsule on a waterbath until its weight remains constant. The amount of thiosinamine thus obtained from 3 g. of the oil should be not less than 3.25 g., nor more than 3.5 g. After cooling, thiosinamine forms a brownish, crystalline mass, fusing at 70° C. (158° F.) and having a leek-like, but no pungent, odor. The mass should be soluble in 2 parts of warm water, forming a solution which should not redden blue litmus paper, and which possesses a somewhat bitter, not persistent taste."

Attention should be called to the fact that, by following this test, carbon disulphide is included in this estimation since it also unites with ammonia according to the following equation:

$$\underset{\text{Carbon disulphide}}{CS_2} + \underset{\text{Ammonia}}{4NH_3} = \underset{\text{Ammonium sulphocyanate}}{NH_4SCN} + \underset{\text{Ammonium sulphide.}}{(NH_4)_2S}$$

Inasmuch as the products of the reaction remain behind upon evaporation, the residue may weigh more than 3.5 g. If this is the case and especially if the residue has the odor of ammonium sulphide an unwarranted amount of carbon disulphide was probably present.

Kremel[1] (1888) has suggested the use of ammonia of definite strength and to determine the excess with N/2 acid. As yet no practical test of this method seems to have been made.

According to Gadamer[2] (1899) mustard oil can be assayed titrimetrically according to a method that is also applicable to spirit of mustard.

1) Pharm. Post, 21, p. 828.

2) Archiv d. Pharm., 237, pp. 110, 372.

A solution of 2 p. of mustard oil in 98 p. of alcohol (*Spiritus sinapis* Ph. G.) is first prepared. 5 cc. of this spirit (4.2 g.) are transferred to a 50 cc. measuring flask, and 25 cc. of N/10 silver nitrate V. S. and 5 cc. ammonia are added. The flask is well stoppered and set aside for 24 hours. The liquid is then diluted to the 50 cc. mark and filtered. 25 cc. of the filtrate, after the addition of 4 cc. of nitric acid and a few drops of ferric chloride T. S., should not require more than 4.5 cc. and not less than 4.1 cc. of N/10 ammonium sulphocyanate to produce a permanent red color. These numbers—4.1 and 4.5—correspond to the requirements of the Ph. G., viz. a content of 92.6 to nearly 100 p. c. of allyl mustard oil.

Grützner[1] converts the allyl isothiocyanate into thiosinamine and oxidizes this with sodium peroxide. The resulting sulphuric acid is estimated as barium sulphate, either gravimetrically or volumetrically.

DETERMINATION OF CARBON DISULPHIDE IN MUSTARD OIL. For the detection of not too small amounts of carbon disulphide, i. e., in case of adulteration, it can be converted into copper ethyl xanthate[2] and determined quantitatively.

20—25 g. of mustard oil are heated on a water bath while a slow current of air is passed through the oil. The vapors of carbon disulphide are thus carried over, cooled by passing through a condenser and conducted into alcoholic potassa where they are converted into potassium ethyl xanthate. After the alkaline solution has been neutralized N/10 copper sulphate V. S. is added until a drop produces a reddish-brown color with potassium ferrocyanide, i. e., until a slight excess of copper sulphate has been added and all potassium ethyl xanthate has been converted into the cuprous salt. From the amount of the consumed copper solution (1 cc. corresponds to 0.0152 g. of carbon disulphide) the percentage of carbon disulphide present can be ascertained. This volumetric process can be supplemented by a gravimetric one. The precipitate of cuprous ethyl xanthate can be collected on a filter, washed, dried and heated to a red heat in a crucible, and the residue of cupric oxide weighed. 1 g. of the oxide corresponds to 1.918 g. of carbon disulphide.

For the quantitative estimation of traces of carbon disulphide which occur in every mustard oil, the method of Hofmann[3] can be used according to which it can be converted into a compound with triethyl phosphine, $P(C_2H_5)_3 + CS_2$, and weighed.

162. Oil of White Mustard.

Sinalbin is the glucoside from the white mustard (*Sinapis alba* L.) corresponding to the sinigrin from the black mustard. Robiquet and Boutron-Charlard[4] (1831) first extracted it with boiling alcohol from the seed deprived of its fatty oil. Its properties, including its hydrolysis

[1] Archiv. d. Pharm., 237, p. 185.
[2]) Zeitschr. f. anal. Chem., 21, p. 188.
[3]) Berichte, 13, p. 1732.
[4]) Journ. de Pharm., II, 17, p. 279.

by myrosin, were examined by Will and Laubenheimer[1] (1879) and recently by Gadamer[2] (1897). The identity of the white mustard oil with para hydroxy benzyl isothiocyanate was established by Salkowski[3] (1889).

Sinalbin mustard oil, $C_6H_4.OH^{[1]}.CH_2NCS^{[4]}$, is but sparingly volatile with water vapor and for this reason cannot be prepared from white mustard by distillation. It is an oily liquid of a burning taste, which draws blisters when in contact with the skin but much slower than allyl mustard oil. The pungent mustard oil odor is noticeable only when heated; when cold it possesses only a faint anise-like odor. It is soluble in dilute alkalies. It is prepared by the hydrolysis of sinalbin, dextrose and sinapine sulphate resulting as by-products.

$$\underset{\text{Sinalbin}}{C_{30}H_{42}N_2S_2O_{15}} + H_2O = \underset{\text{Sinalbin mustard oil}}{C_7H_7ONCS} + \underset{\text{Dextrose}}{C_6H_{12}O_6} + \underset{\text{Sinapine acid sulphate.}}{C_{16}H_{24}N_2O_5HSO_4}$$

Artificially it is prepared by the action of carbon disulphide on p-hydroxy benzylamine and by treating the resulting product with mercuric chloride.

163. Oil of Water-cress.

The oil of *Nasturtium officinale* L. was examined by Hofmann[4] in 1874, who extracted the aqueous distillate (600 k.) from 600 k. of fresh herb with petroleum ether. Upon evaporation of the ether he obtained 40 g. (=0.0066 p. c.) of oil. The oil no longer had the odor of water-cress, sp. gr. 1.0014 at 18° and distilled between 120 and 280°. The principal fraction came over at 261° and was shown to be the nitrile of phenyl propionic acid, $C_6H_5.CH_2.CH_2.CN$. Moreigne found raphanol[5] in the oil.

164. Oil of Radish.

The roots and seeds of *Raphanus sativus* L., (Family *Cruciferae*) yield upon distillation with water vapor a small amount of a colorless, sulphuraceous oil, heavier than water, which possesses the taste but not the odor of the radish.[6]

Upon the distillation of the roots of *Raphanus niger*, Moreigne[7] obtained besides a small amount of oil, 0.0025 p. c. of a substance which crystallized in laminae melting at 62°. It was called raphanol

---

1) Liebig's Annalen, 199, p. 150

2) Archiv d. Pharm., 235, p. 83.

3) Berichte, 22, p. 2143.

4) Berichte, 7, p. 520.

5) See under radish oil.

6) Pless, Liebig's Annalen, 58, p. 40. Comp. also Bertram & Walbaum, Journ. f. prakt. Chem., II, 50, p. 560.

7) Journ. de Pharm. et Chim., VI, 4, p. 10; Bull. Soc. chim., III, 15, p. 797.

or, because it possessed the properties of a lactone, raphanolid. It is free from nitrogen and sulphur, elementary analysis and molecular weight determination indicating the formula $C_{29}H_{58}O_4$.[1] When boiled with acetic acid anhydride, it yields an acetyl derivative melting at 122—123°. The liquid portion of the oil was found to contain sulphur but was free from nitrogen and did not combine with ammonia.

165. Oil from Mignonette Flowers.

Upon steam distillation of the fresh flowers of *Reseda odorata* L. (Family *Resedaceae*) an oil yield of but 0.002 p. c. is obtained. Remarkable is the strong development of carbon disulphide[2] during the process of distillation. The oil is of a dark color, and solid at ordinary temperature, having the consistency of orris oil. When greatly diluted it has the odor of fresh mignonette.[3]

In order to obtain the oil in a more suitable form for use it is distilled with geraniol (1 k. of geraniol to 500 k. of flowers), the product being brought into the market as "Reseda Geraniol."[4]

166. Oil from Mignonette Root.

The fresh roots yield upon distillation with water vapor 0.014—0.035 p. c. of oil. It is a light brownish liquid which has the odor of radish, sp. gr. 1.010—1.084, $\alpha_D = +1°30'$. It begins to boil at 255° with decomposition. Even under diminished pressure it does not boil without decomposition.

The oil was first examined by Vollrath[5] in 1871, who recognized its character as that of a mustard oil. His surmise that it was identical with allyl mustard oil was, however, shown to be erroneous by Bertram and Walbaum[6] (1894), who identified it as phenyl ethyl mustard oil, $C_6H_5CH_2.CH_2NCS$.

The odor of this substance is that of the mignonette root oil. Upon heating with ammonia, the well crystallizing thiourea, $NH_2.CS.$-$NHC_2H_4C_6H_5$ is formed, which melts at 137°. If this thiourea is treated with silver nitrate and baryta water, silver sulphide and phenyl ethyl urea, crystallizing in long needles and melting at 111—112°, are formed. If phenyl ethyl mustard oil is heated with concentrated hydrochloric acid

1) Raphanol has also been isolated by Moreigne from the following plants: the radish, turnip, water-cress, spoonwort and gilly-flower.

2) Bericht von S. & Co., Oct. 1891, p. 40.

3) Ibidem, Oct. 1898, p. 48.

4) Ibidem, Oct. 1894, p. 69.

5) Archiv d. Pharm., 198, p. 156.

6) Journ. f. prakt. Chemie, II, 50, p. 555.

in a sealed tube, crystalline laminae of phenyl ethylamine chlorhydrate melting at 217° are obtained. With oxalic acid ethyl ester, phenyl ethylamine combines to form diphenyl ethyl oxamide melting at 186°.

These derivatives were prepared by Bertram and Walbaum from the oil of the root as well as from the synthetic phenyl ethyl mustard oil[1] obtained by the action of phenyl ethylamine on carbon disulphide, and were found to be identical.

167. Oil of Storax.

Oleum Styracis. — Storaxöl. — Essence de Styrax.

Origin and History. The storax, *Liquidambar orientale* Miller (Family *Hamamelidaceae*) is a tree sometimes as high as 30 m. which resembles the plane and is indigenous to southern Asia Minor. The tree sheds its bark annually. In the tissue of the bark it secretes a fragrant balsam which hardens on exposure to the air. This balsam is obtained by boiling the bark with water and thus prepared constitutes the *Styrax liquidus* of commerce. This is a dirty greyish or greenish, resinous mass of peculiar aromatic odor and a pungent, spicy taste.

The *Styrax calamita*, the present solid storax of the market is storax balsam mixed with sawdust so as to render it almost dry. Storax is already mentioned among the aromatics by the ancients, e. g. by Herodotus, Theophrastus and Dioscorides. Mediaeval literature mentions storax varieties of different sources, some of which were used medicinally at times.

The volatile oil from the storax was distilled by Ryff, Gesner and by Porta who moistened the resin with alcohol. Upon distillation with water, storax yields 0.5 p. c. of volatile oil, when distilled with superheated steam about 1 p. c. of oil.

Properties. Oil of storax is a light yellow to dark brown liquid of a pleasant odor. The specific gravity varies from 0.89 to 1.1, according as the hydrocarbons or the cinnamic esters predominate. It is laevogyrate, $\alpha_D = -3$ to $-38°$. It boils between 150—300° with partial decomposition, cinnamic acid remaining behind.

Composition. The peculiar odor of the oil, reminding somewhat of petroleum, is due to the presence of styrene, $C_6H_5.CH:CH_2$, or phenylethylene (Simon,[2] 1839). This hydrocarbon boils at 146°, is optically inactive and can be identified by means of its dibromide, $C_6HC_5.CHBr.$-CH_2Br. (Comp. p. 103). According to van't Hoff[3] (1876), the optical

1) Berichte, 19, p. 1824.
2) Liebig's Annalen, 31, p. 265.
3) Berichte, 9, p. 5.

activity of the oil is referable to an oxygenated constituent, the styrocamphene, $C_{10}H_{16}O$ or $C_{10}H_{18}O$. The oil also contains more or less of the cinnamates of ethyl,[1] benzyl,[2] phenyl propyl[1] and cinnamic alcohols.[1]

168. Oil from American Storax.

Origin and History. *Ocosotl*, an aromatic balsam from Mexico and Central America is one of the drugs from the new world which attracted interest in Europe. In almost all of its properties it clearly resembled the *Styrax liquidus*, the storax coming from the levant and known since antiquity. Its origin, like that of the levant article and especially that of the other American balsams (Tolu, Peru, Copaiba etc.) was for a long time uncertain and created much confusion. The first description of American storax is found in the works of Monardes, Garcia ab Orto and Matthiolus, who lived during the first half of the sixteenth century.

Liquidambar styraciflaum L. (Family *Hamamelidaceae*) is a tree found in the middle and southern United States, in Mexico and Central America. The resinous balsam of the consistence of honey exudes from the splintwood beneath the bark of the trunk and branches. The excretion is either spontaneous or may be caused by incisions, but occurs abundantly only in hot climates. For this reason it enters the market principally from the Central American countries. According to the method of preparation, the balsam varies in appearance and consistency. In North America it is known as sweet gum and is used as a chewing gum and in the preparation of popular remedies.[3]

The first examination of American storax was made by Bonastre[4] in 1830 and 1831. Upon distillation of an apparently fresh balsam, he obtained 7 p. c. of oil. Further investigations were made by Proctor[5] in 1857 and by Harrison[6] in 1874.

Properties and Composition. According to v. Miller,[7] the oil from American storax differs from that of the levant by being dextrogyrate ($a_D = +16°33'$), and contains styrene (bromide, m. p. 73°) and an optically active substance having the odor of turpentine oil, which

1) Liebig's Annalen, 188, p. 184.
2) Liebig's Annalen, 164, p. 289.
3) Pharm. Rundschau, 13, p. 57.
4) Journ. de Pharm., II, 16, p. 88; II, 17, p. 338; Trommsdorff's Neues Journ. d. Pharm., 21, II, p. 242, and 24, II, p. 286.
5) Am. Journ. Pharm., 29, p. 261; 38, p. 33.—Proc. Am. Pharm. Assoc., 18, p. 160.
6) Am. Journ. Pharm., 46, p. 161; Arch. d. Pharm., 206, p. 541.
7) Archiv d. Pharm., 220, p. 648.

was not further examined. American storax contains the cinnamates of cinnamyl (styracin) and phenyl propyl alcohols, but not of ethyl and benzyl alcohols.

The leaves of the American storax have a peculiar terebinthinate odor. Upon distillation[1] they yielded 0.085 p. c. of a mobile oil, which was greenish-yellow; sp. gr. 0.872; $\alpha_D = -38° 45'$; saponification number 5.9; acetylization number 25.2. The odor of the oil reminds of the pine needle oils. It probably contains borneol and bornyl acetate in addition to terpenes.

169. Oil from Rasamala Wood.

Upon distillation of a wood coming from the Dutch East Indies and designated *Rasamala*, Schimmel & Co.[2] obtained 0.17 p. c. of a volatile oil. At ordinary temperature, this oil is a light brownish. crystalline mass, which melts between 30—40° and the odor of which reminds alike of cinnamon and rhubarb. The principal constituent of the oil is a crystalline substance which melts at 54—55°. It is probably a ketone, since it combines with hydroxylamine to a crystalline compound melting at 106—107°. The other constituents of the oil are liquid.

According to a communication by Dr. van Romburgh of Buitenzorg the term *Rasamala* is not only applied to the genuine but rare rasamala tree, *Altingia excelsa* Nor. (*Liquidambar altingia* Bl., Family *Hamamelidaceae*) but also to different Indian drugs, namely to the liquid storax from *Liq. orientale* (*Getah Rasamala*) as well as to other fragrant balsams and to the fragrant wood from *Canarium microcarpum* Willd. (*Kaju Rasamala*).

Whether the oil distilled by Schimmel & Co. was derived from any one of the above named trees is uncertain. Possibly it may have been the so-called aloe or eagle wood (Ger. *Aloeholz*) from *Aquillaria agallocha* Roxb. which is sold in the Indian markets as *Kaju lakka* and the characteristic rhubarb-like odor[3] of which corresponds with that of the wood distilled by Schimmel & Co.

170. Oil of Spiraea.

Oil from the flowers. In 1835 Pagenstecher[4] obtained upon distillation of the flowers of *Spiraea ulmaria* L. a small amount of oil

1) Bericht von S. & Co., Apr. 1898, p. 58.

2) Bericht von S. & Co., Apr. 1892, p. 48.

3) Bericht von S. & Co., Apr. 1892, p. 48. According to a recent investigation by Prof. Möller (Pharm. Post, 1898) aloewood is odorless and developes a peculiar odor only when burnt.

4) Repert. f. d. Pharm., 49, p. 337.

heavier than water which he gave to Loewig[1] for examination. Dumas[2] (1839), to whom the oil was shown, recognized its similarity with the salicylic aldehyde obtained from salicin by Piria shortly before. Ettling[3] (1839) verified Dumas' surmise by showing that salicylic aldehyde is one of the two or three volatile substances of which the oil is composed. Ettling had obtained a yield of 0.2 p. c. of oil upon distillation of the flowers. According to Wicke[4] (1852) the cultivated variety produces a larger yield than the wild plant. The oil is heavier than water and congeals completely at —18 to —20°.

The investigations of Schneegans and Gerock[5] (1892) have shown that besides salicylic aldehyde (formerly known as spiroyl hydride, spiraeic acid, salicyl hydride, spiroylic or spiric acid) the oil also contains methyl salicylate, also traces of heliotropin (piperonal) and vanillin. Ettling had also found a small amount of white, crystalline matter of a pearly lustre (paraffin?) and an oil of the composition C_5H_8 (terpene or sesquiterpene?). According to Schneegans and Gerock, the flowers do not contain salicylic aldehyde as such, but an unknown substance (not salicin, however, as supposed by Buchner[6]) which is decomposed during the process of distillation by a ferment. In addition to methyl salicylate, the flowers contain free salicylic acid.

Oil from the roots. The statement of Wicke, that the roots of *Spiraea ulmaria* L. contain salicylic aldehyde, is incorrect. According to Nietzki[7] (1875) the oil consists principally of methyl salicylate with traces of another substance, probably a hydrocarbon.

Oil from the herb. In the distillate from the herb, Wicke has shown the presence of salicylic aldehyde.

According to Wicke the herb of *S. digitata*, *S. lobata*, *S. filipendula* and the flowers of *S. aruncus* yield salicylic aldehyde upon distillation. The herb of *S. aruncus*, the leaves of *S. japonica*, the herb and flowers of *S. sorbifolia* yield hydrocyanic acid but no salicylic aldehyde. Neither aldehyde nor acid were found in the oil from *S. laevigata*, *S. acutifolia*, *S. ulmifolia*, and *S. opulifolia*.

---

1) Poggend. Ann., 36, p. 383; Pharm. Centralbl., 1839, p. 129.
2) Liebig's Annalen, 29, p. 306.
3) Liebig's Annalen, 29, p. 309; 35, pp. 1, 24.
4) Liebig's Annalen, 83, p. 175.
5) Journ. d. Pharm. f. Elsass-Lothr., 19, pp. 3 and 55. Abstr. Jahresb. f. Pharm., 1892, p. 164.
6) Liebig's Annalen, 88, p. 284.
7) Archiv d. Pharm., 204, p. 429.

171. Oil (Otto) of Rose.

Oleum Rosarum. — Rosenöl. — Essence de Rose.

History. Since the earliest periods the charm and fragrance of the rose has led to its appreciation and use. This is shown by the entire older literature, and of all the flower perfumes that from the rose has always received preference. In Chinese and Sanskrit writings the fragrance of the rose is much praised. Fats and oils saturated with the rose perfume have been used since earliest antiquity in perfumery. Thus Aphrodite anointed the dead body of Hector with rose oil. The Greeks and Romans celebrated annually a rose festival, at which the graves of the dead were decorated with roses and their tombstones were anointed with rose oil. Of the various flower cults that of the roses has been the most eminent since antiquity.

The earliest description of the method of preparation of the oil of rose of the ancients is found in the writings of Dioscorides. It was an aromatised fatty oil as were the majority of the rose oils of the middle ages, such as *Oleum rosarum*, *O. rosatum* or *O. rosaceum*, etc.

Aside from apocryphal Persian and oriental traditions, the earliest definite directions for the preparation of roses and the use of the distillate is found in the writings of the Arabian historian Ibn Chaldun. He mentions that during the eighth and ninth centuries rose water was an important article of commerce, being carried as far as China and India. In a codex of ceremonies of 946 by the East Roman emperor Constantine VII., Persian rosewater is mentioned as a toilet water. At the beginning of the tenth century, Nonus Theophanes, the physician of emperor Michael VIII., recommended and used rose water as a medicament. Avenzoar, the physician of the calif Ebn Attafir of Morocco, who lived at the beginning of the twelfth century; also his contemporary, Joannes Actuarius, a physician of Constantinople, used rose water as an ophthalmic, and rose sugar as an internal remedy.

During this period, Persia seems to have supplied most of the rose water. During the fourteenth century it was also exported from Mesopotamia. After the prime of the levant commerce, the Portuguese and Dutch were the principal carriers of goods between Aden, the ports of the Persian bay, India and the occident. Rose water constituted one of their principal articles of merchandise. During the tenth century the distillation of roses was introduced into Spain by the Arabs.

Throughout the middle ages, the distillation of rose water seems to have been an important industry of Persia. Unless strongly alcoholic wine was used in the process, one would expect that in the distillation of large amounts of rose water the separation of oil of rose at low temperatures in the form of a butyraceous mass had been noticed at an early date and probably used to perfume fats and fatty oils.

The first statements concerning rose oil, which possibly refer to the distilled oil, is found in the writings of Mesuës, and in the almanac of Harib for the year 961, which mentions the time suitable for the preparation of rose water and a rose preserve. In his Compendium aromatariorum written about the middle of the fifteenth century, Saladin of Asculi, the body physician of a prince of Tarentum, describes the distillation of roses for the preparation of rose water and rose oil.

According to a statement by Langles, distilled rose oil is twice definitely mentioned in Mohammed Achem's history of the great moguls of 1525 to 1667; also in the annals of the Mongolian empire written by Manucci, a Venetian physician who lived 40 years in India. Unquestioned mention of the butyraceous oil of rose is made in 1574 by Hieronymus Rubeus, body physician of Pope Clemens III.; also in the writings of Porta of the year 1563 and again in 1604.

In the apothecary tax-ordinances of Worms of 1582 and of Frankfurt-on-the-Main of 1587, *Oleum rosarum verum* is mentioned in the list of distilled oils. About the same time Angelus Sala describes the distillation of rose oil and designates it strikingly as *candiscente pinguidine, instar spermatis ceti.* In his Pharmacopœia of 1641, Schroeder enumerates the oil under the *Olea destillata usitatoria.*

Up to the seventeenth century and beyond, however, Persia principally seems to have supplied the market with rose water and rose oil. In the course of the century, however, the cultivation of the rose and the oil industry spread to India, Arabia, Tunis, Algiers, and Morrocco to the south, also to Asia Minor, Turkey and Bulgaria to the north. On the island of Chios also considerable rose oil was distilled at the beginning of this century, which entered commerce via Smyrna.

The cultivation of roses in Bulgaria, which became of such importance in later years, was begun about the beginning of the seventeenth century. It seems to about coincide with the founding of Kezanlyk, a city on the southern slope of the Balkan mountains in East Roumelia. It was not until the nineteenth century, however, that the rose industry of Bulgaria became a dangerous competitor of the Persian rose distillation. In recent years Bulgaria, in turn, has found successful competitors in Germany and France.

Since the fourteenth century, rose water and with it small quantities of rose oil, have been distilled for popular and medicinal use, also for perfumery, in the north European countries, especially in France, Germany and England. The amount of oil, however, was so small that rose oil was mostly bouhgt from the orient and later from the Balkan states. The cultivation of roses for the purpose of distilling rose oil on a large scale was begun in France about the middle of this century, in Germany in 1883.[1]

The high price of rose oil and the ease with which it can be adulterated seems to have brought about adulteration in Persia in the course of the seventeenth century. Engelbert Kaempfer from Lemgo, who traveled in Persia in the years 1682 to 1684, mentions that rasped sandalwood is added to the roses in the process of distillation. This observation was verified in 1787 by Archibald Keir in Chatra in the Ramgur, whereas Polier observed in Cashmere during the same year that in this country not sandalwood, but the fragrant Indian grass (Andropogon) is added to the roses for distillation.

Aside from its use during antiquity, the use of *Andropogon schoenanthus* L. for the purpose of adulteration of rose distillates, dates back more than a century. As a more convenient adulterant, palmarosa oil is more recently used in place of the grass from which it is distilled in India.

At an early period oil of rose was used as a perfume and, filled in fancy flasks, became a much sought for article in the bazaars of Constantinople, Smyrna, the levant and the entire orient. The demand being greater than the supply, both, manufacturers as well as dealers, early learned to increase the supply in a manner profitable to themselves. The former added palmarosa oil to the roses in the process of distillation, the latter still further diluted it with indifferent oils and spermaceti, the latter being necessary to maintain the proper congealing point.

ORIGIN. Only a few of the 7,000 cultivated varieties of roses are used in the production of the oil. For the purpose of distillation, handsome appearance is of less importance than hardiness and a rich yield of flowers. Both these qualities are possessed by the *Rosa damascena* Miller, which is cultivated in the Balkan provinces and in recent years also in Germany for the production of rose oil. This variety does not occur wild, but is a product of cultivation and originally may have been a hybrid between *R. gallica* and *R. canina*.

1) Pharm. Rundschau, 12, p. 92.

The Bulgarian rose-plant is armed with numerous not strongly recurved prickles. The glabrous, obtuse three paired lateral leaflets and the terminal leaflet are pure bright green, the calyx smooth and slightly primrose. In many cases the ordinarily racemose clusters—of as many as twenty-seven flowers—are almost cymes. In the fully expanded condition the roses attain a maximum width of 7 cm. and, though double, they are nevertheless provided with numerous stamens with large yellow anthers. In many flowers the outer petals are almost white, becoming redder and redder toward the center and under the most favorable conditions chiefly pure rose-red.

Fig. 74.
Bulgarian Rose Oil Distilling Apparatus.

In Bulgaria and in Germany the roses are grown in fairly dense hedges, about the height of a man. In Bulgaria a white rose, *R. alba* L. is planted to indicate the divisions of the rose fields. It is said to yield an oil of poorer quality which is richer in stearoptene.

In southern France *R. centifolia* L. is principally cultivated for the production of rose water and rose pomade. It is planted in rows but forms less dense and lower hedges than the Bulgarian. It is uncertain which species of rose was cultivated in the famous rose gardens of Schiras. Possibly it was *R. gallica*, the dried petals of which are even to-day exported from Persia in large quantities.

In India where rose oil distilleries have existed for two centuries in Gazipour on the Ganges and in other places of Bengal, *R. damascena* is likewise used. The oil distilled there, however, is never pure, but

Fig. 75.
Rose Fields at Miltitz near Leipzig.

always contains sandalwood oil. As already stated, the roses are generally distilled with sandalwood.

Production of Bulgarian Rose Oil. The rose stills used in Bulgaria are very simple, and are to-day much the same as the ones described by Baur[1] 30 years ago.

The copper still (*Lambic*) of 110 liters capacity rests on a fire-place built of stone (fig. 74) which is heated by wood from the near forests of the Balkan mountains. The conical still is 1.1 m. high and provided with handles with which it can be removed from the fire-place. The middle diameter of the still is 0.8 m., at the neck it is 0.25 m. The helmet which is 0.3 m. high and which has the form of a toad-stool fits closely to the neck of the body of the still. The joints are made tight with clay and strips of cloth. The helmet is provided with an exit tube which is inclined to the ground at an angle of 45° and which connects with the condenser tube proper. The latter is straight, about as thick as a thumb, 2.5 m. long and passes through a tub of oak or beech wood filled with water. The cooling water is conveyed by means of a wooden gutter.

A number of these stills are usually mounted under one shed. Each apparatus is charged with 10 k. of freshly picked roses and 75 l. of water. The distillation is continued until two five liter flasks are filled with aqueous distillate. The water remaining in the still is used for the next charge, a proceeding that is irrational inasmuch as the salts and extractive matter which accumulate cannot be without influence on the delicate perfume.

When a sufficient amount of rose water has accumulated, 40 l. are transferred to a still and 5 l. are distilled off. This second distillate is at first a white, turbid liquid, which becomes clear upon standing, the oily constituents separating at the surface. For the separation of the oil, a small funnel-shaped instrument of tin is employed in Bulgaria. The tube has but a very fine opening through which the water will pass but not the semi-congealed oil. According to unreliable and improbable statements 3,000 k. of flowers yield 1 k. of oil in Bulgaria. In all probability a much larger quantity of roses is necessary, though 3,000 k. of flowers with a sufficient amount of palmarosa oil may yield 1 k. of Bulgarian rose oil.

After the oil (Bulgarian *Güljag*) has received another addition of palmarosa oil, which for this purpose is exposed to the sun in shallow dishes, by the broker, it is transferred to the well known tinned copper flasks (*Estagnons*) and is brought into the market as Bulgarian rose oil. The designation Turkish is no longer correct, for in Turkey no rose oil is distilled.

Production of German Rose Oil. The first attempts by Schimmel & Co. to distill rose oil on a large scale were made in 1883. At first *Rosa centifolia* L. from the neighborhood of Leipzig was used.

---

1) Neues Jahrbuch für Pharmacie und verwandte Fächer, 27, p. 1. Abstr. Jahresb. f. Pharm., 1867, p. 350.

In 1888 the firm obtained a considerable number of rose bushes from Bulgaria which, by skillful treatment were rapidly multiplied. At the present time 35 hectares (=86.4 acres) are being cultivated with this rose variety near Miltitz, a station 12 km. from Leipzig on the Thuringian R. R. The detrimental influence of the transportation on the freshly picked roses, made their distillation on the spot a necessity. Now there stands in the midst of the rose plantation a large factory building equipped with the best modern apparatus. The freshly picked roses are at once taken to the large copper stills, which have a capacity for 1,500 k. of roses in addition to the requisite amount of water (fig. 50, p. 81). From 5,000 to 6,000 k. yield 1 k. of oil.

It goes without saying that here the crudities of the Bulgarian process are not tolerated. The stills are not heated with direct fire but with steam. For every new charge of roses fresh water is used and the oil is collected in a cascade-like series of Florentine flasks like all other oils. Owing to the greater care exercised, the odor of the German oil is by far superior to that of the Bulgarian oil. Although the stearoptene content is much higher, the intensity of its odor is again as great as that of the latter. A product, similar in intensity of odor and stearoptene content to the Bulgarian rose oil, is the rose geraniol of Schimmel & Co. which is made by distilling 2,500 k. of fresh roses with 1 k. of pure geraniol.

PROPERTIES. The commercial Bulgarian rose oil is light yellow in color, and sometimes has a greenish tint. At 21—25° it is of the consistency of sweet almond oil, has a strong odor of fresh roses, and a pungent, balsamic taste. At about 18—21° acicular crystals or shining crystalline laminae separate out, which, on account of their lighter sp. gr. collect in the upper portion of the oil and coat the surface with a thin film that readily parts when the oil is disturbed. When cooled the oil congeals to a translucent, soft mass, which is again liquified by the warmth of the hand.

The sp. gr. varies as a rule between 0.855—0.870 at 20°, being lowered by a higher stearoptene content. The oil is slightly laevogyrate, a_D = up to $-4°$.[1] On account of the difficultly soluble paraffins which rose oil contains, it yields only turbid mixtures with even very large amounts of 90 p. c. alcohol. The liquid portion, the so-called oleoptene forms a clear solution with 70 p. c. alcohol. The

1) Baur, whose statements, however, are not always reliable (comp. p. 430), found $a_D = +4°$.

saponification number is 10—17. Rose oil has a slightly acid reaction, its acid number being 0.5—3. For this reason the test of the German Pharmacopœia, that the chloroform-alcoholic solution should not redden blue litmus paper, is untenable. The temperature at which the rose oil begins to congeal, the congealing point, lies between 15 and 22°, mostly between 17 and 21°. The stearoptene content varies from 10—15 p. c.

On account of its larger stearoptene content, German rose oil from *Rosa damascena* Mill. is at ordinary temperature a greenish mass intermingled with crystals. The odor is much stronger and more persistent than that of the Bulgarian oil. The congealing point lies between 27 and 37°, the sp. gr. between 0.845—0.855 at 30°; $\alpha_D = +1$ to $-1°$. The stearoptene content varies between 26—34 p. c.

An oil distilled in Leipzig from *Rosa centifolia* L. had the following properties: sp. gr. 0.8727 at 25°; $\alpha_D = +0° 49'$; congealing point +28°; saponification number 7.8.

According to Dupont and Guerlain[1], two French oils distilled in 1895 and 1896 had the following properties: sp. gr. 0.8225—0.8407 at 30°; α_D at 30° = − 6° 45′ to − 8° 3′; stearoptene content 35 and 26 p. c.

COMPOSITION. Shortly after the introduction of elementary analysis the first combustion of rose oil and its stearoptene was made by Saussure[2] in 1820. A second analysis was made by Blanchet[3] in 1833. The analysis of the oil merely indicated that it contained oxygen, that of the stearoptene that it was a hydrocarbon. Blanchet supposed the latter to be a terpene, $C_{10}H_{16}$, whereas Flückiger[4] in 1869 first recognized in it a representative of the paraffin series. The vapor density determination made by Power[5] in Flückiger's laboratory indicated the formula $C_{16}H_{34}$. Baur,[6] who has made numerous observations on the distillation of rose oil in Bulgaria and who made detailed reports in 1867 and 1872, claimed that the odorless stearoptene could be converted into fragrant oleoptene by oxidation, and also that the reverse reaction could be made to take place. Later investigations, however, have not confirmed these statements. About the same time (1872) Gladstone[7] examined the liquid portion of the oil and found its boiling

1) Compt. rend., 123, p. 750.
2) Ann. de Chim. et Phys., II, 13, p. 337.
3) Liebig's Annalen, 7, p. 154.
4) Pharm. Journ., II, 10, p. 147.
5) Pharmakognosie, 3rd ed., p. 170.
6) Neues Jahrb. f. Pharm., 27, p. 1; 28, p. 193.
7) Journ. Chem. Soc., 25, p. 12.

point to be 216°. The first detailed investigation was made by Eckart[1] in 1890, who showed that the principal constituent of both Bulgarian and German rose oil was an alcohol $C_{10}H_{18}O$, to which he gave the name rhodinol. Although he ascertained the great similarity between the new alcohol and geraniol, he declared the two to be distinct and suggested two formulas indicative of their difference. Like Gladstone, he found the boiling point of the rhodinol to be 216° (instead of 229°) and it is, no doubt, due to this reason that Eckart did not recognize the geraniol as such.

Shortly afterward (in 1893) Markownikoff and Reformatzky[2] claimed that the principal constituent of Bulgarian rose oil, to which they gave the name roseol, had the formula $C_{10}H_{20}O$. At the same time, however, Barbier[3] arrived at the conclusion that Eckart's formula $C_{10}H_{18}O$ was correct. This was further confirmed by Tiemann and Semmler[4] who recognized the identity of Eckart's rhodinal and citral, the aldehyde, $C_{10}H_{16}O$, corresponding to geraniol. These differences led Bertram and Gildemeister[5] to take up the investigation of rose oil. They ascertained that the principal constituent of German as well as Bulgarian rose oil is geraniol, the alcohol $C_{10}H_{18}O$ (b. p. 229—230°) discovered by Jacobsen in 1870 in palmarosa oil, and that "rhodinol" is impure geraniol.[6]

However, geraniol is not the only alcohol contained in rose oil. Hesse[7] (1894) expressed the surmise that the alcohol "reuniol," $C_{10}H_{20}O$, found by him in the pelargonium oils, might also be contained in rose oil. Tiemann and Schmidt[8] demonstrated the identity of "reuniol" with citronellol, the reduction product of citronellal.

---

1) Archiv d. Pharm., 229, p. 355; also Berichte, 24, p. 4205; Poleck, Berichte, 23, p. 3554.

2) Journ. f. prakt. Chem., II. 48, p. 293.

3) Compt. rend., 117, p. 177.

4) Berichte, 26, p. 2708.

5) Journ. f. prakt. Chem., II. 49, p. 185.

6) As to whether this alcohol is to be designated "geraniol," the name assigned to it by its discoverer, or "rhodinol," the name suggested by Eckart for the impure compound, has given rise to a lively controversy. Without entering upon this discussion, a mere reference to the literature will here be made: H. Erdmann & Huth: "Zur Kenntniss des Rhodinols und Geraniols." Journ. f. prakt. Chem., II, 53, p. 42; Bertram & Gildemeister: "Ueber Rhodinol und Geraniol." Journ. f. prakt. Chem., II, 53, p. 225; A. Hesse: "Ueber die vermeintliche Identität von Rhodinol und Geraniol." Journ. f. prakt. Chem., II, 53, p. 238; H. Erdmann: "Untersuchungen über die Bestandtheile des Rosenöls und verwandter ätherischer Oele." Journ. f. prakt. Chem., II, 56, p. 1; Bertram & Gildemeister: "Die Bestandtheile des Rosenöls und verwandter ätherischer Oele." Journ. f. prakt. Chem., II, 56, p. 506; Th. Poleck: "Zur Rhodinolfrage." Journ. f. prakt. Chem., II, 56, p. 515; Berichte, 31, p. 29; Bertram & Gildemeister: "Zur Rhodinolfrage." Berichte, 31, p. 749.

7) Journ. f. prakt. Chem., II, 50, p. 472.

8) Berichte, 29, p. 922.

As already stated, the greater portion of rose oil consists of geraniol. If the stearoptene is removed by dissolving the oil in dilute alcohol or by vacuum distillation, there remains an oil which distills under ordinary pressure between 228—232°. When treated with carefully dried and very finely powdered calcium chloride, this oil solidifies to a solid mass of geraniol calcium chloride. This is freed from oil by washing with ether, petroleum ether or benzene, and then decomposed with water, yielding chemically pure geraniol.[1] For the identification of geraniol the well crystallizing diphenyl urethane derivative (m. p. 83—84°) discovered by Erdmann and Huth[2] can be employed. Properties and derivatives of geraniol are described on p. 132.

The second alcohol, the l-citronellol, is quantitatively of less importance. Tiemann and Schmidt[3] estimate the amount in Turkish rose oil to be about 20 p. c. For the identification and isolation of citronellol phosphorus trichloride is allowed to act on a well cooled ethereal solution of the alcohols of rose oil. The geraniol is hereby converted partly into geranyl chloride, partly into hydrocarbons, whereas the citronellol is converted into a citronellyl chlorophosphorous acid, which can be removed from the ethereal solution by shaking with caustic soda solution. The aqueous solution of this citronellyl acid ester is shaken out with ether, saponified with strong caustic soda solution, and the free citronellol distilled with water vapor.

The two alcohols, geraniol and citronellol, occur in rose oil principally in the free state and only in small part as ester. The oleoptene of Bulgarian rose oil contains about 90 p. c. of alcohols (calculated as $C_{10}H_{18}O$). Normal Bulgarian oil, including the stearoptene, contains on an average 2.5—3.5 p. c. of ester (calculated as geranyl acetate); a sample of German oil contained 3 p. c.

According to Dupont and Guerlain[4] (1896) the esters deviate polarized light more than the underlying alcohols. The liquid portion of a French oil which deviated the ray of polarized light 10° 30′ to the left, after saponification had an angle of only —7° 55′. The acids which are probably partly combined with geraniol, partly with citronellol, have not yet been investigated. Dupont and Guerlain are of the opinion that the esters play an important role in the production of the rose perfume.

---

1) Bertram & Gildemeister, loc. cit.

2) Journ. f. prakt. Chem., II, 53, p. 45; also Tiemann & Schmidt, Berichte, 29, p. 920; further H. Erdmann, Journ. f. prakt. Chem., II, 56, p. 6.

3) Loc. cit.

4) Compt. rend., 123, p. 750.

As shown by Flückiger[1] the stearoptene of oil of rose belongs to the series of paraffin hydrocarbons. It is not, however, a single hydrocarbon, but consists of at least two, possibly of a number of homologous paraffins. This is shown by the fact that on proper treatment of a large amount of stearoptene, fractions melting at 22° and 40—41° respectively are obtained.[2]

The ethyl alcohol found by Eckart results, according to Schimmel & Co.,[3] only when the roses have become heated on their way from the fields to the stills and have thus undergone fermentation.

Inasmuch as neither geraniol and citronellol, nor their esters either alone or mixed possess the characteristic honey-like odor of the rose oil, it must be assumed that other substances are present in very small amounts which assist in producing the fine aroma of the rose. The different odors of the various varieties would seem to indicate chemical differences.

Examination. Since the danger of adulteration of as expensive an oil as rose oil is very great, a number of empirical tests have been suggested by means of which it was supposed the presence of foreign additions, especially of palmarosa oil (so-called Turkish geranium oil) could be detected. Though these tests were never received with much confidence, their value became even more problematic when it was shown that geraniol constituted the principal constituent of rose oil as well as of palmarosa oil. It is true that some characteristic constituent of the palmarosa oil not contained in rose oil might produce a particular reaction with the proposed reagents and thus betray the presence of the adulterant. A recent investigation, however, of palmarosa oil by Gildemeister and Stephan[4] has not been productive of any result in this direction. In addition, it should be remembered that the Bulgarians subject the palmarosa oil to special treatment before they use it as adulterant. Thus e. g. by shaking the oil with lemon juice and by exposing it to the sun, they try to make the palmarosa oil resemble the rose oil in odor and other properties as far as possible.

Besides palmarosa oil, the true geranium oil is used as an adulterant of rose oil. If the adulteration is at all skillful not only do the numerous suggested color reactions with iodine, sulphuric acid, fuchsin sul-

---

1) See footnote 4, p. 430.
2) Bericht von S. & Co., Oct. 1890, p. 42.—Eckart, Dupont & Guerlain, loc. cit.
3) Bericht von S. & Co., Oct. 1892, p. 36.
4) Archiv d. Pharm., 234, p. 321.

phurous acid,[1] etc. fail, but even the rational physical and chemical examination proves of no avail. This is caused partly by the great similarity of rose oil and its adulterants, partly by the fact that rose oil varies especially in paraffin content according to climate, atmospheric conditions, soil and method of distillation to such an extent that a comparison of physical properties usually proves of no avail in the attempt to detect adulteration.

For this reason there is no positive foundation for the claim that the commercial Bulgarian rose oil is the pure distillate from the roses. Several indications render it improbable. First of all, the large imports of palmarosa oil into Bulgaria create suspicion. Further, the enormous differences between the Bulgarian and German distillates is very striking and not to be explained by mere reference to climatic differences. Rather startling is also the fact that Bulgarian manufacturers have repeatedly exhibited as especially fine products, oils that agreed closely with the German distillate in odor, congealing point and stearoptene content.

On the other hand the oil which is obtained by distilling 2,500 k. of roses with 1 k. of geraniol,[2] cannot be distinguished from the Bulgarian oil of commerce. Intensity of odor, congealing point and stearoptene content are alike.

In order to guard against gross adulteration, specific gravity, optical rotation, congealing point, stearoptene content, saponification number, and eventually the amount of alcoholic constituents should be determined. If the data thus obtained correspond with those of a good average oil, and if the odor is delicate and rich, it may be pronounced unsuspicious. No chemist, however, can guarantee an oil on the strength of its physical and chemical properties.

Specific gravity. Inasmuch as the oil is partly solidified at 15°, the determination must be carried out at 20, 25 or 30°. Palmarosa oil has but little effect on the density; alcohol, which has been observed occasionally, lowers it; sandalwood oil, which is added during the distillation in India, increases it.

Optical rotation. The angle of rotation is scarcely influenced by palmarosa oil. It is remarkable that the rotation of the French oil is much greater than that of the German or Bulgarian.

---

1) Panajotow, Berichte, 24, p. 2700; Bericht von S. & Co., April 1892, p. 32. Comp. also Jedermann, Zeitschr. f. analyt. Chemie, 36, p. 96; Bericht von S. & Co., Apr. 1897, p. 39.

2) Bericht von S. & Co., Oct. 1896, p. 66.

Congealing point. With reference to rose oil, it is that point at which the first crystals separate when the oil is slowly cooled. Raikow,[1] who considers the congealing point as the point of supersaturation of the oleoptene with stearoptene, determines it in the following manner:

About 10 cc. of rose oil are introduced into a test tube of about 15 mm. diameter. A thermometer is so inserted that it touches neither the bottom nor the sides of the tube, but floats freely. The oil in the tube is warmed 4—5° above the saturation point by means of the hand and well shaken. The tube is then securely supported and the oil allowed to cool spontaneously until the first crystals appear. At this point the temperature is read, and the operation repeated.

The congealing point of good commercial Bulgarian oil lies as a rule between 18 and 21°, but deviations upward as well as downward occur. Formerly rose oil was valued according to its congealing point alone. Although the odorless paraffin was valueless the oil demanded a higher price according to its higher congealing point. It was originally and correctly considered that the addition of palmarosa oil would lower the congealing point. Later, when the congealing point was artificially raised by the addition of spermaceti, the test lost its significance and value. It has already been stated that the genuine, normal Bulgarian oil of rose probably contains much more stearoptene than the commercial oils from that source.

Stearoptene assay.[2]

50 g. of oil are heated to 70—80° with 500 g. of 75 p. c. alcohol. When cooled to 0° the stearoptene separates almost quantitatively. It is separated by filtration, again treated with 200 g. of 75 p. c. alcohol and the operation repeated until it has become perfectly odorless. As a rule two such treatments of the crude stearoptene suffice.

Test for spermaceti in the stearoptene.

3—5 g. of the stearoptene are boiled for a short time with 20—25 g. of a 5 p. c. alcoholic solution of potassa. The alcohol is evaporated and the residue treated with hot water. Upon cooling most of the stearoptene separates as a solid crystalline mass on the surface. The alkaline solution is removed, the stearoptene melted with some hot water, again allowed to cool, the water again removed, the operation being repeated until the wash water is neutral. The united aqueous liquids are twice extracted with ether in order to remove any suspended stearoptene. The alkaline aqueous solution is then acidified with dilute sulphuric acid and the acid solution extracted again with ether. Upon evaporation, the ether should leave no residue (fatty acids). By way of control, the recovered stearoptene, dried at 90°, is weighed. Allowance should be made for a slight loss resulting from evaporation while drying.

---

1) Chemiker Zeitung, 22, p. 149.

2) Bericht von S. & Co., Apr. 1889, p. 37.

The spermaceti content can be determined in a simpler manner by saponifying the separated stearoptene with alcoholic potassa of known strength and titrating back with N/2 sulphuric acid. The saponification number of spermaceti is 108.

Saponification. The saponification number of a good commercial oil varies from 10—17, the acid number from 0.8—2.7. Palmarosa oil according to Gildemeister and Stephan[1] has a saponification number of 30—50, the genuine geranium oils 45—100. Additions of the latter may, therefore, become noticeable.

Acetylization. In order to make the examination of the oil complete, it will often be advisable to ascertain the alcohol content (geraniol and citronellol) by acetylization. The amount present will be inversely proportional to the percentage of stearoptene present. Umney[2] found 70—72.5 p. c. (calculated as geraniol) in good rose oil. Palmarosa oil contains 76—92 p. c. of geraniol.[1]

As latest adulterant, guaiac wood oil from *Bulnesia sarmienti*, which has an agreeable tea-rose-like odor, has been employed in Bulgaria.[3] It can be recognized by the microscopic examination of the form of the crystals of guaiol which separate from the oil upon cooling. Guaiol forms needle-shaped crystals which are characterized by a channel-like middle line. The crystals of the rose oil paraffin are smaller and thinner and possess less sharply outlined forms (Dietze[4]).

The positive presence of guaiac wood oil in rose oil should be established by the isolation of guaiol melting at 91°. The new adulterant increases the specific gravity and the optical rotation, and raises the congealing point of the oil; it lowers the saponification number but very little, and upon evaporation leaves a resinous residue.[4]

172. Oil of Bitter Almond.

Oleum Amygdalarum Amararum.—Bittermandelöl.—Essence d'Amandes Amères.

Origin and History. *Prunus amygdalus* Stokes (*Amygdalus communis* L.) which belongs to the family of the *Rosaceae* is cultivated in Europe, Asia and northern Africa, recently also in California. In the course of time several cultivated varieties have been formed, which are distinguished by larger or smaller fruits and seeds. The trees which produce the bitter almond do not reveal any permanent botanical

1) Archiv d. Pharm., 234, p. 326.
2) Chemist and Druggist, 49, p. 795.
3) Bericht von S. & Co., Oct. 1898, p. 43.
4) Süddeutsche Apoth. Zeitung, 38, pp. 672 and 680.

differences from those that produce the sweet almond. It is possible that the wild almond tree originally produced none but bitter almonds which gradually became sweet upon cultivation. Both varieties have been known since antiquity.

Bitter almond oil is first mentioned in the writings of Saladin of 1488 and those of Sancto Amando in the sixteenth century. The distillates from bitter almonds and of other *Prunoideae*, seem to have received but little attention during the period of general use of distilled waters. It is also uncertain whether the poisonous character of bitter almond oil was generally known. Even Scheele, when he discovered hydrocyanic acid in 1782 does not seem to have fully realized its poisonous properties. Murray appears to have been the first in 1784 to have emphasized its very poisonous character.

The presence of hydrocyanic acid in bitter almond oil was first surmised by Remler, an apothecary in Erfurt, in 1785. Its presence was, however, first established by Bohm, an apothecary in Berlin, in 1803. Bitter almond oil and its hydrocyanic acid content were investigated by Schaub, Schrader, Ittner, Gay-Lussac[1] (1831), Robiquet und Vogel,[2] Boutron-Charlard[3] (1837), Liebig and Wöhler[4] (1837), Winckler[5] (1839), and others.

The separation of hydrocyanic acid from the benzaldehyde was first accomplished by Vogel in 1822 by shaking the oil with baryta water.[2] Liebig and Wöhler[6] (1837) shook the oil with ferric sulphate or chloride and milk of lime, a method that is still in use, and thus first prepared pure benzaldehyde. Bertagnini in 1853 suggested the use of sodium bisulphite.[7] The separation of benzoic acid from bitter almond oil upon exposure to air was observed by Stange in 1823.[8]

Preparation. Only a very small amount of the bitter almond oil of commerce is prepared from bitter almonds. For the manufacture of the oil the seeds of the apricot, *Prunus armeniaca* L., serve almost exclusively and the oil thus obtained does not appear to differ in any respect from the oil obtained from bitter almonds. The seeds of the apricot are brought from Asia Minor, the home of the tree, into

1) Poggendorff's Annalen der Physik. New series. 28, pp. 1 and 138.
2) Journ. de Pharm.. II, 8, p. 298; Ann. de Chim. et Phys., 15, p. 29; 21, p. 250.
3) Ann. de Chim. et Phys., 44, p. 352.—Liebig's Annalen, 25, p. 175.
4) Liebig's Annalen, 22, p. 1.
5) Repert. f. d. Pharm., I, 17. p. 156.—Pharm. Centralbl., 1839. p. 684.
6) Liebig's Annalen, 3, p. 252.
7) Ibidem, 85, p. 183.
8) Repert. f. d. Pharm.. I, 14, pp. 329, 361; 16, p. 80.

European commerce as "peach kernels." The seed of the peach, *Prunus persica* Jess. likewise yields an oil that may be regarded as the equivalent of bitter almond oil.

Previous to the distillation of the volatile oil, the seeds must be deprived of their fatty oil. They are ground to a coarse powder and the oil is expressed by means of hydraulic presses, a pressure of 350 atmospheres being applied. Upon cold expression, bitter almonds yield about 50 p. c., apricot seeds about 25—38 p. c. of fatty oil.

The press-cakes are finely ground and are then ready for the manufacture of the volatile oil. This is not present as such in the seeds but is formed by a process of fermentation similar to that producing mustard oil and oil of wintergreen. In the presence of water the glucoside amygdalin which is contained in these seeds is decomposed by a ferment known as emulsin into benzaldehyde, hydrocyanic acid and dextrose according to the following equation:

$$\underset{\text{Amygdalin}}{C_{20}H_{27}NO_{11}} + \underset{\text{Water}}{2\,H_2O} = \underset{\text{Benzaldehyde}}{C_6H_5CHO} + \underset{\text{Hydrocyanic acid}}{CNH} + \underset{\text{Dextrose.}}{2\,C_6H_{12}O_6}$$

Inasmuch as the emulsin coagulates at the boiling temperature of water, and thereby loses its activity, the fermentation must be completed before the distillation is begun. The powder is therefore mixed with 6—8 parts of water of about 50—60° and the mixture set aside for about 12 hours. The oil formed is then distilled over with water vapor.

According to the directions of Pettenkofer[1] (1862) 12 parts of the ground seeds, deprived of their fatty oil, are added to 100—120 parts of boiling water while stirring. The mixture is kept at this temperature for about 15—30 minutes and is then set aside to cool. To the cool mixture 1 part of fresh bitter almond powder, mixed with 6—7 parts of water, is added and allowed to macerate for 12 hours.

By treating the larger portion of the almond powder with boiling water, a more complete solution of the amygdalin is supposed to be accomplished. In order to hydrolyse the amygdalin of 12 parts of the powder, the emulsin of 1 part will suffice.

During the process of distillation, care is to be exercised not to allow the vapors of the very poisonous hydrocyanic acid to escape into the room. Not only should good condensation be provided, but the receiver should be joined air-tight to the condenser by means of bladder or parchment paper. All escaping gases should be conducted into the open atmosphere.

Inasmuch as benzaldehyde is rather readily soluble in water, especially in such containing hydrocyanic acid, the bulk of the oil is obtained only after cohobation of the aqueous distillate. The yield of oil from bitter almond seed varies from 0.5—0.7 p. c., that from apricot seeds from 0.6—1 p. c.

---

[1]) Liebig's Annalen, 122, p. 81.

Inasmuch as the hydrocyanic acid is objectionable in many instances, a part of the oil is deprived of this poisonous acid.

For this purpose the oil is shaken with milk of lime and ferrous sulphate whereby the hydrocyanic acid is precipitated as calcium ferrocyanide. The unchanged benzaldehyde is rectified by means of water vapor. If the operation has been carefully conducted no trace of hydrocyanic acid remains, as may be shown by the tests described under Examination.

In place of the oil deprived of its hydrocyanic acid, the cheaper artificial benzaldehyde is largely used. Inasmuch as the latter article usually contains chlorinated products which possess an unpleasant odor and taste, it can be used only in the manufacture of the cheaper grades of soap, but not in perfumery and in liquors. Artificial benzaldehyde is made by boiling benzyl chloride with lead or copper nitrate, or by heating benzylidene chloride with soda lye or milk of lime. Benzaldehyde obtained in this manner can be detected by means of its chlorine content as described on p. 442.

Properties. Bitter almond oil containing hydrocyanic acid is at first a colorless liquid which later becomes yellow. It is strongly refractive and posesses the well-known odor of bitter almonds. In smelling of the bitter almond oil great care should be exercised on account of the strongly poisonous hydrocyanic acid. The sp. gr. of the normal oil is 1.045—1.06. A higher specific gravity may be due to an abnormally high hydrocyanic acid content or of phenyl oxyaceto-nitrile. (Comp. under Composition on p. 440.) Freshly prepared oil is neutral. Upon standing it acquires an acid reaction owing to the oxidation of benzaldehyde to benzoic acid. The oil is optically inactive.

In water it is relatively soluble: 1 part of oil requires somewhat over 300 parts of pure water for solution. In water containing hydrocyanic acid, however, it is more soluble. The oil is soluble in 90 p. c. alcohol in all proportions, of 70 p. c. alcohol it requires 1½—2 parts. Nitric acid dissolves bitter almond oil at ordinary temperature without the generation of nitric oxide vapors.

Upon distillation of the oil over a direct flame an oil rich in hydrocyanic acid comes over first, later a weaker distillate is obtained. On account of the hydrocyanic acid vapors given off, special care must be taken in conducting this operation. The residue contains benzoin which results from the polymerization of the benzaldehyde under the influence of the hydrocyanic acid.

Bitter almond oil deprived of its hydrocyanic acid, or benzaldehyde, is a colorless, optically inactive liquid; sp. gr. 1.050—1.055; b. p. 179°. It is much more readily acted upon by atmospheric oxygen than the oil containing hydrocyanic acid, which seems to act as a preservative. Atmospheric oxygen quite readily oxidizes benzaldehyde to benzoic acid. Compare under Preservation on p. 442.

COMPOSITION. Bitter almond oil consists of benzaldehyde. hydrocyanic acid and phenyl oxyaceto-nitrile (benzaldehyde cyanhydrin or the nitrile of mandelic acid). Upon the hydrolysis of amygdalin, benzaldehyde and hydrocyanic acid result which, upon prolonged contact, unite to form phenyl oxyaceto-nitrile:

$$\underset{\text{Benzaldehyde}}{C_6H_5CHO} + \underset{\text{Hydrocyanic acid}}{CNH} = \underset{\text{Phenyl oxyaceto-nitrile.}}{C_6H_5CH(OH)CN.}$$

The proof that bitter almond oil contains this nitrile was brought by Fileti[1] in 1818. Upon reduction of bitter almond oil he obtained phenyl ethyl amine whereas a fresh mixture of hydrocyanic acid and benzaldehyde yielded only methyl amine.

Phenyl oxyaceto-nitrile is readily decomposable, breaking up into its components when distilled with water vapor or in a vacuum. It must, therefore, be formed after the distillation of the oil. It is formed in large quantities when the oily distillate remains in contact with the aqueous distillate containing the hydrocyanic acid for a long time.

Inasmuch as the sp. gr. of the nitrile of mandelic (phenyl glycollic) acid is high, viz. 1.124, the density of an oil will increase with the amount of nitrile present. Whereas normal oils having a sp. gr. of from 1.052—1.058 contained 1.6—4 p. c. of hydrocyanic acid, oils having a density of from 1.086—1.096 were found to contain 9—11.4 p. c. of hydrocyanic acid. To further clear up these relations it was observed that the sp. gr. of pure benzaldehyde changed from 1.054 to 1.074 when allowed to stand two days in contact with a 20 p. c. aqueous solution of hydrocyanic acid.[2]

From the above it becomes apparent that the hydrocyanic acid content of an oil will vary considerably with the method of preparation. For medicinal purposes, therefore, a definite percentage strength should be required.[3]

EXAMINATION. Qualitative test for hydrocyanic acid. In order to distinguish a bitter almond oil containing hydrocyanic acid from one which does not contain this acid the following test is made use of:

---

1) Gazz. chim. ital. 8, p. 446; Berichte, 12, Ref. p. 296.

2) Bericht von S. & Co., April 1898, p. 40.

3) Pharm. Zeitung, 41, p. 780.

10—15 drops of the oil to be examined are shaken with 2—3 drops of a strong (or a corresponding amount of dilute) soda solution. Several drops of ferrous sulphate solution containing some ferric salt are then added, the mixture thoroughly shaken and acidulated with dilute hydrochloric acid. Upon solution of the precipitate of ferrous and ferric oxide a precipitate of Prussian blue results if hydrocyanic acid was present. The reaction is so delicate that even minute traces of hydrocyanic acid can be detected in this way.

Assay of hydrocyanic acid. The best results are obtained by the gravimetric method according to the following directions.

1 g. of oil carefully weighed is dissolved in 10—20 g. of alcohol and 10 g. of alcoholic ammonia (free from chlorine) are added. After standing for a short time 1 g. of silver nitrate is added and the mixture acidulated with nitric acid. After the liquid has become clear, the silver cyanide is collected on a dried and weighed filter, carefully washed with water and dried at a temperature of 100°.

The silver precipitate obtained in this way represents all of the hydrocyanic acid contained in the oil. If the oil is not previously treated with ammonia only the free hydrocyanic acid is estimated. The ammonia decomposes the phenyl oxyaceto-nitrile.[1]

More convenient is the volumetric method of Vielhaber[2] but it is less accurate because the end-reaction cannot be recognized with certainty. Liebig's method for the estimation of hydrocyanic acid in bitter almond water is not at all applicable to the oil. The following directions are given by Kremers and Schreiner:[1]

1 g. of oil is carefully weighed in a small Erlenmeyer flask, and 10 cc. of a mixture of freshly prepared magnesium hydroxide and several drops of potassium chromate solution are added. Sufficient N/10 silver nitrate V. S. is added with constant shaking until the formation of the red silver chromate indicates the end of the reaction. In order to obtain the percentage of hydrocyanic acid, the number of cc. of the deci-normal solution used are multiplied by 0.0027.

The presence of foreign oils can readily be detected by converting the benzaldehyde into its sodium acid sulphite addition product and thus separated from the non-aldehyde constituents.

In a test tube of fully 100 cc. capacity 5 g. of the oil to be examined and 45 g. of acid sulphite solution are thoroughly shaken. 60 cc. of water are added and the tube placed in hot water. If the oil was pure a clear solution will result, whereas foreign substances will rise to the surface and can be examined.

In order to test for nitro-benzene (oil of mirbane), either the original oil or the oil floating on the sulphite solution is dissolved in 20 times its volume of alcohol and the solution diluted with water until it becomes turbid. Zinc and sulphuric acid are then added and the solution set aside for several hours. The solution is filtered, the alcohol evaporated and the remaining solution

---

1) Pharm. Review. 14, p. 196.

2) Archiv d. Pharm., 213, p. 408.

boiled with a drop of potassium bichromate solution. The presence of aniline, resulting upon the reduction of any nitro-benzene in the original oil, will be recognized by the violet color of the solution.

The most common adulterant of bitter almond oil, whether deprived of hydrocyanic acid or not, is artificial benzaldehyde. Inasmuch as the latter is almost invariably accompanied by chlorinated products the determination of chlorine affords the easiest method of detection in the following manner:[1]

In a small porcelain capsule, placed within a larger one of about 20 cm. diameter, a folded strip of filter paper saturated with oil is placed and ignited. A large beaker of about 2 l. capacity, the inner surface of which has been moistened with distilled water, is immediately placed over the burning oil. The products of combustion condense on the moistened surface of the beaker and with the aid of a little distilled water are washed on to a filter. The filtrate should not be rendered turbid, much less yield a precipitate with silver nitrate T. S. Genuine oil, i. e., oil prepared from almonds or apricot seeds never yields a chlorine reaction. It may be well, however, to make a duplicate test with a known pure oil inasmuch as the water and the utensils employed may at times be contaminated with some chlorine.

Inasmuch as a benzaldehyde free from chlorine has recently been brought into the market, the absence of chlorine is by no means a sign of the absolute purity of an oil. In all cases, however, where chlorine is found the presence of artificial benzaldehyde may be regarded as being established. The presence of alcohol, which is not infrequently found, can be determined in the usual manner.

Preservation. If a small amount of benzaldehyde be exposed to the air in an open dish, crystals are soon formed and after a short time the oil will be converted into a magma of crystals of benzoic acid. The same change takes place if the oil is kept in half-filled bottles. Bitter almond oil should, therefore, be kept in well-filled bottles. Experiments have shown that the addition of 10 p. c. of alcohol will act as a preservative.[2] If but 5 p. c. of alcohol are added, the oxidation will take place even more rapidly than when undiluted. Bitter almond oil containing hydrocyanic acid is not as readily oxidized as the oil free from acid. The hydrocyanic acid evidently acts in a manner similar to the alcohol when 10 p. c. are used.

173. Oil of Cherry-laurel.

Oleum Laurocerasi. — Kirschlorbeeröl. — Essence de Laurier-Cerise.

Origin, History and Preparation. Cherry-laurel, *Prunus laurocerasus* L.. which is indigenous to Persia and the Caucasus and which

1) Bericht von S. & Co., Apr. 1890, p. 29. 2) Bericht von S. & Co., Apr. 1895, p. 47.

is cultivated in countries with a temperate climate, appears to have become known in Europe since the beginning of the sixteenth century. The aqueous distillate from the leaves has been used medicinally since the first half of the eighteenth century and its poisonous properties were repeatedly observed. The distilled oil is mentioned in medical treatises since 1780. The presence of hydrocyanic acid in the oil was observed simultaneously by Schaub and by Schrader at the beginning of this century.

The oil from the leaves closely resembles the oil from bitter almonds but has a slightly different odor. The method of preparation is likewise similar to that of the bitter almonds. The cut leaves are macerated with water and the mash set aside. The oil thus formed is distilled with water vapor. The yield is about 0.5 p. c. (Umney[1]).

The oil is formed by the hydrolysis of laurocerasin, a glucoside closely related to amygdalin, but not identical with it.[2] Being hygroscopic it is not readily obtained in the crystalline state, but results as a thick syrup or an amorphous mass upon evaporation of the alcoholic extract. Laurocerasin, $C_{40}H_{67}NO_{30}$ is made up of one molecule of amygdalin, $C_{20}H_{27}NO_{11}$, one molecule of amygdalic acid, $C_{20}H_{28}C_{13}$, and six molecules of water. In contact with emulsin, the amygdalin of the laurocerasin is broken up into hydrocyanic acid, benzaldehyde and grape sugar. The amount of oil obtainable from the laurocerasin is, therefore, much smaller than that obtained from an equal weight of amygdalin.

Properties. Oil of cherry-laurel can only be distinguished from bitter almond oil by its odor. The difference in odor is emphasized by first uniting the benzaldehyde with an acid sulphite. Its other properties agree with those of bitter almond oil. Its sp. gr. is 1.054—1.066, it is optically inactive and forms a clear solution with 2 parts of 70 p. c. alcohol.

Inasmuch as cherry-laurel oil is very poisonous on account of the hydrocyanic acid contained therein sufficient regard should be had for this property in the use of the oil.

Composition. Like bitter almond oil, cherry-laurel oil contains benzaldehyde, hydrocyanic acid and phenyl oxyaceto-nitrile (Fileti,[3]

1) Pharm. Journ., III, 5, p. 761.

2) Lehmann (1874), Neues Repert. f. d. Pharm., 23, p. 449; and (1885) Pharm. Zeitschr. f. Russl., 24, pp. 353, 369, 385, 401.—Berichte, 18, p. 569 Referate.

3) Gazz. chim. ital., 8, p. 446; Berichte, 12, p. 296.

1878). According to Umney[1] (1869) hydrocyanic acid is present to the extent of 2 p. c. It is probable, however, that the percentage varies as in bitter almond oil.

According to Tilden[2] (1875), traces of another substance are present, to which the peculiar odor of this oil is probably due. If the oil is shaken with sodium acid sulphite solution, a dark colored oil remains which upon oxidation with chromic acid yields benzoic acid. Tilden supposes this substance to be identical with benzyl alcohol.

ADULTERATION AND EXAMINATION. Oil of cherry-laurel is subjected to the same kind of adulteration as bitter almond oil. The tests are the same as those described on p. 440.

174. Oil of Wild Cherry Bark.

The bark of the wild cherry, *Prunus virginiana* Mill. (*P. serotina* Ehrh.), which is indigenous to North America, has been used for a long time by the natives in the preparation of aromatic beverages and household remedies and was made official in the early editions of the U. S. Pharmacopoeia. That the distillate from the bark contains hydrocyanic acid was observed by Procter[3] in 1834. In 1838 he showed that the oil does not preexist in the bark but results by a process similar to that taking place in the formation of bitter almond oil.[4] A more detailed examination of the oil was made by Power and Weimar[5] in 1887.

The powdered bark when macerated yields 0.2 p. c. of oil, which resembles the oil of bitter almonds, consists largely of benzaldehyde, and contains hydrocyanic acid.[6] Sp. gr. 1.045—1.050.

According to Power and Weimar the bark contains no amygdalin, but a substance that behaves like laurocerasin. The ferment does not consist of emulsin, at least no emulsin can be isolated according to the usual method. The leaves of the wild cherry tree also yield an aqueous distillate containing hydrocyanic acid.[7]

Hydrocyanic acid and benzaldehyde have been obtained from parts of other rosaceous plants. Thus, e. g. from the leaves, branches and seeds of the peach, *Prunus persica* Jess., from the seeds and the fleshy

1) Pharm. Journ., II, 10, p. 467.
2) Pharm. Journ., III, 5, p. 761.
3) Americ. Journ. Pharm., 6, p. 8.
4) Ibidem, 10, p. 197.
5) Pharm. Rundschau, 5, p. 208.
6) Bericht von S. & Co., Apr. 1890, p. 48.
7) Pharmacognosie, 3rd ed., p. 765.

pericarp of the cherry, *P. cerasus* L., from the seeds of the plum, *P. domestica* L., from the bark, leaves, flowers and seeds of *P. padus* L., from the young leaves and flowers of *P. spinosa* L.

175. Oil of Copaiba.

Oleum Balsami Copaivae.—Copaivabalsamöl.—Essence de Baume de Copahu.

Origin. Copaiba balsam, which has been used in Europe since the beginning of the sixteenth century, is obtained from a number of species of *Copaifera* (Family *Leguminosae*) which are indigenous to the territory of the Amazon and its tributaries as far north as Guayana, Venezuela and Columbia. The principal species are *Copaifera officinalis* L., *C. guaianensis* Desf., *C. coriacea* Martius, *C. Langsdorffii* Desf., *C. conferti-flora*, *C. oblongifolia* Martius and *C. rigida* Bentham. The balsam is allowed to flow from the cavity made in the trunk of the tree by means of tin tubes into the vessels used for its transportation. Occasionally the trees are so charged with balsam that the receptacles burst and the balsam exudes from the vertical cracks.

Commercially, the balsam is distinguished according to the ports of export. The most important varieties are the Maracaibo balsam (derived principally from *Copaifera officinalis*) and the Para balsam. The latter is more fluid and inasmuch as it yields the larger percentage of oil, abt. 60—90 p. c., is preferred for distillation. Maracaibo balsam which is more viscid yields upon distillation about 40 p. c. of oil. A like yield was obtained from a Maranhão balsam.

Properties. Copaiba oil from Para or Maracaibo balsam is a colorless, yellowish or brownish liquid possessing the characteristic pepper-like odor of the balsam, and a bitter, grating and persistent taste. Sp. gr. 0.900—0.910; $\alpha_D = -7$ to $-35°$. The boiling temperature lies between 250 and 275°. It is not completely soluble in 90 p. c. alcohol. Of absolute alcohol, usually equal parts are requisite to form a clear solution. According to Kremel,[1] two samples of para oil yielded no saponification number. Dulière[2] examined an oil of unknown origin and of a high specific gravity, viz. 0.9155, and found a saponification number corresponding to 4 p. c. of sesquiterpene acetate ($C_{15}H_{25}OCOCH_3$). After acetylization a saponification number was obtained corresponding to 6.5—7.35 p. of acetate.[3]

An oil from Maranhão balsam had the sp. gr. 0.889 and $\alpha_D = -20°$.

1) Pharm. Post, 21, p. 822.
2) Ann. de Pharm. (Louvain), 3, p. 553; 4, p. 11.
3) Bericht von S. & Co., Oct. 1895, p. 42; Apr. 1898, p. 49.

COMPOSITION. The only known and well defined constituent of copaiba oil is the sesquiterpene caryophyllene, $C_{15}H_{24}$, which also occurs in oil of cloves. If fraction 250—270° is treated according to Wallach[1] with glacial acetic acid and sulphuric acid, caryophyllene hydrate $C_{15}H_{25}OH$ is obtained in handsome crystals melting at 96°. By the action of phosphorus pentachloride on caryophyllene hydrate the chloride $C_{15}H_{25}Cl$ melting at 63° is obtained. The nitrosochloride, characteristic of caryophyllene and melting at 161—163° with decomposition, has also been obtained.

Blanchet[2] (1833), also Soubeiran and Capitaine[3] (1840), obtained a solid hydrochloride by passing hydrochloric acid into copaiba oil. One product melted at 77°, another at 54°. The analysis agreed with the formula ($C_{10}H_{16}2HCl$). Later investigators have been unable to obtain a solid hydrochloride from either Para or Maracaibo oil.[4] Neither Posselt[5] (1896) nor Strauss[6] (1868) obtained definite results by oxidizing both oils with nitric acid. Levy and Engländer[7] (1887), who oxidized Para oil with potassium bichromate and sulphuric acid, were more successful. They obtained a crystalline acid melting at 140° which was identified as a symmetric dimethyl succinic acid, $C_6H_{10}O_4$. The small yield of but 1½ p. c. renders it doubtful whether this acid owes its origin to caryophyllene or to some other minor constituent of the oil. According to Brix, Maracaibo balsam oil yields upon oxidation small amounts of terephthalic acid. The same oil when dried over sodium yields a blue oil which has the composition $C_{20}H_{32}+H_2O$.

ADULTERATION. Copaiba oil is more frequently used as an adulterant than adulterated itself. As adulterant gurjun balsam oil only comes into consideration, which, however, can be readily recognized on account of its higher specific gravity and greater optical activity. The oil of the African copaiba balsam (see next oil) is not soluble in an equal volume of absolute alcohol like the genuine oil.

176. African Copaiba Balsam Oil.

Under the above designation there appeared several years ago a balsam from West Africa. Although its botanical origin is not known, its resemblance to the South American balsam renders its derivation

1) Liebig's Annalen, 271, p. 294.
2) Liebig's Annalen, 7, p. 156.
3) Liebig's Annalen, 84, p. 321.
4) Brix (1882), Monatsh. f. Chem., 2, p. 507; Umney (1898), Pharm. Journ., III, 24, p. 215.
5) Liebig's Annalen, 69, p. 67.
6) Liebig's Annalen, 148, p. 148.
7) Liebig's Annalen, 242, p. 189; Berichte, 18, pp. 3206, 3209.

from a copaifera species probable. Upon distillation, Umney[1] obtained 37—40 p. c. of oil. Sp. gr. 0.917—0.918; $\alpha_D = +20°42'$. Upon passing hydrogen chloride into fraction 240—270° no solid hydrochloride was obtained. Neither could caryophyllene hydrate be obtained by treatment with glacial acetic acid and sulphuric acid.

177. Cabriuva Wood Oil.

Cabriuva, *Myrocarpus fastigiatus* Fr. All. (Family *Leguminosae*) is a handsome tree native to Brazil that acquires a height of 8 meters and more. The yellowish-white flowers are very fragrant, the odor resembling that of a mixture of vanilla and tolu balsam. The wood is counted among the most valuable of Brazil.

The oil from the wood was examined by Schimmel & Co.[2] It is of a light yellow color and has a faint, not unpleasant odor. Sp. gr. 0.9283; $\alpha_D = -8°29'$. From 100 k. of sawdust, Peckolt[3] obtained 4.3 g. of oil. Sp. gr. 0.925 at 13°.

178. Carqueja Oil.

Genista tridentata is supposed to be the plant from which this oil is obtained. This leguminous plant grows wild in Brazil and is used as a domestic remedy.[4]

Carqueja oil is yellow, has a somewhat narcotic, camphor-like, not unpleasant odor. Sp. gr. 0.9962; $\alpha_D = -31°15'$ at 17°; saponification number 190.5. Upon distillation three-fourths of the oil distill over between 200—300° while acetic acid is split off. In the first fraction cineol could be identified. A viscid mass remained in the flask.

179. Indigofera Oil.

According to van Romburgh, the leaves of *Indigofera galegoides* D. C. (Family *Leguminosae*) contain a substance (possibly amygdalin or laurocerasin) which with emulsin yields hydrocyanic acid and benzaldehyde. Upon distillation of the fresh, macerated leaves, 0.2 p. c. of oil were obtained.

Indigofera oil is light yellow in color, its odor resembles that of bitter almond oil but differs from it in being herbaceous. Sp. gr. 1.046.[5] Besides benzaldehyde and hydrocyanic acid, the oil contains traces of

---

1) Pharm. Journ., III, 22, p. 452; 24, p. 215.
2) Bericht von S. & Co., Apr. 1896, p. 69.
3) Catalogue of the Nat. Exhib. in Rio, 1866, p. 48.
4) Bericht von S. & Co., Apr. 1896, p. 70.
5) Bericht von S. & Co., Oct. 1894, p. 75.

ethyl alcohol and also small amounts of methyl alcohol. The latter was identified by its conversion into methyl iodide and into the methyl ester of oxalic acid melting at 54°. If the leaves were digested for 24 hours at a temperature of 50° before being distilled, the first distillate contained larger amounts of ethyl alcohol which, however, was formed during the process of maceration.[1]

180. Sappan Leaf Oil.

The leaves of *Caesalpinia sappan* L., a leguminous tree, from which the sappan wood used for dyeing is obtained, yields according to van Romburgh 0.16 to 0.2 p. c. of an almost colorless oil, sp. gr. 0.825 at 28°. It is strongly dextrogyrate, $\alpha_D = +37°30'$ to $+50°30'$ and boils principally at 170°. The odor is pepper-like, reminding of phellandrene. With sodium nitrite and glacial acetic acid so large an amount of phellandrene nitrite was obtained, that the bulk of the oil may be regarded as d-phellandrene. Upon distillation of the leaves, van Romburgh observed the occurrence of methyl alcohol.[2]

181. Oil of Tolu Balsam.

Oleum Balsami Tolutani.—Tolubalsamöl.—Essence de Baume de Tolu.

Origin and History. The statements made on p. 420 concerning the origin and preparation of American storax apply in general to tolu balsam from *Toluifera balsamum* Mill. It is prepared in the southern states of South America and has been used for a long period, in Europe since the middle of the sixteenth century.

Upon distillation with water vapor, the solid balsam yields 1.5—3 p. c. of oil. Slow distillation yields a light oil, whereas rapid distillation with superheated steam yields more oil of a higher density. The sp. gr., therefore, varies between 0.945—1.09. The oil is slightly laevo- or dextrogyrate, $\alpha_D = -0°58'$ to $+0°54'$. The odor is pleasant, highly aromatic and reminds of hyacinth.

Composition. The older, somewhat contradictory statements give but imperfect information concerning the composition of tolu balsam oil.[3] According to Kopp the hydrocarbon boiling at about 170° and having an elemi-like odor is a terpene (possibly phellandrene?). Inas-

---

1) Bericht von S. & Co., Apr. 1896, p. 75.

2) Bericht von S. & Co., Apr. 1898, p. 57.

3) Deville (1841), Ann. de Chim. et Phys., III, 3, p. 151; Liebig's Annalen, 44, p. 304.—Kopp (1847), Journ. de Pharm. et Chim., III, 11, p. 425.—Scharling (1856), Liebig's Annalen, 97, p. 71.

much as Busse[1] (1876) has shown the presence of benzoic and cinnamic esters of benzyl alcohol in the balsam, these esters are probably also contained in the oil. As a matter of fact the oil has a high saponification number (abt. 180) and from the alkaline saponification liquor crystalline acids (presumably cinnamic and benzoic acids) can be precipitated.

182. Oil of Myroxylon Peruiferum.

From the leaves of the leguminous *Myroxylon peruiferum* L. f., a tree closely related to the Peru balsam tree,[2] Peckolt[3] obtained a small amount of oil with a faint but pleasant odor and a sp. gr. of 0.874 at 14°. The bark yielded two oils, one having the sp. gr. 1.139 at 15°, the other 0.924 at 17°. The oil from the wood had a faint sassafras odor and a sp. gr. of 0.852 at 15°.

183. Oil of Rose Geranium.[5]

Oleum Geranii. — Geranium- or Pelargoniumöl. — Essence de Géranium Rose.

Origin and History. The pelargoniums which are indigenous to South Africa and which are now largely cultivated as decorative plants were introduced into Europe in 1690 (Piesse[4]). Recluz[6] in 1819 first obtained a volatile oil from the leaves. However, Demarson of Paris, in 1847 was the first to cultivate pelargoniums for the distillation of the oil. Since then their cultivation throughout France has been largely extended and was introduced into Algiers by Chiris and Monk. The production in the latter colony now exceeds that of any other country. In Spain pelargoniums were cultivated in the vicinity of Valencia by Robillard. Later they were also cultivated in the province of Almeria. The island of Réunion has entered upon the cultivation of these plants since the eighties and now holds the second position in the manufacture of the oil. Of much less importance are the plantations in Corsica.

Preparation. Three species and their varieties are principally cultivated, viz., *Pelargonium odoratissimum* Willd., *P. capitatum* Ait., and *P. roseum* Willd. The last is regarded as a variety of *P. radula* Ait. (Sawer[7]).

---

1) Berichte, 9, p. 830.

2) Inasmuch as the so-called Peru balsam oil cannot be obtained by steam distillation it is not included among the volatile oils in this treatise.

3) Zeitschr. d. öst. Apt.-Ver., 17, p. 49; Jahresb. f. Pharm., 1879, p. 59.

4) The Art of Perfumery, 4th ed., p. 124.

5) This oil is not to be confounded with palmarosa oil from *Andropogon schoenanthus* L. which is misnamed Indian geranium oil.

6) Pharm. Journ., I, 11, p. 825.

7) Odorographia, vol. 1, p. 42.

The oil is to be found in all green parts of the plant, especially in the leaves. The flowers are perfectly odorless. The plants are harvested before they blossom, when the leaves begin to get yellow and after the lemon-like odor has given way to the more rose-like odor. It is reported that the plants grown on dry soil give a smaller yield but a finer oil than those grown in irrigated soil. The yield varies from 0.15 to 0.33 p. c.

PROPERTIES. Geranium oil is a colorless, greenish or brownish liquid of a pleasant, rose-like odor. Sp. gr. 0.89—0.906; $\alpha_D = -6$ to $-16°$. With the exception of the Spanish oil, all varieties are soluble in 2—3 parts of 70 p. c. alcohol. The saponification number varies from 45 to 100 and the percentage of ester calculated as geranyl tiglinate from 19 to 42 p. c.

The constants of the different oil varieties are as follows:

1. French oil. Sp. gr. 0.897—0.905; $\alpha_D = -7°30'$ to $-9°$. Geranyl tiglinate 25—28 p. c. Forms a clear solution with 2—3 parts of 70 p. c. alcohol.

2. African (Algerian) oil. Sp. gr. 0.892—0.90; $\alpha_D = -6°30'$ to $-10°$. Geranyl tiglinate 19—29 p. c. Solubility same as No. 1.

3. Réunion oil. Sp. gr. 0.889—0.895; $\alpha_D = -8°$ to $-11°$. Geranyl tiglinate 27—33 p. c. Solubility same as No. 1. This oil usually has a green color which does not result from the presence of copper but from one of the compounds contained in the highest fractions which when pure is possibly blue in color.

4. Spanish oil. Sp. gr. abt. 0.897; $\alpha_D = -10$ to $-11°$. Geranyl tiglinate 35—42 p. c. As to solubility this oil differs from the others inasmuch as the solution in 2—3 or more parts of 70 p. c. alcohol is rendered turbid by the presence of minute crystals of paraffin which rise to the surface. The separation of oily drops at the bottom of the tube would indicate adulteration with fatty oil.

5. German oil.[1] This oil has been distilled by way of experiment but is not an article of commerce. Yield 0.16 p. c. Sp. gr. 0.906; $\alpha_D = -16°$. Geranyl tiglinate 27.9 p. c.

Of the commercial varieties, the Spanish oil is rated highest. French and African oil are about alike in quality, while the price of Réunion oil is somewhat lower.

The oil is mostly transported in tin cans and often acquires a brownish color and an odor of rotten eggs. This disagreeable odor,

1) Bericht von S. & Co., Apr. 1894, p. 82.

however, is easily removed by exposing the oil to the air in shallow dishes for several days. As soon as possible the oil should be filled into glass containers.

COMPOSITION. The principal constituent is geraniol,[1] $C_{10}H_{18}O$, which is contained in all commercial varieties of the oil. It can be isolated by treating the saponified and dried oil with very fine calcium chloride. The addition product is carefully washed with ether and the pure geraniol separated by decomposing the compound with water.[2]

Besides geraniol, a second alcohol, $C_{10}H_{20}O$, which has been identified as citronellol by Tiemann and Schmidt[3] has been found in the oil. It occurs to a larger extent in the oil from Réunion. Mixtures of these alcohols have been described as "Rhodinol de Pelargonium" by Barbier and Bouveault[4] and as "Reuniol" by Hesse.[5]

The percentage of geraniol and citronellol in the various commercial varieties has been determined by Tiemann and Schmidt:[3] The Spanish oil contains 70 p. c. of alcohols, of which 65 p. c. are geraniol and 35 p. c. citronellol. The African oil contains 75 p. c. alcohols of which 80 p. c. is geraniol and 20 p. c. citronellol. The Réunion oil contains 80 p. c. of equal parts of geraniol and citronellol. In all of these oils the citronellol is a mixture of the dextro- and laevogyrate modifications. The lower fractions of the oil contain a third alcohol, in all probability linalool. (Barbier and Bouveault).[6]

The acids which were separated from the lye after saponification boiled between 100 and 210°, and in part solidified. The solid acid melted at 64—65°. Its silver salt and its dibromide melting at 87° identify it as tiglinic acid.[7] The liquid acid mixture apparently consists of valerianic, butyric[7] and acetic[6] acids.

Jeancard and Satie[8] have found that the amount of free acid in geranium oil increases when the oil is kept in partly filled bottles. An oil which originally showed an acid number of 56, after two months standing had increased to 66.73. The same authors also made a comparative examination of the different commercial varieties of the oil.

---

1) Jahresb. d. Chem., 1879, p. 942.
2) Journ. f. prakt. Chem., II, 49, p. 191.
3) Berichte, 29. p. 924.
4) Compt. rend., 119, pp. 281 & 334.
5) Journ. f. prakt. Chem., II, 50, p. 472; 58, p. 288.
6) Compt. rend., 119, p. 281.
7) Bericht von S. & Co., Apr. 1894, p. 81. It would be of interest to reexamine the aqueous distillate of *Pelargonium roseum* in order to ascertain whether or not the pelargonic acid of Pless (1846) (Liebig's Annalen, 59, p. 54), may reveal itself as a mixture.
8) Bull. Soc. chim. III, 28, p. 37.

Geraniol and citronellol are contained in the oil both free and combined with acids. The esters of citronellol are more strongly optically active than the alcohol. If geranium oil is saponified with alcoholic potassa the angle of rotation is less after the saponification than before.[1]

Of minor constituents l-menthone remains to be mentioned. It was identified by Flatau and Labbé[2] by means of the semicarbazone melting at 179.5°. In the highest fractions a crystalline substance melting at 63° has been found which resembles the stearoptene from oil of rose. Judging from the behavior of the oil to 70 p. c. alcohol, this paraffin occurs in largest amounts in the Spanish variety.

Adulteration and Examination. Geranium oil is adulterated with turpentine oil, cedar wood oil and fatty oils, all of which can be recognized by their insolubility in 70 p. c. alcohol. Fatty oil remains behind upon distillation with water vapor and can be detected without difficulty.

184. Oil of Garden Nasturtium.

The odor of the leaves of *Tropaeolum majus* L. (Family *Tropaeolaceae*) having an odor similar to that of the cress, induced Hofmann[3] in 1874 to examine the oil. 300 k. of flowering herb and unripe seeds were distilled with water vapor. The aqueous distillate was shaken with benzene and upon evaporation of the benzene 75 g. (=0.025 p. c.) of oil were obtained.

The oil boiled between 160—300° leaving a not inconsiderable residue. Only the first fractions contained sulphur and had an unpleasant odor, the others not. The bulk of the oil distilled at about 231.9°. This fraction was a liquid of strong refractive power, sp. gr. 1.0146 at 18°, and when acted upon with caustic potassa gave off large quantities of ammonia. Analysis showed it to be the nitrile of phenyl acetic acid. The principal constituent of the oil, therefore, is the same as that of *Lepidium sativum* L. The same substance was found in the lower and higher fractions of the oil together with a hydrocarbon not farther investigated.

Quite different results were obtained recently by Gadamer.[4] He prepared the oil by extracting the juice from the comminuted herb with ether, also by distilling the carefully comminuted herb and extracting the aqueous distillate with ether. 4 k. of herb yielded 1.3 g. of a

---

1) Compt. rend., 119, p. 281.

2) Bull. Soc. chim., III, 19, p. 788.

3) Berichte, 7, p. 518.

4) Archiv d. Pharm., 237, p. 111.

brownish colored oil the cress-like odor of which became especially marked when warmed. With ammonia the oil yields almost quantitatively benzyl thiourea melting at 162°, thus showing that the oil consists almost entirely of benzyl mustard oil.

This mustard oil owes its origin to the decomposition of a glucoside, the glucotropaeolin, $C_{14}H_{18}KNS_2O_9 + xH_2O$ by means of a ferment. Inasmuch as the glucoside and ferment are contained in separate cells, the formation of the benzyl mustard oil takes place only when the cell walls are ruptured and both substances can act on each other. If the cell wall is not destroyed before distillation, the ferment is rendered inactive by the heat before it can act on the glucoside. This is then decomposed during the process of distillation with the formation of the nitrile of phenyl acetic acid (benzyl cyanide). If it is assumed that Hofmann did not sufficiently comminute his crude material, the results obtained by him are explained.

185. Oil of Coca Leaves.

The presence of an oil in the coca leaves was first observed in 1860 by Niemann[1] and Lossen.[2] The leaves of *Erythroxylon cocà* Lam. var. *spruceanum* Brck. (Family *Erythroxylaceae*) contain variable amounts of oil according to their development. Van Romburgh[3] obtained from young, undeveloped leaves 0.13 p. c., from fully developed leaves only 0.06 p. c. of oil. It consisted principally of methyl salicylate with small amounts of acetone and methyl alcohol.

186. Oil of Guaiac Wood.

Oleum Ligni Guajaci. — Guajakholzöl. — Essence de Bois Gaïac.

Origin and Preparation. *Bulnesia sarmienti* Lor. is, according to Grisebach,[4] a tree 40—60 feet in height, belonging to the family *Zygophyllaceae*, and is indigenous to the Argentine province of Gran Chaco about half way up the Rio Berjemo. The wood, very similar to the ordinary guaiac wood from *Guajacum officinale* L., occurs since the year 1892 as *Palo balsamo* in commerce. It is exceedingly solid and tenacious, and is colored greenish blue on exposure to the air, which allows of drawing the conclusion that guaiac resin is present. By distil-

---

1) De foliis Erythroxyli, Dissertatio. Göttingen 1860.

2) Ueber die Blätter von *Erythroxylon Coca* Lam., Dissertation, Göttingen 1862.

3) Rec. des trav. chim. des Pays Bas, 13, p. 425.—S'Lands Plantentuin te Buitenzorg, 1894, p. 48.—Bericht von S. & Co., Oct. 1895, p. 47; Apr. 1896, p. 75.

4) Abhandl. d. Königl. Ges. d. Wissensch. zu Göttingen. Vol. 24, p. 75.

lation the wood yields 5—6 p. c. of oil, which was first prepared by Schimmel & Co.[1] and brought into commerce as oil of guaiac wood.[2]

Properties. Oil of guaiac wood is a viscous, heavy oil, which at ordinary temperature gradually solidifies to a crystalline mass. When it has solidified, it does not melt again until between 40 and 50°. The odor of the oil is very pleasant, being violet- and tea-like. The specific gravity lies between 0.965 and 0.975 at 30°, the angle of rotation is -6 to $-7°$ at 30°. The oil is soluble in 70 p. c. alcohol. The saponification number found of an oil was 3.9, the ester number 2.4 and the acid number 1.4.[3]

Composition. The crystalline constituent of the oil is guaiac alcohol or guaiol (Wallach[4]), a sesquiterpene hydrate $C_{15}H_{26}O$. Guaiol is an odorless body, crystallizing in large transparent prisms, and melting at 91°.[5] It boils under ordinary pressure at 288°, under 10 mm. pressure at 148°. Its solution in chloroform is laevogyrate. With dehydrating agents a hydrocarbon $C_{15}H_{24}$ is formed, accompanied by an intensely blue substance. On boiling guaiol with acetic acid anhydride a liquid acetyl compound is produced which boils at 155° under a pressure of 10 mm. The odoriferous constituent of the oil has not yet been investigated.

The oil is used in the perfume industry for the purpose of producing a tea-rose odor. Recently it is being employed in Bulgaria as an adulterant for oil of rose (p. 436).

187. Japanese Oil of Pepper.

The fruit of *Xanthoxylum piperitum* D. C. (Family *Rutaceae*) known as *Piper japonicum* or *Sansho* (Japanese), yields upon distillation[6] with water 3.16 p. c. of a yellowish oil of a pleasant odor reminding of lemon. Sp. gr. 0.973; b. p. 160—230°.

The oil was investigated by Stenhouse[7] in 1857, who established the presence of a terpene boiling at 162°, xanthoxylene (pinene or camphene?), as well as a crystalline body $C_{10}H_{8}O_{4}$. The main constituent of the oil is citral,[6] $C_{10}H_{16}O$.

---

1) Bericht von S. & Co., Apr. 1892, p. 42; Apr. 1893, p. 32; Apr. 1898, p. 26; Oct. 1898, p. 30.

2) The fanciful name "Champaca oil" was later given to this same oil although it has not the slightest resemblance to the genuine champaca oil from *Michelia champaca* L. (Bericht von S. & Co., Apr. 1898, p. 33.)

3) Dietze, Süddeutsche Apoth. Zeitung, 38, p. 680.

4) Liebig's Annalen, 279, p. 395.

5) The designation champacol (Chemiker-Zeitung Repert., 17, p. 31) for this alcohol is of course as little justified as the name champaca oil for the oil.

6) Bericht von S. & Co., Oct. 1890, p. 49.

7) Pharm. Journ., I, 17, p. 19. — Liebig's Annalen, 104, p. 236.

188. Oil of Xanthoxylum Hamiltonianum.

The seeds of *Xanthoxylum hamiltonianum*[1] yield according to Helbing[2] 3.84—5 p. c. of volatile oil of the sp. gr. 0.840. It is colorless and has a pleasant persistent odor, reminding of a mixture of geranium and bergamot oils.

189. Oil of Rue.

Oleum Rutae.—Rautenöl.—Essence de Rue.

ORIGIN AND HISTORY. Garden rue, *Ruta graveolens* L., belonging to the family *Rutaceae*, is indigenous to the Mediterranean countries and is cultivated in different countries having a moderate climate. It contains about 0.06 p. c. of volatile oil, which gives to the plant a peculiar aromatic odor and taste which gave rise to its use as a kitchen spice. In ancient times the plant was employed as a remedy, especially for scurvy and snake bites.

The first mention of oil of rue (although possibly of the fatty oil) is found in Saladin's writings. Gesner distilled the oil about the middle of the sixteenth century and the same is mentioned in the price ordinances of the cities of Berlin for the year 1574 and of Frankfurt for the year 1582, and in the Dispensatorium Noricum of the year 1589.

The yield of volatile oil from rue was first determined by Cartheuser in the beginning of the eighteenth century. The oil was investigated by Neumann and Mähl in Rostock in the year 1811,[3] by Will in the year 1840,[4] Cahours in 1845,[5] Gerhard in 1848,[6] Williams in 1858,[7] Hallwachs in 1859,[8] Harbordt in 1862.[9] Giesecke in 1870[10] and Gorup-Besanez and Grimm in 1871.[11]

PROPERTIES. Oil of rue is a colorless to yellow liquid of an intensive persistent rue odor, which is pleasant only when greatly diluted.

Its specific gravity is lower than that of all known volatile oils and lies between 0.833 and 0.840. Oil of rue is slightly dextrogyrate ($\alpha_D = +0°13'$ to $+2°10'$) and solidifies at a low temperature. Its solidification point lies between + 8 and + 10°.[12] During the distillation very little passes over up to 200°, the largest portion boils from

---

1) *Evodia fraxinifolia* was originally given as the source of this oil.

2) Jahresb. f. Pharm., 1887, p. 157; and 1888, p. 128.

3) Trommsdorff's Journ. d. Pharm., 20, II, p. 29.

4) Liebig's Annalen, 35, p. 235.

5) Compt. rend., 26, p. 262.

6) Liebig's Annalen, 67, p. 242.

7) Liebig's Annalen, 107, p. 374.

8) Liebig's Annalen, 113, p. 107.

9) Liebig's Annalen, 123, p. 293.

10) Zeitschr. für Chemie, 13, p. 428.

11) Liebig's Annalen, 157, p. 275.

12) Bericht von S & Co., Oct. 1896, p. 65.

215 to 232° (Umney,[1] 1895). The oil forms a clear solution with 2—3 parts of 70 p. c. alcohol.

Composition. The only constituent which has been positively identified is methyl nonyl ketone,[2] $CH_3CO.C_9H_{19}$. Besides this body, which forms about 90 p. c. of the oil, there is contained in it according to Williams[3] laurinic aldehyde, $C_{12}H_{24}O$ (b. p. 232°). This assertion is, however, too little supported by facts and, therefore, makes the presence of this aldehyde in oil of rue appear still doubtful.

The terpene found by different investigators can in all probability be traced to adulteration with turpentine oil.

For the preparation of pure methyl nonyl ketone the oil is shaken with sodium bisulphite solution, the resulting mixture made fluid with ether, then pressed, and the operation of washing out and pressing repeated several times. The disintegrated press cakes are decomposed by alkali solution. The separated oil is purified by distillation with steam.

Methyl nonyl ketone has the specific gravity 0.8295 at 17.5°; it boils at 224° and melts at + 15°.[4] According to H. Carette[5] it boils at 226° under 766 mm. pressure (corr. 230.65°) and at 121—122° under a pressure of 24 mm. (corr. 122—123°). Carette also prepared the oxime by treatment with hydroxylamine hydrochloride and alkali. M. p. 46°.

Examination. As the value of the oil of rue depends on the amount of the methyl nonyl ketone present, the determination of its solidification point gives a valuable criterion for judging the oil. This determination is conducted as described on page 187.

All foreign additions lower the solidification point which lies at + 8 to + 10° with good oils, they also raise the specific gravity (with the single exception of petroleum).

Petroleum like turpentine oil is detected by it insolubility in 70 p. c. alcohol. Turpentine oil can be further identified by fractional distillation. The portions boiling below 200° are collected separately and tested for pinene. With a pure oil not more than 5 p. c. distill over below 200.[6]

Adulterants may also be separated from the methyl nonyl ketone by the bisulphite method described under Composition and then identified as such.

---

1) Pharm. Journ., III, 25, p. 1044.
2) Liebig's Annalen, 157, p. 275.
3) Liebig's Annalen, 107, p. 374.
4) Zeitschr. f. Chemie, 13, p. 428.
5) Journ. de Pharm. et Chim. VI, 10, p. 255.
6) Pharm. Journ., III, 25, p. 1044.

190. Boronia Oil.

An oil distilled[1] in Victoria (Australia) from *Boronia polygalifolia* (Family *Rutaceae*) had a sweet odor, reminding of estragon and somewhat of rue. Specific gravity 0.839 (!); $\alpha_D = +10°$.

Upon fractional distillation of the oil, there came over from 150—170° 31 p. c., from 170—180° 38 p. c., from 180—190° 15 p. c., and above 190° 16 p. c.

191. Oil of Buchu Leaves.

Oleum Buccu Foliorum.—Buccublätteröl.—Essence de Feuilles de Buco.

Origin and History. The genus *Barosma* belonging to the family *Rutaceae* is distributed in southern Africa in various species. The resin- and oil-containing leaves of these shrubs appear to have been used for a long time by the natives as a medicine. In the year 1820 the leaves were brought from Capetown through London into the European market (Reece[2]). Since the latter twenties they have been made official in most of the pharmacopœias.

In commerce, round and long buchu leaves are distinguished: the first are the leaves of *Barosma betulina* Bartl., and *B. crenulata* L., the second those of *B. serratifolia* Willd. The latter are sometimes adulterated with the leaves of *Empleurum serrulatum* Ait.[3]

The volatile oil of buchu was prepared in 1827 by Brandes.[4] It was investigated by Flückiger in 1880, by Spica in 1884, by Shimoyama in 1888 and by Bialobrzeski in 1896. From the leaves of *Barosma betulina* there are obtained by distillation 1.3—2 p. c. of oil,[5] from which separate even at ordinary temperature crystals of diosphenol. The liquid portion separated from this has the sp. gr. 0.957—0.97 at 15°. The density of a normal oil which had been warmed to 27° in order to dissolve all the diosphenol, was 0.943.

The leaves of *Barosma serratifolia* yield only 0.8—1 p. c. The oil, poor in diosphenol and liquid at the ordinary temperature, has the sp. gr. 0.944—0.961 at 15°.

Oil of buchu leaves is dark in color and has a strong camphor- and mint-like odor and a bitter, cooling taste.

---

1) Imperial Institute Journal, 2, p. 302; Pharm. Journ., 57, p. 199.
2) Monthly Gazette of Health, London, Feb. 1821, p. 799.
3) Prantl's Lehrbuch der Botanik, von Ferd. Pax, p. 336; Pharmacographia, p. 110.
4) Archiv der Pharm., 22, p. 229.
5) Bericht von S. & Co., Apr. 1891, p. 6; Pharm. Journ., III, 25, p. 796.

COMPOSITION.[1] The stearoptene crystallizing from oil of buchu leaves in the cold is a phenol of the formula $C_{10}H_{16}O_2$. Diosphenol, as the body has been called, melts at 82°, boils under 14 mm. pressure at 112°, under ordinary atmospheric pressure at 232° with partial decomposition, and is optically inactive. With ferric chloride it gives a dark green color reaction. Besides its properties as a phenol, diosphenol shows the behavior of an aldehyde as it reacts with hydroxylamine and phenyl hydrazine. Opposed to the supposition that an aldehyde group is present in diosphenol is the fact that by the action of alcoholic potassa and heat an acid $C_{10}H_{18}O_3 + H_2O$, diolic acid m. p. 96°, is produced, whereas from an aldehyde $C_{10}H_{16}O_2$ the acid $C_{10}H_{16}O_3$ and the alcohol $C_{10}H_{18}O_2$ would be expected. Upon reduction diosphenol yields diolalcohol, $C_{10}H_{18}O_2$, crystals melting at 159°; upon oxidation with moist silver oxide or permanganate, a liquid acid $C_{10}H_{16}O_3$ results.

The remaining constituents of the oil boil lower than diosphenol. In the first portion is contained a hydrocarbon $C_{10}H_{18}$ of the boiling point 174 to 176° (at 14 mm. from 65—67°). This hydrocarbon forms a mobile, strongly active liquid of a pinene-like odor. Sp. gr. 0.8647; $[\alpha]_D = + 60° 40'$.

The fraction boiling from 206—209° consists of a slightly laevogyrate ketone ($[\alpha]_D = - 6° 12'$) $C_{10}H_{18}O$, which possesses a pure peppermint-like odor. With hydroxylamine it yields a viscous oxime, slightly dextrogyrate and boiling between 134—135° under a pressure of 15 mm. The ketone is probably identical with laevogyrate menthone, as its reaction with bromine also indicates, with which it forms an oily compound of the composition $C_{10}H_{17}BrOBr_2$.

192. Oil of Empleurum Serrulatum.

The leaves of *Empleurum serrulatum* Ait., which are now and then mixed in with the long buchu leaves, contain 0.64 p. c. of volatile oil,[2] the odor of which reminds of rue and is quite different from that of buchu leaves. Sp. gr. 0.9464. It boils between 200 and 235°, for the greater part between 220 and 230°. From this fraction a crystalline compound is obtained by treating with sodium bisulphite, which seems to indicate the presence of methyl nonyl ketone, the main constituent of oil of rue.

---

1) Flückiger (1880), Pharm. Journ., III, 11, pp. 174 and 219 —Spica (1885) Gazetta chim. ital., 15, p. 195; Abstr. Jahresb. f. Chem., 1885, p. 56.—Shimoyama (1888), Archiv. d. Pharm., 226, p. 403.—Bialobrzeski (1896), Pharm. Zeitschr. f. Russl., 35, pp. 401, 417, 433, 449.—Kondakow (1896), Journ. f. prakt. Chem., II, 54, p. 433.

2) Pharm. Journ., III, 25, p. 796.

193. Oil of Jaborandi Leaves.

Oleum Foliorum Jaborandi.—Jaborandiblätteröl.—Essence de Feuilles de Jaborandi.

The genuine jaborandi leaves from *Pilocarpus jaborandi* Holmes[1] (Family *Rutaceae*) yield upon distillation 0.2—1.1 p. c. of volatile oil,[2] the yield depending upon the freshness of the material. It has a strong odor reminding somewhat of rue and a mild, fruity taste. Sp. gr. 0.865—0.895; $\alpha_D = +3°25'$. It dissolves in 2 parts of 80 p. c. alcohol to a clear solution. It boils from 180—290° and sometimes solidifies on cooling.

By fractionation a hydrocarbon called pilocarpene (Hardy[3]) (sp. gr. 0.852 at 18°, $[\alpha]_D = +1.21°$) can be isolated, which forms with hydrochloric acid a solid dihydrochloride. According to this, pilocarpene appears to be dipentene accompanied by small amounts of an optically active substance. Kleber[4] has also found methyl nonyl ketone and homologous ketones.

The fractions boiling above 260° solidify in the cold and contain a compound melting at 27—28° which is evidently a hydrocarbon as it is capable of taking up considerable amounts of bromine in petroleum ether solution and probably belongs to the olefinic series.[5]

Jaborandi bark contains but very little volatile oil.

194. Oil of Angostura Bark.

Origin and Properties. The genuine angostura bark from *Cusparia trifoliata* Engl. (*Galipea cusparia* St. Hil., *G. officinalis* Hancock) (Family *Rutaceae*) coming from Venezuela and the upper Orinoco districts, yields by distillation with steam 1.5[6]—1.9[7] p. c. of volatile oil of an aromatic odor and taste. It is slightly yellow at first, but later becomes much darker. Sp. gr. 0.93—0.96; angle of rotation $\alpha_D = -36°$ to $-50°$.

Composition. According to an extensive investigation by Beckurts and Troeger[8] the sesquiterpene alcohol, galipol, $C_{15}H_{28}O$, is the aromatic principle of angostura oil. It boils between 260 and 270°, has the

1) Pharm. Journ., 55, pp. 522, 589.
2) Bericht von S. & Co., Apr. 1888, p. 44
3) Bull. Soc. chim., II, 24, p. 497; Abstr. Chem. Centralbl., 1876, p. 70.
4) Private communication.
5) Bericht von S. & Co., Apr. 1899, p. 28.
6) Bericht von S. & Co., Apr. 1890, p. 47.
7) Journ. de Pharm. et Chim., IV, 26, p. 180; Abstr. Jahresb. d. Pharm., 1877, p. 178.
8) Archiv d. Pharm., 235, pp. 518 and 684; ibid., 236, p. 392. — Comp. also Beckurts & Nehring, Archiv d. Pharm., 229, p. 612; also Herzog, ibid., 148, p. 146.

sp. gr. of 0.9270 at 20° and is optically inactive. The highly unstable alcohol when heated splits off water, and is contained in the oil to the extent of about 14 p. c.

A prominent constituent of the oil is cadinene $C_{15}H_{24}$, which is the cause of the laevogyrate rotation; it was characterized by its crystalline hydrochloric acid addition product.

Besides the laevogyrate cadinene and the inactive alcohol there is contained in angostura oil an inactive sesquiterpene, called galipene. This boils at 255—260° and has the sp. gr. 0.912 at 19°. It forms liquid, easily decomposed compounds with the hydrohalogens.

A terpene is found in small amounts in angostura oil, which appears to be pinene.

195. Toddalia Oil.

Toddalia aculeata Pers. (Family *Rutaceae*) is a shrub which grows wild in the Nilgiri mountains (India) and which is there called wild orange tree. All parts of the plant have a sharp, aromatic taste. The root, which was already known in the seventeenth century under the name of *Radix Indica Lopeziania*[1] is called by the natives *Malakarunnay* and is employed as a domestic remedy for indigestion. The bark of the root contains a volatile oil, which is described by Schnitzer[2] as having a cinnamon and melissa-like odor. The ripe berries are used in India as a spice in place of black pepper; the bark and leaves are also said to possess medicinal properties.[1]

The oil of the leaves has been distilled by D. Hooper.[3] It is very fluid and has a pleasant odor, reminding at the same time of verbena and basilicum, and contains considerable quantities of citronellal and an alcoholic constituent boiling above 200°.

The oil would be useful for purposes of perfumery if it could be furnished at a reasonable price.

THE OILS OF THE AGRUMEN FRUITS.[4]

Origin and History. The subfamily *Aurantieae* of the *Rutaceae* is indigenous to central Asia. The large number of varieties of the citrus fruits, known by the collective term of agrumen (acid) fruits is indicative of a very long period of cultivation. As far as oils that are expressed from the rinds of the fruits are concerned, the following species come into consideration:

---

1) Pharmacographia, p. 111.

2) Wittstein's Vierteljahrsschrift f. prakt. Pharmacie, 11, p. 1.

3) Bericht von S. & Co., Apr. 1898, p. 64.

4) Acid fruits.

for lemon oil: *Citrus limonum* Risso,
for bergamot oil: *Citrus bergamia* Risso,
for grape fruit oil: *Citrus decumana* L.,
for cedro oil: *Citrus medica* Risso,
for limette oil: *Citrus medica, var. acida* Brandis, and *Citrus limetta* Risso,
for mandarin oil: *Citrus madurensis* Loureiro,
for orange oil: *Citrus aurantium* Risso, and *Citrus bigaradia* Risso.

During the process of ripening of the fruit the oil is secreted in cells of the outer layer of the rind. When ruptured the oil exudes and can be collected. This oil must have been known early although it does not seem to have found application.

The earliest statements concerning distilled lemon and orange oils were made by Gesner in 1555. Besson followed in 1571, Porta in 1589. The latter described the preparation of the two oils by distilling the fresh, grated rinds. During the sixties of the last century Gaubius recommended the same process.

The method of the mechanical preparation of the agrumen oils by rupturing the oil cells by means of a grate was described by Geoffroy in the beginning of the eighteenth century. Evidently this method was practiced before this date.

In the price ordinances, the oils of lemon and orange appear first in that of Frankfurt-on-the-Main of 1582. Both oils are mentioned in the Dispensatorium Noricum of 1589 and in the Pharmacopoea Augustana of 1613. Oil of bergamot apparently came into use about 1690.

In 1786 Remmler[1] tried to prepare rosin from oil of lemon. About the same time Liphard[2] mentions that the yield of lemon oil is larger when the fruit is allowed to stand until decay sets in. In 1789 the apothecary Heyer of Braunschweig, upon cooling bergamot oil, obtained crystals which he termed bergamot camphor.[3]

PREPARATION. The preparation of the volatile oils from the agrumen fruits is not effected by the otherwise usual method of distillation, but by expressing the rinds of the fruit. The distilled oil is inferior and valueless. The oil is contained in the cells of the outer layer which need only be ruptured in order that they may yield their odorous content.

---

1) Göttling's Taschenb. f. Scheidekünstler, 1786.
2) Crell's Chemische Annalen, 1787, II, p. 250.
3) Ibidem, 1789, I, p. 320.

In southern Italy, i. e. in Sicily and the southern point of Calabria, three different methods are in use for obtaining lemon, orange and bergamot oils, the *Spugna* process, the *Scorzetta* process and the manufacture with the *Macchina*.

In the *Processa alla Spugna*[1] the worker cuts the fruit into quarters, removes the acid fruit pulp and presses the rind of the fruit against a sponge which he holds firmly in his right hand. The oil cells are broken and give their content to the sponge, which is expressed by the hand as soon as it has absorbed a sufficient quantity of oil. The sponges belong to that kind which are usually designated as medium horse sponges; after such a sponge has been used for about 10 days it loses its usefulness, it is now coarse and unable to further absorb any oil.

Somewhat different from the method just described is the *Scorzetta* process.[2] In this the fruit is cut transversely into halves and the content removed by a kind of spoon. Both halves of the rind are then pressed on all sides against the sponge by continually turning them in the hand. This method of obtaining the oil has the advantage that the rinds so treated are not broken, but retain their original appearance and can be salted and exported as so-called *Salato*. Besides, the whole inner part of the fruit remains intact and can be worked more economically into lemon juice than is possible with the fruit sections treated by the *Spugna* process. After a double expression for oil and juice these can only be used as food for cattle.

The *Macchina* is an ingeniously constructed, fairly complicated machine, in which the outer layer of the fruit is first ruptured and then pressed against sponges which absorb the volatile oil. The machines are now almost wholly used in the manufacture of bergamot oil, as the round, uniform shape of the bergamot is especially suited for this method of preparation. For the production of lemon oil the machine is useless for two reasons. First, the irregular form of the lemon prevents all parts of the rind from coming in contact with the sponges, and second, the lemon oil thus obtained is green and is, therefore, not marketable as such. The lemon oil sometimes prepared with the machines is used for adulterating bergamot oil.

The *Ecuelle à piquer* is an instrument employed in Nice for the preparation of the oil. It consists of a bowl, provided with upright brass needles, which is prolonged into a tube at the bottom. The fruit pricked by the needles allows its oil to run into the bowl and collect in the tube from which it is filled into vessels.

---

1) Archiv d. Pharm., 227, p. 1065.

2) Bericht von S. & Co., Apr. 1896, p. 25.

In southern France hardly any oil is at present obtained from agrumen fruits. The *Ecuelle* method is said to be still used in the West Indies for the preparation of limette oil.

Sometimes the unripe fruit which has fallen from the trees, as well as the expressed residues or the filter pulp, are distilled with water. The oil obtained is without exception of a very poor quality and serves for mixing with only the poorest kinds of the pressed oils.

The finished oil, previously purified by filtration, comes into commerce in copper flasks of 50 k. capacity.

PRODUCTION AND COMMERCE. The following figures, taken from the official reports (*Statistica del Commercio Speciale*), give an interesting exposition of the enormous extent and the increase in the production and export of bergamot, lemon, mandarin and orange fruits during the last few years.

The production of Italy amounted to:

1892	3,163	million fruits
1893	3,140	" "
1894	3,320	" "
1895	3,550	" "
1896	3,337	" "

of these were exported:

1892	1,704,628	double hundred weights
1893	1,978,134	" " "
1894	2,148,011	" " "
1895	2,206,870	" " "
1896	2,372,369	" " "

The harvest of the campagne 1895/96 was:

Agricultural regions	Oranges		Lemons		Cedros, Mandarins, Bergamots etc.		Total Agrumens	
	Number of trees	Number of fruits in thousands	Number of trees	Number of fruits in thousands	Number of trees	Number of fruits in thousands	Number of trees	Number of fruits in thousands
Lombardy	755	39	21,242	1,918	3,350	83	25,347	2,040
Venice	144	5	1,210	156	12	0.2	1,366	162
Liguria	1 4,950	12,045	409,652	15,777	38,750	9,131	553,352	36,953
Marken and Umbria	5 ,509	1,452	17,842	237	460	4	70,811	1,693
Tuscany	2,203	320	16,486	819	759	29	19,448	1,168
Latium	11,760	1,478	19,873	2,411	1,370	79	33,003	3,969
South. Adriatic region	343,730	70,436	134,913	23,022	13,482	1,029	492,125	94,486
South. Mediterranian region	2,666,693	541,634	1,031,116	193,588	1,023,450	179,784	4,721,259	915,003
Sicily	4,359,203	544,350	6,243,158	1,661,994	818,779	57,043	10,921,143	2,263,386
Sardinia	187,797	14,649	40,786	3,237	18,132	693	246,715	18,580
Total	7,729,747	1,186,408	7,936,278	1,903,159	1,418,544	247,875	17,084,569	3,337,443

If 800 fruits are calculated to a double hundred weight the year 1896 would give the following result:

Total production	3,337,443,000 fruits
Total export 2,372,360 double hundred weights @ 800 pieces	1,897,888,000 "
Remain for Italian consumption inclusive of the essence manufacture	1,439,555,000 fruits

Among the consumers in 1896 the most prominent are the United States and Canada with nearly 50 percent of the total exports. Then follows Austria-Hungary with about 20 percent, Great Britain with about 14 percent, Russia with 6 percent and Germany with about 5 percent.

The total export of bergamot, lemon and orange essences of the individual centers of production during the last two years was as follows:

1897

Export from Messina	560,788	k. having a value of	7,570,618	lire
" " Reggio	87,095	" " "	1,306,425	"
" " Catania	12,016	" " "	120,160	"
" " Palermo	72,193	" " "	721,153	"
Total essence export 1897	732,092	k. having a value of	9,718,856	lire

1898

Export from Messina	524,099	k. having a value of	6,813,287	lire
" " Reggio	85,069	" " "	1,446,117	"
" " Catania	6,366	" " "	82,758	"
" " Palermo	51,759	" " "	672,867	"
Total essence export 1898	667,293	k. having a value of	9,015,039	lire

During the last 10 years the essence export amounted to:

1889	277,599	k. having a value of	4,206,258	lire
1890	301,879	" " "	5,056,214	"
1891	264,150	" " "	4,954,655	"
1892	359.378	" " "	5,543 358	"
1893	588,334	" " "	9,356,814	"
1894	666,740	" " "	8,308,148	"
1895	554,191	" " "	8,081,870	"
1896	514,067	" " "	7,579.424	"
1897	732,092	" " "	9,719.133	"
1898	667,293	" " "	9,015,083	"

The manufacture of the essences produced by expression is almost entirely carried on in Calabria and Sicily. The only exception is the preparation of the West Indian limette oil at Montserrat.

It is true that now and then samples of expressed essences from the West Indies and from Florida appear, but one cannot yet speak of a

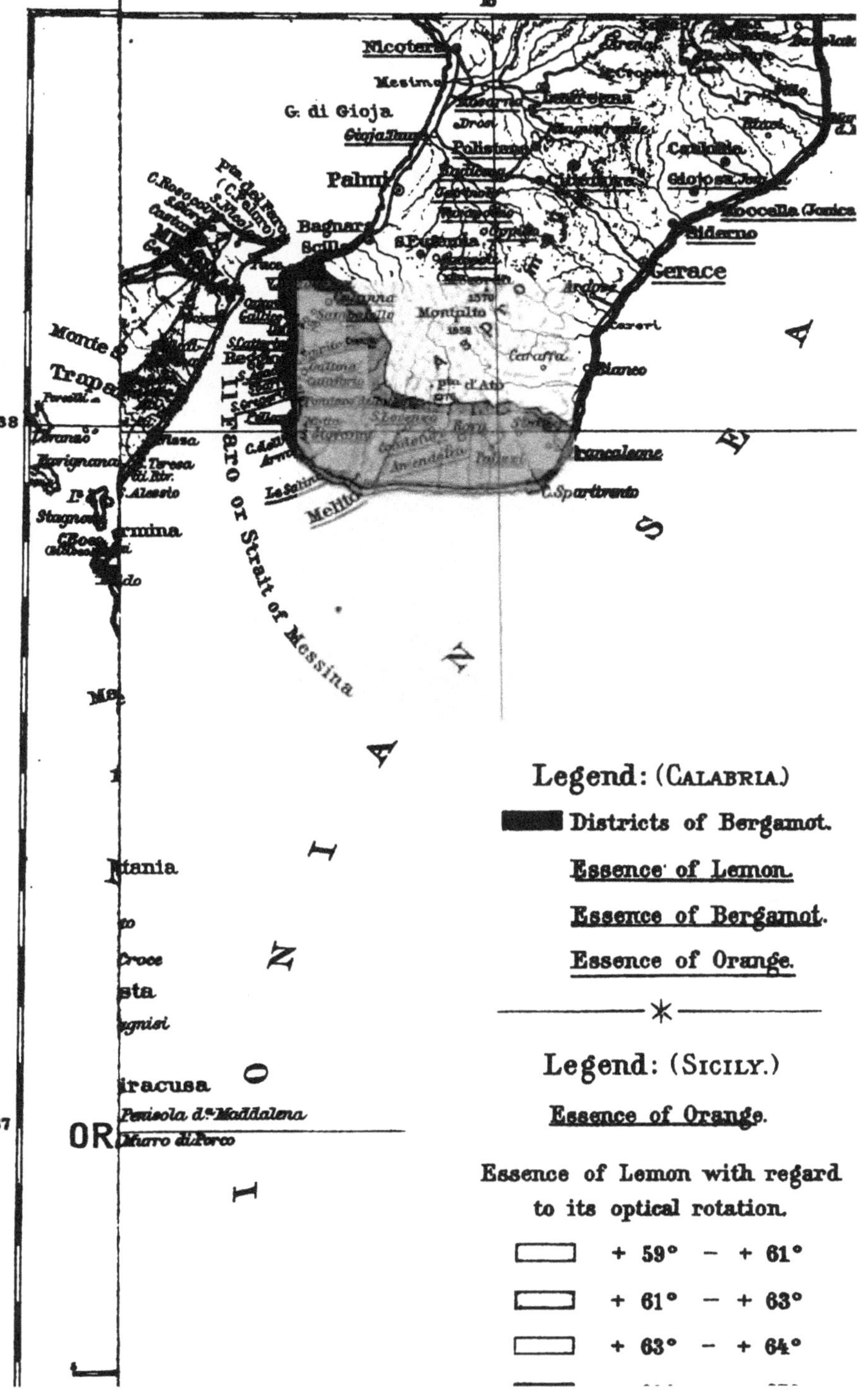

Legend: (Calabria.)
Districts of Bergamot.
Essence of Lemon.
Essence of Bergamot.
Essence of Orange.
Legend: (Sicily.)
Essence of Orange.
Essence of Lemon with regard to its optical rotation.
+ 59° – + 61°
+ 61° – + 63°
+ 63° – + 64°
Il Faro or Strait of Messina
G. di Gioja
Nicotera
Palmi
Bagnara
Scilla
Gerace
Montalto
Siderno
Bianco
Caraffa
C. Spartivento
Melito
Le Saline
Monte S.
Trapani
Siracusa
Penisola d.a Maddalena
Murro di Porco
16
38
37

manufacture in these places. It is hardly to be expected that one will arise, as the high wages there paid cannot compete with the low wages of Italy. The preparation of expressed agrumen oils formerly conducted on a small scale appears to have been entirely given up.

The districts of production can be seen on the accompanying map, which was prepared in 1896 by a representative of the firm of Schimmel & Co. Bergamots are only cultivated in Calabria. The principal center for the oil is Reggio. Lemons and oranges are planted on the continent as well as in Sicily.

In Calabria lemon oil is made in the same places as bergamot oil. In Sicily the principal productive centers of lemon oil are in the provinces of Messina, Catania, Syracuse and Palermo. The oils from the different districts often show quite important variations in their optical rotation. The angle of rotation of the Calabrian oils lies between +59 and +62°30′.

In Sicily the principal places of production, arranged according to the angle of rotation of their oils are as follows:

$\alpha_D = +59°$ to $+61°$. Messina and vicinity, Nizza di Sicilia.
$\alpha_D = +61°$ to $+63°$. Acireale, S. Teresa di Riva, Scaletta, S. Lucia, Patti, S. Agata, S. Stefano.
$\alpha_D = +63°$ to $+64°$. Catania, Giarre, Giardini, Acireale, Leutini.
$\alpha_D = +64°$ to $+67°$. Barcellona, Siracusa.

In the province of Palermo the places colored yellow on the map produce only fruit but no oil, the manufacture of which is restricted to the immediate vicinity of Palermo to which the fruit is transported. The angle of rotation of the oils there produced lies between +59° and +61°; rotations of +61 to 63° are very rare.

196. Oil of Lemon.

Oleum Citri.—Citronenöl.—Essence de Citron.

PROPERTIES. Oil of lemon is a light yellow liquid of the pleasant odor of fresh lemons and an aromatic, mild, somewhat bitter after-taste. Sp. gr. 0.858—0.861. The angle of rotation lies as a rule between 60 and 64° at 20°.[1] Oils with a rotatory power of +64 to 67° are rare.[2]

[1]) It will be well to make the rotation determinations for lemon oil at 20° or to reduce to 20° by calculation, for the same reason as that given under orange oil (remark 6 on page 471). If the rotation is determined at a temperature below 20°, 9 minutes are to be subtracted from the result found for every degree of temperature below 20°. On conducting the investigation at a temperature above 20°, 8.2 minutes are, however, to be added for every degree in temperature in order to find the angle of rotation for +20°.

[2]) The oils obtained in the vicinity of Siracusa and Barcellona (Sicily) distinguish themselves by a high rotatory power. The lemon oils from the province of Palermo are the weakest in rotatory power (+59 to 61°). Compare with above.

The gummy and vegetable wax-like constituents present often prevent the clear solubility of lemon oil in 90 p. c. alcohol. Rectified oil requires about 5 p. of 90 p. c. alcohol to form a clear solution. Absolute alcohol, ether, chloroform, benzol and amyl alcohol dissolve lemon oil in all proportions. The solutions in carbon disulphide and in benzene are usually somewhat cloudy on account of a little water contained in the oil. Oil of lemon, like all essential oils obtained by expression, gives on standing a more or less crystalline deposit.

Inasmuch as rectified and distilled oils decompose rapidly and have an unpleasant, penetrating odor, it is irrational to rectify a lemon oil.

Oil of lemon is rapidly changed by the action of light and air. It loses its color and a thick, brown deposit separates. At the same time the specific gravity and the solubility in 90 percent alcohol increase. This is the same phenomenon that can be observed with old, poorly kept turpentine oil.

Preservation. Oil of lemon is to be kept in carefully closed vessels, filled to the neck, in the dark and in a cool place.

Composition. Although a large number of investigators have worked on oil of lemon since the first quarter of the present century, its composition is for all that not yet satisfactorily cleared up.

A critical examination of the rather extensive literature,[1] the details of which cannot here be considered, shows that a part of the oils investigated were adulterated with turpentine oil. This is not surprising, as it is only now possible, by means of the recent proof[2] that lemon oil contains no pinene, to distinguish pure lemon oil from that adulterated with turpentine oil.

Even by the first analyses of the oil its small amount of oxygen was recognized. Some chemists, however, went so far as to declare it free from oxygen and consisting only of hydrocarbons. This mistake arose from the fact that rectified oils were mostly used in the investigation, in which the much higher boiling oxygenized compounds remained in the residue of the flask.

---

1) Saussure (1820), Ann. de Chim. et Phys., II, 13, p. 262; Liebig's Annalen, 3, p. 157. — Dumas (1833), Ann. de Chim. et Phys., II, 52, p. 45; Liebig's Annalen, 6, p. 255, and ibid., 9, p. 61. — Blanchet & Sell (1833), Liebig's Annalen, 6, p. 318. — Soubeiran & Capitaine (1840), Journ. de Pharm., II, 26, p. 1; Liebig's Annalen, 34, p. 317.—Gerhardt (1848), Compt. rend., 17, p. 314.—Berthelot (1853), Ann. de Chim. et Phys., III, 37, p. 233; ibid. 38, p. 44; ibid., 40, p. 36.—Liebig's Annalen, 88, p. 346. — Oppenheim (1872), Berichte, 5, p. 628. — Lafont (1887), Bull. Soc. chim., II, 48, p 777, and 49, p. 17.

2) Bericht von S. & Co., Apr. 1897, p. 19; Oct. 1897, p. 22.

In reality about 9/10 of the oil consist of hydrocarbons, of which dextrogyrate limonene (Wallach,[1] 1885), (tetrabromide, m. p. 124—125°) is the most important.

Tilden[2] (1877) has called attention to the fact that this terpene is contained in a far less pure state in lemon oil than in orange oil. By the oxidation of the limonene fraction designated by him as "citrene" he obtained paratoluic and terephthalic acids, two acids which do not result by the oxidation of the corresponding fractions from orange oil. "Citrene" behaves likewise different than limonene when treated with concentrated sulphuric acid. Besides resinous polymerization products some cymene is formed. It does not follow, however, that cymene is contained in the original oil. It is much more likely that the cymene resulted from a second terpene by the action of sulphuric acid. As a matter of fact neither cymene nor pseudo-cumene[3] have up to the present been found in lemon oil which had not been treated with sulphuric acid.[4]

The reason for the differences between "citrene" and limonene may probably be found in part in the presence of phellandrene in lemon oil. This terpene has only recently been discovered in the oil and recognized as such by its nitrite melting at 102°.[4]

The behavior of the fraction boiling below 170° observed by Wallach[5] can in all probability be also traced to phellandrene. This fraction gave no solid addition product by direct bromination[6] but after heating from 250—270° it readily yielded dipentene tetrabromide.

Although the terpenes make up by far the greater part of the lemon oil, the odor is largely due to the comparatively small amount of oxygenated compounds. The constituent of greatest importance as to odor is citral, $C_{10}H_{16}O$, an aldehyde, which was found by Bertram in oil of lemon in 1888.[7] The amount of the citral is estimated at 7—10 percent.

Besides citral lemon oil contains a second aldehyde, citronellal. When condensed with pyrotartaric acid and β-naphthylamine this yields

1) Liebig's Annalen, 227, p. 290.

2) Pharm. Journ., III, 8, p. 190; 9, p. 654.

3) Bouchardat & Lafont (Journ. de Pharm., V, 27, p. 49) found, after treating citrene with sulphuric acid, pseudo-cumene besides cymene. They seem to consider these substances as preexisting in lemon oil.

4) Bericht von S. & Co., Oct. 1897, p. 23.

5) Wallach, loc. cit.

6) Oliveri claims to have obtained a tetrabromide of the melting point 81° from the fraction of lemon oil boiling at 170 to 170.5° (Gazz. chim. ital., 21, I, p. 318; Berichte 24, p. 624, Ref.).

7) Bericht von S. & Co., Oct. 1888, p. 17.

according to Doebner[1] citronellal . -naphtho cinchoninic acid, m. p. 225°. Apparently the statements of Ladell[2] who investigated so-called terpene-free lemon oil, refer also to citronellal. He isolated by fractionation a dextrogyrate body $C_{10}H_{18}O$ boiling at 206°.

Tilden[3] in 1877 described as a constituent of lemon oil a dextrogyrate substance of the composition $C_{10}H_{18}O$, boiling a little above 200°, which, with the exception of its rotatory power, agreed in its principal properties with "terpinol." It is probable that Tilden also had citronellal under consideration. Barbier and Bouveault,[4] however, deny the presence of citronellal in lemon oil.

According to Umney and Swinton[5] geranyl acetate is also contained in the lemon oils from Messina and Palermo. They removed from so-called concentrated lemon oil, i. e. from the high boiling fractions rich in oxygen, the aldehyde with hot bisulphite solution, and saponified the non-aldehydes. From the alkaline solution acetic acid was separated and from the oil a fraction was obtained which combined with calcium chloride to form a solid compound, which yielded citral on oxidation, and consequently consisted of geraniol, which must have existed originally in the oil as acetic ester. From Palermo lemon oil, besides geraniol, a fraction of the properties of l-linalool was isolated. Umney and Swinton are of the opinion that the differences in odor which exist between the lemon oils from Palermo and Messina can be attributed to a different ratio of the quantities of citral and citronellal, as well as to the presence of linalyl acetate in the Palermo oil.

A sesquiterpene, $C_{15}H_{24}$, of the boiling point 240—242° is contained according to Oliveri (*loc. cit.*) in the highest boiling fractions of lemon oil.

The statements found in literature[6] referring to the so-called lemon camphor or citraptene, the non-volatile constituent of lemon oil, vary so much both as to composition and properties (m. p. from 45—144°) that the body must at present be considered as a mixture of different substances.

ADULTERATION AND EXAMINATION. Turpentine oil, the adulterant mostly employed, can be readily detected by means of the polariscope.

---

1) Archiv d. Pharm., 232, p. 688; Berichte, 27, p. 852.

2) Pharm. Journ., III, 24, p. 586.

3) Footnote 2 on p. 467: comp. also Wright, Journ. Chem. Soc., II, 12, pp. 2 and 317; Berichte, 6, p. 1320.

4) Compt. rend., 122, p. 85.

5) Pharm. Journ., 61, pp. 196 and 370.

6) Mulder (1839), Liebig's Annalen, 31, p. 69. — Berthelot, loc. cit. — Tilden & Beck (1890), Journ. Chem. Soc., 57, p. 328; Berichte, 23, p. 500, Ref. — Crismer (1891), Bull. Soc. chim., III, 6, p. 30; Berichte, 24, p. 661, Ref.

as it decreases the rotatory power of the lemon oil to a considerable extent. In order to determine approximately the amount of the addition, it must first be determined whether French or American turpentine had been employed. As turpentine oil (pinene) boils lower than the constituents of normal lemon oil, it is to be sought for in the fraction which distills over first. The lemon oil to be investigated is therefore fractionated several times and the rotatory power of the fraction boiling at about 160—165° determined. If the first distillate finally becomes laevogyrate, French turpentine oil is under consideration, but if it remains slightly dextrogyrate, American turpentine oil has been employed.

For the approximate calculation of the amount of turpentine oil from the rotatory power, the mean angle of rotation —30° for French and + 6° for American turpentine oil may be taken.

A lemon oil consisting of half turpentine oil would rotate in a 100 mm. tube in the one case $\frac{+60^\circ + -30^\circ}{2} = +15^\circ$, and in the other case $\frac{+60^\circ + 6^\circ}{2} = +33^\circ$.

According to this it is not difficult to calculate the approximate amount of the addition, providing that it is known whether this addition was French or American turpentine oil.

Detection of Turpentine oil in the presence of orange oil. Lemon oils, in spite of their normal rotatory power, may be adulterated with turpentine oil, when orange oil has been added to compensate the decrease in rotatory power. Even this manipulation is detected in the polariscope by testing individual fractions of the oil.

By numerous experiments it has been determined that by the distillation of pure oils the angle of rotation of the 10 p. c. which distil over first is at most 4—5° smaller than that of the original oil. With oils containing turpentine oil this difference is of course much greater. For the distillation a so-called Ladenburg fractionating flask with three bulbs (fig. 54 on p. 190) is used and from 50 cc. of the oil to be investigated, just 5 cc. are slowly distilled off. At first a few drops of water distill over, which make the distillate turbid and are removed by shaking with some anhydrous sodium sulphate. After filtering, the distillate is investigated in a 50 mm. tube in the polariscope with the proper consideration of the temperature. The result is reduced by calculation to 20° in the manner given in footnote 1 on page 465 and deducted from the rotatory power, likewise reduced to 20°, of the original oil.[1]

In some cases, especially when only small quantities of oil are at command, the modification of Soldaini and Berté[2] is preferable. This consists in that from 25 cc. of the oil one-half is distilled off. The rotation of the 50 p. c. distilled over from pure oil is higher than that of the original oil and higher than that of the residue. When turpentine has been used as adulterant the first half will always show a lower rotatory power.

It should be mentioned that the addition of turpentine oil produces only a very slight increase in the specific gravity.

Fatty oil remains in the residue when the oil is evaporated; the residue of normal lemon oil does not amount to more than 4—5 p. c. For its determination see under bergamot oil on page 476.

Citral Determination. A very dangerous adulterant which up to the present has defied definite detection, consists in the terpenes obtained in the manufacture of the so-called terpene-free lemon oil. As by the addition of these hydrocarbons to the lemon oil the relative amount of the oxygenated constituents, that is of citral and citronellal, is decreased, such an adulteration might be detected by an accurate citral or aldehyde determination. On the other hand it has been claimed that a citral determination is not an absolute standard of purity.[3] Thus, an oil of high citral content might be nothing more than the terpenes resulting from the preparation of so-called concentrated lemon oil, mixed with citral from some other source, for instance, from lemon grass oil. Attempts to establish methods for the estimation of citral have not been wanting, but so far they have all failed on account of insufficient accuracy.

Thus, for instance, it has been tried to determine the amount of aldehyde present in the manner employed for cassia oil,[4] which is indeed very rational, since citral like cinnamic aldehyde forms with bisulphite a compound soluble in water. Citronellal, however, does not possess this property, and its double salt separating partly in the oily and partly in the aqueous layer makes an accurate reading impossible.

---

1) Bericht von S. & Co., Oct. 1896, p. 80.
2) Gazz. chim. ital., 27, II, p. 25.
3) Chemist & Druggist, 58, pp. 120, 292, 949; 54, pp. 495, 545.
4) See Soldaini and Berté. Boll. chim. farm., 38, p. 587; Berichte von S. & Co., Apr. 1900, p. 22.

Garnett[1] tried to convert the aldehydes by reduction with sodium into alcohols, and determine these by acetylization. Unfortunately, as thorough examinations have shown, the reduction does not take place quantitatively and the results are therefore useless.[2]

Walther[3] has published a method for the estimation of citral in lemon oil by treating the latter in alcoholic solution with hydroxylamine hydrochloride and sodium bicarbonate and then titrating back the excess of hydroxylamine. Schimmel & Co.[4] have tested the method and found that the results are too high.

Parry[5] has based a method on the formation of citralidene cyanacetic acid. From 200 cc. of oil 175 cc. are distilled off under 15 mm. pressure. 10 cc. of the residue are shaken in a cassia flask with 5 g. of cyanacetic acid and 5 g. of sodium hydrate in 30 cc. of water. The part insoluble in water is read off on the scale. No determinations were made on known mixtures of citral and terpenes, so that the accuracy of the method is unknown.

197. Oil of Sweet Orange.

Oleum Aurantii Dulcis. — Süsses Pomeranzenschalenöl. Süsses Orangenschalenöl. Apfelsinenschalenöl. — Essence d'Orange Portugal.

PROPERTIES. Oil of orange is a yellow to yellowish-brown liquid of a characteristic orange odor and a mild, aromatic, not bitter taste. Specific gravity 0.848—0.852; $\alpha_D = +96$ to $+98°$ at 20°.[6]

On account of the presence of wax-like, non-volatile substances of unknown composition, which partly separate on standing for some time, the oil as a rule does not form a clear solution with 90 p. c. alcohol. It begins to boil at 175°; up to 180° nine-tenths distill over.

The rectified oil is colorless; its sp. gr. is somewhat lower, the rotatory power slightly higher than that of the original oil. Rectified orange oil is kept with difficulty, it deteriorates rapidly and acquires thereby a stale, grating odor.

1) Chemist and Druggist, 48, p. 599.
2) Bericht von S. & Co., Oct. 1896, p. 32.
3) Pharm. Centralh., 40, p. 621.
4) Bericht von S. & Co., Apr. 1900, p. 20.
5) Chemist & Druggist, 56, p. 376.
6) Since the angle of rotation of orange oil, like that of lemon oil, varies greatly with changes in temperature, decreasing with an increase in temperature, it is necessary, in order to obtain comparable numbers, to ascertain accurately the temperature and to reduce the result to 20° by calculation. As the difference in the angle of rotation between +10° and +20° is 14.5 minutes and between +20° and 30° it is 18 2 minutes for one degree of change in temperature, the reduction to 20° is made by deducting 14.5 minutes for each degree in temperature, when the polarisation was effected at a temperature below 20°. If the determination was made at a temperature above 20°, 18.2 minutes are to be added to the number found in order to find the angle of rotation for +20°.

Composition. Orange oil consists, as Wallach[1] (1884) has shown, of at least 90 p. c. of d-limonene (dihydrochloride, m. p. 50° (Soubeiran and Capitaine,[2] 1840), tetrabromide, m. p. 104—105°[1]). On this account, especially as other hydrocarbons are completely absent, it is well suited for the preparation of this terpene in a pure state. The absence of pinene is of importance for the detection of adulteration with turpentine oil.

Of oxygenated compounds, aldehydes are present in orange oil. By shaking with sodium bisulphite solution crystals of a double compound are formed, which can be isolated by filtration and pressing; by decomposing with soda an oil is obtained which is purified by steam distillation. A part of this boils at 224—228° and consists of citral (Semmler,[3] 1891). The lower boiling fraction contains likewise an aldehyde, the composition of which has not yet been determined.

The assertion made by Wright[4] in 1873 that oil of orange peel contains 0.3 percent of a body boiling at 212—218° identical with the somewhat mystic myristicol ($C_{10}H_{16}O$) of the oil of nutmeg, is too little supported by facts.

The most recent and not yet completed investigation of orange oil is that of Flatau and Labbé.[5] They obtained by shaking orange oil with bisulphite solution a double compound, which yielded besides traces of citronellal, a small amount of a new aldehyde that had a very characteristic orange odor. In addition an acid of possibly twenty-one carbon atoms was isolated. Like the acid, its ethyl ester is difficultly soluble in alcohol and can be precipitated by this solvent from the residue of the oil after 95 p. c. have been distilled off. When purified it melts at 64—65° and has a pleasant and characteristic orange odor. Further communications on this ester are promised.

Parry[6] suspects the presence of the methyl ester of anthranilic acid in sweet orange oil. This observation is confirmed by Schimmel & Co.[7] who have definitely shown its presence.

Of the nature of the orange oil stearoptene, which finds its way into the oil by expressing the peel of the fruit and which remains in the residue when the oil is rectified, nothing is known.

1) Liebig's Annalen, 227, p. 289. Comp. also Völckel (1841), ibid., 39, p. 120; and Wright & Piesse (1871), Chem. News, 24, p. 147.

2) Liebig's Annalen, 84, p. 319.

3) Berichte, 24, p. 202.

4) Chem. News, 27, p. 260; Berichte, 6, p. 148.

5) Bull. Soc. chim., III, 19, p. 361.

6) Chemist & Druggist, 56, p. 462.

7) Bericht von S. & Co., Apr. 1900, p. 18.

EXAMINATION. On account of the low specific gravity and the extraordinarily large rotatory power of orange oil, all kinds of foreign additions can be readily and accurately detected, as there is no adulterant by which these two properties would not be changed.

Formerly, when the polariscope was not so generally used as it is to-day the oil was greatly adulterated with turpentine and sometimes even with lemon oil.

Recently[1] the terpenes remaining from the manufacture of terpene-free lemon oil are used to an enormous extent in Messina for the adulteration of orange oil.

For the detection of turpentine oil the lowest boiling portions of the oil are repeatedly fractionated by employing a dephlegmator, and the pinene may then be recognized by its boiling point, as well as by its rotatory power (strongly laevogyrate with French and slightly dextrogyrate with American turpentine oil). Should such a test be considered as not conclusive, the pinene must be converted into pinene nitrosochloride and into the characteristic pinene nitrolbenzylamine or nitrolpiperidine base.

198. Oil of Bitter Orange.

Oleum Aurantii Amari. — Bitteres Pomeranzenschalenöl. — Essence d'Orange Bigarade.

The oil of bitter orange, which plays only a subordinate role in commerce in comparison with oil of sweet orange, differs from this mainly in its bitter taste. The rotatory power[2] is sometimes slightly lower and varies from + 92° to + 98°.

All other properties are the same as those of the sweet oil, and it is impossible to distinguish between the two oils in any other manner than by their odor and taste.

199. Oil of Bergamot.

Oleum Bergamottae. — Bergamottöl. — Essence de Bergamotte.

PROPERTIES. Oil of bergamot is a brownish-yellow or honey-colored liquid often colored green by the presence of copper.[3] It has a bitter taste and a very pleasant odor. Specific gravity 0.882—0.886. The angle of rotation, which on account of the dark color of the oil can mostly be determined only in a 50 or 20 mm. tube, varies from +8 to +20°.

1) Bericht von S. & Co., Oct. 1899, p. 25.
2) Bericht von S. & Co., Apr. 1896, p. 29.
3) Ibidem, Apr. 1889, p. 16.

The oil yields a clear solution with about ¼ to ½ volume of 90 p. c. alcohol and the solution does not become turbid on the addition of more alcohol. All oils do not dissolve clearly in 80 p. c. alcohol. Many, and especially those of a high ester content, often give turbid mixtures, from which fatty globules separate on the bottom by standing. The reason for this phenomenon has not yet been determined, but can probably be sought in the wax-like constituents, which get into the oil by expressing the peel. This non-volatile substance, which partly separates as a deposit when the oil is kept for some time, consists principally of bergaptene. It remains in the residue when the oil is evaporated on a water bath or by rectification; it amounts to 5—6 p. c.

Rectified oil of bergamot is colorless and has a lower specific gravity (0.870—0.880) as well as a slightly higher rotatory power than the original oil. The rectified oil is as a rule less valuable, because during the steam distillation part of the ester is always decomposed.

The amount of ester present is a measure of the value of bergamot oil, i. e. the oil is the better, the more linalyl acetate it contains. The average content in ester varies somewhat in different years; it usually amounts to between 36 and 40 percent, but sometimes rises as high as 45 percent. Even within the limits of the same harvesting period great variations occur. The oils obtained at the beginning of the harvest from less ripe fruit contain less (down to 30 percent linalyl acetate); with increasing ripeness the ester content increases, for which reason the oil expressed from ripe fruit is the best.

COMPOSITION. As early as 1840 Soubeiran and Capitaine[1] called attention to the presence of different terpenes in bergamot oil. Wallach[2] showed in 1884 that d-limonene was contained in the fraction boiling from 175—180°. When he heated the fraction of the oil boiling from 180—190°, which, indeed, absorbed bromine but yielded no solid bromide, to a higher temperature, high boiling condensation products were formed, and on again fractionating, the portion going over up to 190° gave dipentene tetrabromide melting at 124—125°. It does not follow from this, whether the dipentene detected in this manner is to be considered as an original constituent of the oil, or whether it has been produced by heating the fraction 180—190° which no doubt contained linalool. Likewise, the observation made by Semmler and Tiemann[3] in 1892 according to which the oil boiling 17° higher than the limonene fraction

---

[1]) Liebig's Annalen, 35, p. 318.

[2]) Liebig's Annalen, 227, p. 290.

[3]) Berichte, 25, p. 1182.

yielded dipentene tetrabromide cannot be considered as a proof of the presence of dipentene in bergamot oil.

Our knowledge of the most important constituent as far as the odor of the bergamot oil is concerned, is due to two investigations, published at nearly the same time, by Semmler and Tiemann[1] and by Bertram and Walbaum.[2] By these investigations it was shown that the principal carrier of the bergamot odor is the acetic ester of l-linalool.

In addition to this ester, free l-linalool and possibly substances not yet isolated take part in the formation of the aroma. The properties and derivatives of linalool and linalyl acetate have been described on pp. 128—131.

Charabot[3] has made a comparative study of the oils of bergamot from the green and the ripe fruit. The oil from the green fruit had a sp. gr. of 0.882 at 14°, $\alpha_D = +14°38'$; 0.289 p. c. of free acid, 33.8 p. c. of linalyl acetate. 13.9 p. c. of linalool, and 5.9 p. c. of bergaptene. The oil from the ripe fruit had a sp. gr. of 0.883, $\alpha_D = +20°30'$, 0.283 p. c. of acid, 37.3 p. c. of ester. 5.9 p. c. of linalool and 5.5 p. c. of bergaptene. He draws the conclusion that in ripening the original linalool is changed to the ester and that during this process some of the linalool is dehydrated with the formation of terpenes.

The bergaptene contained in the oil to the extent of about 5 percent is completely odorless. A whole series of investigations has been carried out on this compound.[4] Pomeranz[5] in 1891 succeeded in clearing up its constitution. Bergaptene $C_{12}H_8O_4$ forms soft, white, satin-like, tasteless needles, which are odorless at ordinary temperature, but on heating give off aromatic vapors and melt at 188°. Bergaptene is the monomethyl ether of a dioxycumarin which is traceable to phloroglucin. By treating with methyl iodide and alcoholic potassa methyl bergaptenic acid and its methyl ester result. According to this bergaptene is the inner anhydride of bergaptenic acid.

Examination. The determination of the purity of bergamot oil is not difficult because adulterations of all kinds change the physical constants to a considerable extent. As the specific gravity of pure oils

1) Berichte, 25, p. 1182.

2) Journ. f. prakt. Chem., II, 45, p. 602.

3) Bull. Soc. chim. III, 21, p. 1088.

4) Mulder (1889), Liebig's Annalen, 31, p. 70.—Ohme (1889), ibid., 31, p. 320.—Franke, Dissertation. Erlangen, 1880.—Godefroy (1881), Zeitschr. d. allgem. österr. Apoth. Ver., 19, p. 1; Chem. Centralbl., 1881, p. 872.—Tilden & Beck (1890). Journ. Chem. Soc., 57. p. 828; Chem. Centralbl., 1890, I, p. 719.—Crismer (1891), Bull. Soc. chim., III, 6, p. 80; Chem. Centralbl., 1891, II, p. 879.

5) Monatsh. f. Chem., 12, p. 379; 14, p. 28.

varies within the comparatively narrow limits of 0.882—0.886, the addition of turpentine oil,[1] lemon oil, orange oil, as well as distilled bergamot oil produce a decrease, fatty oil, cedar wood oil or gurjun balsam oil an increase in the density. A part of these adulterants would also change the angle of rotation which lies between +8 and +20° with pure oils.

The solubility determination with 90 percent alcohol gives with bergamot oil results of only slight value, as by it only very extensive adulterations can be recognized. Only a part of pure bergamot oils, as already mentioned, are soluble in 80 percent alcohol. If a bergamot oil dissolves to a clear solution in this solvent it is free from fatty oil, turpentine oil and orange oil. If, however, it does not dissolve, this may be due either to an adulterant, for instance, fatty oil, or also to the presence of large quantities of bergaptene or wax-like constituents.

The detection of fatty oil is effected by weighing the residue left by evaporating the oil at 100°, which with normal oil amounts to 5—6 p. c.

About 5 g. of oil (weighed accurately to 1 cg.) are weighed off in a glass or porcelain dish and heated on a waterbath until that which remains has lost all odor of bergamot oil. After cooling, the dish, previously tared, is weighed with the residue. If this amounts to more than 6 p. c. of the oil used, fatty oil is present. Each additional percent represents one percent of adulterant. Thus e. g. a bergamot oil adulterated with 5 percent of olive oil will leave a residne of from 10 to 11 p. c.

In the oils adulterated with turpentine oil, orange oil or distilled bergamot oil the residue will in certain cases amount to considerably less than 5 or 6 percent.

The determination of the residue is of special importance as fatty oil gives a high saponification number and may therefore easily give rise to mistakes.

ESTER CONTENT. The determination of the ester content[2] which is described in detail on page 193, not only allows of the detection of adulterations, but also furnishes a criterion of the quality of the oil. This is the better, the greater the amount of linalyl acetate.

By the saponification the addition of the essence obtained by the distillation from the press residues or the small rejected fruit,[3] which shows a much smaller saponification number than the expressed oil, is also detected.

---

1) As bergamot oil contains no pinene, the presence of this hydrocarbon can be considered as a proof of the adulteration with turpentine oil.

2) As according to Borntrager (Zeitschr. f. anal. Chem., 35, p. 35) the evaporation residue on saponification gives numbers which correspond to an amount of 2 p. c. of linalyl acetate, this number ought by right to be deducted from the result found. This is, however, not done, as the method would thereby be only made unnecessarily complex.

3) An oil obtained in Messina by distillation from the expressed bergamot peel

200. Oil of Cedro.

The oil offered in commerce under the name of cedro oil or cedrate oil is nothing more than a mixture of lemon oil and other oils.

Genuine cedro oil, obtained by expression from the rind of *Citrus medica* Risso (*Cedro ordinario, Cedratier ordinaire*), is a yellow liquid of a pleasant odor reminding of citral or lemon oil.

In physical as well as chemical behavior cedro oil differs but little from lemon oil. Its specific gravity is 0.871 at 15°, $\alpha_D = +67°8'$.[1]

Upon distillation the larger part of the oil passes over from 177 to 220°. By boiling with an alcoholic solution of β-naphthylamine and pyrotartaric acid citryl-β-naphtho-cinchoninic acid was obtained in yellow crystalline leaflets, melting at 197—200°, by which the presence of citral in cedro oil was shown.

201. Oil of Limette.

Oleum Limettae. — Limettöl. — Essence de Limette.

Two oils coming from different plants and of entirely different properties are designated by the common name of limette oil[2] which according to their source may be called West Indian and Italian limette oils.

West Indian Limette Oil. The West Indian limette or lime, *Citrus medica* L., var. *acida* Brandis,[3] is cultivated on Montserrat, Dominica, Jamaica and Trinidad on account of its acid juice, commercially known as lime juice.

The oil obtained from the peel of the fruit by expression (*oil of limette*) is of a golden yellow color and can hardly be distinguished from a good lemon oil by its odor, if the greater intensity of the limette oil is not considered. Sp. gr. 0.873[4] at 29° to 0.882 at 15°; $\alpha_D = +35$ to $+38°$. The most important constituent of the oil is citral.

Entirely different from the expressed oil is the distilled oil which is obtained as a by-product in the evaporation of the juice and is known in commerce under the name of oil of limes. Its odor is unpleasant, terebinthinate, and no longer reminds of citral. Probably this aldehyde

---

possessed an ester content of only 12 p. c. (sp. gr. 0.865). Two oils distilled from the peel of fallen unripe bergamot fruit contained 6.8 and 23.5 p. c. of ester (sp. gr. 0.868 and 0.889). (Bericht von S. & Co., Oct. 1894, p. 15.).

1) Bericht von S. & Co., Oct. 1895, p. 18.

2) Arch. d. Pharm., 233, p. 174.

3) Bulletin of Miscellaneous Information, Royal Gardens Kew, 1894, p. 118.

4) Pharm. Journ., III, 15, p. 322.

is completely destroyed by the boiling of the acid liquid. Sp. gr. 0.865[1] —0.868;[2] $\alpha_D = +38°52'$. It boils between 175 and 220°.

Italian Limette Oil.[2] The fruit of the South European limette, *Citrus limetta* Risso (*Citrus limetta vulgaris, Lima dulcis, Lima di Spagna dolc., Limettier ordinaire*) distinguishes itself from that of the West Indian by its sweet juice.

The oil obtained by expression from the peel is of a brownish-yellow color, has an odor reminding strongly of bergamot oil, and forms a yellow deposit in considerable amounts on standing. Sp. gr. 0.872; $\alpha_D = +58°19'$; saponification number 75.

The composition of the Italian limette oil is very similar to that of bergamot oil, only the limette oil contains more limonene and less linalyl acetate.

The limonene is the dextrogyrate modification ($\alpha_D = +81°45'$, sp. gr. 0.848) and yields a dihydrochloride[3] melting at 50°, as well as a tetrabromide melting at 105°.

In the oil investigated by Gildemeister 26.3 p. c. of linalyl acetate were present (b. p. 101—103° at 13 mm., sp. gr. 0.898, $\alpha_D = -9°52'$).

After saponification with alkali, acetic acid was found in the alkaline solution, while from the oil l-linalool (b. p. 88.3—89.5° at 13 mm., sp. gr. 0.870, $[\alpha]_D = -20°7'$) was separated by fractional distillation. From it citral was formed on oxidation.[4] Linalool is present in the oil partly in the free state, partly as acetic acid ester.

The limettin which separates from the oil on standing melts according to Tilden[5] at 121—122°. It has the composition $C_6H_8(OCH_3)_2.C_3HO_2$ and yields on melting with potassa, besides acetic acid, phloroglucin.

202. Oil of Limette Leaves.

The oil of the leaves from *Citrus limetta* contains according to Watts[6] (1892) an inactive terpene, boiling at 176—177°, which yields with hydrochloric acid a crystalline hydrochloride melting at 49—50°

1) Pharm. Journ., III, 14, p. 1005.

2) Gildemeister, loc. cit.

3) The same dihydrochloride was obtained by de Luca in 1860 (Compt. rend., 51, p. 258) from the terpene boiling at 180° of an oil, which is designated as coming from *Citrus lumia*, but which in all probability was Italian limette oil. Comp. Gildemeister, loc. cit.

4) To which compound the "Limettsäure," $C_{11}H_4O_6$ obtained by Vohl in 1858 upon the oxidation of the oil (Archiv d. Pharm., 124, p. 16) owes its origin, is uncertain.

5) Journ. Chem. Soc., 61, p. 344; also Tilden & Beck, Journ. Chem. Soc., 57, p. 323.

6) Journ. Chem. Soc., 49, p. 316.

and is, therefore, dipentene or limonene. Fraction 220—230° yielded upon oxidation with chromic acid, acetic acid and pelargonic acid, on account of which it is considered as methyl nonyl ketone.

203. Oil of Mandarins.

Oleum Mandarinae. — Mandarinenöl. — Essence de Mandarines.

ORIGIN. The peel of the pleasant tasting fruit of *Citrus madurensis* Loureiro[1] known as mandarins contains a very pleasant smelling oil, which is obtained like the oils of the other agrumen fruits, by expression.

PROPERTIES. Mandarin oil is a golden yellow liquid with a slight bluish fluorescence, which becomes more prominent when the oil is diluted with alcohol. The odor, although similar to that of lemon oil, is more pleasant and distinctly different from it. Sp. gr. 0.854—0.858; $\alpha_D = +65$ to $+75°$.

Two mandarin oils[2] distilled in Porto Alegre (Brazil) had the following properties: Sp. gr. 0.8515 and 0.8510; angle of rotation $+74°16'$ at 17° and $+74°20'$ at 16°. Both oils were distinguished by a beautiful blue fluorescence.

COMPOSITION. The oil began to boil at 175° and all except a small residue went over up to 179°. The fraction boiling at 175—177° ($\alpha_D = +76°45'$) gave on bromination in glacial acetic acid solution a tetrabromide melting at 104—105° (Gildemeister and Stephan,[3] 1897). By conducting hydrochloric acid into the same fraction dipentene dihydrochloride, m. p. 49°, resulted (de Luca,[4] 1857). According to this the greater part of the mandarin oil consists of d-limonene.

If the portion which did not distill over up to 177° be treated with bisulphite solution, an addition product is obtained, from which an oil is separated by alkali. This behaves like a mixture of citral and citronellal when condensed with pyrotartaric acid and β-naphthylamine. The melting point of the naphtho-cinchoninic acid formed is not constant. At 197° (the melting point of the pure citral compound) the body begins to run together, but does not melt completely until 222° (melting point of the citronellal compound 225°). The positive identification of these two aldehydes in the oil has, therefore, not yet been made.

---

1) According to de Lucca (see footnote 4) the mandarin is obtained from *Citrus bigaradia sinensis* and *C. b. myrtifolia*, according to Sawer (Odorographia, vol. 1, p. 74) and v. Müller (Select Extra-Tropical Plants, 9th ed., p. 129) however, from *Citrus nobilis* Loureiro.

2) Bericht von S. & Co., Apr. 1896, p. 68.

3) Archiv d. Pharm., 235, p. 588.

4) Compt. rend., 45, p. 904.

According to Flatau and Labbé[1] the expressed mandarin oil contains the same ester, difficultly soluble in alcohol, as does orange oil (comp. page 472).

204. Oil of Grape Fruit.

The fruit of *Citrus decumana* L., known as shaddock or grape fruit, contains in its peel a small amount of volatile oil, which can be obtained by expression or by distillation.

An oil[2] expressed from fruit which came from Orlando in Florida possessed an exeedingly pleasant odor, reminding of bitter oranges, the sp. gr. was 0.860 and rotatory power $\alpha_D = +94° 30'$.

205. Oil of Neroli.

Oleum florum Aurantii. — Orangenblüthenöl. — Essence de Néroli.

Origin and History. The oil of neroli is obtained by the distillation with water from the fresh blossoms of the bitter orange, *Citrus bigaradia* Risso.

Oil of orange flowers was known as early as the sixteenth century. Its distillation was described for the first time by Porta. About a century later, in the year 1680, it appears to have been made the fashionable perfume by the Duchess Flavio Orsini, Princess of Neroli, hence the name of essence of neroli. On account of its delicate pleasant odor, the oil has been able to hold its reputation as one of the finest flower perfumes. This is also true of the distilled orange flower water, or *Aqua naphae*, which is used extensively for the aromatizing of food, confections and beverages, also for toilet purposes. The distillation of orange flower oil was described by Benatius in 1806. The oil was investigated in 1825 by Bonastre[3] and in 1828 by Boullay.[4]

Preparation. Oil of neroli is obtained exclusively in southern France from the blossoms of the bitter orange. The oil is not produced in Italy.

Properties. The oil of neroli of commerce is a yellowish, slightly fluorescent liquid, which becomes brownish-red when exposed to light, of an intensive, highly pleasant odor reminding of orange blossoms, and a bitter aromatic taste. Sp. gr. 0.870—0.880. The rotatory power is slightly dextrogyrate, $\alpha_D = +1° 30'$ to $+5°$. The oil is soluble in 1½—2 vols. of 80 p. c. alcohol. On the further addition of alcohol the liquid becomes turbid, and on standing crystalline flakes consisting

[1]) Bull. Soc. chim., III, 19, p. 364.

[2]) Bericht von S. & Co., Apr. 1894, p. 27.

[3]) Journ. d. Pharm. II, 11, p. 529.

[4]) Journ. d. Pharm. II, 14, p. 496.

of paraffin collect on the surface. The alcoholic solution of neroli oil distinguishes itself by a beautiful violet-blue fluorescence, which becomes especially prominent when some alcohol is poured in a layer above the oil.

On strongly cooling, the oil becomes turbid on account of the separation of paraffin. At times it even solidifies to a butter-like mass.

The saponification number of good oils lies between 20 and 52. corresponding to an amount of 7—18 percent of linalyl acetate. Oils with a saponification number higher than 55 are suspicious. Details are mentioned under Examination.

In order to ascertain the properties of oils which were undoubtedly genuine, fresh orange blossoms which were partly preserved with salt, partly with sea water, for transportation, were distilled by the firm of Schimmel & Co.[1] and about 0.1 percent of oil obtained which possessed the following properties:

No.	Specific gravity	α_D	Sap. number	Behavior in a freezing mixture
1.	0.887	inactive	41	solidifies to a butter-like mass.
2.	0.881	—	36	viscous, but not becoming solid.
3.	0.876	−0° 52′	21	
4.	0.872	−0° 40′	21	

The rotatory power of No. 2 could not be determined on account of its dark color. Nos. 3 and 4 are distillates of the same shipment of blossoms. No. 4 consists only of the oil which separated directly in the receiver on distillation. Its preparation corresponds to the method usually employed in southern France, where orange blossom oil is obtained as a by-product in the manufacture of orange blossom water. No. 3 is a normal product, i. e. a mixture of oil separating at once and that obtained by cohobation from the water. The oils obtained by these two methods differ but slightly.

Of still greater importance for the determination of the constants of pure commercial oils than these experiments with preserved blossoms, are the distillations of fresh material by Charabot and Pillet[2] in southern France. The oils distilled in May 1898 in Cannes and Antibes behaved as follows: The sp. gr. was between 0.8720 and 0.8757, the

1) Bericht von S. & Co., Oct. 1891, p. 26; Oct. 1894, p. 40.
2) Bull. Soc. chim., II, 19, p. 858.

angle of rotation, α_D, between $+1.42°$ and $+4.06°$. 1 part of the oil dissolved in 1.3 to 1.6 parts of 80 p. c. alcohol at 20°. The amount of ester ($C_{10}H_{17}OCOCH_3$) present was 13.4—18.0 p. c.

Some interesting quantitative distillations were made by Jean Gras in Cannes during the last harvest (1899). 30 samples were distilled at different periods throughout the harvest. The yield increases as the season advances, from 0.80 to 1.23 p. c. 25 of the 30 samples were investigated by Schimmel & Co.[1] No marked differences were noted. The sp. gr. varied from 0.873—0.877 at 15°, saponification number 35.3—44.8, $\alpha_D = +3°22'$t to $+5°24'$ at 20°. All the oils were soluble in 1¼ and more parts of 80 p. c. alcohol.

Composition. Neroli oil was investigated by Tiemann and Semmler[2] in 1893 and subjected to fractional distillation under diminished pressure at 15 mm. in order to effect a separation of the individual constituents. At 75° a terpene came over which was identified as limonene by means of its tetrabromide melting at 105°.

Between 88 and 94° boiled an optically laevogyrate alcohol corresponding to the formula $C_{10}H_{18}O$ of the sp. gr. 0.8671 at 20°, which possessed the properties of l-linalool.

At 97—104° the acetic acid ester of linalool came over, sp. gr. 0.8972 at 20°, likewise laevogyrate. It was split up into acetic acid and linalool by boiling with potassa.

In the fraction boiling from 110—120° a second alcohol, almost optically inactive and corresponding to the formula $C_{10}H_{18}O$ was contained, which consisted principally of geraniol.

The methyl ester of anthranilic acid occurring in small quantities in the oil and found in 1894 in the laboratory of Schimmel & Co.,[3] plays an important part in the formation of the orange blossom perfume. The fluorescence of the oil is due to the presence of this substance.

The methyl ester of anthranilic acid $NH_2.C_6H_4.COOCH_3$ boils at 132° under a pressure of 14 mm., melts at 25° and in an overcooled state at 15°, it has the sp. gr. 1.168. The odor in the undiluted state is unpleasant. It reminds of the fragrance of the orange blossom only when greatly diluted.

The stearoptene of the orange blossom oil, also called neroli camphor or aurade was first found by Boullay[4] in 1828. It is one of the

1) Bericht von S. & Co. Oct. 1899, p. 42.

2) Berichte, 26, p. 2711.

3) Bericht von S. & Co., Apr. 1899, p. 35. See also Walbaum, Journ. f. prakt. Chemie, II, 59, p. 350; see also Erdmann, Berichte, 32, p. 1213.

4) Journ. de Pharm., 14, p. 497; Trommsdorff's Neues Journ. d. Pharm., 19, 1, p. 227.

paraffins which occur in almost all oils obtained from blossoms and in the pure state is completely odorless and tasteless. Its melting point is 55°.[1]

Inasmuch as the decomposition of the paraffin is wholly improbable, the assertion by Plisson[2] (1829) that the amount of the stearoptene in the oil decreases with age, evidently rests on a wrong observation.

Examination. The most common and most dangerous adulterants are the oils of bergamot and petitgrain. As these for the greater part possess the same constituents as oil of neroli—linalool and linalyl acetate—the detection of small amounts is impossible. Larger additions cause an increase in the specific gravity and the amount of esters, which in pure neroli oil is 7—18 percent (saponification number 20—52), in bergamot oil 35—45 p. c. (s. n. 100—130), in petitgrain oil 38—85 p. c. (s. n. 110—245).

Orange flower oils which show a saponification number higher than 55 are therefore to be rejected as suspicious.

The property of neroli oil to separate paraffin in a freezing mixture has been employed as a test, which is not wholly irrational, as the addition of paraffin free oils might decrease the relative amount of paraffin to such an extent that a separation no longer takes place on cooling. It must, however, be remembered in employing this test that some unadulterated oils may in certain cases be poor in paraffin. When, for instance, at the time of the harvest a large amount of blossoms is to be quickly distilled, it happens, that the distillation is not carried on to its complete exhaustion and that a smaller amount of the difficultly volatile paraffin gets into the oil. For the rest it is necessary in testing to depend on the comparison of the physical properties with those of good oils, especially of the odor.

206. Oil of Neroli Portugal.

Oleum Aurantii Florum Dulce.—Süsses Orangenblütheuöl.—Essence de Néroli Portugal.

Oil of sweet orange blossoms, *Néroli Portugal,* i. e. the oil distilled from the blossoms of the sweet orange does not occur at all in commerce in a pure state. The goods sold under the above designation are always a mixture of different aurantiaceous oils.

An oil distilled in Germany from the fresh blossoms of the sweet orange *Citrus aurantium* Risso, had entirely different properties from

1) Pharmacographia, 2nd ed., p. 127. 2) Journ. de Pharm., II, 15, p. 152.

that procured from southern France.[1] The blossoms used for the distillation were transported from southern Spain in iron caskets from which the air was pumped out after having been filled.

0.154 p. c. of oil was obtained having the sp. gr. 0.893 and the angle of rotation $\alpha_D = +16° 8'$.

207. Oil of Petitgrain.

Oleum Petitgrain.—Petitgrainöl.—Essence de Petit-grain.

Origin and Preparation. Petitgrain oil is obtained from the leaves, twigs and immature fruit of the bitter orange, *Citrus bigaradia* Risso, by distillation with water. Formerly the oil was principally produced in southern France, until toward the end of the seventies French colonists began the distillation in Paraguay. The poor quality of the oils there produced in the beginning improved in the course of the years to such an extent that the South American oil is now generally preferred on account of its greater reliability and uniformity to the often adulterated French product. The market for the Paraguay oil is Asuncion; the principal place of distillation is said to be the little town of Yaguaron.[2]

Properties. The odor of petitgrain oil is similar to that of neroli, but far less delicate, the taste is aromatic and somewhat bitter, the color yellowish. Sp. gr. 0.887—0.900. It turns the polarized ray of light either slightly to the right, or to the left; $\alpha_D = +3° 43'$ to $-1° 22'$. The oil is soluble to a clear solution in 2 parts of 80 p. c. alcohol. Saponification number 110—245 = 38—85 p. c. of linalyl acetate.

Eight different petitgrain oils recently distilled from the leaves of the bitter orange by Charabot and Pillet[3] in Cannes, had the following properties: Sp. gr. 0.8910—0.8934; $\alpha_D = -5° 12'$ to $-6° 15'$. Soluble in 1—1.1 parts of 80 p. c. alcohol. The amounts of ester varied from 51.5—69.6 p. c.

It may be remarked, that oils with so strong a rotatory power as these have not yet been met in commerce. Moreover, as a general rule, petitgrain oil is not distilled exclusively from the leaves, but also from the unripe fruit, by which the deviations of the commercial oils are explained.

Composition. Semmler and Tiemann[4] in 1892 fractionated petitgrain oil under 15 mm. pressure. The first portions contained a

[1]) Bericht von S. & Co., Oct. 1889, p. 38.
[2]) Chemist and Druggist, 51, p. 110.
[3]) Bull. Soc. chim., III, 21, p. 78.
[4]) Berichte, 25, p. 1186.

hydrocarbon which was recognized as limonene. The main fraction, amounting to 70 p. c. of the crude oil, contained oxygen, boiled from 102—106° and had the sp. gr. 0.8988 at 20°. The composition corresponded to the formula $C_{12}H_{20}O_{2}$. By treatment with potassa the oil was decomposed to acetic acid and linalool (aurantiol) and consisted therefore of linalyl acetate.

According to Passy[1] linalool is not the only alcohol $C_{10}H_{18}O$ contained in petitgrain oil, but also geraniol in the free state as well as in the form of the acetic acid ester.

Besides the bodies named other oxygenized, not yet isolated compounds are present in the oil, which take part in developing the specific petitgrain odor.

The amount of esters, linalyl acetate and geranyl acetate, varies between 40 and 85 p. c. and usually amounts to about 50 p. c. The last fractions of the petitgrain oil contain a sesquiterpene.

Charabot and Pillet[2] have recently shown that only the petitgrain oils distilled from the leaves and twigs contain no limonene, so that the occasional presence of d-limonene comes from the accompanying small fruit in the distillation. According to the authors named, petitgrain oil (from the leaves) contains about 70—75 p. c. of l-linalool and 10—15 p. c. of geraniol, as well as small amounts of a sesquiterpene.

Adulteration and Examination. Petitgrain oil is adulterated with orange oil, lemon oil and turpentine oil. These additions are readily recognized by the lowering of the specific gravity, decrease in the saponification number and the solubility, and finally by the change in rotatory power.

Petitgrain citronnier. By this name an oil is designated which is obtained now and then from the twigs, leaves and unripe fruit of the lemon tree. Its odor is similar to that of the petitgrain oil, but the lemon-like odor accompanying it betrays the source of the oil. Sp. gr. 0.868—0.874; $\alpha_D = +22°5'$ to $+34°12'$. Saponification number 14.5—32.2. The odor, also the property of forming a crystalline compound with bisulphite solution caused the presence of citral to be suspected. In fact the citryl β-naphtho cinchoninic acid was obtained, and the presence of this aldehyde thereby proven.[3]

1) Bull. Soc. chim., III, 17, p. 519.

2) Bull. Soc. chim., III, 21, p. 74.

3) Bericht von S. & Co., Oct. 1896, p. 59.

208. Oil of West Indian Sandalwood.

Oleum Santali ex India Occidentali. — Westindisches Sandelholzöl. — Essence de bois de Santal des Indes Occidentales.

The botanical source of the West Indian santalwood was unknown until recently. Through a study of the wood and the leaves Holmes, Kirkby and Petersen came to the conclusion, that the plant must belong to the family of the *Rutaceae*, and by no means be classed with the family of the *Santalaceae*.[1] More than this could not be ascertained on account of the want of blossoms. In compliance with a request by E. M. Holmes of London, Schimmel & Co. finally succeeded in 1898 in obtaining flowering branches of the tree from Puerto Cabello in Venezuela, from whence the wood is brought into commerce. In the investigation Holmes came to the conclusion, that the plant belonged to the family of the *Rutaceae*. He believed it to be a representative of a hitherto unknown genus and named it *Schimmelia oleifera*.[2] Later, however, he agreed with Urban of Berlin, who determined the plant as *Amyris balsamifera* L. The mistake was caused by the fact, that the genus *Amyris* has hitherto been classed by the English botanists with the family of the *Burseraceae*, while in Germany, and according to Holmes' view with perfect right, it is classed with the *Rutaceae*. Now, as no species of the Rutaceae in English botanical literature corresponded to the description of the West Indian sandalwood plant, Holmes had considered it as a new species.

The wood, which has no similarity with the East Indian sandalwood, consists of sticks from a thumb to an arm in thickness. It is white and hard and covered with a gray bark. The anatomical build of the wood has been studied and described by Petersen and Kirkby.

The well comminuted wood yields upon distillation 1.5—3.5 p. c. of a thick, viscous oil of a weak not pleasant odor.

The sp. gr. lies between 0.960 and 0.967, the angle of rotation between $+24$ and $+29^{\circ}$. Of the composition of the West Indian sandalwood oil nothing is known.

209. Oil of Myrrh.

Oleum Myrrhae. — Myrrhenöl. — Essence de Myrrhe.

Origin and History. Myrrh is the dried up emulsion-like juice, originally contained in the parenchyma of the bark, of several species of the genus *Commiphora* which belongs to the family of the *Burseraceae*.

1) Pharm. Journ., III, 16, pp. 757, 821 and 1065.
2) Pharm. Journ., 62, pp. 58, 187 and 205.

These shrubs grow partly wild, and are partly cultivated in the coast districts of the Red Sea especially on the Somali coast of East Africa, and appear to flourish in many parts of Arabia as far as Persia.

The distilled oil of myrrh was well known to Ryff, Cordus and Gesner. In the drug and spice ordinances it is first mentioned in those of the city of Frankfurt-on-the-Main of 1587, and taken up in the Dispensatorium Noricum of 1589.

Observations on the methods of preparation and yield were made in the course of the eighteenth century by Hoffmann, Neumann, Spielmann, Thielebein, and later by Braconnot, Pelletier and Brandes.

The officinal (Ph. G. III.) or Herabol myrrh is derived from *Commiphora abyssinica* Engl. and *C. schimperi* Engl. from southern Arabia and the coast districts of Somali-land. It can be distinguished from other varieties in that the extract made with petroleum ether is colored red by bromine vapors. This color reaction is also peculiar to the volatile oil which is obtained by distillation in yields of 2.5—8. p. c.

PROPERTIES. Oil of myrrh is a thick fluid, of a yellow to greenish color and has a strong myrrh odor.

Sp. gr. 0.988—1.007. Köhler[1] in 1890 found a density of only 0.9624 at 17.5° due possibly to the fact that in the preparation on a small scale the heavier parts are likely to remain in the resin and only the lighter oil comes over. Angle of rotation $\alpha_D = -67°54'$ (Köhler) to $-90°$, Gladstone[2] observed in a 10 inch (25 cm.) tube a rotation of -136.

Oil of myrrh boils according to Köhler from 220—325°, according to Tucholka[3] (1897) from 260—280°. It dissolves to a clear solution in 10 parts of 90 p. c. alcohol.

COMPOSITION. Of the constituents of oil of myrrh not one has been isolated and identified. An elementary analysis made by Ruickhold which had yielded numbers nearly corresponding with the formula $C_{10}H_{14}O$, induced Flückiger[4] (1876) to investigate the oil for carvone. It was found, however, that this body is not contained in oil of myrrh.

Bisabol Oil of Myrrh.[5] Bisabol, or Bissabol myrrh comes from the interior of the Somali-lands and according to Holmes is identical with the opopanax (see this) now occurring in commerce. It does not give the color-reaction with bromine vapors as described for Herabol myrrh.

1) Archiv d. Pharm., 228, p. 291.
2) Journ. Chem. Soc., 17, p. 1.
3) Archiv d. Pharm., 235, p. 298.
4) Berichte, 9, p. 471.
5) Archiv d. Pharm., 235, p. 289.

If 6 drops of the petroleum ether extract of Bisabol myrrh (1:15) are mixed with 3 cc. of glacial acetic acid and this liquid poured in a layer on top of 3 cc. of concentrated sulphuric acid, a rose colored zone is formed at the place of contact of the two layers, and in a short time the entire acetic acid layer will become pink. Herabol myrrh gives under the same conditions only a very slight pink coloration to the acetic acid layer. The zone of contact of the two liquids is green.

Oil of Bisabol myrrh is mobile and light yellow. Sp. gr. 0.8836 at 24°; $[\alpha]_D = -14°20$ at 24°. It boils from 220—270°.

By conducting hydrochloric acid into the ethereal solution of the oil an optically active dihydrochloride $(C_{10}H_{16}2HCl)_2$, melting at 79.3° is formed, the yield being about 6.5 p. c.; $[\alpha]_D = -35°17'$ at +7° in chloroform solution.

By boiling with sodium acetate in glacial acetic acid solution hydrochloric acid is split off, and a hydrocarbon "bisabolene" is formed which is not identical with any of the known terpenes. B. p. 259—260.5; sp. gr. 0.8914; $n_D = 1.4608$.

210. Oil of Opopanax.

The botanical source of the genuine opopanax is still unknown but presumably comes from *Opoponax chironium* Koch (*Ferula opoponax* L.). The opopanax now found in commerce, from which the volatile oil is distilled, is according to Holmes[1] the gum-resin of *Commiphora kataf* Engl. (*Balsamodendron kafal* Kuntlr). In the collection of Chinese drugs it is commonly designated as myrrh and has probably been often mistaken for this. According to Holmes it may probably be the myrrh of the Scriptures.

Upon distillation opopanax yields 6—10 p. c. of oil of a greenish-yellow color and a pleasant balsamic odor. It resinifies readily on exposure to air. Sp. gr. 0.870—0.905; $\alpha_D = -10$ to $-12°$.[2] It dissolves to a clear solution in an equal part of 90 p. c. alcohol.

Oil of opopanax boils with decomposition between 250 and 300° (Baur,[3] 1895). The carrier of the characteristic odor is found in the lower boiling portions.[4]

Recently Knitl[5] has made an investigation of undoubtedly genuine oil. The oil was prepared by shaking out the alcoholic solution of the

---

1) Pharm. Journ., III, 21, p. 838; 25, p. 500; 61, p. 565.

2) An oil distilled in the laboratory of Schimmel & Co., was probably obtained from the oleoresin of another plant. Yield 6.7 p. c.; sp. gr. 0.861; $\alpha_D = +42°30'$.

3) Arch. d. Pharm., 233, p. 235.

4) Bericht von S. & Co., Apr. 1890, p. 34.

5 Archiv d. Pharm., 237, p. 256.

resin with petroleum ether. With bisulphite solution a brownish mass was separated, which on sublimation gave white colorless needles melting at 133—134°. The body corresponded with the formula $C_{20}H_{10}O_7$ and was called oponal. The oil itself, when distilled under diminished pressure gave at first colorless distillate having the odor of lovage oil and later a blue distillate came over. Crystals of oponal separated in the neck of the retort.

211. Oil of Olibanum (Frankincense).

Oleum Olibani. — Weihrauchöl. — Essence d'Oliban.

HISTORY. The distilled oil of olibanum was known to Valerius Cordus, but was seldom mentioned in literature. In the treatises on distillation of the sixteenth century frankincense is mentioned as one of the many constituents used in the distillation of the complex balsams. Oil of frankincense is first found as *Oleum thuris* in the drug ordinances of the city of Berlin for 1574, and of Frankfurt-on-the-Main for 1587; further in the Dispensatorium Noricum of the year 1589.

The older investigations of frankincense as to content of volatile oil as well as the properties of the oil were mostly made in connection with like investigations of oil of myrrh of which the more important have been mentioned on p. 487.

ORIGIN AND PREPARATION. Frankincense is obtained from *Boswellia carterii* Birdw. and other species of the genus *Boswellia* (Family *Burseraceae*) in Somali-land and in south-eastern Arabia. If the bark of the frankincense tree is cut, a white emulsion oozes out, which solidifies after some time and then forms yellow granules (tears or drops) which are separated from the tree trunks or picked up from the ground.

Frankincense yields upon distillation with steam 3—8 p. c. of volatile oil. It is colorless or yellowish and has a pleasant balsamic and slightly lemon-like odor. Sp. gr. 0.875—0.885; $\alpha_D = -11°$ to $-17°$ (Stenhouse,[1] 1840).

COMPOSITION. Oil of olibanum, consisting for the greater part of terpenes, when distilled comes over principally about 162°.[1] After fractionating several times a laevogyrate hydrocarbon of the formula $C_{10}H_{16}$ is obtained, which boils at 157 to 160°. It yields with hydrochloric acid a monohydrochloride melting at 127° (Kurbatow,[2] 1874); with amyl nitrite and hydrochloric acid a nitrosochloride melting at

[1]) Liebig's Annalen, 35, p. 306. [2]) Liebig's Annalen, 173, p. 1.

100—101°. The latter when boiled with alcoholic potassa is readily converted into nitroso pinene $C_{10}H_{15}NO$ melting at 130°, from which it follows, that the terpene boiling at 157 to 160° (the olibene of Kurbatow) is l-pinene (Wallach,[1] 1889).

The fraction boiling between 177 and 179° yields on bromination dipentene tetrabromide.[1] The second constituent of the oil is therefore dipentene.

As oil of olibanum gives with sodium nitrite and glacial acetic acid the phellandrene reaction, phellandrene is the third terpene contained in the oil.[2]

The oxygenized fractions boiling above 175° have not yet been carefully investigated.

212. Oil of Elemi.

Oleum Elemi. — Elemiöl. — Essence d'Elemi.

History. Elemi is a dried up resin juice, comparable with turpentine, of a number of different trees belonging to the family of the *Burseraceae*. At the present time elemi resin comes mainly from the Philippines, especially from Luzon. Besides this, now and then Central American, Brazilian, West African and other varieties occur in the market.

The distilled elemi oil was first mentioned in the ordinance of Frankfurt-on-the-Main of 1587 and included in the Pharmacopoea Augustana of 1613, likewise in the Frankfurt Pharmacopoeia of 1649.

The first determination of the yield of volatile oil was made by Neumann about 1730; Bonastre in 1823 and Manjeau in 1824 repeated the experiment.

Origin and Preparation. The Manila elemi exclusively used for the preparation of elemi oil consists of a thick, turpentine-like, white mass enclosing many small fragments of bark. On keeping for some time it takes on a wax-like appearance on the exterior, due to drying out, by which the outer layer becomes of a yellow color and loses its odor.

When treated with 90 p. c. alcohol in the cold a part dissolves, while a crystalline residue remains. From the solution prepared by boiling, needle-shaped crystals of amyrin separate in large quantity on cooling (Vesterberg,[3] 1887).

---

1) Liebig's Annalen, 252, p. 100.
2) Observation made in the laboratory of Schimmel & Co.
3) Berichte, 20, p. 1242.

The botanical source of the Manila elemi is not definitely known; it is, however, supposed that its source is a species of *Canarium*. Neither *Canarium commune* L. the resin of which forms a dry mass,[1] nor *Canarium muelleri* need be considered in discussing the source of Manila elemi, as the turpentine of the latter plant contains according to Maiden,[2] (1892) no crystalline constituents, and its solutions deposit no crystals on evaporation. Up to 30 p. c. of oil are obtained from Manila elemi by distillation.

Properties. Oil of elemi is colorless or pale yellow and has a decided phellandrene odor. The specific gravity of the oil prepared on a large scale lies between 0.87 and 0.91. Distilled on a small scale terpenes principally come over. Hence the specific gravity is found to be lower, as for instance by Flückiger[3] who determined 0.860.

Manila elemi oil is optically dextrogyrate;[4] $\alpha_D = +44° 3'$. With sodium nitrite and glacial acetic acid it gives a strong phellandrene reaction.

Composition. The fraction of the oil boiling below 175° contains according to Wallach[5] (1888) d-phellandrene. In fraction 175—180° dipentene is present to such an extent that it is very well suited for the preparation of dipentene derivatives. The dipentene was identified by means of the tetrabromide m. p. 125°, and by the nitrosochloride and its conversion into carvoxime melting at 93°.

The elemi oil investigated by Deville[6] in 1849 also contained dipentene, as it yielded a solid hydrochloride.

Besides these terpenes and polyterpenes elemi oil also contains oxygenized products, which readily split off water, even when distilled by themselves, but especially when distilled with potassium acid sulphate.[5]

In the preparation of the oil on a large scale a portion was obtained which sank in water. This boiled in a vacuum of 10 mm. pressure between 153 and 163°, the main portion from 160—161°. Under ordinary atmospheric pressure this fraction boiled at 279—280°, was optically inactive and had the sp. gr. 1.043 at 15°. Upon oxidation with potassium permanganate an acid of the melting point 170° was

---

1) Pharmacographia Indica, vol. 1, p. 321.

2) Pharm. Journ., III, 23, p. 15.

3) Pharmacognosie, 3rd ed , p. 86.

4) An oil examined by Deville in 1849 (Liebig's Annalen, 71. p. 358) was strongly laevogyrate and probably obtained from a different variety of elemi. Comp. also Stenhouse. Liebig's Annalen, 35 p. 304.

5) Liebig's Annalen. 246, p. 233; 252. p. 102.

6) Liebig's Annalen. 71, p. 353.

obtained, the silver salt of which, stable toward light, after being purified by recrystallization from boiling water, was analysed, 29.58 p. c. of silver being found.[1] The composition of the acid, which no doubt represents the direct oxidation product of an oxygenized compound has not yet been determined.

213. Oil of Conima Resin.

The resin known under the name of conima resin or Hyawa gum from *Icica heptaphylla* Aubl. is used in place of frankincense as incense in British Guayana.

It contains small amounts of volatile oil (Stenhouse and Groves.[2] 1876) which can be obtained by distillation with steam. The oil is yellowish and on distillation comes over mainly between 260 and 270°. The highest boiling portions have a bluish-green color. When distilled from sodium a hydrocarbon, conimene, of the boiling point 264° was isolated. Conimene is a sesquiterpene of the formula $C_{15}H_{24}$.

214. Oil of Linaloe.[3]

Oleum Linaloes. — Linaloeöl. — Essence de Linaloé. Essence de Licari.

History. Since the eighteenth century woods with a pleasant odor have come into commerce from Mexico and French Guayana under the name of aloe wood, because they were first considered to be identical with the earlier known aloe wood. Mexican linaloe oil was first introduced into France in the year 1866. Guayana linaloe wood came for the first time to Marseille in the seventies where it was used for making the oil. The oil is distilled in Cayenne itself only since 1893.

Origin.[4] On account of the great similarity of their physical and chemical properties the oils from the wood of entirely different trees are designated by the common name of linaloe oil.

The Mexican linaloewood, called in Mexico *Lignaloë* or *Linalué*, also *Bois de citron de Mexique* comes from *Bursera delphechiana* Poiss. belonging to the family of *Bur.eraceae*, and probably also from *Bursera aloexylon* Engler (*Elaphrium aloexylon* Schiede, *Amyris linaloe* La Llave).

It occurs in commerce in the form of thick trunks, which are usually free from bark and more or less weathered on the surface and of an ash gray appearance. The quite spongy and light wood shows in section a moiré-like figure of brown, close, concentric rings.

1) Bericht von S. & Co., Oct. 1896, p. 95.
2) Liebig's Annalen, 180, p. 258.
3) Also spelled lignaloe.
4) Pharm. Journ., III, 18, p. 132.

Upon distillation, which is carried on in a primitive manner by native Indians in the province of Guerrero, south from the capital of Mexico, the wood of old trunks is said to give a yield of 10—12 p. c. of oil. Wood distilled in Europe gave only 7—9 p. c. of oil.

The linaloewood from French Guayana, or Cayenne linaloewood is called by the natives *Likari*, by the French colonists *Bois de rose mâle.* It is further designated as *Bois de ro e femelle*, *Bois jaune*, *Bois de citron de Cayenne*, *Cèdre jaune* or *Copahu.*

The botanical source has not yet been definitely determined, but according to Moeller[1] *Ocotea caudata* Mez., a tree belonging to the *Lauraceae* may, with considerable probability, be considerd as its source.

The wood is hard and heavy, readily cleavable, on fresh surfaces yellow, on older ones of a reddish color. It is exported from Cayenne in logs deprived of bark, the size of which indicate enormous trees.

PROPERTIES. Although both oils are in general very similar, differences do exist, especially in the specific gravity and optical rotatory power.

The linaloe oil of commerce is almost exclusively the Mexican. This is a colorless to yellowish liquid of pleasant linalool odor.[2] Sp. gr. 0.875 to 0.895; $\alpha_D = -5$ to $-12°$; saponification number 1—10. With two and more parts of 70 p. c. alcohol the oil yields a clear solution.

Cayenne linaloe oil can be readily distinguished from the former by its odor. Sp. gr. 0.870—0.880; $\alpha_D = -15°$ to $-20°$. Its solubility is the same as that of the Mexican oil.

COMPOSITION. Linaloe oil consists almost entirely of an alcohol $C_{10}H_{18}O$, boiling at 198—199°, which was first isolated from the Cayenne oil and called licareol by Morin[3] in 1881. Semmler[4] in 1891 found it in the Mexican oil. He ascertained that it belonged to the aliphatic series and gave it the name of linalool. Barbier[5] (1892) at first considered linalool and licareol as different bodies, but later admitted their identity.[6]

Two further constituents, of subordinate importance, as they are present in only small quantities, and affect the odor but slightly, have

1) Pharm. Post, 29, Nos. 46—48.

2) The odor, which is usually described as lemon- or rose-like, has not the slightest resemblance either to that of rose or that of lemon.

3) Compt. rend., 92, p. 998; 94, p. 733; Ann. de Chim. et Phys., V, 25, p. 427.

4) Berichte, 24. p. 207.

5) Compt. rend., 114, p. 674; 116, p. 883.

6) Compt. rend., 121, p. 168.

been found by Schimmel & Co.[1] in the oil. They are geraniol and methyl heptenone, which so often accompanies the aliphatic terpene alcohols. According to Barbier and Bouveault[2] linaloe oil contains also 3 percent of a sesquiterpene, 0.1 percent of a monatomic and a like amount of a diatomic terpene.

Examination. The adulteration of linaloe oil with fatty oil, which has been observed several times, can be readily determined by its insolubility in 70 p. c. alcohol, by the increase in the specific gravity and by the high saponification number.

215. Oil of Cedrela Wood.

About 12 different species of the genus *Cedrela* (Family *Meliaceae*) indigenous to America, yield a pleasantly odorous wood, which is used for making cigar and sugar boxes. It is incorrectly called cedar wood.

Volatile oils have been distilled from a number of these woods, the botanical sources of which are unknown.

1. Cedrela wood from Corinto (Nicaragua).[3] Yield 2.3 p. c. Yellow oil of the sp. gr. 0.906 and the angle of rotation $\alpha_D = -17°23'$.

2. Cedrela wood from Cuba.[3] Yield 1.75 p. c. Slightly yellow oil of the sp. gr. 0.923 and the angle of rotation $\alpha_D = +18°6'$. It contains large amounts of cadinene, the laevogyrate hydrochloride of which, melting at 118° was prepared.

3. Cedrela wood from La Plata.[3] Yield 0.59 p. c. Optically inactive oil of a light blue color and the sp. gr. 0.928.

4. Cedrela wood from Punta Arenas (Costa Rica).[3] Yield 3.06 p. c. Light blue oil, boiling from 265—270°. Sp. gr. 0.915; $\alpha_D = -5°53'$. Consists principally of cadinene, as was shown by the preparation of the dihydrochloride melting at 118°.

5. Cedrela wood from *Cedrela brasiliensis* (*Cedrela odorata* L.?) from Porto Alegre.[4] From the sawdust of this wood only 0.5 p. c. of a light blue oil of the sp. gr. 0.9348 and the angle of rotation $\alpha_D = -0°22'$ were obtained. The oil may perhaps be identical with the La Plata cedrela wood oil mentioned above.

216. Oil of Senega Root.

The senega belonging to eastern North America, *Polygala senega* L. (Family *Polygalaceae*) contains according to Reuter[5] in its root 0.25

---

1) Bericht von S. & Co., Apr. 1892, p. 24; Oct. 1894, p. 35.

2) Compt. rend., 121, p. 168.

3) Bericht von S. & Co., Apr. 1892, p. 41.

4) Bericht von S. & Co., Apr. 1896, p. 69.

5) Archiv d. Pharm., 227, p. 313.

to 0.33 p. c. of volatile oil, which consists of a mixture of methyl salicylate and an ester of valerianic acid.

Quite a number of other species of the genus *Polygala* also yield methyl salicylate when distilled. Van Romburgh[1] found this ester in the roots of *Polygala variabilis* H. B. K. *β. albiflora* D. C., of *P. oleifera* Heckel and *P. javana*.

Bourquelot[2] showed the presence of methyl salicylate in the roots of *P. vulgaris* L., *P. calcarea* F. Schultz and *P. depressa* Wenderoth.

The investigations of Bourquelot render it probable that the methyl salicylate does not exist as such in the root, but is formed by the action of a ferment, which he designates as gaultherase,[3] on the glucoside gautherin.

217. Oil of Cascarilla.

Oleum Cascarillae.—Cascarillöl.—Essence de Cascarille.

Origin. The cascarilla bark, coming from the Bahama islands and principally used in pharmacy, is derived from *Croton eluteria* Bennet (Family *Euphorbiaceae*). Upon distillation it yields 1.5—3 p. c. of volatile oil.

Properties. Cascarilla oil is yellow to greenish and has a slight aromatic odor and taste. Sp. gr. 0.890—0.925; $\alpha_D = +2$ to $+5°$. In 90 p. c. alcohol it is readily soluble.

Composition. Of the but incompletely investigated oil not a single constituent has been definitely determined. Völckel[4] in 1840 separated the oil by fractionation with water vapor into two portions. The first part was mobile, boiled at 173°, had the sp. gr. 0.862 and was almost free from oxygen. The oil which distilled over later was viscous and contained oxygen. According to Gladstone[5] (1864) the oil consists of two hydrocarbons, of which the one has a lemon-like odor, boils at 172°, and is probably dipentene (Brühl,[6] 1888). The second hydrocarbon is described as being similar to calamus oil and is probably a sesquiterpene.

---

1) Rec. des trav. chim. des Pays-Bas., 18, p. 421.

2) Compt. rend., 119, p. 802; Journ. de Pharm., V, 30, pp. 96, 188, 433; VI, 3, p. 577.

3) Termed betulase by Schneegans in 1896 (Journal der Pharmacie von Elsass Lothringen, 23, p. 17).

4) Liebig's Annalen, 35, p. 307; comp. Trommsdorff, Trommsdorff's Neues Journ. d. Pharm., 26, II, p. 186.

5) Journ. Chem. Soc., 17, p. 1.

6) Berichte, 21, p. 152.

218. Oil of Stillingia.

The purgative root of *Stillingia silvatica* I. Müll. (Family *Euphorbiaceae*) indigenous to North America from Virginia to Florida, and westward to Kansas and Texas, yields upon distillation 3.25 p. c. of a light yellow oil, lighter than water (Bichy,[1] 1885).

219. Oil of Mastiche.

Oleum Masticis. — Mastixöl. — Essence de Mastice.

ORIGIN AND HISTORY. Mastiche is the dried resinous juice of the evergreen tree *Pistacia lentiscus* L. (Family *Anacardiaceae*). This tree grows on several of the islands of the Mediterranean and on the southern coasts of the Mediterranean as far as Morocco and the Canary Islands. The collection of the resin is, however, principally carried on in the island of Chios, where the tree is cultivated for this purpose. The resinous juice oozes out slowly from the slight incisions made in the bark of the trunk, and dries up to form round, colorless granules.

Distilled oil of mastiche, probably obtained by dry distillation, is first mentioned about the middle of the fifteenth century. Such an empyreumatic oil of mastiche is also mentioned in the inventory of the "Rathsapotheke" at Braunschweig of the year 1518. Ryff and Gesner distilled mastiche with wine. In the drug ordinances oil of mastiche is first mentioned in that of Berlin of 1574, in books on medicine in the Pharmacopœa Augustana of 1580 and in the Dispensatorium Noricum of 1589. Later the oil went almost altogether out of use. Recently it is used in Turkey for the preparation of a liquor.

PREPARATION AND PROPERTIES. Mastiche yields upon distillation 1—2 p. c. of volatile oil. This is colorless and smells strongly balsamic like mastiche.[2] Sp. gr. 0.858—0.868; $\alpha_D = +22$ to $+28°$. It begins to boil at 155°[3] and on distillation goes over at 160°. With dilute acids it yields terpin hydrate. According to this the principal constituent of oil of mastiche appears to be d-pinene.

220. Oil of Chios Turpentine.

Chios turpentine, already known to the ancients,[4] is obtained by making incisions in the trunk of *Pistacia terebinthus* L. (Family *Ana-*

1) Amer. Journ. Pharm., 57, p. 581.

2) Bericht von S. & Co., Apr. 1893, p. 64.

3) Archiv d. Pharm., 219, p. 170.

4) Chios turpentine is the *τέρμινθος* of Theophrastus and the *τερέβινθος* of other authors. From this designation the word "Turpentine," at present used for the resinous juice of the different species of *Pinus*, is derived. Flückiger & Hanbury, Pharmacographia, p. 166.

cardiaceae) widely distributed in the orient, especially in the island of Chios. Chios turpentine has the consistency of the ordinary turpentine of the conifers and like it consists of a mixture of resin and volatile oil.

Upon distillation with steam about 14 p. c. of oil are obtained from this turpentine. The oil has a pleasant, mild, turpentine-like odor, which reminds somewhat of mace and of camphor. Sp. gr. 0.868—0.869; $\alpha_D = +12°6'$[1] to $+19°45'$.[2]

When treated with sodium the oil boils at 157° and has, as shown by an elementary analysis, the composition of a hydrocarbon $C_{10}H_{16}$. The distillate saturated with hydrochloric acid yielded a solid compound.[1] From this it follows without a doubt that Chios turpentine oil, like ordinary turpentine oil, consists for the greater part of pinene.

221. Oil of Schinus.

ORIGIN. *Schinus molle* L. indigenous to South America, is often planted in southern Europe on account of its fine feathery leaves and fragrant yellow clusters of blossoms. The aromatic berries serve for the preparation of a wine-like drink. They taste at first sweet, later spicy and at last sharply pepperish and are used quite extensively in Greece in place of pepper. In odor they are similar to elemi, but also remind of pepper and juniper and yield upon distillation 3.35[3] to 5.2 p. c.[4] of volatile oil.

PROPERTIES. A mobile oil, having the odor of phellandrene; sp. gr. 0.850; $\alpha_D = +46°4'$ at 17°, soluble in 3.3 and more parts of 90 p. c. alcohol.

COMPOSITION. By conducting hydrochloric acid into the lowest boiling fractions Spica[5] in 1884 obtained a solid monohydrochloride melting at 115°, the formation of which can probably be traced to pinene. A nitrosochloride could be obtained in only very small quantities from a fraction boiling up to 170° of an oil investigated by Gildemeister and Stephan,[6] from which it follows, that only a small part of the oil estimated, at most ½ p. c., can consist of pinene. The largest amount boils from 170—174° ($\alpha_D = +60°21'$, sp. gr. 0.839) and gave with sodium nitrite and glacial acetic acid a strong phellandrene reaction. As follows from the optical behavior of the fractions of the nitrite, the phellandrene in schinus oil is a mixture of much dextrogyrate and very little laevogyrate phellandrene.

---

1) Archiv d. Pharm., 219, p. 170.
2) Bericht von S. & Co., Oct. 1895, p. 57.
3) Jahresb. der Pharm., 1887, p. 25.
4) Bericht von S. & Co., Apr. 1897, p. 49.
5) Gazz. chim. ital., 14, p. 204.
6) Archiv d. Pharm., 235, p. 589.

With alkali a phenol can be shaken out of the oil, which Spica regarded as thymol[1] on account of the melting point 156° of its nitrite. In the oil investigated by Gildemeister and Stephan, carvacrol, (isocyanate compound, m. p. 140°) was present, but no thymol.

Spica observed in the fraction boiling from 180—185°, which had been exposed for some time to the light, a crystalline body, melting at 160° and supposed it to be identical with pinol hydrate $C_{10}H_{16}O . H_2O$ first found by Sobrero. This, however, is excluded, as pinol hydrate melts at 131°.

222. Oil of Cognac.

Oleum Vitis Viniferae.—Cognacöl.—Huile Volatile de Cognac ou de Li de Vin.

Origin and Preparation. The specific aroma, peculiar to wine and cognac is produced by the ethereal fusel oil. It is also called oil of cognac (Ger. *Drusenöl, Weinbeeröl, Weinöl*) and is a product of the fermentation action of wine yeast. For this reason it is principally found in the yeast deposited from the wine on the bottom after completion of the fermentation. The wine itself contains only minute amounts of it in solution, namely 1 part of oil in 40,000 parts.

For the preparation of oil of cognac either the residues (liquid wine yeast) remaining in the barrels after removing the clear wine, are distilled, or the yeast cakes remaining after expression of the liquid are utilized.

The distillation is conducted in a very primitive manner in the wine districts. Formerly a barrel, lined with lead and provided with a condenser, the contents of which were heated with direct steam, served as distilling apparatus.[2]

According to a more recent description,[3] ordinary whiskey stills are used, which are heated by fire and provided with a stirring arrangement to prevent the burning of the yeast.

Before the distillation additions of various kinds are sometimes made to the yeast mixed with water, the purposes of which are not clearly understood.

According to Rautert[2] to 50 k. of lees 250 g. of sulphuric acid are added, by which the mass is said to become more fluid. In the Palatinate to a mixture of 1 part of lees with 5 parts of water, 2 p. c. of freshly slacked lime, 1 p. c. of potassa and 1 p. c. of salt are added.

---

1) Nitrosocarvacrol melts at 153°.

2) Dingl. polytechn. Journ., 148, p. 71.

3) Chemist and Druggist, 50, p. 188.

The oil collected in the receiver floats on top of the alcohol containing water, from which liquid still further amounts of oil can be obtained by cohobation. The yield of cognac oil from yeast cakes is given as 0.036—0.066 p. c. Schimmel & Co. obtained, however, from the wine yeasts from the Rhine and Palatinate 0.07—0.12 p. c. of cognac oil.

PROPERTIES. The crude cognac oil, usually colored green by copper, contains large amounts of free fatty acid, which, because odorless, may be considered as a worthless constituent. For purification the oil is first freed from copper by shaking with tartaric acid solution, then freed from fatty acids by treatment with soda solution, and in this condition forms the second quality of cognac oil. The finest quality is prepared by the rectification of the oil so purified. In this process a part of the slightly odorous esters of the higher fatty acids are kept back, on account of which the distillate of course becomes stronger. The physical properties of these three varieties, the crude, purified and rectified oil, are of course different. As, however, the observation material at hand is insufficient, general means of differentiation cannot at present be given.

Oil of cognac possesses a stupifying odor, which is unpleasant and even offensive in the undiluted state. Often an empyreumatic odor is perceptable, due to careless preparation.

The specific gravity of different crude cognac oils coming from the Moselle, from Tyrol, from the Rhenish Palatinate and the Rheingau varied between 0.875 and 0.885.[1] Optically they were slightly laevo- or dextrogyrate, $\alpha_D = 0° 3'$ to $+0° 43'$. The oil is but sparingly soluble in 70 p. c. alcohol; of 80 p. c. alcohol 1.5 to 3.5 parts are necessary to form a clear solution. The ester number varies between 140 and 250, the acid number between 50 and over 100.[2]

COMPOSITION. As the result of an investigation made by Pelouze[3] and Liebig in 1836 the oenanthic acid ethyl ester $C_{18}H_{18}O_3$ was considered as the principal constituent of wine fusel oil. Later Delffs[4] asserted that the "oenanthic acid" of the ester was identical with pelargonic acid. Fischer[5] then showed on a Palatinate cognac oil, that the "oenanthic acid" which had been considered as a single substance, was

1) Judging from the low sp. gr. 0.85 at 20° the oil examined by Grimm (see following page) must have contained alcohol.

2) Bericht von S. & Co., Apr. 1899, p. 18.

3) Liebig's Annalen, 19, p. 241.

4) Poggendorff's Annalen, 84, p. 505; Liebig's Annalen, 80, p. 290.

5) Liebig's Annalen, 118, p. 307.

a mixture of much caprinic with a little caprylic acid. In Hungarian cognac oil Grimm[1] (1871) likewise found caprinic and caprylic acid, the caprinic acid being present in larger amount. From the non-acid portions of the saponified oil, Halenke and Kurtz[2] were able to separate only ethyl and amyl alcohol, but no butyl nor propyl alcohol.

According to this, cognac oil consists principally of amyl caprinate and ethyl caprinate. These esters, however, are not its most important constituents, as the specific odor is due to other, not yet isolated compounds.

An investigation made by Morin[3] in 1887 of a genuine cognac, which had been distilled in 1883 in Surgères in lower Charente, is of interest. He found in 100 liters of cognac the following constituents in the stated amounts:

Aldehyde	Traces.
Ethyl alcohol	50,837.00 g.
Normal propyl alcohol	27.17 "
Isobutyl alcohol	6.52 "
Amyl alcohol	190.21 "
Furfurol, Bases	2.19 "
Pleasant smelling oil of wine[4]	7.61 "
Acetic acid	Traces.
Butyric acid	"
Isobutylene glycol	2.19 "
Glycerin	4.38 "

In the investigation of another cognac, Ordonneau[5] in 1886 found 218 g. of normal butyl alcohol in a hectoliter, and considered this as the characteristic product of the fermentation of the elliptical yeast. Later, however, it was found, that the cognac investigated by Ordonneau had probably been prepared from a wine, which had contained the *Bacillus butyricus*, and the presence of the butyl alcohol, which does not form under normal conditions is to be traced to this source.

From these investigations it follows, that the cognac oil, when it is obtained in the manner described above, is far from containing all of the constituents of cognac. It is therefore never possible to obtain with such an oil a product that can be compared to a genuine cognac prepared by distillation from wine.

1) Liebig's Annalen, 157, p. 264.
2) Ibid., p. 270.
3) Compt. rend., 105, p. 1019.
4) Probably identical with the cognac oil obtained from yeast.
5) Compt. rend., 102, p. 217; comp. also Claudon & Morin, Compt. rend., 104, p. 1109; further Sell. Arbeiten aus dem kais. Gesundheitsamte, 6, p. 335.

EXAMINATION. The tests are at present restricted to the determination of a possible presence of alcohol. For this purpose a certain quantity of cognac oil is shaken in a graduated cylinder with an equal volume of water or glycerin. The presence of alcohol is shown by an increase in the water or glycerin layer.

223. Oil of Linden Flowers.

Oil of linden flowers is prepared by treating the water obtained by distilling fresh linden flowers with common salt and shaking out with ether. On evaporating the ether 0.038 p. c. of oil remain. Oil of linden flowers is colorless, very fluid, quite volatile and possesses the odor of the fresh flowers in a high degree. It is soluble in all proportions in ether and alcohol (Winckler[1]).

224. Oil of Ambrette Seeds.

Oleum Abelmoschi Seminis.—Moschuskörneröl.—Essence de Graines d'Ambrette.

Hibiscus abelmoschus L., (*Abelmoschus moschatus* Moench), (Family *Malvaceae*), yields the ambrette seeds, formerly officinal in the German Pharmacopoeia, and no longer used in medicine, but which are still employed in perfumery. The herb-like plant is indigenous to India, but has lately been cultivated in Java and in the West Indies (Martinique).

The oil was first prepared by Schimmel & Co.[2] in 1887. Upon distillation of the comminuted ambrette seeds a yield of 0.2 p. c. of oil is obtained. Furfurol has been isolated from the distillation water.[3]

Oil of ambrette seeds is solid at ordinary temperature; it begins to solidify at about 30—35° and has an agreeable, musk-like odor. Sp. gr. about 0.900 at 25—35°. The oil is inactive or only slightly dextrogyrate. This property can be used to advantage, when it is required to distinguish the mixtures with cedar or copaiba balsam oil found in commerce, from genuine oil.

The compound which causes the solidification of the oil is a fatty acid, most probably palmitic acid. As the result, the oil has a high saponification number lying between 180 and 200.

225. Oil of Tea.

By extracting dried tea leaves, *Thea chinensis* L. (Family *Theaceae*), with ether, 0.6—0.98 p. c. of a citron-yellow extract, solidifying on

---

1) Pharmaceut. Centralblatt, 1887, p. 781.

2) Bericht von S. & Co., Oct. 1887, p. 35; Apr 1888, p. 29; Oct. 1898, p. 45.

3) Bericht von S. & Co., Oct. 1899, p. 36.

cooling, and smelling and tasting strongly of tea, is obtained. Mulder[1] has wrongly called this extract, volatile oil of tea, as it consists for the greater part of non-volatile extractive matter. Upon distillation of dried tea leaves Mulder obtained only a turbid water, but no oil.

The real oil of tea was prepared by van Romburgh[2] by the distillation of fresh fermented tea leaves. The yield was only 0.006 p. c. The oil appears to be formed during the fermentation, probably by the decomposition of a glucoside under the influence of a body similar to the ferment laccase. Oil of tea has the sp. gr. 0.866 at 26°; angle of rotation (200 mm.) —0° 11′.

By repeated fractionation a liquid boiling at 153—154° (740 mm. pressure), of a penetrating fusel-like odor, strongly reminding of tea, is obtained. Its principal constituent is an alcohol of the formula $C_6H_{12}O$, the acetic acid ester of which boils at 160—165°.

Oxidation with potassium bichromate converts the alcohol into an oil of the composition and the properties of butyric acid. By heating the alcohol with hydrochloric acid in a sealed tube to 100°, a chloride boiling at 120° is formed.

The fraction of oil of tea which on distillation passes over between 220—225° is methyl salicylate.

The presence of methyl alcohol (b. p. 66°, nitro methyl meta-phenylenediamine compound and oxalic acid methyl ester, m. p. 51°) in the aqueous distillate was shown. This is, however, not a product of fermentation, but is, as van Romburgh had previously shown, a constituent of the fresh, unfermented tea leaves.

226. Oil of Borneo Camphor.

Origin and Preparation.[3] The Borneo-camphor tree, *Dryobalanops aromatica* Gärtn., is a large and beautiful tree, indigenous to the northwestern coast of Sumatra and to northern Borneo. In the cavities and cracks of the wood of old trees is found the valuable Borneo camphor, also called Baros camphor, Sumatra camphor, or Malay camphor. It is used by the Malays for the embalming of the dead and for ritualistic purposes, and well paid for. Besides the camphor, the trees contain also a volatile oil, which is obtained, either by tapping the tree or by the distillation of the wood.

---

1) Poggendorff's Annalen der Physik, 43, p. 163; Liebig's Annalen, 28, p. 314.

2) Verslag omtrent den Staat van S'Lands Plantentuin te Buitenzorg, 1895, p. 119, and 1896, p. 166; Berichte von S. & Co., Apr. 1897, p. 42, and Apr. 1898, p. 58.

3) For History see under camphor oil, p. 370.

PROPERTIES. Oil of Borneo camphor is a liquid, sometimes colored green by copper, of the sp. gr. 0.882—0.909 (Macewan,[1] 1885).

COMPOSITION. The lack of an investigation along modern lines is probably due only to the fact, that this interesting oil can not be obtained in the market.

From older investigations of Martius[2] (1838), Pelouze[3] (1840) and Gerhardt[4] (1843) it can be seen that borneol[5] is the characteristic constituent of the oil. More recent investigators, however, like Lallemand[6] (1860) and Macewan[1] (1885) deny the presence of this alcohol in the oil.

According to Pelouze[3] the liquid part of the oil has the same composition as the supposed terpene "borneen," which is obtained by the action of phosphoric acid anhydride on borneol. As Wallach[7] has shown (1885), there result by this treatment several hydrocarbons; "borneen" is therefore a mixture of several substances. Although nothing positive can be drawn from this statement of Pelouze, it is, nevertheless, probable also from other considerations that camphene is contained in the liquid parts of the oil of Borneo camphor.

The oil investigated by Lallemand had been brought by Junghuhn from Sumatra, and had been obtained by distilling various comminuted parts of the tree. It turned the plane of polarization to the right (α for red light $= + 7°$) and on distilling began to boil at 180°. The fraction going over at this temperature was relatively small (sp. gr. 0.86, $\alpha_D + 13°$) and yielded after passing hydrochloric acid through it and treating the resulting product with fuming nitric acid, a solid hydrochloride (dipentene?). The principal part of the oil went over from 260—270°, had the sp. gr. 0.90—0.92 and yielded a compound $C_{15}H_{24}2HCl$ melting at 125°. Overlooking the somewhat high melting point the body might be looked upon as cadinene hydrochloride. As already mentioned this Borneo camphor oil contained no borneol.

227. Oil of Gurjun Balsam.

ORIGIN AND HISTORY. Gurjun balsam, called wood oil in India, has been known there for a long time and used as an excellent varnish.

---

1) Pharm. Journ., III, 15, pp. 795 and 1045
2) Liebig's Annalen, 27, p. 63.
3) Compt. rend., 11, p. 365; Liebig's Annalen, 40, p. 326.
4) Liebig's Annalen, 45, p. 38.
5) The borneol from the Borneo camphor tree is dextrogyrate. Flückiger, Pharm. Journ., III, 4, p. 829.
6) Liebig's Annalen, 114, p. 193.
7) Liebig's Annalen, 230, p. 237.

The balsam is obtained in East India, in a manner similar to turpentine, from several species of *Dipterocarpus* (Family *Dipterocarpaceae*), especially from *D. turbinatus* Gaertn. fil., *D. incanus* Roxb. and *D. alatus* Roxb. The *Dipterocarpus* trees belong to the most magnificent trees of the Indian mountain forests, and are so productive, that full grown trees yield up to 180 liters of balsam during a summer.

The collection is as follows: At the end of the dry season a transverse slit 2 inches in length and 6 inches in width is made with an axe through the bark into the wood, just above the ground. At the base of this a deep cavity is made, which is capable of holding at least 1 liter of balsam. This is heated by means of a coal fire which has been started in it, until the walls are charred. The cavity is then emptied and cleaned. The balsam soon begins to flow at such a rapid rate from the slit, that the cavity serving as a reservoir has to be emptied several times a day. When the flow diminishes or is retarded by drying on the walls, the crusts are scraped off or the walls are again heated. When exhausted, the same operation is repeated on the opposite side of the trunk.[1]

The principal places of export of the gurjun balsam are Saigon. Singapore, Moulmein in Tenasserim and Akyab.

Gurjun balsam, according to its source and manner of collection, differs in consistency, color, and behavior to solvents. The better varieties are of a greenish gray color; with reflected light somewhat turbid and slightly fluorescent, with transmitted light it appears entirely clear and reddish brown. Odor and taste remind of copaiba balsam. Its sp. gr. is 0.96—0.97. By steam distillation up to 70 p. c. of oil are obtained.

In Europe attention was called to gurjun balsam in 1811 by Franklin[2] and in 1813 by Ainslie,[3] but its source and manner of collection was first described accurately by Roxburgh about 1827.[4] The similarity of the action of gurjun balsam to that of copaiba balsam was made known in India about the year 1812 by the physician O'Shaughnessy.[5] The balsam also gained in India a considerable reputation as a cure against leprosy and was used later in England in dermatology.[6]

---

1) Pharm. Journ., III, 5, p. 729; III, 18, p. 161; D. Hanbury, Science Papers, p. 118.
2) Tracts on the dominions of Ava. London, 1811, p. 26.
3) Materia medica of Hindoostan. Madras, 1813, p. 186.
4) Plants of the coast of Coromandel, 1828, vol. 3, p. 10.
5) Bengal Dispensatory, Calcutta, 1842, p. 22.
6) Pharmac. Journ., III, 5, p. 729.

PROPERTIES. Gurjun balsam oil is a yellow, somewhat viscous liquid, of the sp. gr. 0.915—0.930. It has at times a very high rotatory power, namely $\alpha_D = -35°$ to $-130°$. Most of the balsams give laevogyrate oils, but strongly dextrogyrate oils have also been observed.[1] The oil is not completely soluble in 90 p. c. alcohol, and even in 95 p. c. alcohol the solubility is limited. It contains no saponifiable constituents, but probably small amounts of alcoholic bodies, for after acetylization the saponification number 9.6 was obtained.

Gurjun balsam oil boils for the greater part at 255—256° and appears to consist almost wholly of sesquiterpene,[2] $C_{15}H_{24}$. With hydrochloric acid this does not give a solid hydrochloride, but only a fine dark blue coloration. Whether this hydrocarbon is identical with one of the known sesquiterpenes has not yet been determined.

Oil of gurjun balsam is on account of its weak odor a very dangerous adulterant for other volatile oils. On account of its high boiling point, its strong rotatory power, and its difficult solubility in alcohol, its detection, however, usually offers no great difficulties.

228. Oil of Ladanum.

ORIGIN AND HISTORY. The ladanum[3] resin, used since antiquity as incense and embalming agent, is an exudation of the bush-like plants *Cistus creticus* L., *C. ladaniferus* L. and others (Family *Cistaceae*) indigenous to Asia Minor, Crete, Cyprus and a few other islands off the coast of Asia Minor. Up to the beginning of this century it was an officinal drug, valued for its pleasant odor, and is often mentioned along with the ancient aromatics storax, myrrh and frankincense in literature and has often, especially in the translations of the Bible, been confounded with galbanum. Lately it has gone almost altogether out of use.

Ladanum oil, distilled at first with wine or spirits of wine (*aqua vitae*) was already known to Ryff, Gesner, Rubeus, and to Porta. It was taken up in medical books, first in the Dispensatorium Noricum of the year 1589 and in the Pharmacopœa Augustana of the year 1613.

PROPERTIES. The oil distilled from the leaves of *Cistus ladaniferus* L. has an unpleasant, narcotic odor. It has the sp. gr. 0.925 and boils from 165—280° with decomposition and splitting off of acetic acid.[4]

---

1) Pharmacographia Indica, vol. 1, p. 193.

2) Pharmacognosie, 2nd ed., p. 102. Comp. also Werner, Zeitschr. f. Chem. und Pharm., 5, p. 588; Jahresb. f. Chem., 1862, p. 461.

3) Also spelled labdanum.

4) Bericht von S. & Co., Oct. 1889, p. 58.

Upon distillation of ladanum resin Schimmel & Co.[1] obtained a golden yellow oil, of a fine strong amber odor. After standing for about half a year, magnificent crystals separated out, which made up about the fourth part of the oil.[2]

229. Oil of Canella.

Origin and History. The evergreen bush *Canella alba* Murray (Family *Canellaceae*), indigenous to the West Indies, has a pleasant aromatic bark, which, when it was brought with other drugs from the new world to Europe, was considered as a variety of cinnamon bark, and later was confounded with other medicinal barks, especially with Winter's bark, from *Drimys winteri* Forst.

The volatile oil of canella bark was probably first distilled in 1707 by Sloane in England, and later, in 1820, by Henry, but appears to have found no application. It was investigated in 1843 by Meyer and von Reiche and later by Bruun and Williams.

Properties. Upon distillation of canella bark, 0.75 to 1.25 p. c. of volatile oil are obtained, the odor of which is similar to that of a mixture of clove and cajeput oil. Sp. gr. 0.920 to 0.935. Optically it is slightly dextrogyrate $\alpha_D = +1° 8'$.

Constituents. By shaking out with alkali, a phenol, eugenol[3] (benzoyl eugenol, Bruun[4]) could be separated.

By the distillation of the remaining portion the lowest fraction went over from 165—170.° (Sp. gr. 0.888; $\alpha_D = -3.38°$). The presence of l-pinene (Williams[5]) was demonstrated by preparing the pinene nitrosochloride and the pinene nitrolbenzylamine base, melting at 122—123°. The next highest fraction consists of cineol[6] (cineolic acid, m. p. 197°).[2] The highest boiling parts of the oil contain caryophyllene,[5] the presence of which was shown by preparing the crystalline caryophyllene hydrate, melting at 92—95°.

230. Oil of Damiana Leaves.

Several species of *Turnera* (Family *Turneraceae*), especially *Turnera aphrodisiaca* Ward and *Turnera diffusa* Ward have been named as the botanical sources of damiana leaves, used medicinally since 1875 in America. The leaves of other plants, however, are also sent into the market as damiana leaves, for instance those of *Bigelovia veneta* Gray

1) Bericht von S. & Co., Apr. 1898, p. 63.
2) Ibidem, Oct. 1893, p. 24.
3) Liebig's Annalen, 47, p. 224.
4) Proc. Wisc. Pharm. Assoc., 1893, p. 36.
5) Pharm. Rundschau, 12, p. 183.
6) Bericht von S. & Co., Oct. 1890, p. 53.

(*Aplopappus discoideus* D. C.). The differences in the material employed no doubt, explain also the differences noticed in the oils obtained by distillation at different times.

Pantzer[1] obtained in 1887 0.5 p. c. of a yellow oil of an aromatic odor and a warm, camphor-like, bitter taste. Schimmel & Co.[2] obtained 0.9 p. c. of a green viscous oil, of a chamomile-like odor. It had the sp. gr. 0.970 and boiled between 250 and 310°. The highest boiling fractions were blue. From another consignment of leaves the same firm obtained[3] in 1896 1 p. c. of oil of the following properties: sp. gr. 0.943; $\alpha_D = -23° 25'$; saponification number 41.8. On standing in the cold, the separation of crystals on the upper surface could be noticed, similar to those which are observed at the beginning of the solidification of rose oil. According to this it is probable, that paraffins are also present in oil of damiana leaves.

231. Oil of Henna.

The blossoms of the henna shrub, *Lawsonia inermis* L. (Family *Lythraceae*) used in the Orient for dyeing, contain according to Holmes[4] a volatile oil, which is said to have an odor closely resembling that of tea rose.

232. Oil of Myrtle.

Oleum Myrti.—Myrtenöl.—Essence de Myrte.

Origin. The myrtle, *Myrtus communis* L. (Family *Myrtaceae*), very widely distributed in the mountainous districts of Spain, Italy and southern France, is distinguished for its fragrant blossoms and leaves. Only the leaves are used for the preparation of the volatile oil[5] giving a yield of 0.3 p. c. The myrtle oil of commerce is usually of French or Spanish origin. The Corsican oil is especially valued.

Properties. Oil of myrtle is a yellow to greenish liquid, of a pleasant and refreshing aroma. Sp. gr. 0.890—0.915; $\alpha_D = +10$ to $+30°$.

Composition. The fraction boiling between 158—160° consists of a dextrogyrate terpene $C_{10}H_{16}$ ($[\alpha] = +36.8°$), the chemical behavior of which pointed to pinene.[6] That the substance under consideration was really d-pinene, was shown by Jahns in 1889 by the preparation of pinene nitrosochloride.[7]

---

1) Am. Journ. Pharm., 59, p. 69.

2) Bericht von S. & Co., Apr. 1888, p. 44.

3) Bericht von S. & Co., Apr. 1897, p. 18.

4) Pharm. Journ., III, 10, p. 685.

5) The fresh fruits likewise contain volatile oil. Raybaud, Journ. de Pharm, II, 20, p. 468.

6) The "Myrtene" of Gladstone (1864), Journ. Chem. Soc., 17, p. 1; 25, p. 1.

7) Archiv d. Pharm., 227, p. 174.

The oil boiling about 176° contains cineol, $C_{10}H_{18}O$. By conducting into it hydrochloric or hydrobromic acid, the characteristic addition products are formed, by the decomposition of which by water the cineol can be readily obtained in a pure form.

The fraction collected at 180° consists of dipentene.[1] This terpene was identified by the preparation of dipentene nitrosochloride and tetrabromide of the melting point 125°.

The fraction boiling from 195—200° acts on sodium with liberation of hydrogen. Jahns suspects it to be a camphor $C_{10}H_{16}O$. According to the boiling point linalool might be contained in this fraction. In a later investigation of oil of myrtle by Bartolotti,[2] to whom the investigations of Jahns and Schimmel & Co. were evidently unknown, not a single constituent was identified.

Oil of myrtle was formerly much used for the preparation of myrtol, as the fraction boiling from 160—180° was called. The reputed antizymotic and deodorizing properties[3] of "myrtol" can no doubt be traced to its content of cineol.

233. Oil of Cheken Leaves.

The cheken leaves, often recommended for medicinal use, the leaves of *Myrtus cheken* Spr. indigenous to Chili, contain about 1 p. c. of an oil very similar to ordinary myrtle oil (Weiss,[4] 1888).

It is quite mobile, of a yellowish green color and has a pleasant odor, reminding of eucalyptus and sage. Sp. gr. 0.8795 at 15°; $\alpha_j = +23.5°$.

The principal part of the oil, about 75 p. c., boils from 155—157° (sp. gr. 0.8635; $\alpha_j = +31.28°$), has the composition $C_{10}H_{16}$ and yields with hydrochloric acid a monohydrochloride melting at 120°. On heating this fraction to 260—270° dipentene resulted (tetrabromide, m. p. 125—126°). From this it follows that the hydrocarbon boiling from 155—157° is d-pinene.

The fraction boiling about 176° (15 p. c.) consists of cineol, as was shown by the formation of the cineol bibromide $C_{10}H_{18}OBr_2$, which separates in red crystals from the petroleum ether solution on the addition of bromine.

About 10 p. c. of the oil pass over, on distillation, from 220—280°. The composition of this fraction has not yet been determined.

1) Bericht von S. & Co., Apr. 1889, p. 28.
2) Gazz. chim. ital., 21, p. 276; Berichte, 24, Referate, p. 572.
3) Pharmaceutische Zeitung, 35, p. 224.
4) Archiv d. Pharm., 226, p. 666.

234. Oil of Pimenta.

Oleum Amomi seu Pimentae.—Pimentöl.—Essence de Piment.

Origin and Preparation. The evergreen shrub *Pimenta officinalis* Lindl. (Family *Myrtaceae*), indigenous to the West Indies, grows on calcareous soil near the coast, in Cuba, Hayti, San Domingo, Trinidad, Antigua, and especially in Jamaica, further in Central America, Mexico, Costa Rica and Venezuela.

For obtaining pimenta valued as an aromatic or allspice,[1] the unripe berries are gathered and dried in the sun. The ripe fruit contains a sweet jelly, but is almost odorless and has therefore no commercial value.

Pimenta yields on distillation, which is at first accompanied by a strong evolution of ammonia, 3—4.5 p. c. of volatile oil.

Properties. Oil of pimenta is of a yellow to brownish color and has a pleasant aromatic odor, similar to clove oil, although sharply differentiated from it, and a piercing, sharp taste. Sp. gr. 1.024[2] to 1.050. Slightly laevogyrate. It is soluble in all proportions in 90 p. c. alcohol, and soluble to a clear solution in 2 parts of 70 p. c. alcohol.

Composition. The experiments of Bonastres,[3] (1825) who prepared the alkali salts of the acid constituents of pimenta oil, already indicated that the similarity of this oil to clove oil was due to the eugenol which was common to both. The proof of the identity of these two phenols was furnished by Oeser[4] in 1864, who separated the eugenol from the oil by means of an alkali solution, and analysed it.

That part of the oil which did not combine with the alkali, boiled at 255° and was slightly laevogyrate. The analysis gave results agreeing with C_5H_8. This is therefore a sesquiterpene.

The odor of oil of pimenta indicates that eugenol and sesquiterpene are not the only constituents. An investigation of the oil according to modern methods might be a profitable and a not too difficult problem.

Oil of pimenta leaves. An oil distilled in Trinidad from the leaves of an undetermined species of *Pimenta*, is a yellow liquid, smelling strongly of lemon, of the sp. gr. 0.882 at 25°; $\alpha_D = -0° 37'$. The lemon-like odor is due to citral, the presence of which was demonstrated by forming the citryl β-naphtho cinchoninic acid.[5]

1) Odorographia, vol. 2, p. 51.
2) Bericht von S. & Co., Apr. 1899, p. 39.
3) Journ. de Pharm., 18, p. 466. Comp. also ibid., 11, p. 187; Trommsdorff's Neues Journ. d. Pharm., 11, I, p. 127.
4) Liebig's Annalen, 181, p. 277.
5) Bericht von S. & Co., Oct. 1896, p. 77.

235. Oil of Bay.

Oleum Myrciae. — Bayöl. — Essence de Myrcia.

Origin and Preparation. As the numerous species of the genera *Myrcia* and *Pimenta* are very variable and very similar to each other, it is highly probable that these small botanical differences are not observed in collecting, and therefore the bay leaves of commerce are not always derived from one and the same plant,[1] but are rather mixtures of the leaves of several species. According to Holmes[2] the botanical source of the genuine bay leaves is *Pimenta acris* Wight (*Eugenia acris* Wight et Arnott) and not *Myrcia acris* D. C. as is given in the U. S. P.

The bay tree is indigenous to the West Indies and flourishes especially on Jamaica, St. Thomas, Antigua, Guadeloupe, Dominica and Barbadoes. The oil is obtained only in part from the fresh material on the islands themselves; the larger part is probably prepared in New York from the dried leaves.

On distilling, part of the oil separates in the receiver on top of the water and another part sinks to the bottom. They must be mixed if a normal oil is to be obtained. This appears not always to be done, as bay oils of a specific gravity greater than water often appear in the market. The yield from dried leaves is 2—2.5 p. c.

Properties. Bay oil is a yellow liquid, soon becoming brown on exposure to the air, of a pleasant odor reminding of clove and a sharp spicy taste. The sp. gr. of the normal oil varies from 0.965 to 0.985 and is as a rule above 0.970; optically it is slightly laevogyrate, $\alpha_D =$ up to $-2°$. The freshly distilled oil is soluble to a clear solution in an equal volume of 90 p. c. alcohol; on standing, however, it soon loses its solubility and then gives only turbid mixtures, a behavior which is due to the polymerization of an olefinic terpene, myrcene, contained in the oil. The content in phenols (eugenol and chavicol) amounts to 59—65 p. c. according to volumetric determinations with dilute potassa solution.

Composition. The identity of the main constituent of greater specific gravity of the oil with eugenol appears to have been first recognized by Markoe[3] in 1877. The first complete investigation of the oil was made by Mittmann[4] in 1889. He found besides eugenol a small amount of methyl eugenol, which statement was later corroborated. By fractional distillation he obtained a small fraction, boiling from

1) Odorographia, vol. 2, p. 56.
2) Pharm. Journ., III, 21, p. 887.
3) Proc. Am. Ph. Assoc., 25, p. 488.
4) Archiv d. Pharm., 227, p. 529.

160—185°, which according to him, consisted of two terpenes, of which he believed the lower boiling to be pinene, and suspected dipentene in the higher boiling portion. According to a more recent investigation by Power and Kleber[1] neither of these terpenes, but a whole series of other bodies, overlooked by Mittmann, are contained in oil of bay.

Upon distillation under 20 mm. pressure of the oil separated from the phenol 80 p. c. went over between 67 and 68°, the remainder at 160°. The first fraction had the sp. gr. 0.8023 and was a colorless liquid of a characteristic odor, but quite different from that of the known terpenes. Analysis and molecular weight determination corresponded with the formula $C_{10}H_{16}$. From the index of refraction, $n_D = 1.4673$, the molecular refraction 47.1 was calculated, according to which the compound contains three double bonds. As with such an arrangement of the atoms a ring configuration is excluded, the atoms must form an open chain, which is also in harmony with the low sp. gr. A so-called olefinic terpene, which was called myrcene was, therefore, under consideration.

Myrcene is very susceptible to change and polymerizes even after standing for only one week to a thick oil. By treating myrcene with glacial acetic acid and sulphuric acid (Bertram's hydration method) there resulted (besides dipentene) an oil of a lavender-like odor, which yielded linalool after saponification. This alcohol was identified by its conversion into citral. Myrcene bears therefore the same relation to linalool, as does camphene to isoborneol, and pinene or dipentene to terpineol. Besides myrcene, bay oil contains l-phellandrene. Other terpenes have not been found by Power and Kleber.

The fractions of bay oil freed from phenols and terpenes, gave a solid compound with sodium bisulphite, the decomposition of which yielded citral (citryl β-naphtho cinchoninic acid).

After separating the citral an oil, smelling like anise, remained, which consisted in part of methyl chavicol. By oxidation it was converted into anisic acid, by treatment with alcoholic potassa into anethol.

The highest boiling fractions of the oil contained methyl eugenol.

By treating the phenols separated from the oil by means of Alkali with methyl iodide, Power & Kleber obtained a mixture of methyl eugenol and methyl chavicol, from which it follows, that besides eugenol, chavicol is also present in oil of bay.

---

[1]) Pharm. Rundschau, 18, p. 60.

The constituents of oil of bay arranged according to the amounts present are as follows:

1. Eugenol, $C_{10}H_{12}O_2$,
2. Myrcene, $C_{10}H_{16}$,
3. Chavicol, $C_9H_{10}O$,
4. Methyl eugenol, $C_{11}H_{14}O_2$,
5. Methyl chavicol, $C_{10}H_{12}O$,
6. Phellandrene, $C_{10}H_{16}$,
7. Citral, $C_{10}H_{16}O$.

EXAMINATION. As oil of bay contains no pinene, any turpentine oil which may have been added, can be detected without difficulty.

From 10. cc. of bay oil 1 cc. is slowly distilled off from a fractionating flask, and the distillate mixed with 1 cc. of amyl nitrite and 2 cc. of glacial acetic acid. To this is added drop by drop with agitation, while the whole is kept well cooled in a freezing mixture, a mixture of equal parts of glacial acetic acid and hydrochloric acid as long as a blue coloration results. If pinene is present a white precipitate of pinene nitrosochloride is formed. In this manner 10 p. c. of turpentine oil in bay oil can be detected.

236. Oil of Cloves.

Oleum Caryophyllorum. — Nelkenöl. — Essence de Girofle.

HISTORY. Oil of cloves appears to have been distilled for the first time in the fifteenth century, but probably, like other aromatics, with wine or the addition of spirits of wine. This method of distillation was described by Ryff, Gesner, Lonicer and others. Gesner also mentions the distillation of the oil, *per descensum*. The pure oil was, however, shortly afterwards distilled by Cordus, by Winther of Andernach and by Porta. In the Dispensatorium Noricum, oil of cloves was not admitted until the edition of 1589. In drug ordinances it was first mentioned in that of the city of Berlin of 1574.

The yield of volatile oil from cloves was determined by Boerhaave, Hoffmann, Neumann and Trommsdorff. Boerhaave remarked that the quite different yield obtained on distillation was sometimes due to the adulteration with cloves which had been exhausted of their oil by distillation and then dried again.

Bonastre[1] in 1827 recognized the acid nature of clove oil and investigated the salt-like compounds of eugenol that were formed with alkalies. Ettling and Liebig[2] in 1834 first showed that besides *Nelkensäure*, eugenic acid, there is also present in the oil an indifferent body.

[1]) Journ. de Pharm., II, 13, pp. 464, 518; Poggendorff's Annalen, 10, pp. 609, 611.
[2]) Liebig's Annalen, 9, p. 68.

Of the older investigations, which were restricted mostly to eugenol, those of Dumas[1] (1833), Böckmann[2] (1838), Stenhouse[3] (1843), Calvi[4] (1856), Brüning[5] (1857), Williams[6] (1858), Hlasiwetz and Grabowski[7] (1866) and Erlenmeyer[8] (1866) may be mentioned.

ORIGIN AND PREPARATION. The ever-green clove tree, *Eugenia caryophyllata* Thunb. (*Caryophyllus aromaticus* L.) (Family *Myrtaceae*), was originally indigenous to the Philippines and is now cultivated on Amboina, Réunion, Mauritius, Madagascar and Malacca (Penang). The plantations on the largest scale are, however, found on the east African islands Zanzibar and Pemba, which taken together produce about ⅘ of the total clove production of the world.

Cloves are the undeveloped blossoms, dried in the air, of the clove tree, which is aromatic in all its parts. Its inflorescence is a perfect cyme containing as many as 35 individual blossoms. Each blossom has a pulpy receptacle nearly 1 cm. in length, which at first is of a light color, later green, and just before blossoming becomes dark red. When this color appears, the blossoms are gathered, because they are then richest in oil, and dried in the air. The flower buds of the cultivated trees contain more oil than those of the trees growing wild.

The berry-like fruit collected just before ripening, formerly came into commerce under the name of *Anthophylli*, mother-cloves.

Only the Zanzibar cloves (under which designation are included the cloves coming from Pemba) are used for distillation. The cloves from Amboina and Réunion are indeed richer in oil, but on account of their better appearance command a much higher price than the difference in the oil content justifies. The most expensive is the Madagascar clove, from St. Marie on the southern point of the island. It yields 18 p. c. of oil, which is considered as finer by the Parisian perfumers, but from other considerations no especial advantages can be assigned to it.[1]

The cloves are distilled either whole or in a comminuted condition; according to the method of distillation (water or dry steam) there is obtained an oil of higher specific gravity and richer in eugenol, or a lighter oil, in which the non-phenol constituents are relatively larger. The oil collecting in the receivers partly sinks in the water and partly floats upon it. By mixing both parts normal oil of cloves is obtained. The yield from Zanzibar cloves amounts to 15—18 p. c.

---

1) Ann. de Chim. et de Phys., II, 53, p. 165; Liebig's Annalen, 9, p. 68.
2) Liebig's Annalen, 27, p. 155.
3) Ibidem, 95, p. 103.
4) Ibidem, 99, p. 242.
5) Ibidem, 104, p. 202.
6) Ibidem, 107, p. 288.
7) Ibidem, 139, p. 95.
8) Zeitschr. f. Chemie, 9, p. 95.

PROPERTIES. Oil of cloves is, when freshly distilled, an almost colorless to yellowish, strongly refracting liquid, which becomes darker with age. The odor is strongly aromatic, the taste persistently burning. The specific gravity varies according to the method of distillation from 1.045 to 1.070. The optical rotation is slightly to the left, α_D = up to $-1° 10'$. The oil dissolves to a clear solution in 2 parts of 70 p. c. alcohol.

The chemical reactions are partly those of eugenol, several color reactions are, however, produced by the traces of furfurol present. Upon distillation oil of cloves goes over between 250 and 260°.

COMPOSITION. Of the constituents of oil of cloves, eugenol, which is present to the extent of 70—85 p. c., was the first to attract the attention of chemists (comp. History, p. 512).

Eugenol, a phenol of the formula $C_{10}H_{12}O_2$, boils under ordinary pressure with slight decomposition at 253—254°[1] (thermometer completely in the vapor). It is the most characteristic and valuable constituent of clove oil; its quantitative determination is described on page 516, its properties and compounds on page 180.

Pure eugenol is obtained by treating oil of cloves with weak (2—5 p. c.) soda solution and after shaking out the alkaline solution several times with ether, decomposing it with dilute sulphuric acid.

Besides free eugenol, there is contained, according to Erdmann,[2] in oil of cloves also 2—3 p. c. of acet-eugenol, which is found in the oil not attacked by the alkaline solution. In this the acet-eugenol can be detected and removed by saponification with alcoholic potassa or by heating with a concentrated aqueous solution of potassa.

The oil remaining after removing the acet-eugenol consists principally, as was already recognized by Church[3] in 1875, of a sesquiterpene, which was more closely investigated by Wallach[4] in 1892 and called caryophyllene.

Caryophyllene, $C_{15}H_{24}$, is a colorless liquid boiling from 258—260°, of a weak odor which does not resemble cloves and the sp. gr. 0.9085. By taking up a molecule of water it is converted into the nicely crystalline alcohol, caryophyllene hydrate, $C_{15}H_{26}O$, melting at 96°. For further information see page 125.

Scheuch[5] in 1863 pointed out the presence of small amounts of salicylic acid. The correctness of this statement was, however, rendered

1) Bericht von S. & Co., Apr. 1892, p. 28.
2) Journ. f. prakt. Chem., II, 56, p. 143.
3) Journ. Chem. Soc., 28, p. 113.
4) Liebig's Annalen, 271, p. 287.
5) Liebig's Annalen, 125, p. 14.

doubtful by Wassermann[1] in 1875. Erdmann[2] definitely showed the presence of salicylic acid in oil of cloves and found that it did not exist free but probably as acet-salicylic acid ester of eugenol. Of other minor constituents of oil of cloves, methyl alcohol must be mentioned, which is best obtained by collecting the first portions in working up the cohobation waters and fractionating this repeatedly. Finally a liquid boiling at 65.5—66° is obtained which can be identified as methyl alcohol[3] by means of its oxalic acid ester, melting at 54°.

By continuing the distillation a liquid boiling at 162° is obtained, which consists of furfurol[3] and was identified by its intense color reaction with aniline and p-toluidine, by converting it into its phenyl hydrazone melting at 96°, as well as by splitting it into pyromucic acid and furfurol alcohol. Furfurol is probably partly the cause of the darkening of oil of cloves.

In the fraction boiling at 150—155°, which cannot be completely separated from furfurol by distillation, is contained methyl amyl ketone,[4] $CH_3.CO.C_5H_{11}$. If from the fraction mentioned, the furfurol be removed by shaking with permanganate solution, the pure ketone, boiling at 151—152° remains. The elementary analysis corresponded to $C_7H_{14}O$. By heating with chromic acid mixture it was gradually oxidized to valerianic acid (together with some capronic acid) and acetic acid. Although the methyl amyl ketone is present in oil of cloves only in a fraction of one percent it, nevertheless, has a considerable effect on the odor of the oil, as it gives rise to the peculiar fruity odor which accompanies the heavy eugenol odor.

A further substance, which has been found in the oil, vanillin,[5] may have been formed by the oxidizing influence of the air on the eugenol. It is worthy of notice that vanillin is already present in the cloves.

Examination. As nearly all adulterants which can be considered, are specifically lighter than clove oil, the determination of the specific gravity is not to be omitted. When it is found less than 1.045 it is to be tested by means of fractional distillation for turpentine and similar low boiling oils. Its solubility in 2 parts of 70 p. c. alcohol is also to be observed. Oils of cedar wood, copaiba and gurjun balsam would show themselves by their difficult solubility and simultaneous increase of the rotatory power. The German Pharmacopoeia gives a test for the highly improbable presence of carbolic acid, in that it directs 1 cc.

1) Liebig's Annalen, 179, p. 369.
2) Journ. f. prakt. Chemie, II, 56, p. 143.
3) Bericht von S. & Co., Oct. 1896, p. 57.
4) Bericht von S. & Co., Apr. 1897, p. 50.
5) Chem. Centralbl., 1890, II, p. 828.

of oil to be shaken with 20 cc. of hot water, and that the filtrate obtained after allowing the mixture to cool, be treated with ferric chloride. A blue coloration would indicate the presence of carbolic acid. Much more common than the adulteration of clove oil, is the substitution of the cheaper and less pleasantly odorous oil of clove stems, the sure detection of which is only possible to a well trained nose. Investigations on a physical or chemical basis are entirely inadequate. Perhaps the recently found fact that only oil of cloves, but not oil of clove stems, contains acet-eugenol, might be used as a means of distinguishing between these oils.

It may often be desirable to know how much eugenol is contained in a clove oil (or clove stem oil). For this purpose H. Thoms[1] has worked out a method of determination which depends on converting the eugenol into its benzoyl derivative, separating and weighing as such. The method is as follows:

In a tared beaker of about 150 cc. 5 g. of clove oil are treated with 20 g. of sodium hydrate solution (15 p. c.) and then 6 g. of benzoyl chloride added. The mixture is well shaken, when a considerable amount of heat will be liberated, until it is uniformly mixed up. After cooling, 50 cc. of water are added, and then heated until the crystalline ester has again become oily and again allowed to cool. The clear supernatant liquid is filtered off and the crystalline mass remaining in the beaker is again treated with 50 cc. of water, heated on a water bath until the ester is melted and filtered after cooling. This operation is repeated with 50 cc. more of water. The excess of soda as well as of sodium salt has then been removed.

After having returned any crystals which may have been washed on to the filter to the beaker, the still moist benzoyl eugenol is at once treated with 25 cc. of alcohol of 90 p. c. by weight, and heated with agitation on a water bath until solution is effected; the beaker is now removed from the water bath and the agitation continued until the benzoyl eugenol has separated in fine crystals. This takes place in a few minutes. The mass is then cooled to 17°, the precipitate transferred to a filter of 9 cm. diameter and the filtrate collected in a graduated cylinder. About 20 cc. of filtrate are usually obtained; the alcoholic solution which is held in the crystalline mass is forced out by adding so much of 90 p. c. alcohol by weight until the total filtrate measures 25 cc. The still moist filter with its precipitate is transferred to a weighing tube (the latter with the filter having previously been dried at 101° and weighed) and dried at 101° until the weight is constant. 25 cc. of 90 p. c. alcohol dissolve at 17° 0.55 g. of pure benzoyl eugenol, which amount must be added to the result obtained.

If a is the amount of benzoic acid ester, b the amount of clove oil used (about 5 g.), and if 25 cc. of the alcoholic solution of the ester are filtered off

1) Ber. der pharm. Gesellschaft, 1, p. 288.

as above described, the percentage content of eugenol in the oil of cloves is found by the following formula:

$$\frac{4100\ (a + 0.55)}{67\ .\ b}$$

This formula results from the two equations:

(Benzoyl eugenol) (Eugenol)

$$268 \quad : \quad 164 = (a + 0.55): \text{Amount of eugenol found}$$

$$\text{Eugenol} = \frac{164\ .\ (a + 0.55)}{268}$$

$$\text{Hence } b: \frac{164\ .\ (a + 0.55)}{268} = 100 : x$$

$$x = \frac{164\ (a + 0.54)\ .\ 100}{268\ .\ b} = \frac{4100\ (a + 0.55)}{67\ .\ b}$$

Mixtures of pure eugenol with caryophyllene yielded by this method results which agreed within 1 p. c., an accuracy which is probably sufficient for most cases.[1]

By Thoms' method only the free eugenol and not that which is in the form of acet-eugenol is determined. If the content of acet-eugenol is also to be found, two determinations must be made; one as before described, and the other after saponifying the oil.[2]

For this purpose the oil is heated with concentrated aqueous potassa solution for a short time to 100°. The saponification must not be made with alcoholic potassa as the alcohol which is removed only with difficulty, would form ethyl benzoate and thus affect the results.

The eugenol content of a good oil of cloves amounts to 70—85 p. c.

Umney[3] has recommended a much simpler and more rapidly conducted eugenol determination by using a Schimmel cassia flask with graduated neck.[4]

A known amount of clove oil is put in the flask, 10 p. c. aqueous potassa solution added, and heated. The oil floating on top (non-phenols) is brought into the neck by filling up the flask, cooled to 15° and its amount read off on the scale.

According to Umney's own statement the results thus obtained by Thom's method are 11—12 p. c. too high, which is explained in that the strong alkali solution dissolves a part of the caryophyllene. By using a more dilute lye (4—5 p. c.) the determinations are more correct and perhaps may then serve for comparative tests as to the eugenol content of several clove oils.

---

1) Bericht von S. & Co., Apr. 1892, p. 28.

2) Journ. f. prakt. Chem., II, 56, p. 148.

3) Pharm. Journ., III, 25, p. 950.

4) Comp. under cassia oil, p. 888, fig. 71.

237. Oil of Clove Stems.

Oleum Caryophyllorum e Stipitibus.—Nelkenstielöl.—Essence de Tiges de Girofle.

Clove stems, the flower stems of the cloves already mentioned, are less aromatic than these, and yield on distillation 5—6 p. c. of oil.

Oil of clove stems was distilled as early as the middle of the sixteenth century. At the beginning of the second decade of this century clove stems were distilled together with cloves for the purpose of cheapening the oil of cloves.[1]

Properties. Oli of clove stems is very similar in its properties to oil of cloves. Its odor, however, is less agreeable, for which reason the oil is not to be used in finer perfumery and in pharmacy; sp. gr. 1.040—1.065; α_D = up to $-1°10'$. It has the same solubility as oil of cloves, i. e. it is soluble to a clear solution in 2 parts of 70 p. c. alcohol.

Composition. The amount of free eugenol contained in oil of clove stems is as a rule somewhat higher than in oil of cloves. The fact that oils of clove stems show a higher content of eugenol by the benzoylchloride method than clove oils of the same specific gravity,[2] is explained in that the specifically heavier acet-eugenol of oil of cloves is absent in oil of clove stems.[3]

Caryophyllene, methyl alcohol and furfurol have been shown to be present in oil of clove stems. Judging from the color, methyl amyl ketone is also present in the oil, however, in much smaller quantity then in oil of cloves.

Tests and eugenol determination are the same as for oil of cloves.

238. Oil of Cajeput.

Oleum Cajeputi.—Cajeputöl.—Essence de Cajeput.

Origin and History. Oil of Cajeput is distilled from the fresh leaves and twigs of various species of *Melaleuca* (Family *Myrtaceae*) especially *Melaleuca leucadendron* L. and the variety known as *M. minor* Smith (*M. cajeputi* Roxb., *M. viridiflora* Gaertn.). These trees, growing to a height of 15 meters, are indigenous to upper India, the islands of the Indian Ocean, to northern Australia, Queensland and New South Wales.

Oil of cajeput appears not to have been brought to Europe until the beginning of the seventeenth century, when the Dutch took possession of the Moluccas. The first accurate account of the source of this oil was made known by the missionary Valentyn and the merchant

---

1) Buchner's Repert. f. d. Pharm., 26, II, p. 278.

2) Comp. Ber. d. pharm. Ges., 1, p. 288.

3) Jour. f. pract. Chem., II, 56, p. 148.

George Eberhard Rumpf of Hanau, both living in Amboina. The latter was an enthusiastic plant collector, and author of the first flora of the island Amboina. According to Rumpf's statement the Malays and Javanese were acquainted with oil of cajeput long before the Moluccas, the Banda and the Sunda islands were taken possession of, and used it as a diaphoretic. In Europe, the oil at first appears to have found no application. The first notice of such is by a physician Lochner in Nürnberg and the apothecary Link in Leipzig. The former mentioned the oil in 1717, the latter had bought the oil about the same time as a novelty from the physician of a ship which had just returned from the East Indies. From this time on cajeput oil was used medicinally in Germany and was introduced into the apothecary shops and mentioned in price ordinances and in medical works. It remained, however, for some time rare and expensive and not until 1730 did larger quantities of the oil come into the European market through Amsterdam. In Germany, it was at first called *Oleum Wittnebianum* after a merchant E. H. Wittneben of Wolfenbüttel, who lived several years in Batavia and had recommended the oil as a valuable remedial agent in German writings. In France and in England oil of cajeput was not used until the beginning of this century.

The first detailed account of the simple methods of distillation of cajeput oil used on the Moluccas, was given by the French traveler Labillardière, who visited the island of Buru in 1792. The use of copper stills and condensers gave rise to a green color due to a small amount of copper in the oil. When perfectly pure, the oil is colorless. The cause of this coloration was first detected by the apothecaries Hellwig in Stralsund in 1786, Westrumb in Hameln in 1788 and Trommsdorff in Erfurt in 1795.

Preparation, Production and Trade. Oil of cajeput is obtained in a primitive manner by the natives of some of the Molucca islands. According to Reinwardt the oil was prepared formerly only on Buru. In 1821 there were only 3 distilling apparatus on this island, in 1855 there were 50. Recently the distillation is also carried on in Ceram. Martin, who in 1891—92 visited Ceram and Buru, described the method of preparation, illustrated in the accompanying figure, as follows:

"Above a crudely mortared fire-place stands a barrel *a*, 1 meter in height, which serves as a distilling vessel; into this are pressed the leaves of the *Melaleuca* and the container is half filled with water. A metallic helm (*b*) which is obtained from Ambon or Java, is mounted on top, and its elongated tube passed through a second, somewhat larger barrel (*c*), serving as a condenser. Water is conducted into the latter from the top by means of a bamboo tube (*d*) from some small channel on the side of a hill. The volatile oil of the plant passes over with the water vapor and separates again after condensation. Water and oil flow into a vessel made of a cocoanut shell, which in turn is connected by means of a short tube with a bottle. Usually one sees a four-cornered brandy flask, as they are frequently seen in India, used for this purpose. This flask is provided at the bottom with a small opening, and stands in a small trough (*e*), filled with water, so that it is likewise filled with

Fig. 76.
Cajeput Oil Distillation on Ceram (Moluccas).

water at the beginning. The distillation product gradually replaces the water in the flask, and the water which has passed over with the oil, likewise flows through the opening into the trough, until finally the entire flask is filled with oil, and can be removed by putting a finger on the opening while under water. The yield of oil as obtained with such an arrangement amounts to about 1½ liters per day. As is well known the light bluish-green liquid is valued in Europe as a stimulant, in Buru it is used as a domestic remedy for all imaginable ills."

The oil is filled into empty wine and beer bottles. 25 bottles are packed at a time into a box made of the stems of the leaves of the sago palm [1] (*Metroxylon*). The exhausted cajeput leaves serve as packing material. Macassar on Celebes is the principal commercial center for oil of cajeput.

---

1) An illustration of such a case is to be found in Tschirch's Indische Heil- und Nutzpflanzen, plate 75 and p. 127.

The total export of this oil from Macassar was:[1]

1894	34,075 kilos.
1895	49,301 "
1896	57,800 "
1897	78,543 "

The greater part of the oil is consumed in the orient, especially in British India. A smaller part comes through Amsterdam, Hamburg and London into the European market.

The import of Amsterdam amounted to:

1894	about 10,700	bottles	= about	6,420 kilos.
1895	" 13,300	"	= "	7,980 "
1896	" —	"	= "	— "
1897	" 3,500	"	= "	2,100 "

The oil is not sold by weight but by bottles. As these, in the course of time, became smaller and smaller, and as considerable loss was experienced by breakage in transporting, Schimmel & Co. tried the more rational proceeding of shipping the oil in iron containers. This practical innovation, which was to establish the trade on a firmer basis, has up to now been unsuccessful on account of the conservatism of the Dutch.

PROPERTIES. Crude oil of cajeput is a green to bluish-green liquid, due to the presence of copper, while the rectified oil is colorless or yellowish. It has the pleasant, camphor-like odor of cineol and an aromatic, somewhat burning, later cooling taste. Sp. gr. 0.920—0.930; $\alpha_D = -0°10'$ to $-2°$. The oil dissolves in 1 part of 80 p. c. alcohol, but often gives clear solutions even with 3—5 parts of 70 p. c. alcohol.

On strongly cooling with liquid carbonic acid and ether, it solidifies to a crystalline mass. The copper can be removed from the oil by shaking with a concentrated solution of tartaric acid. As can be seen from its spectroscopic behavior, chlorophyllan,[2] oxidized chlorophyll, is contained also with the copper in the crude oil.

COMPOSITION. The first chemical investigations dealt almost exclusively with the principal constituent of cajeput oil, the elementary composition of which, $C_{10}H_{18}O$, was correctly recognized as early as 1833 by Blanchet.[3] The identity of this body, which was called cajeputene hydrate by Schmidt,[4] cajeputol by Gladstone,[5] also by Wright and

1) Bericht von S. & Co., Apr. 1898, p. 10; Apr. 1899, p. 8.

2) Pharm. Zeitschr. f. Russl., 27, p. 548; Jahresb. f. Pharm., 1888, p. 817.

3) Liebig's Annalen, 7, p. 161.

4) Journ. Chem. Soc., 14, p. 63; Journ. f. prakt. Chem., 82, p. 189.

5) Journ. Chem. Soc., 25, p. 1; Pharm. Journ., III, 2, p. 746; Jahresb. f. Chemie, 1872, p. 815.

Lambert,[1] with cineol was shown by Wallach[2] in 1884. He prepared the halogen and hydrohalogen compounds. A further proof was furnished by Wallach and Gildemeister[3] by the oxidation of the fraction in question to cineolic acid $C_{10}H_{16}O_5$, melting at 196—197°.

Another important constituent of cajeput oil which in regard to the amount present takes second place, is the solid terpineol,[4] $C_{10}H_{18}O$, discovered by Voiry,[5] which is present in the free state, as well as in the form of its acetic acid ester.

Terpenes are present in the oil to only a small extent; the laevogyrate fraction, boiling at 155—165°, gives with hydrochloric acid a solid laevogyrate monohydrochloride $C_{10}H_{16}HCl$[5] melting at 126—128°. l-Pinene is therefore present.

Several aldehydes are present in the first fraction. Voiry obtained by separation with sodium bisulphite a liquid of the properties of valeric aldehyde. The second aldehyde, smelling like oil of bitter almonds, is probably benzaldehyde.

239. Oil of Niaouli.

Oil of niaouli, the distillate obtained from *Melaleuca viridiflora* Brongn. et Gris, called *Niaouli* in New Caledonia, is very similar to oil of cajeput in its properties and composition. The oil is also called *Gomenol* on account of its preparation in the neighborhood of Gomen. Sp. gr. 0.908—0.922 at 12°. Optically it is either inactive or slightly dextro- or laevogyrate.

Composition.[6] In place of the l-pinene in cajeput oil, niaouli oil contains d-pinene of which a dextrogyrate solid monohydrochloride $C_{10}H_{16}HCl$ was obtained. Cineol is the principal constituent (about 66 p. c.), and is accompanied by a laevogyrate compound of the same boiling point (l-limonene?). Crystallized terpineol, $C_{10}H_{18}O$, and its valerianic acid ester are present to the extent of about 30 p. c.; there are also present traces of acetic and butyric acid esters.[7] Bertrand separated by means of bisulphite two aldehydes from the oil, of which the one had the odor of valeric aldehyde, the other that of bitter

1) Berichte, 7, p. 598; Pharm. Journ., III, 5, p. 284.
2) Liebig's Annalen, 225, p. 315.
3) Liebig's Annalen, 246, p. 276.
4) Bericht von S. & Co., Apr. 1892, p. 7.
5) Compt. rend., 106, p. 1538; Bull. Soc. chim., II, 50, p. 108; Journ. de Pharm., V, 18, p. 149.
6) Bertrand, Bull. Soc. chim., III, 9, p. 432; Compt. rend., 116, p. 1070. — Voiry, Contribution à l'étude chimique des huiles essentielles de quelques Myrtacées. Thèse de l'Ecole de Pharmacie de Paris 1888.
7) Bericht von S. & Co., Apr. 1892, p. 44.

almond oil and boiled at 180° (benzaldehyde?). The unpleasant odor of the crude oil is due to sulphur compounds.

240. Oil of Melaleuca Acuminata F. v. Müll.

Colorless oil, with an odor reminding slightly of juniper berries. Sp. gr. 0.892; $\alpha_D = -15° 20'$. It contains much cineol.[1]

241. Oil of Melaleuca Decussata R. Br.

Distillate of the branches and leaves. Yield 0.037 p. c.; sp. gr. 0.938; b. p. 185—209°. Odor and taste are very similar to cajeput oil (Maiden[2]).

242. Oil of Melaleuca Ericifolia Sm.

The leaves yield 0.033 p. c. of oil which is also similar to cajeput oil. Sp. gr. 0.899—0.902. It boils between 149 and 184°[2] and is dextrogyrate.[4]

243. Oil of Melaleuca Genistifolia Sm.

From the leaves and twigs 0.062 p. c. of oil are obtained. It is yellowish-green in color, and has a mild odor and taste.[3]

244. Oil of Melaleuca Leucadendron var. Lancifolia.

Sp. gr. 0.955; $\alpha_D = -3° 38'$. It consists principally of cineol.[1]

245. Oil of Melaleuca Linariifolia Sm.

Fresh leaves give a yield of 0.17 p. c. of oil which is of a pleasant cajeput-like odor, and has first a mace-like, later a minty taste.[6] It is dextrogyrate,[4] has the sp. gr. 0.898—0.903 and boils from 175—187°.

246. Oil of Melaleuca Squarrosa Sm.

Yield 0.002 p. c. The oil is green and has an unpleasant taste.[6]

247. Oil of Melaleuca Uncinata R. Br.

Sp. gr. 0.925; $\alpha_D = +1° 40'$. The principal part boils between 175 and 180°. The lower boiling fraction has a decided spike- or rosemary-

---

1) Bericht von S. & Co., Apr. 1892, p. 44.
2) The useful native plants of Australia, p. 275.
3) Ibidem, p. 276.
4) Journ. Chem. Soc., 25, p. 1; Pharm. Journ., III, 2, p. 746; Jahresb. f. Chemie, 1872, p. 815.
5) Bericht von S. & Co., Apr. 1892, p. 44.
6) Useful native plants, etc., p. 279.

like odor. The second fraction smells strongly of cineol which is the main constituent of the oil. The highest boiling portions consist probably of terpineol.[1]

248. Oil of Melaleuca Wilsonii F. v. Müll.

The leaves yield 0.024 p. c. of an oil similar to cajeput, of the sp. gr. 0.925.[2]

EUCALYPTUS OILS.

History. The oil of *Eucalyptus piperita* Sm., is the oldest oil of eucalyptus leaves, being mentioned as early as 1790.[3] In 1853 the botanist Ferdinand von Müller recommended to the Province of Victoria the distillation of the leaves of the eucalyptus species.[4] Bosisto,[5] who had distilled experimentally the dried leaves in London, established in 1854 the first factory in Australia,[6] and is, therefore, to be considered as the founder of this extensive industry.

Australian eucalyptus oil came into the German market about 1866 without any botanical reference as to its source. It probably was principally the distillate from *Eucalyptus amygdalina*.

Eucalyptus globulus was discovered in Tasmania in 1792 by Labillardière and introduced into Europe in 1856 by Ramel. The oil of this species was obtained on a large scale, first in southern France, Algiers and California, and is a staple article of commerce only since the early eighties.

Production. The Australian eucalyptus oil industry soon spread from Victoria to South Australia, then to Tasmania and Queensland. The principal places of production are in Victoria, Melbourne and vicinity, the shores of Lake Hindmarsh and Bendigo; in South Australia, Adelaide and Kangaroo Island; in Queensland, Brisbane, Gladstone, Rockhampton, Inghamstown and Wallaroo; on Tasmania, Hobart.

In Australia the leaves of the various species used for distillation are not kept separate with the required care, so that the designations of botanical sources are sometimes unreliable.

According to a report by Maiden[7] the so-called *bulk oil* is obtained

1) Bericht von S. & Co., Apr. 1892, p. 44.
2) Useful native plants, etc., p. 280.
3) Journal of a Voyage to New South Wales by John White, Surgeon-General to the Settlement, published 1790.
4) Eucalyptographia, Melbourne, 1879; Select Extra-Tropical Plants, 9th ed., p. 184.
5) Transact. Royal Soc. Victoria, 1861—64.
6) Bericht von S. & Co., Oct. 1886, p. 13.
7) Bericht von S. & Co., Apr. 1898, p. 27.

in Bendigo (Victoria), where five firms are engaged in the distillation of eucalyptus oil, from a mixture of the leaves of *Eucalyptus sideroxylon* with those of *E. leucoxylon*, *E. melliodora*, *E. polyanthema* and others, while for the preparation of the finer qualities, the leaves of *E. sideroxylon* are exclusively used. The leaves stripped from the felled trees are usually packed into bags by common laborers and transported by them. It is readily seen that with such a system of collection the bags might often contain mixed leaves, and this is also the reason why the quality of the resulting eucalyptus oil often varies so widely.

The distillation is performed in Australia[1] by passing steam through the stills[2] filled with the leaves without the addition of water.

The crude oil from the leaves of *Eucalyptus cneorifolia*, *E. goniocalyx* and *E. incrassata* is of a dark color, that of *E. odorata* is bright yellow. For purification the oil is treated with sodium hydrate solution, which removes the irritating aldehydes and saponifiable bodies, and is then rectified. The removal of the readily volatilized aldehyde is, however, only partially accomplished by this method. The resulting distillate is the ordinary eucalyptus oil of commerce. Ten percent of the original oil remain in the still and form with the sodium hydrate a kind of soap, of a deep dark brown color and syrupy consistency. This distillation residue is used under the name of eucalyptus tar or resin oil as a cheap disinfecting agent or for perfuming common soaps. Frequently, however, it is discarded altogether.

The California eucalyptus oil from *E. globulus* was for a time obtained in considerable quantities as a side product in the manufacture of an anti boiler deposit remedy in Alameda County near San Francisco Bay. This oil no longer occurs in the European market.

The closest competitor of the Australian globulus oil is Algiers, where considerable amounts are produced. In southern France and also in Spain some eucalyptus oil is obtained. What part the oil recently distilled in Portugal[3] will play in the markets of the world cannot yet be estimated. The oil (*E. globulus*) prepared in India in the Nilgiri mountains is exclusively consumed there and not exported.[4]

The eucalyptus oils can, according to their constituents or their odor, be divided into 5 groups.

1) Pharm. Journ., III, 25, p. 501.

2) These distilling apparatus are often constructed of wood, and have a capacity up to 20,000 liters. They are capable of distilling 47 tons of leaves per week. (Bericht von S. & Co., Oct. 1886, p. 18).

3) Bericht von S. & Co., Oct. 1897, p. 24.

4) Bericht von S. & Co., Apr. 1891, p. 18.

First Group: Cineol- (Eucalyptol-) containing Oils.

This group includes primarily all those eucalyptus oils which are of commercial interest. Among these the oil from *E. globulus* is the most important on account of its high content of cineol. The oils from *E. odorata, E. oleosa, E. cneorifolia* and *E. dumosa* which are likewise prepared in larger quantities are to be considered as substitutes for the globulus oil, which they resemble in their properties.

249. Oil of Eucalyptus Globulus.

ORIGIN. Of all the eucalyptus species that of *Eucalyptus globulus* Labillardière is the best known and the most highly valued outside of Australia. On account of its rapid growth and the enormous water evaporation connected therewith, the tree has been planted with success in marshy districts. The beneficial influence which its cultivation exerts in malarial districts, is in all probability principally due to the drying up of the marshes and less to the balsamic exhalation produced by the oil in the leaves.

The tree known in Australia as the blue gum-tree of Victoria and Tasmania requires for its growth about the same climate as the oranges. Its excellent properties are the cause of its cultivation in all parts of the world. It is found in Algiers, southern France, Italy, Spain and Portugal, in California, Florida, Mexico, on Jamaica, in Transvaal and Pretoria, in India and many other places.

According to F. von Müller the fresh leaves yield 0.71 p. c. of oil. Schimmel & Co. obtained a yield of 1.6—3 p. c. from the dried leaves.

PROPERTIES. The oil of *Eucalyptus globulus* is a light yellow, mobile liquid, of a pleasant, refreshing cineol odor, and spicy cooling taste. Aldehydes, (especially valeric aldehyde), make themselves unpleasantly noticeable in the crude oil by their tendency to produce coughing. Well rectified oils do not possess this disagreeable property, or at least only to a very slight degree. Sp. gr. 0.910—0.930; $a_D = +1$ to $+15°$.

The more cineol (eucalyptol) the oil contains, the higher is its specific gravity and the lower its rotatory power. Oils containing much eucalyptol solidify to a white crystalline mass when they are cooled in a freezing mixture of ice and salt. The oil gives a clear solution with 3 parts of 70 p. c. alcohol.

COMPOSITION. The first investigation of eucalyptus oil was undertaken by Cloëz[1] in 1870. He separated from the oil by fractional

[1]) Compt. rend., 70, p. 687; Liebig's Annalen, 154, p. 372.

distillation a body boiling at 175°, which he called eucalyptol. From the low specific gravity (0.905 instead of 0.930) as well as the optical activity of the fraction it can be seen that it was still contaminated with terpenes, on account of which the elementary analysis led Cloëz to the incorrect formula $C_{12}H_{22}O$.

A few years later, in 1874, Faust and Homeyer[1] investigated a eucalyptus oil, the source of which is not mentioned by them. The fact that the fraction of this oil boiling at 171—174° contained no oxygen, shows that it could not have been the oil of *Eucalytus globulus*.

The correct composition of eucalyptol as $C_{10}H_{18}O$ was recognized by Jahns[2] in 1884. He demonstrated its identity with cineol. He employed the method recommended shortly before by Wallach and Brass[3] for the isolation and purification of cineol, by conducting hydrochloric acid into the cineol fraction. For the properties of cineol compare p. 175.

The hydrocarbon accompanying the cineol, and formerly called eucalyptene, is d-pinene. Wallach and Gildemeister[4] obtained solid pinene monohydrochloride by conducting dry hydrochloric acid gas into the fraction boiling about 165°. A nitrosochloride was formed with amyl nitrite and hydrochloric acid, which on boiling with alcoholic potassa was converted into nitrosopinene (m. p. 129—130°), and by heating with piperidine into pinene nitrolpiperidine melting at 116°.

According to a further observation it is probable that eucalyptus oil contains other terpenes besides pinene. Bouchardat and Tardy[5] obtained by the action of formic acid on the terpene fraction of the oil boiling at 156—157°, terpineol, isoborneol and fenchyl alcohol. The formation of the terpineol is explained by the presence of pinene, whereas the formation of isoborneol and fenchyl alcohol can probably be traced to the presence of camphene and fenchene.

The unpleasant, penetrating and irritating odor of the crude eucalyptus oil is occasioned by different aldehydes, principally valeric aldehyde, besides butyric and capronic aldehydes.[6] According to Bouchardat and Oliviero.[7] there are found in the first fraction ethyl and amyl alcohols, and according to Wallach and Gildemeister[8] fatty acids not definitely determined.

1) Berichte, 7, p. 68.

2) Berichte, 17, p. 2941; Archiv d. Pharm., 228, p. 52.

3) Liebig's Annalen, 225, p. 291.

4) Liebig's Annalen, 246, p. 283.—Comp. also Voiry, Bull. Soc. chim., II, 50, p 106.

5) Compt. rend., 120, p. 1417.

6) Bericht von S. & Co., Apr. 1888, p. 18.

7) Bull. Soc. chim., III, 9, p. 429.

8) Liebig's Annalen, 246, p. 288.

The fraction of the oil boiling above 200° is laevogyrate and on distillation splits off an acid (acetic acid?); it contains therefore an ester. After saponifying there is obtained a laevogyrate alcohol ($\alpha_D = -17°$) boiling at 215—220° of the sp. gr. 0.96. This alcohol has not yet been investigated.

Quantitative Determination of Cineol in Eucalyptus Oils. As the value of the eucalyptus oils of this group depends entirely on the amount of cineol present, the quantitative determination of this body is of importance. It is to be regretted that control experiments for the following three methods mostly employed have not been made on mixtures of known cineol content, so that at present nothing can be said as to the degree of accuracy of these methods. As the results of the different methods of determination do not agree, they are only of value when, in stating the cineol content, the method is mentioned by which the latter was determined. The distillation method as well as the phosphoric acid method described in the following are only applicable to oils rich in cineol, while the hydrobromic acid method yields results even when the cineol is present in smaller amounts.

I. The Hydrobromic Acid Method. Dry hydrobromic acid is conducted into the well cooled solution of the eucalyptus oil in an equal or greater volume of petroleum ether as long as a precipitate is formed. The hydrobromide of cineol formed is rapidly filtered off with a suction pump and washed with cold petroleum ether. Hydrobromic acid is again conducted into the filtrate in case a further precipitation takes place. The total hydrobromide of cineol after being dried in a vacuum is decomposed with water, transferred to a burette and the amount of cineol read off on the scale. By multiplying the number of cubic centimeters found by 0.93 (the sp. gr. of the cineol) the weight of the cineol separated from a certain amount of oil is obtained.

II. The Distillation Method.[1] The oil to be investigated is fractionated, and the fractions collected at intervals of 2 degrees, are placed in a freezing mixture and cooled to —15 to —18°. Crystallization is then induced by shaking or by bringing into contact with a crystal of cineol. The portion remaining liquid after standing for one hour in the freezing mixture is removed with a finely pointed pipette. With a little experience an almost dry mass of crystals is obtained from which the last traces of liquid can be removed by a repeated shaking together of the crystals. The melted cineol of all fractions is collected and weighed.

As a certain amount of the cineol remains dissolved in the terpene, it is of course evident, that not all of the compound can be separated in this manner. This method is, therefore, only applicable, when it is desired to know which one of a number of oils is richest in cineol.

---

1) Helbing's Pharmacological Record, VIII, 1892.

III. The Phosphoric Acid Method. This depends on the property of phosphoric acid to combine with cineol to form a compound analogous to the addition products of hydrobromic and hydrochloric acids. According to Helbing and Passmore[1] the method is as follows:

Concentrated phosphoric acid (sp. gr. 1.75) is added drop by drop with constant stirring to 10 g. of eucalyptus oil in a beaker. During this operation the beaker must be kept cold with water, and care must be taken that the contents do not warm up. By the careful addition of the phosphoric acid scarcely a yellow or reddish color is developed. The reaction is complete as soon as a drop of the acid produces a dark red coloration of the mass.

The crystalline mass formed is sharply pressed between filter paper, for which purpose an ordinary copying press may be used. The pressing is repeated with fresh filter paper until no grease spots are visible on the latter. The crystals are then weighed in a beaker and by multiplying by 6.11 the percent of the cineol is obtained.

The factor 6.11 is gotten from the following equation, based on the assumption as yet unproven, that the double compound is an addition product of 1 mol. each of phosphoric acid and of cineol.

$$C_{10}H_{18}O + H_3PO_4 = C_{10}H_{18}OH_3PO_4.$$

The results obtained by the phosphoric acid method are about ¼ higher than those obtained by the freezing method, but are still considerably from the true value. Kebler[2] obtained with pure cineol only 62.14 p. c. instead of 100. He, therefore, recommends the following modification of the method.

4 cc. of phosphoric acid (1.75) are added to 8 g. of oil contained in a beaker cooled with ice water. It is then slowly but thoroughly stirred and the cineol phosphate expressed, decomposed with hot water and the acid titrated with normal alkali. In this manner Kebler found for pure cineol, instead of 100, 103.75 p. c.

Experiments with mixtures of known cineol content were not made, and there is, therefore, no basis for judging the applicability of the method to eucalyptus oils. When it is considered how difficult it is to remove quantitatively the excess of the viscid phosphoric acid, no confidence can be extended even to this modification of the phosphoric acid method.

Allen[3] has recently tested the method by making mixtures of pure eucalyptol with the eucalyptol-free constituents of a globulus oil. In oil containing more than 50 p. c. the experimental result differed from the theoretical by at most 1⅓ p. c. With a content of from 20 to 50 p. c. the highest difference was only 2⅓ p. c. With oils containing less than 20 p. c. the method was found to be inapplicable.

---

1) Helbing's Pharmacological Record, XXIV, 1898.

2) Am. Journ. Pharm., 70, p. 492.

3) Chemist & Druggist. 54, p. 641.

250. Oil of Eucalyptus Odorata.

Eucalyptus odorata Behr. is a tree indigenous to South Australia, Victoria and New-South Wales. The yield of oil from fresh leaves amounts to 1.4 p. c. The light yellow oil has an aromatic camphor-like taste and smells after cineol and Roman caraway oil. Sp. gr. 0.899 to 0.925; slightly laevogyrate. It boils from 157—199° (Maiden[1]) and is often so rich in cineol (identified by the hydrobromic acid compound) that it solidifies in a freezing mixture without being fractionated.

From the rectification residue a considerable amount of cuminic aldehyde was separated by means of its bisulphite compound. It was identified by oxidation to cuminic acid.[2]

251. Oil of Eucalyptus Cneorifolia.

The oil of this bush-like eucalyptus species, growing on Kangaroo island in South Australia has only lately appeared in the market. Its accompanying odor reminds of dill and caraway. Sp. gr. 0.899—0.923; $[\alpha]_D = -4$ to $-14°$ (Wilkinson[3]).

A firm in Adelaide, which had introduced the oil into the market, at first designated it as coming from *Eucalyptus oleosa*. This is due to the fact that *E. cneorifolia* is now considered as a separate species, whereas it was formerly considered as a variety of *E. oleosa*.[4]

252. Oil of Eucalyptus Oleosa.

Eucalyptus oleosa F. v. Müll. likewise belongs to the bush-like eucalyptus species. The yield of oil is 1.25 p. c.;[1] sp. gr. 0.906—0.926: $[\alpha]_D = +4$ to $5°$.[3] It contains cineol and cuminic aldehyde.

253. Oil of Eucalyptus Dumosa.

Eucalyptus dumosa[5] is found in northern Victoria, in southern New South Wales and in South Australia. Yield about 1 p. c.; sp. gr. 0.884 to 0.915; $\alpha_D = +0° 6'$ to $+6° 30'$. It contains large amounts of cineol.[6]

254. Oil of Eucalyptus Amygdalina.

Origin. The tree, known in south-eastern Australia as white and brown peppermint tree, *Eucalyptus amygdalina* Labill., is one of the

1) The useful native plants of Australia, p. 272.
2) Bericht von S. & Co., Apr. 1889, p 19.
3) Proc. Royal Soc. of Victoria, 1898, p. 195.
4) Bericht von S. & Co., Apr. 1892, p. 44.
5) The useful native plants of Australia, p. 267.
6) Bericht von S. & Co., Oct. 1889, p. 26.

tallest of the eucalyptus species. According to F. von Müller it reaches a height of over 400 feet. Its leaves contain more oil than any other species, yielding on distillation over 3 p. c. of oil. Formerly it came in great quantities into the market, but it has recently been replaced more and more by the oils richer in cineol.

According to Baker & Smith[1] the commercial oil of *Eucalytus amygdalina* Labill. is not obtained from this species but from *E. amygdalina* var. *latifolia* Maiden et Deane. The true oil of *E. amygdalina* Labill. contains according to Baker and Smith, when the leaves are distilled at the proper time, about 45 p. c. of cineol.

Properties. The oil is light yellow or colorless. Its cineol odor is almost masked by the terpene odor; sp. gr. 0.850—0.886; $\alpha_D = -25$ to $-70°$. The higher the specific gravity and the lower the rotation, the more cineol and less phellandrene does the oil contain. The phellandrene reaction mentioned below is characteristic for the oil. It is much less soluble in alcohol than oil of *Eucalyptus globulus*. As a rule more than 6 parts of 90 p. c. alcohol are necessary for solution, which in most cases is not even perfect.

Composition. According to Wallach and Gildemeister[2] the oil of *Eucalyptus amygdalina* consists principally of phellandrene.

If the oil diluted with twice its volume of petroleum ether be treated with a concentrated aqueous solution of sodium nitrite, and then acetic acid be added in small portions, the quantities of phellandrene nitrite formed are so large, that often the entire liquid solidifies to a pasty mass.

It may be interesting to note that the phellandrene is laevogyrate and that this optical modification was found for the first time during the investigation of this oil.

Cineol is present only in small quantities in the amygdalina oil. Its presence cannot be shown by the hydrochloric acid reaction; on the other hand, its hydrobromic acid addition product can be readily obtained from the petroleum ether solution of the oil.[2]

255. Oil of Eucalyptus Rostrata.

Eucalyptus rostrata Schlechtd., indigenous to Australia and there distributed from South Australia to northern Queensland, is much cultivated in southern France as well as in Algiers, where the tree is said to be better able to withstand the heat than *E. globulus*. The

1) Chemist and Druggist, 54, p. 864. 2) Liebig's Annalen, 246, p. 278.

yield of oil from fresh leaves amounts to only 0.1 p. c.[1] The oil is yellowish in color and reminds in odor of the oil of *E. odorata*. Sp. gr. 0.912—0.925; $\alpha_D = -1° 8'$ to $+13°$.[2]

It is soluble in 2 parts of 70 p. c. alcohol, boils according to Wittstein and Müller from 137°(?)—181° and contains valeric aldehyde and large quantities of cineol,[3] but no phellandrene.[6]

256. Oil of Eucalyptus Populifera.

Eucalyptus populifera Hook. is distributed over New South Wales, Queensland and North Australia.

The bright red oil smells of cajeput.[1] It contains besides cuminic aldehyde a fair amount of cineol.[4]

257. Oil of Eucalyptus Corymbosa.

Eucalyptus corymbosa Sm. is found in the coast districts of New South Wales and in northern Queensland.

According to Wittstein and Müller the oil smells slightly of lemon and rose (?) and has a bitter, somewhat camphor-like taste.[5] Sp. gr. 0.881. According to Schimmel & Co. it contains much cineol.[4]

258. Oil of Eucalyptus Resinifera.

Eucalyptus resinifera Sm. grows in New South Wales and Queensland. According to Gladstone[6] the oil consists principally of a hydrocarbon smelling of turpentine oil; Schimmel & Co., however, found in it much cineol. An oil, coming from Portugal, and probably obtained from *E. resinifera*, had the sp. gr. 0.893, and the rotatory power $\alpha_D = -17° 8'$. It was not soluble in 70 and 80 p. c. alcohol and contained besides cineol (iodol reaction) also phellandrene.[7]

259. Oil of Eucalyptus Baileyana.

Eucalyptus Baileyana F. v. Müll. occurs in the neighborhood of Brisbane in Queensland. The fresh leaves yield on distillation 0.9 p. c.[5] of oil of the sp. gr. 0.940. It boils from 160—185° and contains about 30 p. c. of cineol.[8]

---

1) Maiden, The useful native plants of Australia, p. 278.
2) Wilkinson, Proc. Roy. Soc. of Victoria, 1893, pp. 197, 198.
3) Bericht von S. & Co., Oct. 1891, p. 40.
4) Bericht von S. & Co., Apr. 1893, p. 28.
5) Maiden, loc. cit., p. 266.
6) Journ. Chem. Soc., 17, p. 1; Jahresb. f. Chemie, 1863, p. 541.
7) Bericht von S. & Co., Oct. 1898, p. 26.
8) Bericht von S. & Co., Apr. 1888, p. 19.

260. Oil of Eucalyptus Microcorys.

Eucalyptus microcorys F. v. Müll. is distributed in the northern coast districts of New South Wales to Cleveland Bay (Queensland). Yield from the fresh leaves 1—2 p. c., sp. gr. 0.896—0.935. The oil boils from 160—200° and contains besides terpenes about 30 p. c. of cineol.[1]

261. Oil of Eucalyptus Risdonia.[2]

Under this name a pleasant and mild smelling eucalyptus oil was introduced in 1874 in London. Sp. gr. 0.915—0.916; $\alpha_D = -4°49'$. It contained cineol and phellandrene.[3]

262. Oil of Eucalyptus Leucoxylon.

Eucalyptus leucoxylon F. v. Müll. (*E. sideroxylon* A. Cunn.) grows in South Australia, Victoria, New South Wales and in southern Queensland. Bosisto[4] reports about 1 p. c. of oil, but remarks that the leaves used for its preparation had lost a part of their oil through heating. Odor and taste are said to be similar to those of the oil from *E. oleosa*. Sp. gr. 0.915—0.927; $[\alpha]_D = +0.5°$ to $+2.7°$.[5]

263. Oil of Eucalyptus Hemiphloia.

Eucalyptus hemiphloia F. v. Müll. is common in eastern South Australia, Victoria, New South Wales, and in southern Queensland. The reddish-brown oil contains cineol and large amounts of cuminic aldehyde.[6]

264. Oil of Eucalyptus Crebra.

Eucalyptus crebra F. v. Müll. (Iron Bark) is indigenous to the coast districts of Queensland and New South Wales. The oil is light yellow, very similar in its odor to the oil of *E. globulus*, and like this is rich in cineol.[7]

265. Oil of Eucalyptus Macrorrhyncha.

The oil of the leaves of *Eucalyptus macrorrhyncha* F. v. Müll. a tree known in New South Wales as red stringybark[8] has been investigated by Baker and Smith.[9]

---

1) Bericht von S. & Co., Apr. 1888, p. 19.

2) According to Maiden *Eucalyptus risdoni* Hook. is identical with *E. amygdalina* Labill. or at least closely related.

3) Bericht von S. & Co., Apr. 1894, p. 29.

4) Maiden, The useful native plants etc., p. 270.

5) Wilkinson, Proc. Roy. Soc. of Victoria, 1893, p. 198.

6) Bericht von S. & Co., Apr. 1892, p. 28.

7) Bericht von S. & Co., Apr. 1893, p. 28.

8) In Australia the following species of *Eucalyptus* are designated as stringybark: *E. obliqa* Herit., *E. baileyana* F. v. Müll., *E. macrorrhyncha* F. v. Müll., *E. capitellata* Sm., *E. eugenioides* Sieb., and *E. fastigata* Deane et Maiden.

9) Journ. and Proc. of the Royal Soc. of N. S. Wales, 32, p. 104.

They obtained by distillation a yield of 0.28—0.31 p. c. of oil. The oil is reddish-brown, has the sp. gr. 0.924—0.927 at 22°, and begins to boil at 172°. It contains a trace of phellandrene, cineol, and the crystallized eudesmol[1] described under oil of *E. piperita* on p. 538.

266. Oil of Eucalyptus Capitellata.

Baker and Smith[2] suggest the name of brown stringybark for *Eucalyptus capitellata* Smith.

The leaves yield on distillation only 0.1 p. c. of a dark red oil of the sp. gr. 0.9153 at 18°. It contains cineol and a trace of phellandrene.

267. Oil of Eucalyptus Eugenioides.

Eucalyptus eugenioides Sieb. is known in Australia as white stringybark. Baker and Smith[2] obtained from the leaves 0.68—0.79 p. c. of oil. The sp. gr. of the oils of two distillations was 0.907 and 0.908; $[\alpha]_D = +3.7$ and $+5.2°$. The oil contains cineol, but no phellandrene.

268. Oil of Eucalyptus Obliqua.

Eucalyptus obliqua Herit. is found in the southern coast districts of New South Wales, principally, however, in Tasmania, Victoria and South Australia. Yield 0.5 p. c. A reddish-yellow oil of mild odor and bitter taste; sp. gr. 0.899. It boils from 171—195°.[3]

An oil obtained in Portugal had the sp. gr. 0.914 and the rotatory power $\alpha_D = -7° 28'$. It was soluble in an equal part of 80 p. c. alcohol and contained cineol (iodol reaction) and phellandrene[4] (nitrite).

269. Oil of Eucalyptus Punctata.

The tree from which the oil was obtained, *Eucalyptus punctata* D. C. (*E. tereticornis* Sm. *var. brachycoris*), is called grey gum in Australia, and yields besides kino a useful hard wood. It is found in the coast districts of New South Wales, of Queensland as far as the border of Victoria.

For the preparation of the oil, the leaves and twigs were used. Altogether nine distillations were made of material collected in different districts. The yield varied between 0.63 and 1.19 p. c. The sp. gr. was between 0.9122 and 0.9205 at 17°. Two oils were laevogyrate

---

1) Journ. and Proc. of the Roy. Soc. of N. S. Wales, 88, p. 86; Bericht von S. & Co., Apr. 1900, p. 24.

2) Journ. and Proc. of the Royal Soc. of N. S. Wales, 82, p. 104.

3) Maiden, The useful native plants etc., p. 272.

4) Bericht von S. & Co., Oct. 1898, p. 27.

($[\alpha]_D$ —0.92° and —2.52°), the other seven were dextrogyrate ($[\alpha]_D$ + 0.54° to 4.44°). A sample mixture of all the distillates had the following properties: sp. gr. 0.915 at 16°, $[\alpha]_D$ = + 0.927°. According to the phosphoric acid method the cineol content of the oil was from 46.4—64.5 p. c. No phellandrene is present in the oil of *E. punctata.*[1]

270. Oil of Eucalyptus Loxophleba.

Eucalyptus loxophleba Benth.[2] is called by the people York gum, on account of its frequent occurrence in the vicinity of the city of York. The oil has a highly unpleasant odor, and produces fits of coughing when inhaled. Sp. gr. 0.8828 at 15.5°; angle of rotation about + 5°. Upon distillation the following fractions were obtained: 168—171° 68 p. c.; 171—176° 14 p. c.; 176—182° 2 p. c.; 182—187° 8 p. c.; residue 8 p. c.

The oil contains phellandrene and cineol. The amount of the latter is estimated at 15—20 p. c. On shaking with bisulphite the oil diminished in volume 20 p. c., which allows of concluding that a considerable amount of aldehydes and ketones is present. Amyl alcohol, of which small quantities were found in the oil of *E. globulus*, and to which, no doubt, are partly due the irritating action of this oil, is not present.

271. Oil of Eucalyptus Dextropinea.

The oil of *Eucalyptus dextropinea* Baker[3] has been prepared by Baker and Smith,[4] as has also the oil of *E. laevopinea* Baker, from the fresh leaves of these trees. Both are indigenous to New South Wales. The yield was in one case 0.825, in another 0.850 p. c. The deep red colored and strongly dextrogyrate oil has the sp. gr. 0.8743—0.8763 at 17°. By distillation the following fractions were obtained: 156—162° 62 p. c.; 162—172° 25 p. c.

The oil consists almost entirely of d-pinene.[4] The main fraction, finally boiling at 156—157° had the sp. gr. $\frac{18^\circ}{16^\circ}$ 0.8629; $[\alpha]_D$ = + 41.2° at 18°.

---

1) On "Grey Gum" (*Eucalyptus punctata* D. C.), particularly in regard to its essential oil, by R. T. Baker F. L. S. and H. G. Smith F. C. S., Technological Museum, Sydney. Journ. and Proc. of the Royal Society of N. S. Wales, 31, p. 259; Bericht von S. & Co., Oct. 1898, p. 27.

2) Pharm. Journ., 61, p. 198. This article by Parry gives the species name as *toxophleba.*

3) The two new species *Eucalyptus dextropinea* and *E. laevopinea* have been described by Baker in Proc. of the Linnean Soc. of New South Wales, 27, p. 414.

4) Journ. and Proc. of the Royal Soc. of N. S. Wales, 32, p 195.

For the identification of the pinene the following derivatives were prepared: pinene nitrosochloride (m. p. 103°) and from this nitrosopinene (m. p. 128—129°), further terpin hydrate, as well as pinene monohydrochloride (m. p. 121—124°).

Besides pinene the oil contains small amounts of cineol, which was recognized by the behavior of the higher boiling fractions toward iodol and bromine.

272. Oil of Eucalyptus Laevopinea.

From the fresh leaves of *Eucalyptus laevopinea* Baker,[1] silver top stringybark, Baker and Smith[2] obtained 0.66 p. c. of a reddish oil having the sp. gr. 0.8732. The following fractions were collected: 157—164° 60 p. c.; 164—172° 28 p. c. Just as the foregoing oil consisted almost entirely of d-pinene, this oil consists almost entirely of l-pinene. The fraction boiling at 157—158°, which can probably be considered as fairly pure pinene, had the sp. gr. 0.8626 at $\frac{19°}{16°}$ and $[\alpha]_D = -48.63°$. The same derivatives of the pinene were prepared as with the foregoing oil. This oil likewise contains only small amounts of cineol.

273. Oil of Eucalyptus Smithii.

The leaves of *Eucalyptus smithii* yield on an average 1.353 p. c. of oil which contains over 70 p. c. of cineol.[4] The oil contains d-pinene, but no phellandrene. The eudesmol, described under the oil of *Eucalyptus piperita* on p. 538 has also been found in this oil.[5]

Second Group: Citronellal-containing Oils.

274. Oil of Eucalyptus Maculata.

The spotted gum-tree, *Eucalyptus maculata*, closely related to *E. citriodora* Hook., grows in New South Wales and Queensland, but has also been transplanted to Ceylon and Algiers. The oil, of citronellal-like odor,[6] has the sp. gr. 0.900, boils from 210—220° and cannot be distinguished from the following oil.

---

[1]) See footnote 3 on p. 585.

[2]) See footnote 4 on p. 585.

[3]) According to Smith almost all of the oils obtained from species closely related to *E. globulus* contain pinene, such as the oils from *E. bridgesiana*, *E. goniocalyx*, etc.

[4]) Proc. of the Linnean Soc. of N. S. Wales, 1899, II, p. 292; Bericht von S. & Co., Apr. 1900, p. 24.

[5]) Journ. and Proc. of the Roy. Soc. of N. S. Wales, 33, p. 86; Bericht von S & Co., Apr. 1900, p. 24.

[6]) Bericht von S. & Co., Apr. 1888, p. 19.

275. Oil of Eucalyptus Citriodora.

Eucalyptus citriodora Hook. is probably only a variety of *E. maculata* Hook. and is for this reason sometimes designated as *E. maculata* Hook. *var. citriodora.* The tree grows best in stony ground. It is indigenous to Queensland and has also been planted with success in India, in Zanzibar and on the Magdalene River.[1] The leaves are distilled in Gladstone (Queensland) and give in the fresh state 1—1.5 p. c., in the dried state 3—4 p. c. of oil.

The oil, distinguished through its pleasant citronellal-like odor, is much used as a perfume for soap. Sp. gr. 0.870—0.905. It is inactive or slightly dextrogyrate (α_D up to + 2°) and is soluble in 4 to 5 parts of 70 p. c. alcohol.

It consists to the extent of 80—90 p. c. of citronellal, $C_{10}H_{18}O$. The remainder of the oil, judging from the odor, consists of geraniol and citronellol. Cineol is not contained in it.[2]

276. Oil of Eucalyptus Dealbata.

Eucalyptus dealbata A. Cunn. (*E. viminalis* Labill.[3]) grows in Tasmania, South Australia, Victoria and New South Wales. The oil has an exceedingly fine, melissa-like odor, due to citronellal. Besides this there is present another body of pleasant odor, reminding of geranium (geraniol?). The oil boils from 206—216° and has the sp. gr. 0.871—0.885.[4]

An oil of *E. viminalis* (no author) described by Wittstein and Müller is so different from that just given, that it is impossible that it should have come from the same plant. The odor was unpleasant; it had the sp. gr. 0.921 and boiled between 159 and 182°, that is, at a temperature at which the oil of *E. dealbata* had not begun to boil.[5]

277. Oil of Eucalyptus Planchoniana.

The fresh leaves of the tree, *Eucalyptus planchoniana* F. v. Müll., occurring in northern New South Wales and in southern Queensland, yield according to Staiger[6] only 0.06 p. c. of volatile oil. It has a peculiar, citronella-like odor and the sp. gr. 0.915.

---

1) v. Müller, Select Extra-Tropical Plants, 9th ed., p. 187.

2) Bericht von S. & Co., Apr. 1888, p. 20; Oct. 1890, pp. 16 and 20; Apr. 1891, p. 19; Apr. 1893, p. 27; Oct. 1893, p. 17.

3) Maiden, The useful native plants etc., p. 527.

4) Bericht von S. & Co., Apr. 1888, p. 19.

5) Maiden, loc. cit., p. 274.

6) Maiden, loc. cit., p. 273.

Third Group: Citral-containing Oils.

278. Oil of Eucalyptus Staigeriana.

Eucalyptus staigeriana F. v. Müll. grows in northern Queensland. Its leaves yield upon distillation 2.75—3.36 p. c. of an oil, smelling pleasantly like lemon and verbena. It has the sp. gr. 0.880—0.901 and boils from 170—230°. The lemon-like odor is due to citral, which besides terpenes forms the principal constituent of the oil.[1]

279. Oil of Backhousia Citriodora.

Backhousia citriodora F. v. Müll. is indigenous to southern Queensland. The leaves yield 4 p. c. of volatile oil, which in odor and properties is similar to the foregoing one, and appears to consist almost entirely of citral. Sp. gr. 0.900; boiling temperature 223—233.[2]

Fourth Group: Oils with a Peppermint-like Odor.

280. Oil of Eucalyptus Haemastoma.

Eucalyptus haemastoma Sm. is distributed from Illawarra (New South Wales) to Wide Bay (Queensland). From the fresh leaves 1.8—1.9 p. c. of oil are obtained. It has a peppermint-like odor, but at the same time reminds of geranium and cumin. Sp. gr. 0.880—0.890; boiling temperature 170—250°. It contains cineol, terpenes and probably also cumin aldehyde and menthone.[1]

281. Oil of Eucalyptus Piperita.

The oil from the leaves of *Eucalyptus piperita* Sm. was already known in 1788. It is mentioned on page 266 of White's "Journal of a Voyage to New South Wales." The name peppermint-tree[3] was given to this plant, on account of the great similarity of the oil of its leaves with that of peppermint (*Mentha piperita*), which grows in England.

The oil has recently been prepared and investigated by Baker and Smith.[4] The leaves and twigs yielded on distillation 0.78 p. c. of oil. It has a light color and a decided peppermint-like odor, which, however, becomes fainter even after several weeks. Sp. gr. 0.909 at 17°; $[\alpha]_D = -2.97°$. It boils between 170 and 272°. In the lower boiling

---

1) Bericht von S. & Co., Apr. 1888, p. 20.

2) Bericht von S. & Co., Apr. 1888, p. 20; Oct. 1888, p. 17.

3) This tree is also designated Sydney peppermint in order to distinguish it from the white and brown peppermint tree, the *E. amygdalina*.

4) Journ. and Proc. of the Royal Soc. of N. S. Wales, 31, p. 195.

fractions phellandrene and cineol were found. From the fractions boiling at 266—272° there separated a compound in well formed crystals, which was called eudesmol. Eudesmol boils at 270—272° and melts at 74—75°. Eudesmol has lately been subjected to a more thorough study by Smith.[1] It crystallizes in white milky needles melting at 79—80°. Its analysis corresponds with the formula $C_{10}H_{16}O$, but contains neither an hydroxyl nor a ketone group. It yields a dinitro compound melting at 90° and a dibromide melting at 55—56°.

Wilkinson[2] found for an oil distilled from *E. piperita* the sp. gr. 0.913 and $[\alpha]_D = +1.6°$.

Fifth Group: Oils Less Known and of Indefinite Odor.

282. Oil of Eucalyptus Diversicolor.

Eucalyptus diversicolor F. v. Müll. grows in south-western Australia, Ceylon and Algiers. The oil has the sp. gr. 0.924; $[\alpha]_D = +9.7°$.[2]

283. Oil of Eucalyptus Fissilis.

The oil from the leaves of *Eucalyptus fissilis*[3] has the sp. gr. 0.928 and is optically inactive.[2]

284. Oil of Eucalyptus Goniocalyx.

Eucalyptus goniocalyx F. v. Müll. is indigenous to Victoria and New South Wales. Yield from the leaves 0.9 p. c. The light yellow oil has a penetrating, quite unpleasant odor and obnoxious taste. Sp. gr. 0.918—0.920; $[\alpha]_D = -4.3°$;[2] boiling temperature 152—175°.[4]

According to Smith[1] the oil contains eudesmol.

285. Oil of Eucalyptus Gracilis.

The oil of *Eucalyptus gracilis* F. v. Müll., growing in Queensland, Victoria and south-western Australia, has the sp. gr. 0.909; $[\alpha]_D = +9.3°$.[5]

286. Oil of Eucalyptus Lehmanni.

Eucalyptus Lehmanni Preiss. occurs in south-western Australia. The oil obtained from it has the sp. gr. 0.923; $[\alpha]_D = +5.9°$.[2]

---

1) Journ. and Proc. of the Roy. Soc. of N. S. Wales, 33, p. 86; Bericht von S. & Co., Apr. 1900, p. 24.

2) Proc. of the Royal Soc. of Victoria, 1898, p. 198.

3) According to Maiden, *Eucalyptus fissilis* F. v. Müll. is synonymous with *E. amygdalina* Labill., the oil of which is strongly laevogyrate. Its specific gravity also does not agree with that obtained by Wilkinson from *E. fissilis*.

4) Maiden, The useful native plants etc., p. 268.

5) Wilkinson, loc. cit., p. 197.

287. Oil of Eucalyptus Longifolia.

Eucalyptus longifolia Lk. is distributed in Victoria and in New South Wales as far as Port Jackson. The oil is viscid, has an aromatic cooling taste and camphor-like odor. Sp. gr. 0.940; boiling temperature 194—215°.[1]

288. Oil of Eucalyptus Occidentalis.

The leaves of *Eucalyptus occidentalis* Endl. growing in south-western Australia give an oil of the sp. gr. 0.9236; $[\alpha]_D = +2.7°$.[2]

289. Oil of Eucalyptus Pauciflora.

Eucalyptus pauciflora Sieb. is found in Tasmania, Victoria and New South Wales. The oil of the leaves has the sp. gr. 0.894—0.920; $[\alpha]_D = +6$ to $+17°$.[2]

290. Oil of Eucalyptus Stuartiana.

Eucalyptus stuartiana F. v. Müll., distributed from Tasmania to Queensland, contains a golden yellow oil, smelling strongly of cymene, but not of cineol.[3] Sp. gr. 0.917—0.932; $[\alpha]_D = -7°$ to $-16°$.[2]

291. Oil of Eucalyptus Tereticornis.

Eucalyptus tereticornis Sm. or red gum is found in Victoria and Queensland, and contains a red oil, of a difficultly definable odor, reminding somewhat of zedoary root. It contains no cineol.[3]

292. Oil of Eucalyptus Tessellaris.

Eucalyptus tessellaris F. v. Müll. (*E. viminalis* Hook. f.), Morton bay ash, grows in South Australia, New South Wales to North Australia. The dark brown oil has an entirely distinct balsamic benzoin-like odor. It contains no cineol.[3]

293. Oil of Eucalyptus Dawsoni.

Eucalyptus dawsoni, called slaty gum, was recognized as a separate species by Baker.[4] The yield of oil amounted to only 0.172 p. c. Its sp. gr. was 0.9414 at 15°. The oil contains no cineol, but much phellandrene. Its principal constituent appears to be a sesquiterpene.

---

1) Maiden, The useful native plants etc., p. 270.

2) Wilkinson, Proc. Roy. Soc. of Victoria, 1898, p. 198.

3) Bericht von S. & Co., Apr. 1898, p. 28.

4) Proc. of the Linnean Soc. of N. S. Wales, 1899, II, p. 292; Bericht von S. & Co., Apr. 1900, p. 24.

294. Oil of Eucalyptus Camphora.

The leaves of *Eucalyptus camphora*, known as sallow or swamp gum, give an average yield of 0.398 p. c. of oil. Sp. gr. 0.916. It contains pinene, cineol and much eudesmol.[1]

295. Oil of Chervil.

By the distillation of the fresh fruit of garden chervil, *Anthriscus cerefolium* Hoffm. (*Chaerophyllum sativum* Lam., Family *Umbelliferae*) Charobot and Pillet[2] in 1899 obtained 0.0118 p. c. of a light yellow oil of an anise-like odor, reminding of estragon. It consists principally of methyl chavicol. On treating the oil with alcoholic potassa, anethol (m. p. 20—21°) was formed, which was converted into anisic aldehyde by oxidation.

Gutzeit[3] in 1875 had distilled the unripe chervil fruit and obtained from 10 k. 27 g. of oil. He showed the presence of ethyl and methyl alcohol in the distillation water.

296. Oil of Coriander.

Oleum Coriandri. — Corianderöl. — Essence de Coriandre.

Origin and History. The coriander plant, *Coriandrum sativum* L. (Family *Umbelliferae*), cultivated in many countries and in nearly all climates, was used as a kitchen spice even before the Christian Era. As such, coriander fruit is mentioned several times in Sanscrit writings, in the Bible and in later Roman writings. Coriander fruit has also been found in old Egyptian monuments of the tenth century B. C. among other still recognizable offerings.

Coriander is also mentioned among the useful plants recommended for cultivation by Charlemagne, but it appears to have received, as with the Arabians, so also with the Germans in the middle ages, only slight consideration. The fruit is again mentioned in the medical and distilling books of the sixteenth century, although it had been employed now and again as a kitchen spice.

The distilled oil of coriander appears to have been first obtained by Porta in the sixteenth century. In the price ordinances of spices, the oil is first included in that of Berlin of 1574 and Frankfurt-on-the-Main of 1587 and in the 1589 edition of the Dispensatorium Noricum.

---

1) See footnote 4 on p. 540.

2) Bull. Soc. chim., III, 21, p. 868.

3) Liebig's Annalen, 177, p. 382.

Coriander oil was investigated in 1785 by Hasse, in 1835 by Trommsdorff,[1] in 1852 by A. Kawalier[2] and in 1881 by B. Grosser.[3] A true insight into its composition, however, was brought about by the investigations of Semmler (1891) and of Barbier (1893).

PREPARATION. Coriander oil is distilled from the fruit[4] which has been crushed between rollers. For the preparation on a large scale, only the coriander from Moravia, Thuringia and Russia is used, the yield being 0.8—1.0 p. c.[5] Only in cases of necessity is the fruit of other districts, which are all much poorer in oil, used. Of these may be mentioned: French (yield about 0.4 p. c.), Dutch (0.6 p. c.) and Italian (0.5 p. c.) coriander. Still less oil is given by the large grained Morocco fruits, namely 0.2—0.3 p. c., while the East Indian fruit with only 0.15—0.2 p. c. gives the lowest yield.

The exhausted and dried fruit is used as cattle food. It contains 11—17 p. c. of proteids and 11—20 p. c. of fat (Uhlitzsch,[6] 1893).

PROPERTIES. Oil of coriander is a colorless or slightly yellowish liquid of a characteristic coriander odor, and an aromatic, mild taste. Sp. gr. 0.870—0.885; $\alpha_D = +8$ to $+13°$. The oil is soluble in 3 parts of 70 p. c. alcohol at 20°.

COMPOSITION. Through the investigations of Kawalier[2] a body of the formula $C_{10}H_{18}O$ was recognized as the principal constituent of the oil. The alcoholic nature of this compound was later shown by Grosser.[3] Semmler[7] in 1891 designated it as coriandrol and found that it was a chain compound belonging to the so-called "olefinic camphors." Barbier[8] in 1893 showed that linalool and coriandrol had the same physical properties with the exception of being opposite in rotation, and that they were also completely alike in their chemical behavior. Both yield on oxidation the same aldehyde viz. citral, split off water in the same manner, and can be converted by proper treatment into geraniol. For these reasons coriandrol must be considered as the dextrogyrate modification of linalool.

For the investigation of the terpene[9] contained in coriander oil, the more volatile portion coming over on rectification by steam distillation

1) Archiv d. Pharm., 52, p. 114.

2) Liebig's Annalen, 84, p. 351; Journ. für prakt. Chem., 58, p. 226.

3) Berichte, 14, p. 2485.

4) The umbelliferous fruits are commonly, though erroneously designated seeds.

5) A yield of 1.1 p. c., as reported by Eck in 1897 (Pharm. Ztg., 32, p. 428) has never been obtained on a large scale.

6) Die landwirthschaftlichen Versuchsstationen, 42, p. 60.

7) Berichte, 24, p. 206.

8) Compt. rend., 116, p. 1460.

9) Bericht von S. & Co., Apr. 1892, p. 11.

was repeatedly fractionated with the use of a distilling column. A fraction boiling between 156 and 160° was obtained. It has the sp. gr. 0.861 and $\alpha_D = +32°42'$. The nitrosochloride gave with benzylamine pinene nitrolbenzylamine melting at 123—124°. According to this the terpene of coriander oil, the amount of which is about 5 p. c., is d-pinene.

As a mixture of pinene with linalool has by no means the odor of coriander oil, another body, at present unknown, must be contained in the oil, to which is due the specific coriander odor.

Examination. Coriander oil is often adulterated with orange or turpentine oil. These additions are readily recognized by the specific gravity and the rotatory power. The solubility test with 70 p. c. alcohol is also very serviceable for detecting adulterations; oils which do not form a clear solution with this solvent are to be rejected.

In order to make a study of the changes which the oil undergoes in the ripening process of the plant, Schimmel & Co.[1] distilled coriander at different stages of its growth. First of all the fresh flowering herb was distilled with the roots immediately after being collected. The plant in this condition had a highly disagreeable, stupefying bed-bug odor. Later, when the herb was half ripe and had already begun to go to seed, the second distillation was untertaken. The bed-bug odor was less prominent and the true coriander odor was more noticeable. As the herb, when the fruit is ripe, is nearly devoid of odor, only the ripe fruits were used immediately after harvesting.

1) Oil from the fresh, flowering plant.

Yield 0.12 p. c. Sp. gr. 0.853. Not soluble in 70 p. c. alcohol. Highly unpleasant bed-bug odor. After 2½ months the sp. gr. had increased to 0.856. The angle of rotation (which had at first not been determined) amounted to $+1°2'$. The bed-bug odor had almost entirely disappeared; the compound appears, therefore, to have suffered polymerization or to have been changed or inverted in some other way.[1]

2) Oil from the fresh, half ripe coriander herb with fruit.

Yield 0.17 p. c. Sp. gr. 0.866; $\alpha_D = +7°10'$. Soluble in 3 parts of 70 p. c. alcohol. The odor is coriander-like accompanied by a slight bed-bug odor. After one month the sp. gr. had risen to 0.869.

---

[1]) Bericht von S. & Co., Oct. 1895, p. 12.

3) Oil from ripe coriander fruit distilled immediately after the harvest.

Yield 0.83 p. c. Sp. gr. 0.876; $\alpha_D = +10°48'$. Soluble in 3 parts of 70 p. c. alcohol. True coriander odor.

297. Oil of Cumin.

Oleum Cumini.—Cuminöl.—Essence de Cumin.

History, Origin and Preparation. The Roman or mother caraway of the orient, *Cuminum cyminum* L has been used together with the common caraway as a spice in antiquity. Both have been often confused, the one with the other and also with the seeds of the black caraway, *Nigella*, in literature. They were used in England toward the end of the thirteenth century and in Germany in the fifteenth century.

The volatile oil of cumin is included in the price ordinances of Berlin for 1574, of Frankfurt for 1582, and in the 1589 edition of the Dispensatorium Noricum.

Oil of cumin is distilled from the seed of the Roman caraway. Syria, Morocco, Malta and East India are of commercial interest as places of production. The individual varieties gave according to determinations in the manufacture on a large scale the following yields:

Maltese	cumin	3.5	percent
Morocco	"	3	"
East Indian	"	3—3.5	"
Syrian	"	2.5—4	"

Properties. Oil of cumin is at first a colorless, later a yellowish liquid, of the unpleasant bug-like, but characteristic odor of cumin, and an aromatic, somewhat bitter taste.

The specific gravity of the oil distilled from cumin from the levant, is 0.91—0.93. With an East Indian oil, as well as with two oils of unknown source, the density was found to vary from 0.893—0.899; $\alpha_D = +4$ to $+8°$.

The solubility varies as much as the specific gravity, which rises and falls with the amount of cumic aldehyde in the oil. The heavy oils dissolve in 3 parts of 80 p. c. alcohol, the lighter oils require as much as 10 parts of this solvent.

Composition. Cumic aldehyde, or cuminol[1] is the constituent to which cumin oil owes its odor and its more prominent properties. Cymene and a terpene $C_{10}H_{16}$ accompany this aldehyde in the oil.

[1]) In accordance with the aldehyde character of this substance it should be designated cuminal.

Gerhardt and Cahours[1] in 1841 separated the hydrocarbons from the oxygenated constituents for purposes of investigation by distilling the oil with caustic potassa. The aldehyde was thereby changed to cumic acid and cumic alcohol and was at the same time retained by the alkali. Bertagnini[2] in 1853 found that cuminol combined with sodium bisulphite and thus recognized its aldehyde nature. This gave at the same time a rational way for the preparation of the body in a pure form. According to Kraut[3] (1854) the lower boiling portions are distilled off and the residue shaken with sodium bisulphite. After standing 24 hours the thick mass is expressed and decomposed by distillation with soda solution or dilute sulphuric acid.

Cumic aldehyde, $C_{10}H_{12}O$, has the sp. gr. 0.972 at 13° (Kopp,[4] 1855), boils at 109.5° under 13.5 mm. pressure, at 232° under 760 mm.[5] and is optically inactive.

The presence of a terpene in cumin oil boiling at 156° and having the sp. gr. 0.865 at 15° was first shown by Warren[6] (1865) and later confirmed by Beilstein and Kupfer[7] (1873). Wolpian[8] (1896) prepared the terpene in its purest form. He found the boiling point 157—158° under 768 mm. pressure, sp. gr. 0.8604, $\alpha_D = +25°25'$. His attempts to prepare a crystallized derivative and thus to identify the hydrocarbon with one of the known terpenes, were fruitless. He believed, therefore, to have found a new terpene which yielded only liquid compounds and gave it the name of hydrocuminene. In order to justify such assumptions, much stronger proofs are necessary then the almost exclusively negative results of Wolpian.

The second hydrocarbon of cumin oil is cymene, the identity of which with cymene from camphor as well as that obtained from other volatile oils was shown by Beilstein and Kupfer, and also by Wolpian.

298. Oil of Celery Seed.

Oleum Apii Graveolentis Seminis.—Selleriesamenöl.—Essence de Semences de Céleri.

Origin and Properties. The volatile oil which is present in all parts of the celery plant, *Apium graveolens* L., is especially well

1) Liebig's Annalen, 38, p. 70.

2) Liebig's Annalen, 85, p. 275.

3) Liebig's Annalen, 92, p. 66.

4) Liebig's Annalen, 94, p. 317.

5) Anschütz & Reitter, Die Destillation unter vermindertem Druck. Bonn, 1895, p. 78.

6) Zeitschr. f. Chemie, 1. p. 667.

7) Liebig's Annalen, 170, p. 282.

8) Pharm. Zeitschr. f. Russl., 35, pp. 97, 113, 129, 145, 161.

represented in the fruit, somewhat sparingly, but still in quantities sufficient for preparation, in the green herb, while the roots give an aromatic water upon distillation, but no oil.

By distillation with water vapor about 2.5—3 p. c. of a very mobile and colorless oil are obtained from the seed. It smells and tastes strongly of celery, has a sp. gr. of 0.870—0.895 and $\alpha_D = +67$ to $+79°$.

COMPOSITION. Celery oil consists of about 90 p. c. of hydrocarbons. In an investigation in the laboratory of Schimmel & Co. a fraction boiling constant at 176—177° was obtained, the rotation of which was $\alpha_D = +107°$. By the action of bromine a solid bromide was formed which melted at 105°. According to this, d-limonene[1] is a constituent of celery oil. Other terpenes, it appears, are not present, in any case, pinene is excluded, as nothing came over before 170° when distilled. This is of interest, as adulteration with turpentine oil can now be readily recognized.

The amount of the oxygenated constituents to which are due the celery odor is very small compared with the terpenes. Sometimes a separation is effected during distillation, in that the heavier portions settle at the bottom of the receiver. Usually, however, no heavy oil is obtained. This is only difficultly volatile with water vapor and therefore often partially remains as residue in the still when the oil is rectified.

Such a residue, as well as the so-called heavy oil was used by Ciamician and Silber[2] in 1897 for their investigations. They found the following compounds:

1) Palmitic acid; 2) a phenol possessing the properties of guaiacol; 3) a second phenol, white needles, melting at 66—67°, of the composition $C_{16}H_{20}O_3$; 4) a liquid boiling from 262 to 269°, probably a sesquiterpene; 5) sedanolid, a lactone $C_{12}H_{18}O_2$. It boils at 185° under a pressure of 17 mm. The corresponding oxy acid, sedanolic acid $C_{12}H_{20}O_3$, melts at 88—89° and is easily changed to sedanolid. From the oxidation results it follows that sedanolic acid is o-oxyamyl-Δ^5-tetrahydrobenzoic acid. 6) Sedanonic acid anhydride, $C_{12}H_{16}O_2$. Sedanonic acid, $C_{12}H_{18}O_3$, is an unsaturated ketone acid (m. p. 113°), o-valeryl-Δ^1-tetrahydrobenzoic acid. Sedanolid and the anhydride of sedanonic acid are to be considered as the constituents giving the characteristic odor to celery oil. Their constitution is shown by the following structural formulas:

---

1) Bericht von S. & Co., Apr. 1892, p. 35.
2) Berichte, 30, pp. 492, 501, 1419, 1424, 1427.

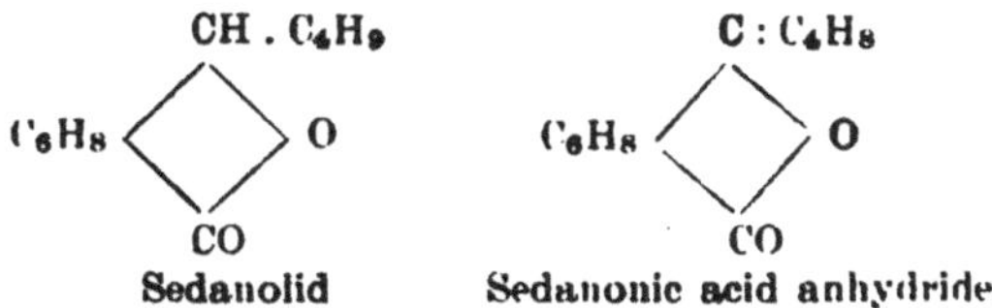

Sedanolid Sedanonic acid anhydride.

299. Oil of Celery Leaves.

Oleum Apii Graveolentis Foliorum.—Sellerieblätteröl.—Essence de Feuilles de Céleri.

The oil prepared from the fresh herb[1] (yield about 0.1 p. c.) has a strong odor of the fresh celery leaves. It is very fluid and of a greenish-yellow color, has the sp. gr. 0.848—0.850 at 15°, the rotatory power $\alpha_D = +48$ to $+52°$ and is soluble to a clear solution in 10 parts of 90 p. c. alcohol.

300. Oil of Parsley.

Oleum Petroselini.—Petersiliensamenöl.—Essence de Persil.

Origin and History. Parsley, *Petroselinum sativum* Hoffm. (*Apium petroselinum* L., *Carum petroselinum* Benth. et Hook.) (Family *Umbelliferae*), was originally indigenous to the Mediterranean countries and to Asia Minor, but is cultivated as a spice in nearly all moderate climates.

Distilled parsley water was during the time of the distilled waters a much used remedy and is often described in the distilling books of the fifteenth and sixteenth centuries.

Distilled oil of parsley does not appear to have come into use until about the middle of the sixteenth century. It is first mentioned in the drug ordinances of Frankfurt-on-the-Main of 1587 and later in the 1589 edition of the Dispensatorium Noricum.

The apiol contained in the oil, and crystallizing out at a low temperature, was observed as early as 1715 by Link, apothecary in Leipzig,[2] and in 1745 by Walther.[3]

The oil from the fresh herb and the fruit was prepared by Pabitzky[4] in 1754. Further, the oil and the crystals separated from it were mentioned by Dehne[5] in 1778, by Bolle[6] in 1829 and by Bley[7] in 1827.

1) Bericht von S. & Co., Oct. 1895, p. 59.
2) Sammlung von Natur und Medicin, wie auch von Kunst- und Litteraturgeschichten. Leipzig and Budissin, 1716.
3) De oleis vegetabilium essentialibus. Dissertatio. Leipzig, 1745, p. 17.
4) Braunschweiger Anzeiger, 1754, p. 1205.
5) Crell's chemisches Journal, 1778, I, p. 40.
6) Archiv d. Pharm., 29, p. 168.
7) Trommsdorff's Neues Journ., 14, II, p. 184.

The first elementary analysis of the "parsley camphor" was made by Blanchet and Sell[1] in 1833. It was further investigated by Löwig and Weidmann[2] in 1839.

Preparation. The volatile oil contained in all parts of the plant is particularly well represented in the fruits. These give upon distillation 2—6 p. c. of oil, which is simply designated in commerce as parsley oil or parsley seed oil.

Properties. A colorless, yellowish or yellowish-green, thick liquid, the odor of which, in spite of any similarity, is quite different from that of the herb. It has the sp. gr. 1.05—1.10 and is slightly laevogyrate. The oil from German seed is usually so rich in apiol, that it is half solid at ordinary temperature. French seeds, however, give an oil much poorer in apiol.

Composition. Apiol can be considered as the principal constituent of parsley oil. It has been studied by v. Gerichten[3] in 1876, by Ginsberg[4] in 1888 and Ciamician and Silber[5] in 1888 and later. The somewhat complicated structure of the compound was almost completely cleared up by the two last named investigators. It only remains to be decided, which of the two formulas proposed[6] belongs to the apiol from dill oil and which to the apiol from parsley oil.

Apiol, $C_{12}H_{14}O_4$, melts at 30° and boils under ordinary pressure at 294°, under a pressure of 33—34 mm. at 179°. It contains an allyl group, which is changed to the propenyl group by boiling with alcoholic potassa. The resulting isoapiol melts at 55—56° and boils at 304°.

By the oxidation of isoapiol is formed: 1) Apiolaldehyde, $C_{10}H_{10}O_5$, small needles melting at 102°. 2) Apiolic acid, $C_{10}H_{10}O_6$, m. p. 175°. 3) Apiolketonic acid, $C_{11}H_{10}O_7$, crystallizing in long, yellow needles, which melt at 160—172°.

The terpene fraction of parsley oil boiling at 160—164°[7] has, according to v. Gerichten the sp. gr. 0.865 at 12° and the angle of rotation in a 100 mm. tube —30.8°. By conducting into it hydrochloric acid no solid hydrochloride was obtained immediately; by diluting with alcohol and pouring on a large surface one was obtained in small quantity melting at 115—116°. According to this it is probable that l-pinene is a constituent of parsley oil.

---

1) Liebig's Annalen, 6, p. 301.

2) Liebig's Annalen, 32, p. 283.

3) Berichte, 9, pp. 258, 1477.

4) Berichte, 21, pp. 1192, 2514; ibid., 23, p. 323.

5) Berichte, 21, pp. 913, 1621; 22, p 2481; 28, p. 2288.

6) These are mentioned under dill oil.

7) Grünling (Dissertation, Strassburg, 1879) found the boiling point of the terpene at 158°.

301. Oil of Parsley Root.

The root of parsley contains only very little oil. According to Schimmel & Co.[1] the yield on distillation from the dry root was 0.08 p. c., from the fresh 0.05 p. c.

The oil had the sp. gr. 1.049 and separated crystals even at the ordinary temperature (probably apiol).

302. Oil of Parsley Leaves.

The yield of oil obtained by distillation from fresh parsley herb amounts to only 0.06—0.08 p. c.[2]

Properties. Oil of parsley leaves is very fluid and of a greenish yellow color. It has the full odor of fresh parsley which is only slight in the oil from the seed. Sp. gr. 0.900—0.925; $\alpha_D = +0°16'$ to $+3°10'$.

The oil distilled at 68—170° under a pressure of 12 mm.; the principal part boiled from 78—96°. By means of Zeisel's methoxyl determination made on the fractions which corresponded to the boiling point of apiol, it was ascertained, that only small quantities of this body were contained in the oil.

303. Oil of Water Hemlock.

The fruit as well as the root of the poisonous water hemlock, *Cicuta virosa* L., contain volatile oil.

Oil of the fruit. Trapp[3] obtained from the fruit collected in the autumn and dried, by distillation 1.2 p. c. of an almost colorless, quite fluid oil, lighter than water and having the odor and taste of Roman caraway oil.

By shaking with sodium bisulphite solution a solid compound was obtained which had the composition of cuminal hydroxy sulphonate of sodium. The oil not acted upon by the bisulphite solution consists of cymene; by treatment with fuming sulphuric acid cymene sulphonic acid was obtained.

The oil of the fruit of water hemlock contains therefore the same constituents as the Roman caraway oil from *Cuminum cyminum* L., cumic aldehyde and cymene.

Oil of the root. The roots yield on distillation 0.12[4]—0.36[5] p. c. of an oil smelling of water fennel and celery and having the sp. gr. 0.870

1) Bericht von S. & Co., Apr. 1894, p. 55.
2) Bericht von S. & Co., Oct. 1895, p. 59.
3) Journ. f. prakt. Chemie, 74, p. 428; Archiv d. Pharm., 231, p. 212.
4) Liebig's Annalen, 31, p. 258.
5) Journ. f. prakt. Chemie, 105, p. 151.

at 18°. It is entirely different from that of the fruit, as it contains neither cymene nor cumic aldehyde. Upon fractional distillation a dextrogyrate terpene boiling at 166° "cicutene" was isolated. By conducting hydrochloric acid into cicutene a hydrochloride was formed, which solidified when cooled. Cicutene is probably not an individual hydrocarbon, but rather a mixture of several terpenes (pinene and phellandrene?).

The oil from the root of the water hemlock was formerly considered as poisonous. That this is not the case was shown by Simon,[1] who experimented on animals.

304. Oil of American Water Hemlock.

The fruit of the poisonous plant, *Cicuta maculata* L., widely distributed in North America, gives on distillation (Glenk[2]) 3.8—4.8 p. c. of an oil smelling like *Chenopodium anthelminticum* and having the sp. gr. 0.840—0.855. The larger part boils between 176 and 183° and consists, as is seen from the elementary analysis of the two fractions 176—178.5° and 178—183°, of terpenes.

305. Oil of Caraway.

Oleum Carvi.—Kümmelöl.—Essence de Carvi.

Origin and History. The caraway plant, *Carum carvi* L. (Family *Umbelliferae*), is cultivated largely in Europe and Asia.

Distilled oil of caraway is first mentioned in the price ordinances of Berlin for 1574 and that of Frankfurt for 1589, as well as in the 1589 edition of the Dispensatorium Noricum.

Preparation. If all the oil contained in the caraway is to be extracted, it, like all other seeds, must be crushed between rollers before distillation. The crushed fruit must be distilled at once, as a great loss of oil is experienced by exposure to the air. At the beginning of the distillation considerable quantities of sulphuretted hydrogen are liberated. The cause of this has not yet been satisfactorily explained. This phenomenon, which also takes place with a few other umbelliferous seeds has been noticed as early as 1823 by Planche.[3] Schimmel & Co.[4] have found methyl alcohol, furfurol and diacetyl in the distillation water. Acetaldehyde is also liberated in large quantities during the distillation.

---

1) Liebig's Annalen, 31, p. 258; comp. also Pharm. Rundschau, 13, p. 108.
2) Amer. Journ. Pharm., 68, p. 880: Stroup, ibid., 68, p. 286.
3) Trommsdorff's Neues Journal, 7, I, p. 358.
4) Bericht von S. & Co., Oct. 1899, p. 32.

The modern distilling apparatus (fig 48 on p. 79) will hold about 2,500 k. of caraway, which yields its oil completely in 6—8 hours. Formerly it was customary—and is even done now and then at the present time — to distil caraway in the uncomminuted state. The exhausted seeds are dried, and used in cheese manufacture or for purposes of adulteration. Of course, the seeds are not entirely deprived of their oil by this treatment, and the yield of oil is therefore less, which is, however, more than counterbalanced by the sale of the dried seeds. Caraway which has been exhausted by distillation, is distinguished from the fresh seed by its darker color, by the very faint odor and taste, and the shrivelled appearance of the seed. Under the microscope, the broken, empty oil cells, and the torn upper layers of cells can be seen in a cross-section.

The exhausted, comminuted caraway is also dried in special apparatus[1] and then used as a cattle food, which is highly prized for its nourishing qualities. According to a series of analyses made at the Royal Saxon Agricultural station at Möckern,[2] dried caraway fodder contains 20—23.5 p. c. of crude protein (of which 75—85 p. c. is digestable) and 14 to 16 p. c. of fat.

The yield of oil varies according to the source of the seeds. Schimmel & Co.[3] found as an average the following yields for the different commercial varieties:

Bavarian, wild	6.5—7	p. c.
German, cultivated	3.5—5	"
Finnish, wild	5 —6	"
Galician	4.5	"
Hessian, wild	6 —7	"
Dutch, cultivated	4 —6.5	"
Moravian	4	"
Norwegian, wild	5 —6.5	"
East Frisian	5.5—6	"
East Prussian, cultivated	5 —5.5	"
Russian, wild	3.2—3.6	"
Swedish, wild	4 —6.5	"
Styrian	6	"
Tyrolese, wild	6.5	"
Württemberg, wild	5.5—6	"

The Dutch, Norwegian and East Prussian are the commercial varieties principally used for distillation. The caraway cultivated in

1) Muspratt-Stohmann, Technische Chemie, 4th ed., vol 1, p. 69.

2) Uhlitzsch, Die landwirthschaftlichen Versuchsstationen, 42, p. 48.

3) Bericht von S. & Co., Apr. 1897, Suppl. p. 26.

other places of northern Germany, is, in spite of its fine appearance. not suitable for oil distillation, on account of the low yield.

How extremely large the consumption of caraway is in Germany is seen from the following figures.[1]

Imported through Hamburg:

1896.

From the Netherlands	1,661.900 kilos
" Norway	41,000 "
" France	14,500 "
Further import from the sea	13,400 "
By rail and from the upper Elbe	105,700 "
Total	1,836,500 kilos

Add to this:

Import to Germany by rail via Groningen	316,600 kilos
	2,153,100 kilos

Of this came from Holland:

Conveyed by sea	1,661,900 kilos
" by land	316.600 "
Total	1,978,500 kilos

a quantity equivalent to 39,570 bales or fully 60 p. c. of the yield of a normal Dutch harvest. Several thousand bales, which were transported to Austria, are indeed included in these figures. These are, however, more than compensated for by the caraway grown in Germany itself. The consumption of the Leipzig factories is probably not overestimated at 10,000 to 12,000 bales per year.

As to profit, the yield is not the only factor that comes into consideration, but the quality of the oil as well. This depends largely on its carvone content, which is indicated by the specific gravity of the oil. It might happen, that the distillation of a consignment of caraway with a lesser yield of oil of high specific gravity would be more profitable than that of another of higher yield but lighter oil.

For the preparation of carvone (carvol, *Oleum Carvi* of the German Pharmacopoeia), the caraway oil is either fractionated in a vacuum or with water vapor. The fractions from the sp. gr. 0.960 on are collected separately as carvone. The limonene (carvene) obtained as a by-product has the sp. gr. 0.850 and is used as a cheap soap perfume.

Properties. Normal caraway oil is a colorless liquid, becoming yellowish in time, of a caraway odor and mild spicy taste. The sp. gr.

1) Bericht von S. & Co., Oct. 1897, p. 81.

lies between 0.907 and 0.915. Oils of lower sp. gr. rarely occur and are less valuable, as they contain less carvone. Its optical rotation α_D is $+70$ to $+80°$. Specific gravity and angle of rotation are in inverse ratio.

Oil of caraway is but sparingly soluble in 70 p. c. alcohol, but yields clear mixtures with 3—10 volumes of 80 p. c. alcohol as well as with an equal volume of 90 p. c. alcohol. It boils from 175 to 230°.

Carvone or carvol (*Oleum Carvi* of the German Pharmacopoeia) has the sp. gr. 0.963—0.966 and the rotation $\alpha_D = +57$ to $60°$. The advantage of carvone over caraway oil lies, aside from its greater intensity of odor and taste, in its ready solubility in dilute alcohol. It is miscible in all proportions with 90 p. c. alcohol. At 20°, 1½—2 parts of 70 p. c. alcohol, and 16—20 parts of 50 p. c. alcohol are sufficient for complete solution. The solubility in 50 p. c. alcohol is a good criterion of the purity of the carvone, as carvone containing 2 p. c. of limonene will not dissolve clearly in 20 parts of 50 p. c. alcohol.

Caraway oil, and especially carvone, is colored yellow by exposure to the air. The oil becomes thereby more viscous and of a higher specific gravity.

If 1 cc. of such a carvone be dissolved in an equal volume of alcohol and a few drops of a very dilute solution of ferric chloride be added, a reddish-violet coloration will be produced which, however, vanishes on the further addition of ferric chloride.

Freshly distilled oil does not give this reaction. This reaction may perhaps be due to the formation of a phenol by the decomposition of the carvone.[1] The proof, however, that a phenol is indeed formed, is still wanting.

COMPOSITION. Völckel[2] in 1840 recognized that caraway oil was a mixture of two bodies, an oxygenated body and one free from oxygen. The carrier of the caraway odor, and therefore the most important constituent, is oxygenated. It has the composition $C_{10}H_{14}O$ and was formerly called carvol;[3] Wallach,[4] however, in 1893 changed the name to carvone in order to express the ketone nature of the compound.

Pure carvone, liberated from its hydrogen sulphide addition product, has the sp. gr. 0.964 and the specific rotatory power[5] $[\alpha]_D = +62.07°$

1) Archiv d. Pharm., 222, p. 362.

2) Liebig's Annalen, 35, p. 308.

3) Liebig's Annalen, 85, p. 246. Gladstone, Journ. Chem. Soc., 25, p. 1; Pharm. Journ., III, 2, p. 746; Jahresb. f. Chem., 1872, p. 815.

4) Liebig's Annalen, 277, p. 107.

5) Archiv d. Pharm., 221, p. 283.

and boils at 229—230° (thermometer completely in the vapor). Good caraway oil contains 50—60 p. c. of carvone. For the chemical behavior and derivatives of carvone see page 160.

The hydrocarbon of caraway oil boiling at 175° and called carvene by Schweizer,[1] is, according to Wallach's[2] investigation, d-limonene (tetrabromide, m. p. 104—105°).

The odor of the carvene obtained by fractionation is quite different from that of pure limonene. It can be changed to the characteristic lemon-like odor of limonene by removing the last traces of carvone by means of the acetate of phenyl hydrazine and then shaking the preparation repeatedly with a dilute solution of potassium permanganate.

EXAMINATION. Caraway oils are often met with in commerce from which part of the valuable carvone has been removed. On the other hand under the designation "carvone" are found oils from which a part, but not all limonene has been removed.

Such products are readily recognized by their abnormal specific gravity. In judging carvone its solubility in 50 p. c. alcohol is also to be considered.

The carvone content, and thus the value of a caraway oil, can be calculated from its density, by taking the specific gravities 0.964 for carvone and 0.850 for limonene as a basis.

If a is the specific gravity of the oil investigated, b the specific gravity of one of the components (limonene), and c the difference of the specific gravity of carvone (0.964) and limonene (0.850), the amount of the other component (carvone) x in percents is given by the following formula:

$$x = \frac{a - b \,.\, 100}{c}$$

This method of determination, which is sufficient for practical purposes, rests on the assumption, that carvone and limonene are the only constituents of caraway oil.

An adulteration often found is that with alcohol, which, moreover, gives the impression that the consumer has under consideration a readily soluble oil, especially suited for the liquor manufacture. For this reason the specific gravity determination and the test for alcohol should never be neglected.

For the direct determination of carvone in volatile oils, Kremers and Schreiner[3] have recommended a method in which the carvone is

---

1) Journ. f. prakt. Chem., 24, p. 257.
2) Liebig's Annalen. 227, p. 291.
3) Pharm. Review, 14, p. 76.

converted into its oxime, and separated from the terpene by steam distillation. The method is as follows:

To a solution of 10 g. of the oil to be tested in 25 cc. of alcohol is added 5 g. of hydroxylamine hydrochloride (in case more than 50 p. c. of carvone is supposed to be present, the amount of hydroxylamine is correspondingly increased) and 0.5 g. of sodium bicarbonate and the mixture boiled on a water bath in a flask connected with a reflux condenser for ¾ hour. 25 cc. of water are then added and the alcohol, which carries over a large quantity of limonene, is distilled off from a water bath. Steam is then passed slowly through the liquid until traces of carvoxime come over. In order that these are not lost for the determination, the last portions of distillate are collected separately in test-tubes, and when traces of crystals of carvoxime appear on the surface the distillation is interrupted. The tube of the condenser is then washed with a little hot water and this, as well as the last collected distillate, containing some carvoxime, returned to the flask. When thoroughly cold the solidified carvoxime is collected on a force filter, washed, and dried by suction. The air-dried carvoxime is then transferred to a tared glass evaporating dish and heated for one hour on a water-bath, and when cool weighed. To the weight thus obtained 0.1 g. is added as this is about the quantity lost during the heating. The weight of oxime found gives by multiplying by the factor 0.9088 the amount of carvone present.

It is to be regretted that this method, which can, perhaps, be improved, gives only approximate results. In order to determine the extent of the error, Schimmel & Co.[1] prepared mixtures of pure carvone, prepared from its hydrogen sulphide addition product, with carefully prepared limonene, and found differences of nearly 7 p. c. between the found and the real carvone content:

A 50 p. c. mixture gave 43.18 p. c. carvone.
" 25 " " " 19.36 " "
" 12.5 " " " 8.54 " "

The source of error lies in that the exact point at which the distillation is to be interrupted, is difficult to determine. Moreover, a mixture of oxime and limonene, which remains liquid for some time, goes over and is lost for the determination, before crystals are noticed on the distillate.

The details of this method have been more carefully studied and the above results refuted by Kremers.[2]

Walter[3] has also based a method on the action of hydroxylamine on carvone. He treats the oil with hydroxylamine and determines the excess by titration with iodine. The applicability and accuracy of the method can not be judged from the available abstract.

---

1) Bericht von S. & Co., Oct. 1896, p. 49.
2) Pharm. Archives, 2, p. 81.
3) Chem. Zeitung. Repert., 28, p. 264.

In order to see in what succession the formation of the individual constituents of caraway oil takes place in the plant, Schimmel & Co.[1] distilled caraway plants at different stages of growth and examined the oils obtained.

Oil No. 1. From long cut, fresh plants, partly in bloom and partly in seed. Sp. gr. 0.882; n_D at 17° 1.48306; $\alpha_D = +65° 12'$.

Oil No. 2. From fresh plants of the same cutting, but after removing the inflorescence and fruits. Sp. gr. about 0.88 (not accurately determined as the amount was small); n_D at 17° 1.50883; $\alpha_D = +20° 36'$.

Oil No. 3. From fresh plants in an advanced state of development, but in which the seeds were not yet fully ripe. Sp. gr. 0.9154; n_D at 17° 1.48825; $\alpha_D = +63° 6'$.

Oil No. 2 had an odor reminding but slightly of caraway; it contained neither carvone nor limonene in amounts that could be detected.

The small specimen for examination was only sufficient to make a boiling point determination. The boiling began at 195°, the thermometer rose rapidly to 230° and about 65—70 p. c. distilled over between 230 and 270°. A resinous residue remained in the flask.

The oils No. 1 and 3, which are comparable, as they were both distilled from the whole herb, show a considerable difference in their specific gravities. The fractional distillation of both gave the following results:

	No. 1.	No. 3.
175—178°	45.0 percent	24.1 percent. (175—185°)
178—185°	21.0 "	
185—190°	4.5 "	17.8 " (185—220°)
190—220°	5.6 "	
220—235°	4.8 "	46.6 "
235—240°	6.4 "	5.5 "
240—270°	9.2 "	6.0 " (240—270° and residue)
Residue and loss	3.5 "	

The distillation residue of both oils solidified to a crystalline mass and contained besides resinous products a hydrocarbon, probably belonging to the paraffin series, crystallizing from hot alcohol in white scales melting at 64°.

The differences between the oils are at once apparent. While the terpene fraction is the larger in No. 1 and the carvone fraction comparatively small, this is quite large in No. 3. It was possible to obtain from the entire carvone fraction of the first oil only about 0.2 g. of

---

[1]) Bericht von S. & Co., Oct. 1896, p. 47.

pure crystallized carvoxime, which speaks for the low carvone content of the oil. On the other hand the corresponding fraction of No. 3 gave carvoxime in good yield.

The result of the fractional distillation indicates that the carvone content of the oil is the lower, the more undeveloped the caraway plant is when distilled; it is highest in the oil distilled from ripe material. The reverse is true for the terpene content. According to this it seems probable, that the terpene is first formed in the plant and from this the oxygenated constituent.

In both oils, the fraction boiling from 240 to 270° contains a body of quite high specific gravity, which is absent in normal caraway oil. This compound, which does not possess the properties of a phenol, and gives with ferric chloride no color reaction, has not been prepared in a pure state, and its relations to the other constituents of the oil are unknown.

306. Oil of Ajowan.

Origin and History. *Carum ajowan* Benth. et Hook. (*Ptychotis ajowan* D. C.) is an annual belonging to the *Umbelliferae*, and is cultivated in India from Panjab to Bengal and the South Decan. It also grows in Egypt, Persia, and Afghanistan. The grayish-brown fruits are similar to those of parsley but are distinguished from these by their rough surface and different odor. The ajowan seeds found in Europe are almost exclusively of Indian source and enter commerce from Bombay. Marwar in Rajputana is reported as the principal Indian market. The plant is called in India *Ajwan, Ajwain* or *Omam.* The thymol crystals separating from the oil are known in the bazaars as *Ajwan Ka-phul*, that is, flower of ajowan. The thymol as well as the distillation water (*Omum* water) are used medicinally in India, especially in cholera.

Ajowan oil mixed with fatty oil is used medicinally like ajowan water in India.

Preparation. The comminuted fruit yields 3—4 p. c. of oil by distillation. The exceedingly large amount of fat in the distilled and dried seeds, makes them very good fodder for cattle. They contain 15—17 p. c. of protein and 25—32 p. c. of fat (Uhlitzsch[1]).

Properties. Ajowan oil is an almost colorless or brownish liquid of a decided thyme odor and a sharp, burning taste. Sp. gr. 0.900—0.930; it is slightly dextrogyrate.

---

[1]) Die landwirthschaftlichen Versuchsstationen, 42, p. 52.

Composition. In Europe, ajowan oil is distilled exclusively for the preparation of its principal constituent, thymol, of which it contains 45—55 p. c.

Thymol, $C_{10}H_{14}O$, was found in the oil by Haines[1] and Stenhouse[2] almost simultaneously in 1855—56. It crystallizes in part from the oil and can be completely separated by shaking with sodium hydrate solution.

The remaining part of the oil, about one half, consists of hydrocarbons, which are sold in commerce under the name of thymene as a soap perfume.

Thymene is a mixture of cymene,[1] and a terpene boiling at 172°, but not further investigated.[2] Thymene is the cheapest source for the preparation of cymene.

307. Oil of Anise.

Oleum Anisi. — Anisöl. — Essence d'Anis.

Origin and History. The anise plant, *Pimpinella anisum* L. (Family *Umbelliferae*) comes originally from the orient, but is now cultivated in nearly all parts of the world. The European market is principally supplied by Russia, Germany, Scandinavia, Bohemia, Moravia, France, the Netherlands and Spain. The greater part of this anise seed is used for distillation for the preparation of anise oil and anethol.

Distilled oil of anise has, on account of its property to solidify, no doubt been noticed as long as anise has been used for the preparation of anise water. The distillation of the oil has, however, been first described in the works of Brunschwig, Lonicer, Ryff, Gesner, Rubeus and Porta. Valerius Cordus in 1540 called attention to the ready solidification of the oil. Nearly a century later Robert Boyle again described the "butter-like" solidification of anise oil.

Anise oil is mentioned in medical books and ordinances first in the Pharmacopœa Augustana of 1580, the Dispensatorium Noricum of 1589, and the Berlin ordinance of Matthaeous Flacco of 1574.

The first accurate investigations of anise oil were undertaken by Saussure[3] in 1820, by Dumas[4] in 1833, by Blanchet and Sell[5] in 1833,

1) Journ. Chem. Soc., 8, p. 289; Jahresb. d. Chem., 1856, p. 622.

2) Liebig's Annalen, 98, p. 269; 98, p. 309.

3) Ann. de Chim. et Phys., II, 13, p. 280; Schweigger's Journal f. Chem. und Phys., 29, p. 165.

4) Liebig's Annalen, 6, p. 245.

5) Ibidem, 6, p. 287.

Cahours[1] in 1841, by A. Laurent[2] and Gerhardt[3] in 1842. Gerhardt called the stearoptene of anise oil anethol and Cahours again pointed out the identity of the stearoptenes of anise and fennel oils, previously recognized by Blanchet.[4]

PRODUCTION. The anise used for distillation comes at present principally from Russia. The cultivation is particularly carried on in the gouvernements Woronesh, (districts Walcysk, Birjutschensk and Ostroy), Kursk, Charkow, Chersson, Podolia and Taurida.

A part of the anise is used for the preparation of the oil at the place of production, the rest is for home consumption and export. The average export for the last 10 years amounted to 150,032 puds valued at 453,721 rubels per annum. The suburbs Krassnaja and Alexejewskaja (gouvernement Woronesh) are the commercial centers for anise. The cultivation is principally done by the peasants.

Anise oil was produced in Russia in 1893 to the value of 105,500 rubels.[5]

The yearly average of anise exported from Russia during the years 1882—1891 amounted to 137,000 puds[6] at a value of 235,000 rubels. The increase during this interval of time is interesting:[7]

Export	1882,	69,000 puds,	value 169,000	rubels
"	1891,	176,000 "	" 504,000	"

One desjatine[8] usually yields 50 puds and under favorable conditions of weather as much as 100 puds of anise. If, however, the weather is cloudy and rainy at the time of blossoming, the yield sinks to 20—25 puds.

The harvest in Russia was estimated in 1896 at 4,380,000 kilos.

The exports of the different anise cultivating countries in 1896 was as follows:

Russia	Export over Libau	722,600 kilos
	" " Riga	396,000 "
Levant	" from Chios	400,000 "
Spain	" " Cadiz	24,200 "
	" " Malaga	102,000 "
Bulgaria		402,463 "

1) Liebig's Annalen, 41, p. 56; 56, p. 177.

2) Ibidem, 44, p. 313.

3) Ibidem, 44, p. 318; 48, p. 234; Journ. f. prakt. Chem., 36, p. 267.

4) Liebig's Annalen, 41, p. 74.

5) W. J. Kowalski, Die Produktivkräfte Russlands. German edition by E Davidson. Leipzig, 1898, p. 322.

6) 1 Pud = 16.375 k.

7) From "Odoriferous plants and volatile oils," a book in the Russian language by A. Bazaroff and N. Monteverte.

8) 1 Desjatine = 109.25 Ar = 2.68 acres.

PREPARATION. The odor and taste of anise seed is due to its volatile oil. They yield their oil only completely when they are distilled in a comminuted condition. The yield obtained from the different commercial varieties is seen from the following table:

Chilian	anise.........Yield	1.9—2.6	p. c.
Italian (Bologna)	" .... "	3.5	"
" ("puglieser")	" .... "	2.7—3.0	"
Macedonian	" .... "	2.2	"
Moravian	" .... "	2.4—3.2	"
Mexican	" .... "	1.9—2.1	"
East Prussian	" .... "	2.4	"
Russian	" .... "	2.4—3.2	"
Spanish	" .... "	3.0	"
Syrian	" .... "	1.5—6.0	"
Thuringian	" .... "	2.4	"

The rather marked differences in yield are not always due to the differences in the seeds themselves, but more often to the accidental or intentional admixture with stems, other seeds, earth and small stones, sometimes amounting to 30 p. c. In Russia and Moravia, earthy particles, very similar to anise in size and color are said to be especially manufactured. They are readily recognized and separated when a sample of the seeds is treated with chloroform or a concentrated solution of salt in a test-tube. The seeds will rise to the surface, whereas the earth particles settle at the bottom. The amount of inorganic admixture can be determined by means of an ash determination. Pure anise gives 7—10 p. c. of ash.

The distilled anise is dried in special apparatus[1] and sold as fodder for cattle, for which purpose it is highly valued on account of its high protein and fat content. According to analyses made in the Royal Saxon Experiment Station at Möckern, the anise residues contain 17—19 p. c. of protein and 16—22 p. c. of fat (Uhlitzsch[2]).

It might be metioned that during the distillation, sulphuretted hydrogen is given off.

PROPERTIES. Anise oil is, at medium temperature (above 20°), a colorless, highly refractory liquid of characteristic odor, and pure, intensely sweet taste. In the cold it solidifies to a snow-white, crystalline mass, which begins to melt at 15° and becomes completely liquid at 19—20°. The oil can under certain conditions be cooled far below its solidification point, without becoming solid and can be kept

1) Muspratt-Stohmann, Technische Chemie, 4th ed., vol. 1, p. 69.

2) Die landwirthschaftlichen Versuchsstationen, 42, p. 29.

for a long while in this over-cooled condition. The falling in of a particle of dust, touching with a crystal of anethol, a sharp agitation, or scratching of the walls of the flask with a glass rod, produces a sudden solidification of the entire mass, with considerable rise in temperature, the thermometer rising up to the true solidification point of the oil. This is, as it is dependent on the anethol-content of the oil, a good indicator of the quality of the oil. In normal oils it lies between 15 and 19°. The method for determining the solidification point is described in detail on p. 187.

The sp. gr. of the liquid oil is 0.980—0.990 at 15°. The plane of polarized light is slightly turned to the left, $\alpha_D = -1° 50'$. (Different from fennel oil and fennel stearoptene which turn the light toward the right.)

For complete solution 1½—5 vol. of 90 p. c. alcohol are required. Anise oil can be distinguished from star anise oil only by the odor and taste. The reaction with alcoholic hydrochloric acid, recommended for this, which was discussed under star anise oil on p. 359, does not give reliable results.

If anise oil be exposed for some time to the light, in contact with air (especially in the liquid state), its crystallizing tendency is diminished and finally disappears altogether.[1] The phenomenon is due to the formation of oxidation products (anisic aldehyde and anisic acid) as well as polymers[2] (metanethol and photo anethol?). At the same time the sp. gr. is increased, which may go so far as to make the oil heavier than water. The oil becomes also more readily soluble in 90 p. c. alcohol.

When anise oil, star anise oil, or anethol (about 2 g.) are evaporated in a small dish on a water bath, a comparatively large amount, 9—10 p. c., of a non-volatile residue remains. This is viscous, odorless, no longer tastes sweet, and propably consists mainly of photo anethol, a polymer, which, according to the investigations of de Varda[3] is formed by the action of light on anethol.

COMPOSITION. Anise oil consist principally of two isomeric compounds $C_{10}H_{12}O$, namely, of the anethol, solid at ordinary temperature, and of the liquid methyl chavicol.[4]

---

1) This was known at the beginning of this century and is mentioned by Hagen, Lehrbuch der Apothekerkunst, 6th ed. (1806), vol. 2, p. 411.

2) According to Grimaux anethol loses its tendency to crystallise by prolonged heating (Bull. Soc. chim., III, 15, p. 778).

3) Gazz. chim. ital., 21, I, p. 183; Berichte, 24, Ref., p. 564.

4) Bericht von S. & Co., Oct. 1895, p. 6. Bouchardat & Tardy later substantiate the occurrence of methyl chavicol (estragol) in Russian anise oil (Compt. rend., 122, p. 624).

The characteristic properties of the oil are due to the anethol, present to the extent of 80—90 p. c., which is therefore its most valuable constituent. It forms snow-white laminae and melts at 21.5° to a colorless, highly refractory, optically inactive liquid of pure anise odor, and an intense sweet taste. The sp. gr. is 0.986 at 25°.

Methyl chavicol is likewise optically inactive and has an anise-like odor, without, however, possessing the sweet anise taste. The properties and characteristic derivatives of anethol and methyl chavicol are described on page 179. In the first runnings of the anise oil, there are contained besides acetaldehyde, bad smelling sulphur compounds, and possibly small amounts of terpenes. In how far these compounds are due to the impurities always accompanying the seeds, cannot be stated.

Attention may briefly be called to a recent observation by Bouchardat and Tardy.[1] It was undoubtedly made on an anise oil adulterated with fennel oil, and might therefore give rise to a false notion of the composition of pure anise oil. The oil investigated by the authors cited was 9/10 solid at 10° and turned the ray of polarized light to the right. The dextrorotation was almost entirely due to a body, $C_{10}H_{16}O$, which could only be separated from the anethol by oxidation of the latter, and which was identical with the fenchone of Wallach. From the liquid, freed as much as possible from anethol by repeated freezing, two further substances were obtained with bisulphite, anisic aldehyde and a ketone, $C_{10}H_{12}O_2$. The "anise ketone" boiled at 263° and was oxidized by permanganate to anisic acid.

Schimmel & Co.[2] remark on this, that the presence of fenchone in anise oil has never been observed by them, although they had used up many thousand kilos of the oil in making anethol, and that it would have been impossible to overlook so stable a body of so penetrating an odor as fenchone. Furthermore it would follow, from the statements of Bouchardat and Tardy (melting point and rotation) that the anise oil had been adulterated with fennel oil to a marked extent.

In regard to the "anise ketone" it must also remain an open question as to whether this body came from the anise oil or the fennel oil. This can only be decided by repeating the experiment on unobjectionably pure material.

Adulteration and Examination. The cruder adulterations observed so far consisted in the addition of turpentine oil, cedar wood oil, copaiba- and gurjun balsam oil, alcohol, spermaceti and fatty oil. All

1) Compt. rend., 122, p. 198; Bull. Soc. chim., III, 15, p. 612.

2) Bericht von S. & Co., Apr. 1896, p. 7.

these adulterants are detected by the determination of the physical properties, as specific gravity, optical rotation, solubility, and solidification point (see p. 187).

The admixture with fennel oil or fennel stearoptene is frequently practiced. Even small amounts can, however, be recognized by the dextrogyration in the polariscope. Dextrogyrate anise oils are, therefore, in all cases to be rejected. In taking the sample for investigation, care must be taken that all the oil is melted and mixed to a homogeneous liquid.

Attention is here again called to the changes described on p. 561 under the heading "Properties," which a normal oil may suffer when carelessly kept and which must be considered in certain cases when the genuineness of an oil is to be determined.

The best criterion for the quality of anise oil is its solidification point, which lies normally between 15 and 19°, usually, however, at + 17°.

308. Oil of Pimpinella Root.

The oil of the white pimpinella root of *Pimpinella saxifraga* L. (Family *Umbelliferae*) is a golden yellow liquid of a penetrating and unpleasant odor, reminding somewhat of parsley oil, of a disagreeable bitter, scratching taste.[1] Sp. gr. 0.959 at 15.° It begins to boil at 240°, then rises to about 300°, a part even passing over above 300°, with considerable decomposition.[2]

The oil of the black pimpinella root, *Pimpinella nigra* Willd. (yield 0.38 p. c.) is light blue, floats on water and smells less penetrating than the preceding. It is changed in the sunlight even in closed vessels to a green color.[3]

309. Oil of Fennel.

Oleum Foeniculi.—Fenchelöl.—Essence de Fenouil.

History. The oil of fennel has no doubt been known since the time of the preparation of the distilled waters. In the sixteenth century it appears to have been introduced along with fennel water as a remedy and its preparation described by Brunschwig and by Porta. In the municipal ordinances of drugs and spices it is first mentioned in that of Berlin of 1574 and of Frankfurt-on-the-Main of 1582; also in the Pharmacopœa Augustana of 1580 and the Dispensatorium Noricum of 1589.

1) Trommsdorff's Neues Journ. d. Pharm., 12, II, p, 68.
2) Bericht von S. & Co., Apr. 1890, p. 87.
3) Trommsdorff's Neues Journ. d. Pharm., 18, II, p. 48.

Early investigations of fennel oil were made in 1779 by Heyer of Braunschweig,[1] in 1792 by Gertinger of Eperies in Hungary,[2] and in 1793 by Göttling of Jena and Giese of Dorpat. Further observations, which like the above deal mainly with fennel camphor (anethol), were made by Buchner[3] and by Goebel. Blanchet and Sell[4] recognized in 1833 the identity of the stearoptenes from fennel and anise oils. This was corroborated in 1842 by Cahours.[5] Wallach[6] investigated fenchone, a body characteristic for fennel oil, which possesses considerable theoretical interest on account of its similarity to camphor.

ORIGIN. Fennel, *Foeniculum vulgare* Gaertn. (*F. capillaceum* Gilib., *Anethum foeniculum* L.) is a fine umbelliferous plant, which is cultivated in Germany (vicinity of Lützen), Moravia, Galicia, Roumania and Macedonia, in France and Italy, also in India and Japan, partly on account of its edible root, but principally on account of its much used fruit. The fruits of the fennel cultivated in the various countries differ not only externally,[7] in form, size and color, but the oils distilled from them show greater differences than is the case with the different varieties of any other plant.

The oils of the fennel from Lützen, Roumania, Galicia, Moravia and Japan, are distinguished by the bitter tasting fenchone, which, together with anethol, produces the characteristic fennel odor. Fenchone is not found in the sweet Roman (French) and Macedonian fennel, anethol, on the other hand, is either entirely absent, or present only in traces in the wild bitter fennel.

In the individual fennel oils the greatest variety of terpenes is found. The oil of the fennel from Lützen contains pinene and dipentene, wild bitter fennel phellandrene and Macedonian fennel limonene. From this it follows that under the name of fennel oil are understood oils of entirely different properties.

In the following the term "fennel oil" applies to the ordinary fennel oil of commerce, the *Oleum Foeniculi* of the Pharmacopœia, as it is obtained by distillation from the fennel from Lützen, Roumania, Moravia and Galicia.

---

1) Crell's Chem. Journal, 8, p. 102.

2) Göttling's Almanach für Scheidekünstler und Apotheker, 14, p. 149.

3) Buchner's Repert. f. die Pharm., 15, p. 163.

4) Liebig's Annalen, 6, p. 287.

5) Liebig's Annalen, 41, p. 74; Journ. für prakt. Chem., 24, p. 359.

6) Liebig's Annalen, 259, p. 324; 263, p. 129.

7) In an article on the commercial varieties of fennel and their essential oils (Pharm. Journ., 58, p. 225) Umney in 1897 described and illustrated the different varieties of commerce.

Upon distillation of crushed seeds, the principal varieties give the following yield:

Saxon	fennel (Lützen)..............	Yield	4.4—5.5 p. c.
Galician	"	"	4.5—6 "
Moravian	"	"	4 "
Roumanian	"	"	4.6 "

The distillation residues, which form a valuable cattle food (comp. under caraway oil, p. 551) contain after drying 14—22 p. c. of protein and 12—18.5 p. c. of fat.[1]

PROPERTIES. At a medium temperature oil of fennel is a colorless or slightly yellow liquid of the peculiar fennel odor, and a taste, which is at first bitter and camphor-like, but having a sweet after taste. Sp. gr. 0.965—0.975; $\alpha_D = +12$ to $+24°$. The solidification point of the normal fennel oil, determined as described on p. 187, lies between $+3$ and $+6°$. As with anise oil, so also here, the oil of the highest solidification point is the best.

Fennel oil is soluble in an equal volume of 90 p. c. alcohol, while 5—8 vol. of 80 p. c. alcohol are required to dissolve 1 vol. of fennel oil.

COMPOSITION. The constituent of fennel oil longest known is anethol, crystallizing out in the cold. (See p. 179.) Good oils contain about 50—60 p. c. of this body. A second compound is also characteristic of the oil.[2] It is found in the fraction 190—192°, and has an intensely bitter, camphor-like taste. According to the investigations of Wallach and Hartmann[3] it is a ketone, which was first called fenchol, later fenchone. Fenchone, $C_{10}H_{16}O$, is isomeric and closely related to camphor, and gives a series of analogous derivatives. It boils at 192—193°, has the sp. gr. 0.9465 at 19°, is strongly dextrogyrate, $[\alpha]_D = +71.97°$, solidifies at a low temperture and melts again at $+5$ to $+6°$. By reduction with sodium it is converted into fenchyl alcohol, $C_{10}H_{18}O$.[4] For further information see p. 166.

Of terpenes, d-pinene and dipentene are contained in common fennel oil. Pinene was detected in the fraction boiling at 157—160° ($\alpha_D = +41° 58'$) by converting it into pinene nitrosochloride and pinene benzylamine, melting at 122°. The fraction boiling at 180° after ten

1) Die landwirthschaftlichen Versuchsstationen, 42, p. 86.

2) Bericht von S. & Co., Apr. 1890, p. 20.

3) Liebig's Annalen, 259, p. 324; 263, p. 129.

4) Laboring under the supposition that this reaction is a quantitative one, Umney has elaborated an assay method by converting fenchone into fenchyl alcohol and this into the acetate. Inasmuch, however, as it requires repeated treatment with sodium to completely reduce the fenchone to fenchyl alcohol, Umney's results must be much too low (Pharm. Journ., 58, p. 225).

fractionations gave, on shaking with hydrobromic acid in glacial acetic acid solution, dipentene dihydrobromide, melting at 94°; by bromination, dipentene tetrabromide, melting at 123—124°, was obtained.

The presence of pinene, fenchone and anethol is corroborated by a more recent investigation by Tardy[1] on an oil from French, cultivated bitter fennel. Newly found were methyl chavicol, as well as anise ketone. Anise ketone combines with bisulphite and boils between 260 and 265°. Upon oxidation with potassium permanganate, acetic and anisic acids result. These meager results, together with an elementary analysis, the results of which do not agree very well, hardly justify the assumption of a new body of the formula $C_6H_4<^{OCH_3}_{CH_2.CO.CH_3}$.

The product obtained by Tardy by conducting dry hydrochloric acid gas into fraction 176—177° could be separated by distillation in a vacuum into two portions, of which the one consisted of cymene, the other of dipentene dihydrochloride. Although the original fraction yielded no phellandrene nitrite, Tardy, nevertheless, proclaims it as a mixture of phellandrene and cymene, assuming that the latter prevents a positive phellandrene reaction. In order to make this explanation more plausible, it would first have to be proven that the cymene was originally in the oil, and that it did not result by the treatment with hydrochloric acid.

The oxidation products present in all anethol-containing oils, namely anisic aldehyde and anisic acid, were also found in this fennel oil.

Examination. The principal test is that for oils deprived of a part of their anethol by fractionation or freezing. The solidification of such oils is then below + 3°, which can be considered as the lowest allowable limit. The addition of alcohol, which has sometimes been observed, is recognized by the lowering of the specific gravity. The same is true for turpentine oil.

The fennel oils described in the following paragraph have more scientific than practical interest.

Oil from the Sweet or Roman Fennel.

The variety of fennel cultivated in southern France, and formerly designated as *Foeniculum dulce* D. C., yields on distillation 2—3 p. c. of oil. It distinguishes itself by its high anethol content and by the absence of fenchone. Sp. gr. 0.976—0.980;[2] $\alpha_D = + 7° 50'$ to $+ 16° 30'$;[2] solidification point + 10 to + 12°.

---

1) Bull. Soc. chim., III, 17, p. 660.

2) Umney, Pharm. Journ., 58, p. 226.

Oil from Macedonian Fennel.

The Macedonian fennel oil is very similar to the oil from the sweet fennel. It has a purely sweet taste and high anethol content. Yield 3.4—3.8 p. c. Sp. gr. 0.970—0.980; $\alpha_D = +5$ to $+12°$; solidification point $+7$ to $+12°$.

Fenchone is entirely absent in Macedonian fennel oil. The terpenes boil from 170—180°. The fraction 170—175° ($\alpha_D = +57°53'$) gave with glacial acetic acid and sodium nitrite a faint, but pronounced phellandrene reaction. The fraction 175—180° ($\alpha_D = +64°33'$) gave by bromination, limonene tetrabromide melting at 104—105°. According to this the oil contains d-phellandrene and d-limonene.[1]

Oil from the Wild Bitter Fennel.

The bitter fennel growing wild in France, Spain and Algiers gives by distillation about 4 p. c. of a volatile oil. Sp. gr. 0.905—0.925; $\alpha_D = +48°$.

The principal constituent of bitter fennel oil is a terpene, found by Cahours[2] and more closely studied by Bunge.[3] Wallach[4] found this terpene to be identical with the d-phellandrene found in water fennel oil by Pesci. The higher boiling fractions are slightly bitter, and appear therefore to contain some fenchone. Anethol is either entirely absent or present only in small amount.

Oil from the Indian Fennel.

Distilled from the Indian variety, *Foeniculum panmorium* D. C. Yield 0.72 (Umney[5]) to 1.2 p. c. Sp. gr. 0.968—0.973; $\alpha_D = +21°$; m. p. 8.2°.[5] It contains fenchone and anethol.

Oil from the Japanese Fennel.

The Japanese fennel is recognized by its small seeds, and therefore sometimes designated in commerce as Japanese anise. The oil is very similar to the German fennel oil. Yield 2.7 p. c. (Umney[6]); sp. gr. 0.975[7]—0.976; $\alpha_D = +10$[7] to $+16°$; solidification point $+7°$. Fenchone and anethol are constituents of the Japanese fennel oil.

---

1) Observation in the laboratory of Schimmel & Co.

2) Liebig's Annalen, 41, p. 74.

3) Zeitschr. f Chemie, 5, p. 579.

4) Liebig's Annalen, 289, p. 40.

5) Pharm. Journ., 58, p. 226.

6) Pharm. Journ., 57, p. 91.

7) Bericht von S. & Co., Oct. 1898, p. 46.

Oil from the Sicilian Ass Fennel.

The sharp tasting fruit of the *Foeniculum piperitum* D. C. (*Finocchio d'asino*) used in southern Italy as a spice, yields upon distillation 2.9 p. c. of oil. The sp. gr. of the oil is 0.951. It can contain only traces of anethol, as on cooling to —5° no separation of anethol took place.

Oil of Fennel from Asia Minor.[1]

Yield 0.75 p. c.; sp. gr. 0.987.

Oil from the Syrian Fennel.[1]

Yield 1.6 p. c.; sp. gr. 0.972.

Oil from the Persian Fennel.[2]

Yield 1.7 p. c.; sp. gr. 0.977; $\alpha_D = +14°$; m. p. 11.2°.

Oil from Russian Fennel.[2]

Yield 4.8 p. c.; sp. gr. 0.967; $\alpha_D = +23°$; m. p. 4.4°.

310. Oil of Meum Athamanticum.

Bärwurzöl.

The dry roots of *Meum athamanticum* Jacq. (Germ. Bärwurz), gave on distillation 0.67 p. c. of a dark yellow oil, very similar in odor to lovage oil. Its sp. gr. at 21° is 0.999. It began to boil at 170° and above 300° greenish-blue fractions, smelling like celery, came over. By distillation from a glass flask about half the oil resinified.[3]

311. Oil of Silaus Pratensis.

Silauöl.

The fruit[4] of *Silaus pratensis* Besser (Family *Umbelliferae*), growing wild in Germany, gives by distillation 1.4 p. c. of oil, the odor of which reminds strongly of estragon. A stearoptene separates in fine needles when exposed to the cold. Sp. gr. 0.982; $\alpha_D = +0° 7'$. Saponification number 20.8.

312. Oil of Water Fennel.

Oleum Phellandrii Aquatici. — Wasserfenchelöl. — Essence de Fenouil d'Eau.

The fruit of the water fennel, *Oenanthe aquatica* Lam. (*Oenanthe phellandrium* Lam., *Phellandrium aquaticum* L.), contains 1—2.5 p. c. of

1) Bericht von S. & Co., Apr. 1897, p. 20 of Suppl.
2) Pharm. Journ., 58, p. 226.
3) Bericht von S. & Co., Apr. 1889, p. 48.
4) Bericht von S. & Co., Oct. 1895, p. 59.

volatile oil. Oil of water fennel is at first a colorless to wine yellow liquid, becoming darker with age. It has a strong penetrating odor and burning taste. The sp. gr. is 0.85—0.89; $\alpha_D = +12°42'$ to $+15°30'$. It begins to boil at about 170° and about 50—60 p. c. go over up to 172°. On further heating the thermometer rises gradually to 300°, and finally a black resin remains in the flask (Bauer[1]).

According to an investigation by Pesci[2] in 1886 the oil consists to the extent of 80 p. c. of a terpene, characterized by a nitrite melting at 103°, which has been called phellandrene, after the name of the plant yielding the oil.[3] For the properties and derivatives of this hydrocarbon see p. 121.

Haensel[4] has observed the separation of a small amount of a heavy oil on the bottom of the Florentine flask during distillation.

313. Oil of Lovage.

Oleum Levistici.—Liebstocköl.—Essence de Livèche.

Origin and History. The original habitat of the umbelliferous plant *Levisticum officinale* Koch (*Angelica levisticum* Baillon, *Ligusticum levisticum* L.), now much cultivated as a kitchen spice, has not been determined, nor has the plant been definitely found growing wild.

The distillates from lovage root are often mentioned in the later treatises on distillation.

The oil distilled from the root appears to have come into use about the middle of the sixteenth century. It is mentioned as *Oleum Levistici* in the price ordinance of Frankfurt-on-the-main of 1587 and as *Oleum Ligustici* in the 1589 edition of the Dispensatorium Noricum.

All parts of the lovage plant contain volatile oil. Formerly only that from the root was prepared, but lately the oil from the fruit and herb is also made.

Preparation. The aromatic, fresh roots of lovage yield upon distillation with water vapor 0.3 to 0.5 p. c., the dried roots 0.6—1 p. c. of oil, the odor of which resembles that of angelica oil. According to whether the fresh or dried roots are used in the distillation, a yellow or a brown oil is obtained. These oils show but a slight difference in specific gravity, but behave differently during the process of distillation.[5] When dry lovage root is distilled there appears together with the oil,

---

1) Ueber das ätherische Oel von *Phellandrium aquaticum*. Inaug. Dissertat. Freiburg, 1885.

2) Gazz. chim. ital., 16, p. 225.

3) Liebig's Annalen, 239, p. 40.

4) Pharm. Zeitung, 48, p 760.

5) Bericht von S. & Co., Apr. 1895, p. 9.

from the very beginning of the distillation, but especially toward the end, a yellow, sticky, resinous body, which in part separates in the outflow tube of the receiver, but the larger part of which remains dissolved in the oil. When the green root is distilled this resin is scarcely noticeable. If the root has just been collected this resin is altogether absent. When the oil from the fresh root is rectified, almost the entire oil is volatile; the oil from the dry root, however, leaves behind a large amount of resin.

PROPERTIES. Sp. gr. 1.000 to 1.040. The oil is inactive (Braun[1]) or slightly (up to + 5°) dextrogyrate.[2] It dissolves to a clear solution in 2—3 p. of 80 p. c. alcohol.

COMPOSITION. If the oil saponified with alcoholic potassa be distilled with water vapor and then fractionated there will be obtained, according to Braun,[1] a fraction boiling at 176° under ordinary atmospheric pressure (sp. gr. 0.8534, $\alpha_D = +5°$) and having the composition of a terpene. Schimmel & Co. isolated as principal fraction a liquid boiling between 107 and 115° under 15 mm. pressure. From this could be separated at ordinary pressure a fraction boiling at 217—218°, which solidified to a crystalline mass and had all the properties of solid d-terpineol;[3] α_D (in over-cooled condition) $= +79° 18'$ at 22°. M. p. of terpinyl phenyl urethane 112°, of terpineol nitrolpiperidine 151—152°. The dihydriodide prepared from the terpineol melted at 77—78°.

314. Oil of Lovage Fruit.

The fruit of lovage yields upon distillation 1.1 p. c. of a distillate very similar to that obtained from the root. Sp. gr. 0.935.[4]

315. Oil of Lovage Herb.

The fresh herb and flower stems of *Levisticum officinale* yield 0.05—0.15 p. c. of oil, which is very similar in odor to that of the root. It has the sp. gr. 0.904—0.940, $\alpha_D = +16$ to $+46°$. It is soluble in an equal part of 90 p. c. alcohol.

316. Oil of Angelica.

Oleum Angelicae.—Angelicawurzelöl.—Essence d'Angélique.

ORIGIN AND HISTORY. Angelica (Ger. *Engelwurz*), *Archangelica officinalis* Hoffm. (*Angelica archangelica* L.) (Family *Umbelliferae*), grows

---

1) Archiv d. Pharm., 235, pp. 2 and 18.

2) Flückiger (Pharmacognosie, 3rd ed., p. 460) found an oil, the source of which is not mentioned, to be laevogyrate.

3) Bericht von S. & Co., Apr. 1897, p. 27; Oct. 1897, p. 9, footnote 3.

4) Bericht von S. & Co., Apr. 1890, p. 48.

here and there throughout northern Europe as far as Siberia and is much cultivated as a drug and for the manufacture of liquors. The plant is the finest of the north European umbellifers and contains in all parts a peculiar aromatic volatile oil, especially in the root and fruit. The fresh green parts of the plant serve as a much liked vegetable in the northern parts of Europe, Sweden and Finnland, and on Iceland and Greenland.

Angelica appears to have first come into use as a spice plant during the fifteenth century, and was no doubt first used for the preparation of the distilled angelica water, the preparation of which is described in Brunschwig's and in later treatises on distillation.

The distilled oil of the roots was not prepared until the second half of the sixteenth century and first mentioned in the price ordinance of Frankfurt in 1582 and in the Dispensatorium Noricum of 1589.

Recently the oil from the fresh roots, the stems and leaves and that of the fruit, have also come into use.

In composition the oils of the various parts, and of the fresh and dried plants, are on the whole identical. Marked differences exist, however, in the quality of the aroma. The aroma, as with so many plants, appears to be affected by the place of growth, the moisture of the atmosphere and the intensity of the light during the time of development of the plant.

Oil of angelica root was investigated by Buchner in 1842,[1] by Beilstein and Wiegand in 1882,[2] by Naudin in 1883,[3] and by Ciamician and Silber in 1896.[4]

Preparation. The oil of the roots as well as that of the fruit (oil of angelica seed) finds practical application. The distillation material comes mostly from Thuringia and Saxony (Erz-Gebirge), although sometimes the fruit from France, from Moravia and the Harz are used for the preparation of the oil.

The yield from the dry root is 0.35—1 p. c.; from the fresh root, which gives a finer oil, 0.2—0.37 p. c.

Properties. The oil from the root is, when freshly distilled, an almost colorless liquid of a pleasant balsamic odor. It becomes yellow to brownish when kept, due to the action of light and air. The odor is very aromatic, pepper-like with a slight admixture of musk. The taste is spicy. Sp. gr. 0.857—0.918; $\alpha_D = +16$ to $+32°$.

---

1) Buchner's Repert. f. d. Pharm., 76, p. 167.
2) Berichte, 15, p. 1741.
3) Bull. Soc. chim., 39, p. 114.
4) Berichte, 29, p. 1811.

Composition. The oil investigated by Beilstein and Wiegand[1] gave on distillation a principal fraction at 160—175°; a small part boiled between 175 and 200°, and still less above 200°. The lowest fraction had after repeated fractionation the constant boiling point 158° and the composition $C_{10}H_{16}$. The terpene absorbed 1 mol. of hydrochloric acid gas, but did not separate a solid hydrochloride. The fraction going over between 170 and 175° formed the principal product and likewise corresponded to the formula $C_{10}H_{16}$. From this fraction a hydrochloride melting at 127° crystallized out, after passing hydrochloric acid gas into the oil. After repeated treatment with sodium the fraction boiling originally at 175—200° boiled constant at 176°. The analysis indicates a mixture of terpenes and cymene. From the difficultly volatile portions a hydrocarbon boiling at 250°, probably a sesquiterpene, was obtained.

Naudin[2] obtained by fractionation a pepper-like smelling terpene of the boiling point 166°, which he called β-terebangelene.

Schimmel & Co.[3] showed the presence of phellandrene in the oil by the preparation of the nitrite. As the solution of the phellandrene nitrite in chloroform turns the ray of polarized light to the left, and as the rotation of the nitrite is in the opposite direction to that of the hydrocarbon, it follows that d-phellandrene is contained in angelica root oil.

From the investigations of Beilstein, Wiegand and Naudin it can be concluded that other terpenes (probably pinene) are also present. No doubt the β-terebangelene of Naudin can be considered as a mixture of phellandrene with a lower boiling terpene.

The high boiling fractions, in which the carrier of the musk-like odor is present, were investigated by Ciamician and Silber.[4]

From the fraction obtained last during a steam distillation, fine crystalline leaflets of the melting point 74—77° separated. The amount was too small for a detailed investigation; it was probably the anhydride of an oxy acid. The oil distilled in a vacuum was saponified with alcoholic potassa. The non-saponifiable part had the characteristic odor of the sesquiterpenes and boiled between 240 and 270°. Sulphuric acid separated two acids from the alkaline solution; 1) a valerianic acid, methyl ethyl acetic acid, the calcium salt of which crystallizes with five mol. of water; 2) oxy penta decylic acid, $C_{15}H_8.O_8$, which when crystallized from ether, forms star grouped needles of the m. p. 84°.

1) Berichte, 15, p. 1741.
2) Bull. Soc. chim., II., 39, p. 407.
3) Bericht von S. & Co., Apr. 1891, p. 3.
4) Berichte, 29, p. 1811.

Of the derivatives of this acid were prepared the barium salt, the acet-penta decylic acid of the m. p. 59°, the brom-penta decylic acid, m. p. 65°, and finally the iod-penta decylic acid, m. p. 78—79° and crystallizing in pearly scales. It may be mentioned that the next lower homologue of oxy penta decylic acid, oxy myristinic acid, is contained in the oil from angelica seed.

317. Oil of Angelica Seed.

Angelikasamenöl.

Angelica seed yields upon distillation 1—1.2 p. c. of oil.

Properties. The oil is very similar to that obtained from the root. Sp. gr. 0.856—0.890; $\alpha_D = +11$ to $+12°$.

Composition. Neither of the investigations for the terpenes of the oil from the seed made by Müller[1] in 1881 and by Naudin[2] in 1882 yielded positive results.

The only hydrocarbon definitely determined is phellandrene.[3] It is probable that, as in the case of the oil of the root, phellandrene is not the only terpene.

Of oxygenated constituents Müller found after saponifying the oil with alcoholic potassa two acids. 1) A valerianic acid, namely methyl ethyl acetic acid, as was shown by the properties of its barium salt. 2) Oxy myristinic acid, pearly leaflets of the m. p. 51°. It is found in the highest boiling and in the non-volatile portions. A number of salts were prepared and analysed. Benzoyl oxy myristinic acid, crystallizing in small white leaflets and melting at 68°, was also prepared.

318. Oil of Angelica Herb.

Angelikakrautöl.

Fresh angelica herb yields about 0.1 p. c. of oil[4] on distillation. In odor it differs but slightly or not at all from the root oil. Sp. gr. 0.870—0.890; $\alpha_D = +8$ to $+21°$.

319. Japanese Oil of Angelica.

Japanisches Angelikaöl.

Two species of angelica are cultivated in Japan for their roots. They are *Angelica refracta* Fr. Schmidt (Japanese *Senkiyu*) and *Angelica anomala* Lall. = *Angelica japonica* A. Gray (Japanese *Biyakushi*) (Rein[5]).

---

1) Berichte, 14, p. 2476.

2) Bull. Soc. chim., II, 37, p. 107; Compt. rend., 93, p. 1146.

3) Bericht von S. & Co., Apr. 1891, p. 8.

4) Bericht von S. & Co., Apr. 1895, p. 10.

5) Japan. Leipzig, 1886, vol. II, p. 159.

The Japanese angelica root[1] is somewhat poor in oil, containing only 0.07—0.1 p. c. Sp. gr. 0.910 at 20°. At +10° the oil separates crystals and at 0° it solidifies. The crystal mass obtained by freezing and suction filtering has the property of a fatty acid, the melting point of which, after several recrystallizations, was 62—63°. (Impure oxy penta decylic acid?)

The boiling point of the oil lies between 170—310°. The fractions coming over last are of a fine bluish-green color. The residue solidifies upon cooling and consists mainly of the non-volatile acid.

The odor of the Japanese oil is exceptionally intensive and persistent, also sharper than that of the German oil. It also has the characteristic admixture of musk odor.

From the fruit of the Japanese angelica Murai obtained 0.67 p. c. of oil.[1]

320. Oil of Asafetida.

Oleum Asae Foetidae. — Asantöl, Oel von Asa foetida. — Essence d'Ase Fétide.

Origin and History. Asafetida (Ger. *Asant*, Stinkasant or Teufelsdreck) is the dried-up milk sap of several species of *Ferula* and *Peucedanum*, especially of *Ferula asa foetida* L. growing in Persia, Afghanistan and the table lands of northwestern Asia.

In the mediaeval distilling books, asafetida is not used by itself, but as an addition in the distillation of alcoholic balsams. The volatile oil of asafetida is apparently first mentioned in the price ordinance of Strassburg for 1685.

Asafetida yields upon distillation 3—6.7 p. c. of volatile oil, which possesses to a high degree the unpleasant odor of the drug, reminding of onions and garlic. Its color is yellow to brown. Sp. gr. 0.975—0.990; optical rotation[2] $\alpha_D = -9° 15'$ (single observation).

Composition. According to an investigation by Hlasiwetz[3] in 1849 the oil is free from oxygen and nitrogen and contains $(C_6H_{11})_2S$, hexenyl sulphide, and $(C_6H_{11})_2S_2$, hexenyl disulphide.

Semmler,[4] however, in 1891 reached entirely different conclusions. He found a small amount of oxygenated compounds and an entirely

1) Bericht von S. & Co., Apr. 1889, p. 4.

2) Flückiger (Pharmacognosie, 3rd ed., p. 59) observed +13 to +19° which led Semmler to suppose that the oils examined by Flückiger consisted only of the lower dextrogyrate fractions. This would also explain the low sp. gr. (0.9515 at 25° = abt. 0.96 at 15°) of Flückiger's oils.

3) Liebig's Annalen, 71, p. 23.

4) Archiv d. Pharm., 229, p. 1; Berichte, 23, p. 3530; 24, p. 78.

different composition for the sulphides. He separated from the lowest boiling fractions, by repeated distillation from metallic potassium, two terpenes:

A hydrocarbon apparently identical with pinene. Sp. gr. 0.8602 at 10°; $\alpha_D = +32° 30'$. It formed a liquid dibrom addition product $C_{10}H_{16}Br_2$. A second terpene in smaller amount, which gave a solid tetrabromide $C_{10}H_{16}Br_4$.

From the higher boiling portions the following compounds were isolated:

A disulphide $C_7H_{14}S_2$, b. p. 83—84° at 9 mm.; sp. gr. 0.9721 at 15°; $\alpha_D = -12° 30'$; it is present to the extent of 45 p. c. in the crude oil.

A disulphide $C_{11}H_{20}S_2$, which makes up 20 p. c. of the oil. Sp. gr. 1.0121 at 14°; b. p. 126—127° at 9 mm.; $\alpha_D = -18° 30'$. The repulsive odor of asafetida oil is due principally to this compound.

A body $(C_{10}H_{16}O)_n$. Sp. gr. 0.9639 at 22°; b. p. 133—145° at 9 mm.; $\alpha_D = -16°$. It is contained to the extent of 20 p. c. in the crude oil. By treatment with sodium, cadinene $C_{15}H_{24}$, results.

A compound $C_8H_{16}S_2$. B. p. 92—96° under 9 mm. pressure.

A disulphide $C_{10}H_{18}S_2$. B. p. 112—116°.

321. Oil of Galbanum.

Galbanumöl.

Origin and History. Galbanum, a gum resin, is the dried milky juice exuding from the trunks and larger branches of the umbelliferous plant *Ferula rubricaulis* Boissier, *F. galbaniflua* Boissier et Buhse (*Peucedanum rubricaule* H. Baillon, *P. galbanifluum* H. Baillon and probably also of *Ferula* (*Peucedanum*) *Schaïr* growing in Persia.

Distilled oil of galbanum was prepared by Ryff, by Gesner, and by Rubeus. It was included in the 1589 edition of the Dispensatorium Noricum and in the Pharmacopoea Augustana of 1580, and mentioned in apothecary and spice ordinances about 1560. Early cursory investigations of the oil were made by Neumann about 1728, by Walther in Leipzig about 1744, by Fiddichow in 1815[1] and by Meissner in 1816.[2]

Preparation and Properties. The aromatic, not unpleasant odor of galbanum is due to its large content of volatile oil. The yield on distillation is different according to the age of the drug and varies between 14 and 22 p. c.

1) Berl. Jahrbuch der Pharmacie, 1816, p. 280.
2) Trommsdorff's N. Journ. d. Pharm., 1, I, p. 3.

Oil of galbanum is yellowish, has the sp. gr. 0.910—0.940 and turns the ray of polarized light either to the right or to the left, $\alpha_D = +20°$ to $-10°$. According to Hirschsohn[1] Persian galbanum yields a dextrogyrate, levant galbanum, however, a laevogyrate oil.

COMPOSITION. An oil investigated by Mössmer,[2] which boiled almost completely between 160 and 165° cannot be considered as normal, as the distillation with water from a glass retort will yield only the lower boiling portions and not the difficultly volatile, higher hydrocarbons. The dextrogyrate hydrocarbon $C_{10}H_{16}$, boiling at 160—161°, gave with hydrochloric acid a crystalline compound, which agreed completely in its properties with the corresponding body from turpentine oil. Mössmer did not succeed in obtaining terpin hydrate with nitric acid, but Flückiger[3] was successful in preparing it in this manner. The terpene of galbanum oil, is therefore, d-pinene.

Fraction 270—280° contains, according to Wallach,[4] cadinene, $C_{15}H_{24}$, the presence of which was shown by the preparation of its dihydrochloride, melting at 117—118°.

322. Oil of Sumbul.

Oleum Sumbuli.—Moschuswurzel- oder Sumbulwurzelöl.—Essence de Sumbul.

In East India the roots of several aromatic plants are designated as sumbul. The root of *Nardostachys jatamansi* D. C. is known as *Sumbul Hindi*, the root of *Valeriana celtica* L. as *Sumbul Ekleti, Sumbul Ekelti, Sumbul Kumi* and *Sumbul italicus*.[5] The root of *Dorema ammoniacum* Don. often used for adulterating the genuine sumbul root, is known as Bombay sumbul or *Boi*.[6]

The genuine sumbul root comes from *Ferula sumbul* Hooker fil. *Euryangium sumbul* Kauffmann) and was first brought into Europe in 1835. On distillation it yields 0.2—0.4 p. c. of a viscous, dark colored oil of musk-like odor, and a sp. gr. of 0.954—0.964. The saponification number of the oil is 92 (single determination). It is not soluble in 10 vol. of 80 p. c. alcohol but in an equal volume of 90 p. c. alcohol.

Nothing is known as to the composition of the oil.

323. Oil of Gum Ammoniac.

Ammoniakgummiöl.

ORIGIN AND HISTORY. The flow of a gum resin from the umbelliferous plant *Dorema ammoniacum* Don. (*Peucedanum ammoniacum* H. Baillon),

---

1) Jahresb. f. d. Pharm., 1875, p. 118.
2) Liebig's Annalen, 119, p. 257.
3) Pharmacognosie, 3rd ed., p. 65.
4) Liebig's Annalen, 238, p. 81.
5) Pharm. Journ., I. 7, p. 546.
6) Pharmacographia, 2nd ed., p. 318.

rich in milky sap, is produced by insects. The gum resin, hardened by contact with the air, was used in antiquity on account of its pleasant odor, as incense, and together with other resins for embalming and probably also as a remedy. The plant grows on sterile ground in many parts of anterior Asia and in eastern North Africa. The botanical source of the gum resin was determined in 1829 by David Don of London.[1]

The oil of gum ammoniac was distilled by Ryff, Cordus and Gesner and is included in the Frankfurt tax for the year 1587 and in the Dispensatorium Noricum of 1589.

Earlier investigations were made by Buchholz[2] and by Calmeyer[3] in 1808, by Braconnot[4] in 1809, and by Hagen[5] in 1814.

PROPERTIES. The ammoniacum of commerce gives on distillation 0.3 p. c. of a dark yellow oil, smelling strongly of the crude material, and reminding vividly of angelica.[6]

It is slightly dextrogyrate,[7] has the sp. gr. 0.891 at 15° and boils principally from 250 to 290°. Between 155 and 170° only a small portion goes over.[8] It is free from sulphur.

324. Oil of Peucedanum Oreoselinum.

Bergpetersilienöl.

Schnedermann and Winkler[9] in 1844 prepared from the fresh herb of *Peucedanum oreoselinum* Moench (*Athamanta oreoselinum* L.) a volatile oil by water distillation. It had a strong, aromatic odor, somewhat similar to juniper. Sp. gr. 0.843. It boiled at 163° and consisted, as is seen from the analysis, almost entirely of terpenes. With hydrochloric acid a liquid monohydrochloride $C_{10}H_{16}HCl$ was obtained.

Which terpene or terpenes are the basis for this compound cannot be decided from the meagre results given.

325. Oil of Peucedanum Ostruthium.

From the dry root of *Peucedanum ostruthium* Koch (*Imperatoria ostruthium* L.) 0.2—0.8 p. c. of oil are obtained by distillation. It has

1) Buchner's Repert., 37, p. 115; Trans. of the Linnean Soc. of London, 16, p. 601.
2) Taschenb. f. Scheidekünstler u. Apoth., 1809, p. 170.
3) Trommsdorff's Journ. d. Pharm., 17, II, p. 82.
4) Ann. de Chim., 68; Trommsdorff's Journ. d. Pharm., 18, I, p. 202.
5) Berl. Jahrbuch d. Pharmacie, 1815, p. 95.
6) Bericht von S. & Co., Apr. 1890, p. 47.
7) Pharmacognosie, 3rd ed., p. 73.
8) Archiv d. Pharm., 233, p. 553.
9) Liebig's Annalen, 51, p. 336.

an odor reminding strongly of angelica oil, and a biting aromatic taste. It has a sp. gr. of 0.877 [1] and boils from 170 to 190°.

The investigations [2] made up to the present have thrown no light on the composition of the oil.

326. Oil of Dill.

Oleum Anethi. — Dillöl. — Essence d'Aneth.

ORIGIN AND HISTORY. The fruit of the dill, *Peucedanum graveolens* Benth. et Hook. (*Anethum graveolens* L.), an umbelliferous plant, indigenous to the Caucasus and Mediterranean countries, but now cultivated in many other places, has been known since antiquity.

In the treatises on distillation of the fifteenth and sixteenth century the distillation of dill is often mentioned. In German apothecary and spice ordinances, dill oil is first mentioned in that of Frankfurt-on-the-Main for 1587.

PREPARATION. Dill, which on account of its aromatic seed is cultivated in Bavaria, Thuringia and Roumania, yields on distillation 3—4 p. c. of oil. The dried distillation residues contain 14.5—15.6 p. c. of protein and 15.5—18 p. c. of fat and are used as cattle food.[3]

PROPERTIES. The oil, though colorless at first, soon becomes yellow on keeping. The odor reminds strongly of caraway oil, yet can be distinguished from this by its peculiar dill aroma. The taste is at first mild, later sharp and burning. Sp. gr. 0.895 (as a rule above 0.900) to 0.915; $\alpha_D = +75$ to $+80°$. Dill oil is soluble in 5—8 parts of 80 p. c. alcohol to a clear solution.

COMPOSITION. Carvone,[4] present to the extent of 40—60 p. c. is the most important constituent of dill oil. As dill carvone has the same rotatory power, as well as the other properties of the carvone from caraway oil, the two are to be considered as entirely identical (Beyer [5]). (Com. p. 160.)

As far as the terpenes in the oil are concerned, Wallach [6] has shown the presence of limonene in fraction 175—180° by preparing its tetrabromide melting at 104—105°.

1) Bericht von S. & Co., Oct. 1887, p. 85.

2) Hirzel (1849), Journ. f. prakt. Chem., 46, p. 292; Pharm. Centralbl., 1849, p. 37. Wagner (1854), Journ. f. prakt. Chem., 62, p. 280.

3) Die landwirthschaftlichen Versuchsstationen, 42, p. 62.

4) Journ. Chem. Soc., 25, p. 1; Jahresb. f. Chem., 1872, p. 816.

5) Archiv d. Pharm., 221, p. 283.

6) Liebig's Annalen, 227, p. 292.

Limonene is the principal hydrocarbon of dill oil. Other terpenes are, however, present. Nietzki[1] in 1874 isolated 10 p. c. of a hydrocarbon, $C_{10}H_{16}$, boiling at 155—160°, which yielded on treatment with dilute acid terpin hydrate, but gave no crystalline hydrochloride. Schimmel & Co.[2] report on an oil distilled in England, in which a decided phellandrene reaction was obtained with sodium nitrite and glacial acetic acid. On repeating this test on German oil, the reaction for phellandrene failed completely, using the oil as a whole. By using the first fraction, however, the phellandrene could also be detected by the formation of the nitrite. Spanish oil, like the English, readily gives the reaction for phellandrene.[3]

Dill-apiol, boiling at 285° and found in the East Indian dill oil, does not occur in the German oil. In a fractional distillation, made for the purpose of detecting this body, it was found that the oil contains no constituents boiling higher than carvone.[2] In the flask remained a slight residue, solidifying to a solid mass, which crystallized from petroleum ether in colorless laminae, melting at 64°. As must be concluded from its indifference to concentrated sulphuric acid, it consisted of paraffin.

327. East Indian Dill Oil.

The oil obtained from the East Indian and Japanese dill of *Anethum sowa* D. C., differs from that of *Anethum graveolens* L. not only physically but also in its chemical compostion.[4]

The East Indian dill oil (yield 2—3 p. c.) has the sp. gr. 0.948[5] to 0.970[6] and the rotatory power $\alpha_D = +41°30'$[7] to $+47°30'$.[5] For the Japanese oil, Umney observed the sp. gr. 0.964 and the angle of rotation $\alpha_D = +50°30'$.

In the distillation a part heavier than water separated at the bottom of the receiver, a phenomenon which has never been noticed with the common dill oil.[6] This specifically heavier portion consists,

---

1) Archiv d. Pharm., 204, p. 317.

2) Bericht von S. & Co., Apr. 1897, p. 13.

3) Bericht von S. & Co., Oct. 1898, p. 20.

4) Although Flückiger and Hanbury (Pharmacographia) do not regard East Indian dill as a separate species, more recently scientists are more inclined to regard the Indian (and Japanese) dill as a distinct species (*Anethum sowa* D. C.) on account of the botanical as well as on the decided chemical differences. (Comp. Umney, Pharm. Journ., 61, p. 176; also Bericht von S. & Co., Oct. 1898, p. 18.)

5) Pharm. Journ., III, 25, p. 977.

6) Bericht von S. & Co., Oct. 1891, p. 12.

7) Berichte, 29, p. 1799.

according to an investigation of Ciamician and Silber,[1] of a body closely related to and isomeric with the apiol from parsley oil, called dill-apiol.

Dill-apiol, $C_{12}H_{14}O_4$, is a thick oily fluid which boils at 162° under 11 mm. pressure, and at 285° under atmospheric pressure. By the action of bromine in excess the dibromide of monobrom-apiol, crystallizing in colorless prisms and melting at 110°, is formed.

Dill-isoapiol is obtained, analogous to isoapiol, by heating dill-apiol with dry sodium ethylate to 150—170°. It forms colorless, shining prisms melting at 44° and boiling at 296° with slight decomposition.

By oxidation of the dill-isoapiol with potassium permanganate the following substances were obtained: 1) Apiol aldehyde, $C_{10}H_{10}O_5$, white needles, m. p. 75°; 2) apiolic acid, $C_{10}H_{10}O_6$, m. p. 151—152°. 3) apiol ketonic acid, $C_{11}H_{10}O_7$, light yellow laminae, m. p. 175°.

By fusing the dill-apiolic acid with potassa, dimethyl apionol carbonic acid is formed, which by dry distillation yields quantitatively dimethyl apionol, boiling at 283°. This can be readily changed by methyl iodide into tetramethyl apionol melting at 89° and identical with that obtained from parsley-apiol.

Apionol.	Tetramethyl apionol.	Apiol I.	Apiol II.
$C_6H_2\begin{array}{l}OH[4]\\OH[3]\\OH[2]\\OH[1]\end{array}$	$C_6H_2\begin{array}{l}OCH_3[4]\\OCH_3[3]\\OCH_3[2]\\OCH_3[1]\end{array}$	$C_6H\begin{array}{ll}\begin{array}{l}O\\O\end{array}>CH_2 & \begin{array}{l}[5]\\[4]\end{array}\\OCH_3 & [3]\\OCH_3 & [2]\\C_3H_5 & [1]\end{array}$	$C_6H\begin{array}{ll}OCH_3 & [5]\\\begin{array}{l}O\\O\end{array}>CH_2 & \begin{array}{l}[4]\\[3]\end{array}\\OCH_3 & [2]\\C_3H_5 & [1]\end{array}$

Which of these two formulas belongs to the apiol from parsley oil and which to that from dill oil is still to be determined.

328. Oil of Peucedanum Sativum.

Pastinaköl.

The dried fruit of *Peucedanum sativum* (*Pastinaca sativa* L.) yields by distillation with water vapor 1.5 to 2.5 p. c. of oil.[2] The oil is yellowish and has a penetrating, persistent odor. Sp. gr. 0.87—0.89; $\alpha_D = -0° 15'$ to $-0° 30'$; saponification number about 170.

Composition. The chemical composition of the oil appears to change with the ripening of the fruit.

Van Renesse[3] in 1873 found that the greater part of an oil, distilled from the ripe fruit, boiled between 244 and 245° and consisted of the

1) Berichte, 29, p. 1799.

2) Wittstein in 1839 (Buchner's Repert. f. d. Pharm., 68, p. 15) obtained a yield of but 0.7 p. c.

3) Liebig's Annalen, 166, p. 84.

octyl ester of normal butyric acid. The acid contained in the portion distilling over below 244° gave on the analysis of its silver salts figures which point to a mixture of butyric acid with an acid of less carbon content. Whether this is propionic acid, as van Renesse suspects, or some other acid, cannot at present be decided.

An oil distilled by Gutzeit[1] from ripe and half ripe fruit was less simple in composition: It consisted for the greater part of the following three fractions:

195—210°......................58.6 p. c.
233—240°......................29.3 "
240—270°......................12.1 "

The presence of ethyl alcohol was shown by Gutzeit in the distillation water.

329. Oil from the Fruit of Peucedanum Grande.

The oil from the fruit of *Peucedanum grande* C. B. Clarke has an exceedingly strong, spicy odor,[2] reminding of the carrot oil. Sp. gr. 0.9008 at 15.5°; $\alpha_D = +36°$. It boils from 185—228°, leaving a quite large residue.[3]

330. Oil of Peucedanum Root.

The dry root of *Peucedanum officinale* L. gives 0.2 p. c. of a yellowish-brown oil[4] of an intensive, persistent odor, hardly pleasant and reminding most of senega root.

Sp. gr. 0.902; $\alpha_D = + 29° 4'$; saponification number 62. Upon standing in the cold the oil separated a solid body, which when twice recrystallized from alcohol formed somewhat yellowish laminae melting at 100°.

331. Oil of Heracleum.

Bärenklauöl.

The oils obtained from the fruits of different species of *Heracleum* are of no practical importance, but are very interesting from a scientific point of view. On the one hand because the esters of the fatty acids of alcohols of the paraffin series, otherwise difficult to obtain, are contained in them, and on the other hand, because the predominant constituents in different stages of the development of the fruit allow of drawing certain conclusions in plant physiology. It has namely been

1) Liebig's Annalen, 177, p. 372.
2) Bericht von S. & Co., Apr. 1891, p. 50.
3) Pharmacographia Indica, vol. 2, p. 126.
4) Bericht von S. & Co., Apr. 1895, p. 78.

determined that not only the oil content, but also the chemical composition of the oil is dependent on the stage of ripeness of the seed. In the oils distilled from half ripe fruit, compounds of low carbon content are found which are absent from the oil of the ripe fruit, and which seem to disappear in the course of the ripening. From this it appears that the bodies of low carbon content are first formed in the fruit, and are then used for the building up of the compounds of higher carbon content.

The yield of oil varies[1] with the fruit of *Heracleum sphondylium* L. from 0.3[2]—3 p. c.[3] The oil is a yellowish, acid reacting liquid, of a penetrating and persistent odor and a sharp taste. Sp. gr. 0.80 to 0.88. The rotatory power observed on two oils was $+0°15'$ and $+0°16'$; saponification number 260—290.

COMPOSITION. An oil investigated by Zincke[2] in 1869 boiled from 190—270° and consisted principally of acetic and capronic acid esters of normal octyl alcohol. The free octyl alcohol also shown to be present is probably not a normal constituent of the oil, but a decomposition product of the ester, formed during the steam distillation of the oil.

Two oils distilled by Möslinger[4] in 1876 were somewhat more complex in composition. They boiled from 110—291° and contained the following constituents: 1) Ethyl butyrate; 2) an hexyl compound, probably hexyl acetate; 3) esters of octyl alcohol, principally the acetate, caprinate and laurinate and possibly also esters of the acids between capronic and laurinic acids; 4) methyl and ethyl alcohol, as well as ammonia, were found in the distillation water.

332. Oil from the Fruit of Heracleum Giganteum.

The oil investigated by Franchimont and Zincke[5] in 1872 had been distilled from a foreign heracleum species, most probably from *H. giganteum* L. Its principal constituents were hexyl butyrate and octyl acetate.

By the distillation of the not fully ripe fruit of *H. giganteum* L., Gutzeit[6] in 1875 obtained 0.56 p. c., from the ripe and partly dried fruit, however, 2 p. c. of oil. Upon fractional distillation it boiled from 130—250°. Fraction 130—170° contained ethyl butyrate.

---

[1]) The great variation in the yield obtained by different investigators is not so much due to differences in the oil content of the fruit as to differences in the percentage of water, i. e., differences in the stage of drying.

[2]) Liebig's Annalen, 152, p. 1.

[3]) Bericht von S. & Co., Oct. 1886, p. 38.

[4]) Berichte, 9, p. 998; Liebig's Annalen, 185, p. 26.

[5]) Liebig's Annalen, 163, p. 193.

[6]) Liebig's Annalen, 177, p. 344.

Methyl and ethyl alcohol were found in the distillation water. The methyl alcohol was more prominent in the distillation of the ripe fruit, whereas with the unripe fruit the ethyl alcohol was present in larger quantity.

333. Oil of Daucus Carota.

Möhrenöl.

The oil distilled from the fruit of the wild carrot, *Daucus carota* L. is colorless to yellow and has a pleasant, carrot-like odor. Yield 0.8—1.6 p. c.; sp. gr. 0.870—0.923; $\alpha_D = -13$ to $-37°$.

According to Landsberg[1] (1890) the principal constituent of the oil is a terpene, $C_{10}H_{16}$, boiling at 159—161°, sp. gr. 0.8525 at 20°, $\alpha_D = +(?)\,32.3°$. By direct bromination it yields a liquid dibromide, after heating to 280°, dipentene tetrabromide, melting at 123—125° results. It is therefore to be considered as pinene.

Upon fractionation of the parts boiling above 200°, the splitting off of water and acetic acid was noticed. The chemical nature of the compounds giving rise to this phenomenon has not been determined.

Carrot root gives only 0.0114 p. c. of a colorless oil by distillation. Sp. gr. 0.8863 at 11.2° (Wackenroder,[2] 1831).

334. Oil of Osmorrhiza Longistylis.

The root of the umbelliferous plant *Osmorrhiza longistylis* Rafinesque, known in North America as sweet cicely, sweet root or sweet anise, smells distinctly of anise and fennel (Green,[3] 1882). Induced by this, Eberhardt[4] in 1887 subjected the root to distillation and obtained 0.63 p. c. of oil having a sp. gr. of 1.0114 at 10°. It solidified at 10—12° and again became liquid at 16°.

On distillation the oil began to boil at 189°; the thermometer then rose rapidly to 225°, the main portion going over from 225—230°, while only a small part went over from 230—280°. The fraction 226—227° consisted, as its properties show, of anethol. By oxidation anisic acid melting at 184° was obtained.

The fraction boiling about 250° gave with bromine a bromide crystallizing in rhombic plates, melting at 139°. The nature of the compound forming the basis of this bromide was not determined.

---

1) Archiv d. Pharm., 228, p. 85.
2) Magaz. d. Pharm., 33, p. 145.
3) Am. Journ. Pharm., 54, p. 895.
4) Pharm. Rundschau, 5, p. 149.

335. Oil of Monotropa Hypopitys.

The volatile oil of *Monotropa hypopitys* L. (Family *Pirolaceae*), often found parasitic on roots in forests, was first prepared in 1857 by Winckler.[1] From the plant almost in full bloom, Winckler distilled an oil which was identical with wintergreen oil from *Gaultheria procumbens.* More recently (1894) Bourquelot[2] again showed the identity of this oil with that of wintergreen. From his investigation it follows, that the oil does not exist in the plant as such, but is present as a glucoside, which is probably identical with the gaultherin isolated by Schneegans and Gerock[3] from the bark of *Betula lenta.* This glucoside splits up under the influence of a ferment contained in the plant, or by dilute sulphuric acid, into methyl salicylate and glucose. Neither the emulsin of almonds, nor the diastase of malt, nor the ferment of saliva are capable of splitting up the gaultherin.

336. Oil of Labrador Tea.

Oleum Ledi Palustris. — Porschöl. — Essence de Ledon.

All parts of *Ledum palustre* L. (Family *Ericaceae*) yield on distillation 0.3—2 p. c. of volatile oil, which is usually so rich in stearoptene that it solidifies at ordinary temperature. Sometimes, however, no crystalline separation can be effected, even by placing separate fractions into a freezing mixture.[4] The reason for this fluctuation in yield, as well as in composition, of the oil, lies according to Trapp in the different stages of development of the plant parts subjected to distillation. If a large yield and an oil rich in stearoptene is desired, the twigs must be distilled before, during, or immediately after blossoming. Hjelt on the other hand could notice no decided influence of the season or growth on the amount of volatile oil.

Properties. Oil of Labrador tea is a greenish or reddish viscid liquid of a penetrating narcotic odor and sharp, unpleasant, persistent taste. Sp. gr. 0.93—0.96. The portions of the oil not consisting of ledum camphor boil between 180—250°.

Composition. The oil first prepared by Rauchfuss[5] in 1796 has

---

1) Neues Jahrb. d. Pharm., 7, p. 107; Vierteljahrsschrift f. prakt. Pharm., 6, p. 571; Jahresb. f. Chemie, 1857, p. 520.

2) Journ. de Pharm. et Chim., V, 30, p. 435; and VI. 8, p. 577; Compt. rend., 119, p. 802; and 122, p 1002.

3) Archiv d. Pharm., 232, p. 437.

4) Bericht von S. & Co., Oct. 1887, p. 35.

5) Trommsdorff's Journ. d. Pharm., 3, p. 189.

since been investigated by numerous chemists.[1] The chemical nature of ledum camphor, however, has only lately been cleared up by Rizza[2] and Hjelt.[3]

Ledum camphor crystallizes from alcohol in fine long needles melting at 104—105°. It boils at 282—283° and is slightly dextrogyrate in alcoholic solution, $[\alpha]_j = +7.98°$. Ledum camphor is a sesquiterpene hydrate $C_{15}H_{26}O$, the hydroxy group of which is so labile, that its alcoholic nature cannot be directly proven. By treatment with benzoyl chloride or sulphuric acid ledene, $C_{15}H_{24}$, results with splitting off of water. Ledene is a sesquiterpene boiling at 255°. Oxidation with nitric acid produces oxalic acid. Potassium permanganate does not act upon it, which indicates that ledum camphor is a tertiary alcohol.

According to experiments by Sundvik, ledum camphor is a poison, acting strongly on the central nervous system.

337. Oil of Wintergreen.

Oleum Gaultheriae. — Wintergrünöl. — Essence de Gaultheria.

Origin and History. Wintergreen, *Gaultheria procumbens* L. (Family *Ericaceae*) grows from the New England States to Minnesota and south as far as Georgia and Alabama. On account of the peculiar pleasant odor and taste which develops when the plant is chewed, it was early used by the natives. The distillation of the oil was probably begun in the first decades of this century along with that of sassafras bark (p. 395) and birch bark (p. 331) in the states of Pennsylvania, New Jersey and New York. At first these aromatics were used for chewing, later for the preparation of refreshing beverages and home remedies. and especially for the much used, supposed blood purifiers. When the preparation of the volatile oils was successful, these were often used instead of the aqueous extraction of the drug. This use is of considerable importance to the history of the introduction of wintergreen and sassafras oils, as both of these were used as popular remedies in the United States since the beginning of this century under the title of patent medicines. The preparation and use of these remedies soon

---

1) Meissner (1812), Berl. Jahrb. d. Pharm., 13, II, p. 170. — Grassmann (1831), Rep. f. d. Pharm., 38, p. 53. — Buchner (1856), ibid., p. 57, and Neues Rep. f. d. Pharm., 5, p. 1. — Willigk (1852). Wiener Academ. Berichte, 9, p. 302. — Fröhde (1861), Journ. f. prakt Chem., 82, p. 181. — Trapp (1869), Zeitschr. f. Chem., 5, p. 350; Berichte, 8 (1875), p. 542; Pharm. Zeitschr. f. Russl., 34 (1895), pp. 561 and 661. — Ivanov (1876), Pharm. Zeitschr. f. Russl., 5, p. 577. — Hjelt & Collan (1882), Berichte, 15, p. 2500. — Rizza (1888), Berichte, 16, p. 2311.

2) Zeitschr. der russ. phys-chem. Ges., 19, I, p. 319. Chem. Centralbl., 1887, p. 1257.

3) Berichte, 28, p. 3087.

became general and with it came a greater demand for the oils. Wintergreen oil was especially in demand for the preparation of one of the oldest popular remedies in the United States, namely Swaim's Panacea, introduced in 1815, which at that time had an enormous sale and in the efficiency of which great confidence was placed.

Wintergreen oil appears not to have been used at that time for other purposes. The first mention of it in literature is found in a botanical work[1] by Bigelow, a physician of Boston, published in 1818. In it, gaultheria oil is mentioned as a stable article of the drug stores, and also that this oil occurs, besides in *Gaultheria*, also in *Spiraea ulmaria*, the root of *Spiraea lobata*, and especially in the bark of *Betula lenta*.[2] In pharmacopœias, the oil was first taken up in that of the United States of 1820. The medicinal use of the oil did not become general until after 1827, when the New York Medical Society made known its use in the preparation of the popular specific mentioned above.[3]

Although the similarity of the volatile oil from *Gaultheria procumbens* L. with that from the bark of *Betula lenta* L. was known before 1818,[1] the identity of their principal constituent was shown scientifically about the same time by Wm. Procter jr.[4] of Philadelphia in 1842 and Cahours[5] in 1844. From that time on the oil was no longer distilled exclusively from wintergreen, but often from this together with birch bark, or only from the latter. The oil came more and more into use as an aromatic for pharmaceutic and cosmetic preparations, for beverages and medicinal remedies,[6] and thus became an important article of commerce. In recent times, however, it was often adulterated with kerosene and alcohol.[7]

Besides the investigations mentioned above, the oil has been more recently examined by Pettigrew[8] in 1883, by Power and Werbke[9] in

---

1) American Medical Botany, vol. 2, p. 28.

2) According to the investigations of Bourquelot (Compt. rend., 119, p. 802, and 122, p. 1002), Schneegans and Gerock (Archiv d. Pharm., 232, p. 439), methyl salicylate is produced by the decomposition of the glucoside gaultherin by the ferment betulase. This glucoside seems to occur not only in wintergreen and sweet birch, but also in a number of other plants, e. g., in the roots of various polygala species, in those of *Spiraea ulmaria*, *S. filipendula*, *S. salicifolia*, *S. lobata*, and the flowers of *Azalea*, in *Monotropa hypopitys*, and many other plants. As to *Gaultheria procumbens*, a part of the oil seems to occur as such in the leaves, for by crushing them a strong odor of methyl salicylate is developed.

3) Pharm. Review, 16, p. 179; Am. Journ. Pharm., 3, p. 199.

4) Am. Journ. Pharm., 14, p. 211.

5) Annal. de Chim. et Phys., III., 10, p. 327; Liebig's Annalen, 48, p. 60; 52, p. 327.

6) New York Medical Record, 22, p. 505; Squibb's Ephemeris, 3, p. 950.

7) Pharm. Rundschau, 7, p. 286.

8) Am. Journ. Pharm., 55, p. 385; and 56, p. 266.

9) Pharm. Rundschau, 6, p. 208; 7, p. 283; 8, p. 38; 10, p. 7.

Fig. 77.
Wintergreen Oil Distillery in Virginia.

1888—1892 and finally by Power and Kleber[1] in 1895, who determined the nature of the constituents other than methyl salicylate.

Methyl salicylate is prepared on a large scale and brought into the market as artificial oil of wintergreen since 1886 by the firm of Schimmel & Co. It is official in the U. S. Pharmacopoeia.

Preparation. The preparation of oil of wintergreen has always been carried on in a primitive manner, the distillation being conducted by smaller farmers at the place of growth. This was first done in the New England states[2] and later in the mountain and forest districts of the states of New York, New Jersey, Pennsylvania, Virginia, and Maryland. Usually old copper whisky stills of various sizes, mostly from 200 to 400 gals. capacity, serve as stills.[3] Sometimes the distillation is done in boxes of oak wood about 8 ft. long, 4 ft. high, and 4—5 ft. broad, mostly, however, in larger alcohol barrels, held together by strong iron hoops, the perforated bottom of which is placed as tightly as possible into a suitable cast iron kettle, which is filled with water for distillation. On the upper part of the barrel is placed a copper helm, which is connected with a condensing worm in a large wooden tub.

In the distillation, which is carried on for only a few months in the year, the still, barrel or box, is filled with finely chopped, well wetted plants. The charge is allowed to stand over night and firing begun in the morning. The distillation is usually complete in eight hours. About 90 p. c. of the oil pass over during the first 2—3 hours, the remaining 10 p. c. in the course of the next 3—4 hours. The crude oil is colored dark by the iron of the condenser. The small producers sell the crude oil obtained to wholesale druggists, who purify it by rectification.[4]

Properties. Gaultheria oil is a colorless, yellow or reddish liquid of a characteristic, strongly aromatic odor, which is distinctly different from that of the oil from *Betula lenta*. The sp. gr. is 1.180—1.187, the boiling temperature lies between 218 and 221°.

Gaultheria oil differs from the inactive betula oil in being slightly optically active, $\alpha_D = -0°25'$ to $-1°$. It yields a perfectly clear solution with 6 p. of 70 p. c. alcohol at 20°. This property, together with the sp. gr. and the optical rotation, allows of the ready detection of most of the adulterants, especially the frequent one with petroleum. Mixtures with the otherwise equal oil of *Betula lenta*, or with artificial methyl salicylate, can only be detected by the depression of the rotatory power below $-0°25'$.

---

1) Pharm. Rundschau, 13, p. 228.

2) Proc. Amer. Pharm. Assoc., 28, p. 269; 30, p. 184.

3) Am. Journ. Pharm., 51, p. 489.

4) Am. Journ. Pharm., 56, p. 264.

The details for further tests have been described under oil of birch bark, page 334.

Composition. The principal constituent of wintergreen oil, the methyl salicylate, was recognized by Cahours and Proctor[1] at the beginning of the forties. Quite contradictory views existed as to the nature of the other constituents. Power and Kleber[2] in 1895 cleared up the nature of the constituents other than methyl salicytate in wintergreen and also in birch bark oil. In order to prevent any possible decomposition, these chemists did not employ the saponification method, which had been used so far in the investigation of this oil, for the separating of the unknown compounds. They took advantage of the property of the methyl salicylate to form a solid salt with potassa, potassium methyl salicylate. The total amount of salicylic ester was removed from the oil by shaking repeatedly with 7.5 p. c. potassa solution. Only 1.05 p. c. of the oil did not combine with the potassa and remained as a semi-solid mass at ordinary temperature. This consisted of the following compounds: 1) A paraffin, C_nH_{2n+2}, which on account of its melting point, 65.5°, must be considered as tricontane, $C_{30}H_{62}$. 2) An aldehyde or ketone, which when separated in a pure form from its bisulphite compound, has an odor like oenanthic aldehyde, and on oxidation with potassium permanganate yields an acid, the silver salt of which corresponds to the formula $C_6H_9O_2Ag$. 3) An alcohol, $C_8H_{16}O$, boiling between 160 and 165°, which corresponds to the ketone or aldehyde mentioned. 4) An ester, $C_{14}H_{24}O_2$, boiling at 230—235° (135° at 25 mm.). This on saponification splits up into the alcohol, $C_8H_{16}O$, and the acid, $C_6H_{10}O_2$, resulting by oxidation from the ketone. The alcohol and also the ester possess the very penetrating characteristic odor, by which wintergreen oil is distinguished from artificial methyl salicylate.

338. Oil of Gaultheria Punctata.

The leaves of *Gaultheria punctata* Blume (*G. fragrantissima* Wallich, *G. fragrans* Don, *Arbutus laurifolia* Hamilton,[3] give upon distillation 1.15 p. c. of a volatile oil, which seems to be identical with that of wintergreen. De Vrij[4] in 1859 prepared the oil from plants which he found in the exhausted volcano Patoea on Java, and in 1871 ascertained that it consisted almost exclusively of methyl salicylate.

---

1) See p. 586, footnotes 4 and 5.

2) Pharm. Rundschau, 13, p. 228. Comp. also Kremers and James, Pharm. Review, 16, p. 105.

3) Sawer. Odorographia, vol. 2, p. 340.—

4) Pharm. Journ., III, 2, p. 503.

The same oil distilled by de Vrij was again investigated in 1879 by Köhler[1] apparently without the knowledge of the statements of its composition made by de Vrij in 1871. The inactive oil boiled at 221—222°; on saponification an amount of salicylic acid, melting at 155—156°, corresponding to the pure methyl ester, was obtained.

Andromeda leschenaultii, growing profusely in the Nilgiri mountains in India, in the leaves of which Broughton[2] in 1867 found methyl salicylate, is probably identical with *G. punctata* Bl.

339. Oil of Gaultheria Leucocarpa.

According to de Vrij[3] the fresh leaves of *Gaultheria leucocarpa* Blume, indigenous to Java, yield on distillation 0.012 p. c. of an oil consisting chiefly of methyl salicylate. Köhler[1] investigated the same oil and confirmed the result obtained by de Vrij. It was inactive and boiled at 221—223°.

340. Primula Root Oil.

The root of *Primula veris* L. (Family *Primulaceae*), contains, besides cyclamin, a glucoside probably identical with saponin, the so-called primula camphor (Mutschler,[4] 1877) which can be obtained by distillation. It separates from the distillate in fine, shining, hexagonal laminae, or as a semi-solid mass, has a fennel- or anise-like odor, and a taste at first burning, then sweetish, fennel-like.

Primula camphor is only difficultly soluble in water; the aqueous solution is colored deep violet to violet-blue by ferric chloride. Its elementary composition corresponds to the formula $C_{22}H_{24}O_{10}$. Upon oxidation with chromic acid, carbon dioxide and salicylic acid resulted; by boiling with potassa there resulted, besides salicylic acid, a very small amount of a second acid, which gave a beautiful dark blue coloration with ferric chloride.

341. Oil of Jasmine.

The general method of obtaining the volatile oils by water or steam distillation is not applicable to a large number of blossoms, as at best only traces of the volatile oil are obtained by it. To these belongs the jasmine blossom of *Jasminum grandiflorum* L. (Family *Oleaceae*). In

---

1) Berichte, 12, p. 246.

2) Pharm. Journ., III, 2, p. 281.

3) Pharm. Journ., III, 2, p. 508.

4) Liebig's Annalen, 185, p. 222.—Comp. also Hünefeld (1836), Journ. f. prakt. Chem., 7, p. 57, and 16, p. 111.

order to isolate the odorous principles of the jasmine blossoms, Verley[1] shook the jasmine pomade obtained by *Enfleurage à froid* with vaseline oil and exhausted the oil from this by treatment with acetone. On evaporation of the acetone a reddish colored oil smelling intensively of jasmine remained behind. Its principal portion came over at 100—101° under 12 mm. pressure and had the composition $C_9H_{10}O_2$. Upon oxidation the compound yielded benzaldehyde, benzoic acid and formaldehyde. By boiling with oxalic acid solution, styrolene alcohol, $C_6H_5.C_2H_3(OH)_2$, was formed. This led to the conclusion that the body was the methylene acetal of phenyl glycol $C_6H_5.C_2H_3{<}^{-O}_{-O}{>}CH_2$. Verley called the compound jasmal and considered it as the odorous principle of the jasmine blossoms.[2]

Jasmal can be prepared synthetically by heating phenyl glycol with formaldehyde and sulphuric acid. It boils at 101° under 12 mm., at 218° under atmospheric pressure. Sp. gr. 1.1334 at 0°.

Hesse and Müller[3] reached entirely different conclusions in their investigation of the oil, which had been prepared in a manner similar to Verley's and purified by steam distillation. They obtained from each kilo of jasmine pomade 4—5 g. of oil having the following properties: sp. gr. 1.007—1.018; $\alpha_D = +2°30'$ to $+3°30'$. The ester-content was 69.1 to 73 p. c., calculated as benzyl acetate, and 90.3 to 95.4 p. c., calculated as linalyl acetate.

According to Hesse and Müller jasmine oil contains large amounts of benzyl acetate, as well as linalool and its acetate. According to a quantitative determination made by a method worked out by them for this purpose they found the following approximate composition for jasmine oil:

65.0 p. c. Benzyl acetate.
7.5 " Linalyl acetate (including possibly other alcohol esters).
6.0 " Benzyl alcohol.
5.5 " other odorous principles.
16.0 " Linalool (eventually also other constituents).

Benzyl acetate, $C_6H_5CH_2OCH_3CO$, boils at 215—216° and has the sp. gr. 1.069.[3]

---

1) Compt. rend., 128, p. 314; Bull. Soc. chim., III, 21, p. 226.
2) According to Verley the homologues of jasmal, the ethylidene, isobutylidene and amylidene acetals of phenyl glycol have a jasmine-like odor.
3) Berichte, 32, pp. 565, 765.

Benzyl alcohol, $C_6H_5CH_2OH$, boils at 88° under 9 mm. pressure. For identification the phenyl urethane melting at 77—79° is well suited.[1]

The jasmal of Verley was not present in the oil investigated by Hesse and Müller. As the phenyl glycol methylene acetal is stable toward alcoholic potassa, it ought to be possible to detect its presence after the saponification of the oil. This, however, was not the case, nor did so much as a trace of styrolene alcohol result by boiling the oil with oxalic acid solution. On the ground of these observations Hesse and Müller came to the conclusion that the volatile oil of jasmine blossoms contained no phenyl glycol methylene acetal.

Schimmel & Co.[2] in their investigation of jasmine oil in 1895 also found benzyl acetate, benzyl alcohol and linalyl acetate, but no compound of the properties of jasmal.

More recently, Hesse[3] has found three further constituents:

2.5 p. c. Indol, C_8H_7N.

0.5 p. c. Anthranilic acid methyl ester, $C_8H_9NO_2$.

3.0 p. c. Jasmone, $C_{11}H_{16}O$.

Jasmone, $C_{11}H_{16}O$ has an intensive, pleasant jasmine odor when diluted. It boils at 257—258° at 755 mm. and has a sp. gr. of 0.945 at 15°. It yields an oxime, by means of which it was separated from the oil, melting at 45°, and a mixture of semicarbazones melting at 200—204°, from which by recrystallization a compound was obtained melting at 204—206°.

312. Oil of Rhodium.

Oleum Ligni Rhodii.—Rosenholzöl.—Essence de Bois de Rose,[4] ou de Rhodes.

The so-called rosewood, i. e. the wood of the roots of the shrubs *Convolvulus scoparius* L. and *C. floridus* L. (Family *Convolvulaceae*), indigenous to the Canary Islands, is generally reported as the material used for the preparation of oil of rhodium.

The oil of rhodium at present in the market is often nothing more than a mixture of rose oil with sandalwood or cedar wood oil.

Oil of rhodium was investigated by Gladstone[5] in 1864. The source of this oil can naturally not be be determined at present. It was viscid, had the sp. gr. 0.906 at 15.5° and the rotatory power

[1]) Bericht von S. & Co., Apr. 1899, p. 27.

[2]) Bericht von S. & Co., Apr. 1899. p. 27.

[3]) Berichte, 32, p. 2611.

[4]) The wood designated *Bois de rose femelle* or *mâle* by the French is Guayana linaloe wood. Comp. p. 498.

[5]) Journ. Chem. Soc., 17, p. 1; Jahresb. f. Chemie, 1863, p. 546.

— 16° in a 250 mm. tube. Four fifths of it consisted of a hydrocarbon $C_{10}H_{16}$ (more probably $C_{15}H_{24}$) boiling at 249°, which had an odor of sandalwood and rose.

The source of an oil distilled by Schimmel & Co.[1] in 1887 is also unknown. The oil had a fine golden-yellow color, with a pleasant rose-like odor. It solidified at + 12° to a mass of needle shaped crystals.

A rose wood from Tenerife[2] recently used by the same firm for distillation, corresponded fairly well with the description of the root wood of *Convolvulus scoparius* L. The odor of the oil did not fulfill the expectations. Sp. gr. 0.951 at 15°; $\alpha_D = +1°30'$; saponification number before acetylization = 0, after acetylization = 151.3. The oil dissolved with a slight turbidity in 10 p. of 95 p. c. alcohol.

343. Oil of Verbena.

The true verbena oil from the leaves of *Verbena triphylla* L. (*Aloysia citriodora* Ort., *Lippia citriodora* Kth., Family *Verbenaceae*) cultivated as an ornamental plant in Spain, northern France and Central America is, as its high price stands in no relation to its true value, no regular article of commerce. It can in most cases be replaced by the much cheaper lemon-grass oil which is similar in odor, and has therefore been called East Indian verbena oil. As genuine verbena oil is so difficult to obtain the statements concerning it have to be taken with some reserve.

Verbena oil has a fine, very pleasant, lemon-like odor, resembling that of lemon-grass oil. An oil distilled in Grasse, according to the statement of the manufacturers, from the leaves of *Lippia citriodora* (yield 0.09 p. c.), had the following properties: Sp. gr. 0.900; $\alpha_D = -12°38'$; soluble in 1—5 volumes of 90 p. c. alcohol, on the further addition of alcohol it became turbid. It contained 35 p. c. of aldehyde, which, as was shown by the preparation of β-citryl naphtho cinchoninic acid (m. p. 195—197°), was citral (Schimmel & Co.)

An oil described as "Oil of true Vervain,"[3] the sp. gr. of which is given as 0.902 and α_D as —12.7°, agrees very well with the above. It was not soluble in 10 p. of 80 p. c. alcohol, but in equal parts of 90 p. c. alcohol, and contained 28 p. c. of aldehyde. After acetylization a decided odor of linalyl acetate was noticed. By saponification of the acetylized oil an alcohol content ($C_{10}H_{18}O$) of 30 p. c. was determined.

1) Bericht von S. & Co., Apr. 1887, p. 28.
2) Bericht von S. & Co., Apr. 1899, p. 41.
3) Chemist and Druggist, 50, p. 218.

Umney[1] reports on an oil distilled in Dunolly (Victoria, Australia) "probably" from *Lippia citriodora.* The sp. gr. after removing the alcohol which had been added, was 0.894, $\alpha_D = -16°$. It contained 74 p. c. of citral.

An oil early investigated by Gladstone[2] (1864) and said to have come from *Aloysia citriodora* was much lighter, its sp. gr. at 15.5° being 0.881; its angle of rotation in a 250 mm. tube was —6°.

344. Oil of Lantana Camara.

The volatile oil of *Lantana camara* L. (Family *Verbenaceae*), widely distributed as a weed in Java, has been distilled in the botanical garden at Buitenzorg. Its odor is not especially pleasant, has a sp. gr. of 0.952 and turns the ray of polarized light 0° 24′ to the left in a 100 mm. tube.[3]

345. Oil of Vitex Trifolia.

The leaves of *Vitex trifolia* L. (Family *Verbenaceae*) are used in India for bathing purposes and as a remedy against various diseases. They contain a volatile oil, which was prepared in the botanical garden at Buitenzorg.[4]

Its odor is pleasantly aromatic, somewhat camphor-like. The latter property is due to cineol, which was shown to be present by the iodol-cineol reaction.

346. Oil of Rosemary.

Oleum Rorismarini.—Rosmarinöl.—Essence de Romarin.

Origin and History. The rosemary bush, *Rosmarinus officinalis* L. belonging to the *Labiatae*, grows in the Mediterranean countries and islands from Greece to Spain.

The first mention of the distillation of rosemary is found in the writings of Arnoldus Villanovus of the thirteenth century. He distilled, probably for medicinal purposes, turpentine oil and rosemary oil. An alcoholic distillate of both or only of rosemary was in use for centuries as the first popular perfume under the name of Hungarian water. The distillation of the oil is described more fully by Raimund Lullus, a disciple of Villanovus.

1) Pharm. Journ., 57, p. 257.
2) Journ. Chem. Soc., 17, p. 1; Jahresb. f. Chem., 1863, pp. 546 and 549.
3) Bericht von S. & Co., Oct. 1896, p. 77.
4) Bericht von S. & Co., Oct. 1894, p. 74.

Oil of rosemary was a much used oil in the middle ages and is often mentioned in the writings of that period. It is described in the index of the Compendium of Saladin at the end of the fifteenth century and in the works of Brunschwig, Ryff, Gesner, Porta, and others, and is mentioned in drug and spice ordinances of the fifteenth century. In several of the treatises on distillation and in medical works of the fourteenth and fifteenth centuries, an empyreumatic oil of rosemary is also mentioned.

One of the first investigations of rosemary oil was made by the Parisian apothecary Geoffroy[1] in 1720. A century later it was more fully investigated by Saussure[2] and in 1837 by Kane.[3] Cartheuser in 1734 determined the yield of oil.[4]

The so-called rosemary camphor was first noticed by Kunkel of Berlin in 1865[5] and a century later (1785) by Arezula of Cadix.[6] Proust prepared it in 1800.[7]

Preparation.[8] In commerce, two rosemary oils are principally distinguished, the Italian and the French.

The Italian or more correctly the Dalmatian rosemary oil is obtained on the islands of Lissa, Lesina and Solta on the Dalmatian coast, where the rosemary grows wild and covers large tracts of land. It is surprising that rosemary grows neither on the mainland nor on the neighboring islands Brazzo, Curzola, Melada and Lagosta. The most pleasant smelling rosemary is that on Solta, where it is, however, more and more suppressed by vineyards, so that distillers come but seldom to this island for the preparation of the oil.

Most of the oil is produced in Lesina, which is also the commercial center for rosemary oil, Lissa is second in importance. Both islands are covered with bushes about a meter in height. The "Rosemary forests" are the property of the community, which gives the license for distillation to the highest bidder at auction. The use of the forests is regulated by law, so that a full harvest is made only once in three years; in the two following years very little oil is distilled. The distillation takes place in July and August, consequently after the flowering period, which lasts from February to April. After the cut twigs have been

1) Mem. de l'Acad. de sc. de Paris, 1721, p. 168.
2) Ann. de Chem. et de Phys., II, 18, p. 278.
3) Trans. of the Roy. Irish Acad., 18, p. 135. — Journ. f. prakt. Chem., 15, p. 156.
4) Elementa Chymiae, II, pp. 88 & 106.
5) Probierstein. p. 397.
6) P. 89; Resultado de las experiencias, p. 8.
7) Trommsdorff's Journ. d. Pharm., 8, II, p. 221.
8) Dingler's Polytechn. Journal, 229, p. 466.—Bericht von S. & Co., Oct. 1896, p. 69.

dried for eight days in the sun, the leaves are stripped off and distilled with water vapor. The distilling apparatus is very primitive, old copper whisky stills being mostly employed.[1] From the port Cittavechia the oil is shipped in tin cans to Triest. From here, often adulterated with turpentine oil, it enters into the world's commerce.

No statistics of the production of the Dalmatian rosemary oil exist. According to inquiries made at the place of production, 20,000 k. of oil are said to be obtained in those years when a full harvest is made. This estimate agrees with a statement made by Flückiger[2] in 1884.

As far as the yield of oil is concerned, it has probably been but seldom determined in Dalmatia, as it may be assumed that the distillers do not weigh the material before putting it into the still. Schimmel & Co.[3] obtained from Dalmatian rosemary leaves 1.4—1.7 p c. of oil.

The French rosemary oil is finer in odor and correspondingly higher in price. It is principally obtained in the Départements du Gard, du Hérault, de la Drôme, des Alpes Maritimes and des Basses Alpes. Here the erect shrub grows to a height of 2 m. and together with *Thymus vulgaris* forms the brushwood of clearings.[2] The distillation is done in itinerant stills in the same manner as described under lavender oil (p. 602). Schimmel & Co.[4] obtained by distillation from the dry French rosemary leaves 2 p. c., from the flowers 1.4 p. c. of oil. Whether the better quality of the French oil depends on the plant itself or on the more careful selection of the material is difficult to state.

Inasmuch as rosemary is also found in Spain, the oil is now and then distilled in that country.[5] It appears, however, to be almost always adulterated with turpentine oil. The properties of the pure Spanish oil are the same as those of the Dalmatian or French oil, as was shown by the recent examination of a Spanish oil in the laboratory of Schimmel & Co.

The English rosemary oil, which is obtained in Mitcham and Market Deeping in small amounts from cultivated plants, is no more important commercially than is the Spanish oil.[6]

Properties. Rosemary oil is a colorless or slightly greenish-yellow liquid of a penetrating camphor-like odor and an aromatic, bitter,

---

1) See p. 67.
2) Archiv d. Pharm., 222, p. 476.
3) Archiv d. Pharm., 235, p. 586; Bericht von S. & Co., Oct. 1897, p. 54.
4) Bericht von S. & Co., Oct. 1898, table in appendix, p. 84.
5) Bericht von S. & Co., Oct. 1889, p. 55.
6) Comp. Holmes, Pharm. Journ., III, 12, p. 288; III, 20, p. 581.— Sawer, Odorographia, vol. 1, p. 370.

cooling taste. The sp. gr. of the French as well as the Italian oil is always higher than 0.900 and sometimes rises to 0.920. The ray of polarized light is always rotated to the right, $\alpha_D = +0°45'$ to $+15°$. The first 10 p. c. obtained by fractional distillation are likewise dextrogyrate. One part of oil gives a clear solution with ½ and more parts of 90 p. c. alcohol, and sometimes with 2, but often as many as 9—10 parts of 80 p. c. alcohol are required. Saponification number 12—20.

The oils distilled by Schimmel & Co. from five different shipments of Dalmatian rosemary leaves had the following properties: Sp. gr. 0.904—0.913; $\alpha_D = +3°40'$ to $+8°52'$.[1] Two oils were distilled in Leipzig from French material.[2] 1) From flowers: sp. gr. 0.920; $\alpha_D = +1°38'$. 2) From leaves: sp. gr. 0.914; $\alpha_D = +14°35'$.

The content of bornyl acetate in two authentic oils was found to be 5.4 and 5.8 p. c. The amount of borneol determined by acetylization was 16.8 and 18.8 p. c.

A Spanish oil recently investigated in the laboratory of Schimmel & Co. had the sp. gr. 0.905 and $\alpha_D = +7°40'$. It was soluble with slight turbidity in 10 parts of 80 p. c. alcohol. The angle of rotation, α_D, of the first 10 p. c. going over on distillation was $+4°1'$. Another sample,[3] investigated later, had the sp. gr. 0.932, $\alpha_D = +17°37'$ at 20°. Spanish distillates which had been investigated at an earlier time[4] were lighter in sp. gr. and laevogyrate, and probably were adulterated with turpentine oil.

The English oil obtained from cultivated plants shows differences in properties, especially in rotatory power. The sp. gr. of three English oils was found by Cripps[5] (1891) to be 0.901, 0.911 and 0.924, the rotatory power, α_D, determined on one of these oils was $-9°35'$. Symes[6] (1879) determined on another, doubtless adulterated English oil the sp. gr. 0.881 and the angle of rotation $\alpha_D = -16°47'$. As the English oils are commercially unimportant, all laevogyrate oils found in commerce can safely be declared as adulterated.

Composition. The following compounds have so far been found in rosemary oil: 1) pinene, 2) camphene, 3) cineol, 4) camphor, 5) borneol.

Up to the present all investigations have been directed toward individual constituents. A thorough investigation of the oil as a whole may therefore result in finding still other substances.

1) Bericht von S. & Co., Oct. 1897, p. 56.
2) Ibidem, Apr. 1891, p. 41.
3) Ibidem, Apr. 1900, p. 40.
4) Ibidem, Oct. 1889, p. 55.
5) Pharm. Journ., III, 21, p. 937.
6) Pharm. Journ., III, 10, p. 212.

Pinene. According to the statement of Bruylants[1] (1879) oil of rosemary contains 80 p. c. of a laevogyrate hydrocarbon $C_{10}H_{16}$ boiling at 157—160°. From the low sp. gr. 0.885 and the laevo-rotation, as well as the large amount of the terpene found, it is evident that the oil investigated was adulterated with French turpentine oil. On account of the frequent adulteration with turpentine oil it became a matter of interest to know whether pure rosemary oil contained pinene or not. Gildemeister and Stephan,[2] therefore, investigated the lower boiling fractions of an oil distilled by themselves from Dalmatian rosemary leaves. The portion boiling at 156—158° after repeated fractionation (sp. gr. 0.867, $\alpha_D = +2°30'$) gave on treating with amyl nitrite and hydrochloric acid in glacial acetic acid solution a nitrosochloride. This yielded with benzylamine pinene nitrolbenzylamine (m. p. 122—123°), from which it follows that pinene—probably a mixture of d- and l-pinene—is a constituent of rosemary oil.

Camphene. The fraction boiling at 160—162° (sp. gr. 0.875, $\alpha_D = -0°45'$) was treated with glacial acetic and sulphuric acids. After saponification of the product of the reaction, isoborneol, of the melting point 211—212°, was obtained. The latter was reconverted into camphene, boiling at 159—160° and melting at about 50°, by boiling with zinc chloride in benzene solution. According to this a second terpene, camphene, is contained in rosemary oil, and judging from the slight rotation of the fraction, it is inactive camphene.

A commercial rosemary oil distilled in Dalmatia gave the same results; pinene and camphene were identified. The first runnings (from 40 k.) of this oil had an odor of acetaldehyde, boiled below 150°, decolorized fuchsin sulphurous acid and was partly soluble in water. The insoluble oil was hydrated with glacial acetic and sulphuric acids, whereby a decided odor of linalyl acetate was developed. This behavior indicates the presence of olefinic terpenes, similar to those occurring in bay, hops, and origanum oils.[2]

Cineol. In the fraction boiling at 176—182° Weber[3] in 1887 found cineol, $C_{10}H_{18}O$, and isolated this body by means of its hydrochloric acid addition product. The identification was made by preparing the dipentene tetrabromide, $C_{10}H_{16}Br_4$, (m. p. 123.5—124°) as well as dipentene dihydriodide, $C_{10}H_{18}I_2$, (m. p. 78.5—79°).

---

1) Journ. de Pharm. et Chim., IV, 29, p. 508; Pharm. Journ., III, 10, p. 327; Jahresb. f. Chemie, 1879, p. 944.

2) Archiv d. Pharm., 235, p. 585.

3) Liebig's Annalen, 238, p. 89.

By conducting hydrochloric acid gas into the ethereal solution of fraction 171—176° Weber obtained dipentene dihydrochloride melting at 49—50°. This compound owes its formation either to cineol, or to dipentene, but as the fraction mentioned contained according to the analysis considerable terpene, its formation is probably due to dipentene (terebene of Bruylants?).

Camphor and borneol. Camphor was first observed in rosemary oil by Lallemand[1] in 1860. Montgolfier[2] in 1876 showed that the rosemary camphor was a mixture of the dextro- and laevogyrate modifications. Bruylants[3] found that the stearoptene crystallizing from the high boiling fractions of the oil consisted not only of camphor, but also of borneol. In the oil, which, as mentioned above, was strongly adulterated, Bruylants determined 4—5 p. c. of borneol and 6—8 p. c. of camphor. Pure oil must therefore contain considerably more of these compounds. Gildemeister and Stephan[5] determined the borneol content by acetylization and found in two cases 16.8 and 18.8 p. c. It was assumed that rosemary contained no other alcohols, such as linalool or geraniol.

According to Haller[4] (1889) camphor is separated from borneol by converting the latter into bornyl acid succinate by heating with succinic acid. The borneol contained in rosemary oil is, as was shown by Haller, like the camphor, a mixture of both optical modifications.

Judging from the saponification number obtained by boiling with potassa, rosemary oil contains small amounts of esters, presumably of borneol. The acid combined with the alcohol has not yet been determined.

Examination. In testing rosemary oil, especial attention should be paid to the specific gravity and rotatory power which should agree with those mentioned under Properties. Turpentine oil, no matter of what source, will cause a decrease in the specific gravity. French turpentine oil is recognized, providing the amount is not insignificant, by the inversion of the angle of rotation to the left. In order to recognize small additions, which do not cause inversion to the left, and do not lower the sp. gr. below 0.900, from 100 cc. of oil 10 cc. are slowly distilled off in the flask described on p. 190 and the distillate tested in the polariscope.[5] With pure rosemary oil the lowest boiling 10 p. c. will always be found dextrogyrate, while if only small amounts of French turpentine oil are present it will be laevogyrate.

---

1) Liebig's Annalen, 114, p. 197.
2) Bull. Soc. chim., II, 25, p. 17.
3) See p. 598, footnote 1.
4) Compt rend., 108, p. 1308.
5) Archiv d. Pharm., 235, p. 585.

Fractions of camphor oil are also often used for adulteration.[1] These, however, usually affect one of the characteristic properties of the rosemary oil. Either the rotatory power (which is usually increased) or the specific gravity, or the solubility in 80 p. c. alcohol is changed.

347. Oil of Lavender.

Oleum Lavandulae.—Lavendelöl.—Essence de Lavande.

Origin and History. The species of lavender used since antiquity on account of their pleasant odor, are indigenous to the northern Mediterranean countries, but flourish especially on the slopes of the French Cevennes and Sea Alps. Whereas formerly the oils of the different species of lavender were known by the general name of spike oil, they have, since a little more than a century ago, been separated into lavender oil and spike oil.

The different lavender species were formerly gathered indiscriminately for oil distillation. In the eighteenth century they were more accurately determined and distinguished botanically. The species determined as *Lavandula spica* $_\alpha$ by Linné, (*L. officinalis* Chaix, *L. angustifolia* Mönch, *L. vulgaris* $_\alpha$ Lamarck, *L. vera* De Candolle) yields the lavender oil.

The spike oil, which is decidedly different in aroma and constituents, is obtained from the more tender *L. spica* D. C. (*L. vulgaris* β Lamarck, *L. latifolia* Villars) growing in the lower hilly and coast districts.

The history of the lavender species and their distillates is given under the earlier known spike oil (p. 607).

Preparation. The French lavender oil is obtained in the higher mountainous regions of southern France in the Départements des Alpes Maritimes, des Basses Alpes, du Hérault, de la Drôme, du Gard and de Vaucluse. As the lavender blossoms cannot be transported, the distillation is carried on as near as possible to the place of collection, as a general rule in portable stills. At the beginning of the flowering period in July, the distillers take their apparatus to the mountains on mules. In the neighborhood of a spring or creek the still is erected on a fire place constructed of stones, and heated with wood, which is abundant. The collection is always begun in the lowest regions where the lavender blooms first. When all obtainable material has been collected and distilled in the neighborhood of the place of distillation the apparatus is carried to a higher point. The oil distilled in the lower regions is considered inferior in value to that distilled toward the end of September at an altitude of 1,500 m.

[1]) Bericht von S. & Co., Apr. 1898, p. 46.

Fig. 78.
Distillation of Lavender Oil near Escragnolles (Département des Alpes Maritimes), Southern France.

The accompanying illustration (fig. 78) shows such a *Distillerie ambulante*[1] in the neighborhood of the village Escragnolles, on the road from Grasse to Paris, thirty kilometers from Grasse. The two copper stills are supported on a fire place made of stones laid in the form of a horse shoe. The condenser consists of an old petroleum barrel, from which the head has been removed, and into which the condensing worm is fitted. The condensing water is furnished by the neighboring spring. In the foreground a laborer is busy turning the pile of lavender flowers with a hay fork, to prevent fermentation. The exhausted blossoms are spread out behind the still. After completion of the distillation the laborer removes the still from the fire, empties it by turning it over, and replaces it by another which had meanwhile been filled.

H. Laval[2] in 1886 made an interesting and exhaustive report on the distribution, collection and distillation of lavender in the Département de Vaucluse. Although not of very recent date, some of his figures may here be mentioned, in order to give an idea of the importance of this industry in southern France.

On the southern slope of the Ventoux mountains lavender is found at a height of 700—1,150 m., together with thyme and satureja. On the northern slope the lavender begins at 450 m., but rises here no higher than 900 m. The ground covered by the plants (*lavandières*) on the Ventoux is approximately 11,000 hectares (27,000 acres); of these, 7,000 hectares (17,200 acres) are property of the communities, the remainder is of private ownership. At the time of flowering, in July and August, the men, women and children go up the mountains. The flower clusters are cut off above the leaves with a sickle and are carried on the head in bundles to the places of distillation. The collection of the lavender is commonly free on the properties of the communities, some, however, make an annual charge of one franc for each family. The amount of fresh lavender flowers collected annually in the Ventoux is estimated at 1,700,000 k. Of these 1,200,000 k. are used for the distillation of the oil (about 6,000 k.) the balance is dried.

Laval describes from personal inspection three different methods of distillation. In a small factory at Sault the distillation was effected by means of steam which was generated in a separate vessel and fed four cylindrical, copper stills. Each had a capacity for 150 k. of fresh

1) See pp. 26 and 67.

2) Journ. de Pharm. et de Chim., V, 13, pp. 593 and 649.

flowers. The warm water from the condenser was used for feeding the boiler. A distillation lasted 1¼ hours. Each still was filled 14 times in 24 hours and thus 8,400 k. of material were distilled daily.

In a second factory in Villes the distillation was also with steam from two egg shaped stills, each of which held a charge of 100 k. of fresh flowers.

A third factory in Bédoin had cylindrical stills, 3 m. high and 1½ m. in diameter, which were heated by direct fire; the flowers were supported on a false bottom, so that they were not in contact with the water but only with the steam. 75 k. were placed on the lowest false bottom, then a second false bottom was put in which carried the same amount of flowers. Alternately a false bottom and flowers followed until the still contained 250 k. Five distillations were made daily.

In the third kind of distillation the flowers are put into the still with water and without sieve bottoms. This method is common to the itinerant stills, and is the one most generally used.

The most rational method of preparation is probably the one with the false bottoms, as this is the most suitable for delicate oils. By the distillation with steam as well as with the water without false bottom no doubt more of the valuable ester is decomposed than in the method used at Bédoin. According to Laval in order to obtain 1 k. of oil, 200 k. of fresh flowers are necessary, which corresponds to a yield of 0.5 p. c. From dried French flowers Schimmel & Co. obtained 1.2 p. c., from German flowers as much as 2.8 p. c. of oil.[1]

The English lavender oil industry is quite insignificant in comparison with the French. Whereas in France only the wild plant is used for distillation in England the cultivated plant is used exclusively. The lavender plantations[2] are found in Surrey county in Mitcham, Carshalton and Beddington, further in Canterbury (Kent), Hitchin (Hertfordshire) and in Market Deeping (Lincolnshire). The distillation, which usually begins in the first weeks of August, is done in the same stills used for the peppermint. The yield from the fresh flowers is given as 0.8 to 1.5 p. c. Both figures appear to be rather high.

---

1) Bericht von S. & Co., Oct. 1898. Table in the appendix, p. 24.

2) More detailed accounts of the cultivation and distillation of English lavender can be found in the following journals and works: Holmes, Pharm. Journ., III, 8 (1877), p. 801; Flückiger and Hanbury, Pharmacographia, p. 477; Sawer, Pharm. Journ., III, 20 (1890), p. 659, and Odorographia, vol. I, p. 356; Chemist and Druggist, 39 (1891), p. 898; Pharm. Journ., 58 (1897), p. 52; Brit. and Colon. Druggist, 34 (1898), p. 388.

Properties. French oil of lavender is a yellowish or yellowish-green[1] liquid of the pleasant, characteristic odor of the lavender flowers. and of a strong aromatic, slightly bitter taste. Sp. gr. 0.885—0.895; $\alpha_D = -3$ to $-9°$. It is soluble to a clear solution in three and more parts of 70 p. c. alcohol. The linalyl acetate content is as a rule 30—40, seldom 45 p. c. According to the amount of ester the lavender oils can be divided into two classes. First, those oils with 36 and more p. c. of ester. They come from the highest regions of the southern French Alps, possess the finest and most intense aroma, and consequently bring the highest prices. Second, those with 30—36 p. c. of ester. Oils with less than 30 p. c. of linalyl acetate are mostly adulterated, much more seldom is the low ester content due to imperfect distillation by which a part of the ester was decomposed.

The English lavender oil differs from the French by a camphor- or cineol-like odor which accompanies it, as well as by its low ester content. Sp. gr. 0.885 to 0.900; $\alpha_D = -1$ to $-10°$. The solubility is the same as that of the French oil, the amount of linalyl acetate is only 5 to 10 p. c.

Composition. According to the earlier works of Saussure[2] (1817—1832), Proust and Dumas[3] (1833), the ordinary laurus camphor, $C_{10}H_{16}O$, had to be considered as a normal constituent of lavender oil. As recent investigations have shown that the oil from *Lavandula vera* D. C. contains no camphor, these statements refer no doubt to oils of other lavender species.[4]

The results of Lallemand[5] (1860) and of Bruylants[6] (1879) in part do not agree with the most recent investigations. Lallemand found a laevogyrate hydrocarbon $C_{10}H_{16}$ (?) boiling from 200 to 210°, and was the first to observe the presence of esters in lavender oil. According

---

[1]) The finest oils with a high ester content mostly have a green tint, whereas rectified oils are colorless. To rectify lavender oil is irrational, because this process decomposes in part the principal constituent, viz. the linalyl acetate. As a result the flavor of the rectified oil is inferior (Bericht von S. & Co., Oct. 1894, p. 80).

[2]) Ann. Chim. et Phys., II, 4, p. 318; II, 13, p. 273; II, 49, p. 225; Liebig's Annalen, 3, p. 168.

[3]) Liebig's Annalen, 6, p. 248.

[4]) In the literature quoted above, Spain is several times mentioned as the country from which the examined oil was obtained. As is shown in an article published by Charabot in 1897 (Bull. Soc. chim., III, 17, p. 378) the Spanish oil, of which the origin is not known, has different properties and is different in composition. Sp. gr. 0.912—0.916; $\alpha_D = +18°20'$ to $+16°25'$; ester content 3.15—3.4 p. c. The oils contained 44.5 to 50.5 p. c. of alcohols as determined by acetylisation. In the higher boiling fractions borneol (m. p. 204°) was found. The Spanish oil, therefore, resembles oil of spike more closely than the oil of lavender.

[5]) Liebig's Annalen, 114, p. 198.

[6]) Journ. de Pharm. et Chim., IV, 30, p. 139.

to Bruylants the oil contains a terpene, $C_{10}H_{16}$, boiling at 162°, which yields a solid monohydrochloride. The principal part (45 p. c.) is said to be a mobile oil, the composition of which corresponds to a mixture of borneol. $C_{10}H_{18}O$, and camphor, $C_{10}H_{16}O$. On cooling to — 25° no solid constituents separated out; on oxidation with potassium dichromate and sulphuric acid, ordinary camphor was obtained. This would seem to show conclusively that Bruylants could not have had genuine lavender oil for examination.

The most recent investigation of French lavender oil was made by Bertram and Walbaum.[1] According to it, the principal constituent is l-linalyl acetate, which is present to the extent of 30—45 p. c. Besides linalyl acetate the ester of butyric acid and perhaps also those of propionic and valerianic acids are present in small amounts. The presence of formic acid could not be shown.

Linalool is present in lavender oil not only as ester, but also in the free state. After acetylization, 67 p. c. of ester were found in an oil which originally contained only 34 p. c. As linalool cannot be quantitatively determined by acetylization, the results always being too low, it might be assumed that the amount of linalool present in lavender oil is about as great in the free state as in the form of ester.

While an oil distilled by Schimmel & Co. from the dried flowers contained no pinene, a small amount of pinene (nitrosochloride, m. p. 102°, nitrolbenzylamine, m. p. 122—123°) was found in the first runnings boiling at 160—170° of a larger quantity of French oil. Cineol is likewise only present in traces. In one case it could only be detected after the linalool contained in the proper fraction had been destroyed by heating with formic acid.[2]

If larger amounts of pinene can be separated from lavender oil, the suspicion of adulteration with turpentine is justified, especially when supported by the ester determination, the sp. gr., the optical rotation, and the solubility in 70 p. c. alcohol. On the other hand a larger cineol content in the French oil indicates adulteration with spike oil.

The linalool is accompanied in lavender oil by a second alcohol $C_{10}H_{18}O$, geraniol. From the fraction boiling at 110—120° under 13 mm. pressure by treatment with calcium chloride, an oil could be separated, from which the diphenyl urethane of geraniol, melting at 82° could be obtained.[3]

---

1) Journ. f. prakt. Chem., II, 45, p. 590.
2) Bericht von S. & Co., Oct. 1893, p. 25.
3) Bericht von S. & Co., Apr. 1898, p. 32.

The English oil of lavender has been investigated by Semmler and Tiemann.[1] From the first runnings of this oil they obtained a tetrabromide melting at 105°, and thus showed the presence of limonene. The fraction boiling at 85—91° under 15 mm. pressure, consisted of l-linalool, that boiling at 97—105° of l-linalyl acetate. In the last boiling fractions a sesquiterpene, $C_{15}H_{24}$, was found but not examined. The English oil differs from the French in its larger content of cineol,[2] as well as its low ester content, amounting to only 5—10 p. c.

Examination and Valuation. The value of lavender oil depends on its content of linalyl acetate. Although the amount of ester alone is not an absolute standard for the quality of the oil, the fineness and value of the oil stands, however, in direct ratio to it, providing, of course, that the oil is a normal distillate, free from any empyreumatic odor due to careless preparation. Such a case is, however, very unlikely to occur, as with an irrational distillation a part of the ester is destroyed. For this reason a carefully prepared oil must always show a comparatively high ester content. If the high linalyl acetate is partly decomposed during the distillation, the free acid acts detrimentally on the linalool and thus influences the odor also in this manner to an appreciable extent.

By adulterations of any kind the ester content is of course also lowered. For these reasons the quantitative saponification is absolutely necessary in testing the value of an oil. The method has been described in detail on p. 193.

In testing for foreign additions, the specific gravity, rotatory power and solubility in 70 p. c. alcohol must be considered.

The more common adulterants are turpentine oil, cedar wood oil, and spike oil. Turpentine oil decreases the specific gravity and the solubility in 70 p. c. alcohol. The presence of pinene can also be readily shown (see under Composition). Spike oil does not influence the solubility, but decreases, as it contains but a small amount of linalyl acetate, the ester content. Besides, spike oil differs by its larger cineol content.

There remains to be mentioned the adulteration of lavender oil with the ethyl ester of succinic acid, which, it is true, has been observed but once.[3] It is, however, worthy of mention, as the addition of comparatively small amounts of this succinic ester to lavender oil, produces a

---

1) Berichte, 25, p. 1186.
2) Bericht von S. & Co., Oct. 1894, p. 31.
3) Bericht von S. & Co., Apr. 1897, p. 25.

high saponification number and thus the amount of linalyl acetate is apparently increased. 8 p. of ethyl succinate give the same saponification number as 18 p. of linalyl acetate. Apart from the large, apparent increase in the ester content, the ethyl succinate is a dangerous adulterant, because it has only a faint odor, which is almost completely covered by that of the lavender oil, and also because it influences but slightly the solubility and rotatory power. The specific gravity of the succinic ester is, however, much higher than that of lavender oil and may thus indicate its presence.

In testing the lavender oil for the esters of succinic, oxalic or similar acids which may here be employed, the property of the acids to form difficultly soluble salts with barium is made use of.

For this purpose about 2 g. of the oil are saponified, the portion insoluble in water separated by shaking with ether and the aqueous solution neutralized with dilute acetic acid. The solution is diluted to 50 cc. and 10 cc. of a cold saturated barium chloride solution added. It is then warmed for two hours on a waterbath and allowed to cool. If a crystalline deposit is formed, the oil is to be considered adulterated, as the acids contained in normal lavender oil, acetic and butyric acids, give soluble barium salts.

348. Oil of Spike.

Oleum Spicae.—Spiköl.—Essence d'Aspic.

ORIGIN. *Lavandula spica* D. C. (*L. spica* β L., *L. vulgaris* β Lam., *L. latifolia* Vill., Ger. *Spiklavendel*), has about the same distribution in the Mediterranean countries as the true lavender. It grows, however, mostly in the lower mountainous regions, not exceeding 700 m., at an altitude where *Lavandula vera* D. C. just begins. The oil is distilled exclusively in southern France in the same manner as lavender oil. 160 k. of flowers yield 1 k. or 0.62 p. c. of oil (Laval[1]).

HISTORY. As was mentioned under lavender oil (p. 600) the distillates of various species of lavender have since antiquity been designated as spike oil. It was not until the end of the sixteenth century that lavender oil and spike oil were differentiated. In antiquity, probably the *Lavandula stoechas* L. indigenous to the Mediterranean coast and distinguished by its aromatic, violet-red blossoms, was commonly used for the preparation of spike oil. The spike or stoechas oil mentioned in the writings of Dioscorides, Pliny, Scribonius Largus and other contemporaries was in all probability only aromatized fatty oil, like the rose and spikenard oil and other aromatic oils much used in antiquity.

---

[1]) Journ. de Pharm. et Chim., V, 13, p. 599.

The distilled spike or spikenard oil was probably known in the fifteenth century. Besides cedar (turpentine) oil it is the only distilled oil mentioned by Brunschwig in his "Destillirbuch" of 1500 as *Oleum de Spica* from "Provinz" (Provence). Saladin also mentioned spikenard oil at the end of the fifteenth century.

Valerius Cordus mentioned in his Dispensatorium Noricum of 1543 only three distilled oils: turpentine oil, juniper oil and spike oil. Ryff described in his treatise on distillation published somewhat earlier, the distillation of "Spik und anderen fürnemen Olen" and added the statement that, "das Spiken- oder Lavendelöl gemeygklich aus der Provinz Frankreich gebracht wird in kleinen glässlin eingefasst und theuer verkaufft." In the sixteenth century the several species of lavender were cultivated in Germany and in England.

Gesner used the name spike oil only and described the distillation of the spike blossoms, whereas Porta at the end of the sixteenth century described also the distillation of lavender blossoms and especially emphasized the superiority of the oil from the French lavender. An interesting description of the preparation of the French spike or lavender oil is contained in the works of Demachy of 1773.

In medical works *Oleum spicae* has been mentioned as early as the thirteenth century. It was included in the first edition of the Dispensatorium Noricum of 1543, and *Oleum lavandulae* is also mentioned together with spike and other essential oils in the 1589 edition. The Pharmacopoea Augustana of Occo, contains up to 1613 only *Oleum spicae*, from that date on also *Oleum lavandulae*.

Spike oil is mentioned in the oldest drug and price ordinances of German cities, *Oleum lavandulae*, however, is not found until 1582 in the Frankfurt ordinance.

The statements of Demachy and other writers of his time agree with the assertions of later authors, that the spike oil found in commerce in the eighteenth century was probably only a distillate or mixture of turpentine and lavender oils.

The yield of oil from the distillation of spike and lavender blossoms appears to have been first mentioned by Lewes,[1] also by Cartheuser.[2] The so-called "lavender camphor" was observed in 1785 by Arezula,[3] and in 1800 by Proust. The first investigations of lavender oil were made by Saussure.[4]

1) The new Dispensatory, 1746.

2) Elementa chymiae, vol. 2, pp. 183 and 149.

3) Resultado de las experiencias, etc.

4) Annal. de Chim. et Phys., 4, p. 318; 18, p. 273; 49, p. 159.

PROPERTIES. Spike oil is a yellowish liquid, of a camphor-like odor, reminding at the same time of lavender and rosemary. The specific gravity lies between 0.905 and 0.915. Spike oil is always dextrogyrate. As a rule the angle of rotation α_D is up to +3°, rarely up to +7°. It is clearly soluble in 2—3 and more parts of 70 p. c. alcohol. The saponification number is about 15, corresponding to 5 p. c. of linalyl acetate (bornyl or terpinyl acetate, resp.).

COMPOSITION. The first constituent identified with certainty in spike oil is camphor (Kane,[1] 1838). This compound does not occur in genuine lavender oil, as was shown in the preceding article, although the older authors claim to have found it. It is, therefore, highly probable that a part of the earlier investigations were made on spike oil and not on lavender oil. Besides camphor, Bruylants[2] (1879) also found borneol in spike oil.

The oil was thoroughly investigated, partly by Bouchardat[3] alone in 1893, partly in co-operation with Voiry[4] in 1888. By repeated fractionation a very insignificant amount (0.2—0.25 p. c.) of an oil boiling about 160° was finally obtained, which yielded a solid hydrochloride melting at 129°. On boiling with alcoholic potassium acetate this was decomposed for the greater part, and gave dextrogyrate camphene, solidifying in the cold. Whether or not d-pinene is present together with the d-camphene, could not be definitely shown. Bouchardat believed that the portion of the solid hydrochloride which was not decomposed by boiling for a short time with alcoholic potassium acetate, was pinene hydrochloride. The fraction boiling about 175°, about 10 p. c. of the oil, solidified in a freezing mixture, and consisted of cineol, $C_{10}H_{18}O$, of which the hydrochloride and dibromide, $C_{10}H_{18}OBr_2$, were prepared.

The portion going over at about 200° was a mixture of l-linalool, d-camphor and d-borneol. The linalool (b. p. 198—199, $\alpha_D = -16° 44'$), was converted into geranyl acetate by boiling with acetic acid anhydride. The camphor oxime prepared from the camphor had all the properties of the oxime obtained from ordinary camphor. The borneol, after it had been separated from the camphor by means of benzoic acid anhydride, likewise turned the ray of polarized light to the right.

Spike oil may also contain terpineol, for Bourchardat obtained dipentene dihydrochloride on conducting hydrochloric acid gas into the

1) Journ. f. prakt. Chem., 15, p. 163. Comp. also Lallemand (1860), Liebig's Annalen, 114, p. 198.

2) Chem. Centralbl., 1879, p. 616.

3) Compt. rend., 117, pp. 53 and 1094.

4) Compt. rend., 106, p. 551.

fraction having the boiling point of terpineol. The formation of this hydrochloride cannot, however, be definitely traced to terpineol, as it might also have been formed from linalool or geraniol. The geraniol also has not been positively identified. Hydrochloric acid yielded with the fraction boiling at 145—160° in a vacuum and smelling like geraniol, a liquid hydrochloride of the properties of geranyl chloride.[1] The analysis of the portion of the oil going over at 250°, gave results agreeing with $C_{15}H_{24}$, which indicates a sesquiterpene.

EXAMINATION. Turpentine oil, with which spike oil is probably most often adulterated, is recognized by the lowering of the specific gravity, the diminution in solubility, and, when French turpentine is present in larger amount, by the inversion of the rotation to the left. If the amount of this oil is so small that only a decrease but no inversion of the angle of rotation takes place, while sp. gr. and solubility of the oil appear suspicious, the first 5—10 p. c. going over on fractional distillation should be tested in the polariscope.

As with rosemary oil (p. 594) the lowest boiling portions of spike oil are always dextrogyrate, while even a small addition of French turpentine oil produces laevorotation. Moreover, with genuine spike oil the amount of the terpenes boiling at about 160° is very small, so that the oil can also be declared adulterated (American turpentine oil, camphor oil) when at this temperature larger amounts of dextrogyrate terpenes distill over.

The addition of rosemary oil is more difficult to recognize. This is, however, much more rarely used as adulterant, as the difference in price is not great. Sp. gr., rotation and boiling temperature of spike and rosemary oils are about alike, only the solubility in 70 p. c. alcohol is different. Mixtures of these two oils often dissolve in 2—3 parts of 70 p. c. alcohol to form a clear solution, but it becomes turbid on the further addition of 70 p. c. alcohol.

Schimmel & Co.[2] have shown that spike oil contained more than 30 p. c., rosemary oil only up to 15 p. c. of alcoholic constituents which could be determined by acetylization. Umney,[3] therefore, suggests for judging the purity of the oil and for detecting the presence of rosemary oil, to acetylize the suspected oil, and to reject oils which show a content of less than 30 p. c. of alcohols as adulterated with

---

[1]) According to Tiemann (Berichte, 31, p. 832) geranyl chloride prepared in this way is a mixture of different isomeric chlorides and not a definite chemical unit.

[2]) Bericht von S. & Co., Oct. 1894, p. 65.

[3]) Chemist and Druggist, 52, p. 166.

rosemary oil. It must here be remembered that the alcoholic constituents of spike oil consist for the greater part of linalool, and that this body cannot, as is well known, be quantitatively acetylized, as it is partly decomposed with the formation of terpenes. Inasmuch as the acetylization result depends on the length of time the acetic acid anhydride reacts on the oil, the results obtained can be used for judging the purity of spike oil, only when the exact conditions under which the acetylization gives uniform results have been determined. Furthermore, the limits between which the apparent alcohol content thus determined varies with normal spike oil, ought to be ascertained by a series of experiments.

349. Oil of Lavandula Stoechas.

Lavandula stoechas L. is known in Spain as *Romero santo* (holy rosemary) (p. 607). Its volatile oil and also that of *Lavandula dentata* L. is there obtained for domestic use, by hanging the fresh, flowering plants with the flower heads downward in bottles, which are sealed and exposed to the sun. On the bottom a mixture of water and volatile oil collects, which is used as a styptic for washing wounds, and also for eruptions.

The odor of the oil does not in the least remind of lavender, but is rather more similar to rosemary, having the camphor-like properties of the latter. The sp. gr. of a sample of oil was 0.942. It boiled from 180—245°; the lower boiling portions contained cineol.[1]

350. Oil of Lavandula Dentata.

The oil of *Lavandula dentata* L. is very similar to that of *Lavandula stoechas* L. Its odor reminds strongly of rosemary oil and camphor. It has the sp. gr. 0.926, distills almost completely between 170—200° and contains cineol.[1]

351. Oil of Lavandula Pedunculata.

An oil of *Lavandula pedunculata* Cav. coming from Portugal, is described by Schimmel & Co.[2] as follows: "The oil has a difficultly definable, not pleasant odor, and is, therefore, useless for practical purposes. Sp. gr. 0.939; $\alpha_D = -44° 54'$. It is soluble in an equal part of 80 p. c. alcohol. The high saponification number 111.7 corresponds to a content of 39 p. c. of an acetic acid ester of an alcohol $C_{10}H_{18}O$.

---

1) Bericht von S. & Co., Oct. 1889, p. 54. 2) Ibidem, Oct. 1898, p. 38.

On distilling the saponified oil with steam a light yellow liquid came over. The first fraction contained cineol, as was shown by the cineol-iodole compound. The odor of this fraction suggests the presence of thujone, together with the cineol."

352. Oil of Catnep.

The oil of *Nepeta cataria* L. indigenous to North America and used as a domestic remedy, has a not pleasant, mint- and camphor-like odor. Sp. gr. 1.041.[1]

353. Oil of Ground-ivy.

From the dried herb of the ground-ivy, *Nepeta glechoma* Benth. (*Glechoma hederacea* L.), Schimmel & Co.[2] obtained 0.3 p. c. of volatile oil. It had a difficultly definable, not pleasant odor, and was dark in color. Sp. gr. 0.925.

354. Oil of Sage.

Oleum Salviae.—Salbeiöl.—Essence de Sauge.

Origin and History. The somewhat shrub-like sage, *Salvia officinalis* L., is indigenous to the northern countries of the Mediterranean, and is cultivated in many countries of moderate climate as a garden plant for medicinal purposes. The plant will grow as far north as the northern part of Norway.

Sage appears to have been used as a medicinal herb at the time of the Romans and was one of the plants recommended by Charlemagne for cultivation. In the "Destillirbuch" by Brunschwig of 1500 a distinction is made between large and small sage for the distillation of sage water

The distilled oil of sage is first mentioned in the price ordinances of Worms of 1582 and of Frankfurt of 1587 and is included in the 1589 edition of the Dispensatorium Noricum. The distillation of the oil has been described by Begnini in 1688, and the yield of oil from the leaves was determined by Wedel in 1715 and Cartheuser about 1732.[3]

In 1720 Geoffroy observed a stearoptene which had crystallized from the oil and which he termed sage camphor.[4] The same substance was again observed by Arezula in 1789 and described by him.[5] The first

1) Bericht von S. & Co., Oct. 1891, p. 40.

2) Bericht von S. & Co., Apr. 1894, p. 55.

3) Elementa chymiae, vol. 2, p. 87.

4) Mem. de l'Acad. roy. des sciences de Paris, 1721, p. 168.

5) Resultado de las experiencias hechas sobre el alcanfor de Murcia non licencia. Segovia, 1789, p. 8.

examination of the leaves appears to have been made by Ilisch[1] in 1810; whereas the oil was investigated by Herberger[2] in 1829 and by Rochleder[3] in 1841.

PREPARATION. For the distillation of the oil the wild Dalmatian herb, which grows abundantly and is brought into the market in large compressed bales, alone seems to be used. The yield from Dalmatian leaves is 1.3—2.5 p. c. German leaves upon distillation yielded 1.4 p. c. of oil.

PROPERTIES. Oil of sage is a yellowish or greenish-yellow liquid possessing the peculiar odor of the herb while at the same time reminding somewhat of tansy and camphor. Sp. gr. 0.915—0.925; $\alpha_D = +10$ to $+25°$. The oil is soluble in two and more parts of 80 p. c. alcohol. Saponification number 107.

COMPOSITION. The presence of the following substances has been definitely determined: pinene, cineol, thujone and borneol.

1) Pinene. From the laevogyrate fraction boiling at 156—158°, Tilden,[4] also Muir and Sugiura[5] obtained a nitrosochloride. Wallach converted the nitrosochloride into nitrosopinene[6] ($C_{10}H_{15}NO$, m. p. 130°) and the original fraction into dipentene[7] (tetrabromide, m. p. 124—125°) and thus established the identity of the hydrocarbon present with pinene. With regard to the optical rotation, the statements vary. Muir and Sugiura found the fraction at one time laevogyrate, at another dextrogyrate; Wallach found it optically inactive. The fraction, therefore, seems to consist of a mixture of the two optical isomers, in which at times the one, at times the other isomer predominates.

2) Cineol. Fraction 174—178° yielded no nitrosochloride. By means of the hydrobrom addition product, Wallach was able to isolate pure cineol.[8]

3) Thujone (tanacetone, salviol, salvone), $C_{10}H_{16}O$. Fraction 198—203° was regarded by Muir and Sugiura as having the composition $C_{10}H_{16}O$, later $C_{10}H_{18}O$ and was termed salviol. Semmler in

---

1) Trommsdorff's Journ. der Pharm., 20, II, p. 7.

2) Buchner's Repert. f. d. Pharm., 34, p. 181.

3) Liebig's Annalen, 44, p. 4.

4) Journ. Chem. Soc., 1877, I, p. 554; Abstr. Jahresb. f. Chem., 1877, p. 427.

5) Philosoph. Magaz. and Journ. of Science, V, 4, p. 336.—Pharm. Journ., III, 8, pp. 191, 994; Journ. Chem. Soc., 1877, II, p. 548; Abstr. Jahresb. f. Chem., 1877, p. 957. — Chem. News, 37, p. 211; Journ. Chem. Soc., 33, p. 292; Abstr. Jahresb. f. Chem., 1878, p. 980. — Journ. Chem. Soc., 37, p. 678; Chem. News, 41, p. 228; Abstr. Jahresb. f. Chem., 1880, p. 1080.

6) Liebig's Annalen, 252, p. 103.

7) Liebig's Annalen, 227, p. 289.

8) Liebig's Annalen, 252, p. 103.

1892 declared salviol, which constituted 50 p. c. of the oil of sage, as identical with tanacetone.[1] Two years later, however, he expressed his doubt as to their identity.[2] By a comparison of the physical properties of the compounds regenerated from their acid sulphite addition products, Schimmel & Co.[3] established the identity of the corresponding ketones from the oils of sage, thuja, tansy and wormwood. Wallach[4] also obtained identical derivatives from thujone, tanacetone, salvone and absinthol. For the properties and derivatives of thujone see p. 167.

4) Borneol. Upon oxidation of oil of sage, Rochleder[5] had obtained camphor which evidently had existed in the oil as such or had been derived from borneol by oxidation. Sugiura and Muir had observed that the higher boiling fractions of the oil separated crystals upon cooling, which resembled camphor but did not agree with it in all of its properties. In order to decide whether camphor is contained in the oil or not, Schimmel & Co.[6] fractionated an oil which they had distilled from Dalmatian herb, and submitted to a freezing mixture those fractions which ought to have contained the camphor if present. No solid substance separated. Acetylization, however, indicated the presence of an alcohol (abt. 8 p. c. calculated as $C_{10}H_{18}O$). A part of the fraction was, therefore, treated with benzoyl chloride. The benzoate was separated and saponified and a substance corresponding with borneol in all of its properties resulted. After repeated crystallization it melted at 204°. In 10 p. c. alcoholic solution it deviated the ray of polarized light 0° 23′ to the right. The borneol of oil of sage is, therefore, a mixture of the dextrogyrate and laevogyrate modifications. Camphor could not be found in the oil examined. This does not, however, exclude its presence in other than the Dalmatian oils.

According to Muir, English oil of sage contains much cedrene, b. p. 260°, some terpene and only traces of oxygenated constituents (?).

355. Oil of Salvia Sclarea.

This oil is distilled from the dry herb, or better from the fresh herb of the muscatel sage *Salvia sclarea* L. Its odor is pleasant, lavender-like and, after evaporation of the more volatile portions, reminds of ambra. Sp. gr. 0.907—0.928; $\alpha_D = -19° 22'$ to $-24° 1'$; saponification number 144, corresponding to 50.4 p. c. of linalyl acetate. Judging

1) Berichte, 25, p. 3850.
2) Berichte, 27, p. 895.
3) Bericht von S. & Co., Oct. 1894, p. 51.
4) Liebig's Annalen, 286, p. 93.
5) Liebig's Annalen, 44, p. 4.
6) Bericht von S. & Co., Oct. 1895, p. 40.

from the odor, this ester is a constituent of the oil,[1] but its presence has not yet been chemically established.

356. Oil of Monarda Punctata.

The American labiate, *Monarda punctata* L., commonly known as horsemint, yields upon distillation about 3 p. c. of volatile oil which is yellowish-red or brownish in color and of a pungent thyme-like and minty odor. Sp. gr. 0.930—0.940; slightly dextrogyrate. Upon prolonged standing thymol separates in large crystals or crusts.

COMPOSITION. Thymol was discovered in the oil by Arppe[2] in 1846. It is present in such quantity that the oil has served for its preparation on a large scale.[3] According to Schröter[4] (1888), the oil is reported to contain 50 p. c. of a laevogyrate hydrocarbon, b. p. 170—173°, 25 p. c. of a dextrogyrate (!), non-crystallizable thymol, and a substance $C_{10}H_{18}O$ (?), b. p. 240—250°. These statements, which do not inspire confidence, throw doubt both upon the results and the material.

From material, identified by botanical authority, Schumann and Kremers[5] obtained an oil which contained as much as 61 p. c. of thymol when assayed according to the iodine method described under oil of thyme. The non-phenol portion of the oil contained cymene, $C_{10}H_{14}$, identified by the oxypropyl benzoic acid, m. p. 155—156°. Fraction 186—202° contains oxygen and consists possibly of linalool. In addition to thymol, Hendricks and Kremers[6] found a phenol fraction which would not solidify upon cooling and possibly contains carvacrol. They also found traces of d-limonene identified by means of its nitrolbenzylamine base melting at 94°.

357. Oil of Monarda Fistulosa.

Wild bergamot or *Monarda fistulosa* L. yields according to Kremers upon distillation an oil similar to that of *M. punctata*.[7] The sp. gr. of a number of oils distilled from the entire plant at different periods from June to September varied from 0.916 to 0.941. Optical rotation is slightly to the left. The amount of phenol varies from 52—58 p. c. The phenol of this species, however, is not thymol but carvacrol. This isomer was identified by means of dicarvacrol (m. p. 147—148°), carva-

1) Bericht von S. & Co., Apr. 1888, p. 44; Oct. 1894, p. 88.
2) Liebig's Annalen, 58, p. 41.
3) Bericht von S. & Co., Oct. 1885, p. 20.
4) Am. Journ. Pharm., 60, p. 118.
5) Pharm. Review, 14, p. 223.
6) Pharm. Archives, 2, p. 73.
7) Pharm. Rundschau, 18, p. 207; Pharm. Review, 14, p. 198.

crol sulphonic acid (m. p. 58—59°), nitroso carvacrol (m. p. 153—154°) and dinitro carvacrol (m. p. 117—119°). Other constituents are cymene, identified by means of oxypropyl benzoic acid (m. p. 156°), and limonene, identified by means of its nitrolbenzylamine base (m. p. 93°).[1]

Upon steam distillation of the phenol separated from its alkaline solution by acid, in one instance a red substance crystallized in the cooler. Purified by sublimation, it melted at 256—266° and behaved toward alkalies like alizarin.

358. Oil of Monarda Didyma.

This oil is presumably similar in composition to that of the two previous oils. Concerning it Flückiger[2] makes the following statement: "In 1796 Brunn, apothecary in Güstrow, observed a crystalline deposit (evidently thymol) in the oil from *Monarda didyma* L., which it seems had been imported from America."

359. Oil of Balm.

Oleum Melissae. — Melissenöl. — Essence de Melisse.

Origin and History. The labiate, *Melissa officinalis* L. is indigenous to the northern Mediterranean countries from Spain to the Caucasus. and is cultivated as a garden plant and for medicinal purposes in Europe and North America.

On account of its fragrance, balm was cultivated by the Greeks, Romans and Arabs, and during the middle ages in Italy, Germany and Scandinavia. During the period of distilled waters, from the fifteenth to the seventeenth centuries, balm water was a current article. Oil of balm appears to have come into use about the middle of the sixteenth century. It is first mentioned in the ordinance of Frankfurt-on-the-Main for 1582 and in the Dispensatorium Noricum of 1589.

Comparable to the distillate from rosemary of the sixteenth century which was the precursor of the *Eau de Cologne* of the eighteenth and nineteenth centuries, the fragrant distillate from balm, lemon peel and lavender of the seventeenth century developed later into a very popular perfume. It was first prepared by the Carmelite monks of Paris in 1611 and became famous as *Eau de Carmelites* (Ger. *Karmelitergeist*). Later the alcoholic distillate was made officinal as *Spiritus Melissae compositus*.

The earlier investigations of the oil were made by Schultz in 1739, by Hoffmann about the same time, and by Dehne in 1779.

1) Pharm. Archives, 2, p. 76.

2) Archiv d. Pharm., 212, p. 488.

The balm oil of commerce is no pure distillate from the balm. It may be an oil of lemon distilled over balm (*Oleum melissae citratum*), or a citronella oil treated in the same way, or merely a fractionated citronella oil. The yield from the herb *Melissa officinalis* L. is so small that the price of the oil would have to be extremely high. An oil distilled from the dry herb by Schimmel & Co. was solid at ordinary temperature and contained citral[1] as was shown by Doebner's reaction. Later the same firm subjected the fresh herb of two varieties of balm to distillation.[2]

1) Fresh herb, just beginning to blossom. Yield 0.014 p. c.; sp. gr. 0.924; $\alpha_D = +0° 30'$. The oil had a very pleasant balm odor.

2) Fresh herb in full blossom. Yield 0.104 p. c.; sp. gr. 0.894; optically inactive. The odor was less pleasant than that of the first oil and distinctly indicated the presence of citral and citronellal.

The attempt to prove the presence of these two aldehydes by means of Doebner's reaction yielded no positive result. The resulting acids from both oils began to melt at about 208° and were completely liquified at 225°. Evidently the mixture consisted of the citral compound (m. p. 197—200°) and the citronellal compound (m. p. 225°).

360. Oil of Pennyroyal.

Oleum Hedeomae.—Pennyroyal- oder amerikanisches Poleiöl.—Essence d'Hedeoma.

Origin and History. The oil from the American labiate, *Hedeoma pulegioides* Persoon is so similar to the oil from the European *Mentha pulegium* L., that it is frequently substituted for the latter. Pennyroyal is found from the Atlantic states to the Rockies. For its distillation very simple apparatus are used and the operation is, therefore, conducted in various regions. The bulk of the oil is reported as being distilled in North Carolina (Harris[3]) and in the southern and eastern sections of Ohio.[4] The primitive stills, like those used for the distillation of sassafras and wintergreen (pp. 397 and 588), consist of a barrel resting on a kettle used as boiler. The boiler is placed over a fire-place dug into the ground and provided with a chimney. The barrel is connected with a tube condenser resting in a trough through which

1) Bericht von S. & Co., Oct. 1894, p. 37.

2) Bericht von S. & Co., Oct. 1895, p. 58.

3) Pharm. Journ., III, 17, p. 672.

4) Comp. Kremers, Proceed. Am. Pharm. Assoc, 35, p. 546.—J. F. Patton, Proc. Penn. Pharm. Ass., 1890; and Proc. Am. Pharm. Assoc., 39, p. 548.

cool water flows. The fresh herb is distilled, a ton yielding about 10—12 pounds of oil. The dry leaves[1] yield about 3 p. c., the dry stems and leaves only 1.3 p. c. oil.

PROPERTIES. American pennyroyal oil is a light yellow liquid, of a characteristic minty and sweetish odor and an aromatic taste. Sp. gr. 0.925—0.940; $\alpha_D = +18$ to $+22°$. It is soluble in two or more parts of 70 p. c. alcohol. By means of this property, adulterations with petroleum, turpentine oil and other essential oils can readily be detected.

COMPOSITION. The principal constituent of pennyroyal oil is pulegone, the presence of which was established by Habhegger[2] by means of the hydrated pulegone oxime, $C_{10}H_{16}NOH.H_2O$, m. p. 147°, and by the benzoyl ester of the latter melting at 141°.

Kremers[3] found in the oil heated with potassa two ketones, $C_{10}H_{18}O$. The one boiled at 168—171° and yielded an oxime melting at 41—43°; the other boiled at 206—209° and yielded an oxime melting at 52°. The latter is possibly identical with menthone. Formic, acetic and isoheptoic acids were also found.

361. Oil of Hyssop.

Oleum Hyssopi. — Isopöl. — Essence d'Hysope.

Hyssop oil was formerly used medicinally and is mentioned in the drug ordinances of Berlin for 1574 and of Frankfurt-on-the-Main for 1582. It is distilled from the herb of *Hyssopus officinalis* L.[4] The yield from the dry herb is about 0.3—0.9 p. c. The oil has a pleasantly aromatic odor reminding of male fern. Sp. gr. 0.925—0.94; $\alpha_D + -17$ to $-23°$. It forms a clear solution with 2—4 parts of 80 p. c. alcohol, while with 70 p. c. alcohol even 10 parts will render only a turbid solution.[5] The saponification number of one oil was 1.4, after acetylization 45.

The oil begins to boil at about 170°; but little passes over below 175°, the bulk distills between 200 and 218° and has a strong odor of thujone or thujyl alcohol.

An examination by Stenhouse[6] was restricted to the analysis of various fractions from which no conclusions as to the composition of

1) Bericht von S. & Co., Oct. 1893, p. 33.

2) Proc. Wis. Pharm. Assoc., 1893, p. 51; Am. Journ. Pharm., 65, p. 417.

3) Proc. Am. Phar. Assoc., 35, p. 546; Am. Journ. Pharm., 59, p. 585.—Pharm. Rundschau, 9, p. 130.

4) Comp. also oil from *Origanum vulgare*.

5) Chemist and Druggist, 50, p. 218.

6) Liebig's Annalen, 44, p. 310; Journ. f. prakt. Chemie, 27, p. 255.

the oil could be drawn. Besides, the oil examined by Stenhouse appears to have been adulterated with turpentine oil. One of his fractions began to boil at 160°, whereas pure oil begins to boil at 170°.

Several French oils had properties that deviated materially from those of pure oils. The sp. gr. of one of these oils was 0.95 and the angle of rotation + (!) 45°. They had a strong odor of fenchone, thus leading to the supposition that they consisted principally of the first runnings of fennel oil.

362. Oil of Satureja Hortensis.

Origin and History. *Satureja hortensis* L. (Ger. *Bohnen-* or *Pfefferkraut*) owes its odor and pungent taste to a volatile oil obtained by steam distillation. The yield from the fresh herb is about 0.1 p. c. The oil is enumerated in the Frankfurt ordinance of 1582 among the medicinal volatile oils.

Properties. The oil possesses the strong aromatic odor of the plant and a sharp, biting taste. The sp. gr. of the oil distilled from the fresh herb lies between 0.900 and 0.925; $\alpha_D = +0°4'$ (determined on a single oil); phenol content 38—42 p. c. The oil forms a clear solution with 10 parts of 80 p. c. alcohol.[1] An oil examined by Jahns[2] in 1882 and distilled from the dry herb had a sp. gr. 0.898; $\alpha_D = -0.62°$; phenol content 30 p. c.

Composition. According to Jahns, this oil contains carvacrol, also traces of a second unknown phenol which gives a blue color reaction with iron salts.

Of the hydrocarbon fractions, about one-third consists of cymene (b. p. 173—175°; cymene sulphonate of barium). The other two-thirds (b. p. 178—180°; sp. gr. 0.855; $\alpha_D = -0.2°$) consist of a terpene or terpenes as shown by analysis.

363. Oil of Satureja Montana.

The oil of *S. montana* L. has much the same properties as that of *S. hortensis*. The two oils cannot be distinguished by their odor.

Fresh, cultivated herb[1] yielded upon distillation 0.18 p. c. of oil. Sp. gr. 0.939; $\alpha_D = -2°35'$; soluble to a clear solution in 4.5 parts of 70 p. c. alcohol, and in 1.5 parts of 80 p. c. alcohol; phenol content 65 p. c.

---

1) Bericht von S. & Co., Oct. 1897, p. 65. 2) Berichte, 15, p. 816.

An oil examined by Haller[1] in 1882 had been distilled from wild herb growing on the Maritime Alps in the neighborhood of Grasse. Sp. gr. 0.9394[2] at 17°; $\alpha_D = -3°25'$. He found 35—40 p. c. carvacrol and a small amount of phenol boiling above 235°. The hydrocarbons boiled between 172—175° and 180—185° and appeared to be terpenes.

364. Oil from Satureja Thymbra.

Satureja thymbra L. was dedicated to Priapos by the ancient Greeks. In Spain it is used commonly as a spice and has the reputation of being a stimulant and disinfectant. These properties are said to be due to its volatile oil. Such an oil was obtained from Spain by Schimmel & Co.[3] and examined by them. It had a sp. gr. 0.905. The oil remaining after the removal of the thymol (abt. 19 p. c.) was fractionated. A fraction boiling abt. 160° contained pinene (nitrolbenzylamine base, m. p. 121°): fraction 175° contained cymene; the following fraction contained traces of dipentene. The fraction above 200° yielded after saponification borneol and acetic acid which are contained in the oil as bornyl acetate. In composition, the oil is therefore closely related to oil of thyme.

365. Oil of Origanum Vulgare.

Origanum vulgare L. is one of the spice plants of antiquity. The hyssop of Luther's translation does not refer to a *Hyssopus* but to *Origanum*. The volatile oil was used during the latter part of the middle ages and is mentioned in the German ordinances of the sixteenth century. The yield from the dry herb is 0.15—0.4 p. c. The oil possesses a strong aromatic odor and a spicy, bitter taste;[4] sp. gr. 0.870—0.910; $\alpha_D = -34.4°$.[5]

According to Kane[6] (1839) the oil contains a stearoptene about which nothing is known. The bulk of the oil is reported by him to boil at 161°. Jahns[5] in 1880 found two phenols in an oil distilled from the fresh herb. One of these gave a green coloration with ferric chloride, and is possibly identical with carvacrol; the other gave a violet color. The oil contained not more than about 0.1 p. c. of combined phenols.

The French oils of commerce are mostly mixtures of a pennyroyal-like odor which possibly contain not a trace of the genuine oil.

---

1) Compt. rend., 94, p. 132.
2) 0.7894 is evidently a printer's error.
3) Bericht von S. & Co., Oct. 1889, p. 55.
4) Bericht von S. & Co., Apr. 1891, p. 49.
5) Archiv d. Pharm., 216, p. 277.
6) Liebig's Annalen, 32, p. 284; Journ. f. prakt. Chemie, 15, p. 157.

366. Oil of Sweet Marjoram.

Oleum Marjoranae.—Majoranöl.—Essence de Marjolaine.

ORIGIN. The fresh, flowering herb of *Origanum majorana* L. yields upon distillation 0.3—0.4 p. c., the dry herb 0.7—0.9 p. c. of oil. The oil of commerce is mostly obtained from Spain.

PROPERTIES. Oil of sweet marjoram is a yellow or greenish-yellow liquid of a pleasant odor reminding of the herb and of cardamom. The taste is spicy but mild. Sp. gr. 0.89—0.91; $\alpha_D = +5$ to $+18°$. The saponification number of a single oil was 21.5. As a rule the oil produces a clear solution with 2 vol. of 80 p. c. alcohol.

COMPOSITION. Oil of sweet marjoram has been repeatedly examined. The stearoptene described and analyzed by Mulder[1] in 1839 possibly was terpin hydrate or pinol hydrate.

According to Bruylants[2] (1879) the oil contains 5 p. c. of a dextrogyrate hydrocarbon $C_{10}H_{16}$, and 85 p. c. of a dextrogyrate mixture of borneol and camphor. This statement has not been confirmed.

Beilstein and Wiegand[3] in 1882 isolated a terpene boiling at 178°, sp. gr. 0.846 at 18.5°, which absorbed one molecule of hydrogen chloride, without yielding a crystalline compound. Fraction 200—220° was analyzed and regarded as a sesquiterpene hydrate, $C_{15}H_{24}H_2O$. The low boiling point, however, renders this conclusion very improbable.

According to a recent investigation by Biltz[4] (1898), the oil contains 40 p. c. of terpenes, principally terpinene (nitrosite m. p. 155—156°); also d-terpineol which, however, could not be obtained crystalline. Fractions 215—218° (sp. gr. 0.930) had the composition $C_{10}H_{18}O$. Oxidized with permanganate it yielded trioxy hexahydro cymene $C_{10}H_{20}O_3$ (m. p. 129—130°).[5] Further oxidation with chromic and sulphuric acids yielded the keto lactone, $C_{10}H_{16}O_3$ (m. p. 61°), obtained by Wallach from terpineol. Fraction 215—218° must, therefore, be pronounced as terpineol. The alcohol is present principally in the free state and only small portions are present in the form of ester. The nature of the substance that produces the peculiar marjoram odor is still unknown.

---

1) Liebig's Annalen, 31, p. 69.—Journ. f. prakt. Chem., 17, p. 108.

2) Journ. de Pharm. et Chim., IV, 30, p. 138; Jahresb. f. Pharm., 1879; p. 160.—Chem. Centralbl., 1879, p. 616.

3) Berichte, 15, p. 2854.

4) Ueber das ätherische Oel von *Origanum majorana*. Inaugur.-Dissertat., Greifswald, 1898; Berichte, 32, p. 995.

5) The substance obtained by Wallach in 1898 from inactive terpineol melted at 121—122° (Liebig's Annalen, 275, p. 152).

367. Oil of Cretian Origanum.

Oleum Origani Cretici. — Spanisch Hopfenöl. — Essence d'Houblon d'Espagne.

The carvacrol-containing oils of several species of *Origanum* growing in Mediterranean countries are known in the German market as *Spanisch Hopfenöl* and *Kretisch Dostenöl.* At present there are two varieties in the market which differ in properties and composition, the Triest oil and the Smyrna oil. The former is dark in color, has a high specific gravity and a corresponding high carvacrol content. The latter is lighter in color, has a milder odor and a lesser carvacrol content in correspondence with its lower specific gravity.

1. Triest Origanum Oil.

Origin. The oil enters the market from Triest. It is uncertain whether the oil is distilled at that place or in the Mediterranean islands which supply the herb. The oil corresponds in all its properties with that distilled in Germany from the dry herb of *Origanum hirtum* Lk. It is probable, therefore, that this is the species yielding the Triest oil. The yield from the dry herb is 2—3 p. c.

Properties. The oil possesses a strong, thyme-like odor and a pungent, persistent taste. Freshly distilled it is of a golden-yellow color, which upon exposure to air is changed to dark brown to greyish-black. The darkening begins at the surface and gradually proceeds downward. Sp. gr. 0.94—0.98. On account of the dark color, the angle of rotation in most instances cannot be observed. In a few instances, when observation was possible, the oil was found inactive or a slight laevorotation, less than 1°, was observed. With 3 p. of 70 p. c. alcohol the oil forms a clear solution. The carvacrol content varies from 60—85 p. c.

Composition. The Triest oil was examined by Jahns[1] in 1879. He showed the presence of carvacrol which previously had been prepared artificially but had not been found in an oil. Properties and derivatives of carvacrol are described on p. 177.

If the oil is treated with dilute soda solution and the solution of the phenylate shaken with ether, all of the carvacrol can be removed. If the alkaline solution which no longer yields anything to the ether is acidulated, a small amount—about 0.2 p. c.—of a second phenol is obtained which produces a violet color with ferric chloride.

The non-phenol portions of the oil, after several rectifications over sodium, boiled principally between 172—176° and consist principally

1) Archiv d. Pharm., 215, p. 1.

of cymene as shown by the formation of cymene sulphonic acid. The fact that the fraction when shaken with sulphuric acid developed the odor of sulphur dioxide and became strongly heated indicates that in addition to cymene other substances, presumably terpenes, are present.

2. Smyrna Origanum Oil.

ORIGIN. The oil is distilled in Asia Minor[1] from *Origanum smyrnaeum* L. and enters the world's market from Smyrna.

PROPERTIES. The oil is of a golden-yellow color and of a mild odor reminding somewhat of linaloe oil or linalool. Sp. gr. 0.915—0.945; $\alpha_D = -3$ to $-13°$. It forms a clear solution with 3 parts of 70 p. c. alcohol and has a carvacrol content of from 25—60 p. c.

COMPOSITION. According to Gildemeister[2] (1895) the Smyrna oil differs from the Triest oil principally as to the linalool which the former contains in appreciable amount. As a result it contains less phenol than the Triest oil. Besides much carvacrol (phenyl isocyanate, m. p. 140°), but very little of the phenol which gives a violet reaction with ferric chloride is present. Fraction 155—163° ($\alpha_D = -3° 28'$) has a remarkably low sp. gr., viz., 0.826 at 15°, thus leading to the supposition that so-called olefinic terpenes may be present.

Fraction 175° contains cymene (oxypropyl benzoic acid, m. p. 156—158°; and isopropyl benzoic acid, m. p. 257—262°); fraction 198—199°, having a sp. gr. of 0.870 and $\alpha_D = -15° 56'$, possessed all of the properties of l-linalool. The presence of this alcohol was proven by oxidizing it to citral and identifying this aldehyde by converting it into citryl-β-naphtho cinchoninic acid, m. p. 198—199°.

EXAMINATION. The solubility of the oil in 70 p. c. alcohol should be tested, by means of which the addition of turpentine oil and other cheap oils can be detected. Of importance is the determination of the carvacrol content as described under oil of thyme. Oils rich in carvacrol command a higher price than those poor in phenol.

368. Oil of Thyme.

Oleum Thymi. — Thymianöl. — Essence de Thym.

HISTORY. The labiate *Thymus vulgaris* L., which is indigenous to the countries bordering on the Mediterranean, is now cultivated in most countries with a temperate climate. During the middle ages the

1) This is possibly the same oil which is prepared in Konia in Anatolia in primitive apparatus and sold in small flasks in the streets of Constantinople as a remedy against rheumatism. Comp. Bericht von S. & Co., Apr. 1891, p. 44.

2) Archiv d. Pharm., 28, p. 182.

distinction between *T. vulgaris* and *T. serpyllum* does not appear to have always been made. Though thyme has always been a rather unimportant remedy, it and the oil of thyme have been officinal since the sixteenth century in most medicinal treatises and in drug and spice ordinances. The oil is enumerated in the Dispensatorium Noricum of 1589. Thyme-camphor was first observed by Neumann in 1719, and by Cartheuser in 1754. It was examined and named thymol by Lallemand in 1853.[1]

Origin. Oil of thyme is distilled principally from the fresh, flowering herb of *Thymus vulgaris* L. which grows abundantly in the wild state in the mountains of southern France. The small knotty and woody stems of the thyme are found in clearings and in the shadeless coast districts of the Riviera, also in the mountainous regions of the Maritime Alps up to an altitude of 1,000 m. It is not definitely known what plant yields the Spanish oil of thyme. Inasmuch as it resembles the Cretian origanum oil in properties and composition, it is not improbable that it is derived from a species of origanum. The yield is known for the oil obtained from the cultivated herb from which the commercial oil, as a rule, is not obtained. Fresh German herb yields 0.3—0.4 p. c., dry German herb 1.7 p. c. of oil; fresh French thyme, cultivated in Germany, yielded 0.9 p. c., dry French herb 2.5—2.6 p. c. of oil.

Properties. Both French and German oil of thyme are of a dirty, dark reddish-brown color, of a pleasant, strong thyme odor and a biting, persistent taste. The sp. gr. of a pure oil is always above 0.900, good French oils having a sp. gr. of 0.905—0.915. Schimmel & Co. observed the sp. gr. 0.909—0.935 on their own distillates. The optical rotation is faintly laevogyrate, but in most instances cannot be observed on account of the dark color of the oil. It is soluble in ½ part of 90 p. c., and in 1—2 parts of 80 p. c. alcohol; of 70 p. c. alcohol 15—30 parts are mostly requisite to form a clear solution. The phenol content of normal oils varies, as a rule, between 20—25 p. c. and in rare instances rises to 42 p. c. The phenol of the French and German oils is mostly thymol, sometimes, however, carvacrol or a mixture of both.

Oil of thyme rectified in the ordinary manner readily resumes the dark color of the crude oil. In order to obtain a light yellow colored oil with full phenol content special precautions are necessary for its rectification. Many consumers lay unnecessary stress on the light color

---

1) Journal de Pharm. et Chim., III, 24, p. 274; Compt. rend., 87, p. 498.

of the oil. In southern France such a "white oil of thyme" is produced by distilling the crude oil with several times its volume of turpentine oil. Such an oil frequently contains less than five per cent, from 1—2 p. c., of thymol. This explains why rectified oil of thyme is frequently quoted for less than the crude oil.

Spanish oil of thyme is quite different from the French and German. Its color is often dark green, sp. gr. 0.93—0.95; phenol content 50—70 p. c. (carvacrol but no thymol). It is also more soluble than either the French or German oils, rendering a clear solution with 2—3 parts of 70 p. c. alcohol. These great differences render it very probable that the Spanish oil is obtained from a different plant.

COMPOSITION. As the amount of cinnamic aldehyde determines the value of cassia oil, eugenol that of oil of cloves and linalyl acetate that of bergamot, so the amount of thymol (or carvacrol) is indicative of the value of oil of thyme. Although thymol was observed as early as 1719 and therefore belongs to those compounds from volatile oils longest known, its composition was first correctly determined by Lallemand[1] in 1853. By analysis he determined the formula $C_{10}H_{14}O$, which took the place of the formula $C_{10}H_{15}O$, found by Doveri[2] a few years earlier.

Occasionally thymol crystallizes from old oils in the cold; it can be removed completely only by shaking with lye. It melts at 50—51° and boils at 232°. Its properties and characteristic derivatives are mentioned on p. 178.

Of other constituents Lallemand found cymene, $C_{10}H_{14}$, b. p. 175°; and thymene, $C_{10}H_{16}$, b. p. 160—165° and laevogyrate. Inasmuch as Schimmel & Co.[3] have shown the identity of this hydrocarbon with l-pinene, the name thymene should be dropped. l-Pinene occurs in the oil in such small quantities only, that its presence cannot be determined without the use of large quantities of oil.

Labbé[4] in 1898 did not succeed in obtaining a solid hydrochloride from fraction 155—158°. With amyl nitrite and hydrochloric acid he obtained a nitrosochloride melting at 106.5°. Inasmuch as the melting point of pinene nitrosochloride is given as 103°, he concludes that pinene is not contained in oil of thyme, whereas the preparation of the benzylamine base would have removed all doubt. In fraction 165—169°

1) Compt. rend., 37, p. 498; Liebig's Annalen, 101, p. 119; Ann. de Chim. et Phys., III, 49, p. 148; Liebig's Annalen, 102, p. 119.

2) Ann. de Chim. et Phys., III, 20, p. 174; Liebig's Annalen, 64, p. 374.

3) Bericht von S. & Co., Oct. 1894, p. 57.

4) Bull. Soc. chim., III, 19, p. 1009.

Labbé claims to have found menthene. He obtained a nitrosochloride, m. p. 113—113.5°, and upon oxidation with permanganate cymene. It is known, however, that the melting points of the nitrosochlorides cannot be safely used for the identification. The supposed oxidation to cymene is also inconclusive, for cymene is already present in the oil and may have been contained in the fraction.

Of the more difficultly volatile fractions, the one boiling between 195—230°[1] had a decided odor of borneol and linalool. Inasmuch as their boiling points lie so close, their separation could not be effected by fractional distillation, but was accomplished by oxidizing the fraction with chromic acid and distilling the products of oxidation in a vacuum. A part of the distillate solidified and was proven to be camphor by means of its oxime melting at 117—118°. The liquid portion yielded with sodium acid sulphite a crystalline derivative from which citral could be regenerated (citryl-β-naphtho cinchoninic acid, m. p. 197°). The formation of camphor indicates the presence of borneol in the oil, that of citral the presence of linalool. The presence of the latter was further demonstrated by Labbé[2] by converting it into geraniol (m. p. of geranyl phthallate of silver 133°). Upon oxidation of the borneol fraction Labbé likewise obtained camphor.

Thymol is not the only phenol in oil of thyme. At times thymol is partly, at times wholly replaced by its isomer carvacrol (see properties on p. 177). What conditions favor the one or the other phenol is not known. Schimmel & Co. have made the following observations: French, dried thyme yielded an oil that contained much thymol, but little carvacrol; French thyme, cultivated in Germany an oil the phenol of which consisted exclusively of carvacrol; both fresh and dried German thyme yielded an oil containing only thymol. In the Spanish oil carvacrol only is found, and this oil has a much larger phenol content than the other oils.

That oil of thyme probably contains a third phenol is indicated by the fact that the oil produces a greenish-black color with ferric chloride.

Examination. The adulteration of oil of thyme is most commonly accomplished with turpentine oil. In the preparation of "white oil of thyme," such an adulteration has become an established custom in southern France. The addition of oil of turpentine reduces the specific gravity below 0.900 and diminishes the solubility in alcohol. A reduction of the phenol content also results from the addition of turpentine oil

---

[1]) Bericht von S. & Co., Oct. 1894, p. 57. [2]) Bull. Soc. chim., III, 19, p. 1009.

thus rendering a phenol assay of the greatest importance. A simple method, which is sufficiently accurate for most practical purposes, is applied in the following manner:

A burette of 60 cc. capacity calibrated into tenths is almost completely filled with 5 p. c. caustic soda solution. 10 cc. of the oil to be examined are then added, the burette stoppered with a well fitting cork, well shaken and set aside for 12—24 hours. Drops of oil adhering to the side of the burette are loosened by tapping and rotating the burette. After the alkaline solution has become clear the amount of non-phenol oil is read off.

In order to determine at the same time whether the oil contains thymol or carvacrol, the alkaline solution of phenylate is separated from the oil, transferred to a separating funnel and acidulated with sulphuric acid. After the phenol has completely separated, the aqueous solution is run off and the oil set aside in a capsule in a cool place. If the oil consists of thymol it solidifies upon standing, or crystallization may be induced by adding a fragment of a thymol crystal. If it consists of carvacrol, the oil remains liquid. If both phenols are present it crystallizes partially.

This method yields approximate results only. On the one hand the alkaline solution of phenols dissolves some of the hydrocarbons, on the other hand a part of the phenol remains dissolved in the oil. Both errors probably compensate each other. To remove the hydrocarbons dissolved in the alkali by shaking with ether, would simply increase the error inasmuch as thymol and still more carvacrol can be partly extracted from their alkaline solutions by ether.

More rational, though more complicated, is the assay method devised by Kremers and Schreiner.[1] It is a modification of the method suggested by Messinger and Vortmann,[2] and is based upon the fact that in alkaline solution thymol combines with iodine to a red insoluble compound and that the excess of iodine can be titrated back by standard thiosulphate after the solution had been acidified. Each molecule of thymol requires four atoms of iodine for precipitation. The process is carried out in the following manner:

5 cc. of the oil to be examined are weighed and transferred to a burette provided with a glass stopper and calibrated into tenths. The oil is diluted with about an equal volume of petroleum ether. 5 p. c. soda solution is added and the two solutions shaken. After the two solutions have separated upon standing, the aqueous solution is drawn off into a 100 cc. flask. This operation is repeated as long as the ethereal solution diminishes in volume. The alkaline phenol solution is diluted to 100 cc. or if necessary to 200 cc.

10 cc. of this solution are transferred to a measuring flask of 500 cc. capacity and N/10 iodine solution added in slight excess whereby the thymol is precipitated as a dark brown iodine compound. In order to ascertain

1) Pharm. Review, 14, p. 221.

2) Berichte, 23, p. 2758.

whether sufficient iodine has been added, a few drops are transferred to a test tube and acidulated with hydrochloric acid. An excess of iodine is indicated by a brown color of the solution, an insufficient amount of iodine by milkiness produced by thymol. An excess of iodine having been established, the entire alkaline solution is acidulated and diluted to 500 cc. In 100 cc. of the filtrate the excess of iodine is ascertained by titration with N/10 hyposulphite solution.

The number of cc. of hyposulphite solution are deducted from the volume of iodine used and the resultant multiplied by five, this product indicating the total amount of iodine used by the thymol.

Each cc. of N/10 iodine solution corresponds to 0.003741 g. of thymol. From the amount of thymol found in the alkaline solution the percentage of phenol in the original oil can readily be ascertained.

The reaction involved is expressed by the following equation:

$$C_{10}H_{14}O + 4I + 2NaOH = C_{10}H_{12}I_2O + 2NaI + 2H_2O.$$

The carvacrol assay should be slightly modified, because the carvacrol iodide separates in a milky state. In order to obtain a precipitate the mixture is thoroughly shaken after the addition of iodine and then filtered. Then the solution is acidulated with hydrochloric acid and the process continued as directed under thymol. The calculation is the same.

369. Oil of Wild Thyme.

Oleum Serpylli. — Quendelöl. — Essence de Serpolet.

Origin. Wild thyme, *Thymus serpyllum* L. (Ger. Quendel, Feldthymian, Feldkümmel) is indigenous to Europe, North America, central and northern Asia and Abyssinia, and yields upon distillation but little oil. The yield from the dry herb is 0.15 to 0.6 p. c., that from the fresh herb less.

Properties. Oil of wild thyme is a colorless or golden yellow colored liquid of a pleasant odor, somewhat balm-like, also reminding faintly of thyme. Sp. gr. 0.890—0.920; $\alpha_D = -10$ to $21°$.

As *Essence de serpolet* mixtures of origanum oil, pennyroyal oil and oil of thyme are frequently sold in southern France which naturally possess quite different properties from those of the genuine oil of wild thyme.

Composition. The bulk of the oil distills between 175—180° and, according to Febve[1] (1881), consists of cymene, $C_{10}H_{14}$, with traces of a dextrogyrate hydrocarbon $C_{10}H_{16}$. With alkalies about 1 p. c. of phenol can be removed. This phenol, however, is not a definite chemical body, but according to Jahns[2] consists of a mixture of carvacrol, thymol and

1) Compt. rend., 92, p. 1290.

2) Archiv d. Pharm., 216, p. 277; Berichte, 15, p. 819.

a phenol which in alcoholic solution produces a violet color with ferric chloride.[1] The higher boiling fractions (200—250°) presumably also contain sesquiterpenes.

An oil examined by Gladstone[2] in 1864, which had a sp. gr. of 0.884 and $\alpha_D = -31.6°$, consisted almost exclusively of a hydrocarbon resembling turpentine oil. The explanation is not difficult, for specific gravity and angle of rotation indicate a very liberal adulteration with French oil of turpentine.

370. Oil of Thymus Capitatus.

An oil, distilled from the fresh herb of *Thymus capitatus* Lk. in the province of Granada in southern Spain, was examined by Schimmel & Co.[3] Its odor was strongly thyme-like, reminding somewhat of origanum. In its composition it closely resembles the oil of *Satureja thymbra*. Sp. gr. 0.901 at 15°; thymol content is small, about 6 p. c. A liquid phenol is also present, the boiling point of which lies close to that of thymol (carvacrol?). Pinene, cymene, dipentene and bornyl acetate are also present.

371. Oil of Bugle Weed.

The dried herb of the American bugle weed, *Lycopus virginicus* Michx. yields upon distillation 0.075 p. c. of an oil with a characteristic but difficultly definable odor. Sp. gr. 0.924 at 15°.[4]

MENTHA OILS.

History. The mints, which are indigenous to temperate climates and some of which are cultivated, yield several valuable and much used oils: viz. oil of peppermint, oil of spearmint, German and American, and oil of pennyroyal, European and American.

The mints have the peculiarity of readily forming varieties by differences in cultivation, soil and climate conditions. These botanical variations have a rather decided influence on the volatile oils. The varieties of *Mentha piperita* L. also of *M. arvensis* D. C. var. *piperascens* Holmes yield peppermint oil, *M. crispa* yields the *Krauseminzöl* of the Germans and *M. viridis* the American spearmint oil.

Although several mints have been in use for culinary or medicinal purposes since antiquity, no well defined distinction is made even in the

1) Archiv d. Pharm., 212, p. 485.
2) Journ. Chem. Soc., 17, p. 1; Jahresb. f. Chem., 1863, pp. 546 and 549.
3) Bericht von S. & Co., Oct. 1889, p. 56.
4) Bericht von S. & Co., Oct. 1890, p. 49.

treatises on distillation of the fifteenth and sixteenth centuries during which period they were extensively used for the preparation of distilled waters. The oils enumerated in price ordinances and older medical treatises are also of uncertain origin. Thus the Berlin ordinance of 1574 mentions *Oleum menthae*, the Frankfurt ordinance of 1582 mentions *Oleum menthae*, *Oleum polemii* and *Oleum pulegii*.

372. Oil of Peppermint.

Oleum Menthae Piperitae.—Pfefferminzöl.—Essence de Menthe Poivrée.

Origin. As peppermint, *Mentha piperita*, are designated a group of botanically unstable species, subspecies and varieties of mint that produce menthol or an oil possessing the properties of peppermint oil. In Europe and North America several varieties are cultivated for the distillation of the oil. The Japanese peppermint is usually not regarded as belonging to *Mentha piperita*. It is supposed to be derived from *Mentha arvensis* D. C. var. *piperascens* Holmes.

Origin and Preparation. Whether or not peppermint was among the mints used during the middle ages can no longer be determined. In the oldest German treatise on distillation the "Liber de arte distillandi" of Brunschwig of the year 1500, the following mints are mentioned as being used in the preparation of distilled waters: *Mentha aquatica*, *M. rubra*, *M. balsamica*, *M. sarcenica* and *M. crispa*, but no distinguishing characteristics are given. Neither is it definitely known whether the kinds of mint used formerly agree with those now in use. As far as known, the only specimens of *Mentha piperita* which are several hundred years old, are found in the herbarium of the British Museum in London. John Ray,[1] the English naturalist, had obtained them from the county of Hertfordshire of southern England in 1696 and described them as *Mentha palustris*, "peper mint." The well preserved specimens correspond in all essential characteristics with the peppermint which is to-day cultivated in Mitcham, county of Surrey, near London (Flückiger[2]). The cultivation of peppermint in Mitcham seems to have begun about the middle of the eighteenth century and was of some importance toward the end of the century. Up to 1805, however, the distillation of peppermint oil was not conducted in Mitcham but in London (Lysons[3]).

The English peppermint industry reached its height about 1850. From that time on American competition caused a decided set-back in

[1]) Historia plantarum, vol. III, p. 284.

[2]) Pharmacognosie, 3rd ed., p. 726.

[3]) Environs of London, 1800, p. 254.

the production.[1] On the continent, peppermint was evidently not cultivated earlier for purposes of distillation than in England. According to the Leyden botanist David Gaubius, it was cultivated for this purpose near Utrecht in 1770. He also mentions the menthol, the *camphora europaea menthae piperitidis*. Meanwhile Linnaeus had named the plant *Mentha piperita*.

About the same time peppermint was cultivated in Germany. Following the example of the London Pharmacopoeia, in which peppermint

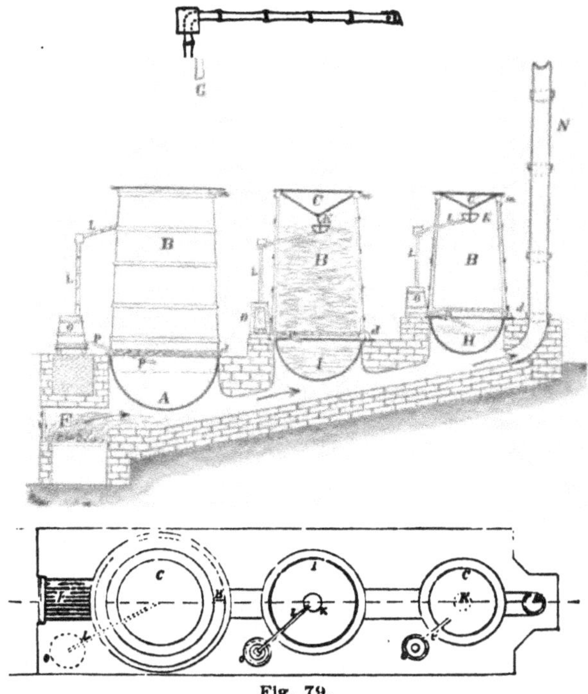

Fig. 79.
Distillation of Peppermint Oil in Japan.

was officinal since 1721 as *Mentha piperitis sapore* it was mentioned in medical and botanical treatises. The treatise by Knigge[2] seems to have made it better known in professional circles.

In Japan, however, the cultivation of peppermint appears to antidate that of any other country. It is reported to have begun before

1) Chemist and Druggist, 1891, p. 405.
2) De mentha piperitide commentatio. Dissertatio. Erlangae. 1780.

the Christian era. Even menthol is supposed to have been known almost as long and to have been used medicinally.[1] The cultivated varieties used in Japan for the distillation of the oil can no longer be traced back to the species from which they have been derived. They differ from the varieties used in Europe and America and in their general characteristics are more closely related to the European *Mentha arvensis* L. and the American *M. canadensis* var. *glabrata* Benth., than to *M. piperita*.

A Japanese outfit was described by E. Marx[2] in 1896. The arrangement, however, is not the same everywhere.

It consists of cast iron kettles with broad rims (*A, I, H*, fig. 79), of wooden casks (*B*) and coolers (*C*). As a rule three apparatus are so arranged that they can be heated by a common fire. The kettles *A, I, H* are first filled with water. The casks *B* with perforated bottoms are placed over the kettles, filled with dry herb and luted with rims of straw and soft clay. The condensers *C* are then placed in position and the fire started in *F*. The vapors saturated with volatile oil condense on the lower surface of the condensers *C* and collect in the cups *K*. From here the mixture of oil and water passes through the bamboo tube *L* into the receiver *O*. The separated water flows through *P* back into the kettles *A, I, H* and replaces in part the evaporated water. As soon as the water in *C* becomes hot it is removed by a bamboo syphon *G*. Sometimes the rims of the kettles are overheated so as to cause a charring of the straw rims. This imparts an empyreumatic odor to the oil which is frequently observable in the Japanese oil.

Another description of the Japanese distillation of peppermint with illustrations is given by T. Asahiva[3] which, however, agrees in the main with the above.

The distillation of peppermint oil on a commercial scale had its origin in Wayne Co., New York, in 1816. It was here conducted on a small scale by individuals and later also in neighboring counties. In 1835 the cultivation of peppermint was begun in Michigan in St. Josephs Co. and was extended to Ohio and northern Indiana.[4]

For a long time the distillation was conducted in copper stills with direct heat. In 1846 several of the larger planters and distillers introduced steam distillation with large wooden stills, experience having shown that the conditions were favorable for this peppermint oil industry. With the older stills but 15 lbs. of oil could mostly be obtained from a charge. Steam distillation from wooden stills allows of the distillation of 75 to 100 lbs. of oil per charge without great increase of cost.

1) Pharmacognosie, 1891, p. 726.
2) Mittheil. d. d. Gesell. für Natur- und Völkerkunde Ostasiens, 6, p. 355.
3) Journ. d. Pharm. f. Elsass-Lothringen, 23, p. 314.
4) Proc. Am. Pharm. Assoc., 34, p. 121.

Wherever modern distilling apparatus are not used, two or three vats are used in connection with each boiler. The plants are packed on a perforated bottom which is lowered by means of chains to within a short distance of the true bottom. For purposes of discharging the false bottom merely has to be raised. The vat is filled by two men, one throwing in the dry mint while the other stamps it down. By admitting some steam from time to time, the herb is moistened somewhat and in better condition to be firmly packed. After the vat is closed steam is admitted from below until the herb is exhausted.

The renewal of the peppermint plantation is accomplished in spring by means of roots and sprouts of the previous year's plants. The largest yield of oil is obtained from the plants cut in September. Formerly it was supposed that a larger yield of oil was obtained by distilling the fresh plant. Experience, however, has taught that it is better to let the herb dry for a short period.[1]

Formerly several labiates and composites growing with peppermint were distilled indiscriminately with the latter and produced a poorer quality of oil.[2] The more important of these plants were *Erigeron canadensis* L., *Erechthites hieracifolia* Raf., *Hedeoma pulegioides* L. and *Ambrosia* species (ragweed). By more careful cultivation and selection these have been largely removed.

Aside from the early cultivation in Japan, the first observations of the crystallization of menthol from the oil (peppermint camphor) were made about 1770 by Gaubius,[3] by Glendenberg about 1784,[4] and by Trommsdorff in 1795.[5] The first examinations of peppermint oil and menthol were made by Dumas[6] in 1832, Blanchet & Sell[7] in 1836, by Walter[8] in 1839 and by Oppenheim[9] in 1861 and 1864.

PRODUCTION AND COMMERCE. In the following list the countries are arranged with reference to the amount of oil produced.

A m e r i c a. The principal places of production are in the states of New York, Michigan and Indiana. Whereas formerly the state of New York controlled the market, Michigan has superseded it during the last ten years and produces at present at least four to five times as much. On the other hand the oil from the former state and especially that from Wayne county with Lyons as the market, is still preferred qualitatively to the Michigan or "Western" oil.

---

1) Proc. Am. Pharm. Assoc., 34, p. 121; Amer. Druggist, Sept. 1886, p. 161, and June 1888; Proc. N. Y. State Pharm. Assoc., 1888; Pharm. Journ., III, 19, pp. 3 and 4.

2) Proc. Am. Pharm. Assoc., 7, p. 449.—Am. Journ. Pharm., 42, p. 120; Archiv d. Pharm., 192, p. 252.

3) Adversariorum varii argumenti liber unus, p. 99.

4) Crell's Chem. Ann., 1785, II, p. 427.

5) Trommsdorff's Journ. d. Pharm., 3, p. 120.

6) Annal. de Chim. èt Phys., 50, p. 282.

7) Liebig's Annalen, 6, p. 293.

8) Ibidem, 32, p. 288.

9) Ibidem, 120, p. 350; 130, p. 176.

Fig. 80.
Peppermint Harvest in Michigan.

Fig. 81.
American Peppermint Distillery.

The largest yearly production of peppermint oil was reached in 1897 and was distributed as follows.

Eastern Michigan	13,000 lbs.
Western "	79,000 "
Northern "	25,000 "
Southern "	55,000 "
Indiana	32,000 "
Various localities	10,000 "
Total	214,000 lbs.
Add to this:	
State of New York	37,000 lbs.
Total	251,000 lbs.

As the result of this enormous production an entirely unexpected drop in price has taken place, which must necessarily lead to restriction in the peppermint culture.

The oldest and best known American trademarks are those of "H. G. Hotchkiss" under which only crude, natural oil is shipped, and "F. S. & Co.," under which rectified American peppermint oil is brought into the market since 1872. Both of these are found in the market exclusively in 1 and 5 lb. glass bottles. Large amounts of crude American oil come into the market in tins or cans. The largest purchasers of these are New York wholesale houses who have agents in the producing districts.

Japan. The peppermint oil production originally had its seat in the vicinity of Yonezawa, but since a few years has also started in the province Bingo Bitchu—both on Hondo—and recently also on the northernmost island Hokkaido (Yezo). The largest production in Japan was probably reached in 1896 with more than 220,000 cätties (1 cätty = 605 g.). This enormous amount was distributed in the different localities as follows:

Uzen with Yonezawa as principal place	abt.	1,200	piculs
Bingo	"	800	"
Bitchu	"	150	"
Bizen	"	50	"
Aki	"	5	"
Yamato	"	7	"
Yamashiro	"	5	"
Shinano	"	5	"
Suruga	"	5	"
Island of Shikoku	"	5	"
Total	abt.	2,232	piculs.[1]

[1]) 1 Picul = 100 cätties = 60.479 k., 1 cätty = 605 g.

Here, likewise, the necessary result was a corresponding decrease in price and a restriction in the peppermint cultivation. The production of peppermint oil decreased as a result in 1897 to about 1,400 piculs, in 1898 to about 1,000 piculs. The principal commercial center for peppermint oil in Japan is Yokohama; Kobe ranks next in importance.

Hamburg has, on account of the splendid steamer connections, become the principal market in Europe for Japan peppermint oil. London and New York are second and third in importance. The great importance of Hamburg for the entire peppermint commerce is seen from the following official statistics:

Import	1897		Average value	1896	
	k.	Mk.[1]		k.	Mk.
From the United States and the Atlantic Ocean	34,190	480.380	14	31,840	542,130
From Japan	30,470	303,860	10	24,000	286,410
" Australia (Mainland)	240	7,480	31	—	—
" China	—	—	—	190	3,110
" Great Britain	7,060	181.250	26	8,020	222.080
" Spain	—	—	—	200	4,700
" France	—	—	—	210	2.950
Other import by water	100	1,200	12	140	1.470
Total by water	72,060	974,170	14	64,600	1,062,850

England. The peppermint culture and oil distillation have their seat in Surrey, Hertfordshire and Lincolnshire, in the vicinity of the towns Mitcham, Waddon, West Croyden, Wallington, Carshalton and Market Deeping. The total area under peppermint cultivation in the Mitcham districts is estimated at 300 acres. About 50 acres are planted with the preferred, so-called white mint. The total production of the districts named is estimated at about 20,000 lbs.; statistics do not exist.

Not entirely unimportant is the peppermint oil production in France, amounting to several thousand kilos. It has its seat exclusively in the Département des Alpes Maritimes. The French oil belongs to the better commercial varieties.

Germany. In Germany the production of peppermint oil has considerably decreased during the last four decades; in Thuringia it has stopped almost altogether. On the other hand the peppermint cultivation has been started on a large scale in the vicinity of Leipzig. A distillery equipped with modern apparatus is in the midst of the fields and, by means of special purifying processes, a product of superior quality is

[1]) Mark = 23.8 cts.

Fig. 82.
Campania Farm of A. M. Todd near Kalamazoo, Michigan.

Fig. 83.
Peppermint Distillery in Decatur, Michigan.

obtained. The annual production has up to the present not exceeded 400 k.

Italy and Russia produce peppermint oils to a similar extent. which are used almost exclusively for home consumption. In Italy the article is protected by a comparatively very high import duty.

An estimation of the annual world production of peppermint oil under normal conditions, leads to the following result:

North America	abt.	90,000 k.
Japan	"	70,000 "
England	"	9,000 "
France	"	3,000 "
Germany	"	800 "
Italy	"	600 "
Russia	"	1,200 "
Various other countries	"	400 "
Total	abt.	175,000 k.

This quantity appears to considerably exceed the world demand. The principal varieties have so decreased in price, that only a restriction in the production can gradually restore a normal state.

PROPERTIES. As already mentioned, the peppermint oils produced in the various countries are not obtained from a single botanical species. This explains why the individual varieties of the oil show such a great diversity in properties and composition. For all practical purposes, the odor and taste give an indication of the quality of the oil. By these properties an experienced connoisseur is able to distinguish between the three principal commercial varieties, the English, American and Japanese. This is of importance on account of the great differences in price. It is to be regretted that the origin of an oil cannot always be definitely recognized by a physical examination, it is usually completely impossible when a mixture of different oils is under consideration.

Peppermint oil is colorless, yellowish or greenish-yellow, of a pleasant, refreshing odor and a cooling persistent taste. It is fairly fluid, but thickens and darkens with age.

Specific Gravity. The specific gravity of the different varieties does not vary greatly, but the differences are, nevertheless, great enough, that sometimes a conclusion as to the origin of the oil can be reached. The lightest oil is the dementholated Japanese oil with a density of 0.895—0.905. The sp. gr. of the English oil lies between 0.900 and 0.910, that of the American mostly between 0.910—0.920. With the German oil a sample as high as 0.930 was found.

Rotation. To a certain extent the rotation of the different varieties is characteristic; it is least with the French (up to —6°) and greatest with the Japanese (up to —42°) oil.

Solubility. The solubility in alcohol of various strengths offers a good means of distinguishing American from English and Japanese oil. English peppermint oil dissolved to a clear solution in 3—5 vol. of 70 p. c. alcohol at 20°. With a further addition of alcohol the solution as a rule remains clear, although sometimes showing a slight opalescence, but with a pure oil never a separation of oil drops. The solubility of the Japanese (dementholated) oil in 70 p. c. alcohol is usually the same as that of the English, often, however, somewhat less. It must be remembered that in this case a normal distillate is not under consideration, but rather a by-product of the menthol manufacture, which varies according to the method employed for the preparation of the menthol. Of the American oils the cheap Western oil shows the same solubility as the English; the better oil from Wayne Co., N. Y., is not clearly soluble in 70 p. c. alcohol. Of 90 p. c. alcohol, ½ vol. is requisite for making a clear solution with the latter oil. If larger quantities of this same strength of alcohol are added, there appears sometimes, especially with unrectified oils, a bluish opalescence, which is undesirable for purposes of liquor manufacture. According to Kennedy[1] this opalescence is due to a decomposition produced by the light. The air appears not to affect the oil detrimentally in this case, as a long continued passage of the air through the oil did not give rise to this phenomenon. Observations which were made during a distillation of Saxon peppermint oil, indicate that the oils are soluble in 70 p. c. alcohol when they have been distilled from the fresh, and insoluble when distilled from the dry, herb. This would explain the differences in solubility of the English and the American oil, inasmuch as it is known that the former is distilled from the fresh, the American (Wayne Co.) from the previously dried, herb. Whether or not other influences play a part, cannot at present be stated.

Menthol Separation in the Cold. The normal Japanese oil is so rich in menthol, that even at ordinary temperature it forms a crystalline mass saturated with oil. American oil solidifies completely in a freezing mixture, while the English as well as the Saxon oil very often shows crystalline separations only after standing for a long while in the freezing mixture. As the two last named oils command the highest

[1]) Proc. of the Texas State Pharm. Assoc., 1888, p. 37.

price, it follows, that the amount of menthol which can be separated from an oil by cooling, is not a criterion of the value of a peppermint oil.

Color Reactions. Numerous color reactions have been suggested for the identification of peppermint oil, of which that produced by acids is the prettiest and most striking.

If 5 drops of American or English peppermint oil are mixed with 1 cc. of glacial acetic acid, a blue coloration will be noticed after several hours, which gradually increases in intensity and reaches its maximum in about 24 hours; the mixture with American oil then shows a deep dark blue with transmitted light, and a fine copper-colored fluorescence with reflected light. With the English oil these phenomena are less intense, often only a light blue coloration with a faint reddish fluorescence appears. Japanese oil[1] does not show this reaction, the mixture remaining colorless.

Slight warming hastens the appearance of the reaction. The color obtained in this manner is, however, not so pure a blue, but rather of a violet shade. Contact with the air is necessary for the reaction. If the air be excluded, no coloration is noticed, even after several days. The reaction is therefore to be considered as an oxidation phenomenon. The reaction is very quickly produced when, according to the U. S. Pharmacopoeia, 2 cc. of oil are mixed with 1 cc. of glacial acetic acid and 1 drop of nitric acid. By this treatment the Japanese oil is also colored slightly violet.

The cause of the color reaction is, according to Polenske,[2] to be found in a nitrogen-free, volatile body accompanying the oil. The colors which this forms with acids show a characteristic spectroscopic behavior. The body itself is decomposed by light, for an oil which has been exposed for some time to the sunlight no longer gives the color reaction.

Other color reactions are produced as follows:

A solution of 1 cc. of oil in 5 cc. of alcohol is heated with 0.5 g. of sugar and 1 cc. of hydrochloric acid. The mixture assumes a deep blue, violet or bluish-green color.[3]

A red color is produced when some chloral hydrate and hydrochloric acid are added to French peppermint oil.[4] German and English oils are said to be colored light brown by this reaction.

The cause of these color phenomena are as little known as the composition of the bodies which produce them. From a practical standpoint but little importance can be attached to these reactions.

---

1) Pharm. Journ., III, 1, p. 682; III, 2, p. 821.

2) Arbeiten a. d. kaiserl. Gesundheitsamte, Berlin (1890), 6, p. 522. — Pharm. Zeitung, 35, p. 547.

3) Chemiker-Zeitung, 13, p. 264.

4) Archiv d. Pharm., 208, p. 29; 205, p. 326.

Composition. Menthol, which is found in all varieties of peppermint oil, must be considered as its characteristic constituent. On account of its ready crystallization it was early observed and repeatedly investigated by the older and the newer chemists. The properties and chemical derivatives of this interesting compound have been described in detail on page 145. Menthol occurs in peppermint oils for the greater part in the free state, in smaller quantities in the form of its acetic and valerianic acid ester. A further constituent supposed to be common to all peppermint oils is menthone, which appears to have been first observed by Beckett and Wright.[1]

The formation of menthol and other constituents of peppermint oil during various stages in the development of the plant has been investigated by Eugène Charabot.[2] The oils examined were derived from plants at three different stages of growth; the first as soon as the inflorescence appeared, and before the formation of flower buds, the second when flower buds were formed, the third when the flowers were fully expanded.

	Before formation of flower buds.	After formation of flower buds		Flowering plants.
		a) leaves.	b) inflorescences.	
Sp. gr. at 18° C.	0.9025	0.9016	0.9081	0.9200
Opt. rot. at 18° C.	—24° 10′	—26°	—20° 15′	—2° 37′
Esters (as menthyl acetate).	3.7 percent	10.3 percent	7.5 percent	10.7 percent
Combined menthol	2.9 "	8.1 "	5.9 "	8.4 "
Free menthol	44.3 "	42.2 "	29.9 "	32.1 "
Total menthol	47.2 "	50.3 "	35.8 "	40.5 "
Menthone	5.2 "	4.2 "	16.7 "	10.2 "

It will thus be seen that at the first stage the plant yields an oil rich in menthol, but containing a relatively small proportion of esters, and in which menthone is only present in small quantity; as, however, the development of the green parts of the plant progresses, the proportion of esters increases, and this esterification takes place in the leaves, for the oil from the inflorescences is less rich in esters. Menthone, however, would appear to be chiefly formed in the flowers, where it increases during the development of the inflorescences, while the proportion of the total menthol diminishes. It is concluded, therefore, that, as in the case of lavender, esterification is confined to the chlorophyll-bearing parts, and that menthone is formed in the flowers by the oxidation of menthol.

1) Journ. Chem. Soc., 1876, I, p. 8. 2) Compt. rend., 130, p. 518.

The more recent investigations are always confined to a single oil of definite origin. Inasmuch as the differences between the various oils depend without doubt on a difference in chemical composition, the results obtained with one oil cannot offhand be applied to another. For this reason the chemical composition will have to be discussed under each individual oil.

American Peppermint Oil.

There are two varieties of American oil, that from Wayne County on Lake Ontario, in New York, which is considered as the finer oil, and the cheaper and less valued oil distilled in Michigan and Indiana, in Wayne, St. Joseph and Van Buren Counties, and known as Western or Michigan oil.

PROPERTIES. The sp. gr. of the peppermint oil from the state of New York lies between 0.91 and 0.92, the angle of rotation α_D between -25 and $-33°$. The oil is not soluble in 70 p. c., but in ½ and more parts of 90 p. c. alcohol, to a clear solution. In a freezing mixture it solidifies quite rapidly to a crystalline mass. It contains a total of 50—60 p. c. of menthol, of which 40—45 p. c. are in the free state and 8—14 p. c. as esters. The amount of menthone present is about 12 p. c.

The Michigan oil has a less fine odor than the preceding oil, but has the advantage, that it is clearly soluble in 4—5 p. of 70 p. c. alcohol. The sp. gr. of the oils so far investigated varied between 0.905 and 0.913, the rotatory power α_D between -18 and $-29°$. A menthol determination made on four oils gave the following results: Free menthol, 43.6—50.3 p. c.; menthol as ester, 4.3—8.5 p. c.; total menthol, 48.6—58 p. c. Two other oils of the same source were, as could be seen from their low total menthol content of 32.6 and 35.8 p. c., adulterated or dementholated.

Power and Kleber[1] have published a very detailed investigation of the constituents of American peppermint oil. The oil had been distilled from the dried herb, free from weeds, collected in Wayne County, New York. The yield was 0.67 p. c.; sp. gr. 0.9140; $\alpha_D = -32° 0'$. Menthol as ester, 14.12 p. c.; free menthol, 45.5 p. c.; total menthol, 59.6 p. c. By distillation from a fractionating flask the following fractions were obtained;

Up to 200°	2.6 p. c.	220—225°	19.6 p. c.
200—205°	2.4 "	225—230°	9.0 "
205—210°	8.6 "	230—235°	3.6 "
210—215°	18.8 "	Residue	12.2 "
215—220°	24.0 "		

1) Pharm. Rundschau, 12, p. 157; Arch. d. Pharm., 232, p. 639.

On account of the greater care taken in the collection of the material, and on account of the more perfect distilling arrangements, the oil had a much purer and more pleasant odor than the ordinary commercial oil.

In American peppermint oil there have been found so far no less than 17 different, well characterized chemical compounds, a number which has up to the present not been found in any other oil. The bodies found can be here given with but a brief mention of their detection. For the details and the course of the analysis the original must be consulted.

American peppermint oil contains:

1) Acetaldehyde, CH_3COH, abt. 0.044 p. c., yielded acetic acid on oxidation.
2) Isovaleric aldehyde, $(CH_3)_2.CH.CH_2.COH$, abt. 0.048 p. c., b. p. 92°, gave valerianic acid on oxidation.
3) Free acetic acid, CH_3COOH.
4) Free isovalerianic acid, $(CH_3)_2.CH.CH_2.COOH$.
5) Pinene,[2] $C_{10}H_{16}$, inactive, perhaps a mixture of d- and l-pinene (pinene nitrolpiperidine, m. p. 118°, pinene nitrolbenzylamine, m. p. 123°).
6) Phellandrene, $C_{10}H_{16}$ (nitrite, m. p. 100°).
7) Cineol, $C_{10}H_{18}O$, b. p. 174—177° (cineol hydrobromide, cineolic acid, m. p. 196°).
8) l-Limonene, $C_{10}H_{16}$, (tetrabromide, m. p. 104°).
9) Menthone, $C_{10}H_{18}O$, (conversion into menthol).
10) Menthol, $C_{10}H_{20}O$, b. p. 215.5°.
11) Menthyl acetate, $C_{10}H_{19}O.C_2H_3O$.
12) Menthyl isovalerianate, $C_{10}H_{19}O.C_5H_9O$.
13) Menthyl ester of an acid $C_8H_{12}O_2$, $C_{10}H_{19}O.C_8H_{11}O$.
14) A lactone $C_{10}H_{16}O_2$, m. p. 23°, of a stale odor, reminding slightly of borneol. The corresponding oxy acid crystallizes from petroleum ether in shining needles, melting at 93°.
15) Cadinene,[1] $C_{15}H_{24}$ (dichlorhydrate, m. p. 118°).
16) Amyl alcohol, $C_5H_{12}O$, (acetate).
17) Dimethyl sulphide,[3] $S(CH_3)_2$.

Besides these there appear to be present in peppermint oil other higher boiling, unstable sulphur compounds, as during the middle of

[1]) Bericht von S. & Co., Apr. 1894, p. 42.
[2]) First observed by Halsey (Proc. Wisconsin Pharm. Assoc., 1893, p. 90).
[3]) Bericht von S. & Co., Oct. 1896, p. 61.

the rectification of the oil, there is frequently noticed a penetrating odor, reminding of putrifying ruta-baga.

The presence of dimethyl sulphide in peppermint oil, is shown as follows:

From 50 cc. of the crude oil about 1 cc. is distilled off, and this poured on an aqueous solution of mercuric chloride. In a short time the formation of a soft skin can be observed at the zone of contact of the two liquids. On account of its volatility the dimethyl sulphide collects in the first portions on rectification of the oil. Oils from which the first runnings of the rectification have been removed, no longer show the reaction.

Menthene, $C_{10}H_{18}$, b. p. 158—160°, which according to Andres and Andreef[1] occurs in Russian peppermint oil, could not, in spite of a diligent search, be found in the American oil. Likewise unsuccessful was the search for a terpene $C_{10}H_{16}$, boiling at 175° and having the properties of Brühl's[2] so-called menthene.

English Peppermint Oil.

The English oil, mostly designated as Mitcham peppermint oil, is very highly valued on account of its fine aroma and pleasant taste.

It is obtained from two varieties, the black and the white mint. The former gives by distillation the larger yield, but the oil from the white mint is considered as superior in quality.

Properties. English peppermint oil has the sp. gr. 0.900—0.910, the rotatory power $\alpha_D = -22°$ to $-33°$. Total menthol, 58—66 p. c.; free menthol, 50—60 p. c.; menthol as ester, 3—14 p. c.; menthone 9—12 p. c. The oils of the two English varieties differ, according to Umney,[3] in their content of menthol as esters. Whereas the oil of the white mint contains 14 p. c. of menthol as esters, the oil of the black mint contains only 7 p. c. The oil of a black mint cultivated in America had, however, a content of 12.2 p. c. of menthol as esters, from which it follows that the difference in the ester content is not decisive. Umney observed great differences on fractionation of the oil of the white and of the black mint.

	Black mint.	White mint.
Below 200°	5 p. c.	24 p. c.
200—205°	27 "	15 "
205—210°	31 "	15 "
210—215°	22 "	15 "
215—220°	7 "	13 "
Above 220°	8 "	18 "

1) Berichte, 25, p. 609.

2) Berichte, 21, p. 157.

3) Pharm. Journ., 56, p. 123; 57, p. 103.

The color reaction with glacial acetic acid is less intense than with the American oil; according to Umney the intensity of coloration increases with the amount of esters present.

Although the menthol content found by acetylization is higher in the English oil than in the American, often only a slight separation of crystals takes place in the freezing mixture. This behavior makes it probable that besides menthol there are present other alcohols (isomeric liquid menthols?) which are estimated as menthol in the acetylization method.

Composition. Flückiger and Power[1] isolated from the English peppermint oil two laevogyrate terpenes, boiling at 165—170° (mixture of pinene and phellandrene?), as well as a dextrogyrate sesquiterpene boiling at 255—260° (probably cadinene). Umney[2] showed the presence of phellandrene by the nitrite reaction and ascertained that the acids combined with the menthol as esters are the same as in the American oil, namely, acetic and isovalerianic acids. He further determined quantitatively the amount of menthone present in the oil.

There is no doubt but that with a thorough investigation a great number of the same substances occurring in the American oil would be found.

Japanese Peppermint Oil.

Properties. The normal Japanese oil is at ordinary temperatures a solid mass of crystals saturated with oil. In commerce occurs the normal oil (*unseparated*), the crude menthol, forming a loose crystal mass (*crystals*), or the liquid oil separated from the latter (*oil*).

The Japanese oil is the cheapest of the peppermint oils, but cannot be used for all purposes on account of its bitter taste. Normal oil has the sp. gr. 0.895—0.900 at 24°; solidification point[3] +17 to +28°; angle of rotation $\alpha_D = -30$ to $-42°$. It is soluble in 3—5 p. of 70 p. c. alcohol. Total menthol, 70—91 p. c.; free menthol, 65—85 p. c.; menthol as ester, 3—6 p. c. The liquid oil obtained in the preparation of menthol has the sp. gr. 0.895—0.905, $\alpha_D = -26$ to $-35°$. It is not always soluble to a clear solution in 3—5 p. of 70 p. c. alcohol. The color reaction with glacial acetic acid described on p. 642 is not given by the Japanese oil, or only to a slight degree.

---

1) Pharm. Journ., III. 11, p. 220; Archiv d. Pharm., 218, p. 222.
2) Pharm. Journ., 56, p. 128; 57, p. 108.
3) Determined as described on p. 187.

COMPOSITION. Japanese peppermint oil was investigated in 1876 by Beckett and Wright.[1] They found in the fraction boiling at 210—215° of the part remaining liquid on cooling, a body $C_{10}H_{18}O$ isomeric with borneol, which in all probability was menthone. In the fraction boiling at 245—255° they suspected a compound $C_{30}H_{50}O$, which, according to their view, was formed by the splitting off of water from 3 molecules of the body $C_{10}H_{18}O$.

$$3C_{10}H_{18}O = C_{30}H_{50}O + 2H_2O.$$

As this formula is not very probable, it must be assumed that the fraction 245—255° consisted of impure menthol, mixed with sesquiterpene.

Saxon Peppermint Oil.

The Saxon peppermint oil is unsurpassed by all other commercial varieties in fineness of aroma and taste. It is the highest priced of all the peppermint oils. As its annual production amounts to only a few hundred kilos, it plays no important part in the world's market. Sp. gr. 0.900—0.915; $\alpha_D = -25$ to $-33°$. The solubility is mostly the same as with the English oil, sometimes, however, a slight opalescent turbidity results on the further addition of 70 p. c. alcohol. Total menthol, 54.7—67.6 p. c.; free menthol, 46.5—61.2 p. c.; menthol as ester, 5.7—8.2 p. c.; menthone, 15.7 p. c. On cooling a crystal separation or solidification takes place only after standing for several days in a freezing mixture.[2]

German Peppermint Oil.

The oil distilled in Gnadenfrei, Silicia, in small amounts, belongs to the best peppermint oils and is similar in its properties to the Saxon oil.

From the refuse, not suitable for medicinal purposes, of the peppermint plants cultivated in Cölleda, Ringleben (Thuringia) and other places, an oil of inferior quality, accompanied by an unpleasant spearmint-like odor, is obtained. Sp. gr. 0.899—0.930; $\alpha_D = -27$ to $-33°$. The oil is usually not soluble to a clear solution in 70 p. c. alcohol.

French Peppermint Oil.

The peppermint oil produced in southern France appears to be principally used in France itself. It distinguishes itself by a high sp. gr. and low rotatory power. Sp. gr. 0.918—0.920; $\alpha_D = -5°54'$ to $-8°20'$. A sample investigated was not soluble in 70 p. c. alcohol. Total

1) Journ. Chem. Soc., 1876, I, p. 8; Jahresb. f. Chem., 1876, p. 897.

2) Bericht von S. & Co., 1896, I, p. 50.

menthol 43.7—46 p. c.; free menthol 35.7—39.4 p. c.; esters 7.1—10 p. c.; menthone 8.8—9.6 p. c. The acids combined with the menthol are the same as those in the American and English oil, namely, acetic and isovalerianic acid.[1]

Charabot,[1] together with C. Ebray, has studied an interesting change in the inflorescence of *Mentha piperita*, which is known to the cultivators and distillers of south-western France as *Menthe basiliquée*. This variety of *Mentha piperita* shows besides the normal inflorescence also some which appear like racemes, similar to the inflorescence of the basilicum after the petals have fallen off. These changed shoots do not blossom; but rather appear as though carrying seed, although in their place are found only bunches of leaves. A microscopic examination of a section through the base of the changed sprout has shown, that the change is brought about by the sting of an insect. The oil obtained from this changed plant has an unpleasant odor and distinguishes itself from the normal oil by a higher sp. gr. and rotatory power, as well as by the lower content of menthol and menthone. For such an oil were determined: sp. gr. 0.924 at 18°; $\alpha_D = +7°$; esters, 8.2 p. c.; total menthol, 41 p. c.; menthone, 3 p. c.

Russian Peppermint Oil.

The Russian peppermint oil likewise plays no part in the world's market, and like the French, is principally distilled for home consumption. Sp. gr. 0.905—0.910; $\alpha_D = -17$ to $-22°$. The amount of total menthol determined in a single case was 50.2 p. c. (free menthol 46.8, menthol as ester 3.4). Crystallization took place only after standing for a long while in the freezing mixture.[2]

Russian peppermint oil has been investigated by Andres and Andreef.[3] Besides menthol, it contains dextrogyrate menthone, probably a mixture of both optical modifications, in which the dextrogyrate modification predominates. The fraction 158—160° gave on analysis numbers which indicate a mixture of a hydrocarbon $C_{10}H_{18}$ with a terpene (pinene?). The authors therefore assume the presence of a menthene,[4] which, however, they were unable to isolate in a pure form. The fraction 173—175° contains l-limonene (tetrabromide, m. p. 102°, nitrosochloride, m. p. 103°, dichlorhydrate, m. p. 49.5—50°.

---

1) Bull. Soc. chim., III, 19, p. 117.

2) Bericht von S. & Co., Apr. 1896, p. 50; ibidem, Apr. 1889, p. 35.

3) Berichte, 25, p. 609; Pharm. Ztsch. f. Russl., 29, p. 341.

4) The menthene, $C_{10}H_{18}$, resulting by splitting off water from menthol, boils at 167—168°.

Italian Peppermint Oil.

The oil distilled in the provinces of Piemont and Padua is not exported, at least not in appreciable quantities. Sp. gr. 0.911—0.926; $\alpha_D = -13$ to $-18°$; b. p. 195—222°. In a freezing mixture none or only a slight menthol separation takes place. Total menthol, 44.1—46.6 p. c.; free menthol, 36.7—41 p. c.; menthol as ester, 5.6—7.4 p. c.

Bohemian Peppermint Oil.

An oil distilled in Bohemia had the following properties: sp. gr. 0.905; $\alpha_D = -27° 22'$. Soluble in 70 p. c. alcohol. Total menthol, 59.9 p. c.; ester menthol, 8.7 p. c.; free menthol, 51.2 p. c.[1]

Chilian Peppermint Oil.

An oil distilled in Osorno (Chili) of the sp. gr. 0.916, had an intense odor of pennyroyal.[2]

Réunion Peppermint Oil.

An oil prepared on the island of Réunion had an odor more of lavender than of peppermint. With iodole it gave Hirschsohn's cineol reaction, had the sp. gr. 0.887, the rotatory power $\alpha_D = -6° 9'$, and was soluble in 4 p. of 70 p. c. alcohol.

EXAMINATION. For the detection and identification of the numerous adulterants of peppermint oil it is necessary, above all, to determine the physical constants, as by them the attention is called to grosser adulterations with alcohol, turpentine oil, and other volatile oils. For distinguishing the different varieties the solubility determination in 70 p. c. alcohol is very useful. All peppermint oils are soluble in equal parts of 90 p. c. alcohol, but only a part of them form clear solutions with 70 p. c. alcohol.

The U. S. Pharmacopoeia requires that peppermint oil solidify at −8 to −20°. If American oil is allowed to stand in a good freezing mixture of ice and salt until thoroughly cooled, and then a small crystal of menthol added, it will crystallize in a short while to a solid mass. With English peppermint oil a good menthol separation usually takes place, but as a rule it does not solidify. The behavior of the other oils in a freezing mixture has been mentioned under their

1) Bericht von S. & Co., Apr. 1896, p. 50.
2) Bericht von S. & Co., Oct. 1894, p. 75.

description. American oil has been repeatedly found in the market from which a part of the menthol had been removed; the solidification test is therefore of importance for this variety. With the English oil the separation of menthol is hardly profitable as the oil commands a higher price than menthol.

Inasmuch as adulterations with other oils decrease the menthol content, a quantitative menthol determination is often of great value. According to Power and Kleber the method is as follows:[1]

20 g. of peppermint oil are heated to boiling with 20 g. of alcoholic normal soda solution (or normal or ½ normal potassa solution) for about an hour, in a flask provided with a reflux condenser (fig. 56, p. 194), in order to decompose the menthol esters. After cooling, the unconsumed alkali is titrated back with normal sulphuric acid, using phenol phthalein as indicator. The saponified oil is repeatedly washed with much water and then heated with an equal volume of glacial acetic acid and 2 g. of anhydrous sodium acetate in a flask provided with a glass-ground tube condenser (fig. 57, p. 195). After cooling, the oil is washed several times with water and dilute soda solution, dried with calcium chloride and filtered. 8—10 g. of this oil are then saponified, as described above, with 50 cc. of alcoholic normal soda solution and the excess of alkali titrated back.

Each cc. of the normal soda solution corresponds to 0.156 g. of menthol or 0.198 g. of menthyl acetate. In order, therefore, to obtain the percentage of menthol in the original oil (not acetylized, but freed from ester) it is necessary to deduct 0.042 g. (the difference between 0.156 and 0.198) for every cc. of normal alkali consumed. If, e. g., s g. of acetylized oil require a cc. of normal soda solution, the total menthol content P (free and ester) can be calculated according to the following formula:

$$P = \frac{a \times 15.6}{s - (a \times 0.042)}$$

The result thus obtained does not express exactly the menthol content, inasmuch as it is assumed for purposes of calculation that all the menthol is present as acetic ester whereas some of it is combined with isovalerianic acid. The resulting error, however, is so small that it can be disregarded.

The amount of menthone is determined in the following manner:

After the menthol content has been determined with a part of the saponified oil, another part is diluted with twice its volume of alcohol and boiled for some time with metallic sodium. Thereby the menthone is reduced to menthol which can be assayed as described above.

373. Oil of Spearmint.

Oleum Menthae Crispae.—Krauseminzöl.—Essence de Menthe Crépue.

Origin and Preparation. Distinction is made between three kinds of spearmint oil, American, German and Russian. The American oil is

---

1) Pharm. Rundschau, 12, p. 162.

distilled in New York and Michigan from the fresh herb of *Mentha viridis* L. The herb is cultivated to a not inconsiderable extent, as much as 12,000 lbs. of oil being obtained in the two states mentioned.[1] In England (Mitcham) also some oil is distilled from the same plant.

The German oil is distilled in small amounts, in Thuringia only, where spearmint is cultivated for medicinal purposes. The waste resulting in the process of drying is used for this purpose. The plant cultivated is the *Mentha crispa* L. which is regarded as a cultural variety of *M. aquatica* L., whereas *M. viridis* L. is probably a cultural variety of *M. silvestris* L.[2]

The botanical source of the Russian oil is not known.

PROPERTIES. American and German spearmint oil resemble each other so closely that no distinction is made in commerce. The oil is a colorless, yellowish, or greenish-yellow liquid and possesses the characteristic, penetrating and disagreeable odor of spearmint. With age and upon exposure to the air, the oil becomes viscid and darker.

The American oil has a sp. gr. of 0.920—0.940 and $\alpha_D = -36$ to $-48°$. It is soluble in equal parts of 90 p. c. alcohol, but the solution is rendered turbid upon the addition of more solvent. An oil distilled by Fritzsche Bros. had somewhat different properties. The spearmint had been cultivated on the factory grounds at Garfield, N. J., and was just in blossom when distilled. The yield was 0.3 p. c. The oil had a sp. gr. of 0.980, consequently higher than that of the ordinary oils; $\alpha_D = -42° 30'$. The odor was quite different from that of the commercial oil, not at all minty or pennyroyal-like, but reminded distinctly of carvone. Upon cohobating the aqueous distillate a considerable amount of oil heavier than water was obtained. It is possible that in the distillation of the commercial oil a part of this heavy oil is lost, thus accounting for the lower specific gravity.[1] After the first harvest toward the close of July, a second was made early in October. The yield from the fresh herb was only 0.18 p. c. The odor of this oil was less delicate, its specific gravity and rotatory power were lower, viz. 0.961 and $\alpha_D = -37° 20'$. Nevertheless, this oil was still heavier than the commercial oils, though no fraction was obtained heavier than water.[3]

COMPOSITION. According to Kane[4] the oil of *Mentha viridis* is supposed to contain a crystalline constituent, but none of the later

1) Bericht von S. & Co., Oct. 1896, p. 45.
2) Pharmacognosie, 3rd ed., p. 727. — See also p. 629.
3) Bericht von S. & Co., Apr. 1897, p. 49.
4) Journ. f. prakt. Chem., 15, p. 168; Liebig's Annalen, 32, p. 286.

investigators observed such a substance. Gladstone[1] found carvone in spearmint oil. With sulphuretted hydrogen he obtained a solid compound which upon treatment with alkalies yielded an oil of the composition $C_{10}H_{14}O$. This turned the plane of polarized light as far to the left as dill carvone turned it to the right. Gladstone gave to this substance the inappropriate name of menthol.

From German spearmint oil Flückiger[2] in 1875 obtained l-carvone of low rotatory power. Beyer,[3] on the other hand, found in 1883 that the angle of rotation of the carvone of the German oil is as great as that of carvone from dill and caraway oils. The amount of carvone in spearmint oil was determined by Schreiner and Kremers[4] as 56 p. c. According to Trimble[5] spearmint oil contains a terpene boiling at 160—167.5°, according to Beyer a laevogyrate hydrocarbon boiling at 168—171°. Brühl[6] concludes from the statements of Gladstone that d-pinene is present. According to Gilman[7] the oil contains l-limonene and probably l-pinene. To which constituent the oil owes its characteristic odor is not yet known.

Russian spearmint oil is reported to be distilled in large quantities but it is principally consumed in Russia. It differs from the American and German oils by its stale odor reminding but slightly of spearmint. Specific gravity and optical rotation are much lower than of the first two varieties, viz., 0.883[8]—0.885; $\alpha_D = -23° 12'$ at 17°. The oil under examination gave a clear solution with 2 parts of 70 p. c. alcohol and had a saponification number of 25.9. It was shown to consist of 50—60 p. c. of l-linalool.[9] Fraction 196—200°, with $\alpha_D = -17° 37'$ at 17°, upon oxidation yielded citral (citryl-β-naphtho cinchoninic acid, m. p. 197°). Fraction 170—175° ($\alpha_D = -24° 54'$), which constituted about 20 p. c. of the oil, yielded the iodole reaction (iodole-cineol, m. p. 113°) for cineol; and also yielded a nitrosochloride melting at 100°, thus indicating the presence of l-limonene. The highest fraction yielded carvone hydrosulphide melting at 210—211° ($[\alpha]_D = -36° 0'$, at 17° and in 5 p. c. chloroform solution). The amount of l-carvone in Russian spearmint oil is estimated at 5—10 p. c.

The difference between American and German spearmint oil on the

1) Journ. Chem. Soc., 25, p. 1; Jahresb. f. Chem., 1872, p. 816.

2) Berichte, 9, p. 473.

3) Archiv d. Pharm., 221, p. 283.

4) Pharm. Review, 14, p. 244.

5) Am. Journ. Pharm., 57, p. 484.

6) Berichte, 21, p. 156.

7) Proc. Wis. Pharm. Assoc., 1893, p. 58. There also seems to be an alcohol $C_{10}H_{18}O$ present as indicated by the $CaCl_2$ compound and analysis of the regenerated oil. Unpublished results of L. Sumner. *E. K.*

8) Bericht von S. & Co., Apr. 1889, p. 23.

9) Bericht von S. & Co., Apr. 1898, p 28

one hand, and Russian oil on the other, lies principally in the linalool content of the latter accompanied by a small percentage of carvone.

EXAMINATION. For the detection of adulterations Kremers and Schreiner[1] suggest the carvone assay of the oil according to the method previously described.[2] As principal adulterants cedarwood and gurjun balsam oils are taken into consideration. Both of them have about the same specific gravity as spearmint oil; both are also laevogyrate. Individually these oils can be added to the extent of 10—15 p. c., combined however in much larger quantity before being detected by the determination of physical constants. To determine the carvone content of oils thus adulterated is not feasable because the boiling points of carvone and the sesquiterpenes are too close together. The oxime method alluded to above also yields results that are too low. This is attributed to the circumstance that the sesquiterpenes when distilled with water vapor carry some of the carvoxime over and retain it in solution.

Qualitatively the presence of the two adulterants can be ascertained by Wallach's sesquiterpene reaction (p. 124). The test is made with the oily distillate obtained in the carvone assay.

374. Oil of Watermint.

The dry herb of *Mentha aquatica* L. yielded upon distillation 0.34 p. c. of volatile oil of a yellowish-green color and pennyroyal-like odor.[3] Sp. gr. 0.880; $\alpha_D = -2° 14'$.

375. Oil of Mentha Arvensis.

From the dry herb of *Mentha arvensis* L. 0.22 p. c. of oil was obtained. Sp. gr. 0.857; $\alpha_D = -2° 44'$.

376. Oil of Wild Mint.

Mentha canadensis L., which grows wild in North America, yielded upon distillation an oil of reddish-yellow color the odor of which reminded of pennyroyal. The yield from the dry herb was 1.23 p. c.[4] Sp. gr. 0.943 at 15°,[4] 0.927—0.935 at 20°;[5] $\alpha_D = +16° 11'$ to $+20° 32'$. It forms a clear solution with twice its volume of 70 p. c. alcohol.[4]

According to Gage,[5] the oil contains pulegone, the presence of which was determined by means of the bisnitroso compound suggested

---

1) Pharm. Review, 14, p. 244.

2) See under caraway oil p. 554.

3) Bericht von S. & Co., Oct. 1889, p. 55.

4) Bericht von S. & Co., Oct. 1898, p. 45.

5) Pharm. Review, 16, p. 412.

by Baeyer. The melting point of the compound, not mentioned by Baeyer, is 81.5°. Besides pulegone, this oil contains small amounts of thymol or carvacrol.

377. Oil of European Pennyroyal.

Oleum Menthae Pulegii.—Poleiöl.—Essence de Pouliot.

Origin and History. European pennyroyal (Ger. *Polei*) is derived from *Mentha pulegium* L. (*Pulegium vulgare* Mill.) or its hybrid varieties, and has been used medicinally since the middle ages and possibly earlier. The distilled *Oleum pulegii* is mentioned in the price ordinance of Frankfurt for 1582 and appears to have been used medicinally like the plant and its distilled water in the sixteenth and seventeenth centuries.

The commercial European oil of pennyroyal is distilled in Spain, southern France and Algiers from the fresh herb. The Spanish oil is favored on account of its greater purity, whereas the oil from the two other countries is less reliable on account of being frequently adulterated.

Properties. European pennyroyal oil is of a yellowish or reddish-yellow color and of a strong, aromatic, minty odor. Sp. gr. 0.930—0.960; $\alpha_D = +17°$ to $+23°$. It forms a clear solution with 2 and more parts of 70 p. c. alcohol. Inasmuch as the addition of turpentine oil renders the oil less soluble, this test is a valuable one. Turpentine oil also lowers the boiling temperature and the specific gravity.

Composition. European pennyroyal oil has a boiling temperature of limited range. Up to 212° about 5 p. c. pass over; the bulk of the oil, abt. 80 p. c., distills over between 212—216° and consists of a ketone, $C_{10}H_{16}O$, named pulegone by Beckmann and Pleissner.[1] The properties and derivatives of this compound, which boils at 221—222° when pure, are described on p. 169.

The oil examined by Kane[2] in 1839 cannot have been a pure oil, neither can the fraction 183—185°, the composition of which agreed with either of the formulas $C_{10}H_{16}O$ or $C_{10}H_{18}O$, have consisted of pulegone.

Russian Pennyroyal Oil.

The only reference concerning this oil is by Butlerow[3] in 1854. According to him it is distilled from the herb of *Pulegium micranthum* Claus, growing on the steppes of southern Russia, especially near

1) Liebig's Annalen, 262, p. 1.
2) Liebig's Annalen, 32, p. 286; Journal f. prakt. Chem., 15, p. 160.
3) Jahresb. f. Chemie, 1854, p. 594.

Sarepta and Astrachan. Sp. gr. 0.934. It begins to boil at 202°, the bulk coming over at 227°. It has the composition $C_{10}H_{16}O$ and presumably consists principally of pulegone.

378. Oil of Patchouly.

Oleum Foliorum Patchouli. — Patchouliöl. — Essence de Patchouli.

ORIGIN AND PREPARATION. Genuine patchouly, *Pogostemon patchouli* Pellet. is cultivated principally in the Straits Settlements, in Penang, also in the province Wellesley. It is either directly distilled or dried and brought into commerce from Singapore. Comparatively little herb is produced on Mauritius and Réunion.[1] Attempts to cultivate it have been made in Paraguay,[2] also on the West Indian islands Dominica, Guadeloupe and Martinique. No appreciable amounts, however, seem to be produced. Formerly a considerable amount of the drug was also supplied by Java, but in recent years this source of supply has given out entirely. The botanical origin of the plant from Java is not yet known.[3]

The drug shipped from Calcutta and Bombay is of poor quality, containing a large amount of stems and yielding an inferior distillate.[4] The herb cultivated in Assam (Silhet and Khasia mountains) is obtained from *Microtaena cymosa* Prain[3] (*Plectranthus patchouli* Clarke) and probably enters commerce via Calcutta. This would explain the difference of this commercial variety. Whether the herb exported from Bombay is derived from *Pogostemon patchouli* is not known.

Concerning the cultivation and the distillation J. Fisher of Singapore makes the following report:[5]

"The variety selected for cultivation is known locally as *Dhelum Wangi*, which was originally obtained from a small island south of Penang, called Rhio (probably one of the Dindings). The soil most suitable is a rather stiff clay, containing only a small percentage of silica. Land of this description is found near the coast (containing traces of marine deposits), and is planted in rows of 4 or 5 feet apart. The plants are propagated by cuttings struck in the open air, which, until rooted, are sheltered from the sun by pieces of cocoanut shell. The harvest is made in dry weather, and when the sun has drawn up the dew from the leaves; the tops and green parts of the plant are taken off, rejecting all yellow and decayed leaves, and as much as possible the woody stems. The selected parts are then dried in the shade, under large sheds (as the sun would draw out the perfume), and to ensure evenness in drying they are spread on bamboo racks, allowing the air to penetrate from beneath.

---

1) Bericht von S. & Co., Oct. 1890, p. 86.
2) Bericht von S. & Co., Oct. 1887, p. 24.
3) Pharm. Journ., 56, p. 222.
4) Bericht von S. & Co., Apr. 1887, p. 25.
5) Odorographia, vol. 1, p. 297.

During this process they are frequently turned over, and when so far dried as to leave just sufficient moisture to permit of a slight fermentation they are piled in heaps and allowed to heat gently;[1] after this they are again spread out and dried — but not to absolute dryness — and are immediately distilled. The addition of about 25 p. c. of the wild herb *Dhelum Outan* is said to increase the fragrance of the distillate. The distillation is effected by passing steam, generated in a boiler apart, through the leaves in the stills. The pressure of steam is not allowed to rise above 20 lbs., the yield under these conditions being about ¼ oz. per pound of leaves; by high-pressure steam the yield would be a little increased, but rank in quality. The stills are sometimes jacketed, and, by passing a separate current of steam into the jacket condensation in the body of the still at the commencement of the operation is avoided."[2]

The bulk, however, of patchouly oil, which is used exclusively for perfumery, is probably distilled in Europe. With the perfect equipment there in use, the yield is as high as 4 p. c.

The purchase of patchouly herb requires considerable care inasmuch as it is frequently adulterated. The most common adulterant is the leaves of *Ocimum basilicum* L. var. *pilosum* (Family *Labiatae*) known as *Ruku* by the Malays. The peculiar odor of these is lost entirely in the presence of patchouly leaves. Frequently the leaves of *Urena lobata* L. var. *sinuata* (Family *Malvaceae*) are also found, which are called *Perpulut* by the Malays. It is a weed common in cocoanut plantations. Other adulterants are the leaves of *Plectranthus fructicosus*, of *Lavatera olbia* and of *Pavonia weldenii* (Paschkis[3]). In addition to these foreign herbs, which sometimes make up 80 p. c. of the bales, as much as 50 p. c. of sand and earth and up to 35 p. c. of moisture have recently been found.

PROPERTIES. Patchouly oil is a yellowish or greenish-brown to dark brown liquid from which crystals occasionally separate upon standing. The odor of the oil is exceedingly intensive and persistent. Sp. gr. of pure oils distilled in Europe varies from 0.970 to 0.995; α_D varies from — 50 to — 68°. The oil renders a clear solution with equal parts of 90 p. c. alcohol which as a rule remains clear upon the addition of more alcohol. Occasionally, however, the solution becomes turbid upon the addition of 2 volumes, but this turbidity disappears when as much as 4—5 volumes have been added.

---

[1]) Inasmuch as the fresh patchouli herb is odorless, it is to be assumed that the oil is formed during this process of fermentation.

[2]) Further interesting communications concerning the cultivation and distillation of patchouly on the Malaccan peninsula from the pen of Wray, curator of the government museum at Perak, are to be found in the Kew Bulletin for June 1889 and are reproduced in the detailed account in Sawer's Odorographia, vol. 1, pp. 298—308.

[3]) Zeitschr. des öst. Apt. Ver., 17, p. 415; Pharm. Journ., III, 11, p. 818.

Of the imported oils many possess the same properties, some, however, differ materially. Of a number of oils imported from Singapore the sp. gr. lies between 0.957 and 0.965; α_D varied from -44 to $-50°$. These oils were soluble in not less then 3 or 7 volumes of 90 p. c. alcohol. Whether these were adulterated or not cannot be determined at present. They must, however, be regarded as suspicious, for by the addition of cedar wood oil or cubeb oil to patchouly oil distilled in Europe, oils of the just enumerated properties were obtained.

COMPOSITION. Concerning the substances that produce the characteristic odor of patchouly oil nothing whatever is known. Thus far, but two constituents, which are inessential as far as the odor is concerned, have been isolated, namely patchouly alcohol and cadinene.

The patchouly alcohol, formerly known as patchouly camphor, occasionally separates upon prolonged standing and therefore, first attracted the attention of chemists. Gal[1] in 1869 proposed the formula $C_{15}H_{28}O$, Montgolfier[2] in 1877, $C_{15}H_{26}O$. The latest investigation is by Wallach[3] (1894) who introduced the more rational term patchouly alcohol.

Patchouly alcohol, $C_{15}H_{26}O$, crystallizes in transparent, colorless, hexagonal prisms which end in six-sided pyramids and melt at 56°. It is strongly laevogyrate, $[\alpha]_D = -118°$.[2] The elements of water are so loosely bound that even weak dehydrating agents such as hydrochloric acid, sulphuric acid and acetic acid anhydride in the cold, or acetic acid, potassium acid sulphate or zinc chloride with the aid of heat, form the hydrocarbon $C_{15}H_{24}$. This patchoulene boils at 254—256°, has a sp. gr. 0.939 at 23° and has a cedar-like odor. By the exchange of halogen for the hydroxy group of the patchouly alcohol very unstable halides result which at once lose the elements of hydrogen halide. The entire behavior of patchouly alcohol seems to indicate that it is a tertiary alcohol (Wallach).

In fraction 270° Gladstone[4] found a sesquiterpene similar to that obtained from oil of cubeb. In the highest fraction he found a portion the vapors of which were blue, the so-called azulene or coeruleine which also occurs in other oils. According to Wallach,[5] the oil is rich in cadinene, $C_{15}H_{24}$, the hydrochloride of which melted at 117—118°.

---

1) Compt. rend., 68, p. 406; Liebig's Annalen, 150, p. 374.
2) Compt. rend., 84, p. 88.
3) Liebig's Annalen, 279, p. 394.
4) Journ. Chem. Soc., 17, p. 3; Jahresb. f. Chemie, 1863, p. 545.
5) Liebig's Annalen, 238, p. 81.

379. Oil of Dilem.

As *Dilem* the Malays designate a number of plants with patchouly odor. The leaves from Java distilled by Schimmel & Co.[1] were identical with the "flowering patchouly herb" of the botanical garden at Buitenzorg and, according to Holmes,[2] are derived from *Pogostemon comosus* Miq.

These dilem leaves upon distillation yielded 1 p. c. of oil, the odor of which resembled that of patchouly oil, but was much finer. It is yellowish-green, rather viscid, sp. gr. 0.960, and boils between 250—300°.[1]

Upon distillation of the flowering patchouly leaves acetone was observed.[3] An oil distilled in Buitenzorg had a green color and an odor similar to that of patchouly, the intensity of which was agreeably subdued by an anise-like by-odor. Sp. gr. 0.961; $\alpha_D = -32° 17'$.[4]

380. Oil of Sweet Basil.

Oleum Basilici. — Basilicumöl. — Essence de Basilic.

History. Oil of sweet basil appears to have been in use during the middle of the sixteenth century. It is enumerated among the oils of the Frankfurt price ordinance for 1582 and in the Dispensatorium Noricum of 1589. The distilled water was used as early as the fifteenth century.

Origin. Oil of sweet basil is distilled in southern France and Spain, occasionally also in Germany from the fresh herb of *Ocimum basilicum* L. The yield from the German herb is only 0.02—0.04 p. c. On Réunion also oil of sweet basil is distilled. The differences in properties and composition of this oil, however, render it probable that some other species or variety of *Ocimum* is used.

Properties. French and German oil of sweet basil are yellowish, of an aromatic, penetrating odor reminding of estragon. The sp. gr. varies from 0.905—0.930; α_D from -6 to $-22°$. It is soluble in 1—2 parts of 80 p. c. alcohol. The Réunion oil differs from the French in having a camphor-like by-odor, also a higher sp. gr., viz. 0.945 to 0.987. Whereas the French and German oils are laevogyrate, the Réunion oil is dextrogyrate, $\alpha_D = +7°$ to $+12°$. The Réunion oil as a rule is soluble in 7 or more parts of 80 p. c. alcohol. In some instances 3 parts will suffice to produce a clear solution.

---

1) Bericht von S. & Co., Oct. 1888, p. 42.
2) Pharm. Journ., 56, p. 228.
3) Verslag omtrent den Staat van s'Lands Plantentuin te Buitenzorg, 1894, p. 48.
4) Ibidem, 1898, p. 55.

COMPOSITION. Bonastre[1] in 1831 found in oil of sweet basil a solid constituent, the so-called basilicum camphor. An analysis made by Dumas and Peligot[2] in 1835 agreed with the formula $C_{10}H_{22}O_3$, i. e. with terpin hydrate. In all probability this substance was formed by the addition of water to pinene or linalool. The French oil was examined more recently (1897) by Dupont and Guerlain[3] who found as principal constituents methyl chavicol and linalool.

Of similar composition is the German oil as shown by Bertram and Walbaum[4] (1897). The lowest boiling fraction contains cineol[5] (cineol iodole, m. p. 112°). Fraction 215° contains methyl chavicol (homoanisic acid, m. p. 85°). Judging from the methoxy determination, the oil contains 24 p. c. of this compound. Fraction 200° was of an alcoholic nature and probably contains linalool which was found in the French oil. Neither the German nor the French oil contains camphor.

In the Réunion oil Bertram and Walbaum found: 1) d-pinene (pinene nitrolbenzylamine, m. p. 123°); 2) cineol (hydro brom cineol, iodole cineol, m. p. 112°); 3) d-camphor (camphor oxime, m. p. 118°): 4) methyl chavicol, which constitutes the bulk of the oil. It was identified by means of its oxidation to homoanisic acid and anisic acid, also by inversion into its isomer anethol. Judging from a methoxy determination according to Zeisel's method, the Réunion oil contains 67.8 p. c. of methyl chavicol. Linalool was not contained in this oil.

From a small sample of Réunion oil Dupont and Guerlain obtained a crystalline substance, m. p. 64—65°, the amount of which was too small for investigation.

The oil from a large leaved variety of *O. basilicum* known to the natives as *Selasih Mekah*, contained 30—40 p. c. of eugenol.[6]

381. Oil of Mosla Japonica.

According to Shimoyama[7] (1892), *Mosla japonica* Maxim. which is indigenous to Japan yields, when dry, 2.13 p. c. of a reddish-brown, laevogyrate oil, the specific gravity of which is 0.820 (?). It contains 44 p. c. of thymol and a fraction 170—180°, probably cymene.

1) Journ. de Pharm., II, 17, p. 647.

2) Liebig's Annalen, 14, p. 75.

3) Compt. rend., 124, p. 300; Bull. Soc. chim., III, 19, p. 151.

4) Archiv d. Pharm., 235, p. 176.

5) The presence of cineol in oil of sweet basil was previously demonstrated by Hirschsohn in 1893 by means of the iodole reaction (Pharm. Zeitschr. f. Russl., 32, p. 419).

6) Annual report of the botanical garden at Buitenzorg, 1898, p. 28; Bericht von S. & Co., Apr. 1900, p. 5.

7) Apt. Ztg., 7, p. 439; Jahresb. f. Pharm., 1892, p. 465, where the plant is named Mosula, whereas Hartwig (Die neuen Arzneidrogen aus dem Pflanzenreiche, p. 220) designates it Morula.

382. Oil of Dittany.

The dry herb of the North American *Cunila mariana* L. yields 0.7 p. c. of a reddish-yellow oil, sp. gr. 0.915, the odor of which resembles oil of thyme. A preliminary investigation shows it to contain 40 p. c. of phenol, probably thymol.[1]

383. Oil of Lophantus.

The oil of *Lophantus anisatus* Forst. has a pure, pleasant anise odor, reminding somewhat of the honey-like odor of *Solidago odora* Ait. Sp. gr. 0.943 at 20°; $\alpha_D = -7° 10'$.[2]

384. Oil of Pycnanthemum Lanceolatum.

The odor of the oil distilled from the herb of *Pycnanthemum lanceolatum* Pursh[3] is scarcely to be distinguished from that of American pennyroyal oil. Sp. gr. 0.918[4]—0.936[5] at 15°; 0.914—0.935 at 20°; $\alpha_D = -0.566°$ to $+11.083°$.[6]

According to Correll[5] it contains 7—9 p. c. of carvacrol (carvacrol sulphonic acid, m. p. 56—57°, dicarvacrol, m. p. 145—147°). The oil deprived of phenol distills between 180—230°. The analysis of fraction 220—230° (sp. gr. 0.922 at 20°, $[\alpha]_D = +14.88°$) corresponded with the formula $C_{10}H_{16}O$ and indicated the presence of pulegone. Its identity with pulegone was established by Alden[6] by means of the pulegone oxime hydrate melting at 151°. The molten oxime, after having been allowed to cool, then melted at 117—118°. The melting point of the pulegone oxime hydrate of Beckmann and Pleissner melts at 156—157°, that of the normal oxime of Wallach at 118°.

385. Oil of Pycnanthemum Incanum.

The dried herb of the North American labiate, *Pycnanthemum incanum* Michx. yields 0.98 p. c. of a reddish-yellow oil[7] with a strongly aromatic odor, sp. gr. 0.935. It produces a clear solution with twice its volume of 70 p. c. alcohol.

---

1) Bericht von S. & Co., Oct. 1898, p. 44; comp. also Millemann, Am. Journ. Pharm., 88, p. 495.

2) Bericht von S. & Co., Apr. 1898, p. 58.

3) *Pycnanthemum lanceolatum* and *Pycnanthemum linifolium* were formerly regarded as one species and described as *Thymus virginicus* L.

4) Amer. Journ. Pharm., 66, p. 65.

5) Pharm. Review, 14, p. 82.

6) Pharm. Review, 16, p. 414.

7) Bericht von S. & Co., Oct. 1898, p. 45.

386. Oil of Fabiana Imbricata.

Kunz-Krause[1] obtained from the ethereal extract of the leaves of *Fabiana imbricata* Ruiz et Pavon, by steam distillation and shaking out the distillate with ether, a small amount of volatile oil. Upon distillation a small fraction came over at 130°, the principal fraction boiling at 275°. Upon analysis this fraction had the composition $C_{54}H_{90}O_4$ and was called fabianol.

387. Oil of Chione Glabra.

Origin. On account of the aromatic odor of its flowers, *Chione glabra*, a tree belonging to the *Rubiaceae*, is known as *Violette* on the West Indian island Grenada. On Porto Rico it is known as *Palo blanco*. The wood and the bark possess an unpleasant, faecal odor which gradually disappears when exposed to the air.

Preparation and Properties. Distilled with water vapor the bark, according to Paul and Cownly[2] yields 1.5 p. c. of a light yellow oil which is heavier than water and which, when cooled to —20°, congeals to a mass of acicular crystals. This oil, which possesses to a remarkable degree the odor of the bark, has been examined by Dunstan and Henry.[3] Its consists principally of a liquid which congeals at low temperatures, boils at 160° under 34 mm. pressure, has a sp. gr. of 0.850 at 15° and the composition $C_8H_8O_2$. Its odor is aromatic and slightly faecal. With acetic acid anhydride it yields an acetic ester melting at 88°. With hydroxylamine and phenyl hydrazine it yields derivatives indicating the presence of a carbonyl group. The oxime melts at 112°, the phenyl hydrazone at 108°. When fused with potassa, salicylic acid and then phenol are obtained. while nitric acid produces picric acid. These reactions reveal the substance to be o-hydoxy aceto phenone, $C_6H_4.OH.CO.CH_3 = 1:2$. As a matter of fact the properties of this substance agree with those of the artificial compound prepared in a round-about manner from o-nitro cinnamic acid.

In addition to o-hydoxy aceto phenone, the oil contains a crystalline substance that melts at 82° and which possibly is an alkyl derivative of the phenol. On account of the small amount it could not be further examined. The oil also contains traces of nitrogenous compounds, but neither indole nor its derivatives, the presence of which are indicated by the odor, could be obtained.

1) Archiv d. Pharm., 237, p. 10.
2) Pharm. Journ., 61, p. 51.
3) Journ. Chem. Soc. 75, p. 66.

388. Oil of Elder Blossoms.

The oil from the flowers of *Sambucus nigra* L. (Family *Caprifoliaceae*) has been prepared from the fresh as well as the dried material.[1] It was usually obtained by saturating the aqueous distillate with salt, shaking out with ether, and allowing the ether to evaporate spontaneously. In this way Pagenstecher obtained 0.32 p. c. of oil. Schimmel & Co. obtained without the use of ether 0.037 p. c. from the fresh flowers, and 0.0027 p. c. from the dried ones.

At ordinary temperature, elder blossom oil as a rule is a butyraceous or wax-like mass of light yellow or greenish-yellow color. It has an intensive elder blossom odor which becomes especially prominent in great dilutions. The oil is occasionally liquid and congeals when cooled.

According to Gladstone[2] (1864) the oil contains a terpene $C_{10}H_{16}$ and a crystalline substance which is difficultly soluble in alcohol, but not in alkalies and which appears to be a paraffin.

389. Oil of Valerian.

Oleum Valerianae.—Baldrianöl.—Essence de Valeriane.

Origin and History. *Valeriana officinalis* L. and several of its varieties are found wild and cultivated in most of the temperate and more northern countries of Europe and Asia. The peculiar oil contained in the roots, presumably has early attracted attention and brought about its medicinal use. For commercial purposes and for distillation, the plant is recently cultivated in Germany (Thuringia), in France (Dép. du Nord), Holland, England and North America.

The treatises on distillation of the sixteenth century contain directions for preparing distillates of valerian with wine or water. The distilled oil was obtained by Hoffmann, Boerhaave and Geoffroy, in part from the dry, in part from the fresh root. Later Graberg (1782) described the oil, and Trommsdorff investigated the root in 1808. In 1830 he named the acid obtained from the aqueous distillate valerianic acid.

Preparation. For purposes of distillation the dried root and more rarely the fresh root is employed. While Trommsdorff claims that the dried root yields relatively more oil, Zeller is of the opinion that the fresh or dry state has no appreciable effect on the yield. From dried

1) Eliason, Trommsdorff's Neues Journ. d. Pharm., 9, I, p. 246; Winckler, Pharm. Centralbl., 1837, p. 781; Repert. f. d. Pharm., 78 (1841), p. 85; Müller, Archiv d. Pharm., 95 (1846), p. 158.

2) Journ. Chem. Soc., 17, p. 1; Jahresb. f. Chem., 1863, p. 545.

Thuringian root, Schimmel & Co. obtained 0.5—0.9 p. c., from Dutch root about 1 p. c. of oil. The strongly acid aqueous distillate contains valerianic acid which presumably is formed from the bornyl valerianate of the oil during the process of distillation.

PROPERTIES. When fully distilled, oil of valerian is a yellowish-green to brownish-yellow liquid which is slightly acid, and possesses a penetrating, characteristic, not unpleasant odor. Old oil is dark brown and viscid. It has a strongly acid reaction and, on account of the large amount of free valerianic acid it contains, has a disagreeable odor. The stearoptene which occasionally separates from old oil consists of borneol.

As a rule the sp. gr. varies from 0.93 to 0.96. Abnormally light were the oils of Oliviero mentioned under "Composition." They were distilled from the fresh roots of plants growing wild in the Départements Vosges and Ardennes. Their sp. gr. is given as 0.880—0.912 at 0°, corresponding to a density 0.875—0.900 at 15°. Oil of valerian is laevogyrate, $\alpha_D = -8$ to $-13°$. Acid number = 20—50; ester number = 80—100; saponification number = 100—150.

COMPOSITION. Although oil of valerian has been examined repeatedly since the beginning of this century,[1] its composition has become known but comparatively recently through the investigations of Bruylants[2] in 1878 and Oliviero[3] in 1893.

Of the constituents valerianic acid, which derives its name from the plant, is longest known. It was also known that upon oxidation of the oil camphor results.

For the systematic investigations of the oil the esters must first be saponified so that the acids which otherwise result during the process of fractionation do not act on the terpenes and terpene alcohols. The saponified oil begins to boil at 155° and up to 160° a liquid passes over which in its physical properties corresponds to l-pinene, but which consists of two terpenes. By passing dry hydrogen chloride into this fraction a solid dextrogyrate hydrochloride results, which when boiled with potassium acetate is in part decomposed to l-camphene. The part not affected consists of laevogyrate pinene hydrochloride.

---

1) Trommsdorff (1809), Trommsdorff's Journ. d. Pharm., 18, I. p. 8; Liebig's Annalen, 6 (1833), p. 176, and 10 (1834), p. 213. — Ettling (1834), ibid., 9, 40. — Gerhardt & Cahours (1841), Ann. de Chim. et Phys., III, 1, p. 60. — Rochleder (1842), Liebig's Annalen, 44, p. 1. — Gerhardt (1843), ibid., 45, p. 29, and Journ. f. prakt. Chem., 27 (1842), p. 124. — Pierlot (1845), Ann. de Chim. et Phys., III, 14, p. 295, and 56 (1859), p. 291.

2) Berichte, 11, p. 452.

3) Compt. rend., 117, p. 1096; Bull. Soc. chim., III, 11, p. 150; 13, p. 917.

The lowest fraction, therefore, consists of l-camphene[1] and l-pinene. Oliviero is of the opinion that "citrene" (?) is also present, but furnishes no positive proof.

When that portion of the oil which distills under ordinary pressure above 180° is distilled under reduced pressure, fraction 130—140° under 50 mm. pressure largely solidifies on account of the l-borneol present. This alcohol is present in the oil as esters of formic, acetic, butyric and valerianic acids, the largest part being combined with valerianic acid. According to Gerock[2] (1892), the oil contains about 9.5 p. c. of valerianic acid ester and 1 p. c. each of the esters of the other three acids.

Fraction 132—140° (50 mm.) probably contains terpineol, which could not be isolated in the crystalline state. The formation of dipentene dihydrochloride, however, renders its presence probable.

Fraction 160—165° (50 mm.) contains a laevogyrate sesquiterpene, $C_{15}H_{24}$, and fraction 190° a substance $C_{15}H_{26}O$ which, judging from its behavior to benzoic acid anhydride and hydrochloric acid, is an alcohol.

From the water obtained in washing the oil after saponification, Oliviero obtained a crystalline alcohol of the formula $C_{10}H_{20}O_2$. It is strongly laevogyrate and melts at 132°. According to Flückiger,[3] the highest fractions (300°) are blue in color.

Basing his claims merely on an elementary analysis, Bruylants[4] supposes fraction 285—290° to contain bornyl ether or bornyl oxide $C_{10}H_{17}.O.C_{10}H_{17}$, an assertion still wanting proof.

Mexican Oil of Valerian.

A Mexican valerian root, probably derived from *Valeriana mexicana* D. C. was distilled by Schimmel & Co.[5] The distillation yielded only a clear water from which no oil separated. Only after cohobation of the aqueous distillate an oily liquid with disagreeable odor and sp. gr. 0.949 at 15° was obtained. It was optically inactive and when shaken with soda solution was dissolved with the exception of a few flakes. Titration with alcoholic potassa yielded as acid number 415, corresponding to 89 p. c. of valerianic acid hydrate $C_5H_{10}O_2 + H_2O$.

---

1) The claim made by Oliviero that the oils of valerian and spike are the first in which camphene has been found is not correct. The occurrence of camphene in citronella oil and ginger oil was reported in the Bericht of Schimmel & Co. of October 1898, pp. 11 and 22. The first publication of Oliviero (also that of Bouchardat on oil of spike) appeared in the Comptes rendus dated December 28, 1898.

2) Journ. d. Pharm. f. Elsass-Lothringen, 19, p. 82.

3) Archiv d. Pharm., 209, p. 204.

4) Berichte, 11, p. 452.

5) Bericht von S. & Co., Apr. 1897, p. 47.

Mexican valerian root, therefore, yields scarcely a trace of volatile oil, but only free valerianic acid. Inasmuch as the root has a strong odor of the acid, it is to be supposed that it is contained as such in the root and is not formed during the process of distillation.

390. Oil of Kesso Root.

ORIGIN AND PREPARATION. Japanese valerian root is not derived, as was first supposed, from *Patrinia scabiosaefolia* Link, but from *Valeriana officinalis* L. var. *angustifolia* Miq., which is known as *Kesso* or *Kanokosô* in Japan. Upon distillation it yields 8 p. c. of oil and is, therefore, much richer in oil than the common valerian root.

PROPERTIES. The odor of the optically laevogyrate oil is barely distinguishable from that of valerian oil. The oil, however, is much heavier, having a sp. gr. of 0.990—0.996.

COMPOSITION. In composition also kesso oil closely resembles valerian oil for it contains almost all constituents found in the latter. In addition it contains kessyl acetate which causes the higher specific gravity (Bertram and Gildemeister,[1] 1890).

When kesso oil is distilled a first fraction is obtained that is strongly acid, has the disagreeable odor of decayed cheese and contains acetic and valerianic acids and probably valerianic aldehyde.

Fraction 155—160° is strongly laevogyrate and like the corresponding fraction of valerian oil contains l-pinene (nitrosopinene, m. p. 101°) and l-camphene[2] (isoborneol, m. p 212°). Fraction 170—180° contains dipentene (tetrabromide, m. p. 123°). It is questionable whether this terpene is an original constituent of the oil or whether it is formed by the action of the acids on pinene or terpineol.

The borneol of kesso oil, like that of valerian oil, is laevogyrate and is present as acetate and isovalerianate. Bornyl formate found in oil of valerian is not contained in kesso oil.

Terpineol was identified by the formation of dipentene dihydriodide (m. p. 76°) from fraction 200—220°. Other reactions, more characteristic for terpineol, could not be applied on account of the borneol which is difficult to remove.

Fraction 260—280° had a decided sesquiterpene odor but did not yield a solid hydrochloride. The highest boiling fractions contain in addition to a blue oil, the acetate of an alcohol $C_{14}H_{24}O_2$, the kessyl acetate.

---

1) Archiv d. Pharm., 228, p. 483. 2) Journ. f. prakt. Chem., II, 49, p. 18.

Kessyl acetate, $C_{14}H_{23}O_2COCH_3$, is a viscid liquid which does not solidify at —20°. Under 15—16 mm. pressure it boils between 178 and 179°; under atmosperic pressure at about 300° with slight decomposition; $\alpha_D = -70° 6'$.

Kessyl alcohol, $C_{14}H_{24}O_2$, crystallizes in large, well formed crystals of the rhombic system. It is odorless, insoluble in water, readily soluble in alcohol, ether, chloroform, benzene and petroleum ether. It melts at 85° and under 11 mm. pressure boils between 155 and 156°, under ordinary pressure between 300 and 302°. Its alcoholic solution is laevogyrate. Upon oxidation with potassium bichromate and sulphuric acid a substance with two hydrogen atoms less is obtained, which crystallizes in thick needles melting at 104—105° and turns the plane of polarized light to the left.

391. Oil of Valeriana Celtica.

The small *Valeriana celtica* L. of the Styrian alps formerly yielded a drug known as *Spica celtica* (Ger. *Alpenspik, celtischer Spik*). The root yields upon distillation 1.5—1.75 p. c. of volatile oil of a strong odor reminding more of Roman chamomile and patchouly than of valerian. Sp. gr. 0.967; boiling temperature 250—300°.[1]

392. Oil of Nardostachys Jatamansi.

The rhizomes of *Nardostachys jatamansi* D. C. (*Patrinia jatamansi* Wallich) and of *N. grandiflora* D. C., (Family *Valerianaceae*) which are indigenous to the Himalaya mountains of northern India have an odor reminding faintly of musk but more strongly of patchouly. On account of its fragrance this root was highly prized during antiquity and was used for perfuming ointments and fatty oils Later other roots were used as substitutes, notably that of *Ferula sumbul* Hook. fil. (p. 576) and still later that of *Valeriana celtica* L. (see above).

According to Kemp[2] the root of *Nardostachys jatamansi* D. C. yields upon distillation 1 p. c. of a volatile oil of light yellow color. Sp. gr. 0.9748 at 22°; $\alpha_D = -19° 5'$.

393. Oil of Dog Fennel.

Eupatorium foeniculaceum Willd. (Family *Compositae*) known as dog fennel on account of its fennel-like leaves, is found along the Virginia coast and in other southern states of North America. An oil

[1] Bericht von S. & Co., Oct. 1887, p. 36.
[2] Pharmacographia Indica, vol. 2, p. 237.

distilled in Florida from the entire plant was light in color and had an aromatic, pepper-like odor not at all resembling fennel. Sp. gr. 0.935; $\alpha_D = +17° 50'$. It contained a large amount of phellandrene.[1]

394. Oil of Ageratum Conyzoides.

This oil was distilled in the Botanical Garden at Buitenzorg. The presence of methyl alcohol in the aqueous distillate was observed.

Sp. gr. 1.015 at 27.5°; optical rotation −5.5° (in 200 mm. tube). It boils at about 260° and probably contains sesquiterpene.[2]

395. Oil of Golden Rod.

In the United States east of the Rocky Mountains there are found about fifty species of *Solidago* known as golden rod, some of which are so common as to be regarded as weeds. Several possess more or less aromatic properties.

The oil of golden rod so-called is derived from *Solidago odora* Aiton. It has a strongly aromatic, but not especially pleasant odor and a sp. gr. of 0.963.[3]

The fresh, flowering herb of *Solidago canadensis* L. yields 0.63 p. c. of oil. It is of a light yellow color and has a very agreeable, sweetly aromatic odor. Sp. gr. 0.859; $\alpha_D = -11° 10'$.[4]

This oil contains about 85 p. c. of terpenes, especially pinene with some phellandrene and dipentene, possibly also limonene. The higher boiling portions consist of borneol (total 9.2 p. c.), bornyl acetate (3.4 p. c.) and cadinene. It is remarkable that this golden rod oil should so closely resemble, as to composition, the pinene needle oils of an entirely different family.[5]

The odor of the oil of *Solidago rugosa* Mill. is said to resemble that of origanum oil.

396. Oil of Erigeron (Fleabane).

Oleum Erigerontis.—Erigeronöl.—Essence d'Erigeron.

Origin. *Erigeron canadensis* L. is a very common weed which is known in America as fleabane, horseweed or butterweed. It is frequently found in peppermint fields. The fresh herb yields upon distillation 0.2—0.4 p. c. of oil which finds limited medical application in the United States and which was made official in the U. S. Pharmacopoeia of 1890.

1) Bericht von S. & Co., Apr. 1896, p. 70.
2) Ibidem, Apr. 1898, p. 57.
3) Ibidem, Oct. 1891, p. 40.
4) Ibidem, Apr. 1894, p. 57.
5) Bericht von S. & Co., Apr. 1897, p. 38.

Properties. When fresh, erigeron oil is a colorless or light yellow, mobile liquid, and possesses a peculiar aromatic, persistent odor and a somewhat prickly taste. Exposed to the air it rapidly resinifies and becomes viscid and darker. Sp. gr. 0.850—0.870; $\alpha_D = +52°$.[1] With an equal volume of 90 p. c. alcohol it forms a clear solution.

Composition. Erigeron oil boils almost completely at 175° (Power[2]) and consists very largely of d-limonene. Vigier and Cloez[3] (1881) prepared a dihydrochloride, which Beilstein and Wiegand[4] (1882) found to melt at 47—48°; Wallach[5] (1885) prepared the tetrabromide melting at 104—105°; Meissner[6] (1893) the nitrosochlorides, and from the α-nitrosochloride the benzylamine base melting at 90—92°.

In fraction 205—210° Hunkel[7] found terpineol which he identified by means of its nitrosochloride and the nitrolpiperidine base melting at 159—160°.

397. Oil of Blumea Balsamifera.

The semi-shrub-like composite *Blumea balsamifera* D. C. is indigenous to India and is found from the Himalayas to Singapore and in the Malay Archipelago. It likewise grows in the Islands Hai-nan and Formosa. In Hai-nan and in the Chinese province Kwang-tung considerable quantities of the so-called Ngai camphor, *Ngai-fên*, are obtained from it by distillation. From Hoi-han on Hai-nan annually 15,000 lbs. are said to be exported. The crude Ngai camphor is rectified in Canton and is then called *Ngai-p'ien* (Holmes[8]). In Burma also this camphor is prepared.[9]

Ngai camphor is used in China for ritualistic and medicinal purposes. It is also added to the better qualities of india ink.[10]

Formerly this camphor was rarely brought to Europe. In 1895 Schimmel & Co. obtained a sample in the form of a yellowish-white, crumbly mass consisting almost entirely of l-borneol.[11] (See p. 143.) The identity of Ngai camphor with l-borneol was first established in 1874 by both Plowman[12] and Flückiger.[13]

---

1) Bericht von S. & Co., Oct. 1894, p. 73.
2) Pharm. Rundschau, 5, p. 201.
3) Journ. de Pharm., V, 4, p. 236.
4) Berichte, 15, p. 2854.
5) Liebig's Annalen, 227, p. 292.
6) Amer. Journ. Pharm., 65, p. 420.
7) Pharm. Rundschau, 13, p. 187.
8) Pharm. Journ., III, 21, p. 1150.
9) Buchner's Neues Repert. f. d. Pharm., 23, p. 321; Hanbury, Science Papers, 1876, p. 394.
10) Pharmacognosie, 1891, p. 158; Pharmacographia, 1879, p. 518.
11) Bericht von S. & Co., Apr. 1894, p. 74.
12) Pharm. Journ., III, 4, p. 710; Neues Repert. f. d. Pharm., 23, p. 325.
13) Pharm. Journ., III, 4, p. 829.

398. Oil of Blumea Lacera.

Blumea lacera D. C. is a perennial plant which is widely distributed in India. It possesses a strong camphor-like odor and is therefore used by the natives against flies and other insects.

Dymock[1] obtained by the distillation of 150 lbs. of fresh, flowering herb about 2 oz. of a light yellow oil, sp. gr. 0.9144 at 26.7° and $\alpha_D = -66°$.

399. Oil of Helichrysum Stoechas.

The oil of the flowering herb of *Helichrysum stoechas* D. C. is used in Spain as a remedy against diseases of the bladder and kidneys. It has the odor of inferior coniferous oils and about three-fourths boil between 155 and 170°, the remaining fourth between 170 and 260°. Pinene is probably the principal constituent.[2]

400. Oil of Elecampane.

Oleum Helenii. — Alantöl. — Essence de Racine d'Aunée.

ORIGIN AND HISTORY. Upon distillation of elecampane, the root of *Inula helenium* L., with water vapor, 1—2 p. c. of a solid, crystalline mass permeated by some liquid oil is obtained. Aqueous distillates were used medicinally during the middle ages. *Oleum radicis helenii* is first mentioned in the Frankfurt ordinance of 1582. Oil of elecampane is used principally for the preparation of alanto-lactone, the helenin of Gerhardt[3] (1839).

COMPOSITION. Oil of elecampane consists almost entirely of alanto-lactone with which are mixed small amounts of alantolic acid, alantol and a substance $(C_6H_8O)_x$, the helenin[4] of Kallen.[5]

Alanto-lactone $C_{14}H_{20}\left\langle\begin{matrix} O \\ | \\ CO \end{matrix}\right.$ was formerly regarded by Kallen as alantolic acid anhydride. Bredt and Posth recognized its character as lactone and changed the name correspondingly.

---

1) Pharm. Journ., III, 14, p. 985.

2) Bericht von S. & Co., Oct. 1889, p. 54.

3) Ann. de Chim. et Phys., II, 72, p. 163; and III, 12, p. 188. — Liebig's Annalen, 34, p. 192, and 52, p. 389.

4) Bredt and Posth (Liebig's Annalen, 285, p. 349) in 1895 suggested to drop the term helenin because it is used to designate three different substances. Gerhardt designated his impure alanto-lactone as helenin; Kallen so designates the substance $(C_6H_8O)x$; finally the inulin from elecampane is sometimes called helenin. The helenin of commerce is almost pure alanto-lactone.

5) Berichte, 6, p. 1506. Compare also Kallen, Ueber Alantolacton und die Anlagerung von Blausäure an ungesättigte Lactone. Inaug.-Dissertation. Rostock, 1895.

Recrystallized from dilute alcohol, the lactone is obtained in colorless, prismatic needles melting at 76°, of faint odor and taste. Under 10 mm. pressure it boils between 205—210°. It is readily soluble in alcohol, ether, chloroform, glacial acetic acid, benzene, petroleum ether, sparingly soluble in water. In sodium carbonate solution it is insoluble in the cold. When heated with solutions of the alkalies, it dissolves to form the salts of the corresponding oxy acid, the alantolic acid,

$$C_{14}H_{20}\left\langle\begin{matrix} O \\ | \\ CO \end{matrix}\right. + KOH = C_{14}H_{20}\left\langle\begin{matrix} OH \\ COOK \end{matrix}\right.$$

Alantolic acid, $C_{14}H_{20}(OH)COOH$, crystallizes in fine needles which melt at 94°, splitting off water. It is readily soluble in sodium carbonate solution, also in alcohol and ether. When heated with water it is largely decomposed. The hot aqueous solution upon cooling separates fine crystals of alantolic acid which are contaminated with the lactone. The easy change from the one to the other explains why the crude oil contains alantolic acid besides the lactone.

The helenin of Kallen, $(C_6H_8O)_x$, crystallizes in four sided prisms which melt at 110°. Under ordinary pressure it does not distill without decomposition; under 14 mm. pressure, however, it distills at 240° without decomposition. It is indifferent toward ordinary reagents. Since it is contained in the root in small amounts only, it has not yet been farther examined.

Alantol is an oil that boils at about 200°, and apparently occurs in small quantities in the fresh root only. It is a yellowish liquid, and an isomer of common camphor, $C_{10}H_{16}O$. Upon distillation with phosphorus pentasulphide it yields cymene, $C_{10}H_{14}$, b. p. 175°.

401. Oil of Osmitopsis Asteriscoides.

An alcoholic infusion of *Osmitopsis asteriscoides* (*Osmites bellidiastrum* Thbg., *Bellidiastrum osmitoides* Less.) is used in Cape Colony as a remedy against lameness.

The volatile oil of the plant was examined by Gorup-Besanez[1] in 1854. It is a mobile, yellowish-green liquid of a not pleasant odor, reminding of camphor and cajeput oil; sp. gr. 0.931 at 16°. The oil began to boil at 176°, two-thirds passed over between 178—188°, the remainder between 188—208°. When the thermometer rose to 206° a sublimate, possibly of camphor, was deposited in the neck of the retort.

1) Liebig's Annalen, 89, p. 214.

Fraction 178—182° had the composition $C_{10}H_{18}O$ and probably consisted of cineol, which is also indicated by the cajeput-like odor and specific gravity.

402. Oil of Ambrosia Artimisiaefolia.

Ambrosia artimisiaefolia is a common weed in North America and is known as ragweed, hogweed, bitterweed and Roman wormwood.

The fresh, flowering herb yielded upon distillation 0.07 p. c. of a dark green oil possessing an aromatic, not unpleasant odor. Sp. gr. 0.870; $\alpha_D = -26°$.[1]

403. Oil of Roman Chamomile.

Oleum Chamomillae Romanae. Oleum Anthemidis. — Römisch Kamillenöl. — Essence de Camomille Romaine.

Origin and History. *Anthemis nobilis* L. occurs wild only in scattered districts of western Europe. In Europe and America it is cultivated as garden plant and for commercial purposes. Occasionally it escapes cultivation.

Owing to the similarity of several species of *Anthemis*, *Chrysanthemum* and *Matricaria* it is not known what plant was designated as *Anthemis* by the writers of antiquity. In the German treatises on distillation the common chamomile seems to have been preferred for medicinal purposes, whereas in England the Roman chamomile was almost exclusively used as chamomile flowers. Distilled oil of Roman chamomile is first mentioned along with *Oleum chamomillae vulgaris* in the price ordinance of Frankfurt for 1587.

Preparation. Up to quite recent times Roman chamomile was cultivated to a considerable extent in the neighborhood of Leipzig, but has been dropped for several years because no longer profitable. At the present time the oil is distilled principally in Mitcham near London, where the plant is cultivated and where it also grows wild. The dried flowers yield about 0.8—1 p. c. of oil.

Properties. Freshly distilled Roman chamomile oil has a light blue color, which under the influence of air and light changes to greenish and brownish-yellow. The odor is strong but pleasant, the taste burning. Sp. gr. 0.905 to 0.915, $\alpha_D = +1$ to $+3°$ (Umney,[2] 1895). The oil has usually a faintly acid reaction, saponification number 250—300. As a rule it forms a clear solution with 6 parts of 70 p. c.

---

1) Bericht von S. & Co., Oct. 1894, p. 78. 2) Pharm. Journ., III, 25, p. 949.

alcohol. A lesser solubility is occasionally due to a larger paraffin content, and does not necessarily indicate adulteration.

COMPOSITION. The oil of Roman chamomile consists principally of a mixture of esters of butyric acid, and the isomeric angelic and tiglinic[1] acids which are present as esters of isobutyl, amyl and hexyl alcohols.

Angelic acid was discovered in the oil by Gerhardt[2] in 1848, and has thus become readily accessible to chemists. The other statements by Gerhardt concerning the presence of angelic aldehyde and a terpene boiling at 175°, chamomillene, have been shown to be wrong.

Demarçay[3] in 1873 showed that angelic acid is not free but combined with butyl and amyl alcohols. He supposed that he had found valerianic acid, but this was not verified by later investigators.

More detailed examinations of the oil were made by Fittig and Kopp[4] in 1876 and by Fittig and Köbig[5] in 1879. Kopp saponified the oil and isolated and identified the acids, whereas Köbig fractionated the oil and examined the various fractions. The two investigations, therefore, supplement each other.

Upon repeated fractionation, five large fractions are obtained: 1) 147—148°; 2) 177—177.5°; 3) 200—201°; 4) 204—205°; 5) above 220° decomposition sets in and about one-third of the oil remains in the residue.

Fraction 1, b. p. 147—148°, contains an ester of isobutyric acid, probably that of isobutyl alcohol, $C_4H_7O.OC_4H_9$, with which is mixed a small amount of a hydrocarbon. In the lowest fraction of the acids, 150—160°, after separation of the alcohols, a white amorphous powder separated which was not isolated in a pure state but which, judging from its properties, consisted of methacrylic acid $CH_2:C(CH_3)COOH$. The first fraction of the non-saponified oil, therefore, may be regarded as an ester of this acid.

Fraction 2 consists of fairly pure angelic acid isobutyl ester, $C_5H_7O.OC_4H_9$. The isobutyl alcohol obtained from it boils at 107—108°, the angelic acid melted at 45°.

Fraction 3, b. p. 200—201°, consisted of a colorless, rather viscid oil having the odor of chamomile. Upon saponification it was resolved

[1]) It is doubtful whether tiglinic acid is contained as such in the oil or is formed from the angelic acid.

[2]) Compt. rend., 26, p. 225; Ann. de Chim. et Phys., III, 24, p. 96; Liebig's Annalen. 67, p. 235; Journal f. prakt. Chemie, 45, p. 235.

[3]) Compt. rend., 77, p. 360; ibid., 80, p. 1400.

[4]) Berichte, 9, p. 1195; ibid., 10, p. 513.

[5]) Liebig's Annalen, 195, pp. 79, 81 and 92.

into inactive amyl alcohol of fermentation, $C_5H_{12}O$, b. p. 129—130°, and angelic acid. It, therefore, consists of angelic acid amyl ester, $C_5H_7O.OC_5H_{11}$.

Fraction 4 consists of a mixture of the amyl esters of angelic and tiglinic acids.

Fraction 5. The portion above 220° was saponified with potassa. Among the volatile acids, tiglinic and angelic acids were found. The oil distilled from the lye was separated into a smaller fraction, 152—153°, and into a larger one, 213.5—214.5°.

Fraction 152—153° upon analysis revealed itself as a hexyl alcohol which upon oxidation yielded a capronic acid. Van Romburgh[1] showed in 1886 that this acid is identical with methyl ethyl propionic acid, and that the alcohol, which was also prepared synthetically, has the formula $C_2H_5(CH_3)CH.CH_2.CH_2OH$. The methyl ethyl propyl alcohol of Roman chamomile oil boils at 154°, has a sp. gr. of 0.829 and is dextrogyrate, $[\alpha]_D = +8.2°$, whereas the synthetic alcohol is optically inactive. The oil contains about 4 p. c. of this hexyl alcohol.

Fraction 213.5—214.5° is a colorless, viscid liquid of camphor-like odor, which does not distill without decomposition under ordinary pressure. It is isomeric with camphor, having the formula $C_{10}H_{16}O$, and was named anthemol. The acetate of this alcohol boils at 234—236°, and upon saponification yields the original alcohol. Chromic acid destroys the alcohol entirely, dilute nitric acid oxidizes it to p-toluic and terephthalic acids, also a more soluble acid, possibly terebinic acid.

There remains to be mentioned that Naudin[2] obtained from the flowers of Roman chamomile, by the extraction with petroleum ether a paraffin, "anthemene", $C_{18}H_{36}$, which melted at 63—64°.

404. Oil of Anthemis Cotula.

Upon distillation of the fresh flowers of *Anthemis cotula* L. and shaking the aqueous distillate with ether, Hurd[3] in 1885 obtained 0.013 p. c. of oil. The entire fresh plant, when treated in like manner, yielded 0.01 p. c. of a reddish oil with acid reaction and bitter taste; sp. gr. 0.858 at 26°. Upon cooling a crystalline acid separated, which melted at 58° after purification. The same acid was obtained upon saponification of the oil. The saponified oil boiled between 185—290° and left a residue of 20 p. c.

---

1) Rec. des Trav. chim. des Pays-Bas, 5, p. 219, and 6, p. 150; Berichte, 20, pp. 375 and 468. Referate.

2) Bull. Soc. chim., II, 41, p. 483.

3) Am. Journ. Pharm., 57, p. 376.

405. Oil of Milfoil.

The fresh flowers of *Achillea millefolium* L. yield upon distillation 0.07—0.13 p. c. of volatile oil of a dark blue color and a strongly aromatic, camphor-like odor, sp. gr. 0.905—0.925. It was first prepared by Bley[1] in 1828. The only known constituent is cineol, the presence of which was demonstrated by Schimmel & Co.[2] The high boiling blue fraction is possibly identical with the so-called azulene or coeruleine of chamomile oil.

From the roots of milfoil, Bley[3] in 1828 obtained 0.032 p. c. of an almost colorless oil possessing an unpleasant taste and a peculiar, faintly valerian-like odor. In the aqueous distillate acetic acid was found. The root contains traces of volatile sulphur compounds.

406. Oil of Achillea Nobilis.

Bley[4] in 1835 prepared the oil from the flowers (0.24 p. c.), dry herb (0.26 p. c.) and the seeds (0.19 p. c.) of *Achillea nobilis* L. The oil from the herb had a sp. gr. of 0.970. It had a strong camphor-like odor similar to that of milfoil, but finer, and a spicy, bitter taste.

407. Oil of Achillea Moschata.

The iva herb used in the manufacture of iva liquor is obtained from *Achillea moschata* L. and owes its agreable odor to a volatile oil of which about 0.5 p. c. are obtained upon distillation.

The oil has a greenish-blue to dark blue color and a strongly aromatic odor and taste. Sp. gr. 0.932—0.934. It boils between 170—260°. The only substance characterized so far is cineol,[5] identified by Hirschsohn's cineol reaction. The so-called ivaol of Planta-Reichenau,[6] $C_{24}H_{40}O_2$, boiling from 170—210°, cannot be an individual substance, as is shown by its boiling temperature.

408. Oil of Achillea Coronopifolia.

Schimmel & Co.[7] obtained from Spain a deep blue, mobile oil of *Achillea coronopifolia* Willd., possessing a strong but pleasant odor reminding of tansy. Sp. gr. 0.924.

---

1) Trommsdorff's N. Journ. d. Pharm., 16, II, p. 96.
2) Bericht von S. & Co., Oct. 1894, p. 55.
3) Trommsdorff's N. Journ. d. Pharm., 16, I, p. 247.
4) Archiv d. Pharm., 52, p. 124.
5) Bericht von S. & Co., Oct. 1894, p. 27.
6) Liebig's Annalen, 155, p. 148.
7) Bericht von S. & Co., Apr. 1898, p. 64.

409. Oil of Achillea Ageratum.

The oil of *Achillea ageratum* L. was distilled by de Luca[1] in 1875 from the flowering plant. Sp. gr. 0.849 at 24°. Upon distillation two fractions were obtained: the first boiling between 165—170°; the second between 180—182° and yielding upon analysis results agreeing with the formula $C_{26}H_{44}O_8$.

410. Oil of German Chamomile.[2]

Oleum Chamomillae. — Kamillenöl. — Essence de Camomille.

Origin and History. The common chamomile, *Matricaria chamomilla* L. (*Chrysanthemum chamomilla* Bernh.) grows throughout Europe with the exception of the extreme north, and has migrated to North America and Australia.

Chamomile was used medicinally by the Greeks and Romans, and was highly valued during the middle ages. Brunschwig describes the distillation of the flowers.

The oil, which attracted attention on account of its blue color, seems to have been known since the middle of the fifteenth century. Saladin mentions it in his list of 1488, and the Nürnberg physician Joachim Camerarius prepared it in 1588. Gesner and Porta distilled the oil after moistening the flowers with *aqua vitae*. In the price ordinances it is first mentioned in that of Berlin for 1574 under the title of *Oleum matricariae* and of Frankfurt-on-the-Main for 1587. In the Dispensatorium Noricum of 1589 it is mentioned as *Oleum chamomillae vulgaris* besides *Oleum matricariae*.

The blue color of the oil was attributed to a copper content due to the distilling vessels, until Pauli and Herford of Copenhagen showed in 1664 that the oil distilled from glass vessels likewise possessed a blue color.[3] On a large scale by steam distillation, chamomile oil was first distilled in 1822 by the apothecary Franz Steer of Kaschau in Hungary.[4] The oil was first examined by Zeller[5] in 1827.

Preparation. For purposes of distillation the more unsightly blossoms, which are not as serviceable for pharmaceutical purposes, are employed. The drug used for distillation comes from Hungary. The

---

1) Ann. de Chim. et Phys., V, 4, p. 182; Jahresb. f. Chem., 1875, p. 849.

2) In the United States and England the term "oil of chamomile" is also used to designate the oil of Roman chamomile, *Anthemis nobilis*.

3) Quadripartitum botanicum, p. 282.

4) Buchner's Repert. f. d. Pharm., 61, p. 85.

5) Buchner's Repert. f. d. Pharm., 25, p. 467.

yield varies from 0.2 to 0.36 p. c. On account of the small yield of oil, chamomile was formerly distilled with lemon oil and thus the *Oleum chamomillae citratum* of the apothecary shops was obtained.

PROPERTIES. At middle temperature, chamomile oil is a rather viscid liquid of a dark blue color, which when carelessly exposed to light and air, passes into green and finally into brown. The odor of the oil is strong and characteristic, the taste bitter and aromatic. Sp. gr. 0.930—0.940 at 15°, at which temperature the oil begins to deposit crystals. When cooled the oil becomes butyraceous and at 0° has congealed to a solid mass. The saponification number averages about 45. On account of the large paraffin content, the oil yields but turbid mixtures with 90 p. c. alcohol.

COMPOSITION. Very little is known about the composition of oil of chamomile for, with the exception of the odorless paraffin, not a single constituent has been isolated. The older investigations of Bornträger[1] (1844) and Bizio[2] (1861) contain nothing worth mentioning. The examination by Gladstone[3] (1864) was restricted principally to the blue fraction, the "coeruleine." The statements made concerning the properties and composition of this substance differ greatly from those of Piesse[4] (1863) and Kachler[5] (1871).

An oil distilled by Kachler himself began to boil at 105°. Up to 180° about 4.5 p. c. of a faintly bluish fraction was obtained which possessed a strong chamomile odor; 8.3 p. c. boiled between 180—250°; between 255—295° 42 p. c. were obtained, distilling with blue vapors. The fraction obtained above 295° (25 p. c.) was very viscid, the vapors becoming violet at last; a residue of 20 p. c. of a pitchy, brownish mass remained. The distillation was accompanied by decomposition, for all of the fractions had an acid reaction. By shaking with aqueous potassa an acid was extracted the silver salt of which gave figures agreeing with caprinic acid, $C_{10}H_{20}O_2$. After a second fractionation of the two lower fractions an oil was obtained between 150—165°, the composition of which accidentally agreed with the formula $C_{10}H_{16}O$. A fraction boiling within a range of 15 degrees cannot represent a chemical individual. The analyses of the following fractions are of no interest for the same reason.

1) Liebig's Annalen, 49, p. 243.

2) Wiener akadem. Ber., 48, 2nd part, p. 292; Jahresb. f. Chemie, 1861, p. 681.

3) Journ. Chem. Soc., 17, p. 1; Jahresb. f. Chem., 1863, p. 550.

4) Compt. rend., 57, p. 1016; Chem. News, 8, pp. 245, 273; Chem. Centralbl., 1864, p. 320.

5) Berichte, 4, p. 36.

The blue oil named azulene by Piesse, boiled between 281—289° and was regarded by Kachler as a polymer of camphor of the formula $(C_{10}H_{16}O)_3$. By treatment with potassium a hydrocarbon boiling between 250—255° was obtained. The formula $C_{30}H_{48}$ assigned to it is not very probable if the boiling point is taken into consideration. (The sesquiterpenes $C_{15}H_{24}$ boil at about the above temperature). Anhydrous phosphoric acid acts on the blue substance with the formation of a hydrocarbon $(C_{10}H_{14})_n$. In its general properties, the blue fraction of chamomile oil agrees with the blue oil obtained by Mössmer[1] (1861) upon the destructive distillation of galbanum. Possibly the high boiling, blue fractions of other oils, such as those of milfoil, kesso, valerian, Roman chamomile, wormwood and others are identical with the blue fractions of the oil of German chamomile. None of the others, however, have such a deep blue color and it would, therefore, seem probable that the substance involved is contained in the purest condition in chamomile oil. It certainly is worthy of careful investigation.

The congealing of chamomile oil at low temperatures is due to its paraffin content. The paraffin, which remains in the flask when the oil is fractionated, as a dark colored mass, is difficultly soluble in alcohol, but readily in ether and persistently retains the blue color of the oil. In the pure state it is white as snow, melts at 53—54°, and shows all the properties of the paraffin hydrocarbons.[2] Whether it is an individual hydrocarbon or a mixture of several homologues, has not yet been determined.

Examination. It is reported that oil of milfoil is used for the adulteration of chamomile oil, it being specially suited to this purpose on account of its blue color.[3] Larger quantities, however, are excluded on account of its totally different odor. More frequently oil of cedarwood[4] seems to be used which has no marked odor of its own. The addition of this oil, however, or of other oils reduces the congealing point. Genuine oil begins to get viscid at + 15° and becomes solid at 0°. Adulterated oils remain liquid.

411. Oil of Feverfew.

Origin. *Matricaria parthenium* L. (*Pyrethrum parthenium* Sm., *Chrysanthemum parthenium* Pers., Ger. *Mutterkraut*) is frequently cultivated in Germany and also occurs wild. It supplied the *Herba Matricariae* formerly officinal in Germany. Upon distillation of the

1) Liebig's Annalen, 119, p. 262.
2) Bericht von S. & Co., Apr. 1894, p. 18.
3) Chem. Ztg., 8, p. 268, and 19, p. 385.
4) Bericht von S. & Co., Apr. 1895, p. 19.

fresh, flowering plant 0.02—0.07 p. c. of a yellow to dark green oil is obtained. Sp. gr. 0.908—0.960. At times the oil deposits crystals at ordinary temperature. According to Chautard the budding herb yields more oil than the flowering herb, but that of the former contains less stearoptene.

COMPOSITION. Oil of feverfew is of special interest because the rare l-camphor[1] was first found in it by Dessaignes and Chautard[2] in 1848. In order to obtain this substance the oil is fractionated, and fraction 200° is subjected to the temperature of a freezing mixture, and the crystals separated by means of a force filter. l-Camphor melts at 175°, boils at 204° and deviates the ray of polarized light as far to the left, ($[\alpha]_D = -33.1°$) as ordinary camphor does to the right.

Feverfew cultivated in the neighborhood of Leipzig yielded 0.068 p. c. of oil, sp. gr. 0.960. At ordinary temperature a considerable quantity of hexagonal crystals separated, which, when recrystallized from petroleum ether, melted at 203—204° and consisted of l-borneol, $[\alpha]_D = -36°$. Camphor was not to be found in the oil.[3]

It seems probable that the oil examined by Dessaignes and Chautard also contained borneol besides camphor, for the mother liquor from the crystals, after treatment with nitric acid, yielded more camphor. It is but rational to assume that the further amounts of camphor obtained after the treatment with nitric acid owed their origin to the borneol present.

According to Chautard[4] the oil further contains a terpene boiling below 200°, which did not yield a solid derivative with hydrogen chloride; and a liquid dextrogyrate constituent boiling above 200° which contained more oxygen than camphor. In connection with the oil mentioned above, Schimmel & Co. obtained the saponification number 131 which indicates a rather high ester content.

412. Oil of Tansy.

Oleum Tanaceti. — Rainfarnöl. — Essence de Tanaisie.

ORIGIN AND HISTORY. *Tanacetum vulgare* L. (*Chrysanthemum tanacetum* Karsch) is found wild in most European countries, likewise in the Atlantic states of North America, and is also cultivated. It is one of the strongly aromatic composites. The oil, which is especially contained

1) l-Camphor occurs also in one other oil, namely in oil of tansy.
2) Journ. f. prakt. Chem., 45, p. 45.
3) Bericht von S. & Co., Oct. 1894, p. 71.
4) Journ. de Pharm., III, 44, p. 18; Jahresber. f Chemie, 1863, p. 555.

in the flowers, has been used as an anthelminthic since the middle ages. In the United States the leaves und flowers are official.

The distilled water from the flowers and leaves was a common remedy during the sixteenth and seventeenth centuries. The oil is first mentioned in the price ordinance of Frankfurt for 1582 and in the Dispensatorium Noricum of 1589. The oil was first examined by Persoz[1] in 1841.

PREPARATION. Fresh, flowering tansy yields upon distillation 0.1 to 0.2, dry herb 0.2—0.3 p. c. of oil.[2] The commercial oil is obtained principally from North America. It should be noted that the oil from different sources varies somewhat.

PROPERTIES. Oil of tansy is a yellowish liquid, which becomes brown under the influence of air and light. The sp. gr. of the oil from fresh herb is 0.925—0.940, that from dried herb 0.955; $\alpha_D = +30$ to $+45°$. The American oil, when pure, forms a clear solution with 3 parts of 70 p. c. alcohol. An oil distilled in Germany gave no clear solution with 70 p. c. alcohol.

An English oil from cultivated tansy[3] had but little resemblance, as to odor, to ordinary oil of tansy but resembled oil of rosemary and had a decided odor of camphor; upon evaporation of the oil a delicate ambra odor was finally developed. Its camphor content was so large that a part of it crystallized out at 0°. Very remarkable was its optical behavior for it was strongly laevogyrate, $\alpha_D = -27°$.

COMPOSITION. The bulk of oil of tansy consists of thujone or tanacetone to which the oil owes its characteristic odor. It was first isolated in a pure state by Bruylants[4] in 1878. If the oil is shaken according to his directions with an equal volume of sodium acid sulphite solution and two parts of alcohol, a compound $C_{10}H_{16}O.NaHSO_3$ is formed upon prolonged standing, from which the pure thujone can be separated by means of soda. That this substance, which Bruylants regarded as an aldehyde and named tanacetyl hydride, is a ketone and identical with the thujone isolated by Wallach from thuja oil, was shown by Semmler.[5] Properties and derivatives of thujone are described on p. 167.

1) Compt. rend., 18, p. 486; Journ. für prakt. Chem., 25, pp. 55, 60.

2) The yield mentioned by Leppig (Berichte, 15, p. 1088, Ref.), viz., 1.49 p. c. from the flowers and 0.66 p. c. from the herb can not well refer to the oil obtained by distillation.

3) Bericht von S. & Co., Oct. 1895, p. 35.

4) Berichte, 11, p. 449.

5) Berichte, 25, p. 3343.

Fraction 203—205° upon oxidation with chromic acid yielded a small amount of camphor. Bruylants supposed that this fraction contained the alcohol $C_{10}H_{18}O$ corresponding to thujone and regarded the camphor as its oxidation product. Inasmuch as it is now known that thujyl alcohol does not yield camphor, this ketone must be regarded either as a constituent of the oil or as a product of oxidation from some substance other than thujyl alcohol. Persoz[1] (1841), also Vohl[2] (1853) had observed that the oil, after having been oxidized with chromic acid, contained camphor. Persoz did not answer the question whether this camphor was contained in the oil as such or whether it was a product of the oxidation.

In order to decide this question Schimmel & Co. removed the thujone as completely as possible by shaking with bisulphite solution, and fractionated the remaining oil. At 205° the distillate congealed partly in the receiver. The crystalline mass was separated by suction and crystallized from 80 p. c. alcohol. Odor and other properties indicated a mixture of camphor and borneol. For their separation Haller's method[3] was employed and a relatively large amount of camphor with little borneol obtained. The camphor was identified by means of the oxime melting at 116°. The optical properties showed that it was not the common dextro camphor, but the very rare laevo camphor. The amount of borneol was too small to determine its rotation.

It remains to be determined whether thujyl alcohol is contained in the oil, its presence being probable according to Bruylants; also whether the terpene of Bruylants, boiling about 160°, is pinene or camphene.

According to Peyraud,[4] oil of tansy is exceedingly poisonous. In animals it produces a condition similar to rabies, *la rage tanacétique.*

413. Oil of Tanacetum Balsamita.

The fresh, flowering herb of *Tanacetum balsamita* L. yielded upon distillation 0.064 p. c. of oil.[5] The odor is agreeably balsamic, but little characteristic, reminding of tansy. Sp. gr. 0.943—0.949; $\alpha_D = -43°40'$ to $-53°48'$; sap. number = 21. In the cold, paraffin-like crystals are formed at the surface. The oil was not soluble in 80 p. c. alcohol. With 1—2 parts of 90 p. c. alcohol, however, it gave a clear solution, but this was rendered turbid upon the addition of more alcohol, white flakes (paraffin?) separating. The oil boiled between 207—283°.

---

1) Compt. rend., 13, p. 436; Liebig's Annalen, 44, p. 318; Journ. f. prakt. Chem., 25, p. 55.

2) Archiv d. Pharm., 124, p. 16.

3) Compt. rend., 108, p. 1308.

4) Compt. rend., 105, p. 525.

5) Bericht von S. & Co., Oct. 1897, p. 66.

414. Kiku Oil.

Kiku oil is distilled in large quantities in the western part of Japan from the leaves and flowers of *Pyrethrum indicum* Cass. (Jap. *Abura-Kuku*). The production in 1887 amounted to about 1,400 k. It is used as a popular remedy.[1]

The oil of the leaves is colorless and has a camphoraceous odor reminding somewhat of eucalyptus. Sp. gr. 0.885; boiling temperature 165—175°.

The oil of the flowers is likewise colorless and has an unpleasant odor. The first fraction distilling at about 180° has a pleasant odor. the higher fractions are camphoraceous but not agreeable.[2]

415. Oil of Artemisia Vulgaris.

Artemisia vulgaris L. (Ger. *Beifuss*) is a common weed found along hedges and roadsides and contains volatile oil in all its parts, the root yielding 0.1 p. c., the herb 0.2 p. c.

The root oil is greenish-yellow, butyraceous, crystalline.[3] It has a nauseating, bitter taste, first burning then cooling.

The oil from the herb is but little characteristic as to odor. Sp. gr. 0.907. According to an observation made in the laboratory of Schimmel & Co. it contains cineol.

416. Oil of Estragon.

Oleum Dracunculi.—Esdragonöl.—Essence d'Es ragon.

Origin. Estragon oil, which is used in the preparation of aromatic vinegars, is distilled from the flowering herb of *Artemisia dracunculus* L. The dry herb yields 0.25—0.8 p. c.; the fresh herb 0.1—0.4 p. c.

Properties. Estragon oil is a colorless to yellowish-green liquid of peculiar, anise-like odor and strongly aromatic, but not sweetish taste. The sp. gr. varies greatly, 0.900—0.945. An oil distilled from the dry Thuringian herb had a sp. gr. of but 0.890. The oil is dextrogyrate, $\alpha_D = +2$ to $+9°$. It is soluble in 10 parts of 80 p. c. alcohol.

Composition. Estragon oil was first examined by Laurent[4] in 1842. Upon oxidation he obtained an acid melting at 175° which he termed dragonic acid. Gerhardt[5] (1844) identified this acid with anisic acid

1) Bericht von S. & Co., Apr. 1888, p. 46.
2) Bericht von S. & Co., Apr. 1887, p. 37.
3) Bretz & Elieson, Taschenbuch für Chemiker und Apotheker, 1826, p. 61.
4) Liebig's Annalen, 44, p. 313.
5) Compt. rend., 19, p. 489; Liebig's Annalen, 52, p. 401.

and showed that estragon oil behaved toward reagents like anise oil, and concluded that the two oils were identical. Since then anethol was regarded as the principal constituent of estragon oil. When in 1892 Schimmel & Co. found methyl chavicol (p-methoxy allyl phenol), the isomer of anethol (p-methoxy propenyl phenol) for the first time in a volatile oil, viz. oil of anise bark, the estragon-like odor attracted attention. This suggested a new investigation of estragon oil which showed that the bulk of this oil consists of methyl chavicol.[1]

Upon energetic oxidation, methyl chavicol like anethol yields anisic acid, milder oxidation with permanganate, however, yields homoanisic acid. The statement by Gerhardt that the acid obtained by the oxidation of anise oil was anisic acid was correct, but his conclusions were wrong.

Somewhat later (1893) Grimaux[2] likewise found methyl chavicol in estragon oil and named it estragol.

The properties and derivatives of methyl chavicol are described on p. 179. In addition to the inactive methyl chavicol, estragon oil contains several other constituents to which the optical activity of the oil is due, but these have not yet been examined.

417. Oil of Levant Wormseed.

Origin. The unexpanded flowers of *Artemisia maritima* var. *stechmanni* are used as a popular anthelmintic. The bulk of the levant wormseed is used in the manufacture of santonin, whereby the oil was obtained as by-product. Since the manufacture of this by-product has been dropped the price of the oil has risen and its consumption decreased. To this circumstance should be added that the principal constituent of the oil, the cineol or eucalyptol, can be had pure and at a low price. The yield varies from 2—3 p. c.

Properties. Oil of levant wormseed is a yellowish liquid possessing the camphoraceous odor of cineol, but also the unpleasant by-odor characteristic of the drug. Sp. gr. 0.915—0.940. The oil is slightly laevogyrate.

Composition. Up to 1884 oil of levant wormseed had been repeatedly examined but the results were unsatisfactory because they were more or less contradictory, particularly those pertaining to the dehydration products of the constituent $C_{10}H_{18}O$. Whereas Völckel[3]

---

1) Bericht von S. & Co., Apr. 1892, p. 17.

2) Compt. rend., 117, p. 1089.

3) Liebig's Annalen, 88, p. 110; 87, p. 312; 89, p. 358.

(1841, 1853 and 1854), Hirzel[1] (1854), Kraut and Wahlforss[2] (1864) and Graebe[3] (1872) claimed that phosphoric acid anhydride produced a hydrocarbon $C_{10}H_{16}$, Faust and Homeyer[4] (1874) showed that this hydrocarbon "cynene" was identical with cymene, $C_{10}H_{14}$. These contradictory points, however, were cleared up by two contributions published at almost the same time, viz. those of Wallach and Brass[5] and Hell and Stürcke.[6]

Wallach and Brass were the first to isolate the oxygenated constituent in a pure state, by making its hydrogen chloride addition product and decomposing this with water. This substance, which they called cineol, is dehydrated by hydrogen chloride or benzoyl chloride yielding a terpene, cynene, which later was termed dipentene. They also showed that dipentene under the influence of concentrated sulphuric acid or phosphorus pentachloride yields cymene. This explains the contradictory statements of earlier chemists who according to the strength and character of the dehydrating agents obtained a mixture in which either dipentene or cymene predominated.

In addition to cineol, the oil contains an unknown hydrocarbon, either a terpene or cymene, the boiling point of which lies very close to that of cineol; also an oxygenated substance boiling higher than cineo and optically laevogyrate.

418. Oil of Wormwood.

Oleum Absynthii. — Wermutöl. — Essence d'Absynthe.

Origin and History. *Artemisia absinthium* L. is indigenous to many European countries and has been introduced into North America. For commercial purposes the plant is frequently cultivated.

The distilled oil of wormwood was known to Porta about 1570, who called attention to its blue color. It was first examined by Hoffmann in 1722 and recommended by him for medicinal purposes. It was also examined by Geoffroy (1721), Kunzemüller (1784), Buchholz (1785), and Margueron (1798). In the price ordinances it occurs in that of Frankfurt for 1587, also in the Dispensatorium Noricum of 1589.

Preparation. Whereas the French oil formerly controlled the market, it is now replaced more and more by the cheaper but less

1) Jahresb. f. Chemie, 1854, p. 591; 1855, p. 655.

2) Liebig's Annalen, 128, p. 298.

3) Berichte, 5, p. 680.

4) Berichte, 7, p. 1429.

5) Liebig's Annalen, 225, p. 291.

6) Berichte, 17, p. 1970.

prized American oil from New York (Wayne Co.), Michigan, Nebraska and Wisconsin. Similar in quality to the French are the Spanish and Algerian oils, also the oil prepared in small quantities on Corsica. The Russian oil which for a time was to be had, has again disappeared from the market. The consumption of wormwood oil has decreased considerably, due possibly to the toxic properties of the oil to which attention has been directed.

From the fresh herb cultivated in Germany ½ p. c. of oil at most is obtained, which at first is dark brown, the color changing to green after prolonged exposure to the air.

PROPERTIES. Oil of wormwood is a somewhat viscid liquid of a dark green or occasional blue color. It has the not pleasant odor of the plant and a bitter, grating and persistent taste. Sp. gr. 0.925—0.955. On account of the dark color the angle of rotation cannot be determined. Inasmuch as the principal constituent of the oil, the thujone, is strongly dextrogyrate, the oil itself must be dextrogyrate. The oil is soluble to a clear solution in 2—4 p. of 80 p. c. alcohol.

COMPOSITION. The first chemical examination of the oil was made by Leblanc[1] in 1845. After repeated rectification over lime, he obtained a principal fraction boiling at 205° which had the composition $C_{10}H_{16}O$. By treating this with phosphoric acid anhydride he obtained a hydrocarbon $C_{10}H_{14}$. These results were confirmed by the later investigations of Cahours[2] (1847), Schwanert[3] (1863) and Gladstone[4] (1864). Beilstein and Kupffer[5] (1873) called the substance $C_{10}H_{16}O$ absynthol and identified its dehydration product with cymene. They also verified Gladstone's statement that the high fraction 270—300° is identical with the corresponding fraction of oil of chamomile.

By washing the various fractions with aqueous potassa, Beilstein and Kupffer obtained an acid, the barium salt of which indicated acetic acid.

The identity of absynthol with tanacetone or thujone was pointed out by Semmler[6] (for properties see p. 167).

Fritzsche Bros.[7] recently examined an oil distilled in their Garfield laboratories. The thujone was removed as completely as possible by

1) Compt. rend., 21, p. 379; Ann. de Chim. et Phys., III, 16, p. 333; Chem. Centralbl., 1846, p. 62.

2) Compt. rend., 25, p. 725.

3) Liebig's Annalen, 128, p. 110.

4) Journ. Chem. Soc., 17, p. 1; Jahresb. f. Chem., 1863, p. 549.

5) Liebig's Annalen, 170, p. 290.

6) Berichte, 25, p. 3350. Comp. also ibid., 27, p. 895, and Bericht von S. & Co., Oct. 1894, p. 51; and Wallach, Liebig's Annalen, 286, p. 93.

7) Bericht von S. & Co., Apr. 1897, p. 51.

means of sodium acid sulphite solution and alcohol, and the remaining oil saponified with alcoholic soda. The saponified oil was distilled with water vapor, but a viscid residue remained which was only partly soluble in ether. The portion not soluble in ether proved to be the sodium salt of an acid which was identified with palmitic acid by means of its melting point and the analysis of its silver salt. Of the volatile acids acetic and isovalerianic acids were identified.

The saponified oil was distilled in vacuum and the lower boiling portions fractionated under ordinary pressure. A small fraction 158—168° resulted in which the presence of phellandrene was demonstrated. An attempt to prepare pinene nitrosochloride resulted in the formation of the characteristic blue color and a few crystals were obtained, but the amount was too small to make possible the identification of pinene.[1]

Fraction 200—203° was found to consist of thujone which cannot be removed completely from the oil by means of the bisulphite. It was characterized by means of the bisulphite addition product, also by its tribromide melting at 121—122°. Neither of these compounds could be obtained from the following fraction 210—215° which was shown to be of alcoholic nature by means of acetylization and subsequent saponification. Careful oxidation with chromic acid converted it almost completely into thujone, thus demonstrating the presence of thujyl (tanacetyl) alcohol.

In the higher, fairly large fraction 260—280° the presence of cadinene was demonstrated by means of the hydrochloride melting at 117—118°.

By saponification of the original and of the acetylized oil it was shown that the oil contains 17.6 p. c. of thujyl acetate (=13.9 p. c. of thujyl alcohol) and 24.2 p. c. of total thujyl alcohol (free and combined).

Oil of wormwood, therefore, consists of the following substances: 1) thujone, $C_{10}H_{16}O$; 2) thujyl alcohol, $C_{10}H_{18}O$, free and combined with acetic, isovalerianic and palmitic acids; 3) phellandrene and possibly pinene; 4) cadinene; 5) blue oil, the elementary composition of which has not yet been definitely determined.

Examination. Oil of wormwood is principally adulterated with turpentine oil. Inasmuch as oil of wormwood itself contains but little of this terpene, its presence is readily ascertained. If a first fraction of 10 p. c. is distilled over, this should be soluble in 2 p. of 80 p. c. alcohol.

[1]) Wright (1874) (Pharm. Journ., III, 5, p. 233) found two hydrocarbons in the oil, the one boiling at 150°, the other between 170—180°. Brühl (1888) (Berichte, 21, p. 156) is of the opinion that the physical properties mentioned by Gladstone indicate d-pinene.

419. Oil of Artemisia Gallica.

According to Heckel and Schlagdenhauffen[1] (1885), *Artemisia gallica* Willd., which is widely distributed throughout France, contains besides santonin 1 p. c. of volatile oil. In its preparation a small amount of a crystalline substance (camphor?) was obtained. Concerning the composition of the oil nothing is known.

420. Oil of Artemisia Barrelieri.

The oil distilled from the flowering herb of *Artemisia barrelieri* Bess. is used in Spain as a popular remedy against colic, and hysteric and epileptic attacks. It is also said to be used in the preparation of Algerian absynthe.

The odor is pleasant, strong, aromatic and reminds strongly of tansy.[2] Sp. gr. 0.923. It boils between 180—210° and consists almost entirely of thujone,[3] thus explaining its similarity to wormwood and tansy. If it could be obtained in large quantities it would be admirably suited for the preparation of pure thujone.

421. Oil of Artemisia Glacialis.

The dry herb of *Artemisia glacialis* L. (*Genepi des Alpes*) (Ger. *Alpenbeifuss, Genepikraut*) yields upon distillation 0.15—0.3 p. c. of volatile oil with a strongly aromatic odor. Sp. gr. 0.964 at 20°. It boils between 195 and 310°. At 0° it solidifies to a butyraceous mass due to the presence of a fatty acid melting at 16°.[4]

422. Oil of Fireweed.

Erechthitis hieracifolia Raf. is frequently found wild in burned forest districts, hence the name. According to Todd[5] its sp. gr. is 0.845—0.855, according to Power[6] 0.838 at 18.5°. It is either dextro- or laevogyrate, $\alpha_D = -2$ to $+2°$.[6]

According to Beilstein and Wiegand[7] (1882) the oil consists principally of a terpene boiling at 172°, sp. gr. 0.838 at 18.5° which absorbs a molecule of hydrogen chloride without forming a crystalline derivative. Fraction 240—310° also has the elementary composition $C_{10}H_{16}$. The fractions boiling above 190° are, according to Power, polymerization products due to boiling.

---

1) Compt. rend., 100, p. 804.
2) Bericht von S. & Co., Oct. 1889, p. 58.
3) Bericht von S. & Co., Oct. 1894, p. 51.
4) Bericht von S. & Co., Apr. 1889, p. 48.
5) Am. Journ. Pharm., 59, p. 312.
6) Pharm. Rundschau, 5, p. 201.
7) Berichte, 15, p. 2854.

423. Oil of Petasites Officinalis.

Petasites officinalis Moench, (Ger. *Pestwurz*), yields upon distillation 0.1 p. c. of volatile oil which does not form a clear solution with even 10 parts of 90 p. c. alcohol. Sp. gr. 0.944; $\alpha_D = +2° 18'$.

424. Oil of Arnica Root.

Origin and Properties. This oil is obtained from the freshly dried root of *Arnica montana* L., with a yield of 0.5—1 p. c. It is at first light yellow in color and darkens with age. It has an odor reminding of radish, and a pungent aromatic taste. Sp. gr. 0.990—1.000; $\alpha_D = -1° 58'$.

Composition. Arnica root oil was first examined by Walz[1] in 1861, who claimed to have found capronic and caprylic esters in the oil and capronic and caprylic acids in the aqueous distillate. Entirely different results were obtained by Sigel[2] in 1873. He found isobutyric acid in the aqueous distillate, also traces of formic acid and a small amount of an acid the silver salt of which gave results agreeing with either angelic or valerianic acids.

The oil boiled between 214 and 263° with decomposition, leaving a brown resinous residue.

For the investigation of the individual constituents the oil was saponified with alcoholic potassa, and the alkaline solution neutralized with dilute sulphuric acid. A phenol, b. p. 224—225°, and sp. gr. 1.015 at 12° was thus obtained, the elementary analysis of which agreed with the formula $C_8H_{10}O$. Sigel, therefore, regards the substance as phlorol (ethyl phenol) without, however, bringing any proof. The ethyl ether of this phenol is a colorless liquid of agreeable aromatic odor, sp. gr. 0.9323 at 18° and b. p. 215—217.° The alkaline liquid also contained isobutyric acid, thus rendering the presence of isobutyric acid phloryl ester in the original oil probable.

The oil separated from the alcoholic alkaline solution by means of water boiled between 224—245° and upon oxidation with chromic acid yielded thymoquinone, $C_{10}H_{12}O_2$. When heated with hydriodic acid in a sealed tube, methyl iodide, thymohydroquinone, $C_{10}H_{14}O_2$, and a phenol boiling at 225—226° and of the composition of phlorol, $C_8H_{10}O$, were obtained. These results indicate the presence of hydrothymoquinone methyl ether and phloryl methyl ether in the oil.

---

1) Neues Jahrb. der Pharm., 15, p. 329; Archiv d. Pharm., 158, p. 1; Jahresb. f. Chemie, 1861, p. 752.

2) Liebig's Annalen, 170, p. 345.

According to Sigel the oil of arnica root consists one-fifth of isobutyric acid phloryl ester, the remaining four-fifths consisting largely of the methyl ether of hydrothymoquinone and of a small amount of the methyl ether of a phlorol.

425. Oil of Arnica Flowers.

ORIGIN AND PROPERTIES. The flowers of *Arnica montana* L. give upon distillation a small yield, 0.04—0.07 p. c., of oil. It is of a reddish-yellow to brown color and has a strong aromatic odor and taste. At normal temperature it is usually a butyraceous mass, sometimes, however, it congeals only when cooled. In one instance[1] the sp. gr. was found 0.906 at 15°; in another[2] 0.900 at 25°; a third oil had the sp. gr. 0.8977 at 15°. Acid number 75.1, saponification number 29.9. It was not completely soluble in 10 parts of 90 p. c. alcohol.

COMPOSITION. The oil has not yet been examined. In the petroleum ether extract Börner[3] found lauric and palmitic acids, also hydrocarbons of the paraffin series. Inasmuch as these three substances are volatile with water vapor they must be contained in the oil, and are undoubtedly the cause of its congealing. Schimmel & Co. isolated an acid melting at 61° from the oil.

426. Oil of Costus Root.

ORIGIN. The root of *Saussurea lappa* Clarke (*Aplotaxis lappa* Decaisne, *A. auriculata* D. C., *Aucklandia costus* Falconer) was known to the Greeks as *κόστος* and was used in the preparation of a fragrant ointment. The plant is indigenous to the northwestern Himalaya mountains and grows at an altitude of 7,000 to 13,000 ft. The root is fully developed in fall and is collected in September and October. In Cashmere as much as 2 million lbs. are said to be collected annually. They are used principally to protect the Cashmere shawls against insects. Large quantities of the root are also exported to China where they are used as incense under the name of *Putchuc*. Upon distillation the root yields 0.8—1 p. c. of volatile oil.

PROPERTIES. The oil of costus root is viscid and of a light yellow color.[4] The odor reminds first of elecampane[5] and later of violet. The oil from old roots sometimes has an unpleasant odor. Sp. gr. 0.982 to 0.987; $a_D = +15$ to $+16°$. The oil begins to boil at 275°. About

1) Bericht von S. & Co., Oct. 1889, p. 5.
2) Bericht von S. & Co., Oct. 1891, p. 4.
3) Inaug.-Dissertation. Erlangen, 1892.
4) Bericht von S. & Co., Apr. 1896, p. 42.
5) Elecampane was used during antiquity as adulterant of costus root.

half passes over by 315° when complete decomposition sets in. By treatment with soda lye a part of the oil is dissolved and is again set free from the salt solution by means of acids.[1]

427. Oil of Carline Thistle.

The root of *Carlina acaulis* L. (Ger. *Eberwurz*) yields upon distillation 1.5—2 p. c. of an oil with a narcotic odor, and sp. gr. 1.030 at 18°. Upon distillation under ordinary pressure, one-half passes over between 265—300°, then decomposition and complete resinification sets in.[2]

By distillation over sodium in a vacuum, Semmler[3] isolated a fraction 139—141° (20 mm.) which was a hydrocarbon of the composition $(C_5H_8)_x$ (b. p. under ordinary pressure 250—253°). According to its boiling point it might be a sesquiterpene, but for such its sp. gr., 0.8733 at 22.8°, is low (sp. gr. of all known sesquiterpenes 0.90—0.92).

The principal constituent of the oil is an oxygenated, specifically heavy substance boiling at 169—171° under 21 mm. pressure. The oil also contains a substance which separates in white, shining laminae upon cooling.

428. Oil of Sphaeranthus Indicus.

The composite *Sphaeranthus indicus* L. with its rose-like odor is used in India as a medicine. According to Dymock[4] (1884) it yields upon distillation a dark red, viscid oil which is rather soluble in water. From 150 lbs. of the fresh herb about ½ oz. of oil was obtained.

OIL OF UNKNOWN BOTANICAL ORIGIN.

429. Oil of Anise Bark.

As anise bark Schimmel & Co. obtained in 1891 a bark from Madagascar,[5] which resembled massoy bark in exterior appearance, but differed materially as to odor. The aroma was more anise- than estragon-like, and reminded somewhat of safrol. The supposition of the importer that it is the bark of the star-anise tree is not probable, inasmuch as *Illicium verum* does not occur on Madagascar. It may be the bark of another species of *Illicium*, possibly of *I. parviflorum* Michx., which is reputed to have a sassafras-like odor. Upon distillation the

1) Bericht von S. & Co., Apr. 1892, p. 41.
2) Bericht von S. & Co., Apr. 1889, p. 44.
3) Chemiker-Zeitung, 18, p. 1158.
4) Pharm. Journ., III, 14, p. 985.
5) Bericht von S. & Co., Apr. 1892, p. 40.

anise bark yielded 3.5 p. c. of a light yellow oil of anise-like odor but without a sweet taste. Sp. gr. 0.969; $\alpha_D = -0°46'$. In addition to small amounts of anethol, the oil consisted largely of methyl chavicol, which had been prepared artificially by Eykman.

430. Oil of Quipita Wood.

This Venezuelan wood is of a light color, dense in texture but not very hard, and enters commerce in billets several meters in length and 5—20 cm. in diameter. The thicker stems have a thin, white outer bark and somewhat resemble birch stems; the bark of the younger stems is grayish-brown.

Upon distillation of the rasped wood, Schimmel & Co.[1] obtained 1 p. c. of a light yellow oil, with terebinthinate odor. Sp. gr. 0.934; $\alpha_D = -34° 31'$. Saponification number of the original oil 2.9; after acetylization 40.2. This shows that in addition to a small amount of ester there are alcoholic constituents in the oil.

1) Bericht von S. & Co., Oct. 1896, p. 75.

BIBLIOGRAPHIC NOTES.

Those who desire further information than that here given concerning the writings mentioned in the historical chapters of this work, or about their authors, may consult the historical appendix to Flückiger's Pharmacognosie des Pflanzenreiches and the works there mentioned. The translator desired to make these bibliographic notes more complete, but time did not permit to do so at present.

Aëtius. A Christian physician, born in Ămid, now Diarbekr, on the upper Tigris, who was educated in Alexandria. Between 540 and 550 he wrote a medical treatise in sixteen books or section. The following editions are referred to in the historical chapters of this work:

Aetii medici graeci ex veteribus medicinae tetrabiblos. Editio Aldina. Veneti 1547.

Aëtius "*βιβλία ἰατρικὰ ἑκκαιδεκὰ*". Libri medicinales sedecim. Editio Aldina 1533.

Anschütz, R. and H. Reitter. Dr. Anschütz was in 1895 Professor of Chemistry at Bonn, when Dr. Reitter was laboratory assistant.

Die Destillation unter vermindertem Druck im Laboratorium. Zweite neu bearbeitete Auflage. Ein Bd., pp. IV, 86. Verlag von Fr. Cohen, Bonn, 1895.

Arabian Physicians. The writings of the Arabian physicians and scientists are preserved principally in a collective edition printed in Venice in 1502. The following works are contained in this volume:

Uni Joannis Mesuë Liber de consolatione medicinarum simplicium et correctione operationem earum canones universales: cum expositione preclarissimi medici magistri Bondini de lentiis felicitur incipiunt.

Additiones Petri Apponi medici clarissimi, et Francisci de Pedemontium.

Joannis Nazareni filii Mesuë Grabaddin medicinarum particularium incipit.

Antidotarium Nicolai cum expositionibus, et glossis clarissimi magistri Platearii.

Expositio Joannis de Santo Amando supra antidotarrii Nicolai incipit feliciter.

Tractatus de synonymis quid pro quo.

Liber Servitoris seu libri XXVIII Bulchasin Ben-aherazern: translatus a Simone Januensi: interprete Abraamo Judeo Tortuosienzi.

Uni Saladini de esculo Servitati principis Tarenti physici principalis compendii aromatiorum opus feciliter incipit.

Quae omnia supradicta hic finem habent ad laudam dei. Veneti impressa anno Domini 1502, die 28 Junii.

The oldest single editions of these works date up to 1471, about the time of the introduction of the printing of books.

Askinson, George Wm. Manufacturer of perfumes. The following works are of A. Hartleben's chemisch-technische Bibliothek.

Die Fabrikation der ätherischen Oele. Anleitung zur Darstellung der ätherischen Oele nach den Methoden der Pressung, Destillation, Extraction, Deplacirung, Maceration und Absorption, etc. Zweite vermehrte und verbesserte Auflage. Ein Bd., pp. VIII, 215, mit 36 Abbildungen. A. Hartleben's Verlag, Wien, 1887.

Die Parfumerie-Fabrikation. Vierte, vermehrte und verbesserte Auflage. Ein Bd., pp. 376, mit 35 Abbildungen. A. Hartleben's Verlag, Wien, 1895.

Avenzoar. (Averrhoës, Averroës.) Abul-Welid Muhammed Ben Ahmed Ibn Roschd el-Maliki was born (1126) and educated in Cordova. A famous Arabian philosopher and physician. Of his numerous works the following is referred to on p. 19.

Liber Theizir Dahalmodana Vahaltadabir prooemium Averrhoi Cordubensis ab Jacobo Hebraeo. Anno 1281. Colliget Veneti 1553.

Ayurvedas. See Susruta.

Bergmann, Torbern. Swedish chemist, born 1735, died 1784.

De primordiis chemiae. Upsala 1779. Editio Hebenstreit. Lipsiae 1787.

Historiae chemiae medium seu obscurum aevum. Editio Hebenstreit. Lipsiae 1787.

Bessonius.

Jacobi Bessonii, De absoluta ratione extrahendi aquas et olea ex medicamentis simplicibus a quodam empirico accepta et a Bassonio locupletata, experimentis confirmata. Tiguri 1559. — French edition, Paris 1573.

Boerhaave, Hermann. Born 1668 near Leiden. Since 1709 Professor of medicine, botany and chemistry at Leiden. Died there 1738.

Elementa chemiae, quae anniversario labore docuit, in publicis, privatisque scholis, Hermannus Boerhaave. Tomus primus, qui continet historiam et artis theoriam. Tomus secundus, qui continet operationes chemicas. Lugduni Batavorum 1732 — Londini 1732 et 1735 — Parisii 1732, 1733, 1753 — Lipsiae 1732 — Basileae 1745 — Veneti 1745 et 1759.

BORNEMANN, GEORG.

Die flüchtigen Oele des Pflanzenreichs, ihr Vorkommen, ihre Gewinnung und Eigenschaften, ihre Untersuchung und Verwendung. Ein Bd., pp. XII, 441, mit einem Atlas von 8 Foliotafeln, enthaltend 88 Abbildungen. Verlag von B. V. Voigt. Weimar 1891.

BRUNFELS, OTTO. 1488—1534. Carthusian friar, then teacher in Strassburg, died as city physician in Basel.

Spiegel der Arznei. Strassburg 1532.—Reformation der Apotheken. Strassburg 1536.

BRUNSCHWIG, HIERONYMUS. About 1450—1530. Born in Strassburg, physician.

Liber de arte distillandi. For the complete title see title pages of both volumes on pp. 23 and 25, reproduced half size.

BURGHART, G. H.

Die zum allgemeinen Gebrauch wohl eingerichtete Destillirkunst. Auch die Bereitung verschiedener destillirter Wässer und Oele. Von G. H. Burghart. Breslau 1736. — Neue Auflage mit vielen Zusätzen von J. Christian Wiegleb. 1754.

Das Brennen der Wasser, Oele und Geister. Wohleingerichtete Destillirkunst und neue Zusätze. Von G. H. Burghart. Breslau 1748.

CARTHEUSER, JOH. FRIEDR. 1704—1769. Professor of medicine, botany and chemistry at the University of Frankfurt-on-the-Oder. His contributions on volatile oils are contained in the following works:

Fundamenta materiae medicae. Francofurt. ad Viadr. 1738 and Paris edition 1752.

Elementa Chymiae dogmatico-experimentalis, una cum synopsi Materiae medicae selectioris. Halae 1736. Editio secunda priore longe emendatior.

Dissertatio chymico-physica de genericis quibusdam plantarum principiis hactenus neglectis. Francof. ad Viadr. 1754. Editio secunda 1764.

Dissertatio physico-chemica medica de quibusdam Materiae medicae subjectis exarat. ac publice habet nunc iter. resus. Francof. ad Viadr. 1774.

Dissertationes nonnullae selectiores physico-chemicae ac medicae. varii argumenti post novam lustrationem ad prelum revocat. Francof. ad Viadr. 1778.

Pharmacologia theoretico-practica praelectionibus academicis accommodata. Berolini 1745.

CHARAKA. See Susruta.

CORDUS, VALERIUS. Born 1515 near Erfurt, died 1544 in Rome. See pp. 27 and 30. His lectures on Dioscorides were published after his death and reedited by Conrad Gesner with the following title:

In hoc volumine continentur Valerii Cordi Simesusii Annototiones in Pedacii Dioscoridis Anązarbei de medica materia libros quinque longe aliae quam antea sunt hac sunt evulgatae.

Ejusdem Val. Cordi Historiae stirpium libri quatuor posthumi nunc primum in lucem editi, adjectis etiam stirpium iconibus et brevissimis Annotatiunculis. Sylva qua rerum fossilium in Germania plurimarum. Metallorum, Lapidum et Stirpium aliquot rariorum noticiam brevissime persequitur, nunc hactenus visa.

De artificiosis extractionibus liber. — Compositiones medicinales aliquot non vulgares. — Hic accedunt Stockhornii et Nessi in Bernatium Helvetiorum ditione montium, et nascentium in eis stirpium, descriptio Benedicti Aretii, Graecae et Hebraicae linguarum in schola Bernensi professoris clarissimi. Item Conradi Gesneri De Hortis Germaniae liber recens una cum descriptione Tulipae Turcarum, Chamaecerasi montani, Chamaeopiti, Chamaenerii et Conizoidis. — Omnia summo studio atque industria doctissima atque excellentis viri Conr. Gesneri medici Tigurini collecta et praefationibus illustrata. — 1561 Argentorati excudebat Josias Ribelius.

An earlier edition referred to on p. 65 bears the following title:

Valerii Cordi Annotationes in Pedacei Dioscoridis de Materia medica libros quinque. Liber de artificiosis extractionibus. Liber II. De destillatione oleorum. Anno dei 1540.

The title of the pharmacopoeia compiled at the request of the Nürnberg council and published 1546 is:

Pharmacorum omnium, quae quidem in usu sunt, conficiendorum ratio. Vulgo vocant Dispensatorium pharmacopolarum. Ex omni genere bonorum authorum, cum veterum tum recentium collectum, et scholiis utilissimis illustratum in quibus obiter, plurium simplicium, hactenus non cognitorum vera noticia traditur. Authore Valerio Cordo. Item de collectione repositione et duratione simplicium. De adulterationibus quorundam simplicium. Simplici aliquo absolute scripto, quid sit accipiendum, Ἀντιβαλλόμενα, id est, Succedanea, sive Quid pro Quo. Qualem virum Pharmacopolam esse conveniat. Cum indice copioso. Norimbergae, apud Joh. Petreium.

Other editions of this Dispensatorium Noricum used in the compilation on p. 32 are enumerated on p. 51, footnote 3.

Conringius.

Hermannus Conringius, De hermetica Aegyptiorum vetere et Paracelsiorum nova medicina libri duo. Helmstadt 1648.

Crato von Kraftheim.

Conciliorum et epistolarum libri vii. Francofurti 1589.

C. C. Cunrathii,

Medulla destillatoria et medica, oder Bericht, wie man den Spiritus vini zur Exultation bringen soll. Leipzig 1549.

Demachy.

Laborant im Grossen oder die Kunst die chemischen Producte fabrikmässig zu verfertigen. Aus dem Französischen übersetzt, mit Zusätzen versehen von Samuel Hahnemann, der Arzneikunde Doctor und Physikus des Amtes Gommern. Leipzig 1784.

DEJEAN.

Traité raisonné de la destillation, ou la destillation reduite en principes avec un traité des odeurs. Par Dejean. Paris 1753.—German edition, Altenburg 1754.

Traité des odeurs, Suite du traité de la destillation. Par Dejean. Paris 1764.

DESTILLIRBÜCHER. See p. 24. Also Brunschwig, Ulstad and Reiff. For more recent works see Burghart, Dejean, and Demachy.

DIOSCORIDES. Pedanius Dioscorides is the first medical writer of importance of the Christian era. He was born about the beginning of the first century in Anazarbus in the southeastern part of Asia Minor. In the capacity of a physician he traveled with the Roman army. His "Materia Medica" written in the second half of the first century is the most thorough work of its kind produced in antiquity. It was regarded as authority throughout the middle ages. Even during Luther's time it was made the basis for lectures and printed commentaries, e. g. by Melanchthon and Valerius Cordus at the University of Wittenberg about the middle of the sixteenth century.

Dioscorides' most important writings are his five books "De materia medica," and his "Alexipharmaca et theriaca," remedies against vegetable and animal poisons, which were added as books six and seven. These and others, more or less apocryphal writings, have been frequently published and commented upon in many languages. Some of the oldest editions are to be found in the library at Leyden. They are a manuscript in the Arabic language, written about the year 940; a very rare Greek edition printed by Aldum Manutium in Venice in 1499; and a Latin edition by J. Allemanum de Medemblich printed in Colle in 1503.

Some of the better translations and commentaries are:

Pedanii Dioscoridis Anazarbensis: de materia medica libri quinque. Jano Coronario medico physico interprete. Basiliae 1529.

Valerii Cordi Simesusii Annotationes in Pedacei Dioscoridis Anazarbei de medica materia libros quinque, longe aliae quam ante hac sunt emulgatae. Ejusdem Historiae stirpium libri quatuor, et de artificiosis extractionibus liber. Tiguri 1540.

Valerii Cordi Simesusii Annotationes in Pedanii Dioscoridis Anazarbei de materia medica libros quinque, longe aliae quam antea sunt haec sunt emulgatae. Ejusdem historia stirpium libri quatuor, et de artificiosis extractionibus liber etc. Translatio Rüllii. Francofurtum ad Moenum 1549. Editio Gessnerii 1561.

Pedanii Dioscoridis Anazarbei de medicinale materia medica libri sex, Joanno Rüllio Suessionensi interprete. Accesserunt priori editioni Valerii Cordi Simisusii Annotationes doctissimi in Dioscoridis de medica materia libros Euricii Cordi judicium de herbis et simplicibus medicinae; ac eorum quae apud medicos controverruntur explicatio. Francofurti 1543.

Petri Andreae Matthioli Opera quae extant omnia. Commentarii in sex libros Pedacei Dioscoridis de materia medica. Veneti 1554.

Petri Andreae Matthioli, Medici Caesarii et Fernandi Archiducis Austriae, Opera quae extant omnia: hoc est Commentarii in sex libros Pedacei Dioscoridis Anazarbei de medica materia. Post diversarum editionem collationem infinitis locis aucti: De ratione destillandi aquas ex omnibus plantis; et quomodo genuini odores in ipsis aquis conservare possint. Veneti 1544. — Basilae 1565.

Πεδακίου Διοσκορίδου 'Αναζαρβεῶς περὶ ὕλης ἰατρικῆς βιβλία or Pedacei Dioscoridis Anazarbei opera quae extant omnia. Ex nova interpretatione. Jani-Antonii Saraceni, Lugduni Medici, Francofurti 1578 and 1598.

A Latin translation of the Dioscorides' "Materia Medica" had appeared as early as 1478, and a Greek edition about the same time in Cologne.

A later edition of Dioscorides' "Materia Medica" which has been used in this work is the edition of Prof. Curtius Sprengel. It constitutes volume 25 of Kühn's collection: "Medicorum graecorum opera quae extant," and consists of two parts. The first part contains: "De Materia medica libri quinquae"; the second part: "Liber de venenis eorumque precautione et medicamentione" (pp. 1—338), and "Commentarius in Dioscoridem" (pp. 340—675).

Dispensatorium Brandenburgicum. The edition of 1698 was consulted. See p. 31, footnote 3.

Dispensatorium Noricum. See Cordus and Gesner.

Flückiger, F. A. Born May 15, 1828, in Langenthal, Switzerland, died Dec. 11, 1894, in Bern. Professor of Pharmacy and Pharmacognosy at Strassburg from 1873—1892.

Pharmacognosie des Pflanzenreiches. Dritte Auflage. Ein Bd., pp. XVI, 1117. Verlag von Hermann Heyfelder, Berlin, 1891.

Flückiger and Hanbury.

Pharmacographia, a history of the principal drugs of vegetable origin met with in Great Britain and British India. Second edition. One vol., pp. XX, 803. Macmillan & Co., London, 1879.

Fourcroy, A. F. Born 1755, died 1809. An influential French teacher of chemistry.

Système des connaissances chimiques, et de leur applications aux phénomènes de la nature et de l'art. Paris 1801.

Fuchs, Leonhard. 1501—1566. Professor of medicine in Ingolstad and Tübingen.

De componendorum miscendorumque medicamentorum ratione. 1549.

FUCHS, REMACLIUS. Born 1510 in Limburg, died 1587 in Brussels.

Remaclii Fuchsii, Historia omnium aquarum, quae in commune hodie practicantium sunt usu, vires et recta destillandi ratio. Veneti 1542. — Parisii 1542.

GALENUS. Born 131 in Pergamon, educated in Smyrna, Alexandria and Rome as physician. Died between 201 and 210. His numerous writings have been edited by Kühn under the title:

Claudii Galeni Opera omnia, in 20 Bänden, Lipsiae 1821—1833. Special mention may here be made of De simplicium medicamentorum temperaturis et facultatibus libri XI.

GEBER. Abou Moussah Dschafar al Sofi, known in western countries as Geber, was active as Arabian physician and scientist during the second half of the eighth century.

In addition to the original writings of Geber which were written in Arabic, a number of works which evidently were written later in the Greek and Latin languages, have been attributed to him up to a comparatively recent date. The apocryphal character of these later works has recently been proven by M. Berthelot ("Introduction à l'étude de la chimie des anciens et du moyen-âge." Paris, 1889; also "Revue des deux mondes," Sept. 15 and Oct. 1, 1893).

Some of the works written by Geber or at least attributed to him and referred to in this book are:

Gebri "de alchimia libri tres." Argentorati arte et impensa. Io. Griegningeri anno 1529.

Gebri "Summa perfectionis magisterii." Ex bibliotheca vaticana exemplari. Gedani 1682.

Alchimiae Gebri Arabis libri excud. Joh. Petrius, Nürembergensis. Bernae 1545.

GESNER, CONRAD. (Euonymus Philiatrus.[1]) Born 1516 in Zürich, died there in 1565. Studied in Bourges, Paris and Basel. Was city physician and later Professor of the natural sciences in Zürich. Edited the works of Cordus (see Cordus). More important than these "Annotationes" is the following:

Thesaurus Euonymi Philiatri, de remediis secretis; liber physicus, medicus et partim etiam chymicus et oeconomicus in vinorum diversi sapores apparatur: medicis et pharmacopolis omnibus praecipue necessarius. Tigur. 1552. Liber I-De destillatione ejusque differentiis in genere. Auctor est Conradus Gesnerus. Tiguri.[2]

---

1) The pseudonym of Gesner is possibly derived from *Euonymus* (Ger. Pfaffenhütchen) and *philiatros*, (φίλος ἰατρός) friend of medicine.

2) Another reference contains after the sentence ending with "necessarius" the following data: Tiguri 1552 — Lugduni 1557—1566 — Francof. 1578.

A German edition appeared in 1555 under the title:

"Ein kostlicher theurer Schatz des Euonymus Philiatrus, darinnen behalten sind vil heymlicher gütter stuck der arzney. fürnemmlich aber die art und eygenschafften der gebrannten wasseren und ölen, wie man dieselbigen bereiten sölle: desgleychen yeder wasseren und ölen art und eygenschafft, nutz und brauch. Item alles mit schönen lieblichen figürlinen angezeigt unnd Item wie man mancherley weyn bereiten sölle, auch den abgestandenen durch hilff der gebrannten wasseren, gewürtzen unnd anderley materi widerumb helffen möge für die augen gestellt, ganz lustig, nutzlich und güt allen Alchemisten, haushalten, insbesonders den Balbiererern, Apothekern und allen liebhaberen der Arztney.—Erstlich in Latin beschrieben durch Euonymum Philiatrum, unn newlich verteutscht durch Johannem Rudolphum Landenberger zu Zürich: vormals in Teutsche sprach niemals gesähen. Getruckt in Zürich bei Andrea und Jacobo den Gessneren gebrüder im jar als man zalt von Christi unsere Heylands geburt 1555."

For more than a century this work was frequently reprinted and translated. The English translation by Moroyng appeared in 1559 under the title:

New book of distillation called the treasure of Euonymus, London 1559, 1564—1565.

A French translation appeared in 1555 in Lyon.

In 1583, after the death of Gesner, this volume was reprinted with a second volume (also written by him and published by Caspar Wolf in the original Latin in 1565) translated into German by Jacob Nüscheler. The title of this second volume is:

Ander Theil des Schatzes Euonymi von allerhand künstlichen und bewerten ölen, wasseren und heymlichen Arzneyen, sampt ihrer ordentlichen bereytung und dienstlichen Figuren. Erstlich zusammen getragen durch Herrn Doctor Cunrat Gesner, Demnach von Capar Wollfen der Arzneyen Doctor. Zürich; in Latin beschrieben und in Truck gefertiget, jetzt aber newlich von Johann Jacobo Nüscheler Doctoren, in Tütsche Sprach vertolmetschet. 1583.

Glauber, Johann Rudolf. 1604—1668. A representative of technical chemistry during the iatrochemical period.

Johanni Rudolphi Glauberi Furni novi philosophici oder Beschreibung der neu erfundenen Destillirkunst. Amsterdam 1648 — Leiden 1648 — Prag 1700.

Gren, F. A. C. 1760—1798, studied pharmacy, was professor of chemistry at Halle. Founded with Gilbert the Annalen der Physik in 1798.

Gren's Grundriss der Chemie nach den neuesten Entdeckungen entworfen und zum Gebrauch akademischer Vorlesungen eingerichtet. Halle 1796.

Hanbury, D. Born Sept. 11, 1825, died March 27, 1875. Pharmacist and writer on numerous pharmacognostical subjects.

Science Papers. Edited with memoir by Joseph Ince. One vol., pp. XI, 543. Macmillan & Co., London, 1876.

Pharmacographia—See Flückiger.

HEUSLER, FR. For several years Professor Wallach's assistant at Göttingen, later Privatdocent at Bonn.

Die Terpene. Braunschweig 1896. Reprint from the new "Handwörterbuch der Chemie." Published by Vieweg und Sohn, Braunschweig, 1896.

HIRZEL, HEINRICH.

Die Toiletten-Chemie. Leipzig 1864.

HOEFER.

Histoire de la chimie. 2nd ed. 1866.

HOFFMANN, FR. Born 1660 in Halle, died 1742. Professor of medicine in Halle. A representative of the phlogistic school of chemistry.

Frederici Hoffmannii, Opera omnia physico-medica. Denuo revisa. correcta et aucta. In sex tomos distributa. Genevae 1740—1761 — Veneti 1745, 17 Volumina — Neapel 1753, 25 Volumina.

Fr. Hoffmannii Opera omnia physico-medica. Supplementum secundum. Genève 1760.

IBN KHALDUN.

Notices et extraits des manuscripts de la bibliothèque impériale à Paris. 1862.

JOANNI RHENANI,

Medici, Solis e puteo emergentis: sive dissertationis chymia technice practica, materia lapidis philosophici et clavis operum Paracelsi, qua abstrusa implicantur deficientia supplentur. Francofurti 1613. Pars 1. Theoremata chymio technica.

KLIEMONT, J. M. Vienna.

Die synthetischen und isolirten Aromatica. Verlag von Eduard Baldamus, Leipzig) 1899.

KOPP, HERMANN. Born 1817 in Hanau, studied under Liebig, and since 1841 Professor at Heidelberg. Known as physical chemist and author of several historical treatises:

Geschichte der Chemie, 1843—1847.
Entwickelung d. Chem. in der neueren Zeit. 1873.
Beiträge zur Geschichte der Chemie, 1869—1875.

LARGUS. See Scribonius Largus.

LEWIS, WILLIAM.

The new Dispensatory: Containing the theory and practice of pharmacy, a description of medicinal simples, according to their virtues and medicinal qualities, the description, use and dose of each article, etc. Intended as a correction and improvement of Quincy. London 1753.

LONICER, ADAM. 1528—1586.

Adami Loniceri, der Arzney Doctor und weiland Ordinarii Primarii Physici zu Francfurt am Meyn, Kräuterbueh und künstliche Conterfeyungen der Bäumen, Stauden, Hecken, Kräutern, Getrayde, Gewürzen und nützlichen Kunst zu destilliren. . . . — Auf das allerfleissigste übersehen, corrigirt und verbessert durch Petrum Uffenbachium, Ordin. Physicus in Francfurt am meyn. Ulm, anno dei 1551, 1573 und 1589.

Adami Loniceri. Kräuter Buch und künstliche Conterfeyungen sammt der schönen und nützlichen Kunst zu destilliren. Von Petrus Uffenbach in's Teutsche übertragen. Ulm 1703.

LULLUS. Raymundus Lullus, born about 1235 of family of Spanish nobility, alchemist, was killed as missionary in Africa in 1315.

Raimundi Lulli, "Experimenta nova" in Manget's Bibliotheca chemica curiosa. Genf 1702.

MATTHIOLUS. Born 1501 near Siena, studied medicine in Padua, body physician of Emperor Maximilian II, succumbed to the pest in Trient in 1577.

Petri Andreae Matthioli, Opera quae extant omnia. Supplementum: De ratione destillandi aquas ex omnibus plantis: et quomodo genuini odores in ipsis aquis conservari possint. Basilae 1565.

MAIER, JULIUS.

Dr. Julius Maier, Die ätherischen Oele, ihre Gewinnung, chemischen und physikalischen Eigenschaften, Zusammensetzung und Anwendung. Verlag von Paul Neff. Stuttgart. 1867.

MIERŽINSKI, STANISLAUS.

Die Fabrikation der ätherischen Oele und Riechstoffe. Berlin 1872.

Die Riechstoffe und ihre Verwendung zur Herstellung von Duftessenzen, Haarölen, Pomaden, Riechkissen etc., sowie anderer kosmetischer Mittel. Siebente Auflage. Ein Bd., pp. XX, 331, mit 70 Abbildungen. Verlag von B. F. Voigt, Weimar, 1894. (Sechster Band von Neuer Schauplatz der Künste und Handwerke.)

MESUË, THE YOUNGER. Yahyâ ben Mâsawaih ben Hamech ben Ali ben Abdallah, body physician of the Chalifa el-Hâkim in Cairo, died about 1015 when more than 90 years old. His "Antidotarium" was the most renowned pharmaceutical treatise of the middle ages.

Mesuë, Antidotarium seu Grabaddin medicamentorum compositorum libri XII. Editio Veneti 1502.

Mesuë, Simplicia et composita, et antidotarii novem posteriores sectiones adnotationes. Venetiae 1602.

NEUMANN, CASPAR. Born 1683, died 1737. Apothecary in Berlin. His work on distilled oil is contained in the second volume of his

Chymia medica dogmatico-experimentalis, oder Gründliche mit Experimenten bewiesene Medicinische Chemie. Herausgegeben von Christ. Heinr. Kessel. 4 Bände. Züllichau 1749—1755.

NONUS THEOPHANUS. Body physician of Emperor Michael VIII in Constantinople.

Nonus Theophanus. Editio Bernardi. Praefatio ad Synesius de febribus. Amstelodami, 1749.

Comp. Synesius.

OCCO, ADOLPH. Second half of sixteenth century, author of the famous Augsburg Pharmacopoeia.

Pharmacopoea seu Medicamentarium pro Republica Augustana. Author Adolphus Occo. Augusta Vindelicorum 1564.

ORTOLFF. Adolph Megtenberger or Meydenberger, also Ortolph or Ortolff von Bayernland, born 1450, author of the first pharmacopoeia in Germany.

Ortloff von Bayrland. Arzneibuch. Hie fahet an eyn büchelin von manigerley Artzeney. Mainz 1485.

PHARMACOPOEA AUGUSTANA. See Occo. The editions of 1580, 1597 and 1640 were consulted in preparing the list on p. 32.

PHILIPPE.

Histoire des Apothicaires. One vol., pp. VII, 452. A la Direction de Publicité Medicale, Paris, 1853.

Translated into German by Ludwig.

Philippe & Ludwig, Geschichte der Apotheker. 1858.

PIESSE, S.

The Art of Perfumerie. London 1862.

PLINIUS. Cajus Plinius Secundus, born in the year 23 A. D. near Como, died in the year 79 near Stabiae (Castellamare) at the time of the famous eruption of Vesuvius. Compiled the natural scientific information of his time in 47 books. Most references in this work are to the following edition:

Plinii Secundi Naturalis Historiae libri 37. Recognovit atque indicibus instruxit Ludovicus Janus. Lipsiae 1859. Littré. 2 vols. Paris 1877.

PORTA. Giovanni Battista della Porta, 1537—1615. A Neapolitan nobleman, known especially for his researches in physics.

Joh. Baptistae Portae, Neapolitani, Magiae naturalis libri viginti, in quibus scientiarum naturalium divitiae et deliciae demonstrantur. Iam de novo, ab omnibus mendis repurgati, in lucem prodierunt. Romae 1565. Antwerp. 1567. Editio: Hanoviae 1619. Liber decimus: Destillat, destillata ad fastigia virium sustollit.[1]

POWER, F. B. See Schimmel & Co.

---

1) Other editions are Romae 1563, Antwerpiae 1564 and 1567, Ravennae 1565, Hanoviae 1619.

PRICE ORDINANCES. See also p. 31.

"Ita sunt nomina medicinarum simplicium sive materialium quae ad apothecam requirentur. In genere et in specie." Prof. F. A. Flückiger under the title "Die Frankfurter Liste."

Register alles Apothekischen Simplicien und Compositen, so in den beiden Messen zu Frankfurt am Main durch Materialisten, Kauffleut, wurzelträger, Kräutler und durch die Apotheker daselbst verkauft werden. Frankfurt a. M. 1582.

Reformatio oder erneute Ordnung der heilig Reichsstadt Frankfurt a. M., die Pflege der Gesundheit betreffend. Den Medicis, Apothekern und Materialisten zur Nachrichtigung gegeben. Darneben den Tax und Werth der Arzneien, welche in den Apotheken allda zu finden. 1587.

RHASES. Abu Bekr Muhammed Ben Zakerijja el-Râze. Born and educated in Raj, at one time director of the hospital in Bagdad, author of numerous works, called the Galen of his time. Died 923 or 932.

Das Buch der Geheimnisse des Abû Bekr Ben Zakarîjâ Er-Râzî. Fleischer's Catalog No. 266. Leipziger Stadtbibliothek. Codex K. 215.

Extracts of Rhases' writings and the unimportant illustrations of several Arabian distilling apparatus were published in 1878 by Prof. E. Wiedemann in vol. 32 (p. 575) of the Zeitschrift der deutschen morgenländischen Gesellschaft.

REIFF. Walther Hermann Ryff, during the first half of the sixteenth century surgeon in Strassburg.

H. Gualtherus Ryff, New gross Destillirbuch, wohl gegründeter künstlicher Destillation, sampt underweisung und bericht, künstlich abzuziehen oder Separiren die fürnembste destillirte Wasser, köstliche aquae vitae, Quintam essentiam, heilsame oel, Balsam und dergleychen vielgüter Abzüge. Recht künstlich und viel auff bequeme art dann bisher, auch mit bequemerem Zeug der Gefäss und Instrument, des ganzen Destillirzeugs von Kreutern, Blümen, Wurzeln, Früchten, Gethier unnd anderen stucken, darinnen natürliche feuchte unnd Elementische krafft, einfach oder mancherley gestalt vermischt und componirt; durch H. Gualtherum Ryff, Medicum & chirurgum Argentinensem, getruckt zu Frankfurt a./m. bei Christian Egenolff's seligen Erben im jar 1556.

Reformirte deutsche Apothek. Frankfurt a./M. 1563.

SALADINUS ASCULANUS. Italian physician. Wrote, probably between 1442 and 1458, a rather remarkable pharmaceutical treatise entitled:

Compendium aromatiorum Saladini, principis Tarenti dignissimi, medici diligenti, correctum et emendatum. Bononae 1488. Editio Veneti 1471, 1488 and 1502.

SANCTO AMANDO.

Expositio Joannis de Sancto Amando supra Antidotarium Nicolai incipit feliciter. With the edition of Mesuë's works. Veneti 1502.

SAWER, J. CH.

Odorographia, a natural history of raw materials and drugs used in the perfume industry. One vol:, pp. XXIII, 883. Gurney, Jackson, London 1892. Second series 1894.

Rhodologia. A discourse on roses and the odor of rose. One vol., pp. 93. W. J. Smith, Brighton 1894.

SCHMIEDER, CHR. G.

Geschichte der Alchemie. Halle 1832.

SCRIBONIUS LARGUS. Roman physician of the first century of the Christian era, who in the year 43 accompanied the Emperor Tiberius Claudius to Britannia.

Scriboni Largi, Compositiones medicamentorum. Editio Schneider.

SCHIMMEL & Co. Since January 1877 the firm of Schimmel & Co., Leipzig, has published a report on volatile oils, at first annually, since 1880 semi-annually. The character of this report was at first commercial, but soon became scientific as well until it became a semi-annual repertory of everything pertaining to volatile oils. The title of the German edition is:

Bericht von Schimmel & Co. (Inhaber Gebr. Fritzsche) in Leipzig. Fabrik äther. Oele, Essenzen und chemischer Präparate. April and October.

Since Oct. 1890 an English translation is also published:

Semi-Annual Report of Schimmel & Co. (Fritzsche Brothers). Leipzig and New York.

The title of the French edition which appears since Oct. 1896 is:

Bulletin Semestriel de Schimmel & Cie (Fritchzse Frères). Leipzig and New York.

In 1893 this firm published:

The factories of Schimmel & Co., Leipzig—Prag and Fritzsche Brothers, New York—Garfield. Text by Professor Dr. F. A. Flückiger—Bern.

An elegant work with 32 heliogravure plates. The text gives a brief account of the history of the volatile oils. A year later the New York branch published a

Descriptive catalogue of essential oils and organic chemical preparations compiled by Frederick B. Power, Ph. G., Ph, D., Director of the laboratories of Fritzsche Brothers, at Garfield, N. J.

SCHOLTZ, MAX.

Die Terpene. Sonderabdruck, pp. 189—246.

SUSRUTA. The name, possibly pseudonym of the author[1] of a book on health *Ayurvedas* of sanskrit literature. It was formerly supposed that this book had been written centuries before Christ, but it is now not

[1]) Hippocrates has been suggested.

placed back farther than the twelfth century of the present era. It was translated into German by Hessler between 1844 and 1855 in Erlangen. The *Charaka* is a similar older treatise possibly of the eighth century.

Susrutas Ayur-vedas, id est medicinae systema a venerabili D'hanvantare demonstratum a Susruta discipulo compositum. Nunc primum ex Sanscrita in Latinum sermonem vertit, introductionem, annotationes et rerum indicem adjecit Dr. Fr. Hessler, Erlangae 1844.

The Susruta, or System of medicine, taught by Dhanvantari and composed by his disciple Susruta. Published by Sri Madhusudana-Gupta, Prof. of medicine at the Sanscrit College at Calcutta. Calcutta 1835. 2 vol.

With regard to the age of these works consult:

Lassen, Indische Alterthumskunde. 1. Aufl., Band 2, p. 551.

J. F. Royle, An essay on the antiquity of Hindoo medicine. London 1837. Deutsche Ausgabe von Wallach und Heusinger, Das Alterthum der indischen Medicin. Cassel 1839, p. 45.

Allan Webb, The historical relations of ancient Hindoo with Greek medicine. Calcutta 1850, p. 45.

Zeitschrift der Deutsch. Morgenländ. Gesellsch. Bd. 30 (1876), p. 617; also Bd. 31, p. 647.

Synesius. Born 375 in Cyrene, a disciple of Hypatia in Alexandria, elected Bishop of Ptolemais in 410, alchemist, died 415.

Synesii Tractatus chymicus ad Dioscoridem. In Fabricii biblia graeca. Tom. 8.

Theophrastus. Born 370 or 392 B. C. in Eresos on the island of Lesbos, disciple of Aristotle, died between 288 and 286 in Athens.

Theophrasti Eresii opera, quae supersunt omnia. Historia plantarum. Editio Wimmer. Parisii 1866.

Ulstad, Philipp. Professor of medicine in Nürnberg during the first half of the sixteenth century.

Philippi Ulstadii, patris nobilis Coelum Philosophorum seu liber de secretis naturae, id est: quomodo non solum e vino, sed etiam ex omnibus metallis, fructibus, radicibus, herbis etc. Quinta essentia, sive aqua vitae, ad conservationem humani corporis educi debeat. Argentor. 1526 und 1528 — Lugduni 1540 und 1553 — Parisii 1543 — August. Treboc. 1558 — Francofurti 1600.

The title of the German edition is:

Dess Edlen und Hocherfahrenen Herrn Philippi Ulstadii von Nürmberg Büchlein von Heimligkeiten der Natur, jetzund verdeutischt. Frankfurt am Mayn 1551.

That of the French edition which appeared in 1547:

Le Ciel des philosophes ou secrets de la nature. Paris 1547.

VICTORIUS FAVENTINUS. The Bologna physician and professor Bennedetto Vettori was born 1481, died 1561.

Victorii Faventini, Practicae magnae de morbis curandis ad tirones. tomi duo. Veneti 1562. Tom. 1, cap. 21. fol. 144.

VILLANOVUS. Little is known about Arnaldus Villanovus as to nativity. Was physician in Barcelona during the second half of the thirteenth century, suffered shipwreck on his way to Avignon to Pope Clemens V about the year 1313. Alchemist.

Arnoldi Villanovi Opera omnia. Veneti 1505. Liber de vinis.

Arnoldi Villanovi Breviarium practicae, proemium in operis omnibus cum N. Taurelli in quos libros annotationibus. Basiliae 1587.

WINTHER, JOHANN. Born 1487 in Andernach, died 1574 as professor of medicine in Strassburg.

Guintheri Andernaci Liber de veteri et nova medicina tum cognoscenda tum facienda. Basiliae 1571.

ZEISE, H. Born 1793 in Holstein, died 1863, apothecary in Altona.

Beiträge zur Nutzanwendung der Wasserdämpfe. Pamphlet. Altona 1826. — Arch. d. Pharm. Bd. 16 (1828), p. 69.

ZELLER, G. H. Apothecary.

Studien über die ätherischen Oele. I. Heft. Des chemischen Theils erster Abschnitt. Landau 1850. — II. Heft. Die physischen und chemischen Eigenschaften der officinellen ätherischen Oele. Stuttgart 1855. — III. Heft. Die Ausbeute und Darstellung der ätherischen Oele aus officinellen Pflanzen. Stuttgart 1855.

ZOSIMOS of Panopolis, an encyclopaedic writer of the fourth century and one of the principal authorities of the alchemists.

"Et quid plura moramur? Unus Zosimos Panopolites libro *περὶ ὀργάνων καὶ καμίνων* loculente ad oculos nobis sistit antiquorum illa vasa destillationibus accommodata; postquam enim jussisset candidatos artis id agere ut ipsis ad manus esset *βίκος ὑέλικος σωλὴν ὀστράκινος λοπὰς καὶ ἄγγος στενόστονον*, mandassetque *ἐπὶ ἄκρα τῶν σωλήνων βίκους ὑέλου μεγάλους παχεῖς ἐπιθεῖναι, ἵνα μὴ ῥαγῶσιν ἀπὸ τῆς θέρμης τοῦ ὕδατος*, tandem, ut clarius sese explicit, ipsas vasorum figuras appingit, quarum nonnullas licet rudiori manu exaratas ex bibliotheca regis christianissimi, et illa D. Marci Venetiis, libuit hic in gratiam curiosorum adjicere." (O. Borrichius "Hermetis Aegyptiorum et chemicarum sapientia" ab Hermanni Conringii animadversionibus vindicata. Hafniae 1674, p. 156.)

A detailed account of Zosimos' discussion on distillation is found in Höfer's Histoire de la chimie. 2. Edit. 1866. Tom 1, pp. 261—270.

INDEX.

www.ingramcontent.com/pod-product-compliance
Lightning Source LLC
Chambersburg PA
CBHW020239290326
41929CB00045B/412